Introduction

The *Networking Fundamentals* textbook is beneficial to anyone wishing to pursue a career in information technology. It provides the skills and basic knowledge required before pursuing studies in specific networking fields, such as network administration, network design, and the support of specific network operating systems. This textbook and the accompanying *Laboratory Manual* and *Study Guide* meet all the required knowledge for the CompTIA Network+ Certification exam and provide the basic skills and knowledge needed to successfully enter the field of networking.

While many books on the market prepare students to pass the Network+ Certification exam, they do not take the time to explain in detail how the various network technologies work. You may be prepared for the exam, but lack the technical foundation to understand the networking technologies underlying everyday operations. In contrast, the *Networking Fundamentals* textbook contains detailed explanations about concepts, rather than just presentations of key points and specifications.

Many concepts may seem confusing or incomplete when first presented. As you progress through the textbook, you will learn other knowledge that will help you grasp previously presented concepts. This is why it is imperative that you constantly review previously learned subjects. Many of the concepts covered in early chapters will not become completely clear until later in the textbook. For example, throughout the textbook, references are made to protocols and the OSI model. These concepts are difficult to fully master until other concepts are presented and learned. Later in the textbook, an entire chapter is devoted to the OSI model and how it is directly related to protocols and equipment. Learning and reviewing the material presented should be a continuous process.

Be sure to visit all the recommended Web sites and perform the *Suggested Laboratory Activities*. A *Laboratory Manual* and a *Study Guide* have been designed to accompany this textbook. I strongly recommend that you complete all laboratory and study guide activities. The *Laboratory Manual* will provide you with a guided opportunity to perform many of the jobs required in a typical network setting. The *Study Guide* will provide you with a valuable means of review and practice essential for mastery of basic network knowledge and skills. By completing all activities and related class work, your opportunity for success on the Network+ Certification exam and in the workplace will be greatly enhanced.

I want to take this opportunity to wish you success in your future.

Sincerely,

Richard Roberts

The Author

For the past 35 years, Richard Roberts has been designing curriculum, teaching Electricity and Electronics as well as Computer Technology, and supervising technical teachers. Mr. Roberts is an accomplished programmer and computer technician. He has experience as the system administrator for Novell NetWare, Microsoft NT, and IBM Token Ring networking systems. He possesses a Bachelor's degree in Technical Education and a Master's degree in Administration/Supervision. He also has current CompTIA A+, Network+, and iNet+ certifications.

His computer experiences started as early as 1974, when he began programming and teaching the Motorola 6800, which eventually evolved into the Motorola 68000—the core processor of the Apple Macintosh computer system. Since then, Mr. Roberts has maintained his teaching status to both instructors and students as the technology has evolved, and he has remained at a state-of-the-art technical level through research, teaching, and applications. He is currently an adjunct instructor at South Florida Community College. He has authored the **Computer Service and Repair** textbook and its ancillaries and coauthored the **Electricity and Electronics** textbook as well as designed and programmed the accompanying interactive CD-ROM.

In addition to his current position, Mr. Roberts has taught at Erwin Technical Center and Tampa Bay Technical High School, and he has taught adults in the military service. His time is now divided between computer consulting and applications, teaching students and instructors, and writing textbooks and other ancillary instructional materials. Occasionally, he goes fishing, but not too often.

Using This Text

Each chapter begins with a number of learning objectives. These are the goals you should set to accomplish while working through the chapter. In addition to your objectives, each chapter begins with a list of new terms, which are important for you to learn as you move through the chapter. When these new terms are introduced in the text, they are printed in a *bold italic* typeface. At that point in the text, you will find these terms defined in the margin.

As you read this text, you will also notice some other words or phrases that stand out. File names that you encounter will appear like notepad.exe, student.txt, or io.sys. Any data you must enter, be it by typing at the DOS and Run prompts or button/tabs/menus that you will click with your mouse are set out like **dir C:** or **Start | Programs | Accessories | System Tools**. Any Internet addresses within the text are in the traditional Web style and in blue, such as www.g-w.com. Internet address listed under *Interesting Web Sites for More Information* at the end of each chapter are in the traditional Web style, underlined, and in blue, such as www.g-w.com.

Be sure to read any Network+ Notes, Tech Tips, or Warnings that you encounter. Network+ Notes contain tips that will help you study for the CompTIA Network+ Certification exam. Tech Tips are useful tidbits that might come in handy in the field. Take heed when you see a Warning. Warnings alert you when an act may damage your computer, computer components, or yourself. Damaging a computer component or device through electrostatic discharge is the most common danger you will encounter with computers and networking. You may also encounter some dangerous voltages, especially when dealing with monitors. Most of those repairs should be left to special technicians.

Each chapter concludes with a summary of some of the key information you should take from the chapter, a large number of questions, a list of useful Web sites, and laboratory activities for you to try. Each chapter has two sets of questions. The first set of questions tests your general comprehension of the material in the chapter. The second set of questions mimics the style of the CompTIA Network+ Certification exam. The questions asked here are on topics that the exams commonly probe.

Hands-on experience is the only way to become proficient in networking, so be sure to attempt the activities at the end of each chapter. If you can complete the activities in this text and in the accompanying laboratory manual, you should have no problem passing the Network+ Certification exam. Each chapter concludes with a complete *Laboratory Activity*. Be sure to work through each of these activities. *Suggested Laboratory Activities* are also included. These activities are loosely structured proceedings that you can attempt on your own or if you have free time in class.

Never forget, the world of networking changes rapidly. Consequently, network administration and the CompTIA Network+ Certification exam must change with it. Each chapter includes a list of Web sites where you can find the latest information on the topics covered. Be sure to check the CompTIA Web site (www.comptia.org) frequently for the latest information on what subjects are being added to the exam and what subjects are being dropped. Also, check the author's Web site (www.RMRoberts.com) for text updates, interesting links, and bonus laboratory material.

Acknowledgments

I would like to thank the following people who helped make this textbook possible by supplying information, details, photographs, artwork, and software.

Aimee Leclerc, American Power
   Conversion
Ana M. Bakas, PCTEL, Maxrad Product
   Group
Beverly Summers, Fluke Corporation
Ed Woodward, Hewlett Packard
   Company
Irene Bammer, Liza Meyers, Ortronics
Sheri Najafi, WildPackets Inc.
Tamara Borg, GFI Software, LTD
Vivian Lien, SOYO Group, Inc.

I would also like to thank the following companies for supplying information, details, photographs, artwork, and software.

3Com Corporation
American Power Conversion
Fluke Corporation
Gemplus
GFI Software, LTD
Ortronics
PCTEL, Maxrad Product Group
Precise Biometrics
Siecor Corporation, Hickory, NC
SOYO Group, Inc.
WildPackets, Inc.

Trademarks

DirectX, Microsoft, MS-DOS, Visual Basic, Windows, Windows 2000 Server, Windows NT, Windows Server 2003, and Windows XP are all trademarks of Microsoft Corporation.

Console One, DirXML, eDirectory, NDS, NetWare, Novell, and Novell iFolder are all trademarks of Novell, Inc.

Zip is a trademark of Iomega Corporation.

Apple, FireWire, Macintosh, MAC OS, QuickTime, and TrueType are registered trademarks of Apple Computer, Inc.

JAVA is a trademark of Sun Microsystems.

IBM is a trademark of International Business Machines.

Linux is a trademark of Linus Torvalds

Other trademarks are registered by their respective owners.

CompTIA Authorized Quality Curriculum

The logo of the CompTIA Authorized Quality Curriculum (CAQC) program and the status of this or other training material as "Authorized" under the CompTIA Authorized Quality Curriculum program signifies that, in CompTIA's opinion, such training material covers the content of the CompTIA's related certification exam. CompTIA has not reviewed or approved the accuracy of the contents of this training material and specifically disclaims any warranties of merchantability or fitness for a particular purpose. CompTIA makes no guarantee concerning the success of persons using any such "Authorized" or other training material in order to prepare for any CompTIA certification exam.

The contents of this training material were created for the CompTIA Network+ exam covering CompTIA certification objectives that were current as of 2005.

How To Become CompTIA Certified

This training material can help you prepare for and pass a related CompTIA certification exam(s). In order to achieve CompTIA certification, you must register for and pass a CompTIA certification exam(s).

In order to become CompTIA certified, you must:

1. Select a certification exam provider. For more information please visit http://www.comptia.org/certification/general_information/exam_locations.aspx.

2. Register for and schedule a time to take the CompTIA certification exam(s) at a convenient location.

3. Read and sign the Candidate Agreement, which will be presented at the time of the exam(s). The text of the Candidate Agreement can be found at http://www.comptia.org/certification/general_information/candidate_agreement.aspx.

4. Take and pass the CompTIA certification exam(s).

For more information about CompTIA's certifications, such as its industry acceptance, benefits, or program news, please visit www.comptia.org/certification.

CompTIA is a not-for-profit trade information technology (IT) trade association. CompTIA's certifications are designed by subject matter experts from across the IT industry. Each CompTIA certification is vendor-neutral, covers multiple technologies, and requires demonstration of skills and knowledge widely sought after by the IT industry.

To contact CompTIA with any questions or comments, please call (1) (630) 678 8300 or email questions@comptia.org.

Chapter Listing

8

Table of Contents

Chapter 2
Network Media—Copper Core Cable 65

Chapter 3
Fiber-Optic Cable 111

Chapter 4
Wireless Technology **139**

Chapter 7
Microsoft Network Operating Systems. 249

Chapter 10
Introduction to the Server 367

Chapter 11
TCP/IP Fundamentals 397

Chapter 15
Remote Access and Long Distance Communications . . . 517

Chapter 16
Network Security 551

Chapter 17
A Closer Look at the OSI Model 595

Chapter 20
Designing and Installing a New Network 707

Chapter 21
Network+ Certification Exam Preparation 739

Chapter 22
Employment in the Field of Networking Technology . . . 763

Introduction to Networking

After studying this chapter, you will be able to:

❏ Explain the advantages and disadvantages of a network system.

❏ Identify the three major network classifications: LAN, MAN, and WAN.

❏ Identify the basic network topologies.

❏ Compare and contrast a peer-to-peer network with a client/server network.

❏ Describe how data is packaged and transmitted.

❏ Explain the purpose of a protocol.

❏ List the common networking protocols.

❏ Explain the purpose of general network devices such as a hub, repeater, switch, and gateway.

❏ Identify the major standards organizations.

❏ Identify and explain the purpose of the IEEE 802 standards.

❏ List and explain the purpose of each OSI layer.

Network+ Exam—Key Points

The following subject areas are frequently tested by the CompTIA Network+ Certification exam:

■ Network topologies.

■ Network protocols.

■ Network models.

■ Network connectivity.

■ LAN, MAN, and WAN.

■ Network standards.

■ Network equipment.

■ OSI model.

These subject areas are introduced in this chapter. Many of them appear in more detail later in the textbook. Begin studying them now as simple concepts, and you will be better prepared to comprehend the information presented in later chapters.

Key Words and Terms

The following words and terms will become important pieces of your networking vocabulary. Be sure you can define them.

active hub
access port
administrative server
American National Standards Institute (ANSI)
application layer
ATM (Asynchronous Transfer Mode)
attenuation
backbone
bridge
brouter
bus topology
cell
centralized administration
CERN
client/server network
concentrator
connectionless-oriented
connection-oriented
copper core cable
data layer
database server
decentralized administration
dedicated server
driver
Electronic Industry Association (EIA)
FIR (Fast Infrared)
fiber-optic cable
file server
gateway
hierarchical star
hub
hybrid topology
Institute of Electrical and Electronic Engineers (IEEE)
IPX/SPX (Internetwork Packet Exchange/Sequenced Packet Exchange)
International Organization for Standardization (ISO)
layer 1 device
layer 2 device
layer 3 device
local area network (LAN)
logical identification

MAC (media access code) address
media
mesh topology
metropolitan area network (MAN)
multistation access unit (MAU)
NetBEUI (NetBIOS Enhanced User Interface)
NetBIOS (Network Basic Input/Output System)
network
network administrator
network interface card (NIC)
network layer
network operating system (NOS)
network topology
node
Open Systems Interconnection (OSI) model
passive hub
peer-to-peer network
physical layer
presentation layer
print server
protocol
repeater
ring topology
routable protocol
router
server
session layer
standard
star topology
switch
TCP/IP (Transmission Control Protocol/Internet Protocol)
Telecommunications Industry Association (TIA)
terminating resistor
transport layer
tree
Underwriters Laboratories (UL)
wide area network (WAN)
wireless topology
World Wide Web Consortium (W3C)

This chapter presents a general overview of networking and introduces basic networking concepts. Each topic presented is expanded and explained in detail in later chapters. For full comprehension of the concepts presented throughout the textbook, it is essential that you complete the correlating labs in the lab manual and perform the suggested lab activities at the end of each chapter. The lab manual and suggested lab activities provide for a valuable learning experience.

Definition of a Network

A *network* is an interconnected collection of computers, computer-related equipment, and communications devices. This collection could include a PC, laptop computer, personal digital assistant, tablet PC, printer, cell phone, telephone, pager, fax machine, camera, satellite, satellite dish, or a bar code scanner, **Figure 1-1.** A network can be as simple as two computers in one room or as large as thousands of computers, computer equipment, and communications devices spread across the entire globe.

The purpose of a network is to share computer-related equipment and data. Examples of data are text files, spreadsheets, database information, images, video, music, and voice. Examples of computer-related equipment are printers, Internet access devices, and fax machines.

network
an interconnected collection of computers, computer-related devices, and communications devices.

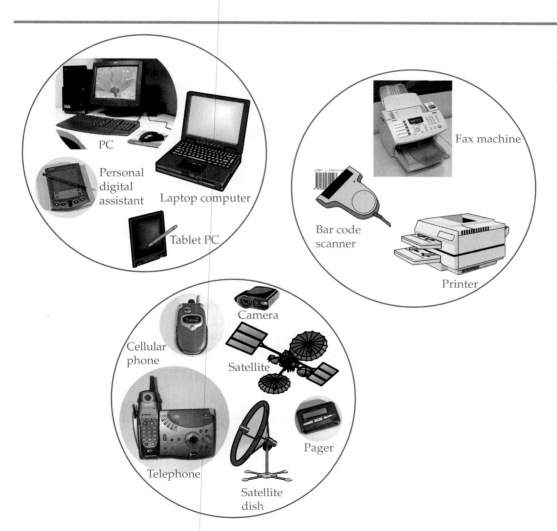

Figure 1-1.
Various devices that can be part of a network.

A network is not the best answer for every situation and should be justified. You must compare the need versus the cost. Next, we will look at the advantages and disadvantages of networking.

Advantages of Networking

Networks provide a means to share software, data, equipment, and communications quickly, easily, and inexpensively. They also provide a means to secure data. A corporate network saves time and money by providing quick access to files and resources throughout the corporation. By using a network system to its full potential, data and communication can be exchanged in the same day in a series of e-mails, phone conversations, and networking conferences. Network meetings can take place quickly and easily between individuals from all over the world, costing only a fraction of what it would cost if the participants had to travel. A small-office or a home-office network can easily share a single printer or data, eliminating the need to transport files on a floppy disk or a CD from computer to computer.

If it were not for networks, the sharing of data would not be quick, easy, or inexpensive. Think about what it would take to share data with another person without the use of a network. You could copy the file to be shared to a floppy disk or CD-RW disc and then physically transport the file to the person. The transportation may involve a short walk or a drive, or it may involve shipping. A walk may not cost much, but a drive or mailing a package can be costly and may not be practical to meet deadlines.

A network can facilitate the delivery of the file to anyone anywhere on the network in a very short time at very little cost. Prior to the use of the Internet, editors and writers were required to exchange work via the US Postal Service. A manuscript would be mailed and delivered in two or three days. The manuscript would be read and edited and then returned within a few days via the public mail system. The writer would review the remarks, changes would be made, and the manuscript would be mailed once again. This process was slow and would typically take a week or more for each chapter. Today, material indicating corrections or clarifications can be exchanged via the network in the same day, saving time and shipping costs.

The ability to share network devices can also save a company a significant amount of time and money. A very expensive poster printer used in commercial advertising and art can be connected to the network. If a person located at another PC wished to print their artwork, they could simply send it there via the network. In fact, everyone in the art department can do the same. There would be no need to buy multiple printers. If the printer was not connected to the network, printing a graphic would be much more difficult and would consume a significant amount of time. The file would have to be copied to a floppy disk or a CD-RW disc and physically transported to the PC to which the printer was attached.

Another advantage of networking is security. Security is a vital part of an organization's structure. It protects information such as customer lists, patient records, legal material, and ideas from being stolen or destroyed. Networks can keep data more secure than a traditional paper system. In an old hard-copy system, data can be photocopied and distributed by anyone having access to the original documents without the company's permission. Customer lists, patient files, financial records, and such can all be copied. The network can be structured to allow users to access only what they need to use for their individual job. Users can be set up as individuals or groups to access particular files, collections of files, and equipment. Allowing specific access is known as assigning user rights or permissions. Security will be covered in detail in Chapter 16—Network Security.

Disadvantages of Networking

Although networking has many advantages, it has its disadvantages as well. These disadvantages include needing additional personnel, losing access to files when the network is down, and vulnerability to hackers, viruses, and disgruntled workers.

Additional personnel are often needed to maintain and modify the network. This adds to the business' operational costs. The number of staff is directly related to the size of the network and the complexity of its design. A small, peer-to-peer network is cost effective and requires few network administration skills for installation and maintenance. A large, client/server-type network can be very expensive and require users to have specialized training prior to its installation and operation. This type of network typically requires a network administrator and additional networking staff. The original cost of installing a network, as well as overcoming the complexity of its usage, can be a disadvantage. However, after long-term use and training, the costs can be justified.

One of the largest disadvantages of a network depends on how it is designed and configured. Often, when a network goes down, files may not be accessible. Unless a backup system is in place, files that are shared from the file server cannot be retrieved until the administrator repairs the network.

Security can be a double-edged sword. A network can provide an excellent means of security, yet fail to provide security because of the many variables involved. A network can be susceptible to hackers, viruses, and disgruntled workers.

Network Classifications—LAN, MAN, and WAN

local area network (LAN)
a network that is usually confined to a single building and is managed by a single entity such as a company.

Networks are classified into three major categories: local area network (LAN), metropolitan area network (MAN), and wide area network (WAN). These categories are based on the physical size of the network, management, and the use of a private or public communication system.

A public communication system can be accessed by anyone who has a telephone. A private system can only be accessed by designated personnel. For example, a telephone line between company offices can be either public or private. A public line is part of the general public telephone system. A private line between company offices is a privately leased line accessible by only designated company employees. An example of a private line is a T1 line. You will learn about T1 lines in Chapter 15—Remote Access and Long Distance Communications. Company networks in which security is a high priority often use private telephone lines rather than public lines.

metropolitan area network (MAN)
a network, under one management, that consists of two or more LANs connected with private communication lines within the same geographic area, such as a city or a university campus.

A *local area network (LAN)* is a network that is usually confined to a single building and is managed by a single entity such as a company. An example of a LAN is a network system connecting computers in an office. A *metropolitan area network (MAN)* consists of two or more LANs connected with private or public communication lines within the same geographic area, such as a city or

wide area network (WAN)
a network that consists of a large number of networks and PCs connected with private and public communication lines throughout many geographic areas.

network topology
the physical arrangement of computers, computer-related devices, communication devices, and cabling in a network.

node
any device attached to the network that is capable of processing and forwarding data.

a university campus. A MAN is also managed by a single entity. The **wide area network (WAN)** consists of a large number of networks and PCs connected with private and public communication lines throughout many geographic areas, **Figure 1-2.** Both public and private telephone companies make communication over a WAN possible. A good example of a WAN is the Internet. The Internet consists of millions of networks spread across the entire globe. A WAN usually requires the services of a telephone company to connect LANs that are under different management.

Network Topologies

The physical arrangement of computers, computer-related devices, communication devices, and cabling in a network is referred to as **network topology.** The four major topologies are bus, ring, star, and mesh. Combining two or more of the four major topologies creates a hybrid topology, **Figure 1-3.**

The most distinctive identifier of network topology is the cable and node arrangement. A **node** is any device attached to the network that is capable of processing and forwarding data. Some of these devices could be a PC, hub, switch, router, server, and repeater, **Figure 1-4.** In a wireless network, cabling is not used and nodes do not have to be arranged in a set pattern; however, each network device in a wireless network is capable of processing and forwarding

Figure 1-2.
A wide area network (WAN) connects networks and PCs together over a large geographic area.

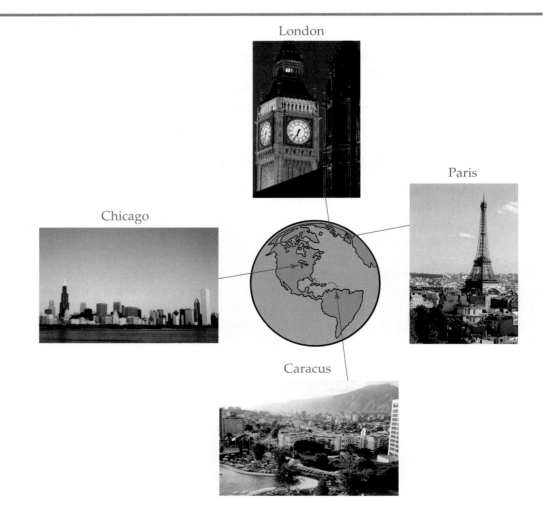

Hybrid			
Bus	Star	Ring	Mesh

Figure 1-3.
A hybrid topology can be formed from two or more of the major network topologies.

data. Wireless technology has become a popular form of linking networks. For these reasons, we will include wireless in our discussion of topologies.

As you read the following sections on each topology, note that each topology is a result of the type of media and network device used to form the network. The following sections briefly mention the type of media and network devices used in each topology. You will learn more about network media and devices later in this chapter and in the following chapters.

Bus Topology

A *bus topology* uses a single cable or conductor to connect all nodes on the network. Look at **Figure 1-5.** Each node is connected to a common cable. A bus topology uses the least amount of cable when compared to other topologies. The bus topology is often used as a backbone to link other topologies. A *backbone* serves as a common path and often employs high-speed network cable such as fiber-optic.

bus topology
a topology that uses a single cable or conductor to connect all nodes on a network.

backbone
a cable that serves as a common path and often employs high-speed network cable such as fiber-optic.

The term *backbone* may change somewhat when applied to specific types of network topology. In the bus topology, it is the common connecting cable. In a star topology, it describes the cable that connects different portions of the network, or different LANs.

Tech Tip

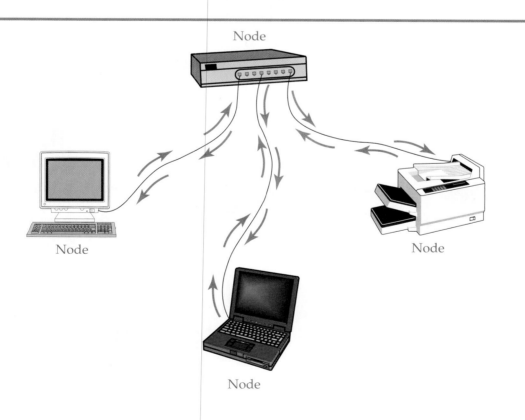

Node

Node

Node

Node

Figure 1-4.
A node denotes a connection point throughout a network system.

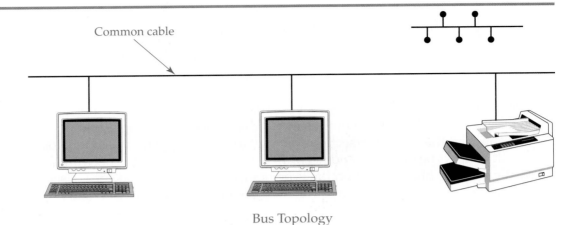

Figure 1-5.
In a bus topology, all
nodes connect to a
common cable.

Common cable

Bus Topology

terminating resistor
an electronic
device, employed
in a coaxial-type
bus topology, that
absorbs transmitted
signals, preventing
the signals from
deflecting and the
data from distorting.

A bus topology that uses coaxial cable requires a terminating resistor at each cable end. A ***terminating resistor*** absorbs the transmitted signals when they reach the end of the bus. Without the terminating resistor, the transmitted signals would deflect, causing the data to be distorted. Terminating resistors are covered in detail in Chapter 2—Network Media—Copper Core Cable.

Star Topology

Cables running from each node to connect to a single point distinguish the ***star topology***, **Figure 1-6.** The center of the star is usually a device known as

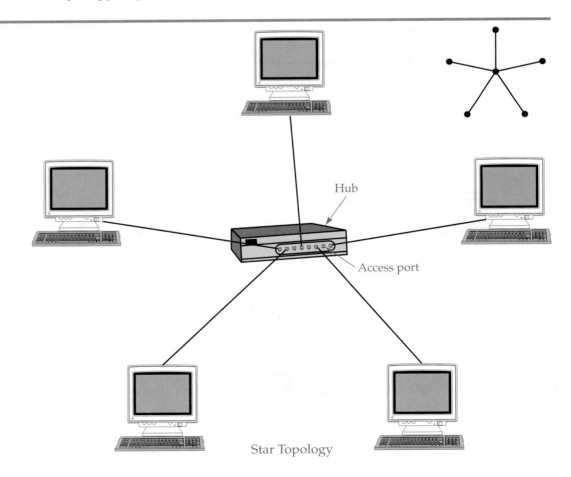

Figure 1-6.
In a star topology, all
nodes connect to a
central point.

Hub

Access port

Star Topology

a hub. A *hub* provides a common electrical connection to all nodes in the star arrangement. Each network cable plugs into a connection called an ***access port*** at the hub. Hubs are discussed in detail in a later section of this chapter.

Ring Topology

The ***ring topology*** consists of a single cable that runs continuously from node to node, **Figure 1-7.** The cable begins and ends at the first node in the network. The ring must remain unbroken.

A ring topology may also resemble a star topology, but because of the equipment used, it is really connected as a ring. When laid out as a physical star, each node is wired directly to a central location called a ***multistation access unit (MAU).*** A MAU is similar to a hub, but does not function exactly like a hub. A MAU allows for the quick connection and disconnection of Token Ring cables while maintaining the logic of the ring topology. In other words, the MAU keeps the circle in tact when one of the Token Ring cables is removed.

Switches at the access ports maintain the ring. See **Figure 1-8.** The unused ports act as closed switches, which maintain the logic of the ring topology. When a cable is plugged into the access port, the switch is opened. The pair of wires inside the cable run to and from the node and MAU. The integrity of the ring is maintained.

star topology
a physical configuration in which cables running from each node in a network connect to a single point, such as a hub.

hub
a network device that provides a common electrical connection to all nodes in a star topology.

access port
a connection into which network cables plug.

ring topology
a physical arrangement that consists of a single cable that runs continuously from node to node.

MAU is also referred to as a MSAU (Multi Station Access Unit).

Tech Tip

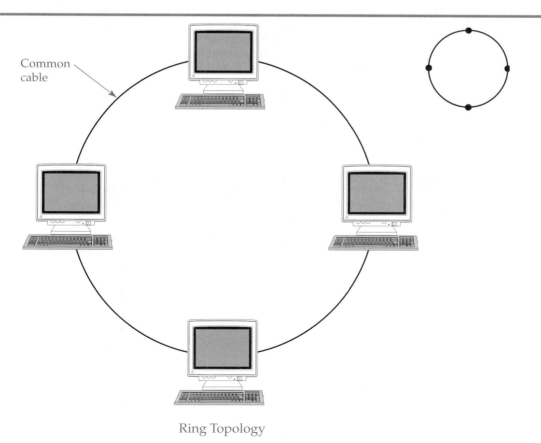

Figure 1-7.
In a ring topology, all nodes connect to a cable, forming a circle or ring.

Common cable

Ring Topology

Figure 1-8.
A MAU maintains ring integrity while cables are connected to or disconnected from the network.

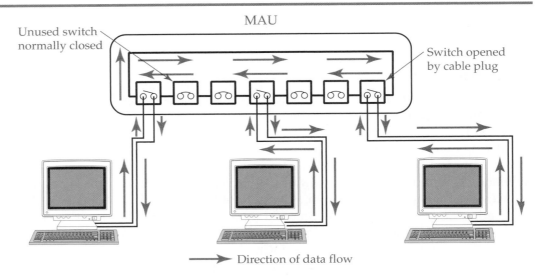

MAU

Unused switch normally closed

Switch opened by cable plug

→ Direction of data flow

multistation access unit (MAU)
a network device that allows for the quick connection and disconnection of Token Ring cables while maintaining the logic of the ring topology.

mesh topology
a physical arrangement in which each node on the network connects to every other node on the network.

wireless topology
a physical arrangement that uses infrared light or radio transmission as a means of communication between nodes.

cell
the area that is served by a radio access tower.

hybrid topology
a physical arrangement that is a mixture of star, bus, and ring topologies.

Mesh Topology

In a *mesh topology*, each node on the network connects to every other node on the network. This is the most reliable network topology and the most expensive because of the additional cost of cabling and equipment. A mesh is only practical when the network is mission critical and cost is not a barrier. A network consisting of multiple servers may use a mesh topology to ensure the reliability of the servers. See **Figure 1-9.**

Wireless Topology

As the name implies, a *wireless topology* does not use cabling. It uses either infrared light or radio transmission to communicate between nodes. Infrared light requires line-of-sight for nodes. This means that nothing that would block the light beam during communications should be placed between the nodes. Radio transmission can experience difficulties caused by building structure and interference generated by other electrical equipment such as radios, motors, welders, and microwave ovens.

Despite these disadvantages, a wireless topology does have many advantages. It is the only answer for communicating with vehicles. The transmission can originate from a building antenna or from a satellite, **Figure 1-10A.** Using cable to connect two buildings could be very expensive and time consuming when compared to installing a wireless network. Two buildings separated by a metropolitan street or a river, or which need immediate connection, can be connected through wireless technology, **Figure 1-10B.** A wireless topology provides a quick way to reconfigure a computer arrangement, whereas moving cables to rearrange computers may not be as easy.

Wireless communications also use cell topology. A *cell* is an area that is served by a radio access tower. The towers are configured throughout an area and divide the area into zones known as cells. A similar technology is used for cellular telephone communication.

Hybrid Topology

As a network grows, it typically changes from a simple network topology, such as a bus or a star, into a hybrid topology. A *hybrid topology* is a mixture of

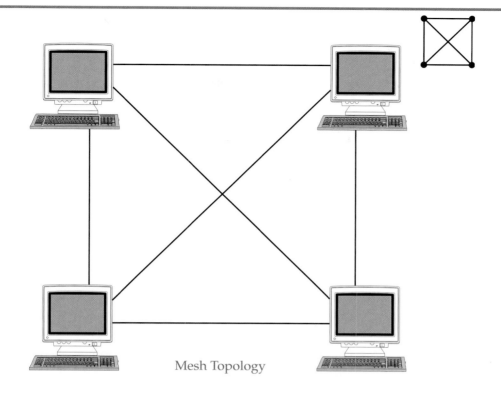

Figure 1-9.
In a mesh topology, each and every node is connected to each other. The number of cable paths rapidly increases as more nodes are added to the mesh topology.

Mesh Topology

topologies. Look at **Figure 1-11.** Notice the different sections of the network. Two star topologies, a bus topology, and a ring topology are connected together.

The design of a hybrid topology changes by the addition of backbones and additional stars or rings. Two subclassifications of the hybrid topology are the hierarchical star and the tree. A *hierarchical star* is created when two or more star topologies are merged using network devices such as hubs, switches, or

hierarchical star
a network topology that is created when two or more star topologies are merged using network devices such as hubs, switches, or routers.

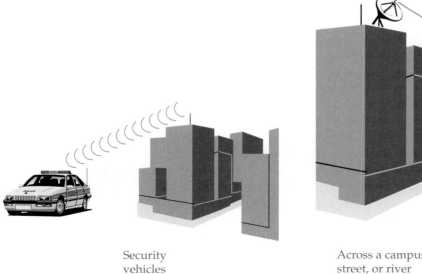

Security
vehicles

Across a campus, busy
street, or river

A

B

Figure 1-10.
A wireless topology allows nodes to connect to one another in areas where cabling would be impossible or difficult. A—A security vehicle receives a transmission from an antenna located on top of a building. B—Wireless technology connecting a network in one building with a network in another building.

Figure 1-11.
This hybrid topology consists of a ring topology, bus topology, and two star topologies.

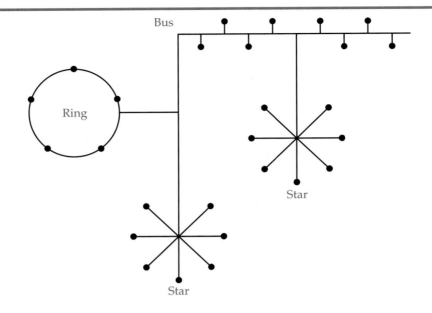

Hybrid Topology

tree
a network topology that is created when two or more star topologies are connected by a common backbone. A tree topology is also defined as a hierarchical star by some network manufacturers.

routers, **Figure 1-12.** A *tree* is created when two or more star topologies are connected together by a common backbone, **Figure 1-13.**

Basic Administration Network Models

The term *network administration model* refers to how a network is designed and administered. There are two basic network administration models—peer-to-peer and client/server. The client/server network uses centralized administration, while the peer-to-peer uses decentralized administration.

Figure 1-12.
Example of a hierarchical star topology.

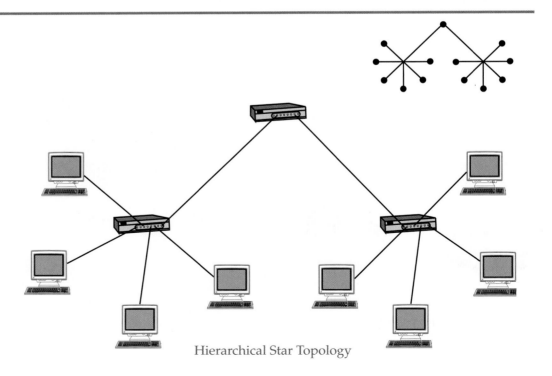

Hierarchical Star Topology

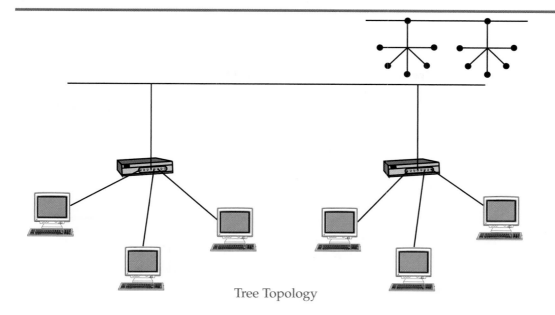

Figure 1-13.
Example of a tree
topology.

Tree Topology

Client/Server Network

The *client/server network* consists of computers connected via a network to one or more servers, **Figure 1-14.** As its name implies, the *server* provides services to networked computers, or clients. Typical services are security, database applications, data storage, Internet access, Web page hosting, and e-mail.

A client/server network has an overall administrator who controls access to the network and to its shares. This person is called the *network administrator.* The methodology used to administer a client/server network is called *centralized administration.* Centralized administration is the only practical solution for controlling a large number of computers. The administrator does not have to go to each server to perform administrative tasks. The administrator can view and access all resource and security information from a central location. Storing this information in a common database or directory makes this possible. An administrator accesses this database to perform administrative tasks such as adding a user or a share. Two of these databases, Microsoft's Active Directory and Novell's Directory Services, are covered in Chapter 7—Microsoft Network Operating Systems and Chapter 8—Novell Network Operating Systems respectively.

A server may provide a multitude of services, such as access to files, the Internet, and mail services, or it may serve a single function. A server that serves a single function is referred to as a *dedicated server.* Some types of dedicated servers are file servers, print servers, database servers, Web page servers, and administrative servers. A *file server* is used to store data files that can be accessed by a client. A *print server* coordinates printing activities between clients and printers. An *administrative server* is used to administer network security and activities. A *database server* contains data files and software programs that query the data.

client/server network
a type of network model that consists of computers connected via a network to one or more servers.

server
a computer that provides services to networked computers or clients.

network administrator
a person who controls access to the network and to its shares.

centralized administration
the methodology used to administer a client/server network.

dedicated server
a server that serves a single function.

file server
a server that stores data files that can be accessed by a client.

Query is a term used to describe extracting data from a database system. Microsoft SQL Server is a typical database query software package.

Tech Tip

Figure 1-14.
A client/server model has a central computer called a server, which controls network security and user access to hardware, software, and programs.

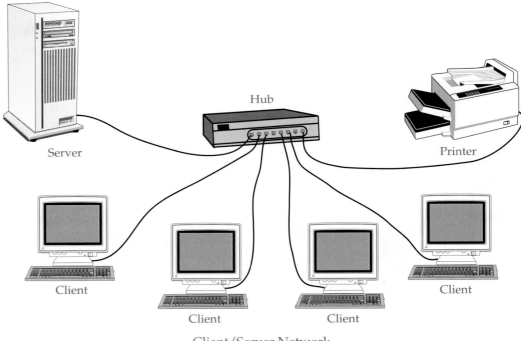

Client/Server Network

The term *server* can be used in two ways, which can be very confusing to a networking novice. The first and most obvious way is when identifying an actual, physical server. The second is when a server is identified by a service such as a proxy server, Web server, or gateway. In other words, a single network server may be referred to by many different names and be used for more than one purpose.

print server
a server that coordinates printing activities between clients and printers.

administrative server
a server that is used to administer network security and activities.

A server is similar to the standard PC in design, and, in fact, many small networks use a standard PC as a server. For large networks, the server is an enhanced PC. It may contain two or more CPUs and ten times or more the normal amount of RAM found in a typical PC. The additional RAM and CPUs allow information to be processed faster, which is useful when the server has many client requests to service at one time. A server is usually equipped with several hard drives. The server may also have one or more duplicate sets of hard drives used to back up the data saved on the first set of hard drives. Backup systems are key to network success. They are covered, along with server hardware, in greater depth in Chapter 10—Introduction to the Server.

Approximately 45% of all business networks are composed of less than 25 computers.

Peer-to-Peer Network

database server
a server that contains data files and software programs that query the data.

All computers are considered as peers or equals in a *peer-to-peer network.* As you can see in **Figure 1-15,** there is no designated server in this model. A computer in a peer-to-peer network can serve as both a client and a server. Each computer is considered equal because each computer's user has equal authority to share their computer's resources with other users on the network.

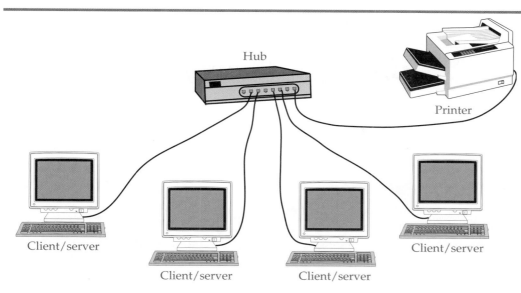

Figure 1-15.
In a simple peer-to-peer network, each computer has equal powers, or administration. Each computer controls the sharing of its data and hardware resources.

Hub

Printer

Client/server

Client/server

Client/server

Client/server

Peer-to-Peer Network

Peer-to-peer networks are often referred to as *workgroups*. Many small networks are constructed as a peer-to-peer network. This type of network is usually inexpensive and easy to install.

In a peer-to-peer network, a computer's user may determine and control which files, drives, and printers will be shared. The computer's user may also determine and control access to the share and designate a password to use the share. Because administration is spread across the entire network, the methodology used to administer a peer-to-peer network is called *decentralized administration*. Administration of this type of network can be very difficult. Without a common database, it is often hard to keep track of all the available users, resources, and access rights. This model is usually used on small networks of less than 25 computers.

The advantages of a peer-to-peer network, however, is that it is inexpensive to install and simple to administer, as long as it remains small. No special, costly networking software is required for operation as it is with the client/server model. A simple peer-to-peer network can be assembled using only Windows 95 or later.

peer-to-peer network
a network in which all computers are considered as peers or equals.

decentralized administration
the methodology used to administer a peer-to-peer network.

Network Operating Systems (NOS)

The *network operating system (NOS)* provides a communication system between the computers, printers, and other intelligent hardware that exist on the network. The most common operating systems today are Novell NetWare, Microsoft Windows Server, UNIX, and Linux. All networks provide communications between different operating systems, different brands of hardware, and various ages of equipment. The network operating system is also composed of software programs that provide for security, user identity, remote access, printer sharing, and other services. In reality, the network operating system *is* the network; everything else is hardware.

network operating system (NOS)
software that provides a communication system between computers, printers, and other intelligent hardware that exist on the network.

Network Communications

Network communication is the exchange of data and commands between network nodes. The data can be simple text to an elaborate multimedia presentation containing sound and full motion video. The network must also communicate commands allowing users to log on, share hardware, and check conditions of the network system. Network communications can be very complicated and are based on network protocols. Network protocols determine how data is exchanged at the packet level.

How Data Is Packaged

Data is transmitted across the network in the form of digital signals, or pulses. Digital signals are rapidly changing levels of voltage often represented in binary terms as 1s and 0s. See **Figure 1-16.** For data to be sent across a network from node to node, a common scheme of packaging must be decided on.

Data is often divided into smaller segments. The data segments are given a sequence number and combined in a packet with other information, such as routing and error-checking code. The packet sequence number allows the receiving computer to reassemble the message in the order it was transmitted. An error-checking code ensures the data was not corrupted. Other information that may be coded into the packet is the packet's destination, the address of the node that originated the packet, and the time it was originated. The type of data and information contained in the packet varies depending on the protocol used. Protocols are discussed in an upcoming section of this chapter and are revisited throughout the textbook.

Network Connectivity

connection-oriented
a type of network connection that maintains a constant and consistent connection between the source and the destination.

Connections between two nodes can be classified as either connection-oriented or connectionless-oriented. *Connection-oriented* requires a direct connection between the source and the destination. For example, a telephone modem is used to make a direct connection between the Internet service provider and your computer. The telephone line usually provides a direct connection from one modem to another. Cable television is also an example of a connection-oriented communication system.

Figure 1-16.
Data is transferred across a network as a series of digital signals, or pulses. These signals are sometimes expressed as binary numbers (1, 0) or as hexadecimal values (1B, F5, etc.). The way the data is presented is determined by the technology used.

12 3C 45 F5 BC 2E AD A2 78 91 9C 2A 34 53 5C FF D3 DC
Hexadecimal code

10100001110010011001010000101101100001100011111000100100
Binary code

Digital pulses

Connectionless-oriented means there is no direct connection between the destination and the source. For example, when a typical Internet connection is established between a PC in Florida and another in California using the Internet, many different paths are used between the destination and source. In fact, the path may change during the process of information exchange between the two PCs. Packets can be exchanged using several different routes. Each packet must be packaged with the source address when a connectionless-oriented transmission mode is used.

connectionless-oriented
a type of network connection in which there is neither a direct connection nor a consistent connection between the destination and the source.

Protocols

For intelligent hardware to communicate across a network, they must use the same protocol. A *protocol* is a group of computer programs that handle packet formatting and control data transmission. A protocol can also be defined as a set of rules governing communication between devices on a network. Some of the things a protocol is responsible for include the following:

- Determining how devices identify each other.
- Determining the method of data exchange.
- Determining the size of each packet.
- Determining the timing for packet transmission.
- Compressing data.
- Determining the signal to be used to end a session.
- Establishing the end of session.
- Providing and establishing error checking.

protocol
a group of computer programs that handle packet formatting and control data transmission. Also, a set of rules governing communication between devices on a network.

The term *protocol* is often used as a label for a software program, a networking standard, or specification. Many times the exact definition depends on the context of the information it is applied to.

 Tech Tip

A protocol determines how to identify other computers that are running the same protocol. For example, a protocol establishes how the two computers will identify each other—by name, such as *Station1* and *Station25,* or by number, such as 192.168.000.010. A protocol also establishes the method for information exchange. Examples of data exchange methods are ASCII code or Bitmap graphic.

Data may be transmitted between two computers by dividing the entire message into smaller packets, or it may be transmitted as a steady stream of data. The protocol determines the size of each packet and the timing of its release over the network. If the packet is too large, the computer hardware may not be able to process it. The protocol divides data into packets of varying lengths. Some packets are as small as 52 bytes and others are several thousand bytes in length. Since only one or two computers may dominate the network usage time, breaking the message into packets and timing its release helps to balance network traffic when many computers are trying to communicate over the network at the same time.

A protocol may compress a packet's contents to allow for faster transmission rates. It also dictates when the communication session will end and the signal that will be sent to denote that the session has ended.

A protocol usually has some error-checking capabilities. Error checking ensures the information received is complete and is reassembled in the correct order. Some network systems, such as the Internet, are very complex. The packets may not be routed to their destination using the same route. Consequently, packets may arrive at different times and would be garbled when reassembled, if not for error checking.

You will study many different protocols and their characteristics throughout the textbook. For now, remember that the protocol is the set of communication rules between two or more devices.

NetBIOS

NetBIOS (Network
Basic Input/Output
System)
software that allows
a computer to com-
municate with other
computers and pro-
vides basic services
for data transfer.

NetBIOS (Network Basic Input/Output System) is a software program or protocol that allows a computer to communicate with many other computers. It provides basic services for the transfer of data between nodes. The NetBIOS software program serves the same function as BIOS software programs do for a single computer. BIOS software allows any computer to communicate with any type of computer hardware. NetBIOS allows network equipment, such as network interface cards, to communicate with each other no matter what hardware brand is used. NetBIOS is at the very core of network communications.

NetBEUI

NetBEUI (NetBIOS
Enhanced User
Interface)
a protocol that is an
enhanced version of
NetBIOS.

NetBEUI is an enhanced version of NetBIOS. NetBEUI is short for NetBIOS Enhanced User Interface. NetBEUI was developed jointly by IBM and Microsoft. It is a simple protocol used for small network systems of a maximum of 100 computers. Ideally, a network that uses the NetBEUI protocol should have a maximum of 25 computers. After the 25th computer, the network starts to significantly slow because of user activity.

routable protocol
a protocol capable of
delivering a packet
to a network differ-
ent from the source
network.

NetBEUI was developed for LANs and not as an Internet protocol. When NetBEUI was developed, the Internet was in its infancy. Internet communications require routable protocols. A **routable protocol** is a protocol that can be routed from one local network to another by use of hardware devices called routers. NetBEUI is not designed to communicate through routers. It is designed to run on a LAN. NetBEUI does not contain a set of rules for delivering data packets to distant locations based on using the Internet addressing scheme. You will learn much more about routing protocols and Internet addressing later in the textbook. At that time, the classification of NetBEUI as not routable will become clear.

TCP/IP

TCP/IP
(Transmission
Control Protocol/
Internet Protocol)
a protocol developed
for UNIX to com-
municate over the
Internet.

TCP/IP (Transmission Control Protocol/Internet Protocol) was developed for UNIX to communicate over the Internet. Because it is robust and routable, it is still the standard protocol used for the Internet and has become a standard protocol for other network operating systems such as Novell NetWare and Microsoft Server 2003.

The name TCP/IP consists of two separate protocol names: TCP and IP. TCP/IP is an entire suite of protocols, which means it consists of many different protocols that serve different functions. For example, TCP/IP network communications consists mainly of three protocols: IP, TCP, and UDP. IP is designed for routing information across a network by the use of an Internet IP address. TCP is designed to send a packet to a destination and wait for a confirmation that the pack arrived intact. UDP is designed to send commands across a network without the need for confirmation.

Another protocol in the TCP/IP suite is the Network Time Protocol (NTP). NTP is used to synchronize time between all network devices. All devices must have the same time reference to ensure security in the network. Time can be examined to determine if a packet has been tampered with. The File Transfer Protocol (FTP) is used to download or upload files to Web servers. SMTP is used to send e-mail messages across a network. To view a complete list of protocols in the TCP/IP suite, visit www.protocols.com.

FIR

FIR (Fast Infrared) is used for transmitting data from laptop computers to desktop computers without the use of cables. Data is transmitted by infrared light. Because FIR is based on wireless technology, a different protocol specification than those developed for wired technology is required. Wireless technology features are discussed in more depth in Chapter 4—Wireless Technology. Then the difference between communication based on wire and wireless devices will become more evident.

IPX/SPX

IPX/SPX (Internetwork Packet Exchange/Sequenced Packet Exchange) was originally developed by Novell to allow NetWare clients and servers to communicate. IPX/SPX is a routable protocol, and it controls how packets of data are delivered and routed between nodes and LANs. SPX guarantees the delivery of the packet; IPX does not.

ATM

ATM (Asynchronous Transfer Mode) is a protocol designed especially for transmitting data, voice, and video. Data is segmented into packets containing 53 bytes each, which are switched between any two nodes in the system at rates ranging from 1.5 Mbps to 622 Mbps. Voice and video data contain thousands more data packets than a typical word document or even a Web page. To transfer massive amounts of data between two points using existing technology such as copper wire systems requires a protocol designed to accomplish this. The ATM protocol is designed specifically to transfer large amounts of data in an uninterrupted stream required by audio and video communications across the Internet. Other existing protocols at the time would not meet this requirement.

There are many other protocols that are designed for specific communication requirements. Some protocols deal with security issues, others with the effective routing of data packets, and some with the remote access of network equipment. Protocol technologies are explored throughout this textbook.

FIR (Fast Infrared) a protocol used for transmitting data from laptop computers to PCs without the use of cables.

IPX/SPX (Internetwork Packet Exchange/Sequenced Packet Exchange) a routable protocol, developed by Novell, that controls how packets of data are delivered and routed between nodes and LANs.

ATM (Asynchronous Transfer Mode) a protocol designed especially for transmitting data, voice, and video.

To learn more about protocols visit www.protocols.com. This Web site provides extensive information for hundreds of protocols. You should bookmark this Web site for future reference.

Tech Tip

Network Media and Devices

A variety of media and devices are associated with networking. This section contains a brief description of each. Some of the network devices presented here may be difficult to distinguish because they are similar in function; however, their differences will become more apparent as protocols and the OSI model are further investigated. Also, each media type and network device is revisited in later chapters, and more detail regarding its use and operation is presented.

Media

media
a general term that identifies the material used to transport packets.

copper core cable
an electrical cable that consists of a copper wire surrounded by plastic or synthetic insulation.

fiber-optic cable
a type of cable that uses a glass or clear plastic core rather than copper.

Media is the general term that identifies the material used to transport packets and data streams between nodes. There are several broad classifications of media. Each type of media has advantages and disadvantages. The most common media is *copper core cable.* Copper core cable consists of a copper wire surrounded by plastic or synthetic insulation. There are many different types of copper core cable used in networking. These types are discussed in depth in Chapter 2—Network Media—Copper Core Cable.

Another type of media, referred to as *fiber-optic cable,* uses a glass or clear plastic core rather than copper. The glass core is thin and flexible. Under normal circumstances, it will not break. Fiber-optic cable carries pulses of light that represent commands and data.

In wireless networks, radio waves carry digital signals between various pieces of equipment in the network. Radio waves are the media and transmission is similar to cell phone technology. Wireless technology is commonly used where cables are impractical, such as to connect a computer installed in a police car to a main network system.

Infrared light is another form of media in which an infrared beam is used to transport a digital signal. Infrared transmission uses a transmitter/receiver unit at each end of the transmission path. A transmitter converts the digital signal into a series of infrared light flashes and sends them to a receiver unit. The receiver unit decodes the signal. Infrared light transmission is similar to the remote control used for televisions.

Network Interface Card

network interface card (NIC)
a network device that contains the electronic components to send and receive a digital signal.

A *network interface card (NIC)* contains the electronic components needed to send and receive a digital signal. A network card is known by many other names such as a network host adapter, network expansion card, and network adapter card. Look at **Figure 1-17** to see what a typical network interface card looks like.

The network interface card fits into one of the motherboard slots inside a PC. This provides a way for the PC to connect to the network media. The network interface card must be equipped with the suitable connector for the type of network media used. Network connectors are discussed in Chapter 2—Network Media—Copper Cable and Chapter 3—Fiber-Optic Cable.

driver
a software program that allows a PC to communicate with and to transfer data to and from computer hardware.

A network interface card must first match the physical communication requirements of the network it is installed for, such as Ethernet, Token Ring, wireless, or high speed FDDI. Next, a driver must be installed so that the computer can communicate with the network interface card. A *driver* is a software program that allows a PC to communicate with and to transfer data to and from computer hardware, such as the network interface card. The network interface card must also be configured with a protocol, such as NetBEUI, TCP/IP, IPX/SPX, or ATM, to support network communications. The network interface card may be configured for a combination of more than one protocol. The exact requirements will become apparent as you progress through the course. Remember that a protocol is a set of rules for communication between two nodes.

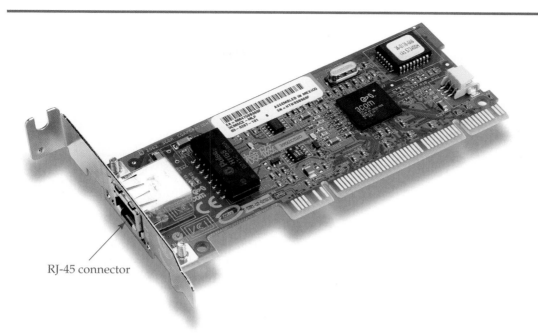

Figure 1-17.
A typical Ethernet
network card equipped
with an RJ-45
connector.
(3Com Corporation)

RJ-45 connector

In general, all network interface cards in the LAN must be configured with the same protocol to communicate with each other.

After the network interface card driver and protocol are installed, the PC must be uniquely identified. There are two types of identification used on a typical network: a physical ID and a logical ID. The network card contains the physical ID, called a *MAC (media access code) address.* The MAC address is a six-byte, hexadecimal number, such as 00 C0 12 2B 14 C5. The first three bytes identify the network interface card's manufacturer, and the second three bytes uniquely identify the card. Together, these two sets of numbers give the network card its own unique, physical identification. The manufacturer stores the MAC address inside an electronic chip on the network interface card.

A *logical identification* is provided by the technician at the time of installation. It is usually a name that uniquely identifies the computer on the network. For example, the name *Station24* would be given to a computer that is the 24th to be added to a group of computers.

Repeater

As digital signals travel across copper wire, the signals lose strength and become distorted. The loss of signal strength is called *attenuation.* To reshape and regenerate the strength of the digital signal, a *repeater* is used. The repeater amplifies or reshapes the weak signal into its original strength and form. See **Figure 1-18.** A repeater allows the network media to exceed its recommended maximum length.

Hub

A *hub* is a central connection point where all network cables are concentrated, **Figure 1-19.** A hub is often called a *concentrator* and is classified as either active or passive. A *passive hub* simply acts as a central connection point for network cables. Packets transmitted from one node are passed to all nodes connected to the passive hub and through the hub to other sections of the network.

MAC (media access code) address
a six-byte hexadecimal number, such as 00 C0 12 2B 14 C5, that uniquely identifies a network card.

logical identification
a name that uniquely identifies a computer on the network.

attenuation
the loss of signal strength.

repeater
a network device that regenerates a weak signal into its original strength and form.

concentrator
a network device that serves as a central connection point for network cables.

passive hub
a network device that acts only as a central connection point for network cables.

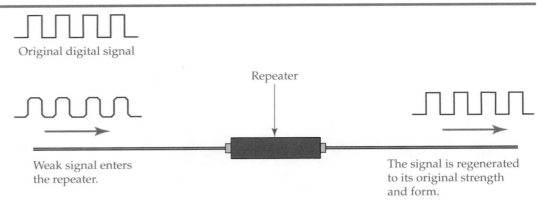

Figure 1-18.
Digital signals lose their strength as they travel along a wire. A repeater is used to regenerate a signal so that it is returned to its original strength and form.

Original digital signal

Repeater

Weak signal enters the repeater.

The signal is regenerated to its original strength and form.

active hub
a network device that acts as a central connection point for network cables and regenerates digital signals like a repeater. The active hub can also determine whether a packet should remain in the isolated section of the network or pass the packet through the hub to another section of the network.

gateway
a network device that connects networks that use different protocols.

Active hubs, sometimes called intelligent hubs or switches, are enhanced passive hubs. An active hub is designed with a power supply. The active hub not only acts as a central connection point for the network cabling, it also regenerates digital signals like a repeater. The active hub can also determine whether a packet should remain in the isolated section of the network or pass the packet through the hub to another section of the network.

Look at **Figure 1-20.** *Station3* attempts to communicate with *Station1.* The intelligent hub does not allow the packets to be transmitted through the hub to the other areas of the network. The hub directs the packets to only those computers that are attached to it. Active hubs are used to reduce excessive data transmission on a network. A network with an excessive number of collisions can be broken into segments by using active hubs. This can reduce the amount of packets transmitted over the entire network. However, this will only reduce the traffic if there are a significant number of transmissions to computers in the same area.

Gateway

A *gateway* connects networks that use different protocols. Gateways make it possible for technologies that use different protocols, such as Token Ring and Ethernet, to communicate. In **Figure 1-21,** a Token Ring network is connected to an Ethernet network via a gateway. The gateway translates the different protocols so that the two networks can exchange information.

Figure 1-19.
A hub is commonly used to connect computers on a network.

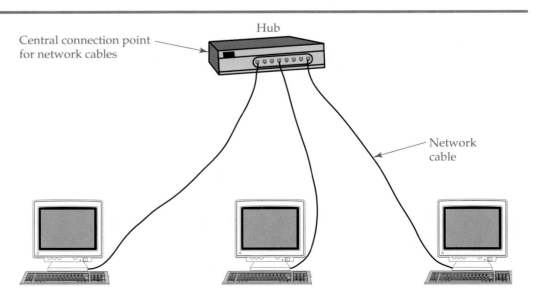

Central connection point for network cables

Hub

Network cable

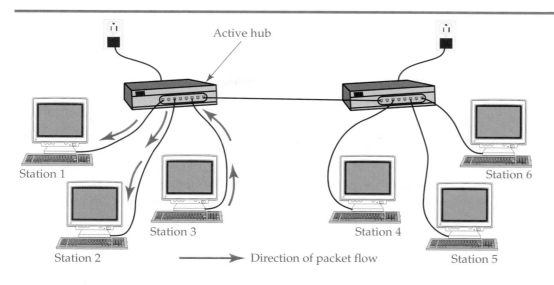

Active hub

Station 1

Station 2

Station 3

Station 4

Station 5

Station 6

Direction of packet flow

Figure 1-20.
In this example, *Station3* communicates with *Station1*. Rather than sending packets to other areas of the network, the active hub directs the packets to only those computers that are attached to it.

Network+ Note:

A server can be classified as a gateway if it provides gateway services to its clients.

Network+

Bridge

A *bridge* can be used to divide the network into smaller segments, reducing the chance of collisions. Networks such as Ethernet broadcast packets throughout the entire segment of the network. When two packets are sent across the network at the same time, a collision occurs, destroying each packet. As a network becomes larger, and more and more packets are transmitted across the network, it is likely that the packets will collide and be destroyed. A bridge controls the flow

bridge
a network device that divides the network into smaller segments, reducing the chance of collisions.

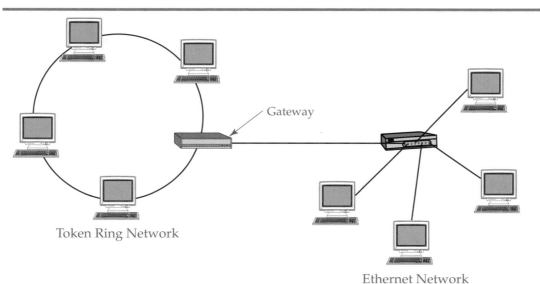

Gateway

Token Ring Network

Ethernet Network

Figure 1-21.
A gateway can connect two dissimilar networks together, such as Token Ring and Ethernet. It can also connect two different network operating systems, such as Microsoft using the TCP/IP protocol and Novell using the IPX/XPX protocol for communications. Often, the piece of network equipment used to connect the LAN to the Internet is referred to as a gateway.

of network traffic between two segments by reading the destination of a network packet. The bridge either allows a packet to pass through to the other segment or restricts the packet to the originating segment.

In **Figure 1-22,** a bridge is used to control network traffic. The network has been divided into two segments: one segment is for the business management department, and the other segment is for the engineering department. The engineering department is responsible for designing new products and for making drawings of the product's parts. The engineers often share the drawings with others in the engineering department. The business management segment contains executives and marketing personnel. By the nature of the arrangement, most of the packets broadcast in the engineering segment are between the engineer's computers. Most of the packets broadcast in the business management segment are between the executives' and the marketing personnel's computers. Only when the business management and engineering departments wish to communicate, packets pass through the bridge.

Switch

switch
a network device that filters network traffic or creates subnetworks from a larger network.

A *switch* filters network traffic or creates subnetworks from a larger network. Some LANs can easily have hundreds or even thousands of nodes. A switch can be used to divide the transmission paths to improve data delivery. When switches form logical networks from a large network, they are called virtual networks or virtual LANs. See **Figure 1-23.**

In the illustration, you can see that all the PCs are connected to one central point, the switch. The switch acts similar to a hub by providing a central connection point for the PCs. A switch also acts like a bridge by limiting traffic on the network. However, a bridge physically isolates segments of a network, while a switch isolates PCs into network segments, thus creating a virtual network. Virtual networks are configured by programming information into the switch, which permits the packets to only flow between designated nodes. By only allowing the packets to flow between the designated nodes, a virtual network within a network is created. The term *virtual*, when applied to networking, means to simulate, rather than a physical network.

Figure 1-22.
A bridge can be used to divide a network into segments, which reduces the amount of overall packet movement on the network media.

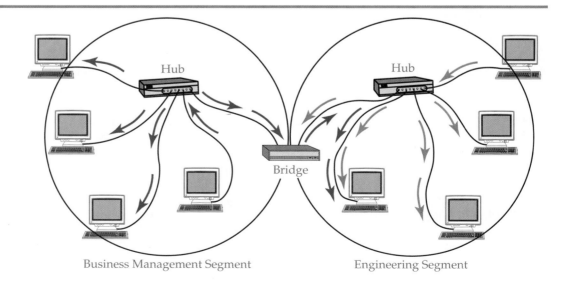

Business Management Segment   Engineering Segment

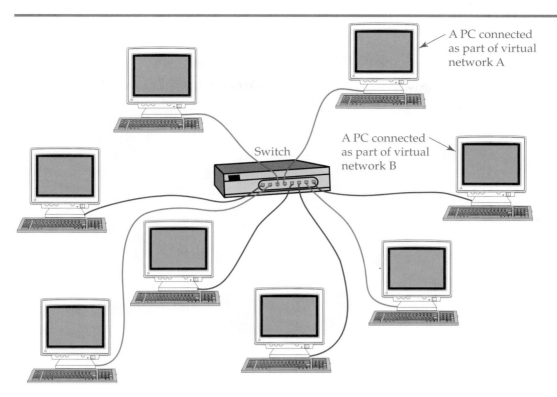

A PC connected as part of virtual network A

A PC connected as part of virtual network B

Switch

Router

A *router* is used to navigate packets across large networks, such as the Internet, using the most efficient route, **Figure 1-24.** A router maintains a table of information containing the location of other routers and their identification. Routers are typically installed between LANs, but may be installed inside a LAN if traffic conditions warrant their installation. Routers "route" data packets across WANs using the TCP/IP protocol addressing scheme. The TCP/IP protocol has become the standard protocol used for WANs and LANs.

router
a network device that navigates packets across large networks, such as the Internet, using the most efficient route.

Router communication is extremely technical. CISCO Systems has their own certification for router technology. Cisco network academies are located all over the United States.

Tech Tip

Brouter

A *brouter* combines router and bridge functions. The brouter functions as a bridge by restricting or passing packets to other sections of a LAN based on the MAC address. It functions as a router by forwarding packets based on the IP address. The main difference between a router and a brouter is that a router typically provides many more functions than a brouter.

brouter
a network device that combines router and bridge functions.

Standards and Organizations

A *standard* is a set of recommendations or practices presented by an organization that defines specific aspects about a technology. For example, an organization may design the specifications for a type of connector that connects to a specific type of cable. Every company who manufactures connectors for

standard
a set of recommendations or practices presented by an organization that help define specific aspects about a technology.

Figure 1-24.
Routers are used to transfer packets across small and large networks. The Internet uses thousands of routers to route packets across the United States and the world.

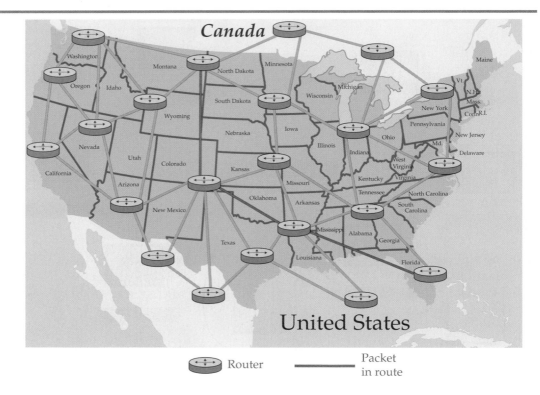

Router ———— Packet in route

that type of cable must know the physical and electrical specifications of the connector to make the connector compatible with the cable. The physical qualities may be length, width, height, and the amount of weight the connector can support. The electrical qualities may specify the amount of voltage and current the connector can handle.

Many times these standards are adopted as law by local jurisdictions. For example, city, county, state, or country may adopt an organization's standards and make them the minimum standard to be followed by contractors or employees. In other words, a company might contract with a contractor to install a network for their business. Since the company wants the network to be installed properly, they will write into the contract "all work shall follow the IEEE standards for networking." Those standards will also serve as the minimum standards for installing the cabling and network devices. The contract will usually specify that all cable and network devices used for the installation must be UL approved. UL stands for Underwriters Laboratories, which is an organization that tests products for safety purposes.

A number of organizations are referred to throughout this textbook. Organization membership is made up of personnel who have an interest in the technology the organization represents. Some of the organizations that are referenced throughout this textbook are IEEE, ISO, UL, ANSI, EIA, and TIA. The following is just a brief explanation of a few of the many organizations that exist. It is recommended that you visit these organizations' Web sites to gain better insight into the scope, membership, and function of each.

IEEE

Institute of Electrical and Electronic Engineers (IEEE)
a professional organization that continually develops standards for the networking and communications industry.

The *Institute of Electrical and Electronic Engineers (IEEE)* (pronounced "I triple E"), founded in 1884, is a professional organization that continually develops standards for the networking and communication industry. This organization consists of scientists, students, commercial vendors, and other interested professionals from the industry. Network standards developed by the IEEE are identified with an 802 prefix. See the chart in **Figure 1-25.**

The specifications outlined in the 802 standards are not laws. They are a set of recommended practices that ensure the quality of a network. They should only be thought of as law if a contract to install a network requires the network to meet the IEEE 802 standards. If a network installation is required to meet the IEEE 802 standards and a problem arises because the standards were not followed, the contractor can be held liable. The 802 standards are referred to throughout your studies of networking. Always check the www.IEEE.org Web site for the latest information on the 802 standards.

ISO

The letters *ISO* are not an acronym, but have been adopted by the organization called the *International Organization for Standardization.* The term *iso* is of Greek origin and means *equal.* The ISO organization was founded in 1946 and comprises 46 different countries and other organizations such as the ANSI (American National Standards Institute) organization. The ISO, interested in the standardization of computer equipment, developed the OSI model for networking technologies. The OSI model is discussed in more detail later in this chapter and again throughout the textbook.

International Organization for Standardization (ISO)
an organization interested in the standardization of computer equipment.

W3C

The *World Wide Web Consortium (W3C)* was established in 1994. The primary concern of the W3C is to provide recommendations for Web page language standards. The organization focuses on the development of programming language standards used for displaying information in the form of Web pages.

Some of the language standards developed and approved by W3C are Hypertext Markup Language (HTML), eXtensible Markup Language (XML),

World Wide Web consortium (W3C)
an organization that focuses on the development of common protocols used for Internet communications.

Standard	Description
802.1	Internetworking
802.2	Logical link control
802.3	Ethernet
802.4	Token Bus
802.5	Token Ring
802.6	Metropolitan area networks
802.7	Broadband
802.8	Fiber optic
802.9	Integrated services
802.10	Security
802.11.1	Wireless networking
802.11.2	Demand Priority Access
802.15	Wireless personal area network
802.16	Wireless metropolitan area network
802.17	Resilient Packet Ring

Figure 1-25.
IEEE 802 standards.

eXtensible Hypertext Markup Language (XHTML), and Cascading Style Sheets (CSS). The Internet uses these programming languages to present text, graphics, and sound. A protocol known as Hypertext Transport Protocol (HTTP) delivers Web pages to a user's computer.

The Internet has become a standard tool for business. An entire chapter in this textbook is devoted to Web servers. It includes an introduction to HTML and Web server communications.

CERN

CERN is an acronym for *Centre European pour la Recherche Nucléaire* (European Center for Nuclear Research). It is located in Geneva, Switzerland. One of the first Hypertext Transfer Protocol servers was placed online in Geneva and is still in operation. This organization is responsible for the original development of the World Wide Web.

UL

Underwriters Laboratories (UL) is a nonprofit organization that tests products and materials for safety standards. Typically, electrical materials and equipment must be UL approved before they can be used in many facilities or as part of a contract specification.

EIA

The *Electronic Industry Association (EIA)* began in 1948 as a trade association concerned with radio communication. It was originally formed as the Radio Manufacturers Association and was responsible for standards such as RS-232. The EIA sponsors many trade shows and conferences throughout the United States.

TIA

The *Telecommunications Industry Association (TIA)* was founded in 1924 by a small group of suppliers who wished to promote the independent telephone industry and to organize trade shows. It soon evolved into a standards maintenance organization. The main areas that TIA is concerned with are fiber optics, user equipment, network equipment, wireless communications, and satellite communications.

ANSI

In 1918, five engineering societies and three government organizations founded the *American National Standards Institute (ANSI)*. ANSI is a private, nonprofit organization that does not develop standards, but rather prompts voluntary conformity and standardization. ANSI is the official representative of many international organizations such as the ISO. ANSI is also one of the founders of the ISO organization. ANSI facilitates meetings and brings together interested parties to resolve issues and create a general census among its members to be presented to other standards committees.

OSI Model

The *Open Systems Interconnection (OSI) model* describes how hardware and software should work together to form a network communications system. It serves as a guide for troubleshooting and designing networks.

CERN
an organization that is responsible for the original development of the World Wide Web.

Underwriters Laboratories (UL)
a nonprofit organization that tests products and materials against safety standards.

Electronic Industry Association (EIA)
a trade association that is concerned with radio communications.

Telecommunications Industry Association (TIA)
an organization that is concerned with fiber optics, user equipment, network devices, wireless communications, and satellite communications.

American National Standards Institute (ANSI)
a private, nonprofit organization that does not develop standards, but rather prompts voluntary conformity and standardization.

Open Systems Interconnection (OSI) model
a model that describes how hardware and software should work together to form a network communications system.

The OSI model consists of seven layers, **Figure 1-26.** Each layer in the OSI model is assigned a specific function. This section provides a brief description of each layer. It is intended as an overview of how networking components work together to send and receive information. The OSI model is referenced throughout the textbook and is discussed in detail in Chapter 17—A Closer Look at the OSI Model.

The OSI model is a joint effort of international members to standardize networking communications systems. Not all software companies follow the strict guidelines of the OSI model. Many systems were in place long before the OSI model was developed and adopted. Some models combine two or more layers into a single unit. It is important to remember that the OSI model simply serves as a guide for future development and illustrates the complexity of transmitting data between network devices.

Tech Tip

Physical Layer

The *physical layer* is the lowest layer of the model. It consists of the cable and connectors used for constructing the network. This layer is only concerned with how digital signals, the binary 1s and 0s, are carried electrically from one networked device to another. When you think of the physical layer, think of cables, radio waves or light pulses, 1s and 0s, and digital signals.

Data Link Layer

The *data link layer* describes how the raw data is packaged for transfer from one network interface card to another. The data link layer also contains information such as the address of the source and destination and the size of the packet, and it provides for error checking. When you think of the data link layer, think of packaging and placing the packet as a stream of digital pulses on the network media.

Network Layer

The *network layer* is responsible for routing packets from one network to another using the IP addressing format. Routing prevents or limits network congestion. It also can prioritize the transmission of packets. As packets are transmitted from one network to another, several different routes may be used. If there is too much traffic on one cable section, the packet may be transmitted along a different route to avoid congestion. When you think of the network layer, think of navigating between networks.

physical layer
the lowest layer of the OSI model, which focuses on the physical characteristics of a network such as cabling and connectors.

data link layer
the layer of the OSI model that describes how raw data is packaged for transfer from one network interface card to another.

network layer
the layer of the OSI model that is responsible for routing packets from one network to another using the IP addressing format.

| Application |
| Presentation |
| Session |
| Transport |
| Network |
| Data link |
| Physical |

Figure 1-26.
The seven layers of the OSI model.

Transport Layer

transport layer
the layer of the OSI
model that ensures
reliable data by
sequencing packets
and reassembling
them into their
correct order.

The main responsibility of the *transport layer* is to ensure reliable data by sequencing packets and reassembling them into their correct order. Packets are transmitted over many miles and may use different routes. The packets may not arrive in the same sequence they were transmitted and may require reassembly into their correct order. Reassembly into the correct order is especially important in transmitting digital images. When you think of the transport layer, think of packet quality.

Session Layer

session layer
the layer of the OSI
model that estab-
lishes a connection
between two com-
puters and provides
security based on
computer and user
name recognition.

The *session layer* establishes a connection between two different computers and provides security based on computer and user name recognition. The two computers may not normally be compatible for data exchange, such as a PC and a Macintosh. The session layer and transport layers are sometimes combined. When you think of the session layer, think of a handshake at the introduction and at the end of the meeting.

Presentation Layer

presentation layer
the layer of the OSI
model that ensures
character code
recognition.

The *presentation layer* ensures character-code recognition. It is responsible for converting character codes into a code that is recognizable by a computer that uses a different character code. For example, Extended Binary Coded Decimal Interchange Code (EBCDIC) is widely used on mainframes, while most PCs use ASCII. EBCDIC uses numbers to represent characters similar to the way ASCII number codes represent character codes. When a PC communicates with a mainframe, the ASCII code must be converted to EBCDIC. When you think of the presentation layer, think of how the data must look for both parties to interpret or understand it.

Application Layer

application layer
the highest layer of
the OSI model that
starts network data
communications
using a program
such as e-mail or a
Web browser.

The *application layer* should not be confused with general software applications. The application layer works with specific networking applications such as Web browser programs, file transfer programs, and e-mail. When you think of the application layer, think of establishing communication with the network.

Network+

Network+ Note:

The Network+ Certification exam by CompTIA requires extensive knowledge of the OSI model and how it relates to network devices, protocols, and network technologies. If the OSI model seems confusing to you, you are not alone. You cannot have a complete understanding of the OSI model until you have a basic understanding of all the networking technologies and how they relate to each other.

OSI Model and Network Devices

It is important to understand how the OSI model and network devices correlate. Look at **Figure 1-27**. In this figure you will see the generally accepted classifications of network devices, such as hubs, switches, and routers, as compared to the OSI model. However, some alterations in the network device's technology may cause a network device to correlate to a different OSI layer.

Network devices are classified according to whether the device makes decisions about how a packet is sent. If the network device makes no decision about where a packet is sent, it is a *layer 1 device.* The device simply moves the packet along the network path such as through cabling, hubs, and repeaters. If decisions are made based on a MAC address or a logical name, it is a *layer 2 device.* If the network device makes a decision about where to move the data based on a protocol such as the Internet protocol, it is a *layer 3 device.*

A network device is often reclassified when its normal function is altered. For example, a switch, normally classified as a layer 2 device, may be reclassified as a layer 3 device if more intelligent decision making is incorporated into it. The decision making of a layer 3 switch is based on some similar information that routers use. A hub, which is normally a layer 1 device, may be reclassified as a layer 2 device when it acts as an intelligent hub. Intelligent hubs make decisions based on MAC addresses and behave more like a switch.

layer 1 device
a network device that makes no decision about where a packet is sent.

layer 2 device
a network device that makes decisions about where a packet is sent based on a MAC address or a logical name.

layer 3 device
a network device that makes a decision about where a packet is sent based on a protocol such as the Internet protocol.

Some switches may be classified as a layer 1 device or layer 3 device. The general classification for the switch is layer 2 device; but if it has some router capabilities, it is usually labeled as a layer 3 device.

Tech Tip

As you progress through the text and learn more about protocols and network communication, the exact placement of certain network devices on the OSI model will become clearer.

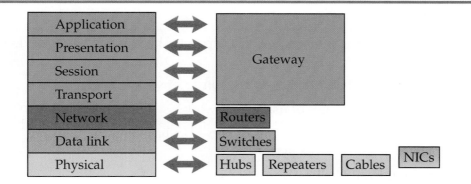

Figure 1-27.
A general correlation between network devices and the layers of the OSI model. This is only a general correlation. Some specific technologies may correlate a network device to a different OSI layer.

Summary

- Networks provide a way to share equipment and data.
- The three classifications of networks based on size, management, and use of private or public communication systems are LAN, MAN, and WAN.
- The four common network topologies are star, ring, bus, and mesh.
- The methodology used to administer a client/server network is centralized administration.
- The methodology used to administer a peer-to-peer network is decentralized administration.
- A network operating system provides a communications system between nodes.
- Data that is to be transmitted on a network is divided into segments.
- A typical packet contains a data segment, source and destination address, error checking, and sequence identification.
- A protocol is a set of rules that control the communication between two nodes.
- In general, two computers need to use the same protocol to communicate with each other.
- A repeater regenerates digital signals, allowing the network media to exceed its recommended length.
- A hub provides a central connection point for a network.
- A bridge is used to create two segments out of a larger network to reduce the chance of collisions.
- A switch serves as a central connection point and can provide network segmentation.
- A router connects various networks together and provides a number of alternate routes for a data packet to travel.
- A gateway connects networks that use different protocols.
- The OSI model serves as a guide for troubleshooting and designing networks.

Review Questions

Answer the following questions on a separate sheet of paper. Please do not write in this book.

1. List some things that might be shared on a network.

2. What are the advantages of using a network?

3. What are the three classifications of networks based on the size and complexity of the network system?

4. The Internet would be best described as a _____.
   A. LAN
   B. MAN
   C. WAN
   D. PAN

5. Rank by size, starting with the smallest: MAN, LAN, and WAN.

6. Six computers connected with a hub in your classroom would *most likely* be classified as a _____ network.

7. Your instructor has a computer that is connected to a powerful computer in another building. This would *most likely* be a _____ network.

8. What are the four major classifications of network topologies?

9. Define *node* and give two examples.

10. What does the term *client* mean in the context of a network?

11. Which type of administrative model uses centralized administration?

12. Name three types of dedicated servers.

13. How does a server differ from a typical PC?

14. Describe how data is transmitted across a network.

15. Data is divided into _____.

16. What does a packet contain?

17. List the layers of the OSI model starting at the application layer.

18. Which layer of the OSI model is mainly concerned with network cables and connectors?

19. Convert the following acronyms into complete words.
    A. MAU
    B. LAN
    C. MAN
    D. WAN
    E. ATM
    F. IPX/SPX
    G. TCP/IP
    H. IEEE
    I. NetBEUI
    J. NetBIOS
    K. FIR
    L. NIC

Sample Network+ Questions

Network+

Answer the following questions on a separate sheet of paper. Please do not write in this book.

1. Which of the following is *not* a typical LAN topology?
   A. Ring
   B. Star
   C. Bus
   D. Square

2. A protocol is best described as a _____.
   A. set of standards that serve as an installation guide for network media and devices
   B. set of rules and procedures that govern how two points on a network communicate
   C. special software utility that determines the amount of traffic on a network
   D. set of rules that assist personnel in using networks to communicate

3. Network cabling is assigned to the _____ layer of the OSI model.
   A. presentation
   B. physical
   C. network
   D. materials

4. Which standard organization is responsible for the 802 standards?
   A. ISO
   B. EIA
   C. ANSI
   D. IEEE

5. Which of the following statements is true about a peer-to-peer network? (Select all that apply.)
   A. It is controlled or administered from one central computer.
   B. Each user controls their personal resources.
   C. It is the best method for controlling a large number of networked PCs.
   D. There is no central file server.

6. What protocol delivers Web page content from a Web server to a user's computer?
   A. XTML
   B. XML
   C. HTTP
   D. HTML

7. Which piece of equipment would most likely be used to provide communications between the local area network and the Internet?

   A. Hub

   B. Bridge

   C. Gateway

   D. Repeater

8. Which protocol is specifically designed to support audio and video communication?

   A. FTP

   B. ATM

   C. HTML

   D. IPX

9. Which item can be shared on a peer-to-peer network?

   A. Hard disk drive

   B. Floppy drive

   C. Printer

   D. Any of the above devices can be shared.

10. What is attenuation?

   A. The loss of network digital signal strength along a cable.

   B. Interference between two network cables.

   C. Network cable resistance measured in ohms.

   D. The term used to describe the procedure for calling attention to a particular node.

Discussion Questions

Answer the following questions on a separate sheet of paper. Please do not write in this book.

1. How many different ways could you deliver a message to a location such as a business, ten miles away from your school location? What is the advantage and disadvantage of each method?

Interesting Web Sites for More Information

www.blackbox.com

www.howstuffworks.com

www.microsoft.com

www.novell.com

www.siemons.com

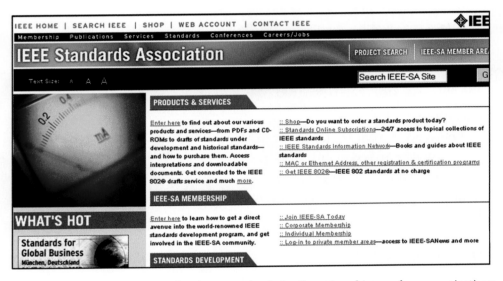

The IEEE Standards Organization develops standards for the networking and communications industry. You will encounter many IEEE standards in your study of networking fundamentals. To learn more about the IEEE organization, visit the www.IEEE.org Web site.

Chapter 1
Laboratory Activity

Identifying a Workstation's IP Configuration Settings

After completing this laboratory activity, you will be able to:
- Explain the purpose of an assigned IP address.
- Explain the purpose of an assigned subnet mask.
- Explain the function of the DHCP service.
- Determine when it is appropriate to use the **winipcfg** command and when to use the **ipconfig** command.
- Identify a workstation's IP configuration settings.

Introduction

This laboratory activity is important because it provides you with the basic information needed to reconfigure the IP configuration settings of your assigned workstation. You will learn how to check your workstation's IP configuration settings using the IP Configuration utility. You will record these settings so that they are available in case you need to restore your workstation to its original condition after you have completed a lab. This lab activity does not require an in-depth understanding of the settings you will record. For now, you only need to know how to reenter the settings. Later in the course, you will learn more about each of the settings in the configuration.

You may need an account set up to access the workstation's IP configuration settings. If this is the case, the instructor will provide you with a user name and password. Some configuration settings require administrative privileges to access and change configuration settings. Administrative privileges refer to a user having an account equal to the system administrator on the local computer or the network. Since the lab area you will be using is set up to train you as a network support technician, you will need administrative privileges for many of the lab activities.

On a separate sheet of paper, record your assigned user name, password, the number of your assigned workstation, and the server you are to log on to. The server name may or may not be required for network logon. You will most likely need to record a server name if there is more than one server used by students. The default server chosen for log on could be a different server than the one used for the network fundamentals class, or there may not be a server specified at all.

User name: _____

Password: _____

Workstation number: _____

Server name: _____

To access the workstation's IP configuration settings, you will issue a command either at the command prompt or in the **Run** dialog box. When using Windows 2000 or Windows XP, you will issue the **ipconfig** or **ipconfig /all** command at the command prompt. The following is a typical result of issuing the **ipconfig** command.

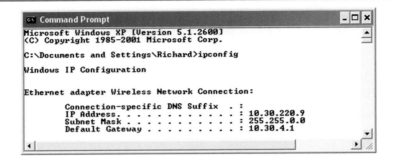

When **ipconfig /all** is issued, more detailed information appears. The following is a typical result of issuing the **ipconfig /all** command.

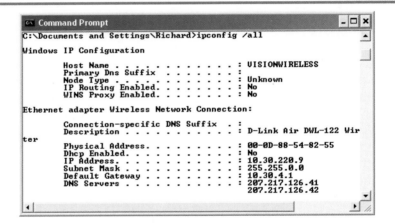

When using a Windows 98 or Windows Me system, you must access the **Run** dialog box through the **Start** menu. In the **Run** dialog box, type and enter **winipcfg**. The following is the result of running the **winipcfg** command.

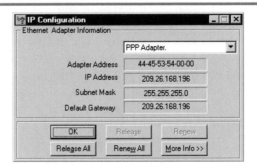

Take special note of the **More Info** button at the lower-right corner of the display. When selected, more information is revealed about the workstation's IP configuration.

The **winipcfg** and **ipconfig** commands can be used as troubleshooting tools to diagnose connection problems on a TCP/IP-based network. In a TCP/IP-based network, computers require an IP address to identify each other. Every computer must have a unique IP address. The IP address is presented as a set of four numbers separated by a period, such as 192.168.000.12. Each computer should have a unique IP address. Each computer must also have a subnet mask. A subnet mask is also a set of four numbers separated by a period, such as 255.255.255.000. Do not be surprised if the subnet mask on your workstation matches the example given. The majority of network subnet masks you will encounter on small networks will be 255.255.255.000. The subnet mask is used to identify the network classification. At this point, you do not need to know any more detail about the subnet mask. The subnet mask will be covered later in the course.

A workstation may be configured as a DHCP client. A DHCP client automatically receives an IP address and other TCP/IP settings from a DHCP server. The DHCP server issues an IP address from a list of IP addresses. There is no guarantee that the same IP address will be issued to a workstation.

To determine if a workstation is configured to receive an IP address from a DHCP server, check if DHCP is set to enabled or if a DHCP server address is listed in the display after running the **winipcfg** command and clicking **More Info** or running the **ipconfig /all** command. In the display, you will also see other assigned IP addresses, such as that for the default gateway, primary and secondary WINS server, and DNS servers. These settings are not always required for the workstation because these services may not be available. It depends on how the network is configured and if the network has Internet access. Detailed information about these services is well beyond the scope of this first laboratory activity. Each of these services is covered in detail later in the course. The main purpose of this laboratory activity is to help you restore the workstation to its original configuration.

Equipment and Materials

■   Assigned workstation—preferably with Windows XP Professional installed as part of a network domain.

Procedures

1. _____   Report to your assigned workstation.

2. _____   Boot the workstation and look for the logon screen. There may not be a logon screen for the workstation at this time. If the workstation is part of a network with a server, you will see a logon dialog box requesting a user name and password and possibly a server to log on to. If the workstation is part of a peer-to-peer network or simply a standalone computer, you will not see a network logon dialog box. The workstation, however, may prompt you for a password if the workstation is set up for multiple users. Your instructor will identify the type of system you will be using.

3. _____ Identify which operating system is installed on the workstation. You will need this information to determine which IP Configuration utility command to use.

4. _____ If the system you are using is Windows 2000 or Windows XP, open the command prompt from the **Start** menu and enter the following command: **ipconfig /all**. If the system you are using is Windows 95, Windows 98, or Windows Me, access the **Run** option from the **Start** menu and enter **winipcfg** in the **Run** dialog box. Click the **More Info** button to reveal more information about the IP configuration settings.

5. _____ On a separate sheet of paper, record the information that has been made available. For example, you may record the following information:

IP address: _____

Subnet mask: _____

Default gateway: _____

DHCP enabled (yes/no): _____

Primary WINS server IP address: _____

Secondary WINS server IP address: _____

DNS server addresses: _____

6. _____ After recording the information, answer the review questions.

7. _____ When you have answered the review questions, shut down your workstation and return all materials to their proper place.

Review Questions

Answer the following questions on a separate sheet of paper. Please do not write in this book.

1. What command is issued from the command prompt on a Windows XP workstation to reveal the assigned IP address?

2. What command is issued from the **Run** dialog box on a Windows 98 workstation to reveal the assigned IP address?

3. What is the purpose of an IP address?

4. What is the purpose of a subnet mask?

5. What is the purpose of the DHCP service or server?

6. Identify what command or program (**ipconfig** or **winipcfg**) will run the IP Configuration utility in the following operating systems. Record your answer (**ipconfig** or **winipcfg**) on a separate sheet of paper.

   A. Windows XP: _____

   B. Windows 2000: _____

   C. Windows Me: _____

   D. Windows 98: _____

2

Network Media— Copper Core Cable

After studying this chapter, you will be able to:

❏ Match the five forms of electronic signals to the media types on which they travel.

❏ Describe the major differences between an analog and a digital signal.

❏ Describe the two methods of data transmission: Broadband and Baseband.

❏ Define simplex, half-duplex, and full-duplex communication.

❏ Define electronic terms such as *impedance, reflected loss,* and *crosstalk.*

❏ List the characteristics of the 802.3 classifications.

❏ List the characteristics of the 802.5 classifications.

❏ Describe the various types of wiring faults.

Network+ Exam—Key Points

To prepare for the Network+ Certification exam, you must be able to specify the characteristics of the IEEE 802.3 and 802.5 classifications. These characteristics include:

■ Cable type.

■ Connectors.

■ Topology.

■ Data rates.

■ Maximum and minimum cable and segment lengths.

The Network+ Certification requirements are constantly changing. Always check the www.comptia.org Web site for the latest information!

Key Words and Terms

The following words and terms will become important pieces of your networking vocabulary. Be sure you can define them.

1000BaseCX
1000BaseT
100BaseT4
100BaseTX
10Base2
10Base5
10BaseT
amplifier
amplitude
analog signal
attenuation
AWG rating
bandwidth
Baseband
Broadband
Carrier Sense Multiple Access with
    Collision Detection (CSMA/CD)
coaxial cable
crossed pair
crosstalk
decibel (dB)
digital signal
Equal Level Far-End Crosstalk
    (ELFEXT)
Far-End Crosstalk (FEXT)

frequency
full-duplex
ground
half-duplex
impedance
industrial field bus technology
interference
latency
magnetic induction
Near-End Crosstalk (NEXT)
network media
noise
open
plenum-rated
reflected loss
resistance
reversed pair
RG-58
RG-8
short
simplex
split pair
terminating resistor
Time to Live (TTL)
twisted pair

Whether you are troubleshooting or designing a network, you must be familiar with the various types of network media. This chapter presents a general overview of network media and its electrical characteristics and then covers in detail copper core cable. In the following chapters, fiber-optic and wireless network media will be covered respectively. To understand the characteristics and limitations of copper core cable, you will first be introduced to some electronic terminology. Later, the IEEE standards covered in Chapter 1—Introduction to Networking will be revisited, specifically the 802.3 and 802.5 standards. You will see how these standards relate to network design and will understand the importance of knowing their specifications.

Network Media

network media
a general term for all forms of pathways that support network communications.

Network media is a general term for all forms of pathways that support network communications. There are two general classifications of network media: cable-based and wireless, **Figure 2-1.** Cable-based network media comprises copper core cabling, such as coaxial and twisted pair, and glass/plastic core cabling, such as fiber-optic. Wireless network media comprises air, or the atmosphere.

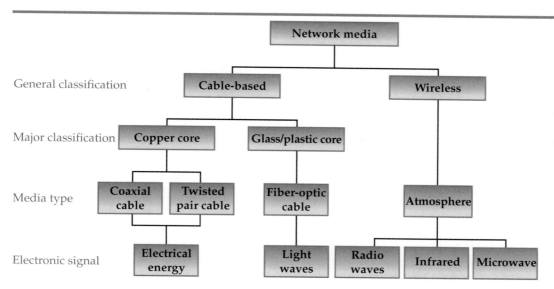

All network communications originate from electronic signals. Electronic signals can take many forms, such as electrical energy, light waves, radio waves, infrared light waves, and microwaves. For example, electronic signals are sent through coaxial and twisted pair cable as pulsating electrical energy; through fiber-optic cable as pulsating light waves; and through the atmosphere as pulsating radio waves, infrared light waves, or microwaves.

To understand the terminology related to network media and network communications, you must first learn some basics about electronic signals. In the following section we will look at the characteristics of two general types of electronic signals: analog and digital.

Analog and Digital Signals

Electronic signals are classified into two general types: analog and digital. An *analog signal* is an electronic signal that varies in values, such as a light that varies in intensity. A *digital signal* is an electronic signal that has discrete values. Discrete values can be on and off, high and low, or 5 volts and 0 volts.

The major difference between analog and digital signals can be seen in the shape of the waveform. Look at **Figure 2-2.** The analog and digital signals of electrical energy are plotted on a simple graph that compares voltage levels to time. Note that the analog signal forms a series of slopes whereas the digital signal rises and falls sharply at right angles. In the analog signal, the degree of the slope is related to the varying degrees of voltage over time, whereas the digital signal holds to discrete voltages (0 volts or +5 volts) over a length of time.

Each type of electronic signal has a cycle, or a pattern of fluctuations. Look again at Figure 2-2 and note the cycle for an analog and digital signal. In this example, a cycle for an analog signal begins at 0 volts, rises to +5 volts, falls to –5 volts, and rises to 0 volts. A cycle for a digital signal begins at 0 volts, rises sharply to +5 volts, and then falls sharply to 0 volts.

Frequency

An electronic signal can be measured by its frequency. *Frequency* is the number of cycles of an electronic signal that occur in 1 second. It is measured

analog signal
an electronic signal that varies in values.

digital signal
an electronic signal that has discrete values.

frequency
the number of cycles of an electronic signal that occur in 1 second, measured in Hertz (Hz).

Figure 2-2.
Analog and digital signals plotted on a simple graph. Note that time affects the degree of the slope in an analog signal and the duration of the voltage level in a digital signal.

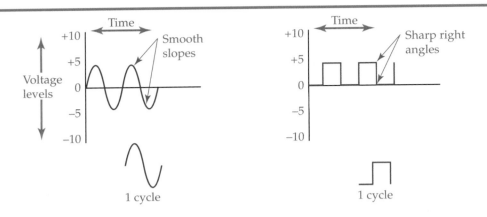

attenuation
the loss of signal strength.

amplitude
the maximum voltage, or height, of an electronic signal.

amplifier
an electronic device designed to raise a signal's amplitude.

interference
an undesired electromagnetic signal imposed on a desired signal that distorts or corrupts the desired signal.

noise
electromagnetic interference.

crosstalk
interference that comes from neighboring conductors inside a wire's insulating jacket.

in Hertz (Hz). Look at the analog and digital signal in **Figure 2-3** as plotted on a graph showing voltage and time. In each graph, the time interval is set to 1 second. The first graph shows an analog signal that achieves 5 cycles in 1 second. The analog signal, therefore, has a frequency of 5 Hz (5 cycles / 1 second). The second graph shows a digital signal that achieves 5 cycles in 1 second. The digital signal also has a frequency of 5 Hz (5 cycles / 1 second). If a cycle for an electronic signal is repeated 100 times in 1 second, the signal has a frequency of 100 Hz (100 cycles / 1 second).

Attenuation

All electronic signals degenerate, or lose amplitude, over long distances. Losing signal strength is referred to as **attenuation**. As an electronic signal travels across the media, the amplitude of the signal is lessened, **Figure 2-4**. *Amplitude* is the maximum voltage, or height, of an electronic signal. The amplitude can be augmented to its original level with an amplifier. An *amplifier* is an electronic device designed to raise a signal's amplitude.

Interference

In addition to attenuation, most electronic signals pick up interference from sources such as motors, fluorescent lights, transformers, radio transmitters, and from other conductors carrying electronic signals. *Interference* is an undesired electromagnetic signal imposed on a desired signal that distorts or corrupts the desired signal. Electromagnetic interference is generally referred to as *noise*. Interference that comes from neighboring conductors inside a wire's insulating jacket is called *crosstalk*. Crosstalk is covered later in detail.

Figure 2-3.
5 Hz frequencies of an analog and a digital signal. On each graph an analog and digital cycle occurs 5 times in 1 second.

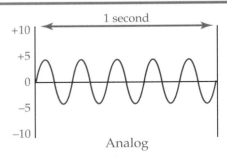

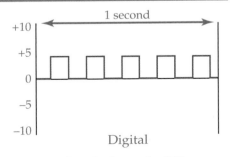

Input signal   Output signal

Figure 2-4.
Attenuation is the loss of signal strength as the signal travels along a conductor.

Noise and crosstalk are unwanted signals that need to be removed from the desired signal, especially before it is amplified. When an analog or digital signal is amplified, the unwanted signals caused by interference are also amplified. To get rid of the unwanted signals, the signal must be filtered. *Filtering* electronic signals is the removal of unwanted signals.

Amplifiers often incorporate filters. The filter removes the unwanted signals before amplification so when the signal is amplified, it is an exact duplicate of the original signal. Networks use repeaters to amplify and filter signals. A repeater is not actually an amplifier. However, in the networking environment, it is often referred to as an amplifier because a repeater counters attenuation by reshaping the signal to its original form.

While both analog and digital signals can be used to express data in network communications, the digital signal is preferred. It is very difficult and sometimes impossible to filter out all of the unwanted signals from an analog signal without changing the analog signal. Amplifying an analog signal not only amplifies the original signal, but also the unwanted signals, which results in a signal that no longer represents the original. However, when a digital signal is amplified, the unwanted signals are more easily filtered to restore the original look of the original signal.

Digital signals have a square or rectangular waveform and are produced at a fixed frequency. When digital signals are regenerated, they are regenerated by digital devices that also produce a square or rectangular waveform. A typical signal from a radio station is an analog signal composed of various amplitudes and a mixture of frequencies inside a set range of frequencies. Because of the various frequencies the radio signal comprises, the interference is sometimes impossible to separate or filter. This is a major drawback of using analog signals. Since interference can be easily removed from digital signals, the digital signal has become the norm for transmitting data over network cables.

Look at **Figure 2-5.** At the top is the original signal in digital and analog form. These signals represent the transmitted data. Below each signal (middle) is a representation of the same signal after attenuation and added interference. The bottom drawings are a representation of the distorted signals after amplification and filtering. Notice that the analog signal is distorted after amplification and is not an exact match of the original signal. The digital signal, however, looks exactly like the original signal after amplification. This is because the repeater is only capable of reproducing the original signal as 5 volt and 0 volt levels. Interference consists of many different levels of analog voltage (amplitude) and cannot be reproduced by the repeater because the repeater only generates two levels of signal strength.

Excessively strong interference, however, can be impossible to filter from a digital signal. Interference strength is directly related to the distance from the digital signal and the power of the source producing the interference. For example, a portable telephone may interfere with a digital signal at a close

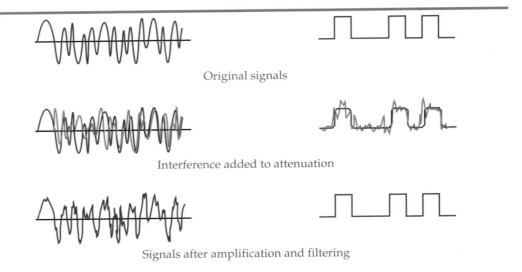

Original signals

Interference added to attenuation

Signals after amplification and filtering

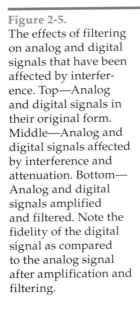

Figure 2-5.
The effects of filtering on analog and digital signals that have been affected by interference. Top—Analog and digital signals in their original form. Middle—Analog and digital signals affected by interference and attenuation. Bottom—Analog and digital signals amplified and filtered. Note the fidelity of the digital signal as compared to the analog signal after amplification and filtering.

proximity of five to ten feet but not generate sufficient signal strength at thirty feet to interfere with the digital signal.

Latency

latency
the amount of time it takes a signal to travel from its source to its destination.

Latency is the amount of time it takes a signal to travel from its source to its destination. A general cause of latency is network equipment along the signal's path. Some networking equipment regenerate, or reshape, a digital signal after it is received. For example, a repeater is used to extend the maximum distance that a signal can normally travel by reshaping the signal. A digital signal may travel through several repeaters before reaching its destination, thus slowing the time it takes the signal to reach its destination.

There is normally a maximum amount of time a packet is allowed to circulate through a network before it is destroyed. If the overall latency is too great, the packet is dropped from the network. The maximum amount of time allowed is referred to as the *Time to Live (TTL)*. If a signal exceeds the allocated TTL, it is removed from the network to prevent it from circulating forever. If signals could circulate forever, they would accumulate and eventually cause the entire system to shut down.

Time to Live (TTL)
a maximum time allowed for an electronic signal to circulate a network.

Data Transmission

Data transmission terminology and concepts are often referenced in network standards and equipment specifications. A good understanding of data transmission terminology will help you master some of the most difficult concepts of networking. Much of the original network terminology has its roots in the field of electronics. Electronic terminology often requires an extensive background in electronic theory to fully grasp the concept being introduced. As networking has evolved over time, network technicians have coined their own terminology and have used electronic terms in some contexts that have slightly changed the original meaning of the term. Several of these terms and their influence on the data rate of data transmission are introduced in this section.

Bandwidth

Bandwidth is a measurement of the network media's ability to carry data. The definition of bandwidth varies according to the context to which it is applied. For example, when referring to a network cable, bandwidth refers to the amount of digital signal the cable can carry based on a given time. Network cable bandwidth is measured in bits per second (bps).

The term *bandwidth*, when applied to network cabling, can be confused with the term *frequency*. Technically speaking, frequency is a measurement of the network media's ability to carry an electronic signal. For example, a network cable can have a frequency rating of 250 MHz but a permissible bandwidth of 1000 MHz. The frequency rating is the rating of the individual wires inside the cable assembly; the bandwidth is the entire cable assembly's (4 pairs of conductors) ability to carry digital signals. In general, the higher the bandwidth, the more data that can be carried. Please take note of the fact that bandwidth does not always match data rates because data rates can be influenced by data compression techniques.

When referring to an analog signal, the term *bandwidth* means a measurement of the maximum frequency of a device or the total range of analog frequencies. Bandwidth in analog transmission is measured in hertz (Hz). A range of frequencies would be all the analog signal frequencies between two specific frequencies. An example would be television or radio analog signals transmitted and identified as channels. A typical television channel is assigned a 6 MHz bandwidth.

In analog transmission, bandwidth can support several different frequencies at once while in digital transmission bandwidth usually supports only one. In other words, when data is transmitted over a cable using a digital signal, the digital signal uses the entire bandwidth. When data is transmitted over a cable using analog signals, many different analog signals, or frequencies, representing data can be transmitted at once. See **Figure 2-6.**

bandwidth
a measurement of the network media's ability to carry data.

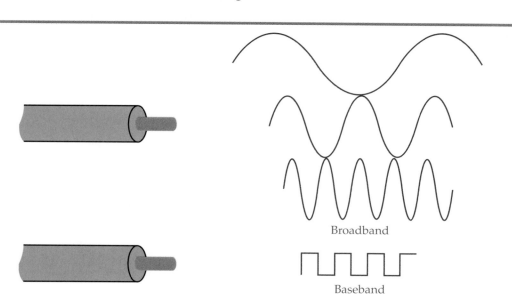

Broadband

Baseband

Figure 2-6.
Baseband transmission allows only one digital signal to be transmitted at a time. The digital signal uses the entire bandwidth of the network media. Broadband transmission allows multiple analog signals to be transmitted simultaneously, each of differing frequencies and amplitudes.

Baseband

Baseband is a method of transmitting data in the form of a digital signal, using the entire bandwidth of a cable. Remember that the bandwidth of a network media is its total capacity to transmit data and that a digital signal uses the entire bandwidth. The network media may be able to support more than one frequency, but in application, it is typically used to carry only one chosen frequency. For example, a copper core cable may be used to transmit several analog signals, or frequencies, at once, but when used as part of an Ethernet network, will transmit digital signals at only one frequency. This frequency will consume the entire bandwidth of the cable.

You cannot send two digital signals on the same wire because they have matching frequencies and voltage levels. Look at **Figure 2-7.** It illustrates the effects of transmitting more than one digital signal on a wire. Each digital signal consists of a pattern of high and low voltage levels which represent data such as letters and numbers. When both signals are placed on one wire, the two signals combine to form a new digital pattern. The new digital pattern of the combined signals cannot be separated into two individual signals, thus all data is distorted and cannot be read at its destination.

Broadband

Broadband is a method of transmitting data in the form of several analog signals at the same time. Certain electronic techniques can be incorporated into communication systems that allow network media to carry more than one frequency at a time. In general, a network conductor that carries a single series of digital signals is Baseband. A conductor that carries two or more analog signals is Broadband.

Wireless media, such as radio, is an example of Broadband transmission. In this case, the atmosphere is capable of carrying many different radio frequencies at the same time. For example, television signals range from 7 MHz to 1002 MHz. Inside this range of frequencies are individual channels specified in smaller ranges. For example, Channel 8 has a frequency range of 180 MHz to 186 MHz. Channel 9 has a frequency range of 186 MHz to 192 MHz. Each of the channels has a bandwidth of 6 MHz inside the allocated television bandwidth of 7 MHz to 1002 MHz. Wireless networks use radio waves as the network medium and use the terms *bandwidth* and *channel* when describing each radio signal path.

Figure 2-7.
Only one digital signal can be carried on a network cable wire. If two signals are sent on the same cable, the two would combine to form a new signal. The new signal cannot be filtered to restore the original signal.

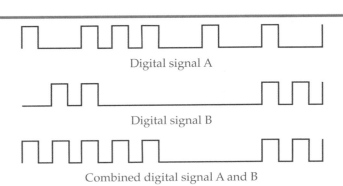

Digital signal A

Digital signal B

Combined digital signal A and B

Simplex, Full-Duplex, and Half-Duplex Communication

Communication between two electronic devices can occur in one of three modes: simplex, full-duplex, and half-duplex. See **Figure 2-8**.

Simplex means that communication occurs in one direction only. An example of simplex communication is the transmission that occurs between a television station and a television. *Full-duplex* communication is the bi-directional communication that occurs between two devices simultaneously. An example of full-duplex communication is communications via telephone. *Half-duplex* communication is also bi-directional communication; however, it can only occur in one direction at a time. Half-duplex communication is used with walkie-talkies. Also, most LANs use half-duplex communication. Ethernet is an excellent example. In an Ethernet network, only one computer or node can successfully transmit data over the network cable at a time. If two computers transmit data at the same time, a collision occurs and the data is destroyed.

Electronic Terms

Networks that incorporate copper cable have limits to the type of cable, length, number of segments, and the number of nodes that can be connected. These limits are based on certain electrical characteristics. As you read about the copper cable types and the IEEE 802 classifications, you will learn what these limitations are. However, there are several basic electronic terms you must first master to fully understand these limitations. These terms are *resistance, impedance, reflected loss,* and *crosstalk.*

Direct Current and Alternating Current

The flow of electrical energy is described as either direct current (DC) or alternating current (AC). The term *current* describes the flow of electrons, which is the actual form of electrical energy. DC means that the current flow is in one

simplex
a communication mode in which communications occur in one direction only.

full-duplex
a communication mode in which bi-directional communication occurs simultaneously.

half-duplex
a communication mode in which bi-directional communication occurs in one direction at a time.

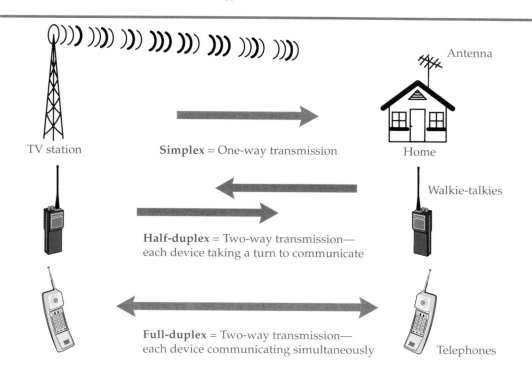

TV station

Simplex = One-way transmission

Antenna

Home

Walkie-talkies

Half-duplex = Two-way transmission—
each device taking a turn to communicate

Full-duplex = Two-way transmission—
each device communicating simultaneously

Telephones

Figure 2-8.
Examples of simplex, half-duplex, and full-duplex communication.

steady direction. One of the most common sources of direct current is a battery. Battery electrical energy flows in a steady direction from the negative (–) terminal to the positive (+) terminal. AC does not flow in one direction. It alternates from a negative charge to a positive charge. The most common source of alternating current is the wall outlet.

Since the two forms of electrical energy flow differently, they produce different electrical characteristics. Because the two systems have different characteristics, different terminology is used to describe their characteristics. Two terms applied to these characteristics are *resistance* and *impedance*. These terms are inherited from the electronics industry and are applied to networking cable.

Resistance and Impedance

resistance
the opposition to direct current (DC) in a conductor.

Resistance is the opposition to direct current (DC) in a conductor. Every conductor has a certain amount of resistance that affects current flow—the longer the conductor, the greater the resistance.

impedance
the opposition to alternating current (AC) in a conductor.

Resistance is a DC term and should not be confused with **impedance,** which refers to the opposition to AC. Impedance increases as frequency increases. In other words, the higher the frequency, the faster the speed of data transmission. However, the quality or integrity of the data will diminish because of impedance. This is why there are limitations to the speed at which data can be delivered across a conductor. The length of a conductor also influences impedance. As the length of a conductor increases, so does the total impedance.

Digital signals change current flow direction just as analog signals do. Digital signals consist of a series of square or rectangular waveforms which sharply change current flow direction after a set time period. For this reason, digital signals are affected by impedance the same way analog signals are affected by impedance. See **Figure 2-9** for a summary of resistance and impedance.

Reflected Loss

In high-speed networks, data is transmitted in full-duplex mode. This means that data can flow in both directions at the same time. The original signal enters the cable and travels to the end. When it reaches the end, part of the signal is reflected. **Reflected loss** is the amount of signal reflected from the end of the cable. If it is of sufficient value, the reflected signal can disrupt communications.

reflected loss
the amount of signal reflected from the far end of the cable.

Figure 2-9.
Comparison of resistance and impedance.

Electrical Term	Definition	Symbol	Comments
Resistance	Opposition to direct current.	Ω (ohm)	Current flow affected by cable length. Resistance is present in battery-powered circuits.
Impedance	Opposition to alternating current.	Z	Current flow is affected by frequency, cable length, induction, and capacitance. Impedance is present in AC-powered circuits.

Crosstalk

Crosstalk is a type of interference that occurs when one pair of conductors imposes a signal on another pair of parallel conductors, **Figure 2-10.** All energized conductors are surrounded by a magnetic field, Figure 2-10A. The magnetic field runs the length of the conductor. The strength of the magnetic field is determined by the amount of current traveling through the conductor. Frequency also affects the strength of the magnetic field. When another conductor is placed in close proximity to a current carrying conductor, the magnetic field encircles the other conductor and induces current flow, Figure 2-10B. This electronic phenomenon is referred to as *magnetic induction,* or *mutual induction.* When a conductor in a cable assembly carries digital pulses, the digital pulses create a magnetic field pattern similar to the digital pulses. The magnetic field induces in a neighboring conductor an electrical current resembling the digital pulse pattern of the original conductor, thus creating crosstalk.

Analog telephone systems also suffer from crosstalk. In fact, the term *crosstalk* originated in telephone communication to describe the effect of hearing an additional telephone conversation from a neighboring telephone line while talking to someone else.

Network conductors, such as twisted pair, are designed to limit the effects of crosstalk by reducing the amount of contact between a pair of conductors, Figure 2-10C. This is why twisted pair must be used in place of the older style of telephone cable that has no twist. However, one place where crosstalk can be generated on twisted pair cable is at the ends near the connectors. When cables are made by hand, the technician may leave too much untwisted conductor near the connector, leaving the cable vulnerable to crosstalk.

Figure 2-11 shows a twisted pair cable properly connected to a connector and a twisted pair cable improperly connected to a connector. Note that the improperly connected twisted pair cable has too much wire exposed and that the pairs are not twisted. The improperly connected cable will produce crosstalk.

Twisted pair cable segments are limited to 100 meters (328 ft.) maximum in length. Longer segment lengths produce crosstalk, as well as a reduction in

magnetic induction
an electrical phenomenon in which the magnetic field encircling a current carrying conductor induces a current flow in a conductor of close proximity.

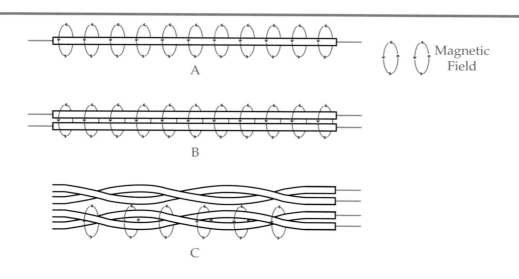

Figure 2-10.
The occurrence of crosstalk. A—An energized conductor is surrounded by a magnetic field. B—Any conductor placed near the energized conductor and in the magnetic field will have an electrical energy transferred to it. C—Twisting the pairs of conductors reduces the amount of electrical energy transferred between pairs of conductors.

Figure 2-11.
Example of a twisted pair cable properly connected to a connector and a twisted pair cable improperly connected to a connector. Note the exposed, untwisted pairs of wire in the improper connection.

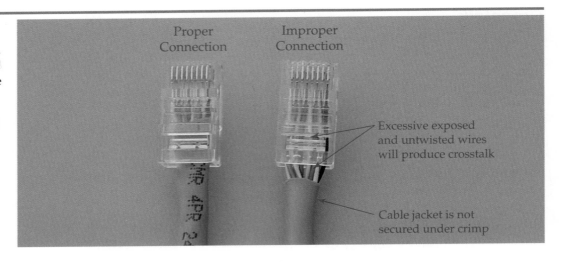

Proper Connection   Improper Connection

Excessive exposed and untwisted wires will produce crosstalk

Cable jacket is not secured under crimp

signal strength. Cable shielding of wire mesh or foil also reduces crosstalk. The shielding absorbs and contains the magnetic field generated by a conductor and protects the pairs from electromagnetic interference generated by an outside source.

Three types of measurement can be taken on twisted pair cable to measure the effects of crosstalk: Near-End Crosstalk (NEXT), Far-End Crosstalk (FEXT), and Equal Level Far-End Crosstalk (ELFEXT). These measurements are taken by a manufacturer to determine cable specs so that the cable can be correctly labeled. They are only taken in the field to see if an existing Category 5 cable could be used as a Category 5e cable.

Near-End Crosstalk (NEXT)
a measurement of the reflected loss at the near-end, or input end, of a cable.

decibel (dB)
a unit of measurement used to express the relationship of power between two electrical forces.

Near-End Crosstalk (NEXT)

Near-End Crosstalk (NEXT) is a measurement of the reflected loss at the near end, or input end, of a cable. Look at **Figure 2-12.**

Losses are typically expressed in decibels. A *decibel (dB)* is a unit of measurement that expresses the relationship of power between two electrical forces. This measurement is often used to show the output-to-input ratio of a signal. In the case of twisted pair cable, it is used to compare the difference between the amount of power in the input signal and the amount of power generated by crosstalk. In the case of networking fundamentals, it is the relationship of the input signal compared to the crosstalk signal.

Figure 2-12.
Near-End Crosstalk (NEXT) is a measurement of crosstalk taken at the near end, or input end, of a cable.

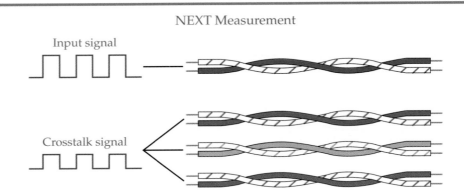

NEXT Measurement

Input signal

Crosstalk signal

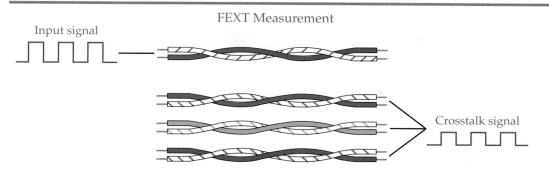

FEXT Measurement

Input signal

Crosstalk signal

Figure 2-13.
Far-End Crosstalk
(FEXT) is a measure-
ment of crosstalk taken
at the far end, or output
end of the cable.

Far-End Crosstalk (FEXT)

As the name implies, *Far-End Crosstalk (FEXT)* is a measurement of reflective loss at the far end, or output end, of the cable, **Figure 2-13.** Far-end losses are also expressed in decibels.

Equal Level Far-End Crosstalk (ELFEXT)

Equal Level Far-End Crosstalk (ELFEXT) is a measurement calculated by subtracting the effects of attenuation from the FEXT. ELFEXT negates much of the attenuation by raising the attenuated signal by a value equal to the attenuation. This brings the attenuated signal back to the desired level. This is what the term *ELFEXT* implies by "equal level." It equalizes the loss with a gain.

Copper Core Cables

Copper core cables are commonly constructed of a center core of copper surrounded by an insulating jacket. All copper core cables have an AWG rating. The *AWG rating* describes the size of a conductor's diameter. The abbreviation AWG represents American Wire Gauge, which is the accepted standard for specifying the size of a conductor. The size of a conductor is expressed as a numeric value such as 22, 20, and 18—the smaller the number, the larger the diameter of the wire.

Figure 2-14 shows a wire gauge. A wire gauge is used to determine the AWG rating or the size of a conductor. Notice that the largest wire measurement on the

Far-End Crosstalk (FEXT)
a measurement of reflective loss at the far end, or output end, of the cable.

Equal Level Far-End Crosstalk (ELFEXT)
a measurement calculated by subtracting the effects of attenuation from the FEXT.

AWG rating
a rating that describes the size of a conductor's diameter.

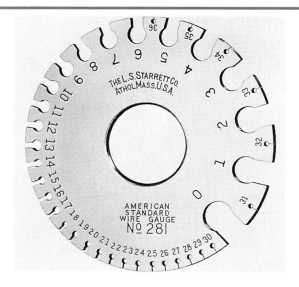

Figure 2-14.
Note the AWG ratings
on this wire gauge.
The AWG rating of 0
indicates the largest
wire size on the gauge,
whereas the AWG rat-
ing of 36 indicates the
smallest wire size on
the gauge.

gauge is marked by the AWG rating 0, and the smallest wire measurement on the gauge is marked by AWG rating 36.

Cables are often identified as plenum-rated. *Plenum-rated* means that the cable has a special type of insulation that will not give off toxic gases should the cable be consumed by fire. The term *plenum-rated* is derived from the plenum in a building. The plenum is the area above the drop ceiling and under a raised floor, **Figure 2-15.** Cables designed to pass through a building plenum must be plenum-rated to ensure they will not give off toxic gases in case of a fire.

Two types of copper core cable are used in networking: coaxial and twisted pair. Twisted pair was derived from the telephone industry. Twisted pair was not originally designed for high frequencies. Voice signals are relatively low frequency compared to other forms of electrical signals, such as radio, and did not require cable designed for high frequencies. Also, the original networks ran at only 1 Mbps to 4 Mbps. Early versions of telephone communications cables worked fine for these applications.

Coaxial cable was designed to carry high frequency signals and was first used in high-frequency radio communications to conduct a radio signal to and from an antenna. Coaxial cable was also used in early networks. It is used to some degree today because of its shielding effect. We will explore more about these two types of cables in the next section.

Coaxial Cable

Coaxial cable, or *coax,* consists of a copper core conductor surrounded by an insulator, **Figure 2-16.** The insulator is covered with a shield. The shield is made of either solid foil or braided wire. The shield protects the copper core conductor from electromagnetic interference, which would corrupt the transmitted data. An insulating jacket protects the shield. Coaxial cable is difficult to work with and relatively expensive when compared to some of the other cable-based media.

A copper core conductor can be either solid or stranded. A solid copper core conductor is stiffer and more difficult to work with than a stranded copper

plenum-rated
a rating given to a cable that has a special type of insulation that will not give off toxic gases should the cable be consumed by fire.

coaxial cable
a cable consisting of a copper core conductor surrounded by an insulator.

Figure 2-15.
The plenum area of a building is located above the drop ceiling and under a raised floor.

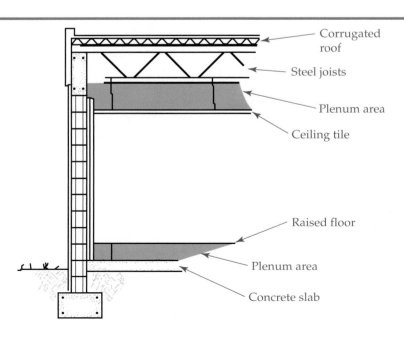

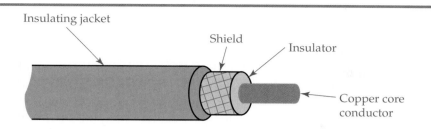

Figure 2-16.
Coaxial cable layers.

core conductor. However, neither is as flexible as today's networking cables. Another interesting fact is a stranded copper core conductor has the ability to carry more electrical energy (current) than a solid copper core conductor of equal dimensions. Stranded copper core coaxial cable, therefore, can contribute to a better quality of signal than solid copper core coaxial cable. Many times, the most important factor in determining which cable type to use is cost. Solid copper core coaxial cable is less expensive than stranded copper core coaxial cable. This is true for all copper cable types manufactured. The exact type used typically varies according to installer preferences or the specifications of the installation.

There are several classifications of network coaxial cable. Coaxial cable used for television is not acceptable for use as a network cable. Its characteristics work well for television transmission but will cause problems if used for computer networks. **Figure 2-17** lists some of the most common types of coaxial cable and their application. The two most common coaxial cable types that have been used for networking are RG-8 (thicknet) and RG-58 (thinnet).

RG-8, or *thicknet*, is a very rigid coaxial cable. It has a solid copper conductor in the center, which adds to the rigidity of the cable and to its overall thickness. RG-8 allows for longer segments than other copper core cables, but because of the difficulty in handling the cable and its wide diameter, it severely limits the number of cables that can be placed inside a conduit. Thicknet is not used today as a network media.

RG-58, or *thinnet*, is smaller in diameter and easier to work with than thicknet coaxial cable but does not carry a signal as far. RG-58 is still permitted but not recommended for use in new network installations.

RG-8
a thick, rigid cable also known as thicknet and used in a 10Base5 network.

RG-58
A thin, flexible cable also known as thinnet and used in a 10Base2 network.

Figure 2-17.
Coaxial cable types and their applications.

Cable	Common Name	Impedance	Actual Size in Diameter	Description
RG-6	Broadband	75 Ohms	.332	Used for cable TV.
RG-8	Thicknet	50 Ohms	.405	Used for Ethernet networks.
RG-11	Thick coax	75 Ohms	.475	Used for cable TV trunk lines.
RG-58	Thinnet	50 Ohms	.195	Used for Ethernet networks.
RG-59	CATV	75 Ohms	.242	Used for cable TV and sometimes used for ARCnet.
RG-62	Baseband	93 Ohms	.249	Used for ARCnet.

Twisted Pair

twisted pair
a type of cable consisting of four pairs of twisted conductors.

Twisted pair consists of four pairs of twisted conductors. Twisted pair cable has been available for many years and was first used by telephone companies to carry voice transmissions. Today, twisted pair is the most common choice of network cable. AWG wire sizes range from 18 to 26, with the AWG wire size of 24 used most often.

There are seven categories of twisted pair cable: Category 1 through Category 7. The categories are based on the physical design, such as the number of pairs or twists per foot, and the capabilities of the cable, such as the maximum frequency rating and the data rate. The maximum frequency rating and data rate are only two of the measurements of a cable's capabilities. Other measurements to consider are crosstalk, NEXT, and impedance.

Twisted pair cable can also be labeled as UTP (unshielded twisted pair) and STP (shielded twisted pair). Shielding can be applied over the entire cable assembly or over individual pairs of conductors. When shielding is applied to individual pairs of conductors, the shielding protects against crosstalk and outside sources of interference. The problem with shielding individual pairs of conductors is it makes the cable much harder to work with when applying connectors. This type of cable is also more expensive than twisted pair cable that does not shield the individual pairs of conductors.

For a complete listing of twisted pair categories, see the chart in **Figure 2-18.** Note that Category 5e is not a separate category but rather an addendum to the Category 5 specification.

Tech Tip

When referring to a category of twisted pair cable, it is common to use the term *Cat* for short. For example, it is common to use the term *Cat 5* instead of *Category 5*.

Category 1

Category 1 cable is limited to low frequency applications, such as voice signals. It consists of two conductors. The two conductors are not twisted. While this design is sufficient for electrical signals representing voice transmission, it is inadequate for computer networks.

Category 2

Category 2 cable consists of four pairs and has a maximum frequency rating of 1 MHz. This design is not acceptable for today's network systems. It was once used in networks that were limited to 4 Mbps. Today's networks run at a minimum of 10 Mbps.

Category 3

Category 3 cable has a maximum frequency rating of 16 MHz. It consists of four twisted pairs and three twists per foot. This cable type can be found on existing networks usually rated at 10 Mbps and 16 Mbps. It may also be found in many existing telephone installations.

Category 4

Category 4 cable has a maximum frequency rating of 20 MHz. It consists of four twisted pairs. This cable has a data rate of 16 Mbps. It is only a slight improvement over Category 3 cable.

Category	Type	Maximum Frequency Rating	Data Rate	Number of Pairs	Comments
Category 1	UTP	None	Less than 1 Mbps	2	Used for electrical signals representing voice transmission.
Category 2	UTP	1 MHz	4 Mbps	4	Used in earlier networks that were limited to 4 Mbps.
Category 3	UTP or STP	16 MHz	10 Mbps 16 Mbps	4	Can be found in existing networks rated at 10 Mbps and 16 Mbps and in some telephone installations.
Category 4	UTP or STP	20 MHz	16 Mbps	4	This cable type was only a slight improvement over Category 3.
Category 5	UTP or STP	100 MHz	100 Mbps 1000 Mbps (using 4 pairs)	4	Commonly used in 10BaseT and 100BaseTX network installations.
Category 5e	UTP or STP	100 MHz	100 Mbps 1000 Mbps (using 4 pairs)	4	This cable type is not a replacement for the Category 5 cable. It is an addendum to the cable classification.
Category 6	UTP or STP	250 MHz	NS	4	Similar in construction to 5 and 5e. No special shielding is required.
Category 7	UTP or STP	600 MHz	NS	4	Each of the four pairs of conductors is protected by foil, and then all pairs are protected by foil or metal braiding.

Figure 2-18.
Twisted pair categories and their characteristics. There are seven twisted pair categories. Category 5e is an addendum to Category 5, not a separate category.

Note: NS = No standard.

Category 5

Category 5 cable has a maximum frequency rating of 100 MHz. It is capable of data rates of 100 Mbps and 1000 Mbps (using four pairs). It is found commonly in 10BaseT and 100BaseTX networks.

Category 5e

Category 5e cable has a maximum frequency rating of 100 MHz. Category 5e is not a replacement for Category 5 cable but rather an addendum to the cable classification. With the growing need for cable that could support higher data rates, Category 5 standards were revised to Category 5e. Category 5e can support data rates up to 1000 Mbps using all four pairs of conductors. The improvements to Category 5 are based on Near-End Crosstalk (NEXT) and Equal Level Far-End Crosstalk (ELFEXT).

At the time of the introduction of Category 5e, any existing Category 5 cable was eligible to be reclassified as Category 5e if it met the requirements of the new standard. This would allow the use of Category 5 cable for newer network installations. Existing Category 5 cable could be tested to see if it met the Category 5e standard. If it met the standard, then it could be used. If not, it would require replacement. Some of the factors that determine the results of the tests are the AWG of the conductors and the insulation.

Category 6

Category 6 has a maximum frequency rating of 250 MHz. It is similar in construction to Category 5 and 5e, but no special shielding is required.

Category 7

Category 7 cable has a maximum frequency rating of 600 MHz. It uses a different cable construction to achieve a higher data rate. Category 7 is constructed of four pairs of twisted conductors with a protective foil or a conductive braid surrounding each pair. In addition to the individually covered pairs, there is an overall protective foil or conductive braid surrounding the complete assembly.

Two different classifications of cable can have the same maximum frequency rating, but as a standard, they will support two different data rates. An example of a cable having the same frequency rating is Category 5 and Category 5e. Both have a frequency rating of 100 MHz, but because Category 5e has a better reflected loss rating based on NEXT and FLEX, it can be used in networks with a higher data rate standard. What makes this section so confusing is that information about cable specifications can change between applications. The two main cable applications are voice (low frequency analog) and data (high frequency digital) signals. Also, different manufacturers sometimes post higher ratings for their cable than the minimum rating.

IEEE 802.3 and 802.5 Standards

In Chapter 1—Introduction to Networking, you learned that the Institute of Electrical and Electronic Engineers (IEEE) develops standards for the networking and communications industry. These standards begin with an 802 prefix, such as 802.1 for Internetworking and 802.2 for Logical Link Control. (See Figure 1-25.)

The IEEE 802 standards describe how a network should be operated and the limitations to be met. For example, the 802 standards specify the allowable data rate, access method, cable type, and topology that should be used. These standards not only help industry develop networking media and hardware that is compatible, they also aid those who are designing a network to select the appropriate hardware, media, topology, and cable lengths.

In this section, you will learn specifically about the 802.3 and the 802.5 standards. These standards describe Ethernet and Token Ring network specifications respectively and specify the use of copper cabling.

IEEE 802.3

The IEEE 802.3 standard is for Ethernet networks. Ethernet networks use a media access method called **CSMA/CD (Carrier Sense Multiple Access with Collision Detection)** to control and ensure data delivery. This is how CSMA/CD works. A workstation listens for data traffic on the network before transmitting data. When the network is silent, the workstation transmits data. If, however, another workstation submits data at the same time, the data collides on the

Carrier Sense Multiple Access with Collision Detection (CSMA/CD) a set of rules that define how two or more devices will access network media and how they respond after a collision occurs.

network. When this happens, the two workstations wait a random period before retransmitting the data. The random period is less than a fraction of a second. A typical network can transmit thousands of data packets in one second. Collisions usually remain unnoticed on a properly installed network. A poorly designed network may operate slowly due to too many collisions. As you see, one inherent problem with Ethernet communications is data collision on the network. When two data packets collide, they become corrupted and cannot be delivered.

The following will help you better understand CSMA/CD as a standard for network communication. The acronym CSMA/CD can be separated into three distinct pairs of letters/words that explain the operation requirements that must be met to match the CSMA/CD standard. Carrier Sense (CS) means that a network interface card can sense when data is transmitted on a network cable. Multiple Access (MA) means that the network cable provides multiple access at the same time to all network interface cards connected to the cable. Collision Detect (CD) means that the network interface card can tell when a collision has occurred and will automatically resend the damaged packet.

The IEEE 802.3 standard comprises various Ethernet classifications. These classifications differ by data rate, topology, and media type and are named with short descriptions, such as 10BaseT and 10Base2. The chart in **Figure 2-19** lists all of the IEEE 802.3 classifications along with the maximum and minimum segment lengths in meters, data rate, cable type, and the topology of each. The 802.3 classifications are divided into three categories: 10 Mbps, Fast Ethernet, and Gigabit Ethernet.

To help you decode the cryptic names of the classifications and to remember the specifications for each, look at **Figure 2-20.** In both examples, the *10* represents the data rate, which is 10 Mbps. The *Base* in these examples stands for Baseband, which means that the digital signal is transmitted at one frequency, and this frequency consumes the entire bandwidth of the cable. The last symbol in a cable classification is usually a number or a letter. A number indicates the

Figure 2-19.
IEEE 802.3 categories and classifications.

Category	Classification	Data Rate	Maximum Segment Length	Minimum Segment Length	Cable Type	Topology
10 Mbps	10Base2	10 Mbps	185 m	0.5 m	RG-58 (thinnet)	Bus
	10Base5	10 Mbps	500 m	2.5 m	RG-8 (thicknet)	Bus
	10BaseT	10 Mbps	100 m	0.6 m	Category 3, 4, and 5	Star
Fast Ethernet	100BaseT4	100 Mbps	100 m	0.6 m	Category 3, 4, and 5	Star
	100BaseTX	100 Mbps	100 m	0.6 m	Category 5	Star
Gigabit Ethernet	1000BaseCX	1000 Mbps	25 m	0.6 m	Category 5	Star
	1000BaseT	1000 Mbps	100 m	.6 m	Category 5e	Star

Note: 1000BaseCX is obsolete and is no longer recognized.

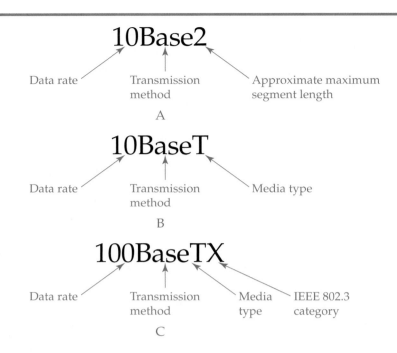

Figure 2-20.
Each IEEE 802.3
classification has an
identifying name
that reveals the
classification's data
rate, transmission
method, and either the
approximate maximum
segment length or
media type. Learning
how to decode the
cryptic names can help
you recall the charac-
teristics of the IEEE
802.3 classifications.

avpproximate maximum segment length in hundreds of meters, and a letter indicates the media type. For example, in Figure 2-20A, the 2 represents the approximate maximum length of a segment in meters. A 10Base2 segment can therefore have a maximum length of approximately 200 meters (2 × 100). The actual maximum length, however, is 185 meters, but the 2 serves as an approximate. The 10Base5 classification specifies a maximum segment length of 500 meters (5 × 100). In this case, 500 meters is the actual maximum segment length.

In Figure 2-20B, the last character is a letter and represents the media type—twisted pair. Sometimes a classification will have two characters at the end, such as *VG* in 100BaseVG or *TX* in 100BaseTX. In the case of 100BaseVG, the *VG* represents voice grade. However, in 100BaseTX, the *T* represents twisted pair and the *X* represents Fast Ethernet, Figure 2-20C.

The IEEE specification for 802.3 is over 1,000 pages long. You can download the complete specification from the IEEE Web site as an Adobe Acrobat file.

---

10 Mbps

Early networks were used primarily to transfer numeric or text data. A 10 Mbps transfer speed was more than adequate for most applications. The three classifications of 10 Mbps that were used are 10Base2, 10Base5, and 10BaseT.

10Base2

10Base2
an IEEE 802.3 classification that specifies the use of RG-58 cable, a data rate of 10 Mbps, a maximum segment length of 185 m, a minimum segment length of .5 m, and a bus topology.

The *10Base2* classification specifies the use of RG-58 (thinnet) cable and a data rate of 10 Mbps. **Figure 2-21** illustrates the physical characteristics of a 10Base2 network. Note that a 10Base2 network is configured in a bus topology and that the maximum segment length is 185 meters. This means that the segment length cannot exceed 185 meters. If it does, the signal strength will deteriorate due to attenuation and impedance. Also note that the minimum segment length between nodes is .5 meters.

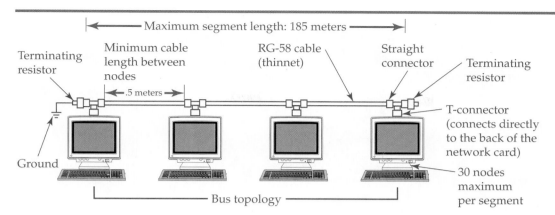

Figure 2-21.
Physical characteristics
of a 10Base2 network.

The network interface cards in each computer on the network attach directly to the segment, or bus. The connectors used in a 10Base2 network are called BNC (British Navel Connector). There are several types of BNC connectors: T-connector, straight connector, and terminating resistor. Look again at Figure 2-21. You will see that the T-connector attaches directly to the BNC connector on the network interface card. The T-connector allows the network to be cabled in a bus topology. Straight connectors attach to the ends of the T-connectors. This forms a common line of communication between all computers on the network. A terminating resistor is installed at each end of the bus topology.

The *terminating resistor* absorbs the electrical signals when they reach the end of the segment. If a terminating resistor were not installed, the signal would reflect, causing the other signals transmitted across the cable to distort. In a 10Base2 network, a 50-ohm terminating resistor is used to match the impedance of the RG-58 cable, which is 50 ohms.

Also, one end of the segment must be electrically grounded. Remember, a fluctuating digital signal traveling along a cable's core produces an electromagnetic wave, which can be absorbed by a neighboring conductor. In the case of a copper core cable, the cable's shielding absorbs the electromagnetic wave. Grounding one end of the cable helps to drain this interference so that the original signal is not affected.

Figure 2-22 shows a typical BNC T-connector, a straight connector (installed on the end of the coax cable), and a terminating resistor. Remember, if there is a break in the cabling or a problem with a BNC connector or a terminating resistor, no communication will take place. This is a major reason why coaxial cable is no longer used in networking.

A 10Base2 network must follow the 5-4-3 rule. This rule states that a 10Base2 network can have a total of 5 segments, 4 repeaters, and 3 segments that are connected to computers. Look at **Figure 2-23.** Note that there are a total of 5 segments and 4 repeaters. Only three of the segments, *Segment 1*, *Segment 3*, and *Segment 5*, are connected to computers. Each of the 3 segments can have a maximum of 30 computers. *Segment 2* links *Repeater 1* and *Repeater 2*, and *Segment 4* links *Repeater 3* and *Repeater 4*.

The 5-4-3 rule was developed to limit signal latency. Remember that signal latency is the time lapse between sending a signal and the arrival of a signal at the destination. All transmissions on a local network have a maximum amount of circulation time allowed. The maximum number of cable segments allowed in a network is directly related to signal latency.

terminating resistor
a BNC-type connector that absorbs electrical signals when they reach the end of the segment.

Figure 2-22.
BNC connectors.

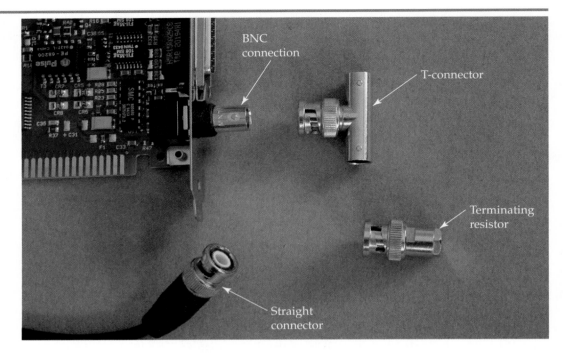

Figure 2-22.
BNC connectors.

10Base5

10Base5
an IEEE 802.3 classi-
fication that specifies
the use of RG-8
cable, a data rate of
10 Mbps, a maximum
segment length of
500 m, a minimum
segment length of
2.5 m, and a bus
topology.

The **10Base5** classification was the first IEEE 802.3 classification developed for Ethernet. The 10Base5 classification specifies the use of RG-8 (thicknet) and a 10 Mbps data rate. The 10Base5 network, like 10Base2, uses a bus topology and follows the 5-4-3 rule. However, the maximum segment length for a 10Base5 network is 500 meters. Each of the three segments that can connect to computers can connect to a maximum of 100. This allows a 10Base5 network to have a total of 300 computers.

A computer attaches to the thicknet coaxial cable through an AUI cable and AUI connectors. AUI is the acronym for Attachment Unit Interface. The AUI connector is a 15-pin D-shell connector that is used to attach a cable to a network interface card or network device. One end of the AUI connector connects to

Figure 2-23.
The 5-4-3 rule as
applied to the 10Base2
network.

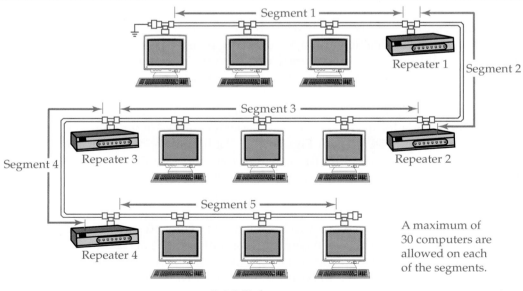

5-4-3 Rule

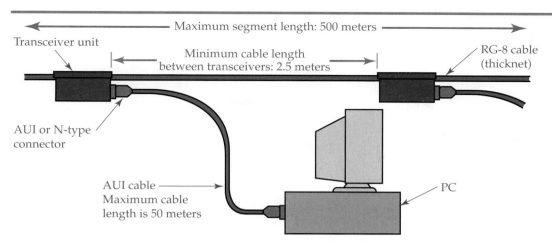

Figure 2-24.
10Base5 connections
using a vampire tap, a
transceiver unit, and
an Attachment Unit
Interface (AUI). Cable
lengths and type are
also indicated.

the network interface card and the other end connects to a transceiver unit, **Figure 2-24.** The transceiver unit has a pair of spikes that pierce the cable's insulating jacket, shield, and insulator and make a connection with the center core conductor. The transceiver unit is often called a vampire tap.

The term *transceiver* is formed from the two electronic terms: transmitter and receiver. They combine to form the new term *transceiver*, which is any electrical device that can both transmit and receive electrical signals. The term *transceiver*, when applied to network systems, generally refers to the early form of transceiver that was a separate piece of equipment used between the network cable and the computer. Today, the transceiver is incorporated into the network adapter card or directly into the circuitry of the network device, such as switches and routers. However, because 10Base5 is expensive and difficult to install, it is no longer used.

10BaseT

The **10BaseT** classification specifies the use of Category 3, Category 4, or Category 5 UTP cable. It is arranged in a star topology and follows the 100-meter rule, **Figure 2-25.** This rule specifies that any cable on the network should not exceed 100 meters. The 100-meter length ensures that the signal strength will be sufficient when reaching the final destination. A repeater, however, may be used to extend the distance to an additional 100 meters. Most hubs encountered are active hubs, which means they also act as a repeater, regenerating electrical signals.

A 10BaseT network can have two types of hub configurations: daisy chain and cascade. Study the hub configurations in **Figure 2-26.** Note that in the

10BaseT
an IEEE 802.3 classi-
fication that specifies
the use of Category 3,
4, or 5 cable, a data
rate of 10 Mbps, a
maximum segment
length of 100 m, a
minimum segment
length of .6 m, and a
star topology.

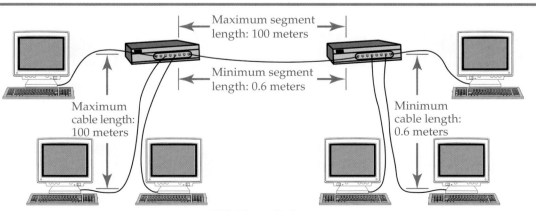

100-Meter Rule

Figure 2-25.
The 10BaseT network
follows the 100-meter
rule. This means that
any cable used in
a 10BaseT network
should not exceed 100
meters in length.

Figure 2-26.
Daisy-chain and cascade hub configurations. A—When hubs are arranged in a daisy-chain configuration, an electronic signal must pass through all hubs to arrive at a computer on the last hub in the chain. B—When hubs are arranged in a cascade configuration, an electronic signal never has to pass through more than three hubs to arrive at its destination.

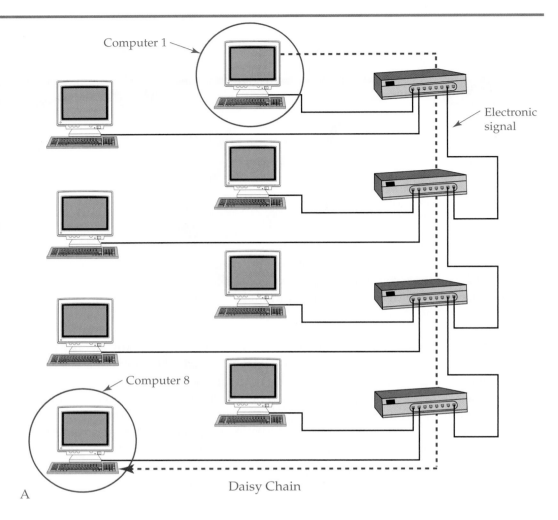

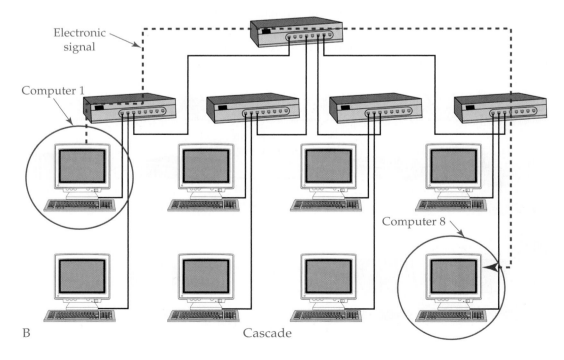

daisy-chain configuration, Figure 2-26A, an electronic signal from *Computer 1* must pass through four hubs before it reaches its destination, *Computer 8*. In the cascade configuration, Figure 2-26B, an electronic signal from *Computer 1* only has to pass through three hubs before it reaches its destination, *Computer 8*. This is a significant detail. Although both configurations have the same amount of hubs, an electronic signal does not have to pass through as many hubs in a cascade configuration as it does in a daisy-chain configuration.

If another hub were added to the daisy-chain configuration, an electronic signal from *Computer 1* would have to travel through five hubs before it reached a destination computer on the fifth hub, thus increasing the latency of the electronic signal. However, if another hub were added to the cascade configuration, an electronic signal from *Computer 1* would still only have to travel through three hubs before it reached a destination computer on the fifth hub. Adding another hub to the configuration would not increase the amount of hubs through which a signal has to travel.

As a general rule, no more than four hubs should be linked in a daisy-chain fashion. However, more than four hubs are allowed as long as an electronic signal does not need to travel through more than four hubs to arrive at its destination. If the signal passes through more than four hubs, the latency of the signal becomes excessive and causes the electronic signal to exceed its TTL setting. The electronic signal is then destroyed.

The 10BaseT classification uses RJ-45 connectors. The RJ-45 connector contains eight connection points, or pins, inside its plastic housing, **Figure 2-27.** To install an RJ-45 connector on a UTP cable, the insulating jacket of the UTP cable is first stripped from the cable, exposing the conductor pairs, **Figure 2-28.** The pairs are untwisted so that each conductor can be inserted into one of the designated pin areas. When the RJ-45 connector is crimped, each conductor makes contact with one of the eight pins, **Figure 2-29.**

When installing an RJ-45 connector, it is important that the conductors, which are color-coded, be inserted into the appropriate pin area. Two standard connections are recognized by industry: 568A and 568B. Long before standards existed, there were two common ways to make an RJ-45 connection. Since the industry was divided, both became standards but are denoted by the letters *A* and *B*. The only real difference in the connections is the color of the conductors inserted into the pin areas. Other than the color of the conductors,

Figure 2-27.
Close-up of an RJ-45 connector.

Plastic housing

Pin

RJ-45 Connector

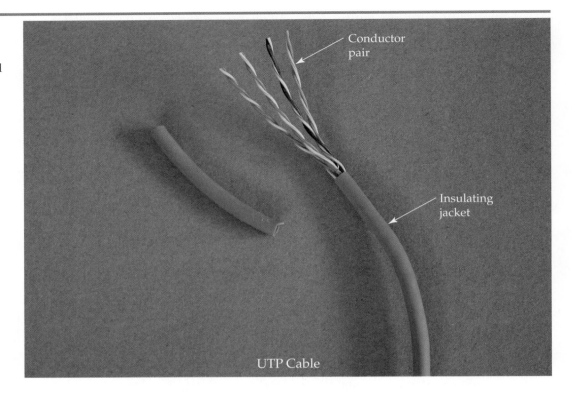

Figure 2-28.
The insulating jacket on this UTP cable has been stripped to reveal the conductor pairs.

Conductor pair

Insulating jacket

UTP Cable

the electrical qualities are the same. Look at **Figure 2-30.** Compare the two standard connections, 568A and 568B, in Figure 2-30A. Notice that for the 10BaseT classification, Figure 2-30B, each connector uses only two pairs of conductors from the cable: the orange pair and the green pair. The other two pairs are not used for communications in a 10BaseT network.

UTP cables have two common classifications of assembly: straight through and crossover. A straight-through cable is used to connect a computer with a

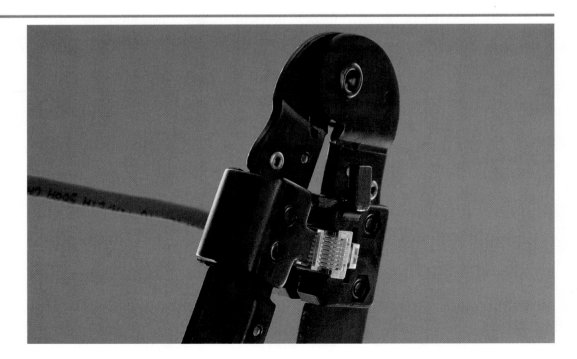

Figure 2-29.
An RJ-45 crimp tool is used to make the connection between the RJ-45 pins and the twisted pair conductors.

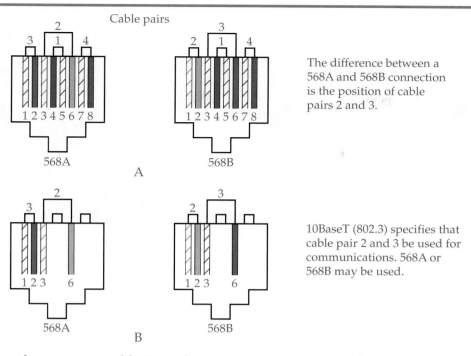

Cable pairs

The difference between a 568A and 568B connection is the position of cable pairs 2 and 3.

10BaseT (802.3) specifies that cable pair 2 and 3 be used for communications. 568A or 568B may be used.

Figure 2-30.
The 568A and 568B connection standards. A—The only difference between the two connection standards is the locations of pairs 2 and 3. B—The 10BaseT classification uses pairs 2 and 3. Either connection standard can be used.

hub, and a crossover cable is used to connect a computer with a computer. The straight-through cable is constructed with each numbered pin connecting to the matching numbered pin on the opposite end of the cable.

Look at the wire map in **Figure 2-31.** Notice that pin 1 connects to pin 1; pin 2 connects to pin 2, and so on. When a two-workstation 10BaseT network is configured, a crossover cable must be used for the network cards to communicate. A crossover cable has two pairs (four individual conductors) that are cross-connected. Cross connecting the pairs allows a transmit signal from one computer to be sent to the receive pins of a network card on the other computer. This allows the computers to communicate without a hub. A crossover cable is not needed when a hub is used because the circuitry at each access port crosses the connection internally. Therefore, in this situation a straight-through cable is used.

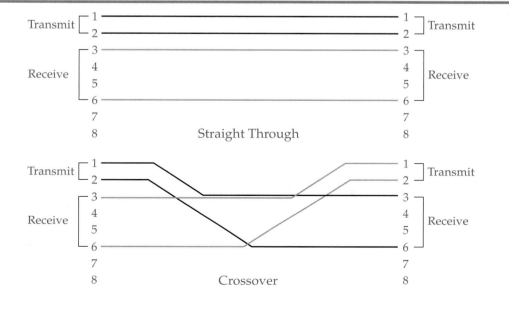

Figure 2-31.
Wire map of a straight-through cable and a crossover cable.

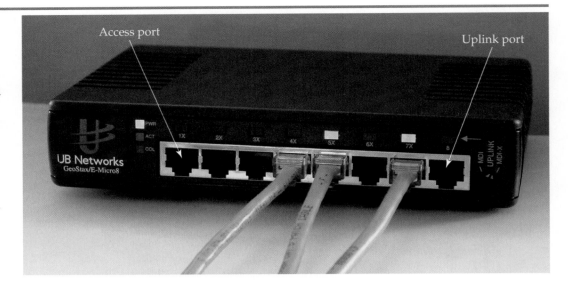

Figure 2-32.
A hub typically has many access ports and one uplink port. The circuitry inside the hub provides a crossed connection at each access port and allows for a straight-through connection at the uplink port.

Look at the hub in **Figure 2-32.** Notice that each access port number on the hub has an *X* beside it. This indicates the access port's internal connection is crossed. Also, notice the uplink port. The uplink port serves a straight-through connection. It can be used to connect via a straight-through cable to an access port on another hub.

A *wire map* is an illustration of the correct circuit path of conductors from one end of a cable run to the opposite end. Wire maps are used to illustrate proper cable connections for a technician who makes cable assemblies.

Fast Ethernet

Networks have rapidly grown in numbers and in applications. As networks and technology evolved, a demand for full motion video and audio was desired. Full motion video and audio requires a great deal of bandwidth because there is a much larger amount of data that needs to be transferred compared to text-based data. In 1998, to accommodate the needs for greater bandwidth, the IEEE standards approved 100BaseT, also known as Fast Ethernet.

The Fast Ethernet classifications differ in the mode of communication and the number of pairs that they use. Look at the chart in **Figure 2-33** for a summary of the two Fast Ethernet types that use copper cable: 100BaseT4 and 100BaseTX. Fast Ethernet works over most existing Category 5, 5e, and 6 cable.

Figure 2-33.
Communication modes used by the Fast Ethernet classifications.

Fast Ethernet Classification	Communications Mode	Pairs Used
100BaseTX	Half-duplex	Pair 1 and 2
100BaseT4	Half-duplex Full-duplex	Pair 1 and 2 Pair 1, 2, 3, and 4

100BaseT4

The *100BaseT4* classification specifies either half-duplex or full-duplex mode. In half-duplex mode, communication occurs on pairs 1 and 2, but in full-duplex mode, communication occur on all four pairs. The 100BaseT4 classification specifies the use of all four pairs of Category 3, 4, or 5 cable.

100BaseT4 was first introduced as a way to upgrade existing 10BaseT networks to 100 Mbps without the need to replace existing cables. Although the existing cable can be applied to the 100BaseT4 standard, existing network interface cards and other networking devices that are designed with a maximum throughput of only 10 Mbps need to be replaced to meet the 100 Mbps required by the 100BaseT4 standard. Remember, a network is only as fast as the slowest device or cable in the network.

100BaseTX

The *100BaseTX* classification specifies the use of Category 5e cable and uses only pairs 1 and 2. Like 10BaseT, it is configured in a star topology, has the same cable limitations, and uses RJ-45 connectors; however, it has a data rate of 100 Mbps. 100BaseTX uses only the half-duplex mode. Remember that in half-duplex mode, transmission occurs in only one direction at a time. All half-duplex communication occurs on pairs 1 and 2.

Gigabit Ethernet

In 1999, 1000BaseT, also known as Gigabit (Gb) Ethernet, was adopted. Higher speeds were accomplished by using cable that supported higher data rates and using all four of the cable pairs to allow for full-duplex transmission. It is important to note, however, that with Fast Ethernet and Gigabit Ethernet, not only must the cable be able to support the higher data rates, the network hardware, such as network cards, hubs, and switches, must also be able to support the higher data rates as well.

Many network cards are manufactured as 10/100 and 10/100/1000, which means that they support 10 Mbps, 100 Mbps, and 1000 Mbps. The network cards and other network hardware automatically detect the data rate of the network and adjust accordingly.

1000BaseCX

The *1000BaseCX* classification specifies a data rate of 1 Gbps and uses Category 5 cable. It has a maximum segment length of 25 m and a minimum segment length of .6 m. It was intended for short runs, but has become obsolete due to the introduction of 1000BaseT.

1000BaseT

The *1000BaseT* classification, also known as Gigabit Ethernet, specifies a 1 Gbps data rate using all four pairs of Category 5e cable. It has a maximum segment length of 100 meters and a minimum segment length of .6 meters.

1000BaseT was originally introduced for network backbone installations, but now it is commonly used to connect workstations. 1000BaseT provides a less expensive alternative to fiber-optic cable when high-speed data rates are desired.

As you can see, the need to create faster network systems and the ability to use existing materials whenever possible has driven the creation of many different networking standards. At times, it can be confusing because many of the standards use the same category of cable, but in a different way. As new standards are created, older standards become obsolete, but can still be found in the field.

100BaseT4
an IEEE 802.3 classification that specifies the use of Category 3, 4, and 5 cable, a data rate of 100 Mbps, a maximum segment length of 100 m, a minimum segment length of .6 m, and a star topology.

100BaseTX
an IEEE 802.3 classification that specifies the use of Category 5e cable, a data rate of 100 Mbps, a maximum segment length of 100 m, a minimum segment length of .6 m, and a star topology.

1000BaseCX
an IEEE 802.3 classification that specifies the use of Category 5 cable, a data rate of 1000 Mbps, a maximum segment length of 25 m, a minimum segment length of .6 m, and a star topology.

1000BaseT
an IEEE 802.3 classification that specifies the use of Category 5e cable, a data rate of 1000 Mbps, a maximum segment length of 100 m, a minimum segment length of .6 m, and a star topology.

IEEE 802.5

The IEEE 802.5 standard describes the Token Ring network. The Token Ring network uses the token passing access method and is configured in a ring topology. Token Ring networks operate at 4 Mbps, 16 Mbps, and 100 Mbps. Token Ring can be used with proprietary cable such as IBM design or UTP and STP. The maximum number of nodes allowed for a Token Ring network varies according to the network distance, speed, cable media, and equipment used. Always check with the manufacturer for the latest specifications.

Tech Tip

Some Token Ring networks do not use UTP as the network media. They use a proprietary cable, but the electrical characteristics are similar.

While a Token Ring network can be wired as a physical ring, **Figure 2-34,** most Token Ring networks use a multiple access unit (MAU) for a center connection point, **Figure 2-35.** When a MAU is used, the network arrangement resembles a star topology. However, because of the circuitry in the MAU, the network is physically connected as a ring. Each port on the MAU is designed to maintain an electrical ring. Look closely at the illustration of the MAU in **Figure 2-36.**

Figure 2-34.
Token Ring network using coaxial cable. When a Token Ring network uses coaxial cable, it resembles a physical ring.

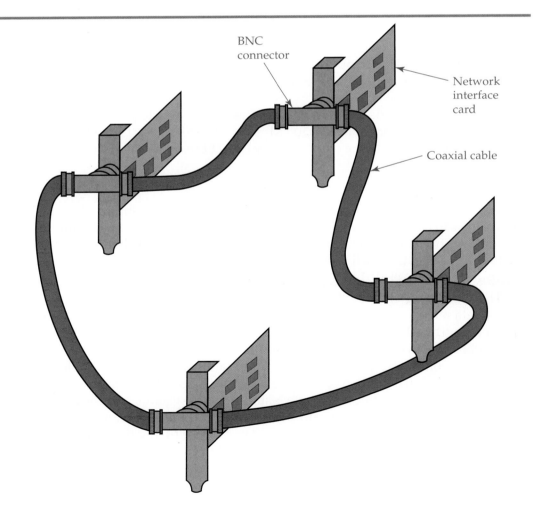

BNC connector

Network interface card

Coaxial cable

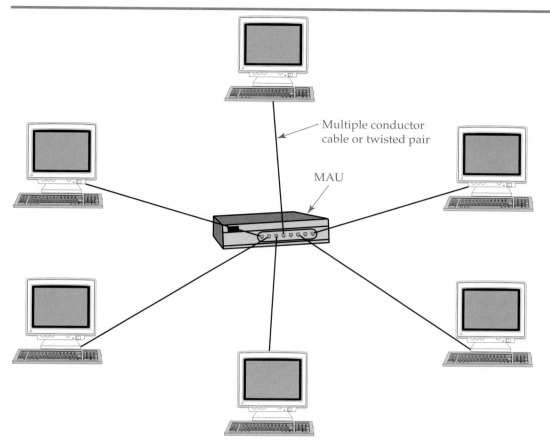

Figure 2-35.
Token Ring network using twisted pair cable. When a Token Ring network uses twisted pair cabling, it resembles a physical star.

Each port located at the MAU is normally closed to ensure the integrity of the ring. When a computer is added by plugging in the network cable, the port opens the normally closed circuit allowing the added computer to join the ring.

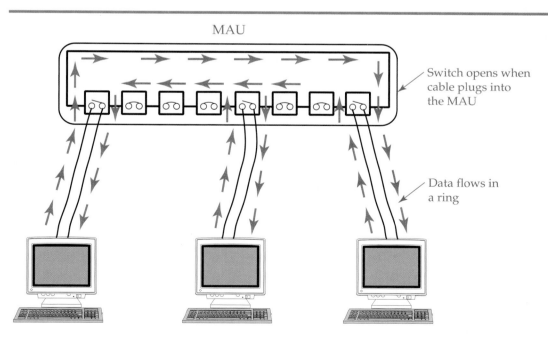

Figure 2-36.
Basic representation of the electronic circuitry inside a MAU. The MAU is designed to allow a continuous loop of data flow in the Token Ring topology. Look closely at the way the switches inside the unit are normally closed and then open when a cable plug is inserted.

A Token Ring network commonly uses several MAUs. When two or more MAUs exist in the ring, integrity is maintained in the same way as with added computers. MAUs, however, connect through special ports called *ring in* and *ring out*.

When a Token Ring network incorporates a MAU, twisted pair cable is used. The twisted pair cable attaches to an RJ-45 connector. Token Ring only uses two pairs for communication and has one standard, 568A. Look at **Figure 2-37.**

ARCnet Standard

ARCnet was designed and introduced by the Datapoint Corporation in the late 1970s. It uses the token passing access method. The acronym/conjunction ARCnet is formed from the term *Attached Resource Computer Network*. It was originally designed as a LAN system for business offices, but quickly became an industrial field bus technology that is still used today. An **industrial field bus technology** is a combination of computer network devices and industrial controller devices sharing the same bus in the industrial environment.

Modern ARCnet systems use twisted pair conductors rather than coaxial cable. This means the system can be more conveniently wired using RJ-45 and RJ-11 connector-based hubs. ARCnet allows a maximum number of 255 nodes per network. The data rate is typically 2.5 Mbps, which might seem slow. If you consider that large packets of data are not transmitted, but rather small packets with commands and monitoring operations such as the opening and closing of switches and timers, the speed is more than adequate. Another interesting aspect of ARCnet is that the transmission speed can be controlled from 19 kbps (.019 Mbps) to a maximum of 10 Mbps. The speed is directly related to the length of the bus. The longer the bus, the lower the rate. There is also a special ARCnet system known as Fast ARCnet that is capable of data transfer rates as high as 20 Mbps.

Wiring Faults

When installing twisted pair cabling, the pairs of conductors inside the cable jacket must be connected correctly at both ends, or the network will not be able to

industrial field bus technology
a combination of computer network devices and industrial controller devices sharing the same bus in the industrial environment.

Figure 2-37.
The 802.5 standard specifies that only the 568A connection standard be used. Note that pairs 1 and 2 are used for communications.

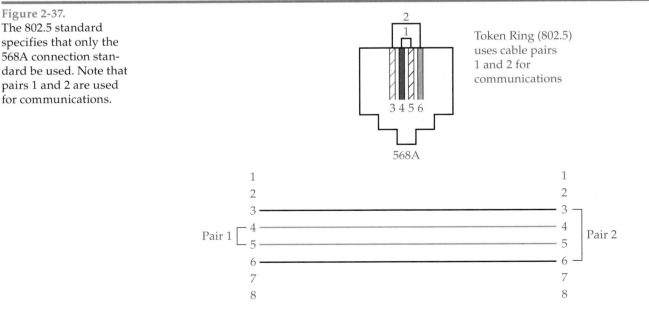

Token Ring (802.5) uses cable pairs 1 and 2 for communications

568A

communicate. The proper connection of individual conductors to pins is outlined in TIA standards: T568A and T568B. The proper connection sequence must be maintained throughout the installation, from the network card to the server room. Incorrect wiring can result in shorts, opens, reversed pairs, crossed pairs, and split pairs.

Short

A *short* occurs in cabling when two conductors are improperly connected, resulting in a shorter circuit path. Look at the lamp circuit in **Figure 2-38.** The lamp circuit is connected with the splice point left open for viewing. Most short circuits and open circuits occur at splice points and connection points. In Figure 2-38A, the circuit is correctly wired. When the switch is closed, electrical energy flows through the lamp, and the lamp lights. In Figure 2-38B, the wires are touching at the splice point. When the switch is closed, the electrical energy cannot reach the lamp because the electrical energy routes through the shortest path, the splice point. When a short occurs in a network cable, data takes a shorter route and does not reach its destination.

short
a wiring fault which results in electrical current flow taking a shorter path.

Open

An *open* is just as the name implies. The circuit has an open spot along the length of the conductor. Look at **Figure 2-39.** In the illustration, you can see the open is at the splice point. When the switch is closed, no electrical energy can reach the lamp. The lamp will not light because the circuit is incomplete. When an open occurs in a network system, no data can travel past the point of the open.

open
a wiring fault in which there is a break in a circuit path.

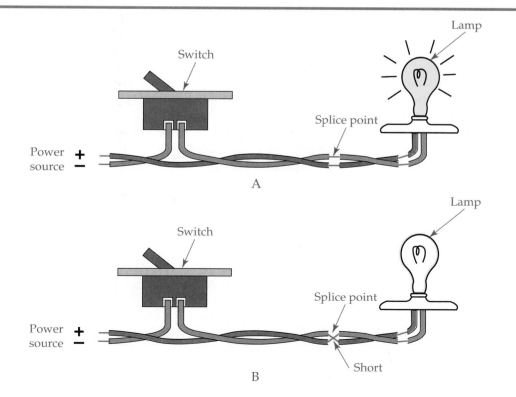

Figure 2-38.
A correctly wired circuit and a shorted circuit. A—This circuit is correctly wired. When the switch is closed, the current flows through the complete circuit, lighting the lamp. B—In this circuit, the wires are shorted (touching). When the switch is closed, the current takes the shortest path, and the lamp does not light.

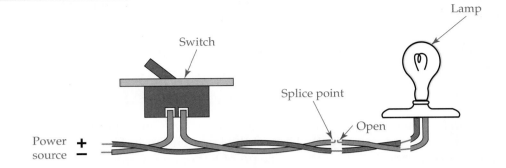

Figure 2-39.
In this circuit, there is a break in the wiring at the splice point. When the switch is closed, the lamp does light. This condition is called an open.

Ground

ground
an occurrence of a conductor connecting to the earth through a continuous path.

A *ground* occurs when a conductor connects to the earth through a continuous path. Grounds typically occur where the conductor's insulating jacket is torn and the copper conductor makes contact with conduit or any other metal surface that is grounded. When a ground occurs in a network cable, all data is prevented from reaching its destination.

Tech Tip

Remember, a short occurs between two conductors, a ground occurs between a conductor and the earth. The connection to the earth is usually provided by a metallic enclosure, such as the metal pipe used to carry network cables, or exposed metal building parts, such as I beams or drop ceiling channels.

Reversed, Crossed and Split Pairs

reversed pair
a wiring fault in which conductors within a twisted pair cable pair reverse their connection at the opposite end of the cable.

crossed pair
a wiring fault in which two pairs within a twisted pair cable have switched positions at the opposite end of the cable.

split pair
a wiring fault in which one conductor from each of two pairs within a twisted pair cable have switched positions at the opposite end of the cable.

Reversed, crossed, and split pairs commonly occur when technicians install connectors on twisted pair cables in the field. It is very easy to misconnect individual wires when installing an RJ-45 connector or when terminating a twisted pair cable in a punch down block. The misconnection may go unnoticed until after the network system is installed and a computer has a problem accessing the network.

Study the examples in **Figure 2-40** of reverse, crossed, and split pairs. A *reversed pair* occurs when two pairs of a cable assembly have reversed two connections, Figure 2-40B. For example, a blue pair has reversed pin connections with a green pair. A *crossed pair* occurs when one of each of the two cable pairs has become part of the other pair's connection. In Figure 2-40C, one of the blue pair conductors is connected as one of the green pair. One of the green conductor pairs is connected as one of the blue conductor pairs. A *split pair* occurs when two pairs of conductors are reversed in connection with another pair. In Figure 2-40D, a complete pair of blue conductors has been reversed with connections intended for the green pair, and the green pair has taken the position intended for the blue pair.

Electronic cable testers are designed for testing a cable for the possibility of opens, crossed pairs, split pairs, and reversed pairs. You will use a cable tester often in the field when troubleshooting network communication problems. **Figure 2-41** shows a cable tester.

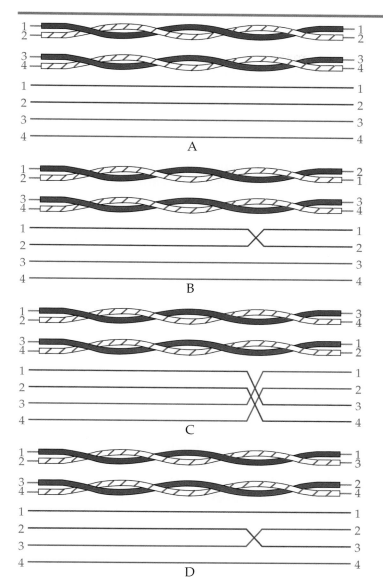

Each pair of conductors are straight through. The connection numbers are the same at each end.

Normal

The blue pair is an example of a reverse pair. Notice how the connections on the right are reversed as compared to the connections on the left.

Reverse Pair

This is an example of crossed pairs. Take special notice of the green and blue pairs. The blue and green have switched positions.

Crossed Pairs

This is an example of a split pair. Each of the two pairs of conductors have had their pair split and connected to the other pair position.

Split Pair

Figure 2-40.
Common wiring errors: reverse pair, crossed pairs, and split pair. The red numbers indicate the errors. A—Normal connection is made. Each conductor is wired to a corresponding connection number. B—Reverse pair. Conductors within the pair reverse their connection at the opposite end of the cable. C—Crossed pairs. Two pairs have switched positions at the opposite end of the cable. D—Split pair. One conductor from each of two pairs have switched positions at the opposite end of the cable.

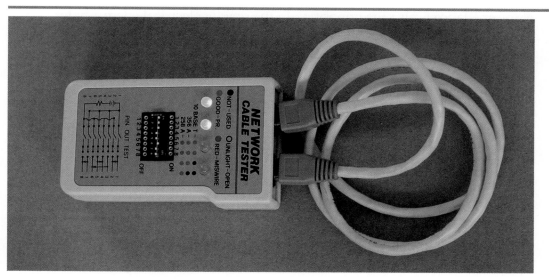

Figure 2-41.
An electronic cable tester is used to check for wiring faults such as crossed pairs, split pairs, and reversed pairs.

Summary

- The four major classifications of network media are coaxial cable, twisted pair cable, fiber-optic cable, and the atmosphere.

- Electronic signals are classified as analog and digital. Analog and digital signals differ by the shape of their waveforms.

- In the analog signal, time affects the degree of the slope. In the digital signal, time affects the duration of the voltage level.

- Both analog and digital waves have a frequency measurement. Frequency is the number of cycles that occur in 1 second and is measured in Hertz (Hz).

- Attenuation is the loss of signal strength.

- Interference is an unwanted signal that combines with and changes the waveform of the original signal.

- Latency is the amount of time it takes a signal to travel from its source to its destination.

- Bandwidth is a measurement of the ability of network media to carry data, whereas frequency is a measurement of the ability of network media to carry an electronic signal.

- Baseband is a method of transmitting a single frequency, using the entire bandwidth of a cable.

- Broadband is a method of transmitting multiple frequencies or data across a cable at once.

- Communications can occur in one of three modes: simplex, half-duplex, and full-duplex. Simplex communication occurs in one direction only. Half-duplex communications occurs in both directions, but not at the same time. Full-duplex communication occurs in both directions at the same time.

- Electrical energy that travels in only one direction is referred to as direct current (DC).

- Electrical energy that constantly reverses its direction of flow is referred to as alternating current (AC).

- Electrical resistance is the opposition to the flow of electrical energy. Resistance is generally used to describe the effect in DC circuits.

- Impedance is also the opposition to the flow of electrical energy, but it expresses the amount of resistance encountered in an AC circuit.

- Reflected loss is the amount of signal reflected from the far end, or receiving end, of a cable.

- The electrical phenomenon of a current-carrying conductor inducing current flow in a cable of close proximity is called crosstalk.

- Near-End Crosstalk is a measurement of crosstalk at the near end, or sending end, of a cable.

- The decibel (dB) is an electrical measurement that is used to express the amount of power or energy gained or lost.

- Far-End Crosstalk is a measurement of crosstalk at the far end, or receiving end, of a cable.

- Equal Level Far-End Crosstalk is a measurement of the negated effects of cable attenuation caused by crosstalk.

- Copper conductor wire size is based on its diameter and is expressed as a size according to the American Wire Gauge AWG—the larger AWG the number, the smaller the wire size.

- Two common coaxial cable types that have been used for networking are RG-8 (thicknet) and RG-58 (thinnet).

- Twisted pair cable is divided into seven categories. The categories are based on the physical design and capabilities of the cable such as the maximum frequency rating, the data rate that the cable is capable of, and number of twists per foot.

- The IEEE 802.3 standard describes the characteristics of various Ethernet technologies such as 10 Mbps, Fast Ethernet, and Gigabit Ethernet.

- The IEEE 802.5 standard describes the characteristics of a Token Ring network.

- Wiring faults include shorts, opens, reversed pairs, crossed pairs, and split pairs.

Review Questions

Answer the following questions on a separate sheet of paper. Please do not write in this book.

1. _____ signals are sent through copper core cable as pulsating electrical energy.

2. Pulsating light waves are commonly sent through _____ cable.

3. Wireless technology commonly transmits _____ waves, _____ waves, and _____ through the atmosphere.

4. What are the two types of electrical signals?

5. Time affects the degree of the slope in a(n) _____ signal.

6. Time affects the duration of the voltage level in a(n) _____ signal.

7. How often an analog or digital cycle repeats in one second is called _____.

8. Frequency is measured in _____ and is represented by the abbreviation _____.

9. How many times does an analog cycle repeat in a 2.3 kHz electronic signal?

10. How many times does a digital cycle repeat in a 10 MHz electronic signal?

11. What is attenuation?

12. Noise and crosstalk are types of _____.

13. What is latency?

14. A method of transmitting data in the form of a digital signal, using the entire bandwidth of a cable is called _____.

15. A method of transmitting data in the form of several analog signals at the same time is called _____.

16. List the three modes in which communication can occur.

17. What is the difference between half-duplex and full-duplex communication?

18. _____ is opposition to current flow in a DC circuit, and _____ is the opposition to current flow in an AC circuit.

19. The size of a conductor's diameter is described by what rating?

20. Which conductor is larger, one that has a 22 AWG rating or one that has a 24 AWG rating?

21. RG-58 is also referred to as _____.

22. List the following characteristics of a 10Base2 network: topology, cable type, connector type, maximum segment length, minimum cable length, and data rate.

23. Explain the 5-4-3 rule and list the 802.3 classifications to which it applies.

24. List the following characteristics of a 10BaseT network: topology, cable type, connector type, maximum segment length, minimum cable length, and data rate.

25. What will happen to an electronic signal in a 10BaseT network if it passes through more than four hubs?

26. What is the term used for a circuit path that is shorter than its intended path?

27. What is the term used for an incomplete circuit path?

28. What is the term used to describe a wire that comes in contact with a metal enclosure that is touching the earth?

29. Name three possible cable problems that might result from improperly wiring twisted cable pairs.

Sample Network+ Exam Questions

Network+

Answer the following questions on a separate sheet of paper. Please do not write in this book.

1. What type of cable is typically associated with a 10BaseT network?
   A. RG-58
   B. UTP
   C. Fiber-optic
   D. RG-8

2. Which of the following connectors is commonly associated with a 10BaseT network?
   A. BNC
   B. RJ-45
   C. AUI
   D. ST

3. What is the maximum segment length allowed for a Category 5 UTP cable?
   A. 100 meters
   B. 185 meters
   C. 200 meters
   D. 500 meters

4. The loss of signal strength from one end of a cable to the opposite end is expressed as _____.

   A. attenuation

   B. crosstalk

   C. NEXT

   D. resistance

5. What is the maximum length of a 10Base2 cable segment?

   A. 100 meters

   B. 185 meters

   C. 200 meters

   D. 500 meters

6. Which of the following is not an 802.3 category?

   A. 10 Mbps

   B. 100 Mbps

   C. Fast Ethernet

   D. Gigabit Ethernet

7. Which one of the following network technologies does the 802.5 standard describe?

   A. Ethernet

   B. 10Base5

   C. Wireless

   D. Token Ring

8. What is the maximum number of segments in a 10Base2 network that can be connected to computers?

   A. 2

   B. 3

   C. 4

   D. 5

9. What is the maximum data rate of a Category 5e cable if four cable pairs are used to transmit data?

   A. 10 Mbps

   B. 100 Mbps

   C. 1000 Mbps

   D. Only two cable pairs on a Category 5e cable can be used to transmit data.

10. A cable with a special type of insulation that will not give off toxic gases should the cable be consumed by fire is _____.

   A. AWG-rated

   B. fireproof-rated

   C. plenum-rated

   D. safety-rated

Suggested Laboratory Activities

1. Make a straight-through patch cable using Category 5e.
2. Make a crossover cable.
3. Use a cable fault detector to verify a cable wire map.
4. Use a wire gauge to size different conductors.
5. Search the Web for various network interface cards to see what cable types they support.

Interesting Web Sites for More Information

www.belden.com
www.belden.com/college/college.htm
www.blackbox.com
www.lanshack.com
www.techfest.com/index.htm

Chapter 2
Laboratory Activity

Making a Straight-Through Patch Cable

After completing this laboratory activity, you will be able to:
■ Construct a patch cable following the 568A and 568B standards.
■ Explain the difference between the 568A and 568B standards.
■ Test a patch cable.

Introduction

In this laboratory activity, you will make a straight-through patch cable for a 10BaseT, 100BaseTX, and 1000BaseT network. This type of cable is commonly used to connect network interface cards to the network wall outlet, network interface cards to hubs, and to connect network segments together using a patch panel located in the network equipment or server room.

There are two recognized standards of straight-through cable pin assignments: 568A and 568B. Both work equally well and provide the same electrical conduction qualities. The differences in the two standards are the arrangement of conductor colors inside the RJ-45 connectors. Look at the following illustration of the 568A and 568B pin assignments.

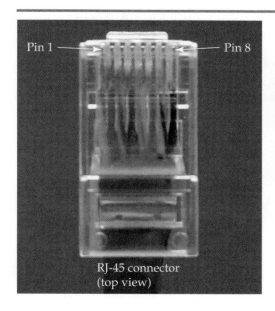

Pin 1   Pin 8

RJ-45 connector
(top view)

568A		568B	
Pin #	Color	Pin #	Color
1	Green/white	1	Orange/white
2	Green	2	Orange
3	Orange/white	3	Green/white
4	Blue	4	Blue
5	Blue/white	5	Blue/white
6	Orange	6	Green
7	Brown/white	7	Brown/white
8	Brown	8	Brown

Notice that the difference between the two standards is the location of the orange/white and orange pair and the green/white and green pair of conductors and their assignments to pins 1, 2, 3, and 6. For a 568A connection, green/white and green are assigned to pins 1 and 2 and orange/white and orange are assigned to pins 3 and 6 respectively. In the 568B standard, the pin assignment is reversed. Orange/white and orange are assigned to pins 1 and 2 and green/white and green are assigned to pins 3 and 6 respectively.

The confusion of the color assignments is a result of independent color codes developed in the communications industry before network standards were introduced. The two most dominant color code assignments used at the time were accepted as a standard. The two standards were simply differentiated by the letters *A* and *B*. Accepting the two pin designations was a compromise to keep different communications companies in agreement. It was the only way a standard could be adopted at the time.

The only conductors used for transmitting and receiving data in a 10BaseT or 100BaseTX cable are located on pins 1, 2, 3, and 6. The other conductors, located on pins 4, 5, 7, and 8, are not used in a 10BaseT or 100BaseTX network. The 1000BaseT standard uses the same wire map as 568A and 568B; however, all four pairs of conductors are used to transmit data.

The type of conductor used for patch cables is Unshielded Twisted Pair (UTP). The cable is constructed of four pairs of twisted conductors, resulting in a total of eight conductors. Category 5, Category 5e, and Category 6 can be used with a standard RJ-45 connector for 10BaseT and 100BaseTX networks. A 1000BaseT network requires Category 5e or Category 6.

Note:
Category 3 is allowed for 10BaseT network patch cables. However, it is rarely used and is not recommended for new installations.

Stranded wire is preferred for patch cables because the cable needs to be flexible. While solid wire cable is also flexible, it tends to break at the connector when flexed excessively. Solid wire is used for horizontal runs. Horizontal runs are the cable runs installed inside building walls. Because the horizontal run is installed inside building walls, the cable is not required to be as flexible as stranded wire.

Equipment and Materials

- 2 (or more) RJ-45 connectors.
- RJ-45 crimping tool.
- UTP cable stripper. (Some crimping tools incorporate a striping tool as part of the assembly.)
- 3-foot length of stranded Category 5e UTP cable.
- Cable tester.

Procedures

1. _____   Gather the required material and report to your assigned workstation.

2. _____   Using the cable stripper, remove approximately 1 1/2″ to 2″ of the cable's insulating jacket. Check closely for nicks in the individual conductors caused by the cable stripper. If the insulation of the individual conductors is damaged, cut off the bad portion and repeat the cable-stripping process.

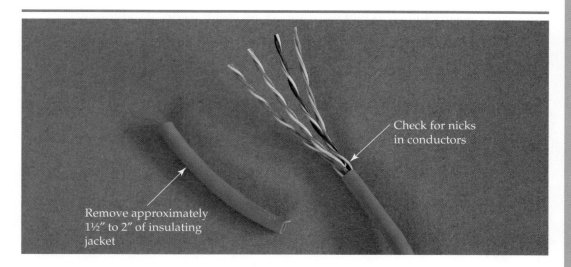

Check for nicks in conductors

Remove approximately 1½″ to 2″ of insulating jacket

3. _____   Study the color arrangement of the assigned standard, 568A or 568B. Separate the conductor pairs and arrange them into the standard's color sequence. After you have arranged the conductors into the proper color sequence, trim the excess cable, leaving approximately 1/2″ of conductors protruding from the insulating jacket. Excessive lengths of untwisted conductors at the connector increase the possibility of crosstalk. Note that the individual conductors do not have their insulation removed.

Neatly trim conductors to approximately ½″

4. _____   Carefully insert the trimmed conductors into the RJ-45 connector. Check the pin 1 position of the RJ-45 connector by comparing it to that in the illustration. It is easy to confuse the top and bottom of the RJ-45 connector and pin 1 location. Be sure the cable's insulating jacket is inserted into the connector. Inspect the wiring arrangement closely after the cable has been inserted into the connector. Be sure each of the conductors is butted against the end of the connector.

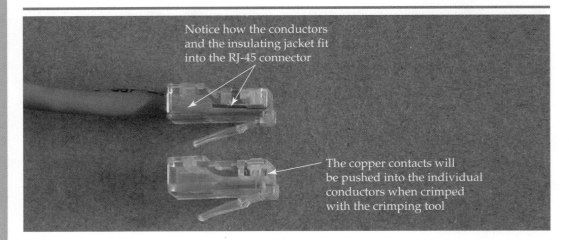

Notice how the conductors and the insulating jacket fit into the RJ-45 connector

The copper contacts will be pushed into the individual conductors when crimped with the crimping tool

5. _____   Place the RJ-45 connector into the crimping tool. Be sure the RJ-45 connector is properly inserted into the crimping tool and that the conductors are firmly in place inside the RJ-45 connector. Notice the series of eight teeth that the crimping tool uses to drive the copper contacts into the conductors.

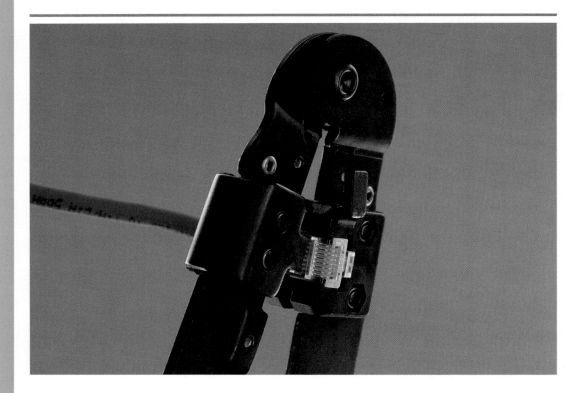

6. _____ Apply firm pressure to the RJ-45 crimping tool until you are sure the copper contacts are fully inserted into the individual conductors.

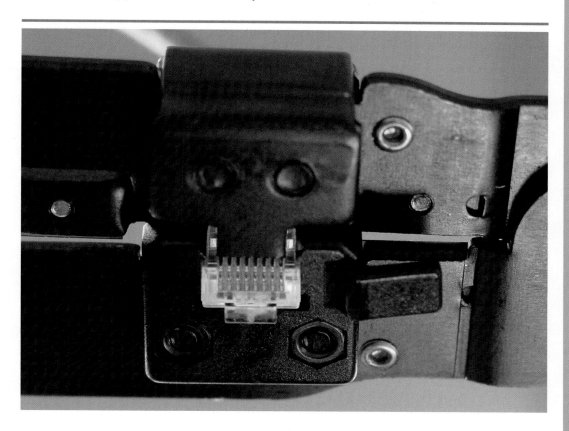

7. _____ Inspect the completed connector for proper color assignments. Also, check that the insulating jacket is secured inside the connector.

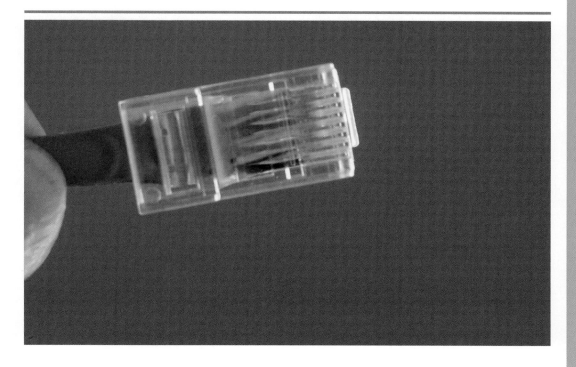

8. _____ Repeat the same sequence of steps for the opposite end of the cable.

9. _____ Test the cable by using a standard UTP cable tester. If the cable tester indicates a fault, have your instructor check your cable.

10. _____ Have your instructor check your completed cable.

11. _____ Repeat steps 2 through 10 using the other cable standard. Once you have constructed a patch cable for both standards, proceed to step 12.

12. _____ Clean up your assigned work area and dispose of all trash. Return the tools and supplies to their storage area.

13. _____ Answer the review questions.

Review Questions

Answer the following questions on a separate sheet of paper. Please do not write in this book.

1. What are the two standards for making a patch cable?

2. Describe the difference between the 568A and 568B standards.

3. For what are straight-through patch cables commonly used?

4. What cable fault may occur if an excessive length of untwisted conductors is left at the connector?

5. How many conductors are used to communicate on a 10BaseT and 100BaseTX network?

6. How many conductors are used to communicate on a 1000BaseT network?

7. Which type of wire (stranded or solid) is used for patch cables?

8. Why is stranded wire preferred for patch cables?

9. What does the acronym UTP represent?

10. What is a horizontal run?

3

Fiber-Optic Cable

After studying this chapter, you will be able to:

❑ List the advantages of fiber-optic cable as compared to copper core cable.

❑ Explain the properties of light associated with fiber-optic cable.

❑ Describe the characteristics of fiber-optic cable transmission.

❑ Describe the difference between multimode and single-mode fiber-optic cable.

❑ List the characteristics and specifications of the IEEE 802.3 fiber-optic standards.

❑ List the characteristics and specifications of the FDDI standard.

❑ Identify and describe the problems associated with fiber-optic cable splices.

❑ Identify common fiber-optic cable test instruments and explain how they are used.

Network+ Exam—Key Points

To prepare for the Network+ Certification exam, you must be able to do the following:

■ List the advantages of fiber-optic cable.

■ Identify the characteristics of fiber-optic cable.

■ Know the differences between single-mode and multimode fiber-optic cable.

■ Identify transmission-related issues.

■ Identify the main features of the FDDI and 802.3 fiber-optic standards, such as the type of cable and connectors used, speed, topology, and cable lengths.

■ Recognize SC, ST, FC, LC, and MTRJ connectors.

Key Words and Terms

The following words and terms will become important pieces of your networking vocabulary. Be sure you can define them.

dispersion
electromagnetic wave
extrinsic losses
Fiber Distributed Data Interface (FDDI)
Fresenel reflection loss
fusion splice
graded-index multimode fiber-optic
    cable

multimode fiber-optic cable
Optical Time Domain Reflectometer
    (OTDR)
scattering
single-mode fiber-optic cable
step-index multimode fiber-optic cable
wavelength

Fiber-optic cable supports transmission rates over 1 Gbps. It is primarily used for network backbones and long-distance runs. Soon it may be economical and common to install fiber-optic cable directly to the PC. Fiber-optic cable has many advantages over copper core cable and follows a different set of network standards such as FDDI and 802.3. This chapter explores the advantages of fiber-optic cable and its construction. It also covers the nature of light, transmission characteristics, the FDDI and 802.3 standards, and cable faults and testing.

Characteristics of Fiber-Optic Cable

Fiber-optic cable consists of a glass or plastic core that carries pulses of light. The pulses of light represent binary data. Look at **Figure 3-1.** A transmitter is located at the source and a receiver is located at the destination. The transmitter and receiver are connected by fiber-optic cable. The transmitter at the source converts the binary data represented by electrical, digital pulses into light pulses. A pulse of light represents a binary one and an absence of light represents a binary zero. The light pulses travel the fiber-optic cable to the receiver at the destination. The receiver converts the light pulses back into an electrical signal consisting of a series of digital pulses.

It may seem like a lot of extra work to convert digital signals into light signals and then back to digital signals. One might question why it would not be more efficient to use copper core cable. There are several reasons fiber-optic

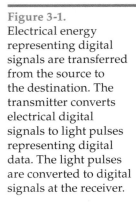

Figure 3-1.
Electrical energy representing digital signals are transferred from the source to the destination. The transmitter converts electrical digital signals to light pulses representing digital data. The light pulses are converted to digital signals at the receiver.

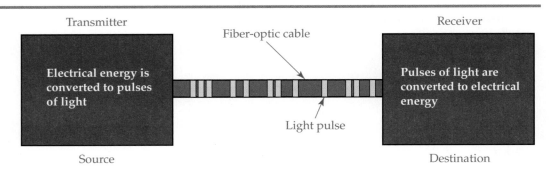

cable might be chosen over copper core cable. These reasons are related to the following beneficial attributes of fiber-optic cable:

- Provides for data security.
- Immune to electromagnetic interference.
- Lightweight and small in diameter.
- Safe from fire and explosion.
- Wide bandwidth.
- Corrosion- and water-resistant.
- Supports data transmission over longer distances than copper core cable.

Security

Information has become a profitable and, in some cases, a priceless commodity. In business, industry, and military applications, it is important to protect this information from unethical people who would seek to gain from using it. Fiber-optic cable is virtually impossible to tap into without being detected. Any cut made to the fiber-optic core disturbs the light signal and thus data transmission. Removing the protective material around the fiber-optic core causes signal loss. Fiber-optic cable is far more secure than copper core cable.

Immunity to Electromagnetic Interference

Fiber-optic cable conducts light waves instead of electrical energy, which makes it immune to electromagnetic interference generated by motors, radio signals, lighting, and other sources of electromagnetic energy. Conventional copper core cable must be shielded to prevent electromagnetic interference. Shielding cable causes cable to be large in diameter and heavy.

The military began research in fiber optics many years ago. Nuclear weapons create a powerful electromagnetic field, which destroys communication systems that consist of copper core cable. The destruction caused by an electromagnetic field could occur many miles from an actual nuclear explosion. Since fiber-optic cable is not affected by electromagnetic fields, military communications that use fiber-optic cable could remain intact.

Weight and Size

Fiber-optic cables are exceptionally lightweight compared to copper core cable. This is unimportant in general wiring applications, but for applications where weight is a critical factor, it is very important. Some of the applications in which weight is a critical factor are aircrafts and ships. The lighter the aircraft or ship, the more cargo it can carry. Fiber-optic cables are approximately 1/10th the weight of a comparable copper core cable.

Fiber-optic cable has a smaller overall diameter than copper core cable. This is extremely important when installing communication lines. Many more communication lines can be provided in the same conduit using fiber-optic cable rather than copper core cable. In a telephone communication system, over 1000 fiber-optic cables can easily fit in the same space as 100 copper core cables.

Safety

Fiber-optic cables are typically installed near explosive vapors and dust, such as in a petroleum factory or food processing plant. Using fiber-optic cable eliminates the possibility of an electrical spark causing an explosion due to a shorted or grounded circuit. In addition, light for illumination can be directly transmitted through fiber-optic cables in place of conventional lamps. When using electrical lamps for illumination in hazardous industries, the lamp fixtures must be the explosion-resistant type. Fiber-optic lamps are safer and less expensive.

Bandwidth

Fiber-optic cable has a greater bandwidth than copper core cable. Light can be transmitted at a much higher frequency through glass core than electronic digital signals through copper core cable. Copper wire suffers from data rate limitations caused by impedance. Fiber optic is not affected by impedance characteristics. Since a higher signal frequency can be used, more data can be transmitted through fiber-optic cable per second than copper core cable. Copper core cable has limitations and losses due to inductive reactance. Copper core cables lose their conduction capabilities at extremely high frequencies. Fiber-optic cable can handle high frequencies with little to no problem.

Corrosion and Water Resistance

The very nature of glass or plastic makes it resistant to most corrosives. Water does not affect the light conduction capabilities of a properly installed fiber-optic system. The fiber-optic cables that have been installed under the world's oceans are expected to last for many, many years.

Greater Distances

Fiber-optic cable can support data transmission over greater distances than copper core cable. Copper core cable used for network systems is typically limited to 100 meters, but fiber-optic cable can be used for long distance applications spanning 20 kilometers or more. Fiber-optic cable is typically used for network backbones of 2000 meters and 3000 meters in a single segment length. For certain applications such as FDDI, distances of 200 kilometers or approximately 124 miles can be supported.

The Nature of Light

Light is a form of energy closely related to electrical energy. While much is known about light, there are still many unanswered questions. One such question is what exactly is light energy? It has been debated whether is it a wave or a particle. For this reason, two main theories exist. The dominant theory, however, is that light is a wave. For the purpose of this textbook, the wave theory will be used to explain the nature of light.

Light is classified as electromagnetic energy and is identified in the electromagnetic wave spectrum chart along with radio waves, X-rays, and Gamma rays, **Figure 3-2.** An *electromagnetic wave* is a form of energy that behaves like a wave and can travel through a vacuum. It needs no other media for support. For example, sound is an energy that travels as a wave, but it must be carried by a gas (air), liquid (water), or solid (steel). Sound cannot travel through a vacuum as can radio waves, X-rays, and light waves. Hence, sound is not a form of electromagnetic energy.

In Chapter 2, Network Media—Copper Core Cable, electrical energy wave patterns were described using the term *frequency*. Remember, frequency is based

electromagnetic wave
a form of energy that behaves like a wave and can travel through a vacuum.

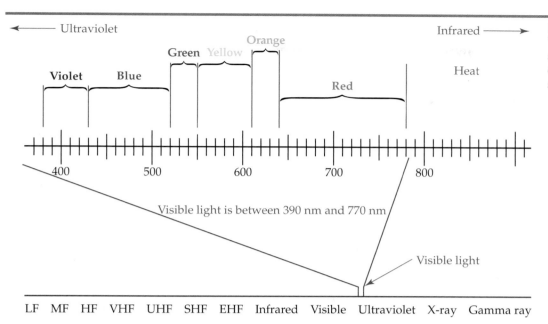

on the number of complete cycles of an electrical wave that occur in one second. Light energy can also be described using the term *frequency,* but the preferred term is *wavelength.* **Wavelength,** in relation to light waves, is the distance of one complete cycle and is measured in meters. For example, visible light is composed of wavelengths between 390 nanometers and 770 nanometers. A nanometer is one billionth of a meter. The nanometer is used to measure the distance of the repeating light wave pattern or the distance of one complete light wave cycle.

wavelength
the total distance an electromagnetic or light wave travels during one full cycle.

The term *wavelength* originated from the field of physics. Many of the terms first introduced in physics have been directly adopted and used to describe the phenomena of fiber-optic cable. It is interesting to note that early on the term *wavelength* was used to describe the fluctuating patterns of radio waves and is still used in the design of antenna systems.

The three common wavelengths for fiber-optic communications are 850 nanometers, 1300 nanometers, and 1550 nanometers. Fiber-optic devices and cable, such as transmitters, receivers, and fiber-optic core diameters, are selected according to wavelength of the signal used. For example, a device such as a transmitter generates light at specific wavelength. This requires that the receiver and fiber-optic core match the same wavelength to achieve the best performance.

Look again at Figure 3-2. Notice that the wavelength of visible light is between 390 nanometers and 770 nanometers. Below visible light are larger wavelengths, such as UHF and VHF television waves and radar waves. Above visible light are much shorter wavelengths, such as X-rays, Gamma rays, and cosmic rays. The actual spectrum of visible light is quite small compared to all other forms of electromagnetic wavelengths.

Most fiber-optic systems use infrared wavelengths between 850 nanometers to 1550 nanometers. At these wavelengths, there is less signal loss. A typical LED has a wavelength that is between 800 nanometers and 900 nanometers or 1250 nanometers and 1350 nanometers.

Fiber-Optic Cable Construction

Fiber-optic cable is composed of a glass or plastic core surrounded by a cladding. See **Figure 3-3.** The glass or plastic core is the medium for transferring light waves. The cladding surrounds the core and causes the transmitted light to remain in the core. Without the cladding on the core, the light would be lost though the sides of the core material. A buffer area surrounds the cladding and core. The buffer is wrapped around the core and cladding to provide physical protection. In addition to the buffer area, waterproofing materials are sometimes inserted under the cable sheath. The sheath is similar to the insulating jacket on copper core conductors. Its composition depends on the environment in which the fiber-optic cable will be installed. Some sheaths are designed to be oil- or water-resistant. Others may be installed in saltwater or buried directly in the earth. Some may be installed in a building plenum area. The size of fiber-optic cable is expressed in micromils (μm).

Fiber-optic cable is generally classified as either loose tube or tight buffer. Loose tube is fiber-optic cable in which the core or cores are loosely fitted inside the sheath. The sheath is typically filled with gel to protect the core(s). The core or cores are free to move inside the jacket. Loose tube is typically used for long cable runs where pulling the cable into conduit could result in breaking the core(s). Tight buffer fiber-optic cable has a sheath that is tightly bound to the core(s). This produces a smaller overall diameter of the cable assembly. The selection of cable type is left to the designer or installer.

When fiber-optic cable is installed in a plenum area, the installation usually falls under the jurisdiction of the National Electrical Code (NEC). The NEC

Figure 3-3.
A typical fiber-optic cable is constructed of a glass or plastic core surrounded by cladding, a buffer, and a sheath. The buffer provides strength for pulling the cable and serves as a soft padding to protect the core and cladding. The sheath protects the cable from physical elements.

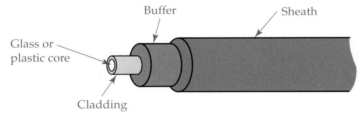

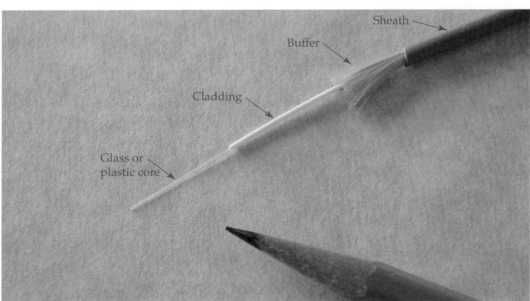

is referred to for building specifications or standards. The standards describe minimal requirements for installing cables and conduits. Many electrical contractors are responsible for installing all of a building's cables, including data cables. Electrical contractors and building codes refer to the NEC as a guide.

The NEC has strict regulations about the composition of the sheath of fiber-optic cables installed in a plenum area. Fumes produced by this covering in a fire may be dangerous to people in the building. In many buildings, the plenum area also contains the air-conditioning system. Fumes and vapors present in the plenum area could be harmful to personnel throughout the building.

Two broad classifications of fiber-optic cable commonly used are single-mode and multimode. To understand the characteristics of these fiber-optic cable classifications, you must first be familiar with the characteristics of fiber-optic transmission.

Fiber-Optic Cable Transmission Characteristics

Since fiber-optic cable carries light and not electrical energy, a new set of terms is needed to describe the characteristics of light carried over fiber-optic cable. While some of the terms might be familiar from the last chapter, some new terminology that is unique to light is introduced. The first characteristic presented is attenuation. Attenuation was discussed and related to copper core media in the previous chapter. The causes for the loss of an electrical signal over copper core cable, however, are different than for the loss of a light signal over fiber-optic cable. The major causes of attenuation in fiber-optic cable can be characterized as scattering, dispersion, extrinsic losses (bends, splices, and connectors) and Fresenel reflection loss.

Attenuation

In Chapter 2, Network Media—Copper Cable, you learned that the loss in transmission of signal power from one end of the cable to the other is called *attenuation*. Attenuation is expressed in decibels (dB), or decibels per kilometer (dB/km). The decibel is a relative measurement for signal strength. It is patterned after our sense of hearing.

Decibel measurement is nonlinear. It does not move in gradual increments, as does an analog speed gauge in an automobile. Look at the table in **Figure 3-4A** and compare the dB rating of 1, 3, 10, 20, and 50 to the power remaining and power loss values. Notice that a 50% power loss is represented by 3 dB. When the dB is doubled to 6, the power loss is only 75%, not the expected 100%. On close examination of the table, you will see that the relationship between power loss and dB is nonlinear. When decibel ratings are plotted as a chart, the result is a curve, not a straight line, **Figure 3-4B.**

The most common use of the decibel chart in fiber-optic cable specifications is to express the attenuation over a given distance. Each cable manufacturer has a slightly different decibel rating of loss for their fiber-optic core. The variation is caused by the amount of impurities left in the core after the refining and manufacturing process is complete. Engineers use these decibel ratings to calculate the expected attenuation of an installation. For example, the engineering staff of many communications carriers calculate the loss of light over a given distance when running fiber-optic cable in long-haul installations. A long-haul installation is the installation of fiber-optic cable between two end points that are

Figure 3-4.
The common unit of measure for power loss or gain is the decibel (dB). Light, sound, and electrical energy are often expressed in decibels. The decibel is based on a complicated mathematical formula. A—Notice that a 3 dB loss equals 50% power loss, but an additional 3 dB loss results in a total loss of 75%. B—The chart plots the power in decibels against power loss. As you can plainly see, the relationship is not linear, but rather curved.

dB	Power Remaining in Percent	Power Loss in Percent	Fraction of Power
1	79	21	
2	63	37	
3	50	50	½ power
4	40	60	
5	32	68	
6	25	75	¼
7	20	80	
8	16	84	
9	12	88	
10	10	90	1/10
11	8	92	
12	6.3	93.7	
13	5	95	
14	4	96	
15	3.2	96.8	
16	2.5	97.5	
17	2	98	
18	1.6	98.4	
19	1.3	98.7	
20	1	99	1/100
30	.1	99.9	
40	.01	99.99	
50	.001	99.999	1/1000

A

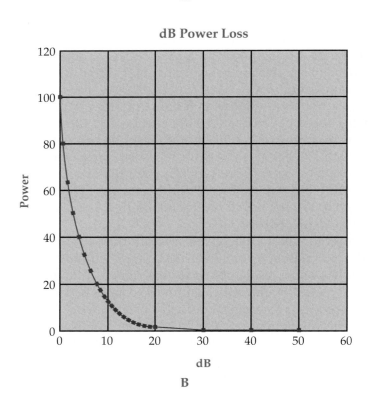

B

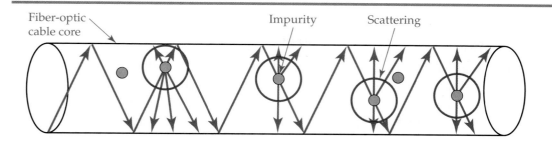

many miles apart. It is not unusual for a long-haul installation to span more than 40 miles. Long distance telephone carriers are a prime example.

Scattering

Scattering is the loss of signal strength due to impurities in the core material. No core material can be made 100% pure. A microscopic amount of impurities will always remain. These impurities cause the light to scatter, **Figure 3-5.** The total amount of impurities along the length of the cable is cumulative—the longer the cable, the greater the signal loss.

Glass core is a better transmitter of light than plastic, but it is more expensive. To reduce attenuation, glass cores are used on very long runs of fiber-optic cable.

scattering
the loss of signal strength due to impurities in the core material.

Dispersion

Dispersion is the distortion of the light wave pattern as it reflects off the core cladding. Dispersion is the main factor that limits the maximum length of a fiber-optic cable used as a communications media.

Light beams transmitted through fiber-optic cable do not travel in a straight line. This causes the light to reflect off the cladding. Look at **Figure 3-6.** Due to dispersion, digital light pulses transmitted into one end of a fiber-optic cable reach the opposite end of the cable at different times. This factor causes the light pulse to be distorted in shape. The light pulse appears flattened and elongated. A receiver can quickly reshape the signal into a square shape. On a short run of fiber-optic cable there is no real problem with dispersion, but on a long run, not

dispersion
the distortion of a light wave as it reflects off the cladding of a fiber-optic cable.

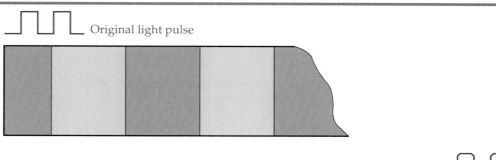

Original light pulse

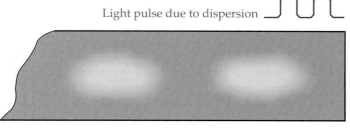

Light pulse due to dispersion

all of the light arrives at the end of the cable at the same time. The slight variation in time can be critical at high frequency rates of transmission.

Extrinsic losses

extrinsic losses
signal losses caused by physical factors outside the normal core such as splices, connectors, and bends in the core.

Extrinsic losses are caused by physical factors outside the normal core. Physical factors can be splices, connectors, and bends in the fiber core. Splices in fiber-optic cable need special attention. Splices, couplings, and connectors are the main reasons for signal loss in normal runs of fiber-optic cable. Splices and connectors are covered in detail later in this chapter. Fiber-optic cables also have a minimum bend radius. Exceeding this radius causes attenuation.

Fresenel reflection loss

Fresenel reflection loss
a type of signal loss that commonly occurs at connection points in fiber-optic cabling and is due to the refraction property differences in the core material, the connector materials used for sealing the connector, and air.

Fresenel reflection loss commonly occurs at connection points in fiber-optic cabling. It is typically due to the refraction property differences in the core material, the connector materials used for sealing the connector, and air. Light waves traveling from one type of core material to another cause signal loss. All glass and plastic core materials, sealing materials, and air have different refraction indices, or properties. Using a sealing material minimizes this type of loss. The sealing material uses a refraction index closely matching the index of the glass or plastic core material. A mismatch of sealing agent to core material can cause an excessive amount of signal loss.

Fiber-Optic Cable Specifications

The two broad classifications of fiber-optic cable based on the diameter of the core are multimode and single-mode. See **Figure 3-7.** In general,

Figure 3-7.
Fiber-optic cable is classified as either single-mode or multimode. Multimode fiber-optic cable has a larger core diameter than single-mode. A large diameter core contributes to dispersion. Single-mode fiber-optic cable is much smaller in diameter and has a more restrictive light path, which results in less dispersion. Single-mode fiber-optic cable is used for long distance applications.

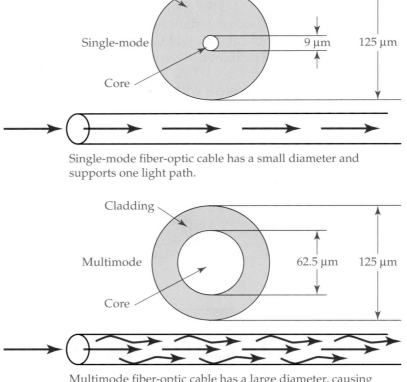

Single-mode fiber-optic cable has a small diameter and supports one light path.

Multimode fiber-optic cable has a large diameter, causing multiple paths of light to be transmitted.

multimode fiber-optic cable has a larger core diameter than single-mode fiber-optic cable. A larger core diameter causes more light loss due to dispersion. *Single-mode fiber-optic cable* is much smaller in diameter than multimode cable. The diameter of single-mode fiber-optic cable is almost equal to the length of the light wave traveling through the cable. Because the core is designed to closely match the wavelength of the light wave, the light wave cannot readily disperse as it does in a large diameter core. The result is single-mode fiber-optic cable can carry light farther than multimode fiber-optic cable.

Look again at Figure 3-7. Note that the core dimensions are measured in micrometers (μm) or millionths of a meter. The size of a fiber-optic cable is expressed in two numerical values separated by a slash. For example, a fiber-optic cable identified as 65.5/125 indicates that the core diameter is equal to 65.5 micrometers, and the overall core and cladding diameter is 125 micrometers. Typical multimode core dimensions are 50/125, 62.5/125, and 100/140. However, 100/140 is an older technology and is not encountered very often. A typical single-mode core is 8.3/125.

Multimode cable is divided into two classifications: step-index and graded-index. These classifications are based on the design of the multimode cable core. *Graded-index multimode fiber-optic cable* is designed with a varying grade of core material. Look at **Figure 3-8.** Note that the core of the graded-index fiber-optic cable is designed with maximum light conduction at the center of the core. The ability of the core to conduct light is gradually reduced toward the cladding. This lessens the light's tendency to disperse. Less dispersion causes less signal distortion and attenuation. The *step-index multimode fiber-optic cable* does not counter dispersion.

IEEE 802.3 Standards

IEEE is mainly concerned with the lower levels of the OSI model and develops standards, which serve as guidelines for hardware applications. The

multimode fiber-optic cable
a type of fiber-optic cable that has a large core diameter and is susceptible to attenuation due to dispersion.

single-mode fiber-optic cable
a type of fiber-optic cable that has a small core diameter and limited dispersion. Single-mode fiber-optic cable can carry light farther than multimode fiber-optic cable.

graded-index multimode fiber-optic cable
a type of multimode fiber-optic cable that has a varying grade of core material. The core is designed with maximum light conduction at its center and gradually diminished light conduction toward its cladding.

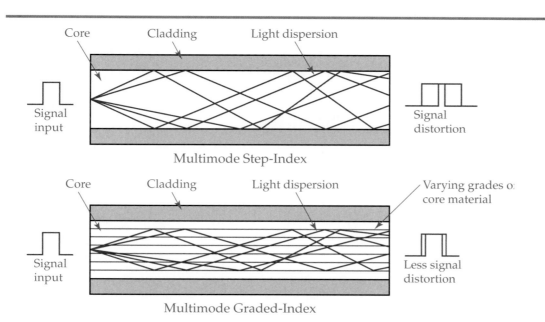

Figure 3-8. There are two types of multimode fiber-optic cable: graded-index and step-index. Graded-index multimode cable is designed with maximum light conduction in the center of the core. Conduction gradually diminishes toward the cladding, which reduces dispersion. Less light dispersion produces less digital signal distortion. No special design is applied to step-index.

step-index multimode fiber-optic cable
a general multimode fiber-optic cable that does not counter dispersion.

original IEEE 802.3 Ethernet standard has evolved as the demand for higher data rates, greater distances, and new technologies are introduced.

When an IEEE standard is modified, it is identified with a lowercase letter. For example, IEEE uses 802.3z to contain the information concerning Gigabit Ethernet. It was adopted in the late 1990s. Gigabit Ethernet identifies special cable applications known as 1000BaseSX, 1000BaseLX, and 1000BaseCX. Take note that 1000BaseCX is a copper core cable and not a fiber-optic cable. 1000BaseCX was limited to 25 meters and used to connect server room equipment. 1000BaseSX uses multimode fiber while 1000BaseLX uses either single-mode or multimode fiber. In **Figure 3-9,** you can readily see that single-mode fiber-optic cable supports much greater distances than multimode fiber-optic cable.

In 2002, the IEEE 802.3 Ethernet standard amended the 10 Gigabit Ethernet standard known as IEEE 802.3ae. The IEEE 802.ae standard only recognizes fiber-optic cables. It does not recognize copper core cables. The 802.3ae standard describes the 10GBaseSR (Short Range), 10GBaseLR (Long Range), and 10GBaseER (Extended Range) classifications. 10GBaseSR supports short distances using multimode fiber-optic cable, mainly 26 meters to 82 meters depending on the type of fiber-optic cable used. 10GBaseLR and 10GBaseER support 10 kilometers and 40 kilometers respectively. Long distance communications use single-mode fiber-optic cable.

Figure 3-9.
Fiber-optic cable classifications.

Standard	Single-mode (S) or Multimode (M)	Core Diameter in Microns	Wavelength in Nanometers (nm)	Cable Distance in Meters	Remarks
10BaseFL	M	62.5 50	850	2000	Early generic fiber-optic standard.
100BaseFX	M	62.5 50	1300	2000	Known as Fast Ethernet.
1000BaseSX	M	62.5 50	850	300	Known as Gigabit Ethernet.
1000BaseLX	M	62.5 50	1300	550	Known as Gigabit Ethernet.
1000BaseLX	S	9	1300	5 k	Known as Gigabit Ethernet.
10GBaseSR	M	62.5 50	850	66-300	Known as 10 Gigabit Ethernet. Distance dependent on bandwidth.
10GBaseLR	S	9	1310	10 k	Known as 10 Gigabit Ethernet.
10GBaseER	S	9	1310	40 k	Known as 10 Gigabit Ethernet.

Note: Companies may calculate greater distances based on manufacturer cable specifications and equipment requirements. Such engineering is common in long haul communication systems such as long distance telecommunications industries.

Media	Horizontal Cross Connect to Main Cross Connect in Meters	Horizontal Cross Connect to Intermediate Cross Connect in Meters	Main Cross Connect to Intermediate Cross Connect in Meters
UTP	800	300	500
62.5/125 50/125 Multimode Fiber-Optic Cable	2,000	300	1,700
Single-Mode Fiber-Optic Cable	3,000	300	2,700

Figure 3-10.
The ANSI/TIA/EIA 568-B 1 standard for backbone distribution.

Note: Single-mode fiber-optic cable can be installed to distances of 60 kilometers or greater based on engineering calculations and product manufacturers specifications.

At times, standards and manufacturers' tables conflict. Look at the ANSI/TIA/EIA 568-B 1 Standard table in **Figure 3-10.** You will see the recommended standard distances for network backbone distribution. This table is solely concerned with the maximum distance of backbone cabling without identifying specific IEEE cable standards. The chart simply categorizes backbone cable into one of three categories: UTP, multimode fiber-optic cable, and single-mode fiber-optic cable. It makes no distinction for cable data rates. The IEEE standards are concerned with data rates and distances based on their developed standards.

Network+ Note:

Be sure to read any test items carefully for reference to ANSI/TIA/EIA or IEEE.

Network+

Fiber Distributed Data Interface (FDDI)

Fiber Distributed Data Interface (FDDI) is a standard developed by ANSI that employs fiber-optic cable over great distances. The distances can reach as far as 40 kilometers in length and support data speeds of 100 Mbps or higher. FDDI is used mainly as a backbone for large network systems such as a MAN or a WAN.

FDDI is structured as a pair of rings, **Figure 3-11.** Two rings of fiber-optic cable connect to each node on the network system. One ring supports data

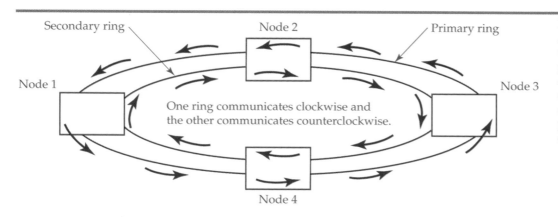

Secondary ring   Node 2   Primary ring

Node 1

One ring communicates clockwise and the other communicates counterclockwise.

Node 3

Node 4

Figure 3-11.
A typical FDDI system uses two fiber-optic paths to provide continuous data communications. If one ring fails, the other provides a path for communications.

Fiber Distributed Data Interface (FDDI) a standard developed by ANSI that employs fiber-optic cable over great distances. The distances can reach as far as 40 kilometers in length and support data speeds of 100 Mbps or higher. FDDI is used mainly as a backbone for large network systems such as a MAN or a WAN.

flow in a clockwise rotation and the other ring supports data flow in a counter-clockwise rotation.

Using FDDI guarantees continuous communication. If one ring fails, the other ring automatically takes over and provides the path for communications. One ring is considered the primary communication ring and the other is the secondary. If a section fails in the primary ring, the secondary ring supports communication. If both the primary and secondary ring is cut or fails, the two rings join at the next node, forming a new ring and continuing communications, **Figure 3-12.**

FDDI uses a token passing method similar to IEEE 802.5. An enhanced version of FDDI called FFDI Full Duplex Technology (FFDT) supports speeds of 200 Mbps by allowing simultaneous communication in both directions. See **Figure 3-13** for a summary of FDDI data rates and cable lengths. FDDI can also use UTP cable.

Fiber-Optic Cable Connectors

There are many different styles of fiber-optic cable connectors. Some are proprietary. It is best if you use connectors from the same manufacturer when installing a new fiber-optic network or extending an old one. The most common styles of fiber-optic cable connectors are SC, ST, FC, LC, and MTRJ.

Figure 3-12.
If the primary ring in an FDDI system fails, the secondary ring provides communication. When both rings break, the nodes closest to the break complete the circuit, allowing communication to continue.

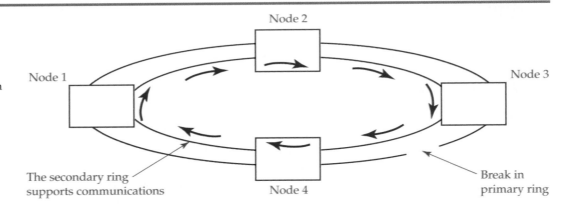

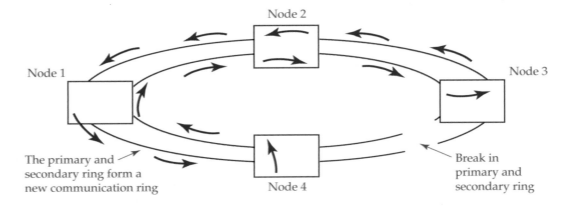

Media	Single-Mode (S) or Multimode (M)	Core Diameter in Microns	Wavelength in Nanometers (nm)	Cable Distance in Meters
FDDI	M	62.5/125	1300	2 k
FDDI	S	9	1300	40 k

Figure 3-13.
FDDI specifications.

Network+ Note:

Although there are many different types of fiber-optic cable connectors, CompTIA lists only five in the Network+ Certification exam objectives: SC and ST, FC, LC, and MTRJ.

Look at the close-up of an ST and SC connector in **Figure 3-14A.** Note that the ST connector is round and uses a push and twist connection similar to the BNC. The SC connector is square and uses retaining clips to hold its position. When two SC or ST connectors are fabricated into a single unit, they are generally referred to as duplex connectors. A single connector is referred to as a simplex connector.

FC connectors are designed to provide a maximum physical connection, **Figure 3-14B.** The FC connector is designed with screw threads as part of the physical design. They are very similar in design to coaxial style screw on connectors.

The LC connector is a relatively new style developed by Lucent Technologies. The LC is classified as a small form factor connector. The term *form factor* refers to the physical attributes of a connector. The LC connector is approximately half the diameter of an SC or ST connector. The LC connector can be converted into a duplex connector by the addition of a clip designed to hold the two separate connectors together. See **Figure 3-14C.**

The MTRJ connector is unique in design because it incorporates two fiber-optic cores into one assembly without the use of a clip, **Figure 3-14D.** It is always considered a duplex connector. The MTRJ is also a small form factor connector.

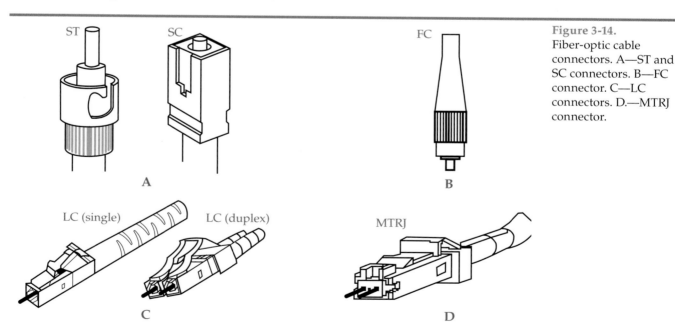

Figure 3-14.
Fiber-optic cable connectors. A—ST and SC connectors. B—FC connector. C—LC connectors. D.—MTRJ connector.

Fiber-Optic Cable Installation and Troubleshooting

Fiber-optic cables are typically not made in the field but are prepared prior to installation in a controlled setting. Installing a fiber-optic connector on a fiber-optic cable requires a great deal of expertise. Typically, a technician who installs fiber-optic cable connectors must be certified. BICSI is an organization that provides certification. Another such organization is the Fiber Optic Association (FOA). Some companies, such as Ortronics, Corning Cable Systems, Nortel Networks, and Belden, oversee their own certification programs. Whether or not a technician requires certification is determined by the vendor or by installation specifications written for a new installation. Usually the purchasing company will prepare these.

Installing Connectors

A supplier may provide the necessary cable for a small, fiber-optic network, such as a LAN. This includes taking measurements and preparing cable splices and connections prior to installation. When fiber-optic cabling is obtained in this way, the highest quality control standards are applied to cable preparation. However, this is not always possible. Cables may have to be installed in the field where factors such as dirt, dust, and chemicals can hamper the cable splicing, termination, and connector installation. Splices, couplings, and connections require special equipment and techniques. Fiber-optic cores must be in near-perfect alignment when splices are made. Misalignment or other improper splicing will cause attenuation. See **Figure 3-15.**

Figure 3-15.
Common causes of attenuation at splice points.

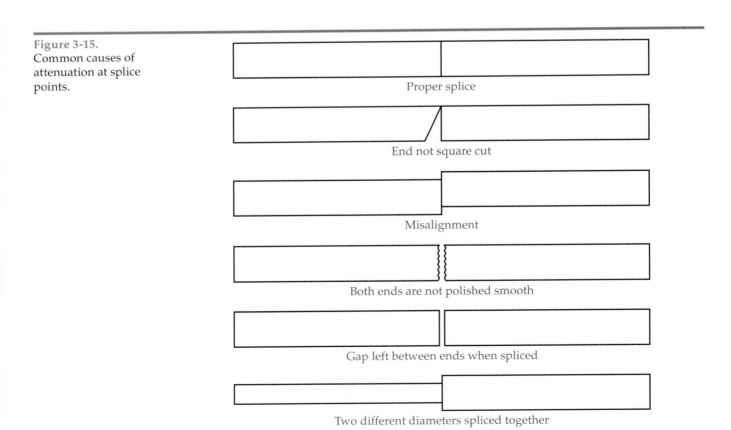

Proper splice

End not square cut

Misalignment

Both ends are not polished smooth

Gap left between ends when spliced

Two different diameters spliced together

The fiber-optic cable is cleaved, not cut like copper core cable. To cleave the fiber-optic cable core, it must be scribed with a sharp, cleaving tool. The edge of the cleaving tool is made of diamond or ceramic material. Once the core has been scribed, pressure is applied to the scribed area until the core breaks. This produces a clean, clear surface at the end of the core. Traditional cutting tools used for copper core cable would badly scar the ends of the core. Marks or scratches on the ends of the core material cause attenuation due to reflection and refraction. The light scatters in many directions rather than transferring directly to the next cable or connector. Plastic core cables do not have to be cleaved. An extremely sharp cutter must be used to obtain a clean, clear cut.

Figure 3-16 shows a high-quality, fiber-optic cleaver and splicer. Note the eyepiece located at the top of the tool. Since the diameter of a fiber-optic core is small, a microscope with at least 30× magnification is needed to inspect the cut and splice. This tool is capable of cutting fiber-optic cable, polishing the ends, and fusing them together for a near-flawless splice.

Making a Fusion Splice

A *fusion splice* is the joining of two fiber-optic cores using heat to fuse, or melt, the materials together. Another, more expensive fusion splicer is shown in **Figure 3-17**. This fusion splicer performs all the functions of the splicer shown in Figure 3-16 and more. It is equipped with video cameras and a liquid crystal display for close, simultaneous inspection of the splice at two different angles.

Before making a fusion splice, the outer, protective sheath and buffer must be removed from the fiber-optic cable end. The splice area of the core must be very clean. Dust and debris can contaminate the splice, resulting in a poor data rate through the finished splice. The cladding must be carefully handled. Damage to the cladding will cause signal loss. Once the fiber-optic cable is prepared, the end of the fiber-optic core is cleaved. A cleaver simply nicks the core rather than cuts through it. The core can then be easily broken.

Next, the two ends of the fiber core must be perfectly aligned. Because of the small diameters of the core, a magnifier is used to aid in the alignment. Some of the more expensive fusion splicing units combine the cleaver magnifier and fuser into one package.

fusion splice
the joining of two fiber-optic cores using heat to fuse, or melt, the materials together.

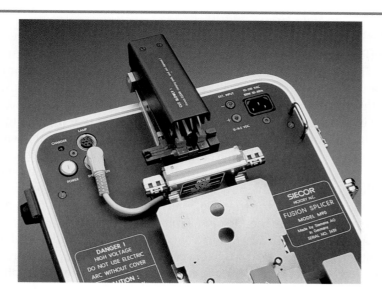

Figure 3-16.
Fiber-optic cable splicer with cleaver. (Siecor Corporation, Hickory, NC)

Figure 3-17.
A multiprocessor-controlled fusion splicer. (Siecor Corporation, Hickory, NC)

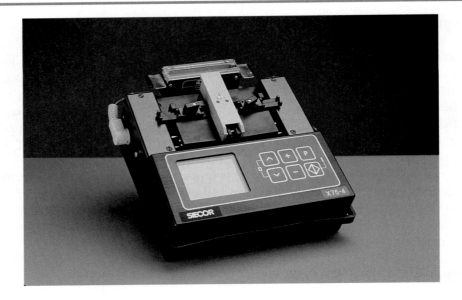

When the two ends are aligned, they are melted together by an electrical arc. The two ends become permanently welded together, or fused. After fusing the cores, the splice must be protected so it will not be damaged through handling or installation. The splice is covered using material such as heat shrink tubing, silicone gel, or mechanical crimps.

Mechanical splices are made in a similar way except for fusing the two cores. Mechanical splices do not use heat but rather a mechanical means of splicing the two cores. The two cores are aligned, and then the mechanical crimper clamps the two cores together. Between the two cores is a gel-like substance that maintains a consistent light path between the two ends by eliminating any air gap between the two cores. An air gap would cause the Fresenel effect, thus reducing the ability of the splice to carry light from one core to the other.

The main difference between the mechanical and the fusion splice is the cost of the equipment and the cost of each splice. The equipment for making a fusion splice is very expensive compared to the equipment for making a mechanical splice. However, the cost of each splice is less expensive when using fusing equipment.

Using Fiber-Optic Cable Meters

Testing fiber-optic cable requires specialized instruments that employ the principles of light to obtain data. The most commonly used meter for testing short runs of fiber-optic cable consists of two items: a power meter and a light source. The light source is attached to one end of the fiber-optic cable, and the power meter is connected to the other end. The light source injects light into the fiber-optic cable, and the light meter measures the amount of light reaching the opposite end. Power loss is calculated and is typically expressed in decibels (dB) or microwatts (μW). The information gathered from the meter can be sent to and stored on a computer. For some new network installations, this test is more than adequate to assure a contractor that the fiber-optic cable installation is acceptable.

Another type of meter is the ***Optical Time Domain Reflectometer (OTDR)***. The OTDR is a sophisticated and expensive meter for testing and troubleshooting long runs of fiber-optic cable. The OTDR accurately measures and records the effects of attenuation. It compares the amount of light injected into the fiber-optic cable core

Optical Time Domain Reflectometer (OTDR)
a meter used for testing and trouble-shooting long runs of fiber-optic cable. The OTDR conducts measurements based on the principle of attenuation.

Figure 3-18.
Fluke OFTM-5612
OTDR. (Reproduced
with permission—
Fluke Corp.)

to the amount of light reflected to the meter. **Figure 3-18** shows the Fluke OFTM-5612 OTDR, which is used to inspect and analyze fiber-optic cable installations.

The OTDR can also accurately measure the total length of a cable run or locate breaks and faults in a cable and at its connections. The typical OTDR presents information in a graphical manner on a liquid crystal display (LCD). A slope is presented that represents the attenuation of the core. Major flaws in splices and such are presented as spikes, **Figure 3-19.** The distance to the fault and other pertinent information, such as signal loss in decibels (dB) or milliwatts (mW), is indicated along the bottom edge of the screen. **Figure 3-20** illustrates how faults and distance are represented on a graph.

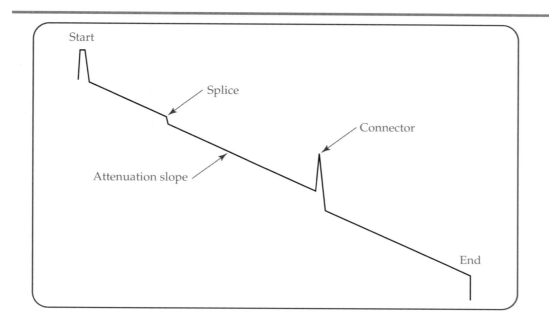

Figure 3-19.
Attenuation and cable
faults as indicated on
an OTDR display. Note
that faults are indicated
by spikes.

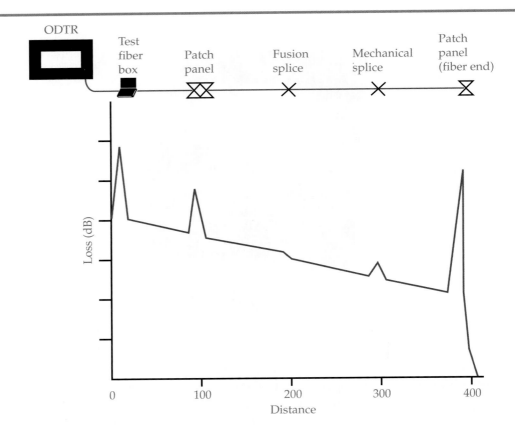

Figure 3-20.
An OTDR can indicate loss in decibels and the exact distance from the OTDR to the fault.

Figure 3-21 shows an OTDR signal trace analysis indicating the attenuation of the cable and the location of a fault of 7.619 dB at approximately 53.481 kilometers from the meter. Using test equipment that can quickly identify the location of a fault can save many hours of testing by other means.

An OTDR is typically equipped with a parallel port for connecting directly to a printer. Screen captures and a full report on the specifications of the fiber-optic cable can be printed. The OTDR is routinely used for the final inspection of a new fiber-optic cable installation. It not only provides information to prove the cable is flawless, but also provides a baseline of information for comparison at a later date when troubleshooting. The typical OTDR is an expensive piece of test equipment; however, it does not need to be purchased. To save money, contractors often rent an OTDR to test and certify a fiber-optic cable system they have installed. OTDR tests are not normally performed on short, fiber-optic cable runs associated with LANs. This type of installation is typically checked with a simple, handheld meter, not an OTDR. The OTDR is best suited for measuring cable runs spanning great distances like those found in network backbones.

Summary

- Fiber-optic cable has the following advantages: provides for data security, immune to electromagnetic interference, lightweight and small in diameter, safe from fire and explosion, wide bandwidth, corrosion- and water-resistant, supports data transmission over longer distances than copper core cable.

- Light is described in wavelengths.

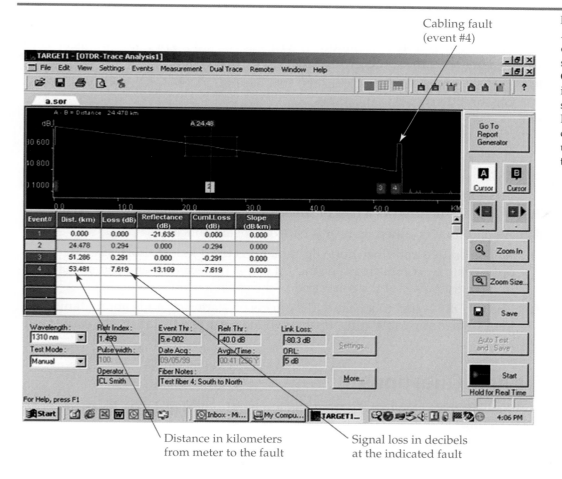

Cabling fault
(event #4)

Figure 3-21.
A trace analysis is
displayed on the LCD
screen of an ODTR.
Cable faults are
indicated by the sharp
spikes along the slope.
Exact losses are also
displayed in decibel
units at the bottom of
the display.

Distance in kilometers
from meter to the fault

Signal loss in decibels
at the indicated fault

- A wavelength is the total distance the electromagnetic wave or light wave travels during one full cycle.

- Wavelengths are measured in nanometers (nanometer), or one billionth of a meter.

- Fiber-optic cable cores are composed of either glass or plastic.

- The glass or plastic core is surrounded by cladding, which restricts the light to the core area.

- Scattering is the loss of light due to impurities in the core material.

- Dispersion is the distortion of light waves caused by the light reflecting from the cladding material and arriving at different times at the far end of the cable.

- Extrinsic losses are caused by physical factors not normally found in the core material, such as at splices and connector locations.

- Two broad classifications of fiber-optic cable based on its ability to carry light are multimode and single-mode.

- Single-mode fiber-optic cable has a smaller core diameter than multimode fiber-optic cable and carries light farther and with less attenuation.

- Two classifications of multimode fiber-optic cable are graded-index and step-index.

■ Graded-index multimode fiber-optic cable is designed with a varying grade of core material that allows for maximum light conduction at the center of the core. Step-index multimode fiber-optic cable does not have a special core design and is therefore greatly affected by dispersion.

■ The 802.3z standard describes the 1000BaseSX, 1000BaseLX, and 1000BaseCX Gigabit Ethernet classifications.

■ The 802.3ae standard describes the 10GBaseSR (Short Range), 10GBaseLR (Long Range), and 10GBaseER (Extended Range) 10 Gigabit Ethernet classifications.

■ FDDI is used mainly as a backbone for large network systems such as a MAN or a WAN.

■ FDDI is structured as a pair of rings.

■ A splice that is joined by heat is called a fusion splice.

■ A light source and fiber-optic light meter is used to test short runs of fiber-optic cable by comparing the amount of light injected in one end of the cable to the light power at the other end of the cable.

■ The OTDR is used to measure the effects of scattering and cabling faults in long fiber-optic cable runs. It can also measure the distance to a cable fault or break.

Review Questions

Answer the following questions on a separate sheet of paper. Please do not write in this book.

1. What are seven advantages of fiber-optic cable?

2. Light is measured in _____.

3. What does the abbreviation nm represent?

4. What is the wavelength range of visible light?

5. What are the three common wavelengths associated with fiber-optic cable?

6. What is *dispersion*?

7. Name three physical factors that contribute to extrinsic losses.

8. What are the two classifications of fiber-optic cable based on the diameter of the core?

9. How does the diameter of the fiber-optic cable core affect the distance light can travel?

10. Fiber-optic cable core diameter is expressed in _____.

11. What are the two most common sizes of multimode fiber-optic cable?

12. How can the light distance carrying capacity of a multimode cable be improved?

13. What are the three classifications that the IEEE 802.ae standard describes?

14. How is FDDI structured?

15. Why does FDDI use two rings of cable?

16. What are the five most common types of fiber-optic cable connector?

17. What are some common causes of attenuation associated with fiber-optic cable splices?

18. Using heat to join two fiber-optic cores is called a _____ splice.

19. What is required to test a short run of fiber-optic cable?

20. What device is commonly used to test long runs of fiber-optic cable?

21. What principle of fiber-optic cable loss does the OTDR use for measurements?

Sample Network+ Exam Questions

Network+

Answer the following questions on a separate sheet of paper. Please do not write in this book.

1. Which of the following network media is typically the most secure from ease dropping devices or security taps?
   A. Fiber-optic cable
   B. Copper core cable
   C. Wireless
   D. Infrared

2. Identify the connector in the following exhibit.
   A. SC
   B. ST
   C. RJ-45
   D. BNC

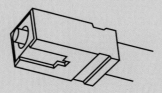

3. Identify the connector in the following exhibit.
   A. SC
   B. ST
   C. RJ-45
   D. BNC

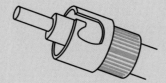

4. Where is fiber-optic cable typically found? (Select all that apply.)
   A. In network backbones.
   B. In network segments that are exposed to electromagnetic interference.
   C. Where security from wiretaps is of great concern.
   D. In network segments requiring slow data transmission rates.

5. Fiber-optic cable losses are expressed in which unit?
   A. Decibels
   B. Ohms
   C. Resistance
   D. Impedance

6. The loss of signal strength in the core of a fiber-optic cable caused by impurities in the core is referred to as _____.
   A. scattering
   B. dispersion
   C. diffusion
   D. reflection

7. Which cable standard supports the longest distances?
   A. 10GbaseER
   B. 10GBaseLX
   C. 1000BaseLX
   D. 100BaseCX

8. According to the ANSI/TIA/EIA standards for backbone distribution inside a premise, what is the maximum recommended distance for multimode fiber-optic cable?
   A. 100 meters
   B. 200 meters
   C. 500 meters
   D. 2000 meters

9. What type of networking media is used in FDDI?
   A. Fiber-optic multimode
   B. Fiber-optic single-mode
   C. UTP
   D. All the above materials can be used for FDDI.

10. Which is a correct description of FDDI topology?
   A. Star
   B. Bus
   C. Single ring
   D. Dual ring

Suggested Laboratory Activities

1. Determine the core size of a fiber-optic cable using a micrometer.

2. Visit the Web sites of several network interface card venders and see how much a fiber-optic network interface card would cost for your home computer.

3. Price a 100 meters of multimode fiber-optic cable and compare to 100 meters of UTP.

4. Conduct an Internet search to locate the steps required to make a fiber-optic patch cable.

5. Visit the BICSI and Fiber Optic Association (FOA) Web site and see what it takes to become a certified fiber-optic technician.

Interesting Web Sites for More Information

http://archive.comlab.ox.ac.uk/other/museums/computing.html

http://bwcecom.belden.com/college/college.htm

www.bell-labs.com/technology/lightwave/

www.bell-labs.com/technology/wireless/

www.cableu.net

www.fiber-optics.info

www.fiber-optics.info/glossary-f.htm

www.levitonvoicedata.com/learning/default.asp

www.lucent.com

www.occfiber.com

www.ortronic.com

www.siemon.com

Check out Belden's Cable College located at www.belden.com/college/college htm. The Cable College includes technical papers, cable installation procedures, a glossary of cabling terms, and other reference materials.

**Chapter 3
Laboratory Activity**

Fiber-Optic Connector Identification

After completing this laboratory activity, you will be able to:
- Identify common fiber-optic connector types.
- Distinguish between common fiber-optic connectors by physical traits.

Introduction

This laboratory activity has been designed to familiarize you with various fiber-optic cable, connectors, and vendors. In this laboratory activity, you will search the Internet for images of common, fiber-optic connectors. Once located, you will copy the images and paste them into a word processing document and then label the image with the appropriate acronym.

To copy a Web page image, right-click the image and then select **Copy** from the shortcut menu. To insert the image into a word processing document, right-click a blank area of the document page and click **Paste**. When you have finished copying the images specified in the laboratory activity, you will print two copies of the assignment, one for your reference and the other to turn in to your instructor.

Some fiber-optic cable manufacturers and distributors include the following: Ortronics, Siemon, Belden, Black Box, Agilent, Fluke, and 3Com. You may wish to visit their Web sites for information about various fiber-optic cables and connectors.

Equipment and Materials

- Workstation with Internet access and word processing software, such as Microsoft WordPad. (WordPad is included in all Microsoft operating systems).
- Printer access.

Procedures

1. _____ Report to your assigned workstation.

2. _____ Boot the workstation and verify it is in working order.

4. _____ Open Internet Explorer.

5. _____ Conduct a search for fiber-optic connectors. Use the following phrases in your search:

   ■ Fiber-optic connectors.

   ■ Fiber optic connectors.

   ■ Fiber-optic cable connectors.

   ■ Fiber optic cable connectors.

6. _____ After locating images of fiber-optic connectors, copy each image from the Web page and paste it into a word processing document. Be sure to label each example. Use the following labels, one for each connector type: Biconic, ESCON, FC, FDDI, MTRJ, SC, SC Duplex, SMA, and ST.

7. _____ When you are finished, print out two copies of your document.

8. _____ Answer the review questions. You may use the Internet to research the answers.

9. _____ Return your workstation to its original condition.

Review Questions

Answer the following questions on a separate sheet of paper. Please do not write in this book.

1. What is the most obvious physical difference between an ST and an SC connector?

2. What is the main physical difference between an LC connector and an ST connector?

3. How many fiber-optic cores are typically associated with an FDDI connector?

4. How many fiber-optic cores are typically associated with an MTRJ connector?

5. How many fiber-optic cores are typically associated with an ST connector?

6. What does the acronym ESCON represent?

7. What does the acronym FDDI represent?

8. In general, what does the term *duplex* mean when referring to fiber-optic connectors?

4

Wireless Technology

After studying this chapter, you will be able to:

❏ Describe the principles of radio wave transmission.

❏ Identify and describe the major antenna styles and the purpose of each.

❏ Describe the three transmission techniques used in radio wave-based transmission.

❏ Identify the characteristics of the U-NII classifications.

❏ Identify the key characteristics of the IEEE 802.11 wireless networking standards.

❏ Describe the CSMA/CA access method.

❏ Identify the key characteristics of the Bluetooth standard.

❏ Explain how cellular technology works.

❏ Describe the two types of microwave networks.

❏ Describe the two types of infrared transmission.

❏ List the advantages and disadvantages of wireless networking.

❏ Explain the purpose of the SSID.

❏ Explain how security is provided in wireless networks.

Network+ Exam—Key Points

To prepare for the CompTIA Network+ exam, you must be able to specify the characteristics of the IEEE 802.11 classification. These characteristics include speed, cable lengths, and access method. You should also be able to define and identify an ad hoc network and an infrastructure wireless network.

To better prepare yourself to answer questions on wireless technology, be sure to install and configure a wireless network card and view all of the configuration properties that are featured.

Key Words and Terms

The following words and terms will become important pieces of your networking vocabulary. Be sure you can define them.

ad hoc mode
carrier wave
cellular technology
channel
demodulation
direct sequencing
directional
Extensible Authentication Protocol (EAP)
frequency hopping
geosynchronous orbit
infrastructure mode
ISM band
modulation
omni-directional
orthogonal frequency-division
   multiplexing (OFDM)

piconet
propagation delay
radio interference
radio waves
receiver
spread spectrum
transmitter
unbounded media
Wi-Fi
Wi-Fi Protected Access (WPA)
Wireless Access Point (WAP)
Wireless Application Protocol (WAP)
Wired Equivalent Privacy (WEP)
working group

unbounded media an unrestricted path for network transmissions.

Wireless media is often referred to as unbounded media. *Unbounded media* means the path for network transmissions is unrestricted. When copper core or fiber-optic cable is used, the transmitted network signals are bound to the medium. When the atmosphere is used, the transmission is spread throughout the atmosphere and is not limited to a single path.

Network signals transmitted through the atmosphere are electromagnetic waves. In Chapter 3—Fiber-Optic Cable, you learned that light waves are a part of the electromagnetic wave spectrum and are categorized as visible light. You learned specifically how light waves travel through fiber-optic cable to transmit data. In this chapter, you will learn about the three categories of electromagnetic waves that are used to transmit data across the atmosphere: radio waves, infrared, and microwaves. You will also learn about wireless transmission techniques and associated standards. To help you better understand wireless networking technologies, electromagnetic waves will be discussed in detail.

Electromagnetic Waves

The atmosphere is full of electromagnetic waves. Electromagnetic waves are categorized according to frequency ranges. For example, common radio waves used for communications start at .5 MHz for AM radio and span to 22 GHz for satellite communications. The frequency of an electromagnetic wave is based on the repeating pattern of its waveform just as it is with electrical energy. One complete waveform is called a cycle, **Figure 4-1,** and frequency is the number of times a cycle occurs in one second.

Electromagnetic waves are produced both intentionally and unintentionally as a by-product of electrical energy. Radio and television stations produce electromagnetic waves intentionally while many household appliances produce electromagnetic waves unintentionally. You should recall from Chapter 2, Network Media—Copper Core Cable that unintentional electromagnetic waves are referred to as interference, or more specifically, noise.

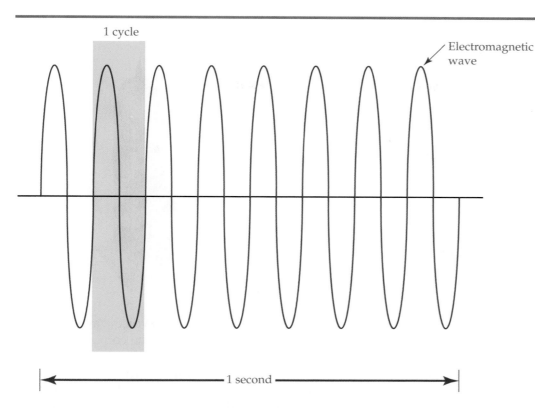

1 cycle

Electromagnetic wave

1 second

Figure 4-1. Electromagnetic frequency is based on the repeating pattern of a waveform. One complete waveform is called a cycle. The frequency is how many cycles occur in one second.

Radio and Microwave Transmission

Radio and microwave transmission works on the principle of producing a carrier wave as the means of communication between two wireless devices. A *carrier wave* is an electromagnetic wave of a set frequency that is used to carry data. It is identified by a frequency number. For example, radio and television stations are assigned specific frequencies on which they must transmit. A radio identification such as 104.5 FM represents a carrier wave of 104.5 MHz. The carrier wave is how individual stations are identified on a radio or television, **Figure 4-2.**

The carrier wave is mixed with the data signal. The mixing of the carrier wave and data signal is known as *modulation.* The technique of modulation is how AM radio, FM radio, and television operate. A simple radio broadcast consists of a *transmitter,* which generates the carrier wave and modulates information into the carrier wave, and a *receiver,* which receives the modulated wave and demodulates it. The transceiver and receiver must both be at or very near the same carrier wave frequency for communication to occur.

Look at the example of a voice wave broadcast in **Figure 4-3.** In the example, a carrier wave of 104.5 MHz, is shown. The human voice produces sound at a much lower frequency, typically in a range from a 400 cycles per second to approximately 4,000 cycles per second. The human voice is converted into electrical energy using a microphone. The microphone produces a pattern of electrical energy in direct proportion to the human voice. The electrical energy is mixed with the carrier wave. The carrier wave has a much higher frequency than the electromagnetic wave produced by the human voice. The two are combined so that the human voice can be transmitted.

The combined wave is transmitted across the atmosphere. When the combined wave reaches the receiver and is accepted, the electromagnetic energy

carrier wave
an electromagnetic wave of a set frequency that is used to carry data.

modulation
the process of mixing a data signal with a carrier wave.

transmitter
an electronic device that generates a carrier wave and modulates data signals into the carrier wave.

receiver
an electronic device that receives a modulated signal and demodulates it.

Figure 4-2.
A carrier wave is the means of communication between two wireless devices. In this example, a radio station assigned the 104.5 MHz frequency transmits data on a 104.5 MHz carrier wave. A radio with its dial set to 104.5 picks up the information broadcast from the radio station.

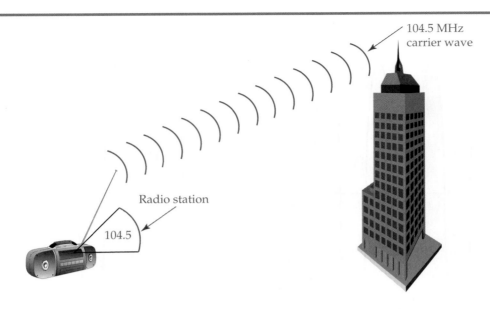

104.5 MHz carrier wave

Radio station

104.5

demodulation
the process of separating a data signal from a carrier wave.

is converted into electrical energy. The receiver separates the voice wave from the carrier wave. This process is known as *demodulation.* After the combined wave is demodulated, the transceiver discards the carrier wave and then amplifies the voice wave and sends it to a speaker. The speaker converts the electrical energy into a voice wave. While this is a simple, nontechnical explanation, it is important to remember that the carrier wave and the voice wave are combined, or modulated, before they are transmitted and are separated, or demodulated, after they are received.

The same principle is used to transmit digital data signals. A carrier wave establishes the transmitter/receiver relationship. The carrier wave is modulated with a wave pattern resembling the digital data signal. The two waves are combined before transmission and then separated at the receiver.

To modulate data, the carrier wave must be at a much higher frequency than the digital data. In the example of the voice wave and carrier wave frequency,

Figure 4-3.
Example of a voice wave broadcast.

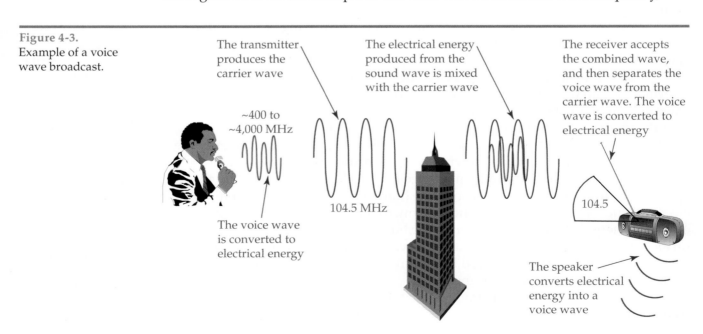

The transmitter produces the carrier wave

The electrical energy produced from the sound wave is mixed with the carrier wave

The receiver accepts the combined wave, and then separates the voice wave from the carrier wave. The voice wave is converted to electrical energy

~400 to ~4,000 MHz

The voice wave is converted to electrical energy

104.5 MHz

104.5

The speaker converts electrical energy into a voice wave

the carrier wave is 104.5 MHz while the voice wave fluctuates between 400 Hz and 4,000 Hz. Based on an average voice frequency of 2,000 Hz, an approximate 500:1 ratio exists between the two frequencies. The carrier wave is only slightly distorted when combined with the voice wave. The same principle applies when a carrier wave is combined with a digital data signal. If the two signals do not have a high ratio, the digital data signal distorts the carrier wave to a point where the transmitter cannot recognize it.

A 104.5 MHz carrier wave is 200 kHz in width. Technically, a 104.5 MHz carrier wave has a bandwidth of 200 kHz. The bandwidth of a carrier wave is referred to as a ***channel.*** Technically, a channel is a small portion of the electromagnetic spectrum and is used to designate a set of frequencies for a particular electronic application. The FCC assigns channels and bandwidths for electromagnetic waves.

channel
the bandwidth of a carrier wave.

A channel is identified by the assigned frequency that represents the starting point of the band. For example, 104.5 is the identification of the channel even though the channel spans the next 200 kHz. Look at **Figure 4-4.** The 145 MHz is the designated channel for the carrier wave. This channel is a single position in the entire radio frequency spectrum represented by the 145 MHz designation. The carrier wave channel is 200 kHz wide starting at exactly 145 MHz and ending at 145.2 MHz. The carrier wave must stay within the 200 kHz band as specified by FCC regulation.

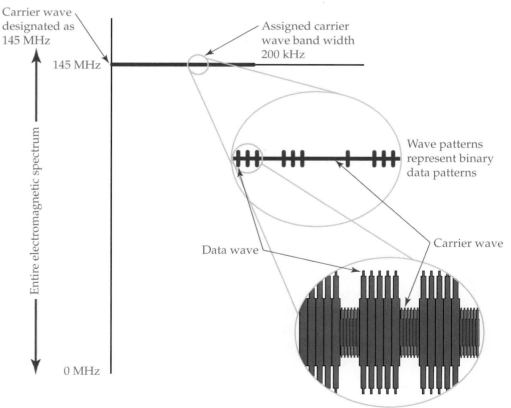

Carrier wave designated as 145 MHz

Assigned carrier wave band width 200 kHz

145 MHz

Entire electromagnetic spectrum

0 MHz

Wave patterns represent binary data patterns

Data wave

Carrier wave

Zoomed close-up of the modulated radio signal, which is a combination of carrier wave and data wave patterns

Figure 4-4.
The carrier wave and the data wave are electronically combined or modulated. The modulated signal must remain inside the radio frequency range or channel assigned by the FCC.

The exact center of the bandwidth area is 145.1 MHz. It is nearly impossible for the carrier wave to remain exactly at 145.1 MHz. The electronic components that are used to create the circuit that produce the exact frequency change directly with temperature. The components heat up because of the electrical energy passing through them and are also influenced by environmental temperatures. Electronic transmitters are enclosed and air-conditioned to keep the components at a predetermined temperature. The enclosure compensates for environmental changes in temperature and the heat effects of the electronic components. A change in component temperature causes a direct change in the carrier wave frequency produced by electronic components.

Infrared Transmission

Infrared uses a series of digital light pulses. The light is either on or off. A typical television remote control uses infrared technology to transmit to the television receiver rather than radio wave technology. There are two distinct disadvantages to infrared technology. First, the two devices that are communicating must be in direct line of sight of each other. This means that the infrared receiver/transmitter of each device must be aimed, or at least positioned, in the general direction of the other.

The infrared beam is weak and must be transmitted through an optical lens to prevent dispersion. The other disadvantage is an infrared beam can travel a relatively limited distance compared to wireless radio wave technologies. Some infrared communication links for networks use a laser to achieve greater distances.

Even with its disadvantages, infrared can be found in many of the same applications as Bluetooth. It is commonly used in some models of personal digital assistants, palm tops, and laptops. Typical data rates for infrared devices are 1 Mbps and 2 Mbps. Data rates can be much higher when lasers are used for the direct connection of two LANs.

Radio Interference

radio interference
interference that
matches the
frequency of the
carrier wave.

Radio interference is interference that matches the frequency of the carrier wave. The Federal Communications Commission (FCC) is responsible for dividing the entire electromagnetic spectrum to prevent electronic equipment from interfering with one another. However, this is not always possible. Think for a minute how many different devices use and produce electromagnetic waves—remote controls for remote control cars, remote control airplanes, garage door openers, television, AM radio, FM radio, satellites, pagers, cellular phones, electrical power lines, radar, motors, fluorescent lights, and such. The list is ever increasing. There are thousands of products that produce electromagnetic waves. The FCC regulates the electromagnetic spectrum and dictates the frequency that is to be used for each group of devices. Even with all these regulations, equipment fails and produces undesirable frequencies that can interfere with a regulated frequency. If you move the tuner of an AM radio across the various stations, you will hear an excellent example of interference.

AM radio is an old technology that was susceptible to radio interference. FM radio is an improvement over AM radio because it is less susceptible to radio interference.

Virtually any type of electrical equipment can produce radio interference even if it is not assigned to the radio frequency spectrum assigned to wireless LAN communications. For example, some other sources of radio interference that could corrupt data packets are fluorescent lighting, electric motors, electrical control systems, welding equipment, portable radios, and such. While not intended to produce radio interference in the assigned wireless spectrum, a defective piece of equipment can produce electrical radio harmonics. The harmonic signal is a multiple of an original signal. For example, a radio frequency of 12,000 can also produce signals of 24,000, 36,000, 48,000, and so on. This means that harmonic frequencies produced by other areas of the electromagnetic spectrum can cause interference with the wireless network.

Another major factor is proximity to the source of the signal. If radio interference is in close proximity to the wireless network system, it need not be at the same frequency. Close proximity and a powerful signal can disrupt the wireless communications. Radio signals can also reflect or bounce off surrounding materials, usually metallic surfaces, **Figure 4-5.** The reflected signal is an exact copy of the original signal. The reflected signal does not always disrupt the original signal. At times, it actually increases the strength of the original signal by combining with it. The problem occurs if there is too much delay between the original signal and the reflected signal. When there is too much delay, the two signals overlap causing the combined signal to be distorted. The difference in time it takes the two signals to travel from the source to the destination determines the level of distortion. The amount of time it takes to reach the destination depends on the angle and distance of the reflection.

There are many sources of radio interference that can affect communications on an IEEE 802.11 wireless LAN. Industrial, scientific and medical devices occupy the frequency band allocated to the wireless network devices specified under the IEEE 802.11 standard. Cordless phones and microwave ovens are also included in the same frequency ranges. The band of radio frequencies associated with industrial, scientific, and medical devices is referred to as the *ISM band,* **Figure 4-6.** The acronym ISM represents Industrial, Scientific, and Medical.

ISM band
the band of radio frequencies associated with industrial, scientific, and medical devices.

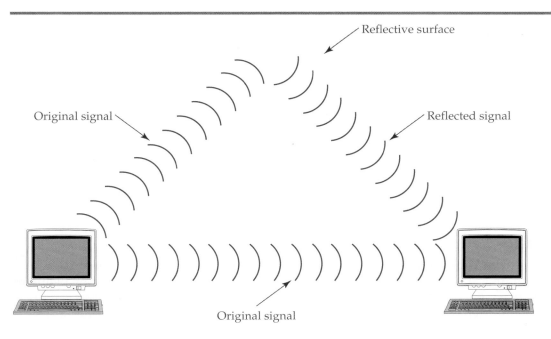

Reflective surface

Original signal

Reflected signal

Original signal

Figure 4-5.
A reflected radio signal can combine with the intended radio signal and either disrupt the intended signal or enhance it.

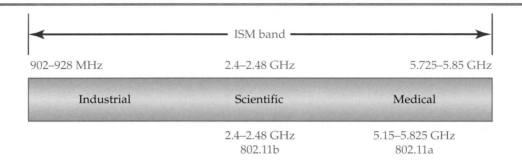

Antenna Styles

Two major classifications of antennae are associated with radio wave- and microwave-based wireless networks: omni-directional and directional. These classifications are based on an antenna's ability to transmit electromagnetic signals. *Omni-directional* is the transmission of electromagnetic signals in all directions. *Directional* is the transmission of electromagnetic signals in a focused or aimed direction. Antennae can be further broken down by their individual style of construction such as omni, dipole, flat panel, Yagi, and parabolic dish. **Figure 4-7** shows each antenna style and the electromagnetic wave pattern it produces. Note that the electromagnetic wave patterns are viewed from overhead.

omni-directional
the ability of an antenna to transmit electromagnetic signals in all directions.

directional
the ability of an antenna to focus or aim an electromagnetic signal in a particular direction.

Omni

The omni is a straight piece of wire. The wire is engineered to match the exact length or a fraction of the frequency's wavelength. For example, a frequency of 2.4 GHz produces a wavelength of approximately 2.19″. An antenna 2.19″ in length would match the wavelength exactly. Matching the antenna length to the wavelength of the radio frequency ensures the best possible reception and reduces the possibility of picking up interference.

The omni antenna is typically used for a wireless transmitter to broadcast in a 360° pattern. This type of antenna is used for the source of a wireless transmission. For example, a wireless Internet service provider would use an omni antenna to facilitate broadcasting in all directions to better serve customers throughout the area.

Dipole

The dipole is one of the most common radio antennae used. What makes it popular is it is relatively inexpensive to manufacture compared to many other antenna styles. The dipole is commonly used as a client or receiver antenna rather than as a broadcast antenna. The dipole antenna is bidirectional, as seen in Figure 4-7. Rotating the antenna until the dipole aligns with the source of the radio wave transmission can enhance the received signal.

Yagi

The Yagi antenna is used for point-to-point links. It is a directional-type antenna. A Yagi antenna is typically designed from many radio antenna elements (tubes). Each element is progressively larger or smaller than the main element by approximately 5%. The way the Yagi antenna enhances the radiation of the

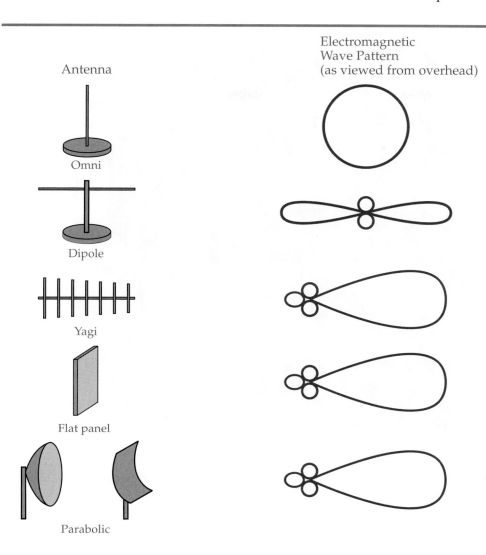

Antenna

Electromagnetic
Wave Pattern
(as viewed from overhead)

Omni

Dipole

Yagi

Flat panel

Parabolic

Figure 4-7.
Basic antenna styles and the electromagnetic wave patterns they produce.

electromagnetic wave is beyond the scope of this textbook. For now, just note and recognize the shape of the Yagi antenna and its electromagnetic wave pattern. Some Yagi antennae are constructed inside a metal tube to further enhance the reception and transmission of electromagnetic waves. See **Figure 4-8.**

Flat Panel

The flat panel is a directional-type antenna used for point-to-point links. The main advantage of a flat panel is the aesthetics. The antenna blends in well with building architecture, and many times it is unnoticeable. However, the disadvantage of a flat panel antenna is the consideration of wind load. High wind areas can catch the flat panel design like a sail on a ship. Flat panels must be rigidly supported to reduce the effects of wind.

Parabolic

The parabolic antenna is used for point-to-point links. It a directional-type antenna typically constructed from a grid of rods or mesh wiring. The parabolic antenna enhances reception by reflecting incoming electromagnetic waves with its curved surface to a horn at its center. When the parabolic antenna transmits

Figure 4-8.
Yagi antenna constructed inside a metal tube. The metal tube enhances reception and transmission. (Photo reprinted with the permission of PCTEL, Maxrad Product Group.)

electromagnetic waves, the horn transmits the signal toward the curved surface of the antenna. The curved surface reflects the electromagnetic waves to produce a beamlike pattern in the same way light is reflected from the curved surface of a flashlight to produce a beam of light. The parabolic antenna is constructed as a simple, curved surface or in the shape of a dish. When it is constructed in the shape of a dish, it is usually called a parabolic disk or simply dish antenna. The parabolic antenna greatly amplifies a weak radio wave signal compared to other antenna types.

Radio Wave Transmission Techniques and Networking

radio waves
electromagnetic waves with a frequency range of 10 kHz to 3,000,000 MHz.

Radio waves are electromagnetic waves with a frequency range of 10 kHz to 3,000,000 MHz. They have the longest wavelength compared to microwave and infrared. Radio waves are used in LANs. Radio wave-based networks adhere to the IEEE 802.11 and Bluetooth standards and operate at 2.4 GHz.

Radio Wave-Based Transmission Techniques

To communicate between radio-based wireless network devices, several transmission techniques are used: single-frequency, spread-spectrum, and orthogonal frequency-division multiplexing. You must become familiar with these techniques because most wireless network technologies are described using these terms. The techniques are based on the technology, the frequency band of operation, and the manufacturer's idea of the best way to achieve a high data rate. A high data rate not only relies on how fast the data can move between two points, but also on how much data has to be retransmitted because of interference.

Transmission techniques divide an allocated frequency band into many separate frequency ranges, or channels. After the frequency band is divided, a carrier wave is generated for each of the channels.

Spread spectrum

Spread spectrum is a transmission technique that uses multiple channels to transmit data either simultaneously or sequentially. The term *spread spectrum* refers to transmission channels *spread* across the *spectrum* of the available bandwidth.

Spread spectrum transmission works in a similar manner that highways work. A highway system consists of several separate lanes to carry vehicles. You can think of spread spectrum as several radio wave paths designed to carry radio waves. In the highway system, trucks carry supplies to a store. Each truck uses a separate lane. If one of the lanes is blocked (interference), the other lanes can still carry the supplies to the store. This is the same method employed by spread spectrum. If one of the channels is blocked by radio interference, the other channels can still carry the radio wave data.

In the spread spectrum technique, data can be transmitted on multiple channels simultaneously or sequentially (one at a time). The spread spectrum technique that transmits data on multiple channels simultaneously is called *frequency hopping.* The spread spectrum technique that transmits data on multiple channels sequentially is called *direct sequencing.*

Spread spectrum is the chosen transmission method of most wireless technologies. Transmitting data on multiple channels decreases the likelihood of interference. Interference is typically limited to only one or two of the channels. The other channels in the frequency band are free to carry data undisturbed. Data that is lost can be easily retransmitted on a channel that is not affected by the interference.

Frequency hopping

Frequency hopping is also referred to as *frequency hopping spread spectrum (FHSS)*. Frequency hopping is used with wireless devices that use the 2.4 GHz radio band. The 2.4 GHz frequency has a bandwidth of 83.5 MHz. Rather than use the entire range as a single channel to carry radio data, the frequency band is divided into seventy-nine, 1 MHz channels. See **Figure 4-9.**

Instead of transmitting the data packets over a single channel, the data packets hop from one channel to another in a set pattern determined by a software algorithm. None of the 79 channels are occupied for more than .4 seconds. Since the data packets switch from channel to channel, or rather frequency to frequency, this transmission technique is called frequency hopping.

Many people assume that because data packets hop to various channels, this transmission technique was designed as a security measure. This assumption is false. The reason that the frequency hopping technique is used is to avoid interference. If any interference matches the same frequency as the wireless networking devices, the data would be corrupted. If only one frequency was used and it encountered interference, then the network would be useless. By using the frequency hopping technique, interference can be tolerated. The interference would likely only affect one or two of the available frequencies in the entire seventy-nine 1 MHz channels. This allows more than an ample number of channels to ensure continuous transmission between wireless networking devices.

spread spectrum
a transmission technique that uses multiple channels to transmit data either simultaneously or sequentially.

frequency hopping
a spread spectrum technique that transmits data on multiple channels simultaneously.

direct sequencing
a spread spectrum technique that transmits data on multiple channels sequentially.

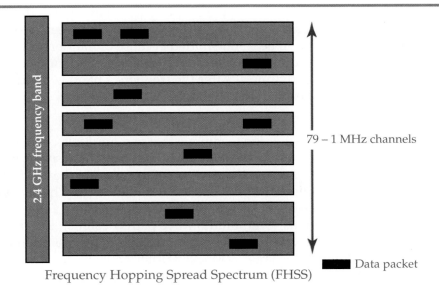

Figure 4-9.
In frequency hopping,
data is transmitted over
79-1 MHz channels.
The data transmissions
continuously use dif-
ferent channels in short
increments of less than
0.4 seconds each.

79 – 1 MHz channels

■ Data packet

Frequency Hopping Spread Spectrum (FHSS)

The frequency hopping technique is limited to a maximum of a 2 Mbps data rate. A much higher data rate can be accomplished using direct sequencing.

Direct sequencing

Direct sequencing divides the 2.4 GHz frequency band into eleven over-lapping channels of 83 MHz each. Within the eleven channels are three channels with a 22 MHz bandwidth. The three channels do not overlap and can be used simultaneously. Using three channels at the same time results in higher data rates than frequency hopping. The data rates for direct sequencing are 11 Mbps and 33 Mbps. The 33 Mbps is a result of using all three 22 Mbps channels at the same time, **Figure 4-10.**

One disadvantage of direct sequencing is that a much larger portion of the transmitted data is affected by electromagnetic interference than with frequency hopping. The data rate of direct sequencing is, therefore, drastically affected by interference.

Direct sequencing is also referred to as *direct sequencing spread spectrum (DSSS)*. Most vendors use DSSS technology at 11 Mbps for wireless network systems.

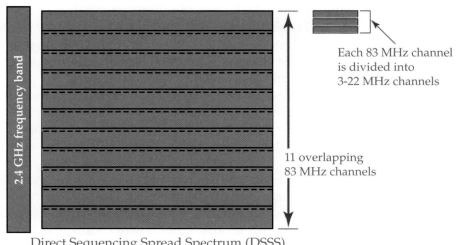

Figure 4-10.
In direct sequencing,
a 2.4 GHz frequency
band is divided into
eleven overlapping,
83 MHz channels.
Each 83 MHz channel
is further divided
into three 22 MHz
channels.

Each 83 MHz channel
is divided into
3-22 MHz channels

11 overlapping
83 MHz channels

Direct Sequencing Spread Spectrum (DSSS)

Orthogonal frequency-division multiplexing

The *orthogonal frequency-division multiplexing (OFDM)* transmission technique is used with wireless devices that use the 5 GHz radio band and can achieve a data rate as high as 54 Mbps. The OFDM transmission technique divides the allotted frequency into channels similar to frequency hopping and direct sequencing. *Orthogonal* means separate side by side over a range of values. In wireless application, the term *orthogonal* means there are multiple separate radio channels side by side within an assigned radio band. *Frequency division* means to divide the assigned frequency range into multiple, narrow sub-frequencies. *Multiplexing* is an electronics term, which means to combine content from different sources and transmit them collectively over a single, common carrier. By combining the three terms, OFDM means to communicate wireless data over several different channels within an assigned frequency range. However, in OFDM, each channel is broadcast separately and is referred to as multiplexed.

OFDM is used in conjunction with the Unlicensed National Information Infrastructure (U-NII) frequency ranges. The FCC divided the 5 GHz radio frequency into three, 20 MHz channels and classified them as the Unlicensed National Information Infrastructure (U-NII). The three classifications are U-NII-1, U-NII-2, and U-NII-3. See **Figure 4-11.** Each of the three U-NII classifications has a frequency range of 100 MHz. Using the OFDM transmission technique, each 100 MHz frequency range is broken into four separate 20 MHz channels. Each of the 20 MHz channels are further divided into fifty-two, 300 kHz sub-channels. Forty-eight of the fifty-two sub-channels are used to transmit data, and the remaining four are used for error correction. It is the large number of channels that provide the high data rates. Additionally, communication is not affected as adversely by interference as it is with the other techniques mentioned. If one or two sub-channels are affected, the overall data rate is not affected.

The FCC U-NII classifications are based on the frequency range of the broadcast, the allowable maximum amount of power allotted to the broadcast, and the location where the device may be used. There is no maximum distance measurement in feet or meters for the different classifications. The maximum distances are controlled by the maximum amount of output wattage that can be generated by the devices. The actual range varies considerably due to influences such as building structures and materials, the electromagnetic environment,

orthogonal frequency-division multiplexing (OFDM) a transmission technique that transmits data over different channels within an assigned frequency range. Each channel is broadcast separately and is referred to as multiplexed. It can achieve data rates as high as 54 Mbps.

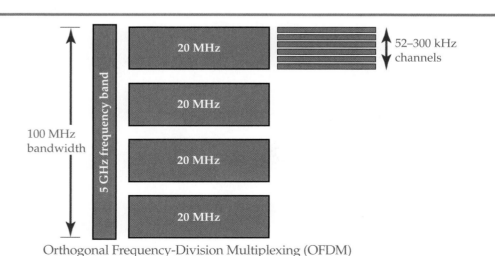

Orthogonal Frequency-Division Multiplexing (OFDM)

Figure 4-11. Orthogonal frequency-division multiplexing (OFDM) is used in conjunction with the U-NII frequency ranges to achieve a data rate as high as 54 Mbps. Each of the U-NII frequency ranges is 100 MHz wide. This bandwidth is divided into three 20 MHz channels, which are further divided into fifty-two 300 kHz channels.

Figure 4-12.
U-NII classifications and
their characteristics.

Classification	Frequency Range	Power	Comment
U-NII 1	5.15 GHz–5.25 GHz	50 mW	Indoors
U-NII 2	5.25 GHz–5.35 GHz	250 mW	Indoors/outdoors
U-NII 3	5.725 GHz–5.825 GHz	1 W	Outdoors

and atmospheric conditions. Use the chart in **Figure 4-12** to get a relative idea of expected maximum distances.

Distances vary by manufacturer and by location conditions such as placement of metal cabinets and building materials. The maximum power output of the device has a direct relationship to data throughput. Packet loss is generally caused by radio interference or excessive distance between two devices. When packet loss increases, the data rate decreases. The data rate is automatically adjusted to a lower rate when an excessive number of packets are lost. The data rate continues to be lowered until an acceptable packet loss is reached. The more powerful the signal, the less interference can disrupt the signal. This means that there will be fewer packets lost. Consequently, data rate is better when the signal is more powerful. The maximum transmission power rating for a wireless device is set by the FCC. Do not attempt to memorize distances because they are not standard.

Radio Wave-Based Networking

Wireless Access Point (WAP)
a wireless network device that provides a connection between a wireless network and a cable-based network and controls the flow of all packets on the wireless network.

Radio wave-based networks are rapidly becoming the choice of many networking systems because of the fast and easy installation and the convenience of no wires. A simple wireless network, such as one designed for home or business use, consists of two or more computers with wireless network adapters. **Figure 4-13** shows a USB wireless network adapter.

While not required, most simple wireless networks contain a Wireless Access Point (WAP). A *Wireless Access Point (WAP)* provides a connection between a wireless network and a cable-based network. Wireless Access Points typically provide access from wireless network devices to needed hardwired network devices, such as printers, modems, and routers.

Figure 4-13.
USB wireless network
adapter.

A typical WAP comes equipped with two omni antennae, **Figure 4-14.** One antenna is used for transmitting, and the other is used for receiving. This allows for full-duplex communication, which makes the WAP more efficient than if a single antenna were used.

When a WAP is present in a wireless network, all communications must go through it. You can think of a WAP as a traffic director for packets. The WAP controls the flow of all packets on the wireless network. When multiple Wireless Access Points are used, the roaming device (laptop, palm top, personal digital assistant) automatically detects and connects to the WAP with the strongest signal, which is typically the closest WAP.

All Wireless Access Points use the same Service Set Identifier (SSID). The SSID is similar in nature to a workgroup name or domain name. The SSID is a name that identifies Wireless Access Points within the same network. Multiple Wireless Access Points within the same network should be configured with the same SSID to support network roaming. When separate network systems are within range of each other, the Wireless Access Points for each separate network should use a different SSID to distinguish themselves from each other. Most Wireless Access Points and wireless network adapters come with a default SSID. If they are installed with minimal or no configuration at all, all Wireless Access Points will have the same SSID. The SSID can be readily reconfigured.

Service Set Identifier (SSID) names are case sensitive.

Tech Tip

Wireless network adapters and Wireless Access Points come with default settings to make configuration easy and, in some cases, automatic. The problem is easy or preconfigured settings make for weak security. Each device by the same manufacturer uses the same default SSID. The SSID should be changed to make the system more secure. By default most manufacturers do not enable encryption. The encryption process slows the overall data rate of the system. It takes time to encrypt and decrypt the packets. Encryption, however, should be enabled to make the network more secure.

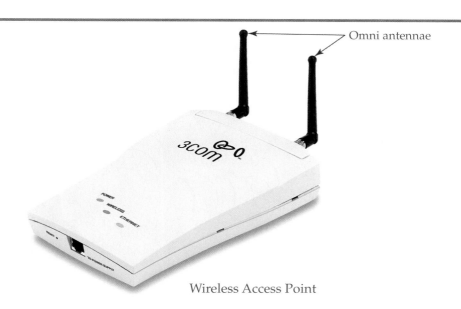

Omni antennae

Wireless Access Point

Figure 4-14.
Wireless Access Point (WAP). Note that the WAP has two omni antennae. One antenna is used for receiving, and the other antenna is used for transmitting.

Figure 4-15.
A wireless network in
infrastructure mode.

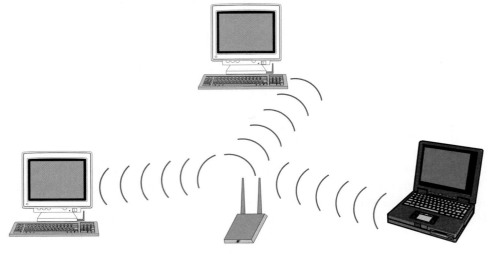

Infrastructure Mode

infrastructure mode
a wireless network
that contains one or
more Wireless Access
Points (WAPs).

ad hoc mode
a wireless network
that does not contain
a Wireless Access
Point (WAP).

A wireless network that contains one or more Wireless Access Points is arranged in an *infrastructure mode*, **Figure 4-15.** When a WAP is not present in a wireless network, it is arranged in an *ad hoc mode*, **Figure 4-16.** The reference to the term *ad hoc* means that the system is put together rapidly and is usually intended to be a temporary rather than a permanent installation. An ad hoc network can contain a maximum of 20 computers. Typically, an ad hoc network consists of a laptop and a desktop computer. An ad hoc network can be compared to a peer-to-peer network where all devices are equal.

The IEEE 802.11 Standard

Radio wave-based networks adhere to the 802.11 standard. The 802.11 standard consists of three classifications of wireless networks: 802.11a, 802.11b, 802.11g. See **Figure 4-17** for a chart of the 802.11 specifications and their characteristics.

Figure 4-16.
A wireless network in
ad hoc mode.

Ad Hoc Mode

802.11 Specification	Radio Frequency	Frequency Range	Data Rate	Range (approximate)	Transmission Method
802.11a	5 GHz	5.15 GHz–5.825 GHz	8 Mbps 12 Mbps 18 Mbps 24 Mbps 36 Mbps 54 Mbps	50 m	OFDM
802.11b	2.4 GHz	2.4 GHz–2.4835 GHz	1 Mbps 2 Mbps 5.5 Mbps 11 Mbps	100 m	DSSS
802.11g	2.4 GHz	2.4 GHz–2.4835 GHz	1 Mbps 2 Mbps 5.5 Mbps 11 Mbps	100 m	DSSS
			54 Mbps	50 m	OFDM

Figure 4-17.
IEEE 802.11 specifications and their characteristics.

802.11a

The 802.11a operates at the 5 GHz frequency and has a maximum data rate of 54 Mbps. Other lower data rates an 802.11a device may use are 48 Mbps, 36 Mbps, 24 Mbps, 18 Mbps, 12 Mbps, 8 Mbps, and 6 Mbps. At the 5 GHz frequency, 802.11a networking devices are not susceptible to interference from devices that cause interference at the 2.4 GHz frequency range. 802.11a devices are incompatible with 802.11b and 802.11g devices. Also, 802.11a devices use a higher frequency than 802.11b or 802.11g devices. The higher frequency cannot penetrate materials such as building walls like the lower frequency devices can. This results in 802.11a devices having a shorter range compared to 802.11b and 802.11g devices.

802.11b

Although the 802.11a and 802.11b classifications were developed at the same time, 802.11b was the first to be adopted by industry. The maximum data rate for 802.11b is 11 Mbps. When the highest rate cannot be achieved because of distance or radio interference, a lower rate is automatically selected. The lower rates are 5.5 Mbps, 2 Mbps, and 1 Mbps. 802.11b can operate over any of the 11 channels within the assigned bandwidth. When communicating between wireless devices, all devices should use the same channel. When using devices from the same manufacturer, the same channel is automatically selected by default. Two wireless networks, one constructed of 802.11b devices and the other constructed of 802.11a devices, can coexist without interfering with each other because they use different assigned frequencies. This allows for two different wireless networks to operate within the same area without interfering with the other.

802.11b networks are often referred to as *Wi-Fi*, which stands for *wireless fidelity*. The use of the term *Wi-Fi* was due to manufacturers forming the Wireless Ethernet Compatibility Alliance (WECA) in an effort to standardize wireless network devices. Devices approved as compatible with the 802.11b standards are given a "Wi-Fi compliant" seal, which means any device bearing the Wi-Fi seal is compatible with any other device bearing the seal, **Figure 4-18.** This process is an attempt to standardize the industry so that users are able to mix and match components from various manufacturers. There are many proprietary devices and software products on the market that are not compatible with other devices and software.

Wi-Fi
a term coined by the Wi-Fi Alliance that refers to wireless network products used in the 802.11b category.

Figure 4-18.
Wireless PCMCIA
adapter card with the
Wi-Fi compliant seal.

802.11g

The IEEE 802.11g standard followed the 802.11a and 802.11b standards. 802.11g operates in the 802.11b frequency range of 2.4 GHz. This makes it downward compatible with 802.11b devices. When communicating with 802.11b devices, the maximum data rate is reduced to 11 Mbps. The maximum throughput for the 802.11g standard is 54 Mbps but the maximum distance is typically much shorter than an 802.11b device. Lower data rates an 802.11g device can use are 48 Mbps, 36 Mbps, 28 Mbps, 24 Mbps, 12 Mbps, 11 Mbps, 9 Mbps, 6 Mbps, 5.5 Mbps, 2 Mbps, and 1 Mbps. Since 802.11g is assigned to the same frequency range as 802.11b, it is susceptible to the same sources of radio interference.

802.11g and 802.11b devices are not compatible with 802.11a devices because they use different frequencies. It must be noted that while the standards are different, there are devices on the market that can communicate with any of the mentioned wireless standards. In other words, there are wireless devices that can communicate with 802.11a, 802.11b, and 802.11g devices.

IEEE 802.11 Access Method

IEEE 802.11 networks rely on Carrier Sense Multiple Access with Collision Avoidance (CSMA/CA) as a media access protocol. Do not confuse this with CSMA/CD, which is used by IEEE 802.3 networks. The Carrier Sense Multiple Access (CSMA) portion of the technology is the same for both the 802.11 and 802.3 networks. The difference is in the Collision Detection (CD) versus Collision Avoidance (CA). Collision Detection (CD) detects a collision on the network after it occurs, while Collision Avoidance (CA) attempts to avoid a collision.

To understand the CSMA/CA process, look at **Figure 4-19.** The laptop equipped with a wireless card first listens for network traffic. If the airwaves are clear, it signals the WAP with a request to send (RTS) message. The WAP returns either a clear to send (CTS) or busy signal to the laptop. The process is repeated until the laptop is cleared to send data. After the data has been sent to the WAP, the WAP sends an acknowledge (ACK) to the laptop. Collision avoidance technology solves the problem of broadcast storms, which are associated with collision detection technology. The collision detection technology is designed to accept network collisions, wait, and then resend the complete transmission.

The CSMA/CA access method was selected over CSMA/CD because of the nature of wireless media. In a typical Ethernet environment that uses cable, a collision can be detected anywhere on a segment by all nodes in that segment.

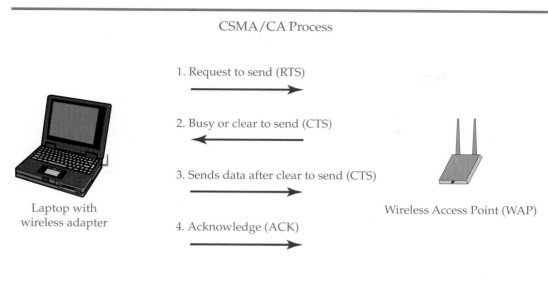

CSMA/CA Process

1. Request to send (RTS)

2. Busy or clear to send (CTS)

3. Sends data after clear to send (CTS)

4. Acknowledge (ACK)

Laptop with
wireless adapter

Wireless Access Point (WAP)

Figure 4-19.
A laptop equipped with a wireless network adapter listens for network traffic. If all is clear, it sends a request to send (RTS) signal to the Wireless Access Point (WAP). The WAP returns a clear to send (CTS) or a busy signal to the laptop. This process is repeated until the laptop is cleared to send the data. After the data is sent to the WAP, the WAP sends an acknowledge (AWK) signal to the laptop.

The same is not true of a wireless network. Look at **Figure 4-20** and compare the two networking technologies. In the cable-based network, all nodes are connected to the same segment via a hub. All nodes can detect any broadcast that takes place. There is no need to broadcast an intention to communicate.

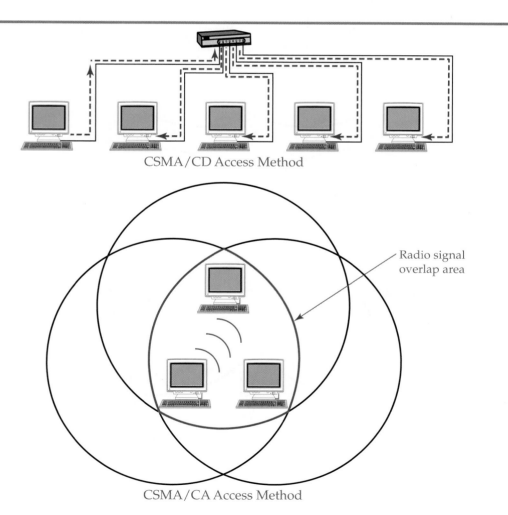

CSMA/CD Access Method

Radio signal
overlap area

CSMA/CA Access Method

Figure 4-20.
In an 802.3 network, which uses the CSMA/CD access method, all computers are wired to each other and each computer can hear traffic on the network. There is no need for a computer to send a signal notifying the other computers it is about to transmit data. In an 802.11 network, which uses the CSMA/CA access method, computers with wireless network interface cards communicate within an overlap area. A Wireless Access Point (WAP) controls all communication giving permission to the individual devices to communicate in an organized fashion.

For nodes to communicate with each other on a wireless network, all nodes must be inside the same broadcast area, called an *overlap area*. See **Figure 4-21.** This is not always possible, especially with mobile devices such as laptops, PDAs, cell phones, automobiles, and such. It is not unusual to have a mobile computer outside of the normal overlap area. There is also the possibility of mobile users moving in and out of the overlap area, further complicating communications. The best logical choice of media access for a wireless network is CSMA/CA.

CSMA/CA is used because of the way wireless networks communicate as compared to wired networks. When communicating on a wired network,

Figure 4-21.
Overlap area of an ad hoc and infrastructure wireless network.
A—For computers in an ad hoc wireless network to communicate with each other, all computers must be in the same overlap area. B—In an infrastructure wireless network, the Wireless Access Point (WAP) must be in the common overlap area of the wireless network computers. The WAP controls all communications.

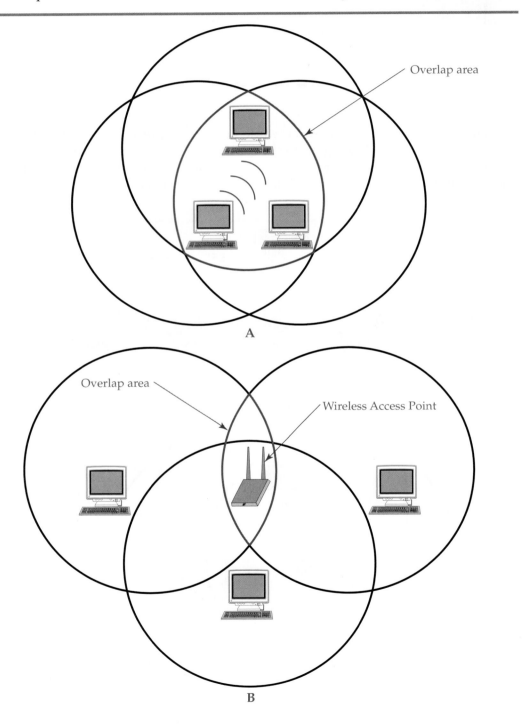

all devices in the same collision domain can hear each other when they are communicating. The idea is that each device waits for when the cable is clear of communications before it communicates on the cable. When a collection of wireless devices are connected as a network, not all devices will always be within the range of all other devices. This causes a problem because a wireless device may not be aware when other devices are communicating. This is why a wireless network system requires CSMA/CA to access the network media, the wireless network area.

CSMA/CA is designed to make the WAP in charge of all communications. The WAP permits wireless devices to communicate or denies wireless devices to communicate. The WAP is centrally located and can communicate with all the devices in the wireless network. The WAP hears all communications. A device sitting at the edge of the wireless network cannot hear the devices farthest from it, but the WAP can. If CSMA/CD was used as the access method, the devices would not take turns as permitted by the WAP and the result would be too many communication collisions on the network caused by two or more devices attempting to communicate at the same time.

Adding more Wireless Access Points can expand the geographic area covered by a wireless network, **Figure 4-22.** Connecting several Wireless Access Points with cable can also enlarge a wireless LAN. **Figure 4-23** shows several Wireless Access Points joined by copper core cable. This connection allows multiple Wireless Access

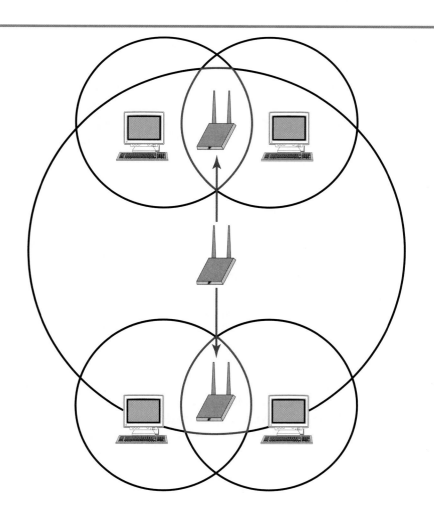

Figure 4-22.
The range of a wireless network can be extended by adding additional Wireless Access Points (WAPs).

Figure 4-23.
A wireless network can be greatly expanded by connecting multiple Wireless Access Points (WAPs) to a cable-based Ethernet network such as 10BaseT.

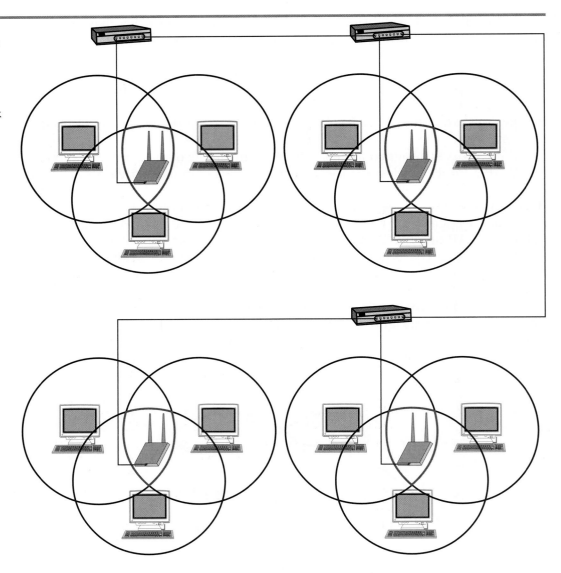

Points to be spread over a large area without the need for overlapping radio signals. This type of arrangement is ideal for mobile users. As the user travels through the network system with a mobile device such as a laptop, they can access the network system from anywhere. For example, a college campus spanning hundreds of acres could incorporate a mesh arrangement of Wireless Access Points throughout the campus area. A professor or student could use their laptop equipped with a wireless PCMCIA card to access the network system from anywhere on the campus such as classrooms, the library, cafeteria, a dorm room, or even from outside on the lawn areas.

New IEEE Wireless Standards

working group
a standard not fully developed and adopted as an official standard recognized by IEEE.

There are two other major IEEE standards that are directly related to wireless technology: 802.15 Wireless Personal Networks (WPAN) and 802.16 Broadband Wireless Access (BWA). While not adopted at this time, they are still worth mentioning since their adoption is anticipated soon. Standards not fully developed and adopted as an official standard recognized by IEEE are referred to as a *working group.* Titles for the standards being developed always use the term *working group* so people will know that they are under development.

802.15 Working Group for Wireless Personal Area Networks is better known as Personal Area Networks (PAN) or Wireless Personal Area Networks (WPAN). Many of the standards developed here coexist with standards already developed and adopted by 802.11 wireless LANs. One of the concerns of 802.15 is the compatibility of Bluetooth standards with other mobile communications devices such as telephones, portable computers, and personal digital assistants.

The 802.16 Working Group for Wireless Access Standard is better known as Broadband Wireless Access (BWA) or Wireless Metropolitan Area Networks (WMAN). The project is concerned with connecting LANs to each other using wireless technologies. The work group is hoping to design data rates as high as 70 Mbps over distances of over 30 miles. The work group is also attempting to expand the radio frequencies assigned by the FCC and radio frequencies dedicated to only wireless mesh-type networks, such as wireless MANs. At the present time, private companies can achieve high data rates over many miles, but there is no single standard that allows devices manufactured by different companies to communicate with the other.

Bluetooth

Bluetooth is a short-range, wireless system that is designed for 30 feet (10 meters) or less. Bluetooth uses 79 channels that use the frequency hopping spread spectrum transmission technique, starting at 2.4 GHz. The Bluetooth standard was developed separately from the IEEE network standards. It was never intended as a networking standard designed to carry massive amounts of information. Bluetooth was designed for appliances such as telephones, laptops, palm tops, digital cameras, personal digital assistants, headsets, printers, keyboards, and mice.

A Bluetooth network is referred to as a *piconet* or a *Personal Area Network (PAN)*. A piconet is a very small network. Bluetooth became recognized by the IEEE organization and was incorporated into the IEEE 802.15 Working Group Wireless Personal Area Networks in July, 2004. Bluetooth suffers from the same radio interference sources as other 802.11 devices, which are part of the ISM band. Bluetooth will not interfere with 802.11b devices when operated in the same area because they use different formats for configuring data. In other words, a wireless keyboard and mouse based on the Bluetooth standard will not interfere with the operation of an 802.11b wireless network.

piconet
a Bluetooth network.

Cellular Technology

Cellular technology is based on radio waves connecting to designated areas referred to as cells. Rather than communicate directly by radio wave from one cell to another, a remote device connects to a radio transmitter/receiver within its cell. The radio transmitter/receiver communicates to remote cells via microwave transmission or telephone lines. In the remote cell, the message is sent to a radio

cellular technology
a technology based on radio waves connecting to designated areas referred to as cells.

transceiver/receiver. The radio transceiver/receiver sends, via radio waves, the message to the remote device within its cell. See **Figure 4-24.**

Cellular technology is responsible for wireless telephone and pager technology. The same technology connects mobile and stationary computer equipment. A text message can be transmitted to a pager by typing in a message using a desktop computer. The message is sent over an Internet connection to a mobile telephone switching office. From there, radio microwaves transmit the encoded message to the distant pager system. See **Figure 4-25.**

Cellular technology supports duplex communications using a device such as a personal digital assistant or a palmtop configured for wireless radio services.

Microwave Transmission and Networking

The term *microwave* is used to describe radio waves in the electromagnetic spectrum that have a wavelength from 1 millimeter (mm) to 30 centimeters (cm) and radio wave frequencies between 1 GHz and 300 GHz. The amount of data carried by a radio wave is directly proportional to its frequency. In short, the higher the frequency, the more data that can be transmitted in a given period of time. The portion of the electromagnetic spectrum identified as microwave is the preferred technology used for wireless networking. Its application is described in the IEEE 802.11 standards. Microwave radio wave transmission can be broadcast directly between wireless devices or routed through satellites orbiting the earth.

Satellites are often used as part of a WAN distribution. Satellites can offer the advantage of providing a wireless network connection to remote or mobile locations that cannot be achieved using conventional methods.

Look at **Figure 4-26.** The satellite in this illustration is positioned 22,300 miles (35,880 km) above the earth's surface and moves at a speed of approximately 68,000 mph. At this distance, it takes the satellite exactly 24 hours to make one revolution around the earth. This time is equal to the time of the earth's rotation. This equality causes the satellite to appear in a stationary position above the earth and is said to be in *geosynchronous orbit.* In other words, the satellite's

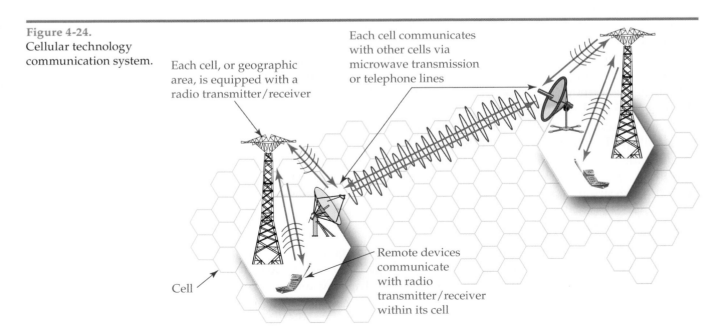

Figure 4-24.
Cellular technology communication system.

Each cell, or geographic area, is equipped with a radio transmitter/receiver

Each cell communicates with other cells via microwave transmission or telephone lines

Cell

Remote devices communicate with radio transmitter/receiver within its cell

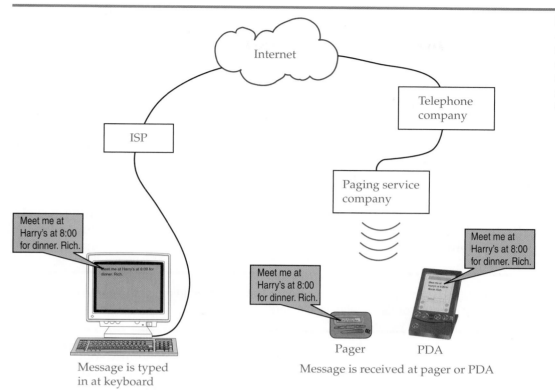

Figure 4-25.
An example of using cellular technology to send a message from a computer to a mobile network device such as a pager or a PDA.

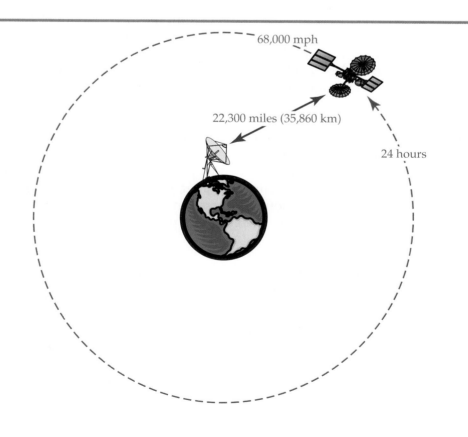

Figure 4-26.
A typical satellite in geosynchronous orbit is approximately 22,300 miles (35,860 km) above the earth's surface. At this distance the satellite can maintain a fixed position above the earth.

geosynchronous orbit an orbit in which a satellite's rotational speed is synchronized with the earth's rotational speed, making the satellite appear to be in a stationary position.

propagation delay the time it takes data to be transmitted from the earth and satellite.

speed is synchronized with the earth's rotational speed. If the satellite were in a lower orbit, it would need to move at a much higher orbital speed. This would result in the need to track the satellite as it moves across the sky. Tracking a satellite in stationary position is easy compared to tracking a constantly moving satellite.

One significant disadvantage to satellite communications is propagation delay. *Propagation delay* is the time it takes for data to be transmitted from the earth and satellite. It is caused by the great distance between the satellite and earth and can be compared to latency. In the past, the Motorola Company attempted to send a large group of 66 satellites into orbit at approximately 460 miles (740 km) above the earth to eliminate much of the propagation delay. At this height, the propagation delay is significantly reduced and is relatively unnoticed by the users. Unfortunately, the project, called Iridium, proved too costly and was abandoned before it could be fully implemented.

Tech Tip

The term *propagation delay* is used in satellite transmission in much the same way that the term *latency* is used for data delay in network systems.

While the propagation delay that normally occurs is generally short—only a fraction of a second—the effect of the delay depends on the type of data transmitted. For example, a delay of 250 milliseconds (ms) for a message consisting of several pages of text would go completely unnoticed by the end user. The same delay for data containing a telephone conversation would show slight pauses in the conversation. A series of frames containing millions of bits of picture or video data could show serious evidence of the delay. Picture or video data needs to be of relatively low resolution to limit the total amount of data transmitted. Transmitting high-resolution picture or video data would only be practical in half-duplex communication. Full-duplex communication for high-resolution picture or video data transmission is not yet practical. There are still very noticeable delays in high-resolution, two-way conference calls using satellite links in a network.

For one-way data transmission, buffering techniques similar to the buffering techniques used for downloading sound on a slow communications link, such as a 56k modem, could be used to prevent jitter in the transmission of the video. But this can only be used in one direction. Two-way transmissions in real time could not take advantage of buffering to eliminate jitter. Delays of 250 milliseconds (ms) or less are tolerable for live telephone conversations and low-resolution graphics. Delays above 250 milliseconds (ms) are generally unacceptable.

Infrared Transmission and Networking

Infrared transmission is used in point-to-point communications, which are also known as direct line of sight. Direct line of sight means that both devices are in direct alignment with each other and that there are no objects between the two devices. The infrared light beam is positioned to aim directly at the receiving device. Each device must be oriented in a position that aims toward the other device. This is one of the main limitations to using infrared in networking applications.

Infrared is typically used for point-to-point communications between two devices such as a personal digital assistant and a desktop computer. Infrared point-to-point communications can also be used in place of radio wave communication where there might be excessive interference, such as in factory or manufacturing settings.

Advantages and Disadvantages of Wireless Technology

Wireless technology can be cost effective compared to cable-based network media when spanning long distances such as continents or oceans. Spanning across a city, a business district, or across a college campus can be difficult and expensive with copper-core cable or fiber-optic cable. The installation of landlines are not only expensive, they can cause major disruption while digging up streets and parking lots for installation. When installing a temporary network, wireless technology can be much more cost-effective than remodeling a building to accommodate wiring. Wireless technology is most appropriate for mobile devices such as pagers, palmtops, communicators, personal digital assistants, and laptops. Hand held scanners used to scan bar codes on product packages also use wireless technology to transmit data to a computer or cash register system.

Certain wireless technologies are also affected to various degrees by atmospheric conditions such as rain, lightning, fog, and sunspots. The greatest disadvantage or concern for wireless technology is security. Network signals are transmitted in the open air and are capable of being picked up by an unauthorized receiver. Of course, data transmission may be encrypted for security, but even the best security can sometimes be compromised.

Wireless Security

One major concern of wireless networks is security. Network infrastructures designed to use cable are inherently more secure than wireless networks. Cable can be installed so that it is physically secured. Cables can be installed inside walls, pipes, and locked server rooms. Because cables can be physically secured, cabled networks are considered more secure than wireless networks. Unauthorized persons cannot readily connect physically to a private, cabled network system. Wireless networks, on the other hand, transmit data through the air, making it possible for anyone with a standard wireless network card to intercept the radio waves.

Radio waves fill the building areas and areas outside the building. One of the most common building materials used today is glass, especially in commercial establishments. Windows do not limit radio wave transmissions. Anyone near a building that uses wireless devices can easily intercept the wireless network signals with a laptop equipped with a wireless network adapter. However, security features can be implemented that will secure the transmission.

This section covers common wireless security features and provides a brief overview of wireless security. A more in-depth exploration of security features, such as encryption methods, authentication, and security standards are covered in Chapter 16—Network Security.

802.1X Authentication

IEEE 802.1x is a draft standard for authentication methods for wireless networking. It is referred to as a draft standard because it is not complete. The

802.1x draft standard provides a means for a client and server to authenticate with each other. Authentication is typically achieved through the exchange of a user name and password based on the Extensible Authentication Protocol (EAP).

Extensible Authentication Protocol (EAP) an authentication protocol used on both wired and wireless network systems.

The *Extensible Authentication Protocol (EAP)* is a protocol used on both wired and wireless network systems. EAP ensures authorized access to the network system and network resources. The improved version of EAP is called PEAP, which represents Protected EAP.

Figure 4-27 shows the **WLAN properties**, **Authentication** page of Windows XP. The **Enable IEEE 802.1x authentication for this network** option has been selected. Note the **EAP type** options, such as **Protected EAP (PEAP)** and **Smart Card or other Certificate**. Also notice that the computer is authenticated, not the person using the computer. A person who has their password compromised could have their password used to access a wireless network. However, when configured to authenticate the computer, an intruder would have to use the computer to access the wireless network.

802.1X Encryption

The second feature of the 802.1x standard is a way to hide the contents of network packets. Since the packets are broadcast through the open air, anyone could capture the packets and inspect the contents using a protocol analyzer or packet sniffer. A wireless network encryption key is used as part of the mathematical equation (algorithm) to encrypt data that is to be transmitted over a wireless network. There will be more about encryption keys in Chapter 16—Network Security.

The packets that are exchanged are encrypted using any one of a number of encryption software protocols. To make the encryption process unique to a particular network system, a key is used. You can think of a key as a string of alpha and numeric characters that feed the random character generator used to encrypt the contents of each packet. The only way to crack the encryption would be to guess or steal the encryption key. The encryption key can be provided by a security service, manufactured into a hardware device, or created by the network administrator.

Figure 4-27.
Windows XP
WLAN properties,
Authentication page.

IEEE 802.1x Authentication is enabled

Computer is selected to be authenticated

Extensible Authentication Protocol (EAP) types

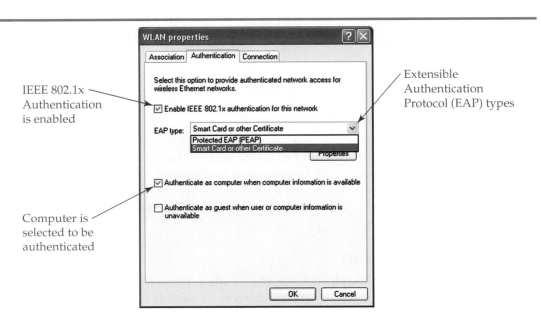

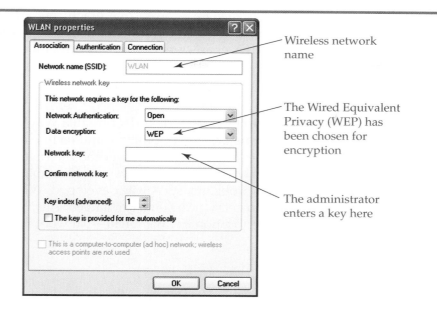

Wireless network name

The Wired Equivalent Privacy (WEP) has been chosen for encryption

The administrator enters a key here

Figure 4-28.
Windows XP
WLAN properties,
Association page.

Figure 4-28 shows the **Association** page of the **WLAN properties** dialog box in Windows XP. Notice that the wireless network has been identified as *WLAN*. The data encryption protocol selected is WEP. The acronym WEP represents Wired Equivalent Privacy. You can see by its name that the *Wired Equivalent Privacy (WEP)* protocol is intended to make a wireless network as secure as a wired network.

Below the **Data encryption** option is an option labeled **Network key**. The administrator enters the network key into the textbox and then enters the key once more in the option labeled **Confirm network key**. This procedure is repeated at other network clients and servers on the wireless network. Each computer must use the same key to be able to exchange encrypted data with each other.

The WEP protocol was followed by an enhanced protocol to improve on the security of WEP. The enhanced protocol was introduced in Windows XP Service Pack 1 and was called Wireless Application Protocol (WAP). *Wireless Application Protocol (WAP)* combines the authentication method with encryption. Both features are incorporated into one protocol. An additional improvement is the encryption was made more difficult to crack. WAP automatically changes the encryption key with each packet exchanged on the network. This significantly improves security by making it harder to crack.

A variation of WAP is WPA-PSK. WPA-PSK represents Wi-Fi Protected Access-Pre Shared Key. WPA-PSK was designed for small-office/home-office (SOHO) networks. It is designed to make it easy to configure encryption and authentication. WPA-PSK was incorporated into the new Wireless Network Setup Wizard that came with Windows XP Service Pack 2. One of the improvements provided by WPA-PSK is the ability to automatically generate a new key after a specified length of time or after a number of packets are exchanged. Changing keys often improves security because the key can be guessed or cracked after an amount of time. The following is a sample of what an encryption key might look like:

- A1D4FFBB
- Opensezzime
- BigDogRunsHere

Wired Equivalent Privacy (WEP)
a data encryption protocol that makes a wireless network as secure as a wired network.

Wireless Application Protocol (WAP)
a protocol that combines the authentication method with encryption.

Keys are often referred to as a *passphrase* by some manufacturers. Also, some keys require all characters to be constructed from HEX characters, as in the first example. A key constructed from HEX characters typically is difficult to guess.

Wi-Fi Protected Access

Wi-Fi Protected Access (WPA) a standard developed to ensure the safe exchange of data between a wireless network and a portable Wi-Fi device, such as a cell phone.

Wi-Fi Protected Access (WPA) is a standard developed to ensure the safe exchange of data between a wireless network and a portable Wi-Fi device, such as a cell phone. As the cell phone grew in popularity and technology was incorporated to allow the cell phone to exchange data over the Internet, security became a concern. To better protect wireless network systems from exposure to hackers, viruses, worms, and other problems, WPA was developed. WPA uses a set of keys to identify a device and to encrypt the data exchanged. Again, there will be more about encryption and authentication in Chapter 16—Network Security.

Summary

- Radio wave and microwave transmissions work on the principle of producing a carrier wave as the means of communication between two network devices.
- Modulation is the mixing of two radio signals, typically a carrier wave with a data signal.
- Infrared is used in line of sight transmissions and is not susceptible to radio interference.
- The FCC regulates the use of the electromagnetic spectrum by assigning radio frequencies and maximum power ratings to devices.
- A common source of interference for wireless networks is industrial, scientific, and medical devices that use frequencies in the same range as wireless networks.
- Two main classifications of antennae based on the shape of the electromagnetic wave pattern it produces are directional and omni.
- Radio waves are electromagnetic waves that cover the 10 kHz to 3,000,000 MHz frequency range.
- Several transmission techniques are used to communicate between radio wave-based network devices: single-frequency, spread-spectrum, and orthogonal frequency-division multiplexing.
- Spread spectrum is the radio transmission technique that subdivides the allocated frequency range into smaller units called channels. The two types of spread spectrum transmission techniques are frequency hopping and direct sequencing.
- Frequency hopping avoids interference by changing channels while transmitting data.
- Another name for frequency hopping is frequency hopping spread spectrum (FHSS).
- Direct sequencing consists of 11 overlapping channels of 83 MHz each in a 2.4 GHz spectrum. Within the 11 channels are three channels with a 22 MHz bandwidth. The three channels do not overlap and can be used simultaneously.
- Another name for direct sequencing is direct sequencing spread spectrum (DSSS).

- Orthogonal frequency division multiplexing is similar to frequency hopping and direct sequencing and employs multiplexing.
- A Wireless Access Point (WAP) provides a connection between a wireless network and a cable-based network. It also serves as the center of communications in a wireless network.
- A wireless network that contains one or more Wireless Access Points is arranged in infrastructure mode.
- A wireless network that does not contain a Wireless Access Point (WAP) is arranged in ad hoc mode.
- The IEEE 802.11a standard describes wireless networks operating at 5 GHz with a maximum data rate of 54 Mbps.
- The IEEE 802.11b standard describes wireless networks operating at 2.4 GHz with a maximum data rate of 11 Mbps.
- The IEEE 802.11g standard describes wireless networks operating at either 2.4 or 5 GHz with a maximum data rate of 54 Mbps.
- IEEE 802.15 standard describes the use of Wireless Personal Area Networks (WPAN).
- Wireless networks specified under the IEEE 802.11 standard use Carrier Sense Multiple Access with Collision Avoidance (CSMA/CA) as the media access method.
- Bluetooth is a standard developed by a group of manufacturers to allow their devices to interoperate. Bluetooth is recognized by the 802.11 standard and is used for short-range data transfer applications using the 2.4 GHz radio band.
- Infrared is commonly used for point-to-point transmission between two devices such as a personal digital assistant and a PC.
- Satellite communications experience propagation delay because of the great distance a signal must travel.
- Wireless technology can be cost effective compared to cable-based network media.
- A disadvantage of employing wireless technology is security.
- A Service Set Identifier (SSID) is similar in function to a workgroup name or domain name and must be assigned to a Wireless Access Point (WAP).
- Wireless networks are not secure when default settings are used for the configuration.
- Wireless encryption keys should be changed on a regular basis to maintain a high degree of security.

Review Questions

Answer the following questions on a separate sheet of paper. Please do not write in this book.

1. A(n) _____ is an electromagnetic wave of a set frequency that is used to carry data in radio wave- and microwave-based networks.

2. What is modulation?

3. A simple radio broadcast consists of a _____, which generates a carrier wave and a _____, which receives a carrier wave.

4. What is a channel?

5. What are the two major classifications of antennae?

6. Which of the following antenna types is not used in point-to-point links?
   A. Flat panel
   B. Omni
   C. Parabolic
   D. Yagi

7. Which antenna style was developed with aesthetics in mind?

8. What is spread spectrum?

9. The spread spectrum technique that transmits data on multiple channels simultaneously is called _____.

10. The spread spectrum technique that transmits data on multiple channels sequentially is called _____.

11. Why is frequency hopping used?

12. The _____ transmission technique uses the 5 GHz frequency and can achieve data rates as high as 54 Mbps.

13. List the frequency range and related power rating of each of the three U-NII classifications.

14. What is an SSID?

15. What frequency does the IEEE 802.11a classification specify, and what is the maximum achievable distance?

16. What frequency does IEEE 802.11b classification specify, and what is the maximum achievable distance?

17. What frequencies does IEEE 802.11g classification specify, and what are the related maximum achievable distances?

18. Which media access method does the 802.11 wireless standard specify?

19. What is the difference between CSMA/CA and CSMA/CD?

20. What is the maximum achievable distance of a Bluetooth device?

21. _____ technology is based on radio waves connecting to designated areas referred to as cells.

22. What is a disadvantage of satellite communications?

23. What is a geosynchronous orbit?

24. What is generally the maximum amount of satellite transmission delay acceptable for telephone conversations?

25. What is an advantage of wireless networking?

26. What is the one major disadvantage of wireless technology?

27. What does the acronym WEP represent?

28. Write an example of an encryption key.

Sample Network+ Exam Questions

Answer the following questions on a separate sheet of paper. Please do not write in this book.

1. Which IEEE 802.11 standard specifies the requirements for a wireless network using the 5 GHz frequency?
   A. IEEE 802.3
   B. IEEE 802.4
   C. IEEE 802.11a
   D. IEEE 802.11b

2. Which media access method is associated with wireless networks?
   A. ARCnet
   B. CSMA/CA
   C. CSMA/CD
   D. Token Ring

3. Which of the following items is used to connect a wireless network to an Ethernet network?
   A. Wireless Access Point
   B. Ad hoc converter
   C. Ethernet converter
   D. Ethernet sequencing device

4. What would *most likely* generate radio interference for an 802.11b network device?
   A. A cordless phone assigned to work at 2.4 GHz.
   B. Another computer using an 802.11g network adapter card.
   C. A wireless keyboard.
   D. An HP LaserJet printer.

5. Who is responsible for regulating the electromagnetic spectrum and dictating the frequency that is to be used for each group of devices?
   A. The IEEE organization.
   B. The telephone company in the local area.
   C. The FCC.
   D. WECA.

6. Which item would *most likely* interfere with the operation of an 802.11b wireless network?
   A. Hub
   B. Cordless phone
   C. Incandescent lights
   D. Copier

7. What is the function of an SSID?

   A. To replace the network MAC address.

   B. To encrypt all wireless communications.

   C. To identify the wireless network by name.

   D. To increase the data throughput of a wireless network.

8. Which protocol is used to encrypt data inside packets on a wireless network?

   A. HTTP

   B. SNMP

   C. FTP

   D. WEP

9. Which IEEE standard is used to describe how to achieve secure authentication to a wireless network?

   A. 802.3

   B. 802.1x

   C. 802.5b

   D. 802.4a

10. Which wireless transmission technique is used to transmit data according to the IEEE 802.11b standard?

    A. DSSS

    B. OFDM

    C. Both DSSS and OFDM

    D. Neither DSSS nor OFDM

Suggested Laboratory Activities

1. Set up an ad hoc network using a PC and a laptop.

2. Set up a two- or three-station wireless network using a Wireless Access Point (WAP).

3. Research a particular brand of wireless network device and list the expected transmission ranges and related speeds.

4. Transfer data between a laptop and a PC by using an infrared port.

5. Observe the effect of a variety of materials on a wireless network. For example, separate two computers on the wireless network with a wooden wall, sheet rock, metal door, or file cabinet. Note the effects on distance and data rate.

6. Observe the effect of a common electric drill being operated near the wireless network, and then observe the effects of a cell phone or walkie-talkie on the wireless network.

Interesting Web Sites for More Information

http://grouper.ieee.org/groups/802/802_tutorials/index.htm

www.cwt.vt.edu

www.palowireless.com/ofdm/tutorials.asp

www.proxim.com

Wireless technology is ideal for college campuses. It can allow students to access the school network, Internet, and mail server from any location on campus, giving students more options on where and how they study.

Chapter 4
Laboratory Activity

Installing an Infrastructure Mode Wireless Network

After completing this laboratory activity, you will be able to:
- Explain the purpose of a Wireless Access Point (WAP).
- Identify typical wireless network configuration requirements.
- Explain the purpose of an SSID.
- Explain why Wireless Access Point (WAP) default settings do not provide security.
- Install an 802.11b USB wireless network adapter.
- Install an 802.11b Wireless Access Point (WAP).

Introduction

In this laboratory activity, you will install an 802.b USB wireless network adapter into a computer and then install and configure a Wireless Access Point (WAP). Wireless devices fall into three categories as defined by IEEE: 802.11a, 802.11b, and 802.11g. The following table lists the assigned frequency and maximum data rate of each.

IEEE Wireless Classification	Assigned Frequency	Maximum Data Rate	Comments
802.11a	5 GHz	54 Mbps	Not compatible with 802.11b or 802.11g.
802.11b	2.4 GHz	11 Mbps	Compatible with 802.11g.
802.11g	2.4 GHz	11 Mbps in 802.b mode 54 Mbps in 802.g mode	Compatible with 802.11b.

These are the maximum data rates designated by the IEEE. Some devices have higher, advertised speeds than those recognized by the IEEE classifications. Maximum data rates can only be produced under ideal conditions. Ideal conditions are when two wireless devices are relatively close to one another, with no partitions or other materials between them or radio interference.

An 802.11g wireless device is compatible with an 802.11b wireless device, and vice versa. Some Wireless Access Points are compatible with all three classifications. An 802.11g wireless device can achieve a maximum data rate of 11 Mbps when communicating with an 802.11b wireless device and a maximum data rate of 54 Mbps when communicating with an 802.11g wireless device.

Windows XP was the first operating system released by Microsoft that came with generic drivers and an application program for wireless devices. When using the generic drivers and the application program, wireless devices do not always work properly or all options are not available.

Microsoft Windows automatically detects new hardware devices when they are installed and automatically configures the device driver if it can match a driver to the new device. This means that Microsoft generic device drivers for wireless devices are automatically installed when a new wireless device is installed. To correct this problem, you should first install the manufacturer's drivers and application program that is supplied on the wireless device installation CD. If you are using an operating system that does not contain generic drivers, you must use the installation CD from the manufacturer.

Wireless devices can be configured in two different modes: ad hoc and infrastructure. When configured in ad hoc mode, the wireless devices are free to communicate at will. When configured in infrastructure mode, a WAP, wireless switch, wireless router, or a wireless gateway controls all wireless device communications. Wireless switches, routers, and gateways typically incorporate the access feature. Wireless Access Points are also referred to as bridges because they bridge two different network technologies: wireless to cable-based.

All wireless devices of the same manufacturer have the same Service Set Identifier (SSID) to make installation easy. The SSID is the name for the wireless network. It is comparable to a workgroup name. There is no one standard set of default settings used by all wireless devices; however, default settings are usually the same within a given manufacturer. For example, D-Link uses *Wlan* or *default*. NETGEAR uses *Wireless*, and 3COM uses *Wlan* or *3Com* as the default SSID. In this laboratory activity, you are directed to use wireless devices from the same manufacturer. Be aware, however, that some manufacturers use more than one default name.

Leaving a wireless device set to the default SSID is a high-security risk. Anyone with a wireless network adapter can successfully attach to a wireless network that is configured with a default SSID. To increase the security, the default SSID should be changed. Changing the SSID is covered in later laboratory activities.

The default wireless channel, or dedicated frequency, also varies according to manufacturer. In the United States, the FCC has designated 11 wireless channels for use in wireless networking. Each wireless network configured in infrastructure mode should use the same channel for communication. If channels differ between wireless devices, the wireless devices will not be able to communicate.

The configuration settings of a WAP can be accessed and changed from a wireless workstation. At the wireless workstation, simply type the assigned WAP IP address into the Web browser address bar. For example, if the default IP address is 192.168.0.1, enter the IP address into the area you would normally type a URL address. A connection should be made automatically. You will then be prompted for a user name and password. Again, the user name and password is set by default and varies according to manufacturer. In the following screen capture, the default administrator user name is *Admin*. The default user name and password can be found in the hardware manual or at the manufacturer's Web site.

All manufacturers incorporate a reset button into their Wireless Access Points. The reset button is used to reset the configuration to its default settings. The reset button resets the SSID, channel, user name, user password, and other configuration values to the manufacturer's original settings. This is an important feature to be aware of when performing laboratory activities. The last student may have changed the configuration, making it impossible for your wireless workstation to access the WAP or connect to the wireless network.

Before changing any configuration settings, remember to write down all changes you plan to make, especially the IP address and subnet mask. After you change the configuration, a wireless network adapter will not be able to connect to the WAP. You must make similar changes to the wireless network adapter configuration. If the wireless network adapter configuration does not match the WAP configuration, you will not be able to attach to the wireless network.

Equipment and Materials

- 802.11b USB wireless adapter with USB cable, installation CD, and installation manual.
- 802.11b Wireless Access Point, installation CD, and installation manual.
- Patch cable.
- Paper clip. (Used to reset the Wireless Access Point if needed.)
- Stand-alone computer running Windows XP SP2. (This lab can also be performed using Windows 2000 or Windows 98.)
- A computer connected to a hub, such as in a peer-to-peer network. The computer should be running Windows XP SP2.

Note:
Both the USB adapter and the WAP should be from the same manufacturer. If not, the default settings may not match and will need to be modified to work correctly.

Note:
All wireless devices should match the same classification, such as 802.11a, 802.11b, or 802.11g. Remember, 802.11b and 802.11g are compatible. Some Wireless Access Points will support all three classifications. Be sure to check the device's specifications before setting up the lab.

You will need the following configuration information before attempting this laboratory activity. Record this information on a separate sheet of paper.

Wireless Network Interface Card

IP address: _____

Subnet mask: _____

Default gateway: _____

Wireless channel: _____

Wireless Access Point

IP address: _____

Subnet mask: _____

Default gateway: _____

User name: _____

User password: _____

Wireless channel: _____

SSID: _____

Note:
You should review the manufacturer's installation procedures before attempting to install any devices or running the installation CD. The manufacturer information can be found either on a disc that accompanies the wireless device or on the manufacturer's Web site.

Procedures

1. _____ Gather all required materials and report to your assigned workstation.

2. _____ Boot the workstation and make sure it is working properly.

3. _____ Review the manufacturer's installation manual for the USB wireless network adapter. Most manufacturers include a copy of the installation manual on the installation CD. Typically, you first install the drivers located on the disc that accompanies the USB wireless device. After inserting the wireless driver disc into a CD or DVD drive, a menu will appear asking if you would like to install the drivers, read the manual, or access the manufacturer's Web site.

4. _____ Install the USB wireless network adapter into the stand-alone PC.

5. _____ After installing the USB network interface card, check if the device appears in the **Safely Remove Hardware** dialog box. To do this, double-click the **Safely Remove Hardware** icon.

Safely Remove
Hardware icon

The **Safely Remove Hardware** dialog box displays all hardware that can be installed and removed while the computer is powered on, such as USB devices and PCMCI cards.

6. _____ You can now proceed to assign the IP address, subnet mask, and default gateway address to the wireless network adapter. You can configure the USB wireless network adapter manually or with a wizard. Typically, the wireless adapter installation CD and the Microsoft XP operating system contain a wizard. The wizard automatically starts after the USB wireless network adapter is detected.

You can manually configure the wireless network adapter by accessing the **Wireless Network Connections** dialog box. This can be accessed through **Start | Control Panel | Network and Internet Connections | Network Connections**, right-clicking the wireless device, and selecting **Properties** from the shortcut menu. After the **Wireless Network Connections** dialog box opens, highlight **Internet Protocol (TCP/IP)** and then select the **Properties** button. In the **Internet Protocol (TCP/IP) Properties** dialog box, enter the IP address and subnet mask settings.

7. _____ Read the installation procedures for Wireless Access Point.

8. _____ Connect the WAP to the hub with a Category 5e patch cable.

9. _____ Install the Wireless Access Point's power cable.

10. _____   View the available Wireless Access Points by opening **Network Connections** in **Control Panel** (**Start | Control Panel | Network and Internet Connections | Network connections**). Right-click the wireless network icon and select the **View Available Wireless Networks** from the shortcut menu. (This feature may not be available if the Microsoft operating system did not configure the wireless network adapter.)

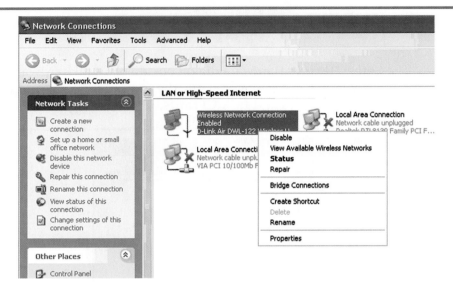

The **Wireless Network Connection** dialog box will display. Notice that the SSID of the wireless network is listed under **Available wireless networks**. This dialog box allows you to choose the SSID to which to connect.

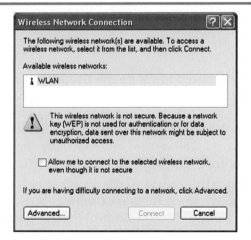

Note:
If an SSID is not listed under **Available wireless networks**, *make sure the WAP and workstation have the same subnet mask. If the default configuration of the WAP has changed, you may need to reset it. Use a paper clip to reset the WAP. You should typically hold the reset with the paper clip for 3–10 seconds. Remember, the power must remain applied to the WAP during this time to enable a reset. Removing the power cord and pressing the reset button will not cause the WAP to return to the original default configuration. If you cannot establish a connection to the WAP, call your instructor for assistance.*

Note the option labeled **Allow me to connect to the selected wireless network, even though it is not secure**. This feature secures wireless networks by not allowing unauthorized access by wireless devices in close proximity. For example, two homes, apartments, or offices, very close together could easily connect to the other's wireless network. If you are having trouble connecting to the WAP, select this option. The Microsoft operating system by default does not automatically allow you to connect directly to a WAP.

When you are finished viewing these settings, close the **Wireless Network Connection** dialog box. In the following procedures, you will examine the WAP configuration screens and features.

Note:
Some Wireless Access Points can be configured through a workstation connected to the wired network. Typically, a software program supplied on the installation CD is required.

11. _____ At the wireless workstation, open Internet Explorer.

12. _____ Type the default IP address of the WAP into the Web browser address bar.

13. _____ Explore the configuration screens and features. Look for where you can change the assigned IP address, assign an SSID, and enable encryption. The following screen captures show the configuration screens for a D-Link WAP. Your WAP configuration screens will not match exactly but will be similar. Most Wireless Access Points have similar features.

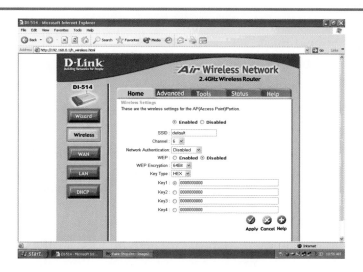

14. _____   If required by your instructor, change the WAP IP address and possibly the subnet mask. Be sure to record the changes. If you do not remember the new IP address and subnet mask, you will not be able to connect to the WAP later to change configurations. Also, change the SSID. Close Internet Explorer when you are finished.

15. _____   Open the **Wireless Connection** dialog box again and click the **Advanced** button. Change the wireless workstation network adapter to ad hoc mode and see if you can connect to the WAP. Change the wireless channel and see the effect on the connection to the WAP. Experiment with other settings. On a separate sheet of paper, list the changes that prevent the wireless network adapter from connecting to the WAP. Settings that can adversely affect the connection include the following:

■ Subnet mask.

■ SSID.

■ Channel.

■ Mode.

Note:
Equipment that uses the same frequencies as the wireless network can affect the connection. Such items are cell phones, microwave ovens, walkie-talkies, and equipment found in electronics laboratories. Keep this in mind if your lab is situated in a school electronics laboratory.

16. _____   After completing the laboratory activity, answer the review questions. Windows XP has a lot of good information about wireless networks in **Help and Support** (**Start | Help and Support**). You may want to review this information at this time. Use the word *wireless* in your search.

17. _____   Return all materials in the condition specified by your instructor.

Review Questions

Answer the following questions on a separate sheet of paper. Please do not write in this book.

1. Why is a manufacturer's installation CD required for the Windows 98 operating system?

2. Why do some manufacturers request that you use their installation CD before installing the wireless device?

3. What does the acronym SSID represent?

4. Why do manufacturers use a default SSID?

5. Why is the default SSID and other default configuration settings considered a security risk?

6. What is a channel?

7. Why must the channels match between the wireless network adapter and the WAP?

8. What configuration settings are changed to the default when the reset button on the WAP is pressed?

9. What other device name can be used when referring to a WAP?

10. What other wireless devices incorporate WAP features into the device?

11. Why might a wireless client fail to establish a connection with a WAP?

12. What is the assigned frequency of an IEEE 802.11a device?

13. What is the maximum data rate of an IEEE 802.11a device?

14. What is the assigned frequency of an IEEE 802.11b device?

15. What is the maximum data rate of an IEEE 802.11b device?

16. What is the assigned frequency of an IEEE 802.11g device?

17. What is the maximum data rate of an IEEE 802.11g device?

18. Can an IEEE 802.11b device communicate with an IEEE 802.11a device?

19. Conduct an Internet search and locate the default settings for the following wireless manufacturers.

	SSID	Channel	Password
3COM			
D-Link			
Belkin			
Cisco			

The palowireless Web site (www.palowireless.com) is a great resource for tutorials on wireless technology.

5

Digital Encoding and Data Transmission

After studying this chapter, you will be able to:

❏ Describe how data is represented as binary and digital signals.

❏ Describe the two modes of transmitting data between two points.

❏ Explain data integrity inspection using a parity check.

❏ Explain data integrity inspection using a Cyclic Redundancy Check (CRC).

❏ Describe the complete data packaging process.

❏ Explain the difference between connection-oriented and connectionless data transmission.

❏ Explain the difference between packet switching and circuit switching.

❏ Describe the various data codes such as ASCII, BCD, UNICODE, and EBCDIC.

❏ Describe the structure and contents of a UDP frame.

❏ Describe the structure and contents of an Ethernet frame.

❏ Relate digital encoding and data packaging to the layers of the OSI model.

Network+ Exam—Key Points

Approximately 20% of the Network+ Certification exam is based on basic networking concepts such as those presented in this chapter. For example, when a protocol is described, it is often referred to as either connection-oriented or connectionless, whether it uses circuit switching or packet switching technology, and which particular layer(s) of the OSI model it is associated with.

The Network+ Certification exam tests the applicant's ability to understand these basic concepts. Review this chapter often as you progress through the course in preparation for the Network+ Certification exam.

Key Words and Terms

The following words and terms will become important pieces of your networking vocabulary. Be sure you can define them.

American Standard Code for
    Information Interchange (ASCII)
asynchronous transmission
Binary Coded Decimal (BCD)
bipolar digital signal
broadcast frame
circuit switching
connectionless communication
connection-oriented communication
Cyclic Redundancy Check (CRC)
data encryption
digital encoding
encapsulation
Extended Binary Coded Decimal
    Interchange Code (EBCDIC)

frame
Hypertext Markup Language (HTML)
Manchester encoding
multicast frame
non-return to zero (NRZ)
one's compliment
packet switching
parity check
port
segment
synchronous transmission
time period
Unicode
unipolar digital signal
virtual network

Basic terminology about data packaging and transmission is presented. Learning this terminology will be beneficial when more specific aspects of networking are covered in later chapters. Also, a good understanding of the OSI model will help you when you further your networking education by studying security concepts, network infrastructure, and IEEE standards.

The chapter ends with a review of the OSI model and its relationship to digital encoding and data packaging. The OSI model is presented using some of the terminology you have learned thus far. The OSI model will be referenced throughout the textbook as you learn more network terminology and concepts. Presenting this material throughout the textbook will enable you to gain a solid understanding of the OSI model.

Digital Encoding and Transmission

Digital signals are used to transmit data and commands across network systems. In the Chapter 2, Network Media—Copper Core Cable, you learned that all network communications originate from electronic signals. Electronic signals take many forms, such as electrical energy, light waves, radio waves, infrared light waves, and microwaves. All of these forms of electronic signals originate from digital signals.

digital encoding takes place when a network interface card converts data into a digital pattern acceptable to the networking media. After the data is encoded into a digital pattern, it is placed on the networking media. So that you can better understand digital encoding and digital patterns, the digital signal will be reviewed.

As you have already learned, a digital signal is an electronic signal that has discrete values. Discrete values can be on and off, high and low, or 5 volts and 0 volts. The waveform of the digital signal is drawn at right angles, resembling a series of incomplete rectangles and squares, **Figure 5-1.** The height, or amplitude,

digital encoding
the conversion of data into a digital pattern acceptable to the networking media.

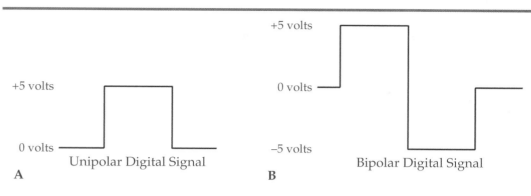

Figure 5-1.
A digital signal has discrete voltage levels. Its waveform resembles a series of incomplete squares and rectangles. A—A unipolar digital signal fluctuates between 0 volts and +5 volts. B—A bipolar digital signal fluctuates between +5 volts and –5 volts.

of the digital signal represents the electrical voltage level. In Figure 5-1A, a positive voltage of 5 volts is illustrated. The digital signal fluctuates between a positive five-volt and zero-volt level. This pattern is similar to a light switch repeatedly turned on and off and is called a ***unipolar digital signal.***

Some digital electronic circuit designs fluctuate between a positive voltage level and a negative voltage level. In Figure 5-1B, the digital signal fluctuates between a positive five-volt level and a negative five-volt level. This pattern is called a ***bipolar digital signal.***

The examples in Figure 5-1 are not meant to imply that all digital systems work at a standard five-volt level. Some systems work at lower voltage levels such as 3.3 volts.

Another important characteristic of the digital signal is the time period. The ***time period*** represents the rate of recurrence of an expected signal level change. The time period used to encode digital data should not be confused with the term *time period* used when talking about frequency or cycles. In the context of encoding a digital signal, the time period represents the time period of an expected digital wave shape. A digital signal may or may not change within a given time period, but a change is expected. In some digital encoding schemes, the time at which the digital signal changes voltage levels is important. The transition of the digital signal's voltage level may occur at the beginning of the time period or at the midpoint of the time period. See **Figure 5-2.**

By changing the transitioning pattern and fluctuation pattern of a digital signal, many digital encoding schemes can be created. Non-Return to Zero (NRZ) is a special classification of digital encoding. Look at **Figure 5-3** and compare the two digital signal patterns. In the first digital signal pattern, the voltage level is restored to zero after the digital pulse. In the second digital signal

unipolar digital signal
a digital signal that fluctuates between a positive five-volt and zero-volt level.

bipolar digital signal
a digital signal that fluctuates between a positive five-volt level and a negative five-volt level.

time period
the rate of recurrence of an expected signal level change.

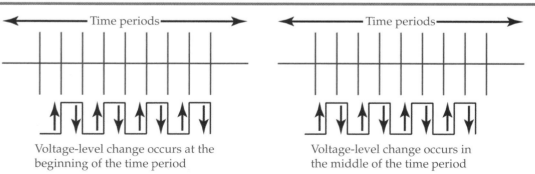

Figure 5-2.
Some digital encoding schemes are marked by the voltage level changing during a time period. The voltage-level change can occur at the beginning of the time period or in the middle of the time period.

Figure 5-3.
The voltage level in a digital signal can either return to 0 volts or not return to 0 volts between transitions. When the signal does not return to zero, it is classified as non-return to zero (NRZ).

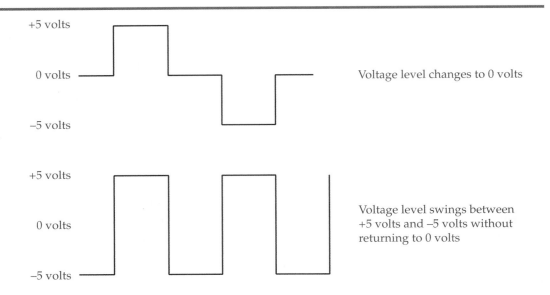

Voltage level changes to 0 volts

Voltage level swings between +5 volts and −5 volts without returning to 0 volts

non-return to zero (NRZ)
a digital signal that fluctuates between a high voltage level and a low voltage level and never returns to zero volts for any measurable period of time.

Manchester encoding
an encoding scheme that is characterized by a digital pulse transitioning during the midpoint of the timing period.

pattern, known as *non-return to zero (NRZ)*, the signal fluctuates between the high voltage level and the lower voltage level and never returns to zero for any measurable period of time.

One of the most popular encoding schemes is called *Manchester encoding.* This encoding scheme is characterized by the digital pulse transitioning during the midpoint of the time period. Look at **Figure 5-4.** Note that a binary one is represented by a transition from five volts to zero volts in the midpoint of the time period. A binary zero is represented by a transition from zero to five volts in the midpoint of the time period.

Digital signal characteristics vary from one encoding scheme to another. The selection of the encoding scheme varies according to the electronic characteristics designed into the network hardware. Further study into the encoding schemes would require more detailed electronics study than this textbook provides. The main point to remember is digital encoding happens at the data link layer of OSI model.

Two modes of transmitting data between two points are synchronous and asynchronous. In *synchronous transmission,* the digital signal is synchronized

Figure 5-4.
The main characteristic of Manchester encoding is the direction of the voltage level during the midpoint of the time period, which is represented by a binary one or zero.

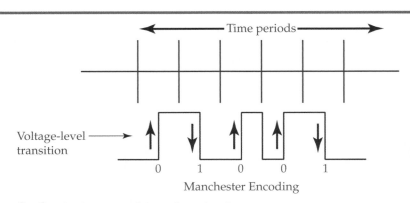

with a reference signal to ensure proper timing. The timing of the digital signal is extremely important for decoding some variations of digital encoding patterns. In *asynchronous transmission,* no reference signal is present.

You may be wondering how a digital signal can be decoded without a reference signal. A reference signal is not needed if there is a way to determine when a byte or stream of data begins or ends. Some data transmission schemes use a series of ones and zeros to identify the beginning and end of a byte or stream of data. A long period of no digital signal can also be used to signify the beginning. Again, there are many different ways the beginning and end can be acknowledged. There is no one universal method. The methods with which data are encoded vary greatly between the computer bus system, network media, telephone signals, radio signals, and such.

Data Packaging and Transmission

To send data from one point to another across a network or the Internet, data is packaged with extra information to ensure its delivery and integrity. Often data is broken into smaller, deliverable pieces. Each piece is called a *segment* and contains information to aid in reassembly.

synchronous transmission
a type of transmission in which a digital signal is synchronized with a reference signal to ensure proper timing.

asynchronous transmission
a type of transmission in which a digital signal is not synchronized with a reference signal.

segment
raw data that is divided into a smaller unit.

Be aware that many references refer to data segments as packets.

Tech Tip

Sometimes, communication is established between a source and destination computer before data is sent. Other times, data is sent without establishing communication and without requiring a confirmation of delivery. Data routing can also differ among data transmissions. This section describes the components involved in data packaging, such as parity checks, segmentation, and encapsulation. It also covers communications and routing.

Parity Checks

Digital signals can become corrupted for different reasons. Some of the causes can be crosstalk, electromagnetic interference, loose connections, and faulty and improperly grounded equipment. A *parity check* is a method of verifying the integrity of transmitted data. Look at **Figure 5-5A.**

parity check
a method of verifying the integrity of transmitted data.

1 0 0 1 0 1 0 0   Original byte

1 0 0 0 0 1 0 0   Corrupted byte

A bit with a binary value of 1 has been changed to 0

A

┌──────One byte──────┐
Data   Parity bit

1 0 0 1 0 1 0   1   Odd and even parity is signified by the last bit in the byte

1 0 0 0 0 1 0   0

B

Figure 5-5.
To perform parity checks, the last bit position in an 8-bit data transmission can be used for the parity code. A value of 1 or 0 in the parity bit may indicate odd or even parity.

In the example, one bit in the entire byte has changed from a binary one to a zero. The meaning of the byte has therefore been altered. Originally, the byte might have represented the letter *A*, but now it represents the letter *R* or a command such as "end of transmission." One way to ensure that data sent from the source to the destination is intact is to include a parity code with the data. For example, rather than eight bits of a byte representing alphanumeric symbols, seven bits could represent alphanumeric symbols and the eighth bit could represent the parity code.

Look at **Figure 5-5B.** In the example, the total number of zeros in the data portion of the byte is used to validate data. An even number of zeros is represented by a parity code of a binary one. An odd number of zeros is represented by a parity code of a binary zero. When the destination receives the byte, it compares the parity code to the number of binary zeros to tell if the data has been corrupted. If the data is corrupted, the destination rejects the data and requests the data be resent.

Cyclic Redundancy Check (CRC)

Cyclic Redundancy Check (CRC)
a sophisticated data integrity check that uses complicated mathematical algorithms to determine if one or more bits are corrupt.

A *Cyclic Redundancy Check (CRC)* is based on the same principle for error detection as the parity check. However, the CRC is a more sophisticated check. The main problem with the parity check is if an even number of bits is corrupt, the corruption goes undetected. The CRC uses complicated mathematical algorithms to determine if one or more bits are corrupt. A CRC can detect the corrupted bit and correct it without requiring the data be retransmitted. The only drawback is a CRC requires more data to be sent. The extra data is the parity code. This slows the data transmission rate. Still, the CRC is a popular method of verifying data. It is incorporated into many different network protocols when data integrity is critical.

Segmentation and Encapsulation

encapsulation
the process of adding information to a segment that identifies such things as the source address, the destination address, the end of the segment, and the size of the segment.

Unless the amount of data transferred between two points is small, the data needs to be divided into smaller units, called segments. After the data is divided into segments, it is encapsulated. *Encapsulation* is the process of adding information to the segment that identifies such things as the source address, the destination address, the end of the segment, and the size of the segment.

Figure 5-6 shows the complete data packaging process, from raw data (text file, sound, or graphic) to digital signals. Raw data is first broken into segments. Segment size is determined by the networking technology, such as Ethernet or Token Ring, and the networking protocol, such as IPX/SPX or TCP/IP. The segmented data is then encapsulated into packets. Packets of data contain additional information needed to deliver the segments. This additional information is called the network header. The network header contains a source and destination address. Other additional information that can be added to the packet is error-checking codes. Error-checking codes ensure the packet is delivered without a distorted data signal.

Packets are limited to traversing only LANs. Since the data will be sent across the Internet, it is further encapsulated, or framed. A packet that is

Figure 5-6.
Data packaging process. Raw data is broken into segments (step 1). The segments are encapsulated into packets (step 2). The packets contain extra information such as source and destination addresses and error-checking codes. Since the data in this example will travel the Internet, the packets are encapsulated into a frame (step 3). The frame contains extra information needed to navigate the Internet. The frame is converted into binary code (step 4). The binary code is converted to digital signals and placed on the network media (step 5).

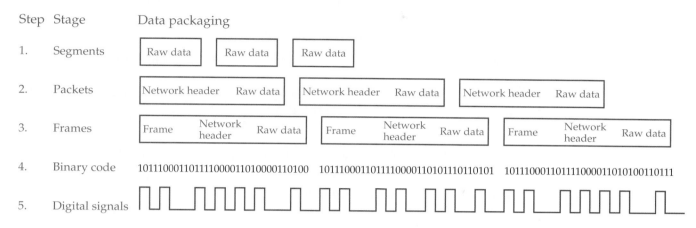

encapsulated with information needed to travel the Internet is called a *frame.* The frame contains additional information such as an Internet address. An Internet address is different from a machine (MAC) address. The machine address is sufficient when delivering data on a LAN.

After the packet has been encapsulated, the frame is converted into binary code. The binary code is then converted to digital signals. The entire series of frames is placed on the network in the form of digital signals and is transmitted to the destination computer. The destination computer removes the information that is no longer needed, such as the destination address, and reconstructs the entire message by placing the packets into their proper sequence. Error checks are also performed to ensure the data has not been corrupted.

This explanation and Figure 5-6 is a generic example of how data is transmitted between two computers. Encapsulation varies depending on which technology and protocols are used. More information will be provided as each of the protocols are explored and explained.

The terms *segment, frame, data gram, protocol data unit (PDU),* and *packet* are often used interchangeably when discussing blocks of data. You can think of all of the terms as units of data transferred between two points. The ISO tends to use the term *PDU* when discussing data packaging. When discussing TCI/IP, the terms *packet* and *frame* are often used with packets described as fitting inside a frame. The term *PDU* is the most technically correct term, but it is awkward to use in conversation. Consequently, technicians tend to use the term *packet* to collectively represent all types of data units.

Many times it is acceptable to use any of the terms when discussing network data, but at other times, it is more important to be precise. Correct terminology is especially important when explaining protocols, OSI layers, and certain types of network equipment.

frame
a packet that is encapsulated with information needed to travel the Internet.

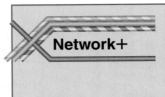

Network+ Note:

Data packaging terminology can vary a great deal from one manufacturer to another. For example, one manufacturer may refer to the entire encapsulation process as framing, while another may break it down into different and distinct stages. When studying for the Network+ Certification exam, remember that CompTIA certifications are vendor-neutral and use generally accepted terms like those used in this textbook to describe various technologies.

If you are planning to take a certification exam for a specific networking technology, such as Nortel, Microsoft, or Novell, it is best to become familiar with the specific vendor's terminology. Many of the mentioned companies provide a vocabulary listing on their Web site.

Connection-Oriented and Connectionless Communication

connection-oriented communication
a type of connection in which communication is first established between the source and destination computers before data is transmitted.

connectionless communication
a type of connection in which data is transmitted to the destination without first establishing a connection.

Communication between two computers can be classified as connection-oriented or connectionless. In **connection-oriented communication,** a connection is first established between the source and destination computers before data is transmitted. After the data is transferred, the connection is terminated. In **connectionless communication,** the data or commands are sent without an established connection. After the data is transferred, there is no need to terminate or release the connection. See **Figure 5-7.**

A protocol can be described as either connection-oriented or connectionless. IP is a connectionless protocol. It simply moves data from point A to point B. IP does not verify the delivery of the data. In contrast, TCP is a connection-oriented protocol. It verifies a connection before transmitting data and then terminates the connection when the data has been transferred. The best of both protocols is used when using TCP/IP in WAN communications. A connection is first established with the TCP protocol. Next, data is transferred with the IP protocol. After all data has been transferred, the connection is terminated with the TCP protocol.

Connectionless protocols are used to transfer audio or video data. Time is very important when conducting a conversation, and the fastest protocols for transferring data are connectionless. A connection-oriented protocol would slow the transfer of voice data since every frame of voice data would have to be

Figure 5-7.
Connection-oriented communication establishes contact with the destination before sending data. Connectionless communication simply sends the data and does not bother to verify if the destination is there or if the message was received.

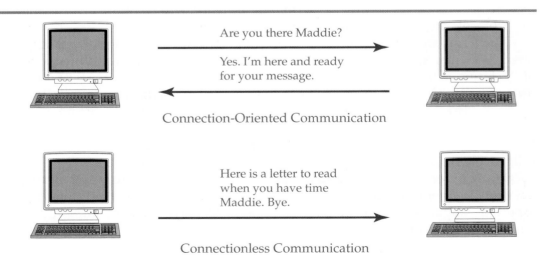

Are you there Maddie?

Yes. I'm here and ready for your message.

Connection-Oriented Communication

Here is a letter to read when you have time Maddie. Bye.

Connectionless Communication

verified. A connection-oriented protocol is used first to verify the destination of the call. A connectionless protocol is used to transfer the voice data. There is no need to verify that each packet of voice data has been received since it is assumed the connection will last through the entire conversation. If a packet is lost and does not reach the destination, a simple "What did you say?" will cause the sender to regenerate the lost data.

Commands can be sent between two computers with the UDP protocol. UDP is a connectionless protocol that requires minimal system resources. Think of a connectionless protocol as one that makes its best effort to send a command or data to a destination but has no way of guaranteeing its arrival. A connectionless protocol typically depends on reliable equipment such as a dedicated line to ensure delivery.

Network+ Note:

It is common to see a question on the Network+ Certification exam that identifies connection and connectionless protocols. Try to remember that IP and IPX are connectionless protocols. TCP and SPX are connection-oriented protocols.

Circuit Switching/Packet Switching

Two main categories describing the way data are routed between two points are packet switching and circuit switching. *Circuit switching* establishes a permanent connection between two points for the duration of the data transfer period. For example, when you use a telephone modem to connect to an Internet service provider, the connection is circuit switching. The telephone line connection is a permanent connection between the telephone modem and Internet service provider. Other types of telephone lines that use circuit switching are ISDN, ATM, and T1. When you think of circuit switching, think of a permanent or dedicated line.

Packet switching does not use a permanent connection. Packet switching breaks the data transmission into smaller parts called packets. Each packet has a source and destination address and a sequence number attached to it. You could say each small unit of data is encapsulated into a packet that contains source and destination addresses and a sequence number. The Internet uses packet switching to send data between two points on the Internet such as a home user and a Web server at a remote location. When a Web page is downloaded from the remote site to the home computer, the Web page is broken into smaller packets. The packets are sent out onto the Internet and may each take a different route to their destination, **Figure 5-8.**

When packets travel across the maze of communication lines, many different factors can delay the packets, such as heavy traffic conditions in the different communication lines. The amount of traffic in each of the many communication lines constantly changes. Also, a communication line could be rendered unusable during transmission due to a lightning strike, maintenance, or equipment failure. The arrival time of the transmitted packets could be out of sequence due to the unforeseen circumstances. Sequence numbers included with the packets help to reconstruct the data. The sequence number also identifies missing packets that may have been destroyed en route to the destination. Some examples of packet switching technologies besides the Internet are FDDI, frame relay, and Ethernet.

circuit switching
a type of transmission which establishes a permanent connection between two points for the duration of the data transfer period.

packet switching
a type of transmission which does not use a permanent connection between two points for the duration of the data transfer period. Packets may travel different routes to the same destination.

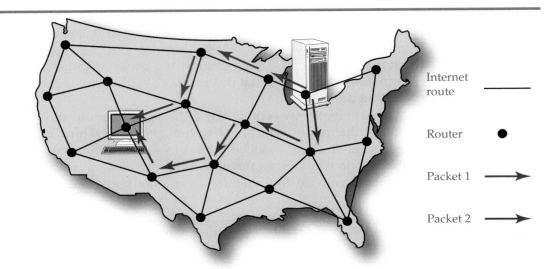

Figure 5-8.
The Internet is an excellent example of how data transfer is based on packet switching. The total amount of data is divided into packets that may take different routes to their destination.

Data Codes

There are many different coding standards, or data codes, for representing the written word. There are also many standards used for representing images, audio sound, and multimedia. Some you may be familiar with include JPEG, BMP, GIF, TIFF, and MPEG. Data codes must be converted for data exchange to take place. In the OSI model, this is addressed at the presentation layer. For example, a typical computer using ASCII and an IBM mainframe computer using EBCDIC cannot understand or decipher an exchange of information. The two encoding schemes must be converted so that the systems can understand the data when it is received. The most commonly encountered data codes are ASCII, BCD, EBCDIC, and UNICODE. Each of these data codes will be briefly introduced. Remember that in the OSI model, the presentation layer is responsible for putting the data into a format that is agreed on by the source and the destination.

ASCII

American Standard Code for Information Interchange (ASCII) a character code that uses eight bits to represent alphanumeric characters.

The ***American Standard Code for Information Interchange (ASCII)*** was an early attempt to standardize data codes. Look at the table of the ASCII character set in **Figure 5-9.** Note that 32 codes are command codes and 96 represent characters found on a standard keyboard. There is an additional 128 symbols not shown in the figure that are operating system dependent. Operating system dependent means that the additional symbols are not standard but rather can only be interpreted by a specific software program. If the data file is opened by the wrong software application, the symbols are incorrectly interpreted and the result is unintelligible. The additional symbols represent such items as fractions, foreign letters, special symbols, lines, and more. The upper 128 characters are not standard. Also note that an ASCII symbol does not indicate font attributes such as size, shape, color, or spacing. ASCII characters interpreted by a software application such as Microsoft Word or Word Perfect can represent these additional attributes.

ASCII files are sometimes referred to as plain text files. They contain no special alphanumeric enhancements such as bold, italic, or underline. The ASCII file format is the most easily exchanged file format used by word processing programs. ASCII uses eight bits to represent individual characters. The bit pattern represents the digital pulses used to encode data as alphanumeric characters.

In the early days of computing, there were many problems because of a lack of standard encoding patterns. Many times an encoding pattern was proprietary. The hardware and software had to be purchased from the same vendor and may have been incompatible with other software programs or hardware devices. Today, some problems are encountered when exchanging text files. Two different versions of the same word processing software may not be compatible. Incompatibility may cause the file not to open when prompted. A good way to ensure a text file can be exchanged between two different systems is by saving the file as an ASCII file or as a plain text file. **Figure 5-10** shows some of the options for saving the text file Chap 5 Encoding and Transmitting Data.doc. Note that the **Text Only (*.txt)** option is selected.

Figure 5-9.
ASCII character set.

Standard ASCII Characters			(Continued)		(Continued)	
0	NUL	Null	43	+	86	V
1	SOH	Start of header	44	,	87	W
2	STX	Start of text	45	-	88	X
3	ETX	End of text	46	.	89	Y
4	EOT	End of transmission	47	/	90	Z
5	ENQ	Enquiry	48	0	91	[
6	ACK	Acknowledgment	49	1	92	\
7	BEL	Bell	50	2	93	]
8	BS	Backspace	51	3	94	^
9	HT	Horizontal tab	52	4	95	_
10	LF	Line feed	53	5	96	`
11	VT	Vertical tab	54	6	97	a
12	FF	Form feed	55	7	98	b
13	CR	Carriage return	56	8	99	c
14	SO	Shift out	57	9	100	d
15	SI	Shift in	58	:	101	e
16	DLE	Data link escape	59	;	102	f
17	DC1	Device control 1	60	<	103	g
18	DC2	Device control 2	61	=	104	h
19	DC3	Device control 3	62	>	105	i
20	DC4	Device control 4	63	?	106	j
21	NAK	Negative acknowledgment	64	@	107	k
22	SYN	Synchronous idle	65	A	108	l
23	ETB	End of transmit block	66	B	109	m
24	CAN	Cancel	67	C	110	n
25	EM	End of medium	68	D	111	o
26	SUB	Substitute	69	E	112	p
27	ESC	Escape	70	F	113	q
28	FS	File separator	71	G	114	r
29	GS	Group separator	72	H	115	s
30	RS	Record separator	73	I	116	t
31	US	Unit separator	74	J	117	u
32	SP	Space	75	K	118	v
33	!		76	L	119	w
34	"		77	M	120	x
35	#		78	N	121	y
36	$		79	O	122	z
37	%		80	P	123	{
38	&		81	Q	124		
39	'		82	R	125	}
40	(		83	S	126	~
41	)		84	T	127	DEL
42	*		85	U		

Figure 5-10.
Options in a word processing program for saving a text file.

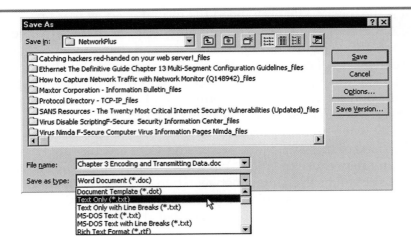

BCD

Binary Coded Decimal (BCD) is a binary number format in which each number is represented as a four-digit binary code. For example, the three groups of four binary digits 0010 0011 1001 represent the decimal numbers 2, 3, and 9. The BCD is used only for numerical applications.

EBCDIC

Extended Binary Coded Decimal Interchange Code (EBCDIC) is an IBM character code similar to ASCII. EBCDIC is pronounced as "ebb si dik." It is widely used on IBM mainframe computers. See **Figure 5-11.**

Unicode

Unicode is similar in principal to ASCII but uses 16 bits to represent individual characters. By using 16 bits for each character, over 65,000 possible characters can be represented. This may seem like an extreme number of possibilities when thinking in terms of the English alphabet, but languages, such as Chinese, that rely on symbols require many more possibilities than the typical 8-bit ASCII code pattern will produce. There are also many characters that are unique to the various languages around the world, such as Russian, Hebrew, Tibetan, Mongolian, and Cherokee, which require extra bits. Unicode is also used in bar codes for scanning merchandise and in Braille for the blind.

Look at the various groups of Unicode listed in **Figure 5-12.** The Unicode groups Thai and Lao represent the Thai language symbols and the Laotian language symbols respectively. The references to CJK represent a joint collection of symbols used by China, Japan, and Korea. The Unicode collection represents over 96,000 symbols. To learn more about Unicode visit the Unicode Web site at www.unicode.org.

HTML

Hypertext Markup Language (HTML) is an authoring language used to create documents that can be downloaded from the Internet and viewed by a Web browser. Part of the HTML language is a standard set of color codes. The color codes are inserted into the coding of a Web page. When a Web browser interprets the page, the color code is translated into the color the code represents.

EBCDIC Characters

Figure 5-11.
EBCDIC character set.

Dec	Hex	Code	Dec	Hex	Code	Dec	Hex	Code	Dec	Hex	Code
0	00	NUL	32	20		64	40	space	96	60	-
1	01	SOH	33	21		65	41		97	61	/
2	02	STX	34	22		66	42		98	62	
3	03	ETX	35	23		67	43		99	63	
4	04		36	24		68	44		100	64	
5	05	HT	37	25	LF	69	45		101	65	
6	06		38	26	ETB	70	46		102	66	
7	07	DEL	39	27	ESC	71	47		103	67	
8	08		40	28		72	48		104	68	
9	09		41	29		73	49		105	69	
10	0A		42	2A		74	4A	[	106	6A	|
11	0B	VT	43	2B		75	4B	.	107	6B	,
12	0C	FF	44	2C		76	4C	<	108	6C	%
13	0D	CR	45	2D	ENQ	77	4D	(	109	6D	_
14	0E	SO	46	2E	ACK	78	4E	+	110	6E	>
15	0F	SI	47	2F	BEL	79	4F	| !	111	6F	?
16	10	DLE	48	30		80	50	&	112	70	
17	11		49	31		81	51		113	71	
18	12		50	32	SYN	82	52		114	72	
19	13		51	33		83	53		115	73	
20	14		52	34		84	54		116	74	
21	15		53	35		85	55		117	75	
22	16	BS	54	36		86	56		118	76	
23	17		55	37	EOT	87	57		119	77	
24	18	CAN	56	38		88	58		120	78	
25	19	EM	57	39		89	59		121	79	'
26	1A		58	3A		90	5A	!]	122	7A	:
27	1B		59	3B		91	5B	$	123	7B	#
28	1C	IFS	60	3C		92	5C	*	124	7C	@
29	1D	IGS	61	3D	NAK	93	5D	)	125	7D	'
30	1E	IRS	62	3E		94	5E	;	126	7E	=
31	1F	IUS	63	3F	SUB	95	5F	^	127	7F	"

(Continued)

Figure 5-13 shows a partial listing of color codes. Note the color codes are expressed in hexadecimal form, such as FFFFFF, FFFFCC, FFFF99, FFFF66, FFFF33, and FFFF00. Web browsers, such as Internet Explorer or Netscape, interpret the color code and create the exact shade of color indicated.

Protocol Frame Structures

In Chapter 1—Introduction to Networking, you learned that a protocol is a set of rules that determine how two nodes will communicate. One part of a

Figure 5-11.
(Continued.)

EBCDIC Characters

Dec	Hex	Code	Dec	Hex	Code	Dec	Hex	Code	Dec	Hex	Code
128	80		160	A0		192	C0	{	224	E0	\
129	81	a	161	A1	~	193	C1	A	225	E1	
130	82	b	162	A2	s	194	C2	B	226	E2	S
131	83	c	163	A3	t	195	C3	C	227	E3	T
132	84	d	164	A4	u	196	C4	D	228	E4	U
133	85	e	165	A5	v	197	C5	E	229	E5	V
134	86	f	166	A6	w	198	C6	F	230	E6	W
135	87	g	167	A7	x	199	C7	G	231	E7	X
136	88	h	168	A8	y	200	C8	H	232	E8	Y
137	89	i	169	A9	z	201	C9	I	233	E9	Z
138	8A		170	AA		202	CA		234	EA	
139	8B		171	AB		203	CB		235	EB	
140	8C		172	AC		204	CC		236	EC	
141	8D		173	AD		205	CD		237	ED	
142	8E		174	AE		206	CE		238	EE	
143	8F		175	AF		207	CF		239	EF	
144	90		176	B0		208	D0	}	240	F0	0
145	91	j	177	B1		209	D1	J	241	F1	1
146	92	k	178	B2		210	D2	K	242	F2	2
147	93	l	179	B3		211	D3	L	243	F3	3
148	94	m	180	B4		212	D4	M	244	F4	4
149	95	n	181	B5		213	D5	N	245	F5	5
150	96	o	182	B6		214	D6	O	246	F6	6
151	97	p	183	B7		215	D7	P	247	F7	7
152	98	q	184	B8		216	D8	Q	248	F8	8
153	99	r	185	B9		217	D9	R	249	F9	9
154	9A		186	BA		218	DA		250	FA	
155	9B		187	BB		219	DB		251	FB	
156	9C		188	BC		220	DC		252	FC	
157	9D		189	BD		221	DD		253	FD	
158	9E		190	BE		222	DE		254	FE	
159	9F		191	BF		223	DF		255	FF	

protocol determines how data and commands are structured into packages such as packets and frames. Many pieces of information have to be built into each packet, such as the source and destination address and the data field size. Many different protocol data structures are used to communicate across networks. The exact protocol data structure is determined by the purpose of the communication (such as commands, text file transfer, and multimedia transfer), the type of network architecture (such as Ethernet, and Token Ring), and the use of specialized equipment and media during the transfer (such as telephone lines, modems, and routers). For example, Ethernet, Token Ring, Wireless, and

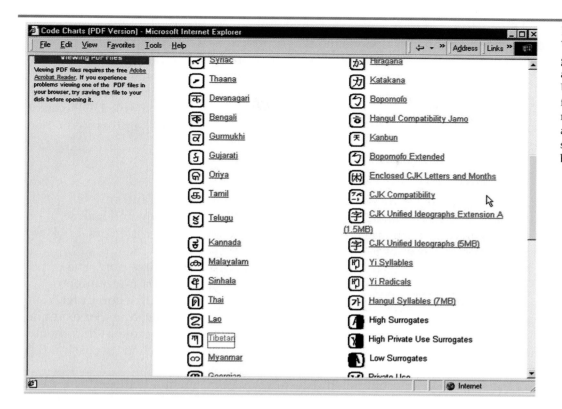

Figure 5-12.
Various Unicode groups. Only languages are shown here, but Unicode also has charts for Braille patterns, mathematical symbols, and drawing symbols, such as arrows and boxes.

Color code	Color	Color code	Color
#FFFFFF		#FFCC33	
#FFFFCC		#FFCC00	
#FFFF99		#FF99FF	
#FFFF66		#FF99CC	
#FFFF33		#FF9999	
#FFFF00		#FF9966	
#FFCCFF		#FF9933	
#FFCCCC		#FF9900	
#FFCC99		#FF66FF	
#FFCC66		#FF66CC	
#FFCC33		#FF6699	
#FFCC00		#FF6666	
#FF99FF		#FF6633	
#FF99CC		#FF6600	
#FF9999		#FF33FF	
#FF9966		#FF33CC	

Figure 5-13.
HTML color codes are written in hexadecimal format and represent various colors.

FDDI use different frame formats to match the technology. The PPP protocol is designed to establish telephone modem connections. Network protocols such as Ethernet and Token Ring can transport other protocols. For example, a TCP or UDP protocol packet is typically carried inside an Ethernet protocol frame.

To help you better understand the concept of protocol data structures and how data is encoded into packets, we will first look at how a simple UDP packet is constructed. Afterwards, a more complicated protocol will be presented as a comparison.

UDP Frame Structure

The User Datagram Protocol (UDP) is relatively compact and simple in design. It contains the data to be transferred, the destination and source information, the length of the packet, and a method to check for errors. Look at the illustration in **Figure 5-14.**

port
a number that represents a logical connection and matches a service with a computer.

The first 16 bits (two bytes) of the packet identify the source port. The source port is the port from which the UDP originated. A *port* is a number that represents a logical connection and matches a service with a computer. A computer uses many different port numbers when communicating. For example, the common port number for the HTTP service is port 80. There will be a much more detailed explanation about ports later in the textbook when discussing the TCP/IP protocol.

The second 16 bits contains the destination port, which identifies the port to which the UDP is being delivered. The third 16 bits indicate the length of the entire UDP packet so that the destination can be sure the entire message or data package was delivered. Next, a 16-bit checksum block of information is included. The checksum is similar to the CRC discussed earlier. It uses a digital-based mathematical calculation called *one's compliment* to check for errors or corruption. The last block of information contains the actual data.

one's compliment
a digital-based mathematical calculation.

Ethernet Frame Structure

Four types of Ethernet frames exist in networking technology: Ethernet II (DIX), IEEE 802.2, IEEE 802.3, and Ethernet SNAP. The Xerox Corporation developed the original Ethernet standard in the early 1970s. In 1982, companies such as Digital Equipment, Intel, and Xerox joined to develop another release of Ethernet known as DIX (Digital, Intel, Xerox). The DIX Ethernet frame format became better known as Ethernet II.

Figure 5-14.
UDP frame.

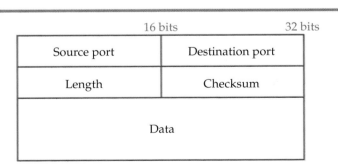

The Institute of Electrical and Electronic Engineers (IEEE) developed and released their first standard for Ethernet in 1985. The standard is called IEEE 802.3: Carrier Sense Multiple Access with Collision Detection (CSMA/CD) Access Method and Physical Layer Specifications. It is interesting to note that IEEE 802.3 CSMA/CD is the technically correct terminology for the type of media access described, and the term *Ethernet* and *802.3* has become the generally accepted term meaning the same as CSMA/CD.

The Ethernet II frame type was developed before the release of the IEEE 802.3 CSMA/CD standard. Consequently, there are two different frame types. While the two frame types Ethernet II and 802.3 are generally compatible, there is no guarantee of full compatibility. Network interface cards are usually set to automatically detect the frame type broadcast, but the network interface card can be configured to accept a specific frame type to improve network performance.

The default frame type associated with the IPX/SPX protocol is 802.3. Ethernet SNAP is a modification of the original 802.3 and Ethernet II frame types. We will now look at two common frame types you will most likely encounter. A study of frame differences will help you better understand the compatibility issues that exist between different systems. Look at **Figure 5-15.**

You will notice that the two frame types, Ethernet II and 802.3, are similar in construction. In fact, both frame types can coexist within the same network. However, there is no guarantee that the two frames can be used to exchange information compatibly. This is due to the slight variation in frame construction. This variation exists in the length of the preamble fields.

The preamble is used to identify and synchronize the start of the frame. It consists of a series of alternating ones and zeros and ends with a series of steady

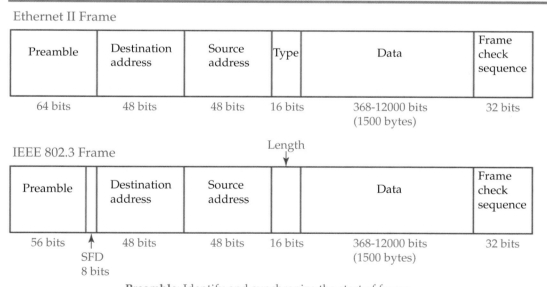

Figure 5-15.
Ethernet II and 802.3 frame comparison.

Preamble: Identify and synchronize the start of frame.
Start Frame Delimiter (SFD): Indicates the start of frame with a one-byte binary pattern in an 802.3 frame.
Destination address: Where the data is being sent.
Source address: Where the data is coming from.
Type: Which upper layer provides guaranteed delivery.
Length: The length of the data field.
Data: The data being transmitted.
Frame Check Sequence: Ensures data integrity.

ones. The start frame delimiter (SFD) in the 802.3 frame follows directly behind the preamble. Prior to Ethernet frame version IEEE 802.3, the SFD and preamble were combined. In the IEEE 802.3 frame, the SFD indicates the actual start of the frame information by using a one-byte field consisting of the following binary pattern: 10101011.

The preamble is followed by the destination address. In the Ethernet II and 802.3 frame types, there is a slight variation of the initial bits in the destination field. The variations in the binary pattern are used to indicate if the intended destination is multicast (a group of computers) or unicast (only one computer). The remaining destination field identifies the MAC address of the computer or device.

It is important to note that a frame containing only the MAC address can only be used for a LAN. To transmit the Ethernet frame to a distant network or across a WAN, additional destination information needs to encapsulate the frame. The Ethernet frame must contain both the source and destination IP addresses. The Internet is a WAN connected via routers. Routers are configured to recognize IP addresses, not MAC addresses. The router selects the proper route for the packet based on the IP address contained in the Ethernet frame. This is in contrast to NetBEUI, which does not use IP addresses to locate the destination. NetBEUI uses the name assigned to the destination and the MAC address. Since NetBEUI does not use IP addresses to identify the destination and source, a router cannot route NetBEUI transmissions.

Another point of interest is that when the destination address is filled with all ones, it becomes a broadcast frame. A *broadcast frame* is intended for every computer on the network; whereas, a *multicast frame* is intended for a preselected number of computers, such as a specific workgroup. Ethernet is also described as broadcasting to every network interface card on the network with each frame it sends. This type of broadcasting is different than that described in the broadcast frame. When described as broadcasting, Ethernet frames are broadcast to every network interface card, but only a network card that matches the MAC address indicated in the destination field receives the frames. If the destination field does not match the MAC address of the network interface card, the frame will not pass through to the computer. The exception is when a broadcast frame is used.

Another difference between the two frame types is the type and length fields. The original Ethernet II frame was designed not to perform error checking at the data link layer. Since the original Ethernet II was not intended to perform data error correction at this level, the type field was used to indicate where error correction would occur. In other words, it identifies the upper layer of the OSI model that is responsible for checking damaged frames.

In the 802.3 frame, the length field exists in place of the type field. The length field indicates the total length of the frame (not including the preamble). This information is used to perform error checking at the data link layer rather than pass the responsibility to upper level layers of the OSI model. This is a fine example of how variations of network technology can occur because of developments in technology.

The data field contains the data, such as text, graphics, or sound, which is to be delivered to the destination. The two frames have identical data field maximum and minimum lengths.

The Frame Check Sequence (FCS) field is used to determine a successful delivery of the frame. The data link layer does not perform the check. Rather, an upper-level layer in the OSI model uses the FCS to verify the frame's integrity.

broadcast frame
a frame intended for every computer on the network to read.

multicast frame
a frame intended for a preselected number of computers, such as a specific workgroup.

As you can see, the two Ethernet frames are very similar, but there are sufficient differences in the way the frames are constructed. These differences can cause compatibility problems due to variations in software programs written to use information contained in the frames. Most network interface cards have an auto-detect function to identify which type of Ethernet frame is being used. Auto detect is usually the default selection. However, you can improve the performance of a network if you know the exact frame format used. The exact frame format can be set in the network card configuration dialog box. See **Figure 5-16.**

The **NWLink IPX/SPX/NetBIOS Compatible Transport Protocol** dialog box only appears in the properties for the network interface card when the network interface card is configured for IPX/SPX compatibility. In Figure 5-16, you can see the four Ethernet frame types associated with Novell Netware IPX/SPX operating system. Also, notice that the **Auto Detect** option is the default setting. When the **Auto Detect** option is selected, the network interface card will automatically detect and configure the correct Ethernet frame type for the network.

Data Encoding, Transmission, and the OSI Model

The OSI model is used as a reference to determine how data can most effectively be transferred between networks. It is specifically designed for communications across WANs and MANs. LANs did not need a complex model when they were first developed. They were, in fact, proprietary in nature since communications were limited to nodes only on the local network. The network operating system could handle all that was required for identification, packet size, media access, security, and such. However, as the Internet grew, it became apparent that many different operating systems, networking devices, and media were going to be used to support communications. Some sort of standard or model was needed to help programmers and manufacturers design systems and computer programs to support the vast array of network technologies.

The OSI model is sometimes referred to as the *protocol stack* because it is, in essence, a stack of protocols. Various protocols are associated and designed to work at different levels of the OSI model. Some protocols are designed to

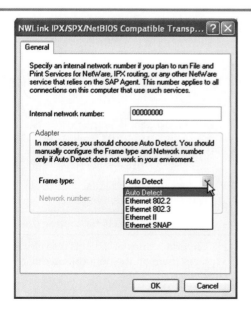

Figure 5-16.
Setting the frame type in the **NWLink IPX/SPX/ NetBIOS Compatible Transport Protocol** dialog box.

work with specific types of media, some are designed for security, and others are designed to encapsulate various types of data such as text, audio, or video. The exact protocol needed is determined by the type of media and networking devices used, the nature of communication, and the type of data sent.

For example, you may use the Internet to access information on a distant server. As the signal travels between your computer and the remote server, it may pass through a maze of telephone exchanges, fiber-optic cables, wireless satellite connections, microwave stations, and such. It may also pass through multiplexers, switches, routers, and such. As the data passes through various network media and devices, the data is packaged to match the technology it encounters. This is all done at remarkably high speeds.

When negotiating a connection, the protocol used for making the initial connection may be different from the protocol used to transfer the data. For example, the Point-to-Point Protocol (PPP) may be used to initially attach to an Internet service provider through a telephone modem. After the connection is established, IP and UDP are used to make a connection at the distant site. After the connection is established, IP and TCP may be used to transfer the data. As you see, no one protocol is used.

As data moves through the layers of the OSI model, each layer is responsible for a specific aspect of communication. In other words, each layer has a different set of responsibilities. The following sections present an overview of how communication takes place at each layer of the OSI model. **Figure 5-17** provides a summary of each layer's function.

Application Layer

The application layer is where the user interfaces with network operating system. The application layer is at the top of the OSI model and is the start and the final destination of all data communications. When a user generates a request, such as to connect to a Web page, the request is packaged, sent down the protocol stack to the physical layer, and sent across network media to the destination computer. At the destination computer, the request enters the physical layer and is sent up the protocol stack to the application layer. The data is finalized in the appropriate form, such as displaying the Web page. The

Figure 5-17.
OSI layer and
functions.

Layer	Function
Application	Interfaces to the network system.
Presentation	Packages data into a universally agreed on form, such as ASCII, BCD, BMP, JPG, and WAV.
Session	Establishes and coordinates communications between two points.
Transport	Ensures accurate delivery.
Network	Encapsulates packets for routing.
Data link	Converts frames or packets into electronic signals and places them on the network media.
Physical	The network media.

application layer is associated with applications such as Web browsers and e-mail. The application layer also starts the file transfer mechanism associated with specific operating systems.

Presentation Layer

Data at the source computer leaves the application layer and enters the presentation layer where the raw data is packaged into a universally agreed on form. The data byte order must be agreed on and presented in the same manner. For example, most computer system CPUs are one of two types: Intel or Motorola. The familiar Pentium processors are manufactured by Intel and the Apple Mac CPU is manufactured by Motorola. Mainframe computers are constructed from both. The two different CPUs write data in opposite order. The order of data written into the packets must be agreed on and presented in the same fashion; otherwise, the data will not be interpreted correctly. If all the hardware and software in the network were exactly the same, there would be no need for a presentation layer. Data encryption also takes place at the presentation layer. *Data encryption* is the encoding of data based on a mathematical formula, which converts the original data symbol into another symbol.

data encryption the encoding of data based on a mathematical formula, which converts the original data symbol into another symbol.

Session Layer

The session layer establishes a dialog between the source and destination computers. It is easier to visualize the session layer as a negotiation rather than as a layer. The session layer negotiates decisions about how data flow is controlled and how a session ends. It decides on whether the destination computer needs to confirm that the data has been received or if the data will be sent without requiring a confirmation of its arrival. For example, video may need to be sent in one continuous flow to preserve the quality of the audio, such as in a telephone conversation. Confirming the arrival of each packet would not be possible. Sending an illustration or digital image, may require that each packet not only be confirmed but also that it be assembled in correct order so that the image is not distorted.

Transport Layer

The transport layer is responsible for the flow of data to and from the destination computer. The source and destination computers decide on an appropriate amount of data that can be sent at one time. They also identify if the data needs to be transmitted in sequential order or if a simple one-packet transmission is enough. For example, a large block of data may need to be segmented. Each segment would need a unique number assigned to it to ensure that it is assembled in the correct order. A simple authentication, which consists of a user name and password to access the destination computer, may only require one frame. Thus, one frame would not need to be assigned a sequence number.

The transport layer is also responsible for overseeing that the complete data package reaches the destination computer. When a large block of data is divided into many segments, it is most likely that a form of response will be needed. A response, such as a UDP packet, is sent from the destination to the source computer confirming that the packets were received and that the next group of packets may be sent or that all of the data has been received.

Network Layer

The network layer provides a means of routing data packets across a WAN or MAN. For example, sending a data packet across a LAN only requires the destination computer's address. This means the name of the computer or the MAC address can be used to identify the destination. When sending a data packet across a WAN or MAN, the packet must be encapsulated with an additional address. The MAC address and computer name is not a practical solution. A more precise address is required to negotiate a route across the Internet, specifically an IP address. Each packet of data must be repackaged according to a protocol standard such as TCP/IP. The protocol used depends on the need for verification of the delivered package. You should recall that the IP protocol does not require verification and is a connectionless protocol. TCP requires verification and is called a connection-oriented protocol.

It is at the network layer that data packets are turned into frames. The frame terminology comes from the fact that the network layer surrounds the data packet with a frame that can navigate across routers. Routers use IP addresses to transfer frames to their destination.

virtual network
a logical network
within a LAN.

The network layer is also responsible for virtual networks. A *virtual network* is a logical network within a LAN. Virtual networks are created to help eliminate collisions, thus increasing the bandwidth capabilities of the LAN. A virtual network is typically created in a LAN to reduce the number of collisions.

TCP/IP was originally designed to communicate over the Internet connecting millions of computers to each other. Originally LANs used proprietary protocols to communicate. Today TCP/IP is the default protocol used to communicate over LANs and the Internet. TCP/IP is covered in detail in Chapter 11—TCP/IP Fundamentals.

Look at **Figure 5-18.** Two virtual networks have been created with a router. The physical network consists of six computers; however, the router creates two separate virtual networks. The router constructs a database of all connected devices by IP address. After all devices have been identified, the router can be programmed to limit packet broadcasts to only the devices identified or grouped together by IP addresses. In the example, two computers are grouped as virtual network A and four computers are grouped as virtual network B. Since broadcasts are limited to the designated virtual networks A and B, communications is limited to these networks. The two networks are not aware of each other. The router can also be programmed to allow communications beyond the virtual network.

This same type of virtual network can be created using a switch. A layer 2 switch creates networks based on MAC addresses. Remember, a layer 2 device operates at the data link layer.

Network switches have become more complex as networking science has evolved. Switches are classified as layer 2, layer 3, and layer 4.

While the intent of the OSI model is to make data packaging appear as a precise sequence of operations, data packaging flows more freely. For example, if data is sent to a destination on a LAN and does not require an IP address, the network layer is not really needed. The data is passed through to the data link layer. The data link layer will have enough information to deliver the data package to the final destination on the LAN.

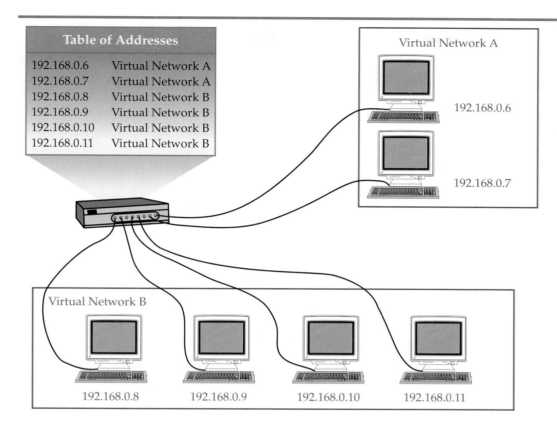

Table of Addresses	
192.168.0.6	Virtual Network A
192.168.0.7	Virtual Network A
192.168.0.8	Virtual Network B
192.168.0.9	Virtual Network B
192.168.0.10	Virtual Network B
192.168.0.11	Virtual Network B

Figure 5-18.
A switch or router can be used to divide a physical network into two or more virtual networks. The router or switch maintains a table of addresses, which it uses as a reference.

Data Link Layer

The data layer converts the data package into electrical pulses and places them on the network media. The data link layer is subdivided into the logical link control (LLC) sublayer and the MAC sublayer. At the MAC sublayer, the data code is converted into electrical pulses.

The MAC sublayer is not only concerned with how the data is encoded, but also with the type of topology and cable used. For example, it decides if the data frame needs to be converted for an 802.11 (wireless) network or an 802.3 10BaseT network.

The LLC sublayer is only interested in getting the data to the destination on the LAN. The LLC sublayer uses the MAC address to address the packet.

In an ARCnet network, a number from *1–255* is assigned to each network interface card. The LLC uses this number to address the packet.

Tech Tip

If data is to be sent across a WAN, the data link layer sends the data to the appropriate router. The router forwards the package based on the IP address inside the package. The router does not need the MAC address to deliver the data package.

The data link layer, like the transport layer, is responsible for error checking. However, at this level, the error checks are based on the binary coded digital patterns. Checks such as parity are made to detect missing binary information. CRC checks are also performed at the data link layer.

Network+ Note:

For the Network+ Certification exam, remember that parity and CRC checks are performed at the data link layer. Checks at the transport layer ensure the entire data block was received intact. The transport layer also checks packet sequencing to ensure all data packets are reassembled in the proper order.

Physical Layer

The physical layer is concerned with the media, hardware, and topology of a network. The media may be infrared signals, copper wire, glass or plastic fiber, or electromagnetic radio waves. The choice of topology is typically a star, ring, mesh, or hybrid. The hardware is the connectors used for the media and unintelligible devices, such as hubs and repeaters.

An unintelligent device makes no decision about where the data will go. It simply passes electrical pulses though the device. An example of an unintelligent device is a repeater or a hub. In Chapter 1—Introduction to Networking, you learned that hubs could be either active or passive. A passive hub is an unintelligent device.

Network+ Note:

For this textbook and for the Network+ Certification exam, a hub is considered unintelligent. It simply provides a connection for the physical media and passes signals without making any choices.

Summary

- Digital encoding is the conversion of data into a digital pattern acceptable to the networking media.
- A digital signal that fluctuates between a positive voltage level and zero-volt level is called a unipolar digital signal.
- A digital signal that fluctuates between a positive voltage level and a negative voltage level is called a bipolar digital signal.
- Manchester encoding is a common digital-encoding scheme used on LANs. It is characterized by the digital pulse transitioning during the midpoint of the timing period.
- A digital signal that fluctuates between a positive voltage level and a negative voltage level and never returns to zero for any measurable period of time is known as non-return to zero (NRZ).
- A synchronous signal is synchronized with a reference signal to ensure proper timing.
- An asynchronous signal is not synchronized with a reference signal and typically uses a binary pattern to determine the stop and start of the signal.
- A parity check is used to check the integrity of the data.

- A Cyclic Redundancy Check (CRC) is an enhanced parity check that not only identifies when an error occurs but also corrects the error.

- Data encapsulation is the process of surrounding the raw data with other information needed for delivery.

- The terms *segment, frame, datagram,* and *protocol data unit (PDU)* are used to describe blocks of data.

- The term *protocol data unit (PDU)* is the most technically correct term used to describe data packaging.

- A connection-oriented protocol establishes a connection with the destination, transfers the data, and then releases the connection after all communication has ended.

- A connectionless protocol simply sends data to the destination.

- A circuit-switching network establishes a physical connection between two points.

- A packet-switching network establishes a logical connection and may use many different, physical paths to send data from the source to the destination.

- ASCII, EBCDIC, and UNICODE are coding standards, or data codes, that represent the written word.

- Binary Coded Decimal (BCD) is a coding standard that uses a four-digit binary code to represent a number.

- Various protocols are designed to package data in particular ways.

- The OSI model is composed of seven layers, which explain the process of encapsulating data for communications between two points on a network system.

Review Questions

Answer the following questions on a separate sheet of paper. Please do not write in this book.

1. Where does digital encoding take place?

2. Describe the Manchester digital encoding scheme and how it is related to binary signals.

3. The type of transmission in which the digital signal is synchronized with a reference signal to ensure proper timing is called _____ transmission.

4. The type of transmission in which no reference signal is present is called _____ transmission.

5. What is a parity check?

6. What is a Cyclic Redundancy Check (CRC)?

7. Describe the complete data packaging process, from raw data (text file, sound, or graphic) to digital signals.

8. A ____ protocol first establishes a connection with the destination computer before transmitting data.

9. A _____ protocol sends data to a destination computer without first establishing a connection.

10. What is the difference between packet switching and circuit switching?

11. Name three encoding standards used to represent written language as binary numbers.

12. What does the acronym ASCII represent?

13. Which standard was designed to represent plain text files?

14. What code is used to represent strictly numerical data?

15. What code format does IBM mainframe computers use?

16. What code can be used to represent over 65,000 different characters?

17. Draw and label a UDP frame.

18. Compare and contrast an Ethernet II frame with an IEEE 802.3 frame.

19. What layer of the OSI model is responsible for the conversion of different coding formats?

20. What layer of the OSI model routes data packages based on the IP address?

21. What layer of the OSI model is concerned with the media, hardware, and topology of the network?

22. What OSI layer is responsible for converting data into digital signals?

23. What OSI layer converts data into a form acceptable to both the source and destination?

24. What OSI layer is concerned with encrypting data?

25. What OSI layer is concerned with establishing and ending the connection to the destination?

26. What OSI layer negotiates the connection between the destination and source?

27. What OSI layer decides if TCP or UDP is to be used as the protocol for transmitting the data?

Sample Network+ Exam Questions

Answer the following questions on a separate sheet of paper. Please do not write in this book.

1. Data integrity of certain protocols is verified by using a technique known as _____.

    A. Cyclic Redundancy Check (CRC)

    B. Data Parity Reference (DPR)

    C. Kerberos Security Check (KSC)

    D. Parity Data Unit (PDU)

2. Which is true concerning a connectionless-oriented protocol? (Select all that apply.)

    A. When establishing a connection between the destination and source, the destination must send a "clear to send" signal to the source before data can be transmitted.

    B. When ending a session, the source must send a termination signal to the destination.

    C. There is no verification from the destination that the data arrived intact.

    D. A connectionless protocol typically relies on the integrity of the network equipment and assumes the packet arrived at the destination.

3. A connection established via modem over a telephone line with an ISP is an example of which of the following technologies?

    A. Packet switching

    B. Circuit switching

    C. Modular switching

    D. Enterprise switching

4. Which layer in the OSI model is responsible for converting data into electrical pulses?

    A. Physical

    B. Data link

    C. Application

    D. Presentation

5. The IEEE 802.3 standard is associated with which media access method?

    A. Carrier Sense Multiple Access with Collision Detection (CSMA/CD)

    B. Carrier Sense Multiple Access with Collision Avoidance (CSMA/CA)

    C. Asynchronous Transmission Mode (ATM)

    D. Wireless Ethernet Protocol (WEP)

6. What is the purpose of a port number?

    A. Identifies the network interface manufacturer.

    B. Identifies a computer operating system service.

    C. Identifies the data rate used by the protocol.

    D. Identifies what type of protocol is used for the transmission.

Network+

7. Which Ethernet frame type is associated with IPX/SPX?
   A. UDP
   B. 802.11
   C. 802.3
   D. TCP/IP

8. Which protocol is used to establish a connection with an ISP when using a telephone modem?
   A. TCP
   B. IP
   C. UDP
   D. PPP

9. Which items are responsible for identifying a destination computer? (Select all that apply.)
   A. IP number
   B. MAC address
   C. Port number
   D. Sequence number

10. What is the purpose of a multicast Ethernet frame?
    A. To broadcast a message to a specifically identified set of workstations.
    B. To broadcast a message to all workstations in the LAN.
    C. To broadcast a single frame to all servers in the WAN.
    D. There is no such thing as a multicast Ethernet frame.

Suggested Laboratory Activities

1. Using the Internet, research various Ethernet frame formats. List the contents of each section of the frame that surrounds the data. See if you can determine where the contents of the frame that is responsible for asynchronous data transmission are located. (Not all frame types will contain an asynchronous marker.)

2. Download a free or limited-time protocol analyzer from the Internet. Use the protocol analyzer to capture frames transmitted from your PC and analyze the contents. Determine where the MAC address, IP address, and data contents are located. See if you can decode the contents of the data frame and addresses.

3. On a Windows 98, Windows 2000, and Windows XP networked computer, locate the properties for the network interface card and identify the various protocols associated with it.

Interesting Web Sites for More Information

www.protocols.com

www.unicode.org

Chapter 5
Laboratory Activity

Ethereal Protocol Analyzer

After completing this laboratory activity, you will be able to:

- Explain the purpose of a network protocol analyzer.
- Describe the common features of a protocol analyzer.
- Explain the purpose of the WinPcap program.
- Define the term *promiscuous* as related to a network protocol analyzer.
- Download and install the Ethereal program.

Introduction

In this laboratory activity, you will download, install, and configure the Ethereal protocol analyzer. The word *Ethereal* is pronounced like two words: *Ether* and *real*. There are many commercial protocol analyzers available that cost a thousand dollars or more. Ethereal is a free, open source product developed under the GNU General Public License. While originally developed for Linux computer systems, it will run on Windows and MAC systems. Many Linux operating systems incorporate a version of Ethereal in the administrative tools.

Note:
While Ethereal will install on most hardware and operating systems, there can still be occasional problems. Check the www.ethereal.com *Web site for more detailed information.*

The understanding of network protocols can be enhanced through the use of a protocol analyzer such as Ethereal. Capturing protocols as they appear in a network system will provide you with an analytical approach to learning how network protocols function—not only the data link layer but also at the higher layers of the OSI model.

This laboratory activity provides a brief introduction to Ethereal. A complete manual designed for Ethereal support and a Frequently Asked Questions (FAQs) list can be found at the Ethereal Web site. You may want to download a copy of the manual at home to learn more about this protocol analyzer.

The latest version of Ethereal can capture and identify over 600 protocols. This number is steadily increasing because of the contributions of so many programmers involved in the project. Ethereal can capture packets and frame information for Ethernet, Token Ring, FDDI, and wireless networks.

Since you will be using a Windows XP operating system on your workstation, it will be necessary to download and install the WinPcap program. WinPcap is an open source library designed for packet capture and analysis of Win32 platform software by Microsoft. WinPcap makes it possible to use Windows 95 and later operating systems with Ethereal. Since you will be using Windows XP, it is a required program. More information is available about WinPcap at the Ethereal Web site.

Ethereal can operate in two different modes: promiscuous and non-promiscuous. In non-promiscuous mode, Ethereal captures only packets intended for the workstation network interface card. In promiscuous mode, all packets on the immediate network segment are captured. The two modes of operation will prove to be very valuable when studying network activity and activity on the network interface card. In promiscuous mode you will be able to capture all activity on the local network segment and see the overall operation of a LAN.

Equipment and Materials

■ Workstation configured with Windows XP and Internet access. (Internet access is required to download a copy of Ethereal and WinPcap.) The workstation must also be equipped with a network interface card.

Procedures

1. _____ Report to your assigned workstation.

2. _____ Boot the workstation and make sure it is in working order.

3. _____ Connect to the Internet and then locate the Ethereal Web site (www.ethereal.com).

4. _____ Read the instructions found on the download site. Failure to read and follow the instructions can cause unpredictable results. Pay particular attention to the requirement for the WinPcap program.

5. _____ Download and install a copy of Ethereal that is compatible with your workstation operating system.

6. _____ After Ethereal is installed, look for a desktop icon similar to the following. If you do not see this icon, call your instructor for assistance.

7. _____ Activate Ethereal by double-clicking the Ethereal icon. The Ethereal graphical user interface should appear. The packet capture areas will be blank.

8. _____ Start the Ethereal protocol analyzer by selecting the **Start** option from the **Capture** menu. The **Ethereal: Capture Options** dialog box will display.

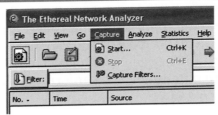

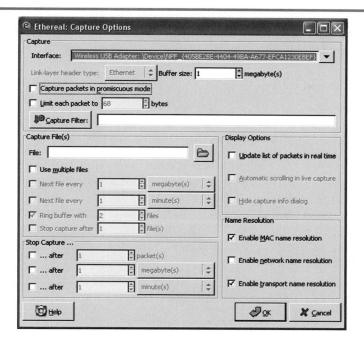

The default options should work fine for your first test run. The only real problem can be if you have more than one network interface card, such as a wired network connection and a wireless network connection. Also, be aware that some modems are detected as a network adapter when connected to an ISP. Pay close attention to the **Interface** textbox located at the top of the dialog box. The **Interface** textbox lists all available network adapters. You can choose which one to use with the Ethereal protocol analyzer. Check it now to verify the correct network adapter is being used and then click **OK**.

9. _____ A new window will appear on the screen similar to the following:

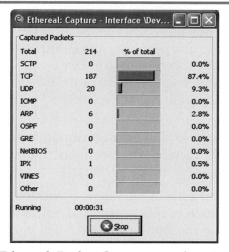

This is the Ethereal: Packet Capture window. It displays the total number of packets captured according to packet type. Notice that in this example, a total of 214 packets are captured among three categories: TCP, UDP, and ARP. When the Ethereal: Packet Capture window first appears, it will be empty. The Ethereal program will capture packets until you click the **Stop** button. Once you've clicked **Stop**, the protocol analyzer will display the number of packets captured.

Note:
If no packets have been captured, you have not established a connection to the Internet or to the local network. Check your connection.

10. _____ Open Internet Explorer and use the Google search engine to find www.RMRoberts.com or enter www.RMRoberts.com into your Web browser address bar. You need to establish a connection with a Web page to generate packets for capture by Ethereal.

11. _____ Once you have established with a Web page, return to the Ethereal: Packet Capture window and click **Stop**. You should see a display similar to the following:

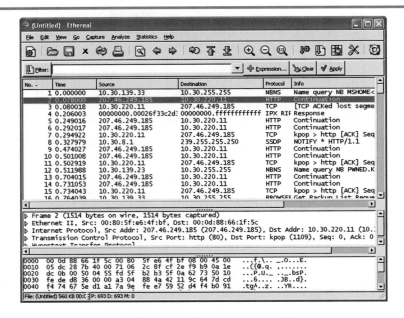

Your display of packets will not be an exact match to that in the screen capture. The protocols and capture sequences listed in your display will vary according to your network setup and Internet access method. The main point of the exercise is to have some packets captured and displayed as they are in the screen capture.

Once you have successfully completed your first packet capture, you can go on to explore some common aspects of Ethereal. If you have not made a successful capture of a series of packets, call your instructor for assistance.

12. _____ Look at the top pane of the Ethereal window. This pane displays a summary of each packet captured. Look at the column labeled **No.** and notice that the summaries are listed in the order in which they were captured. Clicking on the column label will change the listing from descending to ascending numerical order. Click the column label now and observe the effect.

13. _____ Now, click on the **Protocol** column label to see how the information is sorted by protocol rather than by the captured packet sequence.

14. _____ Click on the **Source** column label to see how the source addresses rearrange.

15. _____ Click once more on the **No.** column label to rearrange the packets into sequential order.

16. _____ From the **Protocol** column, select one of the HTTP packets by clicking on the desired summary listing. The contents of the frame and packet are displayed in the middle and bottom panes. Examine the contents displayed. Scroll to see any hidden information.

17. _____ Look at the textbox labeled **Filter** at the top of the Ethereal window. The summary listings in the top pane can be filtered by various categories, such as the type of protocol. Type one of the displayed protocols, such as HTTP, into the **Filter** textbox. Always use lowercase letters for protocols—for example, http. Once the filter has been entered, only the HTTP protocol packets will be displayed. There will be more about filtering in later laboratory activities.

18. _____ Now, look at the middle pane that contains Ethernet frame information. Take note of the series of small boxes with a plus sign (+) inside. The structure is similar to the way Windows Explorer displays directory information. The plus sign in the box means there is more information contained within the line. Simply click on the box to expand the information. Try it now.

Note:
Some versions of Ethereal use an arrow in place of the plus sign.

There is a tremendous amount of information captured in each frame. Do not be overwhelmed by all the information presented at this time. You will use the Ethereal protocol analyzer throughout the laboratory activities. You will acquire more information about how to use the protocol analyzer as you progress through your studies.

19. _____ Exit the Ethereal program by either selecting **Exit** from the **File** menu or clicking the Close button in the upper-right corner of the Ethereal window. A message box similar to the following will appear.

The message box gives you an option to save the series of captured packets or simply exit the program. Do not save the packets at this time unless required by the instructor.

20. _____ Answer the review questions.

21. _____ Return your workstation to the condition specified by your instructor.

Review Questions

Answer the following questions on a separate sheet of paper. Please do not write in this book.

1. How much does Ethereal cost?

2. How many protocols can Ethereal capture and identify?

3. What is WinPCap?

4. What is the difference between promiscuous mode and non-promiscuous mode?

5. How can the sequence of captured packets be changed to and from ascending and descending order?

6. How can you sort protocols by type?

7. What does the box with a plus sign mean at the beginning of an information line?

6 Network Operating Systems and Network Communications

After studying this chapter, you will be able to:

❏ Describe the common traits of all major network operating systems.

❏ Describe the purpose of the data link layer of the OSI model.

❏ Explain the principle of Ethernet communication.

❏ Explain the principle of AppleTalk communication.

❏ Explain the principle of Token Ring communication.

❏ Explain the principle of Token Bus communication.

❏ Explain the principle of ARCnet communication.

❏ Describe the function of NetBIOS.

❏ Describe the function of NetBEUI.

Network+ Exam—Key Points

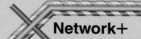

A large portion of the Network+ Certification exam is dedicated to network communication basics such as those presented in this chapter. The Network+ Certification exam will undoubtedly ask questions about the way Ethernet and Token Ring networks behave. Be sure you thoroughly understand how each network technology accesses the network media. Other typical test items focus on NetBEUI and the Universal Naming Convention (UNC).

Key Words and Terms

The following words and terms will become important pieces of your networking vocabulary. Be sure you can define them.

access method

active monitor

AppleTalk

beaconing

broadcast storm

collision domain

command prompt

command syntax

datagram

de facto standard

dumb terminal

graphical user interface (GUI)

LocalTalk

loopback test

monitor contention

ring polling

ring purge

segmenting

session

standby monitor

station address

token

Universal Naming Convention (UNC)

virtual circuit

As you have previously learned, digital data can be encoded in many different ways. It is the job of protocols to create a common language between different network hardware and network operating systems. In this chapter, the relationship between networking technologies, protocols, and network operating systems is explained. You will learn about the features that are common to all major network operating systems and will further explore network protocols and network access methods that make network communications possible. This chapter lays the foundation for the next three chapters, which cover the major network operating systems: Microsoft, Novell, and UNIX/Linux.

Common Network Operating System Traits

Network operating systems share some common features. See **Figure 6-1.** For example, they provide a means of sharing computer resources. Resource sharing, such as sharing files and hardware, is the main purpose of a network system. All network operating systems provide a way to store and manage files in an organized manner. They also provide a means of backing up and encrypting data.

Network operating systems provide security through user authentication. Authentication is accomplished by entering a user name and password into a logon dialog box on a networked computer. Files, folders, and other resources are secured by setting permissions. Permissions are levels of security assigned to a resource. Permissions work by allowing only designated users to use the resource. Each user can also be assigned different access rights to the resource. You will learn more about permissions and access rights in Chapter 7—Microsoft Network Operating Systems.

All network operating systems provide a means of communication over the Internet by making use of the TCP/IP protocol. The TCP/IP protocol can also be used to communicate within the LAN.

Network operating systems provide troubleshooting utilities and services such as e-mail. Each network operating system is also very portable. The term *portability* is a programming term that means the operating system can be ported (installed and run) on all major hardware systems.

Common Features of a Network Operating System	Examples
Internet communications	TCP/IP protocol
Resource sharing	Printer, scanner, storage devices, files
Security	Logon/authentication and resource, file and directory permissions
Services	Web, e-mail, FTP
Storage and file management	File management utilities, backup utilities, encryption
Troubleshooting utilities	Network and server diagnostics
User interface	GUI and command line

Figure 6-1.
Common features of network operating systems.

Networks provide a user interface, both graphical and text-based. The network operating system is the software that interfaces the user with the network hardware. It determines the type of interface, such as a command-line or graphical user interface (GUI), to be used between the user and hardware. A *graphical user interface (GUI)* is a pictorial representation of commands and computer hardware that allow a user to access resources and programs with a click of a mouse button. In **Figure 6-2** a user has right-clicked the **My Network Places** icon to expose a hidden menu (shortcut menu). The menu contains commands, which the user can click to invoke. This is easier than typing commands at a command line.

In a command-line environment, commands are typed and entered at a *command prompt.* A command prompt differs among operating systems and user settings. Compare the command prompts in **Figure 6-3** and **Figure 6-4.**

graphical user interface (GUI)
a pictorial representation of commands and computer hardware that allow a user to access resources and programs with a click of a mouse button.

command prompt
a text-based interface where commands are typed and entered.

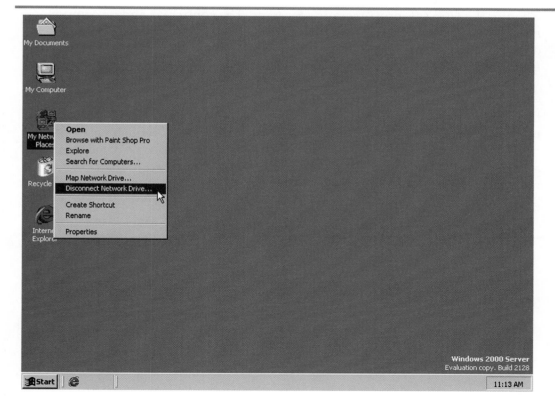

Figure 6-2.
Windows 2003 Server GUI. Right-clicking the mouse reveals a shortcut menu. The shortcut menu contains a list of commands.

Figure 6-3.
Command prompt in Windows Server 2003. A command prompt used in network operating systems is similar to the familiar DOS prompt. A command prompt appears on the screen and text commands are entered to the right of the prompt.

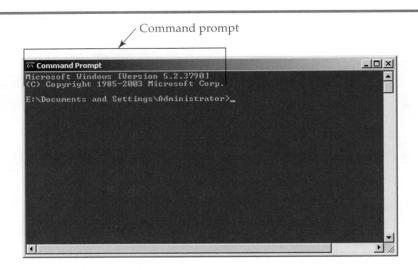

The command prompt in Figure 6-3 is from Windows Server 2003. It consists of a directory path (E:\Documents and Settings\Administrator) and a greater than symbol (>). The command prompt in Figure 6-4 is from the SuSE Server 9 Linux operating system and consists of the user name, server name, colon, current working directory, and the greater than sign. Note that the current working directory is represented by the tilde (~) symbol. In UNIX/Linux operating systems, the tilde (~) symbol represents the user's home directory.

The user can change the command prompt format. Each operating system has a set of command prompt codes from which to choose. For example, a code to display the date and time can be added to the command prompt.

Commands are typed to the right of the command prompt. Many commands consist of several components. You will learn about these components in Chapter 9—UNIX/Linux Operating Systems. The command and its components have to be typed in correctly and in the correct arrangement or an error will occur. The correct manner and arrangement in which a command is to be typed is referred to as the ***command syntax.***

command syntax
the correct manner and arrangement in which a command is to be typed.

Although network operating systems provide similar services to users, such as file sharing and file management, a user interface, and authentication, there are many variations in these services. Network operating systems differ mainly in terminology and procedures, or methods, to accomplish networking tasks. The most confusion among users is due to network service vendors using a different term to name the same device, service, or function in their own networking system. Network operating systems can be compared to automobiles. All automobiles are designed to travel from point to point. They all have starters,

Figure 6-4.
Command prompt in the SuSE Server 9 Linux operating system.

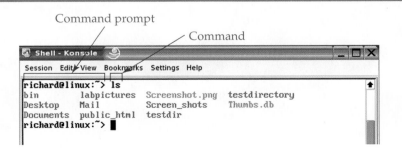

engines, tires, and other components, some of which may be identical, but each automobile brand chooses to maintain a specific identity.

As you will see in the next three chapters, there are similarities between the major network operating systems, and there are also many differences. As you become familiar with one of the operating systems, you will be simultaneously learning some knowledge that can be transferred to another operating system.

All the various network operating systems are excellent systems. Some have better characteristics than others when it comes to certain tasks. There are many opinions among network administrators and technicians about which system is best. There are also strong biases among administrators and technicians, usually based on their own personal training or experiences. Students of networking should not waste valuable time in any negative discussions regarding the best system. Often some network systems are better suited to some purposes than others, but in any given situation, given many different variables, there may be another better-suited system. The best approach may be to know that the best system is the one you are working on at the time, because it may be your job to move from system to system.

Network Operating Systems and Hardware Protocols

As you have learned in previous chapters, networking technologies such as Ethernet, Token Ring, and wireless are tied to certain standards that define the hardware, topology, access method, and other specifications of that technology. Each technology uses a different frame type or way of packaging data and other information for access over the network. This type of access takes place at the data link layer of the OSI model.

The data link layer is responsible for accessing the network media and either placing the data on the network media or receiving the data and passing it up the OSI layers, **Figure 6-5.** Token Ring and Ethernet are actually protocols that operate at the lower level of the OSI model. The method of gaining access to the network media is called an *access method.* Examples of access methods are CSMA/CD, CSMA/CA, and token passing. The access method is designed into the network interface card electronics. This is the reason network interface cards are designed to match the network technology used, such as Ethernet, Token Ring, wireless, and ARCnet.

access method
a method gaining access to the network media.

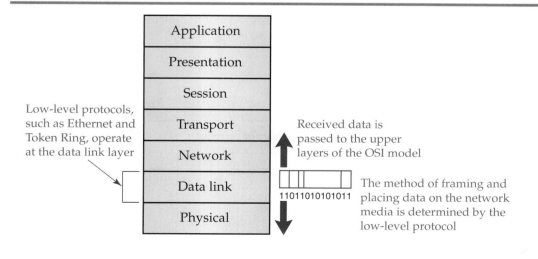

Low-level protocols, such as Ethernet and Token Ring, operate at the data link layer

Application
Presentation
Session
Transport
Network
Data link
Physical

Received data is passed to the upper layers of the OSI model

11011010101011

The method of framing and placing data on the network media is determined by the low-level protocol

Figure 6-5.
The data link layer is responsible for framing the data unit, acknowledging the reception of data, error detection, recovery, and data frame flow control.

The name of the network technology is often used interchangeably with the name of the access method. For example, the term *Ethernet* may be used in place of the term *CSMA/CD*.

The operating system software interfaces with the access method. The network operating system, such as Windows 2000, Novell, and Linux, operates above the data link layer and can run on any of the networking technologies. The access method and the network operating system combine to create the characteristics of the various network operating systems.

Originally, layer 2 of the OSI model was called the data link layer, but later it was subdivided into two distinct sublayers called the logical link control (LLC) sublayer and the media access control (MAC) sublayer, **Figure 6-6.** The LLC sublayer serves an interface between the data link layer and network operating system. It is responsible for communicating with the upper layers of the OSI regardless of the network operating system. The LLC sublayer is described in the IEEE 802.2 standard.

The MAC sublayer is responsible for formatting the data into frames and placing the frames on the physical media. Examples of frame formats are Ethernet, Token Ring, and ATM. You can think of the MAC sublayer as the layer where raw data is converted into binary ones and zeros and packaged into frame formats according to the network architecture or network technology.

Before going on to discuss the high-level protocols associated with the various network operating systems, we will first take a closer look at some of the lower-level protocols and the access methods they use.

Ethernet

Ethernet networks use the CSMA/CD access method to control and ensure the delivery of data. CSMA/CD is a broadcast method of communication. When a computer sends data to another computer, it does so by broadcasting the information to all the computers on the network. This is similar to one person yelling in a room full of people "Bob do you hear me?" Everyone in the room hears Bob's name being called, but only Bob will reply if he is in the room.

An Ethernet system communicates in a similar manner. Each computer on the network has a unique name; no two computers may have the same name.

Figure 6-6.
The data link layer is divided into two distinct sublayers: Logical Link Control (LLC) and Media Access Control (MAC). The LLC sublayer is responsible for interfacing with the network operating system. The MAC sublayer is responsible for formatting the data into frames and placing the data on the network media.

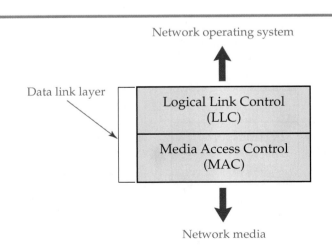

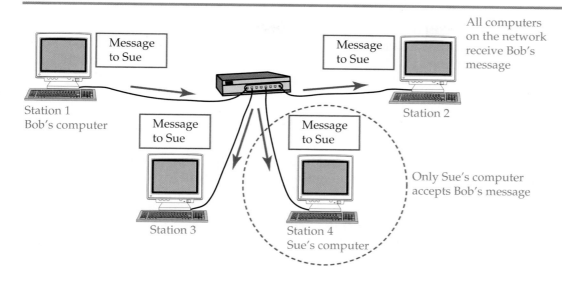

Figure 6-7.
A typical session on
an Ethernet network.
Note that all computers
receive Bob's message,
but only Sue's computer
accepts the message.

Typical computer names are *Station1*, *Accounting3*, *Smith_J*, and *Engineering_Bob*. However, a computer name can be almost anything you desire. Computer names are covered in more detail later in this chapter.

Each computer in the network also has a unique hex address, called a MAC address, programmed into the computer's network interface card. The MAC address is six bytes long. An example of a MAC address is C0 0B 08 1A 2D 2F. No other computer on the network has the same number. Using the hex numbering system to communicate is difficult for humans. Therefore, a database is used to match the MAC address of the network interface card to the unique name given to the computer.

A typical session on an Ethernet network would go something like this. Bob wants to send some data to Sue. Bob sends the data to Sue at a computer named *Station4*. When Bob sends a message to Sue, he is actually sending the message to all computers on the network. All computers on an Ethernet system hear the call from Bob to Sue, but only Sue's computer accepts the message, **Figure 6-7.** Let's take a closer look to see how this happens. First, review the Ethernet 802.3 frame in **Figure 6-8.** Note the frame sections and the size in bytes of each.

After the preamble and the start frame delimiter (SFD), the six bytes that represent the address of the destination computer are transmitted on the network. When the six bytes match Sue's computer, the data packet is accepted. When the data packet is accepted, the next six bytes, which identify the source

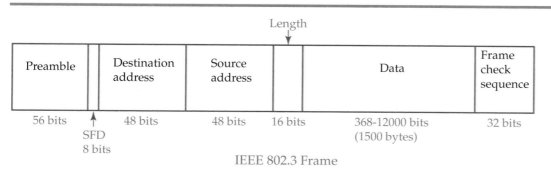

Figure 6-8.
Ethernet 802.3 frame.

computer (Bob's computer), are decoded and stored for a return message. Next, the data portion of the packet is decoded and checked for errors. If no errors are detected, a return message is transmitted to Bob's PC. If a return message is not transmitted to Bob's PC, the data packet will be retransmitted. The data will be retransmitted several more times until a message is received from Sue's PC.

Collisions

Ethernet is an excellent networking system until collisions become excessive. As collisions increase, so does the likelihood of more collisions. Each time a collision occurs, each computer involved in generating the collision ceases broadcasting for a random period of time. After that time, each computer rebroadcasts the packet. This increases the likelihood of another collision. Each collision increases the possibility of additional collisions. If the number of collisions reaches a point that the network is flooded with a continuous number of collisions and rebroadcasts, a *broadcast storm* has occurred. When a broadcast storm occurs, there are so many collisions and rebroadcasts that no significant amount of packets can be exchanged. The only solution to a broadcast storm is to eliminate a significant number of network computers that are causing the problem or to shut down the entire network, and then reboot it. Shutting down the network automatically disconnects the computers that have caused the broadcast storm.

broadcast storm
a condition in which a network is flooded with a continuous number of collisions and rebroadcasts.

Collision domains

The section of a network where collisions occur is referred to as a *collision domain.* A collision domain consists of computers that can directly communicate with each other using broadcasts. The collision domain can be isolated by equipment that controls or limits the broadcasts. Equipment such as switches and routers control broadcasts by limiting the extent or physical boundary of a collision domain. Equipment such as hubs and repeaters do *not* limit broadcasts, and thus do *not* limit the physical size of the collision domain. Repeaters and hubs simply pass the packets and frames. Switches and routers sort packets and frames depending on the destination address. This helps to isolate the network into sections called *segments.* The act of dividing a network into smaller sections to avoid collisions is called *segmenting.* Collision domains can be resized by strategically installing network devices such as switches and routers in a network topology. Look at **Figure 6-9.**

collision domain
a section of a network where collisions occur.

In the illustration, a LAN is segmented by replacing the hubs with switches. The switches isolate communications to a particular segment of the network, reducing the amount of collisions. The network is further segmented into three distinct departments: Engineering, Sales, and Management. By limiting broadcasts to a particular department, the total number of broadcasts across the entire network is reduced. Each of these departments can still communicate with the entire network, but when communications is desired with a member of the department, the packets are limited to that particular segment of the network.

segmenting
the act of dividing a network into smaller sections to avoid collisions.

As more computers are added to a network segment, the likelihood of collision becomes greater. When the number of computers grows to the point of unacceptable collisions, the network segment again needs to be segmented to reduce the number of collisions. It is important to note, however, that as more equipment is added to reduce the number of collisions, latency increases. Each piece of equipment that must analyze or repackage a packet increases the latency of the digital signal.

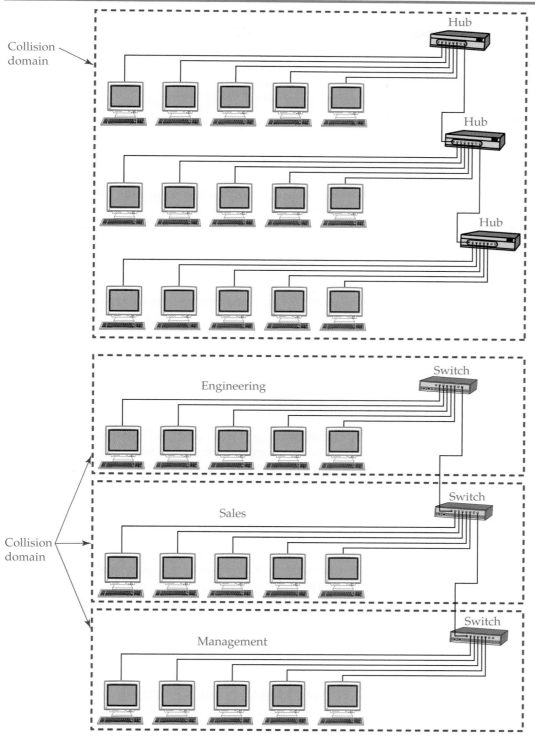

Figure 6-9.
Replacing hubs with switches on an Ethernet network can reduce collisions. Switches limit broadcasts to specific areas of the network by creating virtual networks. In the illustration, the hubs are replaced with switches to create network segments isolating some of the communication by departments, or work groups.

AppleTalk

AppleTalk is a suite of protocols developed to support the Macintosh computers when configured as a network. AppleTalk uses CSMA/CA as its access method. You should recall that wireless LANs also use CSMA/CA. CSMA/CA is similar to the access method used in Ethernet networks. However, CSMA/CA attempts to avoid collisions, rather than detect collisions. In the

AppleTalk
a suite of protocols developed to support the Macintosh computer when configured as a network. AppleTalk uses CSMA/CA as its access method.

Figure 6-10.
The token format consists only of three bytes. A workstation must first capture the token to send a Token Ring frame containing data.

Token Format

SD = Starting delimeter   1 byte
AC = Access control   1 byte
ED = Ending delimeter   1 byte

CSMA/CA access method, a computer not only listens to the media (carrier sense) to detect communications, it also sends a short packet to the destination computer to see if it is ready to receive the data. The destination computer could be offline or busy with some other task. If the destination computer is free to receive the data, it will reply. This short packet also serves as an announcement to other computers on the cable that data transmission is about to take place.

Token Ring

A Token Ring network uses the token passing access method. Before a computer can transmit information on the network, it must seize a token to take control of the network. A *token* is a short binary code that is passed around to the computers on a ring topology and, in some cases, a bus topology. See **Figure 6-10.**

token

a short binary code that is passed to computers on a ring topology and in some cases a bus topology.

In contrast to CSMA/CD, token passing is designed to prevent collisions. Transmission on the Token Ring network is organized around the computer that has possession of the token. A computer must possess the token before it can transmit data across the network.

A token is generated by the network software and is circulated throughout the network. Computers on the network must capture, or accept, the token before they are allowed to communicate. If a network interface card does not accept the token, it regenerates the token and passes it along to the next computer. By using this method, only one computer at a time can control the flow of data.

When a network interface card captures the token, it generates a new type of frame. The new frame contains source and destination addresses and data. See **Figure 6-11.** Although at first it may appear that the token passing method is time-consuming, you must remember that this process only takes a few milliseconds, in contrast to an Ethernet system that may generate many delays due to collisions.

Figure 6-11.
The Token Ring frame format is long in comparison to the token. The Token Ring frame contains data and information such as the destination and source MAC addresses.

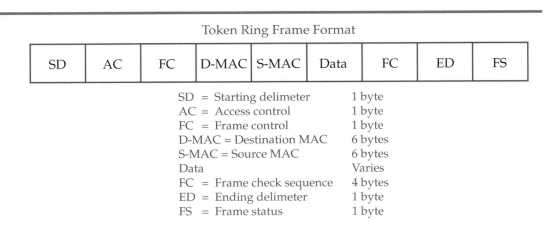

Token Ring Frame Format

| SD | AC | FC | D-MAC | S-MAC | Data | FC | ED | FS |

SD = Starting delimiter   1 byte
AC = Access control   1 byte
FC = Frame control   1 byte
D-MAC = Destination MAC   6 bytes
S-MAC = Source MAC   6 bytes
Data   Varies
FC = Frame check sequence   4 bytes
ED = Ending delimiter   1 byte
FS = Frame status   1 byte

Active monitor

When a Token Ring network is first activated, a specific computer is selected as the active monitor. An *active monitor* is a computer that is responsible for monitoring the necessary administrative functions associated with Token Ring technology. The functions of the active monitor include the following:

- Monitor the presence of the token on the ring.
- Identify added or deleted computers.
- Remove continuously circulating tokens.
- Insert a 24-bit delay between transmissions.
- Synchronize timing of all computers.

The active monitor is selected through a process called *monitor contention* or *monitor election*. Active monitor selection is based on the MAC address of each network interface card on the ring. The network interface card with the highest MAC address number wins the election. Any computer can become an active monitor and all other computers are considered *standby monitors.* If the active monitor fails or logs off the network, one of the standby monitors becomes the new active monitor.

Let's take a closer look at Token Ring communications. As the name implies, the active monitor keeps track of the activities on the Token Ring network. It conducts periodic checks to ensure a token is circulating the ring. If a token is not detected in 10 milliseconds (default), a new token is generated and placed on the ring.

The active monitor conducts a ring poll to identify any new computer logged on to the ring and detects any computers that have logged off the ring. This process is generally referred to as *ring polling* or *polling the ring.* The ring polling occurs every seven seconds. Remember that in Token Ring communications, each computer communicates with the computer immediately upstream and downstream of its location. A computer receives the token and data from its upstream neighbor and then regenerates the token and passes it to the computer immediately downstream of its location.

The active monitor is responsible for removing circulating tokens. If a token becomes corrupt, the token may circulate the ring forever, thus stopping all communications on the ring. The active monitor performs a *ring purge,* which means it removes the defective token and replaces it with a new one.

The active monitor enforces a 24-bit delay between tokens. The 24-bit delay is necessary to ensure that the complete frame (token) is deliverable. The 24-bit space is necessary because it is the exact length of an empty token. The delay is created by holding a frame until sufficient time (equaling 24-bits) lapses.

Lastly, the active monitor is responsible for synchronizing the timing of all computers on the ring. Timing is a critical part of Token Ring operation. Ethernet is random and chaotic, while Token Ring relies on a set timing pattern and orderly procedure for delivering data and monitoring functions. Therefore, all computers must be synchronized with each other so they measure time lapse equally and expect the appearance of a token at a specific time. Remember that any computer on the ring may need to assume the responsibilities of the active monitor and keep the timing sequence going.

active monitor
a computer that is responsible for monitoring the necessary administrative functions associated with Token Ring technology.

monitor contention
the process in which an active monitor is selected.

standby monitor
a computer on a Token Ring network that can become an active monitor if the current active monitor fails or logs off the network.

ring polling
a ring poll conducted by the active monitor that identifies if a computer has logged on or off the ring.

ring purge
removing a defective token and replacing it with a new one.

While synchronized timing is not required for Ethernet communication, it is used in Ethernet systems for security purposes. Security is covered in detail in Chapter 16—Network Security.

Tech Tip

Beaconing

beaconing
a recovery process
used when a hard-
ware failure occurs
on the ring.

Beaconing is a recovery process used when a hardware failure occurs on the ring. The beaconing process is used to remove a defective computer from the ring. When a computer does not receive a token from its upstream neighbor within a set period of time, it begins the beaconing process. First, it generates a beacon frame and sends it to its downstream neighbor. The downstream computer regenerates the beacon frame and sends it to its downstream neighbor. This process continues around the ring. When any computer receives eight beacon frames, it temporarily takes itself off the ring and performs a self-test, which is known as a loopback test. The *loopback test* is simply a signal that tests the network interface card to ensure it is functioning properly. If the loopback test fails, the computer remains offline. If the loopback test indicates no problem with the network interface card, the computer reattaches itself to the ring. This process continues until the defective computer is identified or the network administrator intervenes.

loopback test
a signal that tests the
network interface
card to ensure that
it is functioning
properly.

It may seem like a lot of activity happens on the Token Ring network and that the various methods to ensure frame delivery slow down the transmission rate. Rest assured that data transfer occurs at a blazing speed (unless downloading graphics or sound). The activities previously described typically go unnoticed.

When transferring data, it is not done as one continuous steady stream but as parts of a total sequence of transmissions. When the token is captured, it can be held up to approximately 10 milliseconds (1/100 second). After that, it must be released. If the total transmission is not complete, the computer needs to recapture the token and continue doing so until complete. By incorporating a maximum time the token can be captured and released, the Token Ring system provides an opportunity for other computers to capture the token. Token passing is a well-organized access method that allows each computer to take its turn. If no other computer needs to send data or a command, the token returns to the computer that last sent the token. In less than a second, thousands of bits of data can be transmitted and received. It usually appears to the user that they have had complete control of the ring, when in reality, two, three, or hundreds of users have been communicating over the same network using the same token.

Token Bus

A Token Bus network is defined in the IEEE 802.4 standard and operates similar to a Token Ring network. The difference is in how the token is passed from computer to computer. In a typical Token Ring network, the token is passed to the next computer on the ring. Since the Token Bus topology uses a bus rather than a ring, a variation of token passing must be used. A list of computers is created in a database. Each computer is identified by a MAC address and computer name. A sequential list of addresses is generated which becomes the sequence for the token to follow. The token can also have a priority code set to allow a specific workstation to have the token before any of the other workstations.

ARCnet

ARCnet was originally designed for LANs, but has evolved into an industrial control field bus technology. ARCnet is a popular industrial control technology because of several features:

- The protocol design is simple and compact.
- The protocol uses very short packet lengths, which is critical in industrial manufacturing timing systems.

The ARCnet technology is based on token passing for accessing the media. Token passing ensures equal access to all nodes on the network. This is critical to the design of an industrial manufacturing type application where sharing and timing of machine control is critical. Any one computer on the network cannot be forced to wait too long to transmit data to another on the network. CSMA/CD does provide access to all nodes, but it is does not provide equal access.

The access method of ARCnet is similar to the access method of Token Ring. ARCnet, however, uses a deterministic method of cable access by passing the token to the next, highest assigned node number not necessarily the closest node, like Token Ring.

ARCnet allows a maximum number of 255 nodes per network. Each node on the network must be uniquely identified by numbers ranging from *1* to *255*. Zero is reserved for broadcast messages. Each node uses a network interface module (NIM) not a network interface card (NIC), and it is assigned a MAC ID for identification rather than having one burned in at the factory. Assigning a MAC ID is unique to ARCnet networks.

Many times specific networking technologies use variations on acceptable terminology when identifying a particular type of equipment or technology. One such term associated with ARCnet is *network interface module (NIM)*. The term *NIM* is the equivalent to network interface card (NIC) when talking about the circuit board used to connect the network device to the network medium. ARCnet technology was among the first network systems ever used. Since the ARCnet standards use the acronym NIM not NIC, you should also use the preferred term *NIM*.

Network Operating Systems and Networking Protocols

When networking was in its infancy, there were many different network operating systems. Each of the operating systems was developed with its own way of communicating to network hardware on the LAN. There were no set of standards, no OSI model, and no one set of protocols. Network companies were in direct competition with other companies and many kept secret the way they programmed their software and protocols. As time went by, it became apparent that people purchasing network operating systems and hardware wanted to be able to communicate with other network systems and use the Internet.

Network operating systems had to provide software programs and protocols that would enable the exchange of data between different network operating systems. The exchange of data between different network operating systems was not easy to establish. One of the reasons for the difficulty was many network operating systems used different ways to identify the nodes on a LAN.

The common denominator for communicating between different networks finally became apparent. Each of the various network operating systems would use the same method to identify individual nodes while maintaining a compatibility with its own set of protocols. The result was numerous protocols belonging to the various suites of vendor-specific protocols. However, while Novell, Apple Computer, and Microsoft all have different protocols, they all are capable of being encapsulated at the lower levels of the OSI model in protocols defined by the IEEE, such as Ethernet and Token Ring, **Figure 6-12.**

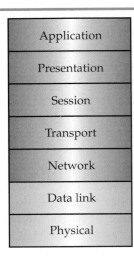
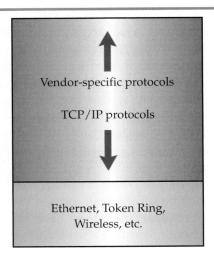

Figure 6-12.
Different protocols such as those from Novell, Apple Computer, and Microsoft are capable of being encapsulated at the lower levels of the OSI model in the protocols defined by the IEEE.

When you install a network interface card, you also automatically install some default protocols. For example, when a network interface card is installed in a Windows 98 computer, specific protocols are also installed as part of the device configuration. If you check the network interface card properties, you will see the NetBEUI and the TCP/IP protocols are installed.

With the introduction of Windows XP, NetBEUI is no longer installed by default, only TCP/IP. If NetBEUI is required to communicate with older operating systems such as Windows 95, it can still be added.

The NetBEUI protocol was developed to communicate across Microsoft LANs. It is not a routable protocol. This means that it has no area in its data package for storing a destination IP address. Therefore, NetBEUI cannot communicate over the Internet and negotiate routes. In contrast, the TCP/IP protocol does contain an area in its data package for a destination IP address. This means that data encapsulated with the TCP/IP protocol can be delivered to a destination over the Internet.

When communicating with a different operating system such as Novell Netware 4.0 or earlier, the data needs to be packaged differently. It is necessary to install additional protocols to support communications over different network media or network operating systems. Look at **Figure 6-13.** It shows

Figure 6-13.
A network operating system can run on many types of protocols. Note the protocols that are available for installation on a Windows 98.

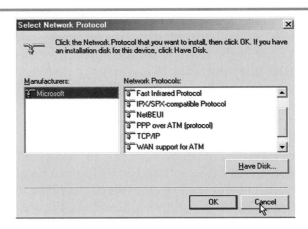

the **Select Network Protocol** dialog box of a Windows 98 computer. You can see some of the many protocols available. As you can see, protocols play an important part in network communications.

NetBIOS

IBM originally developed the Network Basic Input/Output System (NetBIOS) to network IBM PCs. In a typical computer, the BIOS software supports communication between the computer software and hardware. The computer BIOS allows hardware and software programs created by different manufacturers to communicate. NetBIOS functions in a similar fashion to support communications between different computer systems through the use of a network interface card.

Before NetBIOS was developed, the traditional method of networking was connecting a dumb terminal to a mainframe via cabling. A *dumb terminal* consists of only a display and keyboard. The keyboard contains the electronic components needed to support communication between the mainframe and the terminal. No separate microprocessor acts as the brain in the terminal. Consequently, the term *dumb* was used to infer that it contained no CPU. See **Figure 6-14.** In these systems, the mainframe contained the CPUs used to process data and the terminals did not. The mainframe also contained the storage media for retaining data on large magnetic spools of tape. The spools of tape stored data in much the same way as a hard disk drive stores data.

NetBIOS became the de facto standard for communications in a LAN. *De facto standard* means that the software became a standard because of its widely accepted use by industry, not because of the actions or ratification of any standards committee. NetBIOS was also the first protocol designed for communications between PCs rather than through a mainframe. For the first time, a network could be created without needing a mainframe. This made the network inexpensive compared to the mainframe model.

Datagrams and sessions

The NetBIOS protocol uses two common network communication methods: datagram and session. A *datagram* is a short message block that can be sent to a particular computer, sent to a group of computers, or broadcasts to all computers connected to the media, **Figure 6-15.** When the datagram is broadcast to a group of specific computers, it is referred to as multicast. The multicast communication method sends the datagram to a specified group of computers on the network rather than to all the computers on the network. Broadcasts and multicasts are referred to as a connectionless communications. They do not guarantee data delivery.

dumb terminal
a type of computer that consists of only a display and keyboard. The keyboard contains the electronic components needed to support communication between the mainframe and the terminal. A dumb terminal does not contain a CPU.

de facto standard
a standard developed because of its widely accepted use by industry.

datagram
a short message block that can be sent to a particular computer, sent to a group of computers, or broadcast to all computers connected to the media.

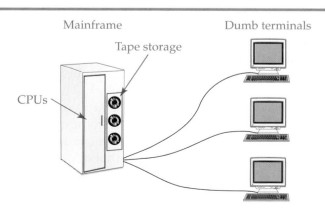

Mainframe   Dumb terminals

Tape storage

CPUs

Figure 6-14.
Dumb terminals are wired directly to a mainframe computer. They communicate directly with the mainframe and cannot process data on their own because they do not contain a CPU. The CPU(s) are located in the mainframe.

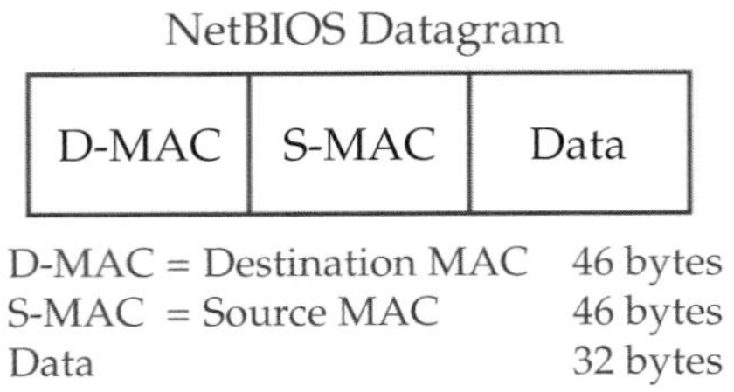

Figure 6-15.
NetBIOS datagram.

NetBIOS Datagram

D-MAC	S-MAC	Data

D-MAC = Destination MAC   46 bytes
S-MAC = Source MAC   46 bytes
Data   32 bytes

When communicating on small network systems, broadcast messages do not interfere with network operation. When a large network is established, the broadcast method of communication is undesirable. Too many broadcast messages fill the network with communication traffic and create a broadcast storm. In this condition, so many broadcast messages occur they collide with one another and never reach their destination. On very large networks, the session-type of communications is much more effective. Network systems use both methods of communications while limiting the amount of direct broadcasts by users.

When communication is limited between two particular computers, it is referred to as a *session.* When computers establish a session, they create a connection-oriented communication known as a *virtual circuit.* A virtual circuit behaves like a direct connection between two or more devices. During a session, communication remains established until one of the participants terminates the session. Communication between two computers in an established session guarantees data delivery. It uses some form of error checking to ensure the delivery of each data packet.

session
communication that is limited between two particular computers.

virtual circuit
a connection-oriented communication.

NetBIOS names, group names, and addresses

NetBIOS can identify a computer by its MAC address, its logical name, or by the group to which it belongs. The MAC address is electronically encoded into the network interface card and cannot be changed. Each MAC address is specifically unique for proper computer identification on a network. Each computer also receives a unique name. When a network interface card is installed in a computer, a default name is automatically assigned. The default name is typically a randomly generated series of alphanumeric characters with no special meaning. The default name can be changed to one with more meaning, such as *Station1*, *PC3*, *Accounting*, or *Billy*. The user name must be unique like the MAC address. If two computers use the same name for identification, communications will not operate properly. The computer can also be assigned a group name to identify it as part of a special group of computers. Similar to the computer name, the group name is assigned a default group name. In the Windows networking environment, the default group name is *Workgroup*.

In older systems, the identification of each computer was originally stored in a text file located on each computer on the network. The text file contained the computer name and the associated MAC address. Any computer on a NetBIOS network could read the text file of any of the other computers on the network. This system worked for earlier and smaller networks, but as networking evolved in size and capability, protocols were developed that allowed for an automated method of communicating identification information across the LAN. We will discuss these protocols in later chapters.

NetBIOS names and UNC

NetBIOS computer names are used to identify computers and servers. NetBIOS names can contain a maximum of 15 characters. NetBIOS uses a standard naming convention called the **Universal Naming Convention (UNC)** to identify shared resources on the network. A typical UNC is \\server_name\share_name. The *server_name* is the name of the computer or server that contains a shared resource, and the *share_name* is the name assigned to the shared resource. The UNC can also include a path to a subdirectory beneath the shared resource such as \\server_name\My Documents\Projects\. In this example, My Documents is the shared resource and Projects is a subdirectory beneath the shared resource. Think of descending a hierarchical structure. The file server contains all shares, so it is the highest level of the hierarchical structure. Next is the share, and last, is the path.

Universal Naming Convention (UNC) a standard naming convention used by NetBIOS.

NetBEUI

The NetBIOS Extended User Interface (NetBEUI) protocol was designed as an enhanced version of NetBIOS. It was also designed by IBM to support their LAN Manager network system. Later, Microsoft adopted NetBEUI into their networking products. It is constantly referenced by technical documentation and can still be used to support LAN systems.

NetBEUI and NetBIOS are often referred to as the same protocol, but they are technically not the same. However, they do share similarities. NetBIOS and NetBEUI are broadcast-type network protocols, and a typical router will not pass broadcast-type protocols. To communicate with a PC in a large network such as the Internet, another protocol must be used such as IP.

A protocol that can transmit packets across a router is referred to as a routable protocol. The most common routable protocol is TCP/IP.

Tech Tip

Network+ Note:

For the Network+ Certification exam, remember that NetBEUI is *not* a routable protocol.

Network+

Figure 6-16 shows the relationship of NetBEUI to the OSI model. The Server Message Block (SMB) protocol is an upper layer protocol that initiates client requests to a server. The request is passed down to the NetBEUI protocol. The NetBEUI protocol passes the request to the data link layer. At the data link layer, the request is sent to the network media. As you can see, SMB and NetBEUI combine to cover five layers of the OSI model.

IPX/SPX

Novell NetWare was first introduced in the early 1980s. Early versions of Novell NetWare used the IPX/SPX protocol suite exclusively. Starting with Novell NetWare 5, TCP/IP was offered as alternative to IPX/SPX. A key point to remember is proprietary networks developed their local network communication system independent of Internet communication. Later, the proprietary networks modified their local network system to be compatible with the TCP/IP protocol

Figure 6-16.
The Server Message Block (SMB) and NetBEUI protocols span five layers of the OSI model. They were developed long before the OSI model was ever designed as a suggested standard.

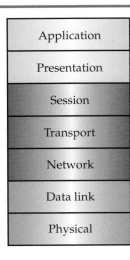
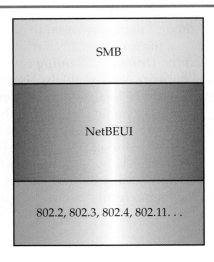

suite to enable communication over the Internet. Today, the TCP/IP protocol suite is selected by default on network operating systems for communications on the LAN and WAN.

Proprietary networks also incorporate methods to exchange data between each other. For example, Windows XP supports communication with IPX/SPX networks by implementing an IPX/SPX, NetBIOS compatible protocol known as NWLink. Interoperability with NetWare network is covered in detail in Chapter 8—Novell Network Operating Systems.

An IPX/SPX address consists of a network number, node address, and socket number. See **Figure 6-17.** The IPX/SPX address is written in hexadecimal form. The network number is the number that identifies the network segment. It is 32 bits, or 4 bytes, long. The node address is the physical address, or MAC address, of the workstation. This address is 48 bits, or 6 bytes, long. The socket number is a number that identifies a process. This number is 16 bits, or 2 bytes long.

When performing diagnostics, you may see the IPX/SPX address written as a station address. A *station address* consists of only a network number and node address. Leading zeros in the network number are often truncated. **Figure 6-18** shows the two ways a station address can be written.

station address
an IPX/SPX-based address that consists of only a network number and node address.

AppleTalk

AppleTalk was originally developed in the 1980s to support Macintosh computers connected as a network. AppleTalk consists of its own set of protocols such as EtherTalk, LocalTalk, TokenTalk, FDDITalk, AppleTalk Address Resolution Protocol (AARP), and more. The original version of AppleTalk simply operated as a local network and was limited to 254 nodes. *LocalTalk* was the commonly used term to describe the implementation of a LAN that supported Macintosh computers. Technically, LocalTalk refers to the network media and AppleTalk refers to a suite of protocols.

LocalTalk
the network media in an AppleTalk network.

Figure 6-17.
An IPX/SPX address consists of a network number, node address, and a socket number.

Figure 6-18.
An IPX/SPX station
address consists only of
a network number and
a node address. It can
be written in two ways:
a colon separating the
network number and
node address or a colon
separating each four
bytes of hexadecimal
numbers.

Soon it was realized that larger network capabilities were needed. In 1989, AppleTalk Phase 2 was introduced. The original Apple Computer network was now referred to as AppleTalk Phase 1. AppleTalk Phase 2 could support up to 16 million nodes.

The ability to increase the number of nodes was due to changing the node address to a more complex system. The new system consists of three parts: a network number, a node number, and a socket number. The network number is a 16-bit (2 bytes) value that identifies the specific network. The node number is a unique, 8-bit (1 byte) value. The socket number is an 8-bit (1 byte) value.

AppleTalk Phase 2 introduced a feature called AppleTalk Zones. AppleTalk Zones is similar to the concept of workgroups and domains found in Microsoft networking systems. An AppleTalk Zone allows for the logical grouping of computers. AppleTalk Zones also allow AppleTalk to support LANs consisting of more than 255 clients on 65,534 networks. Through the use of AppleTalk Zones, the new node addressing technique, and new Apple protocols that support packet routing between networks, the network size increased from 254 nodes to over 16 million nodes. This was a significant advancement for Apple Computer.

AppleTalk is a proprietary protocol that can only be interpreted directly by other AppleTalk computers. To support communication with Microsoft Windows computers and other computers that support TCP/IP communications, AppleTalk over IP was developed. This protocol was modified to support the naming requirements found in the TCP/IP suite.

Microsoft XP does not support direct communication with Apple computers. To communicate with an Apple computer, a Microsoft server must have the Services for Macintosh gateway installed. When a server is configured as a gateway for communication, Microsoft XP clients can access shared files on the Macintosh network. For a Microsoft XP client to access a Macintosh printer, the AppleTalk protocol must be installed on the Microsoft XP client.

TCP/IP

TCP/IP is not a single protocol but rather an entire suite of protocols designed specifically to support communication between computers and network devices connected via the Internet. TCP/IP is native to UNIX. UNIX was the original operating system designed to communicate via the Internet. Other network operating systems were not originally designed to communicate via the Internet, but rather were designed to communicate over a LAN. In these operating systems, two or more LANs could be connected to form a MAN or WAN. These operating systems used their own naming convention to identify nodes on their network. The naming convention did not match the naming convention used by the TCP/IP suite of protocols. Thus, these network operating systems developed software utilities often referred to as protocols to communicate with other operating systems that were based on the TCP/IP naming convention and protocols.

Today, all major network operating systems can communicate across the Internet and adhere to the naming rules of TCP/IP. It does this by encapsulating proprietary protocols inside an Ethernet frame and transmitting the frame across the Internet. This is how computer systems based on different technologies, AppleTalk, DEC, NetWare, and more can communicate with each other and exchange data.

TCP/IP is used by Microsoft Windows XP to communicate on the LAN and has stopped supporting NetBEUI as the default protocol. Windows 98 was the first version of a Microsoft operating system to allow a LAN to be configured for TCP/IP only. While this was only a brief introduction to TCP/IP, Chapter 11—TCP/IP Fundamentals covers TCP/IP in detail.

Summary

- Some common features that a network operating system provides are shared resources, file storage and management, security, troubleshooting utilities, services, and a user interface.
- The data link layer is responsible for placing data on the network media or receiving data and passing it up the OSI layers.
- The data link layer is defined by the IEEE 802.2 standard.
- Ethernet uses CSMA/CD to access the network media.
- CSMA/CD broadcasts a message to all computers in a collision domain. Only the intended computer accepts the message.
- A collision domain is an area where collisions occur in an Ethernet network.
- Segmenting an Ethernet network can reduce collisions.
- A broadcast storm occurs when there is continuous activity on an Ethernet network.
- Macintosh computers use the AppleTalk protocol for communication and the CSMA/CA access method.
- The CSMA/CA access method attempts to avoid collisions through detecting communications on the network media.
- A Token Ring uses the token passing access method. Only one computer may control the token at any one moment.
- Token Bus uses a token passing access method similar to Token Ring. However, the token is passed according to a list of sequential addresses rather than to the proximate computer.
- Token Bus is defined by the IEEE 802.4 standard.
- ARCnet uses token passing for media access.
- ARCnet assigns the MAC address for each network interface module (NIM).
- NetBIOS was originally developed by IBM to network IBM PCs.
- NetBIOS uses two common network communication methods: datagram and session.

- A datagram is a short message block that can be sent to a particular computer, sent to a group of computers, or broadcasts to all computers connected to the media.
- A datagram that is broadcast to a group of specific computers is referred to as multicast.
- A session is communication that is limited between two particular computers.
- NetBEUI is an enhanced version of NetBIOS.
- NetBEUI is *not* a routable protocol.
- Novell Netware allows for the use of either IPX/SPX or TCP/IP as the network protocol.
- AppleTalk Phase 1 could only support LANs.
- AppleTalk Phase 2 could support communications between LANs.
- The Windows XP default network protocol is TCP/IP.

Review Questions

Answer the following questions on a separate sheet of paper. Please do not write in this book.

1. List some common features of network operating systems.
2. What is the purpose of the data link layer?
3. The data link layer is subdivided into what two layers?
4. The Ethernet protocol uses the _____ access method.
5. Describe a typical session on an Ethernet network.
6. When too many packets on a network collide, a _____ storm has occurred.
7. On which type of network do broadcast storms typically occur?
8. Which access method does AppleTalk use?
9. Which access method does Token Ring use?
10. What is monitor contention?
11. What does the term *ring purge* mean?
12. Which computer on a Token Ring network starts the beaconing process?
13. Which IEEE standard describes Token Bus?
14. How does token passing in a Token Bus system differ from token passing in a Token Ring system?
15. Describe ARCnet's access method.
16. What acronym used in ARCnet technology means the same as network interface card?
17. What does the term *de facto standard* mean?
18. Define the term *broadcast*.
19. Define the term *multicast*.

20. Define the term *session*.

21. Which form of communication guarantees delivery of a packet—a session or a datagram?

22. Which form of communication relies on a broadcast-type of transmission—a session or a datagram?

23. Which form of communication is best for large networks—a session or a broadcast?

24. Define the term *virtual circuit*.

25. _____ and _____ are *not* routable protocols.

Sample Network+ Exam Questions

Network+

Answer the following questions on a separate sheet of paper. Please do not write in this book.

1. Which of the following is the UNC format for a shared directory called MyFiles on a server called *Server1*?
   A. //Server1/MyFiles
   B. //MyFiles /Server1
   C. \\Server1\MyFiles
   D. \Server1\MyFiles

2. NetBEUI is native to which operating system?
   A. Apple
   B. UNIX
   C. Windows NT
   D. Novell NetWare

3. Which standard describes the access method for a Token Bus network?
   A. 802.2
   B. 802.3
   C. 802.4
   D. 802.5

4. Removing a workstation from a Token Ring network is referred to as _____.
   A. beaconing
   B. monitoring
   C. purging
   D. ejecting

5. Broadcasting data to a few select workstations is referred to as _____.
   A. beaconing
   B. multicasting
   C. conferencing
   D. multiplexing

6. The IEEE standard 802.3 is associated with which access method?
   A. Carrier Sense Multiple Access with Collision Detection (CSMA/CD)
   B. Carrier Sense Multiple Access with Collision Avoidance (CSMA/CA)
   C. Asynchronous Transmission Mode (ATM)
   D. Wireless Ethernet Protocol (WEP)

7. Which of the following is the default protocol used for Apple clients in an Apple network?
   A. NetBEUI
   B. IPX/SPX
   C. AppleTalk
   D. AppleNet

8. What access method is used by a Token Ring network?
   A. ring polling
   B. token passing
   C. CSMA/CA
   D. CSMA/CD

9. Which network operating system can communicate over Ethernet?
   A. Microsoft NT
   B. Novell NetWare
   C. UNIX
   D. All of the above.

10. Which answer is a correct example of a MAC address?
    A. AB C3 D4 23 54 F2
    B. 192.168.000.123
    C. A12.34E.676.FF2
    D. 1926867721

Suggested Laboratory Activities

1. Install an Ethernet, peer-to-peer network.
2. Install a Token Ring, peer-to-peer network.
3. Install a Token Bus, peer-to-peer network.

Interesting Web Sites for More Information

www.javvin.com/protocolsuite.html

www.protocols.com

www.wildpackets.com

Networks are often used in education to share data and peripherals and to connect to the Internet.

Chapter 6
Laboratory Activity

Installing Client Service for NetWare

After completing this laboratory activity, you will be able to:
■ Install the Client Service for NetWare on a Windows XP client.
■ Remove the Client Service for NetWare from a Windows XP client.
■ Identify the different Ethernet frame types available to support communications with a NetWare server.

Introduction

In this laboratory activity, you will install the Client Service for NetWare on a Windows XP client. This will automatically install the NWLink IPX/SPX/NetBIOS Compatible Transport Protocol. Novell Network servers originally used IPX/SPX as the default protocol suite. Starting with NetWare 5, Novell offered TCP/IP as the native protocol rather than IPX/SPX. However, Novell still provides an option to install the IPX/SPX protocol instead of TCP/IP.

Microsoft and Novell each provide a client service for connecting a Windows XP client to a NetWare server. The client from Microsoft is called Client Service for NetWare (CSNW), and the client from Novell is called Novell Client for Windows NT/2000/XP. In this laboratory activity, you will only be using Client Service for NetWare. One big advantage of using the Novell Client for Windows NT/2000/XP, however, is it supports administrative tools or utilities native to the NetWare system. The Microsoft Client Service for NetWare does not support NetWare administrative tools or utilities native to the NetWare system.

> *Note:*
> *A technical term for each of the NetWare client services is "redirector." A redirector is a software program typically referred to as a protocol or service that acts as a translator between two different network operating systems.*

> *Note:*
> *Microsoft recommends the following selections for Windows XP clients connecting to a Novell NetWare Server 5.x and later. Use Client Service for NetWare and NWLink to connect to NetWare 5.x and later servers running IPX/SPX. Use Novell Client for Windows NT/2000/XP to connect to NetWare 5.x and later servers running TCP/IP.*

Windows XP and Novell NetWare use two different protocols for file and print communications. Windows XP uses the Common Internet File System (CIFS) protocol for file and print services. CIFS is an enhanced version of Service Message Block (SMB), which was Microsoft's original file and print service protocol. NetWare uses the NetWare Core Protocol (NCP) for file and print services. This means that the two operating systems cannot communicate directly because each is based on a different file and print service protocol. This is why a client service is required. A client service installs the needed file and print service protocols.

One of the unique aspects of NetWare is that it can support four different Ethernet frame types. The default frame type selection is **Auto Detect**. When set to Auto Detect, the correct Ethernet frame type will be selected automatically. An administrator may also set the frame type specifically.

Equipment and Materials

■ Workstation with Windows XP.

Note:
A NetWare server is not required for this laboratory activity.

Procedure

1. _____ Report to your assigned workstation.

2. _____ Boot the workstation and verify it is in working order.

3. _____ Access the **Local Area Connection Properties** dialog box for the network adapter through **Start | Control Panel | Network Connections**. Right-click the **Local Area Connection** icon and select **Properties** from the shortcut menu. Your **Local Area Connections Properties** dialog box should look similar to the following screen capture. Pay particular attention to the fact that the only network protocol installed is TCP/IP.

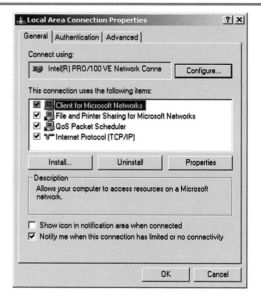

4. _____ Click the **Install** button. The **Select Network Component Type** dialog box will display.

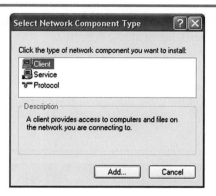

5. _____ Select **Client** and then click the **Add** button. The **Select Network Client** dialog box will display.

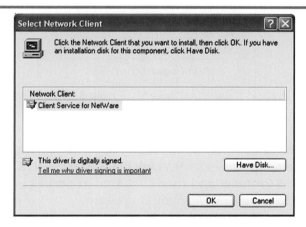

6. _____ Select **Client Service for NetWare** and then select **OK**. This will install the Client Service for NetWare.

7. _____ Now, look in the **Local Area Connection Properties** dialog box for the network adapter to see if the Client Service for Netware and NWLink IPX/SPX/NetBIOS Compatible Transport Protocol has been added to the list. If both entries are listed, you have successfully added the required Microsoft IPX/SPX compatible protocols. If you do not see these entries, call your instructor for assistance.

8. _____ After successfully installing the Client Service for Netware, the computer will display a message informing you that the system must be restarted before the change will take effect. Allow the system to reboot.

9. _____ When the computer reboots, the Windows logon dialog box will appear requesting your user name and password. Your user name should appear by default and your password will be blank unless you use a password to normally access this workstation.

10. _____ The **Select NetWare Logon** dialog box will display. You may select the **Help** button to learn more about the dialog box. When you are through, close **Help** and then click the **Cancel** button in the **Select NetWare Logon** dialog box. You will not change any settings in this dialog box in this laboratory activity.

Note:
The **Select NetWare Logon** *dialog box may not appear if another class has previously used the workstation for this laboratory activity and a NetWare server was not detected.*

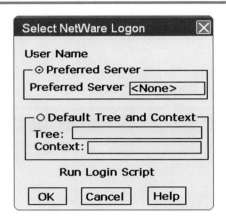

11. _____ Open the **Local Area Connection Properties** dialog box for the network adapter. Select **NWLink IPX/SPX/NetBIOS Compatible Transport Protocol** and then click the **Properties** button. A dialog box similar to the following should appear.

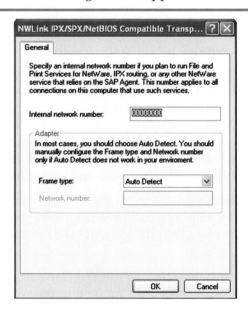

12. _____ Click the arrow of the **Frame** type drop-down list box to see the various frame types available. List the four types on a separate sheet of paper.

13. _____ Click **Cancel** to return to the **Local Area Connection Properties** dialog box.

14. _____ Remove the Client Service for NetWare by highlighting **Client Service for NetWare** and then clicking **Uninstall**.

15. _____ After removing Client Service for NetWare, inspect the **Local Area Connection Properties** dialog box for the network adapter to see if Client Service for NetWare and the NWLink IPX/SPX/NetBIOS Compatible Transport Protocol has been removed. If the Client Service for NetWare and the NWLink IPX/SPX/NetBIOS Compatible Transport Protocol has been removed, click **OK** to exit the **Local Area Connection Properties** dialog box. If they have not been removed, call your instructor for assistance.

16. _____ Now you will remove the **Logon** dialog box(es) from the startup screen. If you don't, you will be required to pass through the Windows logon screen each time the computer is started and possibly the Novell NetWare logon screen. To remove the **Logon** dialog box, open **User Accounts (Start | Control Panel | User Accounts)**. Click **Change the way users log on or off** and then select the **Use the Welcome Screen** option. This will remove the classic Windows **Logon** dialog box. If you have a problem restoring the original screen, call your instructor. Remember, you must leave the workstation the way you found it unless otherwise directed by your instructor.

17. _____ You may repeat the laboratory activity several times until you feel comfortable with adding and removing Client Service for NetWare. Be sure to inspect the **Local Area Connection Properties** dialog box for the network adapter each time you install and uninstall Client Service for NetWare.

18. _____ Answer the review questions.

Review Questions

Answer the following questions on a separate sheet of paper. Please do not write in this book.

1. What is the name of the Microsoft client used to connect a Windows XP workstation to a Novell NetWare 5.x server?

2. What network protocol is installed by default on a workstation running Windows XP?

3. What is the name of the client provided by Novell to connect a Windows XP workstation to a Novell NetWare 5.x server?

4. What is the file and print protocol used by Microsoft Windows XP?

5. What is the file and print protocol used by NetWare?

6. How many Ethernet frame types are available for Novell NetWare?

7. What is the Ethernet frame type default setting for a Novell NetWare client?

7

Microsoft Network Operating Systems

After studying this chapter, you will be able to:

❏ Identify the major differences between a Microsoft peer-to-peer network and a Microsoft client/server network.

❏ Discuss the differences between FAT16, NTFS4.0, and NTFS5.0.

❏ Describe the Windows NT domain model.

❏ Explain the Windows NT authentication process.

❏ Describe the Windows 2000 Server and Windows Server 2003 Active Directory structure.

❏ Explain the Active Directory authentication process.

❏ Explain the purpose of the Microsoft Management Console (MMC).

❏ Explain how a domain user and group account is set up in Active Directory.

❏ Explain the ways to obtain interoperability between clients and servers in networks with different network operating systems.

Network+ Exam—Key Points

The Network+ Certification exam requires knowledge about the interoperability between different networking systems. This is the most important aspect of the entire chapter. The Network+ Certification exam also expects a basic user-level understanding on how to log on to various network operating systems. An example question might be: A user cannot log on to a network. What is the problem, and what is the correct procedure to correct it? The answer would be: Verify the user is logging on correctly, and check if a connection can be established.

A question about Lightweight Directory Access Protocol (LDAP) is most likely to be on the Network+ Certification exam. Remember that LDAP is independent of the operating system.

The CompTIA Network+ Certification is not a certification for Windows Server 2003 or any other network operating system. It is a general certification that measures general knowledge, not specific systems. You will need to know the general concepts associated with Windows networks such as logon procedures, how to add users, common user logon errors, and such. You will not need to know the Windows 2000 Server or the Windows Server 2003 operating systems in depth.

Key Words and Terms

The following words and terms will become important pieces of your networking vocabulary. Be sure you can define them.

access control list (ACL)
auditing
authentication
backup domain controllers (BDC)
basic disk
clustering
complex trust relationship
contiguous namespace
disjointed namespace
disk quota
distinguished name
domain
dynamic disk
EFI partition
forest
gateway service
global security policy
group account
interactive logon
Local Security Authority (LSA)
local security policy
Logical Disk Manager (LDM) partition
logon right
Microsoft Reserved (MSR) partition
member server
multimaster replication

namespace
network authentication
network share
New Technology File System (NTFS)
NTFS permissions
one-way trust relationship
organizational unit
pass-through authentication
permission
primary domain controller (PDC)
security access token
Security Account Manager (SAM)
security policy
share permissions
share-level security
simple volume
snap-in
spanned volume
stand-alone server
striped volume
tree
trust relationship
two-way trust relationship
user-level security
workgroup

This chapter and the next two chapters introduce the dominant network operating systems from Microsoft, Novell, and the UNIX/Linux corporations. As you cover the content of these chapters, you will see the similarities and differences among network operating systems. The material presented in these chapters is limited to an introductory level. Knowledge of these network operating systems fill volumes of textbooks. In fact, a typical resource book for any of these operating systems typically contains over 1000 pages.

Network+ Note:

The Network+ Certification exam only requires introductory-level knowledge of the major network operating systems. After finishing this course on networking fundamentals, you may wish to specialize in one of the dominant network operating systems.

A Brief History of Microsoft Network Operating Systems

Microsoft began developing network operating systems in 1993 when it introduced Windows for Workgroups and Windows NT 3.1. Windows for Workgroups was designed as a peer-to-peer technology, while Windows NT 3.1 was designed as a client/server technology. Both network operating systems became popular because of the GUI.

Windows for Workgroups was one of the first peer-to-peer network models developed. At the time, most networks were the client/server type requiring expensive server software and hardware or a system to access mainframes through dumb terminals. The peer-to-peer model allows a collection of computers to communicate in similar fashion of a client/server model, and also allows them to function as stand-alone computers.

Computers in a Windows for Workgroups peer-to-peer network are grouped as workgroups. A *workgroup* is a group of computers that share resources such as files and hardware, **Figure 7-1.** This type of network structure is limited in scope and is not used for large networks. In the peer-to-peer workgroup model, each computer contains its own database of users. Each user must have a separate account on each computer to use the computer's shared resources. As a peer-to-peer network grows in size, it becomes impractical to manage.

workgroup
a group of computers that share resources such as files and hardware.

Windows NT 3.1 based its network administration on the client/server model. A client/server model stores user and resource information in one location. In a client/server network, the security database is stored on a server. The user logs on to the server and is then cleared to access the network resources.

Computers in a Windows client/server network are grouped together in a domain. A *domain* is a logical grouping of users and equipment as defined by the network administrator, **Figure 7-2.** Domain members share a common security database. It is much easier to manage a large number of users when the network is configured as a client/server.

domain
a logical grouping of users and equipment as defined by the network administrator.

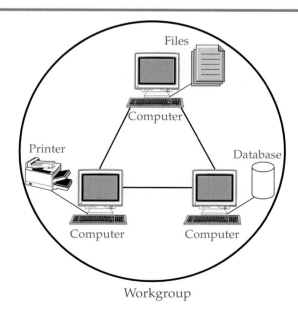

Figure 7-1.
A workgroup is a group of computers that share resources. Some examples of resources are files, directories, databases, and printers.

Figure 7-2.
A domain consists of a collection of users and equipment under one administration.

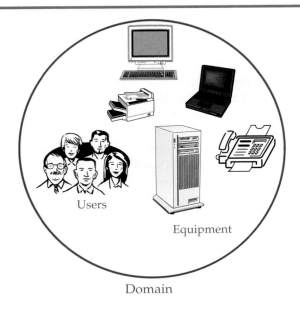

Users

Equipment

Domain

Workgroups can also be implemented on a client/server network. However, the ability of a Microsoft client/server model to organize users into groups makes workgroups obsolete.

It is interesting to note that the first release of Windows NT was called Windows NT 3.1. Typically the first release number of software is labeled as 1.0. Major releases follow in whole numbers such as 2.0, 3.0, and 4.0. Revisions increment in tenths, such as 2.1 and 2.2, and minor revisions generally increment in one hundredths, such as 2.11 and 2.22. The first version of Windows NT, however, was labeled 3.1 to match the version of DOS that was available at the time.

In 1996, Microsoft introduced Windows NT Workstation 4.0 and Windows NT Server 4.0. The workstation version could function as a stand-alone computer or support up to ten computers in a peer-to-peer environment.

Two of the major differences between Windows NT 3.1 and Windows NT 4.0 are the interface and Plug-and-Play capabilities. Windows NT 4.0 was designed to look like Windows 95. The operating system was capable of recognizing and installing Plug-and-Play devices, a feature that NT 3.1 lacked.

In February of 2000, Microsoft introduced Windows 2000 Professional to replace Windows NT Workstation 4.0 and Windows 2000 Server to replace Windows NT Server 4.0. In addition, Microsoft introduced Windows 2000 Advanced Server, which was designed to replace Windows NT 5.0 Server Enterprise Edition and Windows 2000 Datacenter Server.

In 2003, Microsoft released Windows Server 2003. Windows Server 2003 comes in four versions: Standard Edition, Enterprise Edition, Datacenter Edition, and Web Edition. The names reflect the design and administrative tool collection of each. For example, the Standard Edition functions as a network server for controlling a simple network. The Enterprise Edition is designed especially for a large-scale network. The Datacenter Edition is designed to serve as a storage server for a central database. The Web Edition is designed to serve as a Web server with limited capabilities.

Each edition supports IPv6, Remote Desktop, Active Directory, IIS6, .NET Application Services, and the Encrypting File System (EFS). Each has a similar interface. The major differences are the technical features such as how much RAM and how many processors each can support and whether it supports hot-add memory and Cluster services. Look at **Figure 7-3** for a comparison of Windows Server 2003 versions. You can see a complete comparison of features at the Microsoft Windows Server 2003 Web site.

Microsoft also has two versions of Server 2003 for the small business: Small Business Server 2003 Standard Edition and Small Business Server 2003 Premium Edition. These versions are designed to provide a cost-effective, small network server with many of the same features as the Enterprise Edition. The Small Business Server 2003 versions, however, can only support a maximum of 75 workstations.

The Microsoft network operating systems you will encounter most often are Windows 2000 Server and Windows Server 2003. Microsoft has designed many different levels of network operating systems to accommodate peer-to-peer networks to large-scale networks, referred to as "Enterprise" networks.

Common Windows Server Administrative Components

This section provides an overview of the common administrative components and basic terminology of all Microsoft network operating systems. A basic understanding of Microsoft network operating systems will serve as a foundation when studying the specific Microsoft network operating systems that are covered later in this chapter.

	Web Edition	Standard Edition	Enterprise Edition	Datacenter Edition	Small Business Server Standard Edition	Small Business Server Premium Edition
CPU (Minimum)	133 MHz	133 MHz	400 MHz (32-bit) 733 MHz (64-bit)	400 MHz (32-bit) 733 MHz (64-bit)	300 MHz	330 MHz
Maximum number of CPUs	2	4	8	64	4	4
RAM (Recommended minimum)	256MB	256MB	256MB	1GB	384MB	512MB
RAM (Maximum)	2GB	4GB	32GB	64GB	4GB	4GB
Disk space (Minimum for setup)	1.5GB	1.5GB	1.5GB (32-bit) 2GB (64-bit)	1.5GB (32-bit) 2GB(64-bit)	4GB	5GB
Cluster capability	No	No	Yes (8 node maximum)	Yes (8 node maximum)	No	No
Internet Connection Sharing (ICS)	No	Yes	Yes	No	Yes	Yes
Fax service	No	Yes	Yes	Yes	Yes	Yes

Figure 7-3.
Windows Server 2003 version comparison chart.

User Account

Each person who needs to access a Microsoft server and its resources must have a user account. The basic requirement for a user account is a user name and password. A user account can be assigned properties such as rights and restrictions. Rights include logon rights and permissions. A *logon right* is the ability to log on to the network. This ability may be restricted by disabling a user's account or by limiting the time the user is allowed to access the network. A *permission* is the ability to access a network share. Permissions may be limited. For example, a user can be given permission to read a file on the server, but not given permission to write to the file. A user can also be given permission to store personal files on the server's hard drive, yet restricted to using a limited amount of hard drive space.

logon right
the ability to log on to the network.

permission
the ability to access a network share.

When a computer running a Windows server operating system boots, a dialog box appears prompting the user to press [Ctrl] [Alt] [Del] to begin. After the user presses [Ctrl] [Alt] [Del], a logon dialog box displays, prompting the user for a user name and password. After the user name and password are authenticated, the user's default desktop appears.

One of the most common errors made by users is to incorrectly type their user name or password or to forget their passwords. The following is a typical Windows Server 2003 message that is displayed when a user incorrectly enters their name or password:

> The system could not log you on. Make sure your user name and domain are correct, and then type your password again. Letters in passwords must be typed using correct case. Make sure that CAPS LOCK is not accidentally on.

Network+

Network+ Note:

Logon failures are a common question topic on the Network+ Certification exam. Practice logon procedures and create logon failures to observe error messages first hand.

Group Account

To simplify user and resource administration, a network administrator can create a group account and then add users to that account. A *group account* is a collection of users that typically share a common job-oriented goal or similar function. Think of a group as a container or a collection of users who share the same responsibilities, duties, or interests.

group account
a collection of users that typically share a common job-oriented goal or similar function.

The group account may also consist of equipment. A group account can be assigned rights and restrictions like a user. Assigning rights and restrictions to the group level rather than to the user level makes administration simpler. For example, if users of a group need to access a new folder, the administrator can simply change the rights of the group account rather than change each user account.

Some groups are created by default when a network operating system is installed. The following is a list of some typical default groups found in Windows 2000:

- Administrators.
- Backup Operators.

- Guests.
- Power Users.
- Replicator.
- Users.

The system administrator may create other groups. For example, typical groups that may be created for a school setting would be *Teachers*, *Students*, *Science*, *English*, *Math*, *Counseling*, and *Administration*. Once the groups are set up with permissions and restrictions, users can be assigned to the groups. Users may be assigned to more than one group. For example, a teacher may be assigned to the *Teacher* group, the *Math* group, and the default *Backup Operators* group.

Security Policy

A *security policy* is a blanket policy that secures resources on the network. This policy is set before users are added to the network. Users added to the network are automatically affected by these policies unless specific properties are assigned to the users that override the policy's settings. Two types of security policies exist: local security policies and global security policies. A *local security policy* affects local users. A *global security policy* affects users throughout the domain.

Security policies can define password requirements such as whether a user should be allowed to reuse an old password when a password change is required, **Figure 7-4.** Security policies can also be set that affect auditing, **Figure 7-5.** *Auditing* is a service that tracks the events, use, and access of network resources and writes these actions to a log.

Network Share

A *network share*, or *shared resource,* is a resource on the network that is shared among assigned users. Examples of resources include files, directories, hard drives, CD drives, or printers.

security policy
a blanket policy that secures resources on the network.

local security policy
a security policy that affects local users.

global security policy
a security policy that affects domain users.

auditing
a service that tracks the events, use, and access of network resources and writes these actions to a log.

network share
a resource on the network that is shared among assigned users.

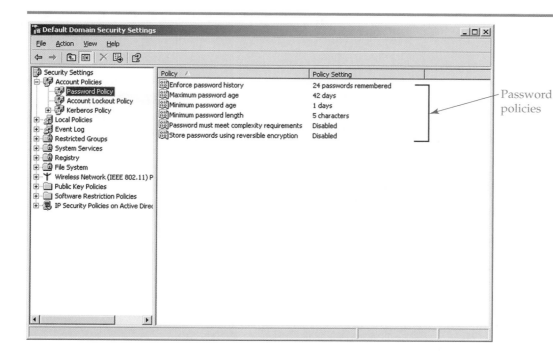

Figure 7-4.
Domain password policies in Windows Server 2003.

Figure 7-5.
Audit policies in
Windows Server 2003.
Audit policies track
certain events, such as
logon and directory
service access, and re-
cord the events in a log.

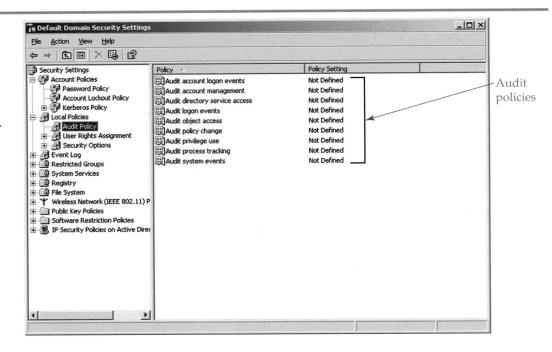

A network share is identified by an open hand icon. Look at **Figure 7-6.** An open hand beneath the folder called **Public ON Adv Server** indicates that this folder is shared. The name of the share indicates that the shared folder is public and is located on the server named *Advanced Server*. An administrator often appends the name Public to a share name when the share is to be used by everyone on the network. Indicating the location of the share is also a good idea. In this example, the location of the share has been indicated by the phrase "ON Adv Server."

Permissions granted to shares delegate who can use the share and in what way. Permissions to a share can be granted at the user or group level. Examples of permissions are Full Control, Modify, Read, and Write.

Figure 7-6.
A shared resource, such
as this shared folder, is
indicated by an open
hand icon.

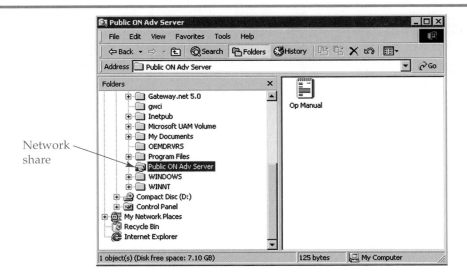

The permissions available for selection will vary according to partition file type (NTFS, FAT, etc.), the type of user account (Local or Domain), and which operating system has been selected, (Windows XP Home Edition, Windows XP Professional, etc.).

Tech Tip

Figure 7-7 shows the permissions assigned to the group *Everyone* for the shared folder, **Public ON Adv Server**. The group *Everyone* is assigned the Full Control permission over the contents of the folder. When a user is assigned to the group *Everyone*, the user will be able to have full control over the shared folder, **Public ON Adv Server**.

Disk Management

The disk management tool that comes with Windows NT, Windows 2000, and Windows Server 2003 is graphical. See **Figure 7-8.** The screen capture shows a hard disk drive that has eight partitions formatted with FAT, FAT32, and NTFS. The size of each partition and its location in the directory structure is displayed. Note the color code used to indicate the primary partition, extended partition, and logical drives. Disk management tools and disk terminology is discussed in more detail later in this chapter.

The *New Technology File System (NTFS)* is the native file format for Windows NT and Windows 2000/2003. NTFS was introduced with Windows NT 3.1. When Windows 2000 was released, the NTFS file system was further improved and was introduced by Microsoft as dynamic disk. Since the release of dynamic disk, it has been renamed NTFS5.0. The original NT file system was renamed NTFS4.0 so the two could be easily distinguished. NTFS4.0 and NTFS5.0 are covered in detail later in this chapter.

New Technology File System (NTFS)
the native file format for Windows NT and Windows 2000/2003.

An excellent feature of the NTFS file system is disk quota. *Disk quota* is the amount of disk space assigned to specific users. A common problem with networks that have shared hard disk drive space is users filling disk space with graphics and audio files. Graphics and audio files take a tremendous amount of storage space compared to database files and document files.

disk quota
the amount of disk space assigned to specific users.

NTFS allows the administrator to allot a specific amount of disk space to a user and a warning level, **Figure 7-9.** As the user approaches the limit (warning level) of

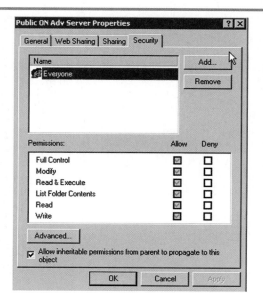

Figure 7-7.
Permissions granted to the group *Everyone* for the Public On Adv Server folder.

Figure 7-8.
The Disk Management utility as shown in the right-hand pane of the Computer Management utility.

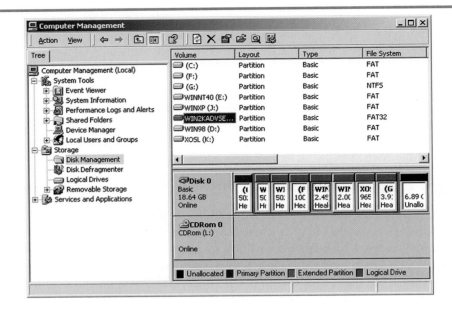

their assigned disk space, a warning message appears on their screen. The message is displayed before the user runs out of space. This is so the user may purge some of their files or request additional space from the system administrator.

Administrative Tools

Administrative tools are designed to make network administration easy and convenient. Some of the tasks administrative tools allow the administrator to do is view logs, manage the domain, manage users and groups, configure the server, set security policies, and add and manage services.

Some examples of administrative tools are User Manager for Domains found in Windows NT and the Microsoft Management Console (MMC) found in

Figure 7-9.
The NTFS disk quota feature allows an administrator to set a limit on the amount of hard disk drive space a user can use. The administrator can also set a warning level. When the user approaches the warning level, a warning message displays. The warning message tells the user that they have almost used their quota of hard disk drive space. The user can then delete unneeded files or ask the administrator to increase their quota.

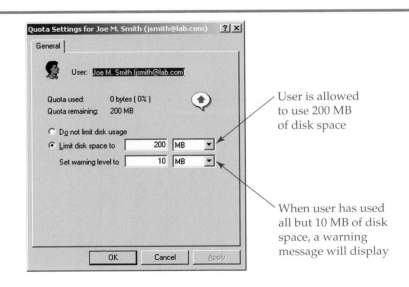

User is allowed to use 200 MB of disk space

When user has used all but 10 MB of disk space, a warning message will display

Windows 2000/2003. User Manager for Domains allows administrators to add users to a domain and manage users. The MMC is a console that can be configured to display the administrative tools that are most used by the administrator. The MMC and other administrative tools are covered in detail later in this chapter. You will find that the features of theses administrative tools are typical and similar to most network operating systems.

You have just received a very brief overview of some of the capabilities of the Microsoft network operating systems. The following sections explore Windows NT and Windows 2000/2003 in more detail.

Windows NT

Windows NT introduced several unique features for networking. One of the most significant improvements was the New Technology File System (NTFS). The original file system, known as File Allocation Table (FAT), or FAT16, has several limitations. It is limited to a 2 GB partition. This is a severe limitation for a server that needs to provide a large amount of storage space. FAT16 is also limited to eight-character file names and a three-character extension. NTFS, on the other hand, can create partition sizes as large as 16 EB. NTFS file names can be as long as 256 characters and have multiple extensions.

share-level security
a type of security that provides password protection and minimal share permissions to network shares. Share-level security applies only to shares that are accessed over the network.

Be aware that many resources refer to any Microsoft server product and network system as simply Microsoft NT even though the Microsoft server product is a later technology belonging to Windows Server 2000 or Windows 2003 Server.

Tech Tip

Not only does NTFS offer larger disk capacity, it also offers many security features that were not available directly for FAT16. The NTFS file system can be set up to share directories and control access to files. FAT16 allows only for share-level security. ***Share-level security*** provides password protection for a share and minimal share permissions. See **Figure 7-10.** Share-level security applies only to shares that are accessed over the network. It does not secure shares that are accessed locally. For example, a user who accesses a shared directory over the

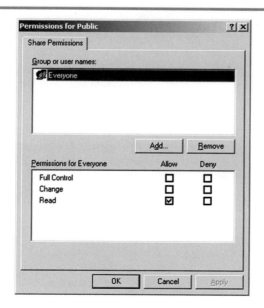

Figure 7-10.
Share permissions assigned to the group *Everyone* for the Public folder. Share permissions only apply when the share is accessed over the network.

share permissions
permissions assigned
to a network share.
Common share permissions
are full control,
change, and read.

user-level security
a type of security
that requires a user to
authenticate through
a security database to
access a share.

network may be asked to provide a password or may be limited to only Read access. If the same user accesses the shared directory locally (at the computer that hosts the share), no security would be enforced. The user would be able to make changes to the file.

Share-level permissions are commonly called *share permissions*. The following is a list of share permissions for the Windows operating systems:

- Full Control.
- Change.
- Read.
- No Access.

Tech Tip

Share permissions for Windows 2000 and Windows Server 2003 differ in that the No Access permission option is omitted.

NTFS permissions
permissions assigned
to directories and
files. NTFS permissions
are effective
locally and over the
network.

NTFS allows for user-level security. **User-level security** requires a user to authenticate through a security database to access a share. During authentication, user permissions are checked and access to the share is given accordingly. User-level security provides security to shares accessed over the network and locally.

User-level security permissions are commonly called NTFS permissions. **NTFS permissions** can be applied to both the directory level and file level. See **Figure 7-11** for a comparison of file and directory permissions for Windows NT, Windows 2000, and Windows Server 2003. As you can see, user-level security allows for more options and better security than share-level security.

Tech Tip

Although both share permissions and NTFS permissions can be applied to a share, it is best to set only the NTFS permissions and to leave the share permissions set to their default. Working with both types of permissions is unnecessary and only complicates administration.

Figure 7-11.
Share permissions for
files and directories in
Windows NT, Windows
2000, and Windows
Server 2003.

Windows NT	
File	**Folder**
Read	Read
Write	Write
Delete	Delete
Change permissions	Change permissions
Execute	Execute
Take ownership	Take ownership

Windows 2000 and Windows Server 2003	
File	**Folder**
Read	Read
Read and execute	Read and Execute
	List folder contents
Write	Write
Modify	Modify
Full control	Full control

Windows NT Network Administrative Models

Microsoft NT can be set up as a peer-to-peer network or as a client/server network. In the peer-to-peer network, all computers have equal administrative powers. Computers that need to share resources with each other are assigned to the same workgroup. Each computer in the workgroup has its own security database and can act as both a client and a server.

The client/server network uses a centralized server and is more secure than a peer-to-peer network. It is easier to administer because of the central location of the security database. It can be more economical than a peer-to-peer network because less time is required for the installation of software and data. By using a client/server model, all application software can be loaded on a central server and made accessible to all computers in the network if desired.

The administrative structure for a client/server-based Windows NT network is divided into domains, groups, and objects. The structure is hierarchical in nature starting with the domain and working its way down to the smallest unit, the object, which can be a shared resource such as file or printer.

The Domain Model

Users and equipment, such as computers, are assigned to a domain for administrative purposes and security. A small company may use a single domain name for the entire business. A large company may divide the company into several distinct domains. For example, a company spanning the United States may divide the company into regions, such as San Francisco, New York, and Orlando. See **Figure 7-12.** The example consists of three distant locations connected via the Internet or a private line. Although each region is represented as a separate domain, it is set up to allow the free flow of data and communications between them. The flow of data and communications can also be restricted.

Another scenario is the company illustrated in **Figure 7-13.** The company spans several distant locations yet is configured into a single domain. Typically multiple domains are used to better organize a large company into logical

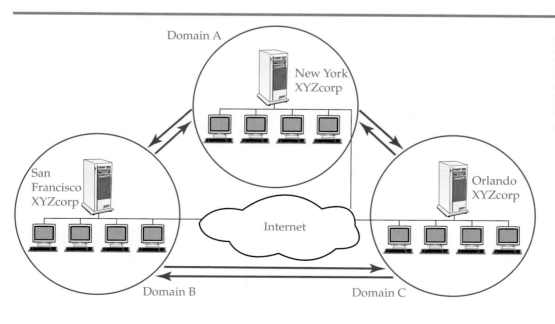

Figure 7-12.
The XYZcorp company has assigned each of its locations to a separate domain. The domains have been configured to allow the free flow of data between them.

Figure 7-13.
A company may span several distant locations yet be configured as a single domain.

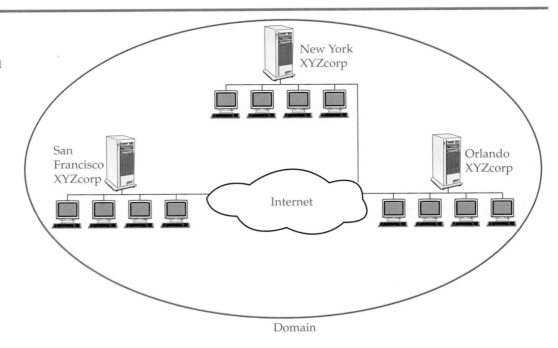

working units such as research, sales, accounting, executive administration, and human resources. There is no one set method to arrange a domain structure. The size and arrangement of a domain structure design is left to the network engineer or administrator.

Tech Tip

The term *domain* as it is used here should not be confused with the term *domain* used in Internet terminology. The term *domain* when used in Internet terminology describes a collection of computers sharing an Internet name or address. A better understanding of the differences will become evident after reading Chapter 11—TCP/IP Fundamentals.

Domain trusts relationships

trust relationship
a relationship between domains, which allows users from one domain to access resources on another domain in which they do not have a user account. The user can then access resources on another domain without having to log on to that domain. The user, however, must have permissions to access the resources.

To allow the free flow of data and communications between domains, a trust relationship between the domains must be established. A *trust relationship* is a relationship between domains, which allows users from one domain to access resources on another domain in which they do not have a user account. The user can then access resources on another domain without having to log on to that domain. The user, however, must have permissions to access the resources.

A trust relationship can be set up to allow domains to have a two-way trust relationship or a one-way trust relationship. In a *two-way trust relationship* both domains are designated as a *trusted domain* and a *trusting domain*. A two-way trust relationship allows both domains in the trust relationship to share its resources with the other. A two-way trust relationship is sometimes referred to as a *full-trust relationship*. **Figure 7-14** shows an example of a two-way trust relationship. *Domain A* trusts *Domain B* and allows *Domain B* to access its resources. *Domain B* trusts *Domain A* and allows *Domain A* to access its resources.

In a *one-way trust relationship*, one domain is the trusted domain and the other is the trusting domain. The trusting domain allows the trusted domain to

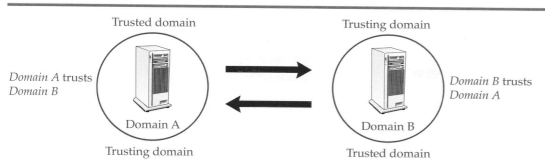

Two-Way Trust Relationship

Figure 7-14.
This is an example of a two-way trust relationship established between two domains. Each domain can access resources on the other domain.

access its resources. Look at **Figure 7-15.** A one-way trust relationship has been established. *Domain B* has been designated the trusting domain and *Domain A* has been designated the trusted domain. In other words, *Domain B* trusts *Domain A* to access its resources. However, since a one-way trust has been established, *Domain A* does not trust *Domain B* to access its resources.

Two-way trust relationships can be implemented to develop a complex trust relationship. A *complex trust relationship* is a trust relationship in which more than two domains have a full-trust relationship. Look at **Figure 7-16.** Notice that a two-way trust relationship is established between all three domains.

Server roles

Servers in a Windows NT network can perform one of the following three roles: primary domain controller, backup domain controller, and server. The *primary domain controller (PDC)* manages user access to the network. The chief function of the primary domain controller is to authenticate domain users as they log on to the network. The PDC hosts the user and security database. Changes to the security database, such as adding users and groups, must be performed through the PDC. There can only be one PDC for the entire Windows NT domain.

As other servers are added to the network, they may be designated as backup domain controllers. *Backup domain controllers (BDC)* are used to store backup copies of the user account and security database. A BDC can perform authentication and can be promoted to a PDC in case the PDC fails. Changes to the security database cannot be made through a BDC.

two-way trust relationship
a trust relationship in which both domains are designated as a trusted domain and a trusting domain. A two-way trust relationship allows both domains in the trust relationship to share its resources with the other.

one-way trust relationship
a trust relationship in which one domain is the trusted domain and the other is the trusting domain. The trusting domain allows the trusted domain to access its resources, but the trusted domain does not allow the trusting domain to access it resources.

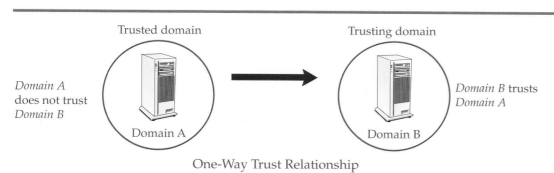

One-Way Trust Relationship

Figure 7-15.
In a one-way trust relationship, a domain may access resources on another domain even though there is no user account established on the other domain. The trust honors the logon authentication of the domain the user is in.

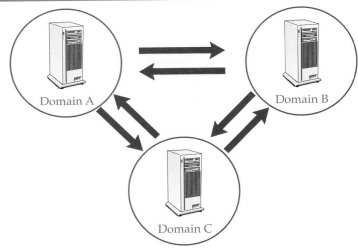

Figure 7-16.
An example of a complex trust relationship. You can think of a complex trust relationship as two-way trusts established between all domains in the configuration.

Complex Trust Relationship

complex trust relationship
a trust relationship in which more than two domains have a full-trust relationship.

primary domain controller (PDC)
a domain controller that hosts the user and security database and manages user access to the network. The chief function of the primary domain controller is to authenticate domain users as they log on to the network.

backup domain controllers (BDC)
a domain controller that stores backup copies of the user account and security database. A BDC can perform authentication and can be promoted to a PDC in case the PDC fails.

member server
a server that belongs to a domain but does not serve as a domain controller. It authenticates users through another server that acts as the PDC or a BDC.

A server that assumes the server role can be designated as a member server or as a stand-alone server. A *member server* is a server that belongs to a domain but does not serve as a domain controller. It authenticates users through another server that acts as the PDC or a BDC. A *stand-alone server* is not part of a domain. Security is handled locally and all security information is stored in its local database.

Authentication

Authentication is the process of verifying a user's identity and allowing the user access to the network. Three components handle authentication in a Windows NT network. These components include the following:

- Security Account Manager (SAM).
- Local Security Authority (LSA).
- Security Resource Monitor (SRM).

The *Security Account Manager (SAM)* maintains the security account database. The security account database contains a list of user and group names, user and group security identifier (SID) numbers, and account information. The *Local Security Authority (LSA)* is responsible for validating local and remote logons and for generating a security access token. The *security access token* contains the user's security identifier (SID) number and the security identifier number of any groups the user belongs to. The Security Resource Monitor (SRM) compares the token information to an access control list (ACL) that is associated with a shared resource. Each shared resource, or object, in the domain has an *access control list (ACL)* associated with it. The ACL contains a list of users and groups that are allowed access to the resource and a list of permissions that are granted to each user and group.

This is how the authentication process works within a domain. When a user logs on to the domain, the user's name and password are sent to the nearest PDC or BDC on the network. The password is sent in an encrypted form so that it cannot be intercepted and read as it travels the network. The SAM compares the user name and password with that in the security account database. If the user name and password matches that in the security account database, the LSA generates a security access token. When the user requests access to a shared resource (object), the Security Resource Monitor (SRM) compares the token information to the permissions associated with the resource. The user is then allowed to access the resource according to their permissions as listed in the ACL. See **Figure 7-17.**

If a user needs to use resources that are located in a different domain and a trust relationship has not been established between the domains, the user has to log on to that domain to use its resources. If a trust relationship has been established between the domains, the user does not have to log on to the other domain to access its resources. A single logon is sufficient. The ability to access all resources throughout the entire network system with only a single logon is call *pass-through authentication.*

Windows NT Administration

Windows NT administration includes managing users, groups, and resources. NT comes with many utilities to help an administrator accomplish this. Three of the most common Windows NT utilities an administrator uses are User Manager, User Manager for Domains, and Disk Administrator.

Managing local users and groups

Local user and group accounts are used in a workgroup environment and for users and groups that need to have local access to the server. Local users and groups are managed through the User Manager utility.

stand-alone server a server that is not part of a domain. Security on a stand-alone server is handled locally and all security information is stored in the local database.

authentication the process of verifying a user's identity and allowing the user access to the network.

Security Account Manager (SAM) a service that maintains the security account database in a Windows NT domain.

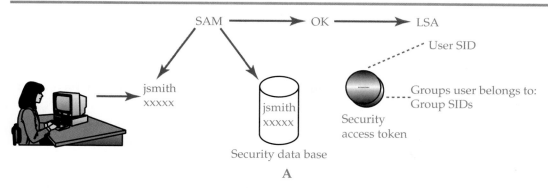

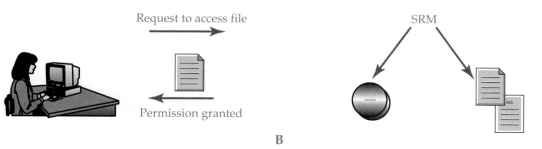

Figure 7-17.
The Windows NT authentication process. A—When a user enters their user name and password, the Security Account Manager (SAM) compares the user name and password to that in the security database. If a match is found, the SAM sends an OK to the Local Security Authority (LSA). The LSA generates a security access token. The security access token contains the user's security identifier (SID) number and the SIDs of the groups the user belongs to. B—When the user request to access a resource, the Security Resource Manager (SRM) compares the token information to the resource's access control list (ACL). If the information matches, the user is granted permission to access the resource.

Tech Tip   Local user accounts are for accessing the local computer. Domain user accounts are for accessing the entire domain.

Local Security Authority (LSA)
a service that validates local and remote logons and generates a security access token.

security access token
contains the user's security identifier (SID) number and the security identifier number of any groups the user belongs to.

access control list (ACL)
a list of users, groups, and permissions that are associated with a resource.

pass-through authentication
the ability to access all resources throughout the entire network system with only a single logon.

To add a local user, open the User Manager utility and select **User | New User** from the menu. The **New User** dialog box displays. Fill in the requested information, such as user name, full name, and password. Select one of the options beneath the **Confirm Password** text box: **User Must Change Password at Next Logon**, **User Cannot Change Password**, **Password Never Expires**, and **Account Disabled**.

Selecting one of the three buttons at the bottom of the **User Properties** dialog box opens other dialog boxes that allow you to assign the user to a group, configure the user's profile, and set up rights for the user to dial into this computer from a remote location.

To add a local group, open the User Manager utility and select **User | New Local Group** from the menu. The **New Local Group** dialog box displays. Enter the name of the group in the **Group Name** text box. To add members to the group, click the **Add** button. The **Add Users and Groups** dialog box displays. You can add a single user by selecting the user's name and clicking **Add**. You can add multiple users by holding down the [Ctrl] key while selecting users and then clicking **Add**. Click **OK** to exit this dialog box and return to the **New Local Group** dialog box. You will see the users you added to the group listed in the **Members** window. Click **OK** to exit and save your changes.

Managing domain users and groups

To add a user domain, open the User Manager for Domains utility and select **User | New User** from the menu. The **User Properties** dialog box displays, **Figure 7-18.** Fill in the requested information, such as user name, full name, and password. Select one of the options beneath the **Confirm Password** text box: **User Must Change Password at Next Logon**, **User Cannot Change Password**, **Password Never Expires**, and **Account Disabled**.

Note the buttons at the bottom of the dialog box. Selecting these buttons opens other dialog boxes that allow you to assign the user to a group, configure

Figure 7-18.
New User dialog box.

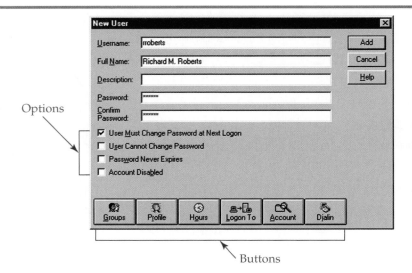

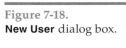

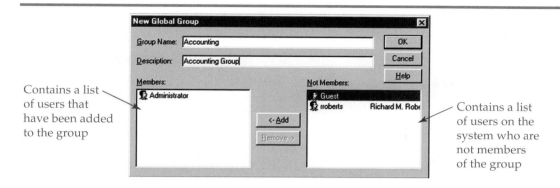

Figure 7-19.
New Global Group
dialog box.

Contains a list of users that have been added to the group

Contains a list of users on the system who are not members of the group

the user's profile, assign the hours the user is allowed to log on to the computer, assign which computers the user can log on to this computer from, and change the account expiration date and the account type.

To add a domain group, open the User Manager for Domains utility and select **User | New Global Group** from the menu. The **New Global Group** dialog box displays, **Figure 7-19.** Enter the name of the group in the **Group Name** text box.

To add members to the group, select members from the **Not Members** list. You can add a single user by selecting the user's name or multiple users by holding down the [Ctrl] key while selecting users. Click the **Add** button. The member or members you selected will display in the **Members** list. Click **OK** to exit and save your changes.

Disk management

The Disk Administrator utility is used to create partition and volumes, format volumes, and to add fault tolerance. Windows NT can be installed on a FAT16 or an NTFS partition. Remember, for added security, you should choose NTFS as the file system.

Some terminology you should be familiar with when working with the Disk Administrator utility is *physical drive, free space, partition, primary partition, logical drive, volume,* and *volume set.* The *physical drive* is the hard disk drive that is inside the computer. The computer BIOS assigns a number to each physical drive in the computer. This number is displayed in the Disk Administrator as *Disk 0, Disk 1,* and so on. See **Figure 7-20.** Displaying the disk number aids the administrator in properly identifying the hard disk drive before making changes to it.

Free space is the space on a hard disk drive that has not been partitioned. A *partition* is an area of the hard disk drive that is to be allocated to an operating system. When a hard disk drive is partitioned, it is assigned a drive letter and becomes a *logical drive.* This term makes sense when you think that a hard disk drive can be divided into multiple partitions and assigned a separate drive letter.

A hard disk drive may contain up to four primary partitions, **Figure 7-21.** A *primary partition* is a partition that stores a bootable copy of an operating system. Although four primary partitions, or bootable partitions, may exist on a single hard disk drive, only one primary partition can be active at a time, Figure 7-21A.

An administrator may want to store an operating system's boot files and system files on the primary partition and create an extended partition with logical drives on which to store data. An *extended partition* is a partition that can contain one or more logical drives. Only one extended partition can exist on a hard disk drive, Figure 7-21B.

Figure 7-20.
The Disk Administrator
utility.

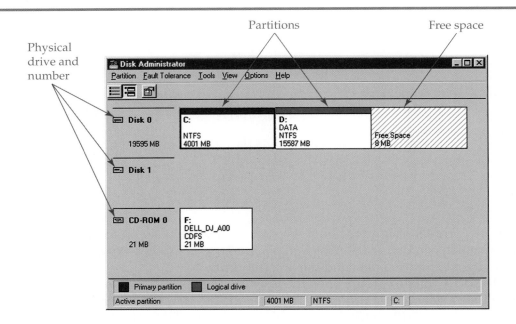

Once the hard disk drive is partitioned, it can be formatted. When a hard disk drive is formatted, it is physically prepared to receive an operating system. This preparation includes dividing the surface of the hard disk drive's platters into tracks and sectors and writing pertinent information to the drive such as a file allocation table (FAT).

A *volume* is an accessible unit of hard disk drive space as seen through the Windows interface, such as in Windows Explorer. A volume may be assigned a single drive letter yet contain multiple partitions. A volume that consists of partitions from two or more hard disk drives is called a *volume set*. See **Figure 7-22.**

The Disk Administrator utility provides tools for applying fault tolerance. The two levels of fault tolerance that can be applied to hard disk drives in a Windows NT system are RAID 1 and RAID 5. Fault tolerance is discussed in detail in Chapter 10—Introduction to the Server.

Figure 7-21.
Hard disk drive partition configurations. A—A hard disk drive may contain up to four primary partitions. A primary partition contains an operating system's boot files. Only one primary partition on a hard disk drive may be marked active. B—A hard disk drive may contain only one extended partition. An extended partition is not assigned a drive letter. It is instead divided into logical drives, which are assigned drive letters.

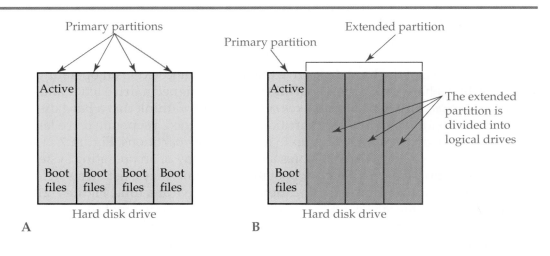

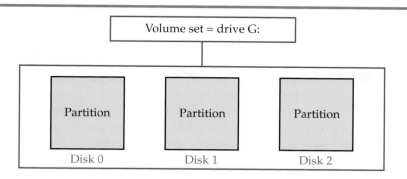

Figure 7-22.
A volume set is assigned a single drive letter and consists of partitions from multiple hard disk drives.

While network systems such as Windows NT are excellent systems and could be used for years to come, they become obsolete because of the lack of technical support provided by Microsoft. Microsoft supports their product line by constantly releasing patches and service packs, which keep the system up-to-date with new hardware and security features. Once the company stops releasing new hardware and security features for a particular operating system, the operating system becomes obsolete. For example, as new hardware products are developed such as the USB port, there is a need for a software patch for existing operating systems that were released before the first USB port was introduced. If a patch that enables the operating system to recognize and use the USB port is not released, the operating system soon becomes obsolete. Normally users do not settle on using older hardware such as a parallel or serial port when newer, faster hardware is available.

Windows 2000 Server and Windows Server 2003

Windows 2000 Server and Windows Server 2003 share many of the same features. The major difference between the two has to do with Internet support. These differences are covered in detail later in this chapter.

Active Directory

As network technology evolves, there are always better ways to design things. When Microsoft developed Windows 2000, they wanted to make it easier to span multiple domains and make accessing and viewing resources across the entire network transparent to the user. The solution was Microsoft Active Directory. Active Directory uses the Lightweight Directory Access Protocol (LDAP) to share its information with network services and other types of networks. LDAP has become a standard protocol among network operating systems for accessing and modifying directory information.

Active Directory uses a hierarchal directory structure that is designed as a database containing information about objects belonging to the entire network. In the Windows 2000/2003 environment, an object is any physical or logical unit that is defined as part of the network. Some examples of objects are users, groups, printers, volumes, directories, and services. Active Directory stores information about each object in its database. For example, for a user object, Active Directory stores information such as user name, password, permissions, and restrictions.

Active Directory can span multiple domains and is replicated on each domain controller. To the user who has the proper permissions, the directory for the entire network is accessible through a single logon procedure, even when spanning thousands of miles. Before Active Directory, the typical user needed to log on to a different server or domain to access its resources.

Active Directory allows the use of more than one domain controller within a domain. The Active Directory database is replicated on all domain controllers within the domain so that each has the current status of all objects in the network. For example, if an administrator disables a user account, the information is instantly transmitted to all domain controllers in the domain.

Active Directory also allows the network administrator to access and modify the entire network from a single logon. There is no need to log on to separate domains. The entire structure of multiple domains is handled as a single unit.

Active Directory uses Lightweight Directory Access Protocol (LDAP) as well as Hypertext Transfer Protocol (HTTP) to transmit directory information between domain controllers. Both LDAP and HTTP are Internet standard protocols. LDAP allows the exchange of directory information between different computer platforms. For example, a server running Windows 2000 can freely exchange directory information with a Novell or UNIX/Linux server as long as both use LDAP. HTTP is the protocol used on the Internet that determines how Web pages are displayed. HTTP is used to display directory information across the network in a form of a Web page. Later in the textbook, you will see the striking similarity of the Active Directory structure and the Novell NetWare directory structure. This is because both are based on LDAP technology.

To demonstrate the naming conventions associated with Active Directory, look at **Figure 7-23.** It shows that the Active Directory domain is named *NetWorkFundClass*. You can see that the domain is listed under **Directory** in the Internet Explorer directory structure. Only Active Directories are displayed beneath the **Directory** label. Look at the top of the screen capture. You will notice that the domain is identified as ntds://NetWorkFundClass.edu. Active Directory follows the Internet domain naming convention plus allows the domain to be identified as *NetworkFundClass*.

Figure 7-23.
An Active Directory domain as seen through Internet Explorer. Note the various ways the domain name can be addressed.

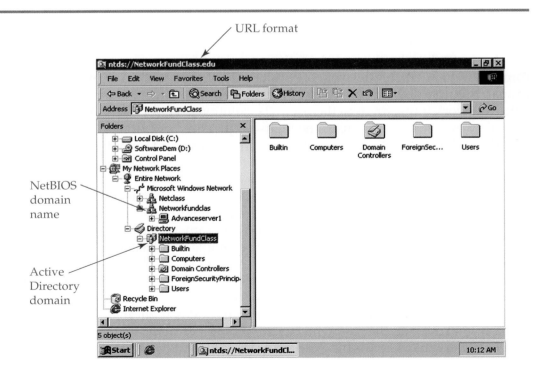

Now, look at domain name as it is displayed under **Microsoft Windows Network**. Note that the word *class* is spelled with one *s*. This is different when compared to the word *class* in the URL address and as listed under **Directory**. It is typical for an Active Directory domain to share two names: one for the Active Directory technology and the other for legacy network technologies such as Windows NT. The use of only 15 letters in the domain name is to allow the domain to be recognized by legacy network operating systems that use the standard NetBIOS naming convention. The NetBIOS naming convention is limited to 15 letters maximum.

The screen capture also reveals how the HTTP protocol relies on the name of the domain as it appears in standard URL format. Note the .edu in the URL format. This name represents the education root domain on the Internet. The Internet domain naming system is covered in detail in Chapter 11—TCP/IP Fundamentals.

Do not confuse HTTP with HTML. HTML is a hypertext markup language, which is an authoring language used to create documents that appear as Web pages. It is not a protocol.

Tech Tip

Active Directory structure

The Active Directory is a hierarchal structure that consists of a forest, trees, domains, and organizational units (OUs), **Figure 7-24**. The definition and an introduction to each is presented briefly in this section. An in-depth knowledge of the Active Directory structure requires in-depth study and more information than can be presented here.

The hierarchical structure is designed like the Internet structure. Look at the **Figure 7-25**. Objects inside a hierarchical structure can be traced to a root, or root domain. Each object in the hierarchical structure is a part of the object above it.

The members of the domain need not be physically located together. They are joined by using the same root domain name and having an account established in the domain. Typically, each domain in the Active Directory structure must conform to the naming convention used by the Domain Name System (DNS). The term *root domain* can be confusing because it is also used

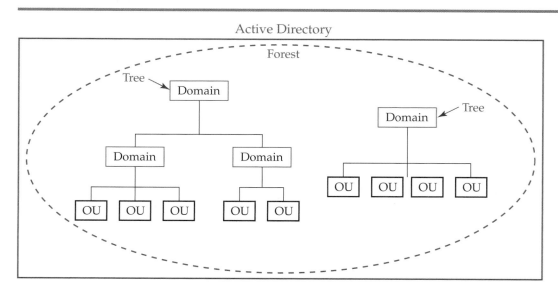

Figure 7-24.
An Active Directory hierarchical structure consists of a forest, trees, domains, and organizational units.

Figure 7-25.
Internet domain structure. A—In an Internet domain structure, objects can be traced back to a root domain. The root domain in this example is com. B—The domain structure in this example has been broken into sections to illustrate how each object in the hierarchy belongs to the object above it.

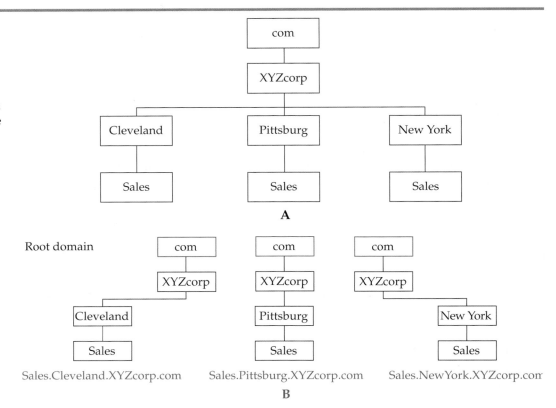

tree
a collection of domains that share a common root domain name and Active Directory database.

forest
a collection of domain trees that share a common Active Directory database.

namespace
the label that identifies a unique location in a structure such as the Internet.

contiguous namespace
a namespace in which the location must use the root domain name as part of its URL or as its complete name.

when describing the Internet structure. In the Internet, the root domain is the highest level of the organizational structure and is composed of a single dot and a name such as .edu and .com. In Active Directory, the root domain is the highest level of the organizational structure but consists of a name followed by a dot and a name such as XYZ.com.

Forest and trees

In the simplest of terms, a *tree* is a collection of domains that share a common root domain name and Active Directory database. A *forest* is a collection of domain trees that share a common Active Directory database. The definitions are very similar, but the most important trait that distinguishes whether a domain is part of a tree or forest is the domain name. For a domain to belong to a tree it must share a contiguous namespace. A forest is a collection of domain trees that have a disjointed namespace. To better understand these definitions, let's look at the terms *namespace, contiguous namespace,* and *disjointed namespace.*

A *namespace* is the label that identifies a unique location in a structure such as the Internet. The uniform resource locator (URL) standard uses a namespace to identify locations on the Internet. For an object or location to conform to a *contiguous namespace* standard, the location must use the root domain name as part of its URL or as its complete name. For example, the root domain name in an Active Directory structure is G-W.com. For a domain in the Active Directory to have a contiguous namespace with G-W.com, the G-W.com must be part of its domain name, for example, Chicago.G-W.com. Any domain name that is a part of the Active Directory, but does not follow the contiguous namespace requirement is considered a *disjointed namespace.* A disjointed namespace occurs commonly when two networks are merged or an existing network is upgraded using

Windows 2000 Active Directory. When two organizations or companies merge, it is unlikely that they will have a contiguous namespace. Look at **Figure 7-26.**

In the illustration, you can see three domains with a contiguous namespace: Chicago.XYZcorp.com, Dallas.XYZcorp.com, and NewYork.XYZcorp.com. These three are in contrast to the other domain name, Chicago.DEFcorp.com, which has a different root domain name. The root domain name is the determining factor that distinguishes whether a domain is part of a tree or forest. Chicago.DEF.com is a disjointed namespace and is therefore part of a forest.

disjointed namespace
a domain name that is a part of the Active Directory, but does not follow the contiguous namespace requirement.

You can install Windows 2000 and not install the Active Directory feature. The Active Directory feature allows for easy administration and communication across the entire forest. Without it, domains may be required to be administrated as separate entities.

Tech Tip

Forest and tree definitions came after the original NT domain structure. Because of the radical changes, an existing NT domain may not meet the criteria to belong to a tree.

Organizational unit

An *organizational unit* is a container that holds objects or other organizational units. Organizational units serve to organize a network into manageable units and often model a company's business structure. Look at **Figure 7-27.** A tree with a root domain name of XYZcorp.com has a domain named Chicago. This domain has an organizational unit for each department in the Chicago office. The Chicago.XYZcorp.com domain has an *Accounting*, *Administration*, and *Warehouse* organizational unit. Each organizational unit contains objects that belong to each department. An object in Active Directory can be any component that is part of the network, such as users, groups, computers, and printers.

organizational unit
a container that holds objects or other organizational units and is used to organize a network into manageable units.

Figure 7-26.
Example of a contiguous namespace and a disjointed namespace. Domains that share a contiguous namespace have a common root domain name. A disjointed namespace does not share a common root domain with the other domains in the Active Directory. A disjointed namespace in an Active Directory indicates that the domains within it are part of a forest.

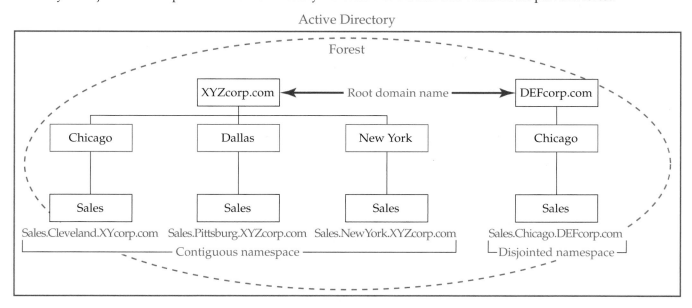

Figure 7-27.
Organizational units are used to organize a network into manageable units. An organizational unit can contain organizational units and objects. Examples of objects are users, groups, computers, and printers.

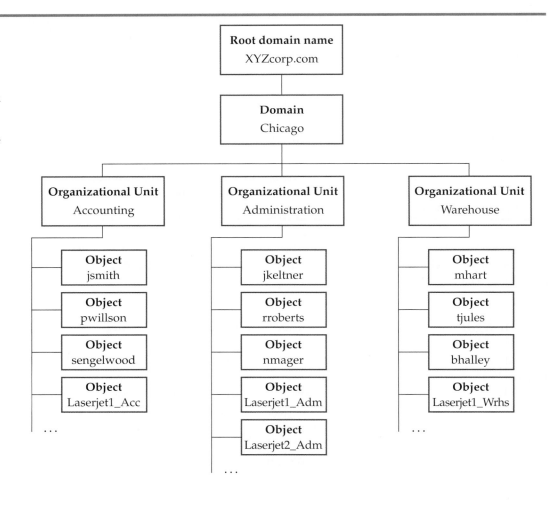

Common name

When an administrator creates a new object in the directory, the administrator assigns the object a name to identify it. This name is called a common name (CN). The object also receives a number called a globally unique identifier (GUID). The GUID links to the distinguished name. A ***distinguished name*** uniquely identifies the location of the object in the Active Directory structure. An example of a distinguished name for an object named *jsmith* located in the Accounting organizational unit in the Chicago.XYZcorp.com domain would be CN=*jsmith*, OU=*Accounting*, DC=Chicago.XYZcorp.com.

The letters in front of the names are referred to as attributes. The following lists defines each attribute:

distinguished name
a naming system that uniquely identifies the location of the object in the Active Directory structure.

- DC=Domain component.
- OU=Organizational unit.
- CN=Common name.

The distinguished name is an LDAP requirement. The attributes are used to decode distinguished names when transferring directory information between two dissimilar network operating system directories. You will see a similar naming structure in Novell NetWare.

Network+ Note:

Be careful when answering a question that contains the term *domain*. Look at the context in which the term *domain* is used. The term *domain* in a Windows NT network content refers to the network, not to the Internet domain. In a Windows 2000/2003 network, the term *domain* means the network name and Internet domain name.

Microsoft Windows 2000/2003 server roles

Servers in a Microsoft Windows 2000/2003 network can perform one of three roles: domain controller (DC), member server, and stand-alone server. Note that a Windows 2000/2003 network does not have a PDC or BDC like Windows NT. The Windows 2000/2003 environment uses a process called **multimaster replication.** In this model all DCs are equal. Each DC stores a copy of the Active Directory database, **Figure 7-28.** Changes to the database can be made at any DC in the domain. When changes to the database are made, the changes are replicated to the other DCs.

A member server is a server that is part of the domain but does not store directory information. A stand-alone server is not part of a domain. It contains its own security database.

Authentication

In a Windows 2000/2003 network, authentication occurs during two types of processes: when a user logs on to the network and when the user requests access to a resource. The first process is called interactive logon, and the second process is called network authentication.

During the **interactive logon,** the user is verified and given access to the Active Directory. When the user accesses a resource, the **network authentication** process occurs. In this process, the security descriptor of the resource the

multimaster replication
a type of security database replication in which all domain controllers store a copy of the Active Directory database. Changes to the database can be made at any DC in the domain. When changes to the database are made, the changes are replicated to the other domain controllers.

interactive logon
an authentication process in which the user is verified and given access to the Active Directory.

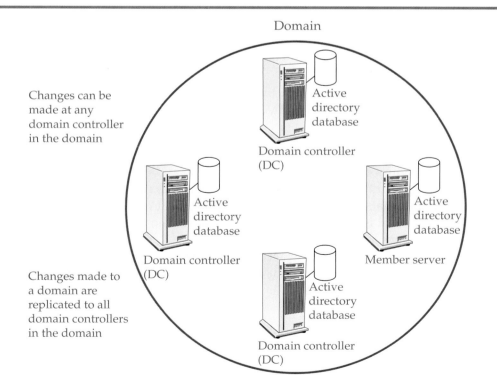

Domain

Changes can be made at any domain controller in the domain

Changes made to a domain are replicated to all domain controllers in the domain

Active directory database

Domain controller (DC)

Active directory database

Domain controller (DC)

Active directory database

Domain controller (DC)

Active directory database

Member server

Figure 7-28.
A Windows 2000/2003 Active Directory can have multiple domain controllers. An administrator may make changes to the Active Directory at any of the domain controllers. Changes are replicated to every domain controller in the domain.

network authentication
an authentication process that occurs when a user accesses a resource.

user wishes to access is checked. The security descriptor contains information regarding users and groups, permissions, auditing, and ownership. The security descriptor is much like the ACL in Windows NT. However, in Windows NT, the ACL applies only to resources, which are called objects in the NT domain. In the Windows 2000/2003, the security descriptor applies to resources, users, and groups, which are called objects in the Active Directory.

The interactive logon and the network authentication processes use the Kerberos version 5 authentication protocol. Kerberos is covered in detail in Chapter 16—Security. A Windows 2000/2003 network can also use the NT LAN Manager (NTLM) protocol to allow for backward compatibility with Windows NT domain controllers.

Windows 2000/2003 network authentication uses a single sign-on method. This means that once the user is authenticated and given access to the Active Directory, they can access any computer in the Active Directory to which they have permission. The user does not have to log on again. Authentication occurs behind the scenes and is handled through the network authentication process.

Windows 2000/2003 Administration

Windows 2000/2003 administration includes utilities for managing objects in the Active Directory and setting security policies. The following is a table of the utilities covered in this section and a description of each:

Utility	Description
Computer Management	Allows an administrator to add local users to a Windows 2000/2003 computer.
Active Directory Users and Computers	Allows administrator to manage users, groups, computers, and OUs in the Active Directory.
Local Security Policy and Domain Security Policy	Allows an administrator to set global security policies for a single computer or for the entire domain.
Disk Management	Allows an administrator to manage a server's hard disk drives.

These utilities can be accessed individually or through the Microsoft Management Console (MMC). The MMC was first introduced with Windows 2000.

MMC

Windows 2000 and later uses the Microsoft Management Console (MMC) as an interface to commonly used utilities. The MMC can be used on any Microsoft system including Windows 95, 98, and NT. The MMC is not a utility itself, but rather a container that holds commonly used utilities. It is set up as a customized Windows interface. It is empty by default and allows the user to select utilities from one convenient location.

snap-in
a tool or utility that is added to the Microsoft Management Console (MMC).

A utility that is added to the MMC is referred to as a *snap-in*. **Figure 7-29** shows some of the many snap-ins that can be added to the MMC. When the snap-ins are selected, they are automatically sent to the MMC. **Figure 7-30** shows a customized MMC.

Managing local users and groups

Local users and groups are managed on the server by using the Computer Management utility or the MMC Computer Management snap-in. To add a new

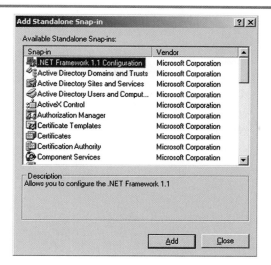

Figure 7-29.
List of snap-ins that
can be added to the
Microsoft Management
Console (MMC) in
Windows Server 2003.

user, right-click the **Users** folder in the left-hand pane of the Computer
Management window or the MMC. Select **New User** from the shortcut menu. The
New User dialog box displays, **Figure 7-31.** Complete the requested information.

You cannot create local user account on a domain controller or view local user accounts
on a server that has been promoted to domain controller status.

Tech Tip

You may select a password option from the options beneath the **Confirm
Password** text box or leave it set to the default. The default is set to the **User Must
Change Password at Next Logon** option. This option means that the user needs to
change their password when they log on to the system the first time. When the
user completes the logon, they will be prompted to change their password. The
administrator will not know the user's password after they change it. As you can
see, security starts with the user password and can be very strict.

Adding more restrictions to passwords can further enhance security. This
can be accomplished in the Local Security Policy utility. Various policies can be

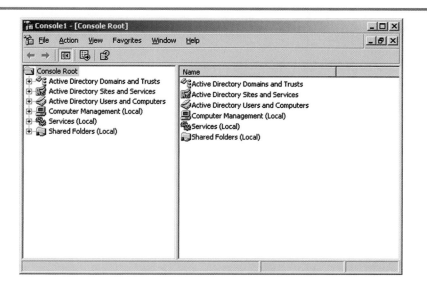

Figure 7-30.
Example of a custom-
ized MMC.

Figure 7-31.
The Windows 2000
New User dialog box
for adding local users.

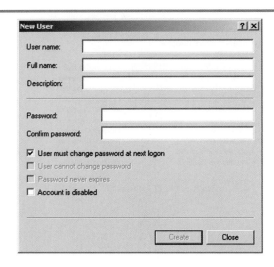

set to increase password security, such as keeping a history of the user's password to prevent reuse, specifying a minimum and maximum password length, and forcing the user to select a password that meets complexity requirements.

Account lockout policies can be set that cause the user account to be locked if the user incorrectly enters their password a number of times, **Figure 7-32.** This prevents unauthorized persons from attempting to guess a user's password. Lockout times can be set to last a few minutes to an indefinite period of time. A lockout time that is set for an indefinite period requires that the system administrator unlock the account. This explanation is a brief overview of how a user account can be set up and some of the security policies that can be established.

Figure 7-32.
Account lockout policy
settings.

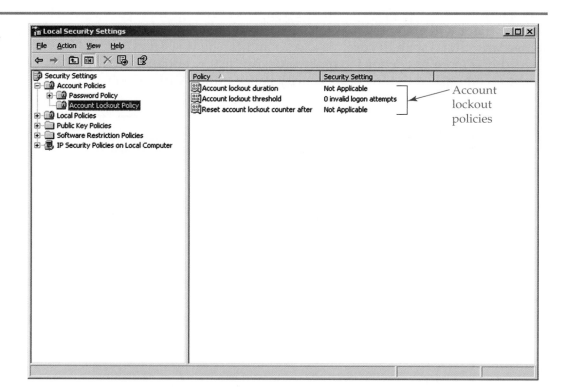

You can add groups by right-clicking the **Groups** folder and selecting **New Group** from the shortcut menu. When the **New Group** dialog box displays, enter a group name and a description. To add members to the group, click **Add**. The **Select Users** dialog box displays. Click **Advanced** and then **Find Now**. You can select multiple users by holding down the [Ctrl] key as you click each user's name. After you have selected the desired users, click **OK**. Click **OK** again to close the **Select Users** dialog box. In the **New Group** dialog box, click **Create** and then click **Close**.

Network security has become such an issue in networking and e-commerce that it has created the demand for highly trained specialists. If you wish to become a certified network administrator you must learn everything there is about a particular network operating system's security.

Managing domain users and groups

Domain users and groups are accounts that are members of a domain. Domain user and group accounts can only be created at a server that is a domain controller. The Active Directory Users and Computers utility is used to add user accounts and group accounts to the domain.

To add a user or a group to the domain, open the Active Directory Users and Computers utility. Right-click the container you wish to add the user to and select **New | User** from the shortcut menu. The **New Object—User** dialog box displays, **Figure 7-33.** Enter the user's first, middle, and last name in the appropriate text boxes. The **Full name** text box automatically fills in as you enter this information. Enter a logon name and then click **Next**. Enter a password in the **Password** and **Confirm Password** text boxes. Change the password options to reflect the security policies set for the network. Click **Next** and review the account settings. Click **Finish** if the settings are correct. Otherwise, click **Back** to review and change the settings in the previous screens.

Password restrictions and account policies for a domain can be further enhanced through the Domain Security Policy utility. The Domain Security Policy utility provides the same policies as the Local Security Policy utility.

To modify the password and account lockout policy settings for the domain, open the Domain Security Policy utility. In the left-hand pane of the Domain Security Policy window, expand the **Account Policies** folder. Select **Password Policy.** In the right-hand pane, right-click the policy you wish to modify. Select

Figure 7-33.
The **New Object—User** dialog box in the Active Directory Users and Computers utility.

basic disk
the term applied to
the old system of hard
disk drive configura-
tion, which includes
creating primary and
extended partitions.

dynamic disk
a new system of disk
configuration and
management intro-
duced with NTFS5.0.
Dynamic Disks are
managed by the sys-
tem as a group and
allow for the creation
of five different vol-
ume types: simple,
spanned, mirrored,
RAID-5, and striped.

simple volume
a dynamic disk
volume that exists on
a single drive.

spanned volume
a single dynamic
disk volume that
spans many drives.

Properties from the shortcut menu. The **Properties** dialog box for the selected policy displays, **Figure 7-34.** Make the desired changes, and then click **OK**.

Disk management

Windows 2000 introduced two new disk management terms: *basic disk* and *dynamic disk*. **Basic disk** is the term applied to the old system of hard disk drive configuration. This includes creating primary and extended partitions.

In Windows 2000/2003, a basic disk cannot contain volumes, volume sets, or any of the fault tolerance levels. These features are reserved for drives configured as dynamic disk. **Dynamic disk** is a new system of disk configuration and management introduced with NTFS5.0. Dynamic Disks are managed by the system as a group. Dynamic Disk is a requirement for Active Directory. The previous versions of disk formats (FAT16, FAT32, and NTFS 4) do not meet the requirements of Active Directory. Dynamic disk allows the creation of five different volume types. These types include the following:

- Simple.
- Spanned.
- Mirrored.
- RAID-5.
- Striped.

A *simple volume* exists on a single drive. A simple volume is similar to a volume in Windows NT. It is an accessible unit of hard disk drive space as seen through the Windows interface, such as in Windows Explorer. A *spanned volume* is similar to a Windows NT volume set in that it is a single volume that combines two or more physical drives. Mirrored and RAID-5 volumes incorporate fault tolerance. Fault tolerance is discussed in detail in Chapter 10—Introduction to the Server. A *striped volume* increases the read/write access speed by spreading

Figure 7-34.
Password policies are
modified by right-
clicking the password
policy, selecting
Properties from the
shortcut menu, and
entering the changes.

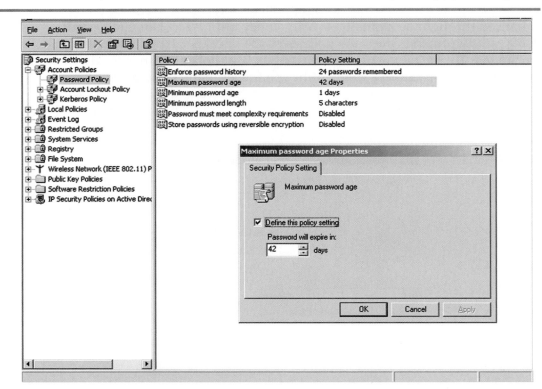

data across multiple hard disk drives. The disadvantage to this is if one of the hard disk drives in the striped set fails, the data will no longer be accessible from any of the hard disk drives in the striped volume. A striped volume, therefore, does not provide fault tolerance.

Windows 2000 and Windows Server 2003 can be installed on a FAT16, FAT32, and NTFS partition. An NTFS partition offers the highest level of security.

Tech Tip

Starting with the introduction of 64-bit processors, such as the Intel® Itanium®, and 64-bit operating systems, such as Windows XP Professional and Windows Enterprise Server 2003, a new partitioning style was developed. The new partitioning style is called globally unique identifier (GUID). It uses a partition table called GUID partition table (GPT).

The GPT partition table is required for a 64-bit operating system to boot. The GPT can support up to 18 exabytes and 128 partitions per disk. You can have both MBR and GPT disks on the same computer. The GPT disk contains three new partitions in addition to any volumes created on the disk: EFI, Logical Disk manager (LDM), and Microsoft Reserved (MSR). The EFI partition was developed as a replacement for BIOS functions. The *EFI partition* contains all programs required to boot the computer in the same way the BIOS boots the computer. The EFI is not a required partition. It can be created on a basic disk or on a GPT disk. At the time of this writing, EFI is an unproven technology and is only available for 64-bit operating systems.

The *Logical Disk Manager (LDM) partition* contains information about dynamic volumes and is created during the conversion from NTFS4.0 to NTFS5.0. There can be more than one LDM partition in a system. The *Microsoft Reserved (MSR) partition* is a required partition on every GPT disk. The MSR reserves disk space for use by system components. On a dynamic GPT disk, the first partition is the EFI if it is present, the second is the LDM, and the third is the MSR partition. Other partitions would be any additional LDM partitions.

Major Differences in Windows Server 2003

Windows Server 2003 was a continuation of the technologies developed in Windows 2000 Server. However, Windows Server 2003 places a greater emphasis on Internet support for enterprise-type networks. It also includes support for Microsoft .NET. (pronounced dot net). The .NET technology is better able to support communication between different Web Programming languages such as HTML, XML, Visual Java, Visual Basic, C++, and the Microsoft software applications. At the time .NET was developed, XML was the latest Web page programming language standard to be released. Microsoft perceived the Web as a media that would rapidly expand the role of user interaction with servers. With .NET, Microsoft introduced new programming technology that better supported Web page development and user interaction based on XML.

striped volume
a dynamic disk volume that increases the read/write access speed by spreading data across multiple hard disk drives.

EFI partition
a partition on a GPT disk that contains all programs required to boot the computer in the same way the BIOS boots the computer.

Logical Disk Manager (LDM) partition
a partition on a GPT disk that contains information about dynamic volumes and is created during the conversion from NTFS4.0 to NTFS5.0. There can be more than one LDM partition in a system.

Microsoft Reserved (MSR) partition
a partition on a disk that reserves disk space for use by system components.

The term *Microsoft Server.Net* was the first name used by Microsoft for the Server 2003 series in an effort to advertise their new .NET technology. The name created a problem with users because they had believed an entire new technology or server system was released, and felt that they were not ready for it. This caused Microsoft to later rename their new server product from Microsoft Server.Net to Microsoft Server 2003. Microsoft Server 2003 was better received.

Tech Tip

The influence of the Internet support design can be seen during a typical Windows Server 2003 installation. During installation, the administrator is prompted for the assigned server URL (domain name) rather than the NetBIOS server name. Active Directory is still the choice of directory structure. Microsoft Server 2003 supports earlier legacy server versions such as Windows NT 4.0. A Windows NT 4.0 server can belong to and be managed by a Windows 2003 server.

clustering
a technology that allows up to eight servers to be connected and act as a single server.

Windows Server 2003 includes support for Internet Information Services (IIS 6.0), which is the Web server support for hosting Web services. Clustering is included in the Windows Server 2003 and the Enterprise Server editions. *Clustering* is a technology that allows up to eight servers to be connected and act as a single server. The servers remain in constant communication offering services to clients. If one or more of the servers fail, the others take over the responsibility of delivering the failed server's services to its clients. This is an extremely important feature for achieving a high level of availability.

There are many other features that are incorporated into the Windows Server 2003 family, but they are beyond the scope of a simple introductory chapter. To learn more, visit the Microsoft Windows 2003 Server home page.

POSIX

Windows NT and Windows 2000 Server support the POSIX standard as defined by the IEEE. The acronym POSIX represents Portable Operating System Interface. The term *portable* is a programming term, which means that the program can be installed and run on various computer hardware systems. The C programming language by nature is very portable. A program written in C language can be run on most hardware systems despite the hardware system's manufacturer.

The POSIX standard covers areas such as file name structure, file access, and file sharing. The POSIX standard was originally designed to ensure compatibility between applications and the UNIX operating system. POSIX had originally stood for Portable Operating System Interface for UNIX. The term has come to mean portability with all operating systems, not just UNIX.

Microsoft Windows Services for UNIX introduced with Windows Server 2003 far exceed the original POSIX standards. Windows XP and Windows Server 2003 can run some Linux programs, such as scripts designed for the C Shell and Korn Shell. Linux shells are covered in Chapter 9—UNIX/Linux Network Operating Systems.

Interoperability

Microsoft supports communication with Novell, Linux, UNIX, Macintosh, and other network operating systems. There are several ways to configure a client or a server to communicate with a server that relies on a different protocol for communication and file sharing. The exact procedure varies according to the following:

■ The type of operating system with which the client needs to communicate.

■ The type of operating system on the client.

■ The services and protocols available to support communication between the different network systems.

Gateways and Services

Generally, to support communications between clients and servers in two different networks with different network operating systems, you would install a gateway or a gateway service. A gateway generally refers to a networking device such as a server, router, or a computer that hosts a gateway service. A **gateway service** is a software program that translates communications between networks that use different network operating systems, such as Novell and Microsoft. For example, a Microsoft server that is running Gateway Services for NetWare would be acting as a translator, or gateway, between the two, different network systems.

Figure 7-35 shows a gateway placed between an IBM mainframe and a Microsoft client. The gateway provides the protocol translation between the two systems. This strategy allows clients to communicate with a network operating system that uses a different protocol for communications and file sharing.

The Windows NT server uses a service called Gateway Services for NetWare (GSNW) to provide access to Novell network file server resources. Gateway Services for NetWare is typically installed on a Windows NT server. Communications between the Windows NT clients and the NetWare server are made through the NT server. When GSNW is used to allow clients access to a NetWare server, the administrator does not have to install a NetWare compatible protocol or client software on each workstation in the Microsoft network. This saves the administrator time. The drawback is, all clients in the Microsoft network use a common account to access the NetWare server. A common account does not allow for individual account policies and permissions. Also, GSNW only supports the SPX/IPX protocol. If the NetWare server is using TCP/IP, you will not be able to use GSNW.

Another way to set up communications is through installing a network service on a server. A network service is a software program that performs a task requested by a client. For example, making a remote connection to a server over a telephone line requires a service known as Remote Access Service (RAS).

Figure 7-36 shows a list of three services provided by a typical Windows 2000 server: File and Printer Sharing for Microsoft Networks, File and Printer Sharing for NetWare Networks, and Service for NetWare Directory Services. File

gateway service
a service that translates communications between networks that use different network operating systems.

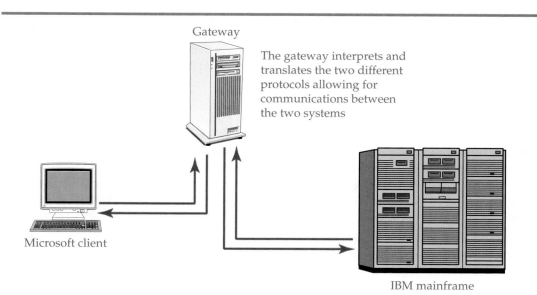

Gateway

The gateway interprets and translates the two different protocols allowing for communications between the two systems

Microsoft client

IBM mainframe

Figure 7-35.
A Microsoft client can communicate with an IBM mainframe by using a gateway service on the server to act as a gateway between the two dissimilar operating systems.

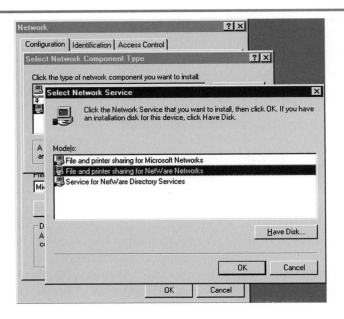

Figure 7-36.
Services provided by
Windows 2000 Server.

and Printer Sharing for Microsoft Networks allows the server to share its file and print resources with clients. File and Printer Sharing for NetWare Networks and Service for NetWare Directory Services provide for interoperability between Microsoft and Novell networks. Because Microsoft and NetWare use two different file systems, the File and Printer Sharing for NetWare Networks service is needed to act as a translator. This is especially important on networks composed of both NetWare and Microsoft servers.

Take special notice of the **Have Disk** button. You can add additional services with a floppy disk or CD provided by a manufacturer. The same feature is true for adding any protocols that are not listed in the **Add Protocol** dialog box.

Microsoft Operating System Client Configuration

A Microsoft-based computer can serve as a client to a server that has a different network operating system than itself. For example, a Microsoft-based computer can serve as a client to a NetWare server or an IBM mainframe. This is accomplished through binding the proper protocol to the network interface card and by using the appropriate client software. **Figure 7-37** shows a lists of protocols provided by Windows 2000. Remember, you can add additional protocols with a floppy disk or CD provided by a manufacturer by clicking the **Have Disk** button.

To access resources on a NetWare server, you must first bind the proper protocol to the network interface card. This is done through the **Network Properties** dialog box on the client. The protocol you will bind to the network interface card depends on the protocol that the NetWare server is using to communicate. NetWare 3.x and 4.x servers use the IPX/SPX protocol, which is its native protocol. NetWare 5.x and later servers typically use the TCP/IP protocol. Be aware that a NetWare 5.0 server can be configured to use IPX/SPX. TCP/IP did not become native to the NetWare operating system until the release of NetWare 6.0.

The second thing you need to do to access resources on a NetWare server is to install NetWare client software. Microsoft includes Client Services for NetWare

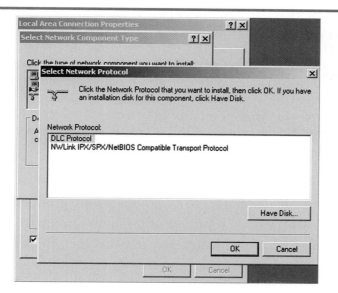

Figure 7-37.
Network protocols
provided by Windows
2000 Server.

(CSNW). Client Services for NetWare is added through the **Network Properties** dialog box on the client. However, CSNW does not support the TCP/IP protocol. If the NetWare server is using the TCP/IP protocol, you must install the Novell Client. The Novell Client is covered in Chapter 8—Novell Network Operating Systems.

When connecting a Microsoft-based computer to an IBM host, you must install the data link control (DLC) protocol on the Microsoft client. The DLC protocol allows the Microsoft client to communicate with and access the resources available on an IBM mainframe.

Microsoft computers can communicate with Macintosh computers by installing the AppleTalk protocol. Look at the **Local Area Connection Properties** dialog box in **Figure 7-38.** The AppleTalk protocol is listed. When the AppleTalk protocol is selected and installed on the Microsoft client or server, communications and file sharing is enabled between the Microsoft client and computers on the Apple network. Macintosh computers generally connect as a workgroup or as clients with access to the file server.

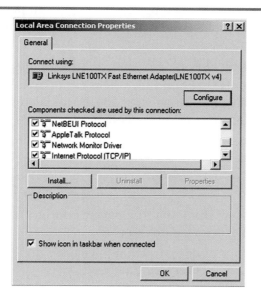

Figure 7-38.
The AppleTalk protocol
is bound to the network
adapter card so that
communications can
occur with Macintosh
computers.

As you can see, there are a number of ways to achieve communications between networks that normally could not exchange information or communicate. The exact method to use is determined by the system engineer or administrator, as well as the company's financial status. The cost of additional equipment, such as a dedicated gateway, and the cost of labor to install protocols across the entire network system on each client must be taken into account. We will discuss more about interoperability in Chapter 8—Novell Network Operating Systems and Chapter 9—UNIX/Linux Network Operating Systems.

Summary

- Computers in a Microsoft peer-to-peer network are grouped as workgroups.
- A workgroup is a group of computers that share resources.
- Computers in a Microsoft client/server network are grouped together in a domain.
- A domain is a logical grouping of users and equipment as defined by the network administrator.
- The New Technology File System (NTFS) is the native file format for Windows NT and Windows 2000/2003.
- The NTFS file system is an improvement over FAT16.
- NTFS can create partition sizes as large as 16 EB. NTFS file names can be as long as 256 characters and have multiple extensions.
- FAT16 is limited to a 2 GB partition and to an eight-character file name with a three-character extension.
- Dynamic disk is an improved version of NTFS and is referred to as NTFS5.0.
- Network shares have assigned permissions, such as no access, full control, and read only.
- A trust relationship in a Window NT domain allows the free flow of data and communications between domains.
- Windows NT domains can have a two-way trust relationship or a one-way trust relationship.
- A Windows NT server that belongs to a domain can perform one of three roles: primary domain controller (PDC), backup domain controller (BDC), and member server.
- There can only be one PDC in a Windows NT domain.
- The three components that handle authentication in a Windows NT network are the Security Account Manager (SAM), Local Security Authority (LSA), and the Security Resource Monitor (SRM).
- Active Directory uses LDAP and HTTP to transfer information between network controllers.
- The Active Directory structure consists of trees, forests, organizational units, and objects.
- A tree is a collection of domains that have a contiguous namespace and share a common Active Directory database.
- A forest is a collection of domains that have a disjointed namespace and share a common Active Directory database.

- An organizational unit is a container that holds objects or other organizational units and serves to organize a network into manageable units.
- The two processes that handle authentication in a Windows 2000/2003 network are interactive logon and network authentication.
- Windows 2000/2003 networks use the Kerberos version 5 protocol for authentication.
- The Microsoft Management Console MMC is a utility used to organize commonly used utilities.
- The Active Directory Users and Computers utility is used to add user and group accounts to a Windows 2000/2003 domain.
- A POSIX compliant operating system can run on numerous hardware platforms.
- A gateway service provides communication between dissimilar network systems such as Microsoft NT and IBM mainframes.
- The DLC protocol installed on a Microsoft computer supports communications between an IBM mainframe and a Microsoft workstation or server.
- Windows NT Server uses the Gateway Services for NetWare (GSNW) service to support communication between a Novell Network and a Microsoft network.

Review Questions

Answer the following questions on a separate sheet of paper. Please do not write in this book.

1. A(n) _____ is a group of computers that share resources such as files and hardware.

2. In a(n) _____ network, each computer contains its own database of users.

3. A(n) _____ network stores user and resource information in one location.

4. A(n) _____ is a logical grouping of users and equipment as defined by the network administrator.

5. Another name for NTFS5.0 is _____.

6. The _____ file system is limited to a 2 GB partition.

7. The _____ file system can create partition sizes as large as 16 EB.

8. The _____ file system can be set up to share directories and control access to files.

9. The _____ file system allows only for share-level security.

10. The administrative structure for a client/server-based Windows NT network is divided into _____, _____, and _____.

11. Explain the Windows NT authentication process.

12. What two Internet protocols are used by Active Directory to exchange directory information with other servers?

13. The Active Directory is a hierarchal structure that consists of _____, _____ _____, and _____.

14. What namespace feature determines if a domain is part of a tree or forest?

15. What is the difference between a disjointed and a contiguous namespace?

16. Explain the Windows 2000/2003 authentication process.

17. What is the Microsoft Management Console (MMC)?

18. What utility would an administrator use to add users and groups to an Active Directory?

19. Define *basic disk* and *dynamic disk*.

20. What is a gateway service?

21. Explain how Gateway Services for NetWare (GSNW) works.

22. How can installing Gateway Services for NetWare (GSNW) save an administrator time?

23. What limitations does using Gateway Services for NetWare (GSNW) impose?

24. Explain how File and Printer Sharing for NetWare Networks works.

25. Which protocol is required to be installed to communicate with a Macintosh computer?

26. Which protocol is used to connect to an IBM host?

Sample Network+ Exam Questions

Answer the following questions on a separate sheet of paper. Please do not write in this book.

1. What will happen on a Windows 2000 Server if a user presses the [Ctrl] [Alt] [Del] keys simultaneously after the server boots?

   A. The system will reboot.

   B. The **Windows Security** dialog box will display.

   C. The **Logon** dialog box will appear.

   D. Nothing will happen. Only the system administrator can invoke [Ctrl] [Alt] [Del].

2. Which protocol is most commonly associated with communicating with an IBM mainframe?

   A. IPX/SPX

   B. AppleTalk

   C. DLC

   D. HyperLink

3. On which file system can Windows 2000 Server be loaded? (Select all that apply.)

   A. FAT16

   B. FAT32

   C. NTFS

   D. OS/2

4. Which network device translates protocols between two different networking systems?

   A. Switch

   B. Router

   C. Gateway

   D. Repeater

5. Which of the following is *not* a classification associated with a Windows 2000 Active Directory structure?

   A. Forest

   B. Tree

   C. Branch

   D. Domain

6. Which of the following authentication protocols can Windows 2000 Server and Windows Server 2003 use to authenticate users? (Select all that apply.)

   A. NTLM

   B. LDAP

   C. Kerberos version 5

   D. TCP/IP

7. Which of the following server roles are available to an NT server that is part of domain? (Select all that apply.)

   A. Domain controller

   B. Primary domain controller

   C. Backup domain controller

   D. Member server

8. Which of the following server roles are available to a Windows 2000 server that is part of an Active Directory? (Select all that apply.)

   A. Domain controller

   B. Primary domain controller

   C. Stand-alone server

   D. Member server

9. How many domain controllers are allowed in an Active Directory?

   A. One

   B. Three

   C. There is no limit.

   D. None. Active Directory uses primary domain controllers.

Network+

10. You are adding a Windows 2000 server to an existing NetWare 4.0 network running IPX/SPX. The workstations on the network need to access files and resource on both systems. What are your options for allowing the workstations to access both systems? (Select all that apply.)

A. Install GSNW on the NetWare server.

B. Install GSNW on the Windows 2000 server.

C. Install a NetWare compatible protocol on each workstation and GSNW.

D. Install a NetWare compatible protocol on each workstation and CSNW.

Suggested Laboratory Activities

1. Log on to a Windows 2000 or 2003 server and set up a new user account.

2. Log on to a Windows 2000 or 2003 server and experiment with a user's password restrictions such as history, length, and lockout policy.

3. Set up a new group on a network and assign users to the group.

4. Make a list of suggested groups you might find in a typical high school network and list the types of objects they might typically share. Do not underestimate this assignment. It can be very involved if done correctly.

5. Using the knowledgebase on the Microsoft Support Web site, research logon errors in Windows 2000 Server and Windows Server 2003 and note the reasons for their generation and their associated error messages.

6. Obtain information from the Microsoft Web site on how to install Windows 2000 Server or Windows Server 2003. Use this information to install the server's operating system.

Interesting Web Sites for More Information

www.microsoft.com

(*Check out knowledgebase articles Q310996 and Q310997 on Active Directory.*)
www.microsoft.com/windowsserver2003

Chapter 7
Laboratory Activity

Adding Users to Windows Server 2003

After completing this laboratory activity, you will be able to:
- Explain the purpose of a user account.
- Describe the basic requirements of a user account.
- Explain the various password options available.
- Create a new user account.

Introduction

In this laboratory activity, you will add a new user to a Windows 2003 Server domain. One of the most common jobs a network technician performs is adding new users and giving users new passwords when the user forgets their password. Adding new users to a network system is also known as creating a user account. A user cannot access system resources on a network without a user account. At minimum, the user account requires a user name and a password.

There are two types of user and group accounts that can be created: domain and local. A local account is created for accessing the local computer or server. A local account is typically used on a peer-to-peer network. The domain user account is created on a server that is a domain controller. With the proper permission, this domain account is good for the entire domain. Once the domain user account is created, a user can log on to the network from any workstation. (The user must have permissions to log on from any workstation.) A local account only allows a user to log on to the computer the user is at, not the entire network.

A domain user account serves two main purposes: authentication and authorization. Authorization is the process of verifying a user's rights to access specific network services, such as printers and shares. Authentication is the process of verifying a user's identify. The user name and password is used for this purpose. This is why it is so important to keep the password secret. Anyone who knows the user password can authenticate as the real user. Active Directory requires user passwords to be more than simple passwords. If you attempt to use a simple password, you will generate an error message similar to that in the following screen capture.

As you can see, the password failed to meet one or more of the following requirements: length, history, and complexity. The requirements are a minimum password length of six characters, a history of 24 passwords, and a complex password, which includes uppercase and lowercase letters, numbers, and at least one special character. You can learn more about user accounts by accessing **Help and Support** located off the **Start** menu. The following list includes a few basic rules for setting up user accounts:

- Passwords can contain a maximum of 127 characters.
- Passwords with over 14 characters cannot be used by Windows 98 or earlier operating systems to access the domain.
- User names must be unique.
- User names should not contain the characters / \ [ ] ; : | = , + * ? < > because they have special meaning to the computer system and can cause an unexpected problem. You are free to use any remaining special characters, such as ! $ # % and &.

Companies generally have a naming convention in place for user names. Typically, they use a last name followed by the initial of the first name. They put the last names in first so that the user will be listed alphabetically in the Active Directory Users and Computers window. It makes it easy to locate a user when all names are listed alphabetically by last name.

For this laboratory activity, the naming scheme you will use for entering user names and group names will be a combination of your name and the name of the object you are creating followed by a decimal number: 1, 2, or 3. For example, a student named Karen will create the following users: KarenUser1, KarenUser2, and so on. This naming scheme will make it easy for your instructor to check your project.

Read the entire laboratory activity before you begin. The laboratory activity contains many examples. Familiarizing yourself with the examples and reading the laboratory activity will give you a clear understanding of what is to be accomplished. After reading the entire laboratory activity, you may begin.

Equipment and Materials

- A computer with Windows Server 2003 or Windows 2000 Server. (The two operating systems are very similar. There will not be a significant difference that will prevent you from completing the laboratory activity.)

On a separate sheet of paper, record any additional instructor notes that will be needed to complete the laboratory activity.

Procedures

1. _____ Report to your assigned workstation.

2. _____ Boot the workstation and verify it is in working order.

3. _____ Check if the server is up and running. You will add the new user accounts at the server.

4. _____ At the server, access the Active Directory Users and Computers utility (**Start |All Programs | Administrative Tools | Active Directory Users and Computers**).

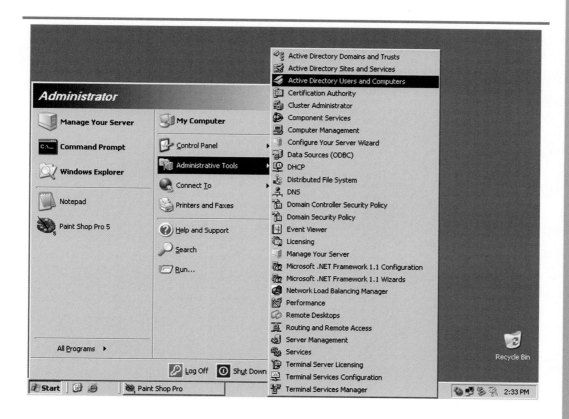

Note:
If Active Directory is not installed, there will be no option for **Active Directory Users and Computers**. *Select* **Computer Management** *instead. Creating a new user in the Computer Management utility is similar to creating a new user in Active Directory Users and Computers.*

Look at the following screen capture of the Active Directory Users and Computers utility. Notice how the right pane of the window lists users and groups currently in the domain. The left pane is displayed in a typical, Windows directory format. At the top of the directory is the name of the Active Directory that contains all the listed users, groups, domain controllers, and more.

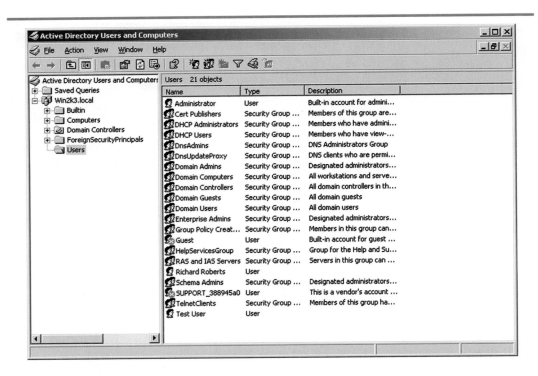

5. _____   From the **Action** menu, select **New | User**. Take note that you can also select **Computer**, **Contact**, **Group**, or **Printer**.

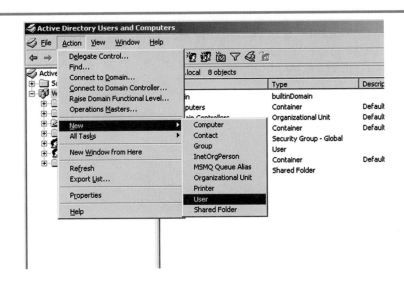

6. _____ The **New Object—User** dialog box will display. You will be required to fill in a series of dialog boxes. The first dialog box should look similar to that in the following screen capture. Fill in the required information and then click **Next**. Remember to create a new user based on your name, such as BillUser1.

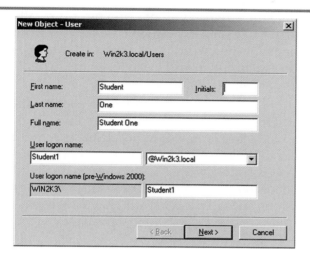

7. _____ In the next dialog box, enter a user password. For this laboratory activity, you will use the word *PwD123#* for the password, unless otherwise instructed by your teacher. This password meets the complexity required by Active Directory.

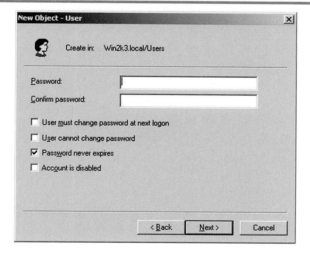

Take special note of the password options available. The first option, **User must change password at next logon**, means the administrator will assign a password. Then, the first time the user logs on to the system, the user will be prompted to change their password. This option allows the user to choose their own password so that only they know what it is, not the administrator. Companies typically use the same password for all first time users.

The second option, **User cannot change password**, means the network administrator will assign a password to the user, and the user must use the assigned password each time they log on to the system. The user cannot change the password, only the Administrator can.

The next option, **Password never expires**, means as stated that the assigned password will never expire. Typically, a network administrator sets an expiration date for passwords, forcing the user to change their password periodically to enhance security. This is a common practice. You will learn to set the expiration date and other features in a later laboratory activity.

The last option, **Account is disabled**, is used to disable a user account. This option is used when personnel go on vacation or take an extended leave. You would not delete the user account because the user would lose all permissions and privileges. The associated permissions and privileges would not be regained by creating a new user account using the previous user name and password. Permissions and privileges would have to be reassigned. It is much easier, therefore, to disable and re-enable an account.

8. _____ After entering the password information, click **Next**. A summary of the user information you entered will display. You can either accept the information by clicking **Finish**, or modify the information by clicking **Back** until you reach the appropriate dialog box to which you want to make changes.

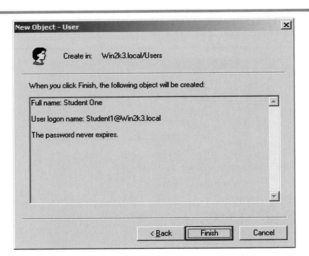

9. _____ After clicking **Finish**, the new user you just created will be automatically added to the list of users. Check the **Users** folder in the Active Directory Users and Computers utility for the user you have just added. If you have successfully created a new account, repeat the steps five through seven until you have created two more users. Remember to create user names based on your name, such as BillUser2 and BillUser3. If you do not see the user account you have just created, call your instructor.

Note:
Active Directory refers to users, computers, groups, and other items listed in the directory as objects.

10. _____ Answer the review questions.

11. _____ If required by your instructor, return the server to its original condition.

Review Questions

Answer the following questions on a separate sheet of paper. Please do not write in this book.

1. What are the minimum requirements for a user account? List two items.

2. What are the two main types of user accounts?

3. What are the two main reasons for a user account?

4. What options are available for the user password?

5. What must a password contain to meet the complexity requirement?

6. What utility is used to create domain users in Windows 2000 Server and Windows Server 2003?

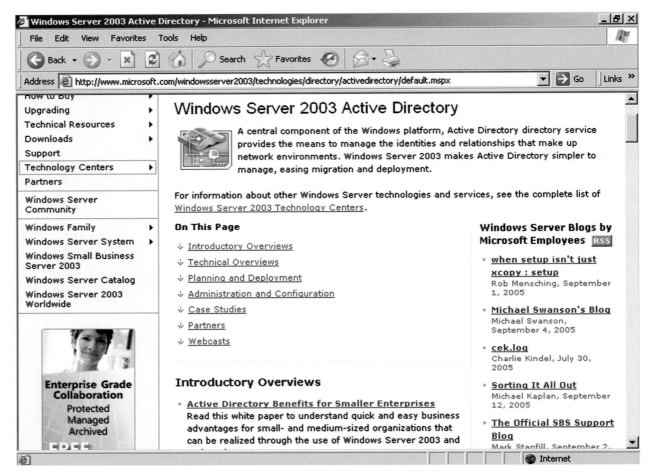

The Microsoft Web site includes information on Windows Server 2003 Active Directory as well as other Windows Server 2003 technologies and services.

8

Novell Network Operating Systems

After studying this chapter, you will be able to:

❑ Define NLM.

❑ Explain the NetWare boot process.

❑ Identify the purpose and features of ConsoleOne.

❑ Define and use the major console commands associated with Novell NetWare.

❑ Describe NetWare's NWFS and NSS file system structures.

❑ Describe the eDirectory structure.

❑ Identify and describe the common NetWare 6 administration utilities.

❑ Identify the three ways a client can access a NetWare server.

❑ Explain how the Native File Access Pack works.

Network+ Exam—Key Points

You do not need in-depth experience with Novell NetWare to be prepared for the Network+ Certification exam. The exam focuses on some introductory-level NetWare terminology as discussed in this chapter. Specifically, the Network+ Certification exam requires knowledge that Novell networks use IPX/SPX as their native network protocol and that the Novell NetWare 5 and later versions can be installed with the IPX/SPX and TCP/IP protocols. Some knowledge of how to log on to a Novell network is essential. You should be aware of the key components that need to be entered in a NetWare logon box. You should also be aware of the various ways to access files on a NetWare server.

Network+

Key Words and Terms

The following words and terms will become important pieces of your networking vocabulary. Be sure you can define them.

browser-based management utility	Novell Storage Services (NSS)
console	object rights
context	overbooking
eDirectory	redirector
explicit rights	screen
file access protocol	server-centric database
inherited rights	servertop
IPX internal network number	storage pools
JAVA-based management utility	trustee
journaling file system	typeful name
NetWare File System (NWFS)	typeless name
network-centric database	Web-based management utility

This chapter is an introduction to the Novell NetWare operating system. It presents the basic concepts unique to NetWare and is intended to familiarize you with the features and terminology associated with NetWare.

A Brief History of NetWare

Novell has been one of the leaders in networking technology since 1983. It was the dominant leader throughout the 1980s and into the early 1990s. Novell began losing its dominance when Microsoft developed a user-friendly network operating system that incorporated a GUI.

Early versions of NetWare were controlled by text-based commands entered at the console prompt similar to the way DOS commands are entered at the command prompt. Text-based command line administration and DOS-based utilities are more difficult to use compared to a GUI. NetWare administration utilities that could be run at a workstation were DOS-based. One such utility was called Syscon. Through Syscon, an administrator could manage network users and groups. Before NetWare 4.0, NetWare used a simple security database, which did not allow for a single logon across multiple servers. A user had to log on to each server to access its resources.

With the introduction of NetWare 4, Novell rebounded by providing a more robust security database called NDS along with the NetWare Administrator (NWAdmin) utility. NWAdmin is a Windows-based utility for managing the network. It was designed to replace Syscon and to work with the new security database. Through NWAdmin's GUI, administrators could monitor and modify the NDS structure and manage users and groups.

A JAVA-based GUI for the server was introduced with NetWare 5. The GUI was referred to as the *servertop* and had some basic utilities to change the display properties and to access files on the server. Another JAVA-based program, ConsoleOne, could be loaded from the command line or the servertop. ConsoleOne was an administrative program that could run on a server or a workstation. However, at that time, it did not offer as much functionality as NWAdmin. The servertop and ConsoleOne required a large amount of server resources and ran very slowly. Because of this, administrators often chose not to run these programs.

A more robust servertop and ConsoleOne were introduced with NetWare 6. Both programs ran faster than their initial versions and offered more functionality. ConsoleOne has come to replace NWAdmin.

Novell made some major changes in their server software with the release of NetWare 6. NetWare 6 incorporated some features from the UNIX/Linux open standard software programs. Open standard means that the code is not proprietary but completely accessible. Open standards can be freely distributed without royalties. There will be more about open standards in the Chapter 9—UNIX/Linux Network Operating Systems.

Some of the new open standard features incorporated were Apache Web Server, MySQL, Perl, and Tomcat. The overall improvements made in NetWare 6.0 were based on the full integration of the Lightweight Directory Access Protocol (LDAP). This led to improvement in NDS, which is now referred as eDirectory. The eDirectory is Novell's equivalent to Microsoft's Active Directory. They both share many similarities.

Some of the latest application features of eDirectory are Novell iPrint, Novell eGuide, Novell iFolder, and Novell Virtual Office. Each is a development based on using Internet protocols, such as TCP/IP and LDAP, for communications. The combined effect of TCP/IP and LDAP allow the new features to communicate over any size network. For example, iPrint allows a client to access a printer located anywhere on the worldwide network and send print jobs to it. The eGuide feature uses LDAP to access user information from an LDAP-based directory in any location. User information such as an address, telephone number, e-mail address, and fax number can be obtained as long as the user accessing the information has the proper permissions. The Novell Virtual Office is a collaboration utility that allows users to conference with each other in real time across the entire network. The term *real time* means that users can conference with one another as if they were together in the same room.

Starting with NetWare 6, NetWare no longer requires the Novell Client to be installed on a workstation to support communications with a NetWare server. Since NetWare 6 uses TCP/IP and LDAP as a communications protocol instead of IPX/SPX, a client can choose other means of accessing a NetWare server, such as using its own file access protocol or using one of Novell's applications that use Internet protocols for communication, such as iPrint and iFolder. Communicating with a NetWare server is covered in detail later in this chapter.

Following the Microsoft lead of incorporating the latest Web page program language and protocol, XML, Novell incorporated DirXML into their server package. DirXML supports the exchange of information with Microsoft Active Directory and Microsoft's Internet Information Server (IIS). As you can see, although the various server operating systems started out as proprietary, they are becoming more similar as they evolve.

After achieving your Network+ Certification, it is normal to pursue a more advanced certification in a network operating system, such as Novell, Microsoft, UNIX, or Linux. This would prove to be a valuable asset when seeking employment in the IT industry.

Tech Tip

Ironically, many senior IT professionals do not have advanced certifications, and some do not even hold minimal certification. This generation of IT professionals typically worked with the technologies before IT certifications existed. These individuals are held in high regard in the industry because of their lengthy and in-depth experience with such technologies. Do not be surprised to meet knowledgeable experts in the field who do not hold any high-level certifications. It is a matter of professional etiquette to respect years of on-the-job training.

Novell Kernel and NetWare Loadable Modules (NLM)

Server.exe is the core component of the NetWare operating system. Additional software programs called NetWare Loadable Modules (NLMs) enhance server.exe. Novell has always designed their network operating systems to be conservative with computer resources such as memory. This is why the network operating system is designed as a series of modules that add functionality to the core. An administrator can load and unload NLMs as needed without having to reboot the server.

There are several types of NLMs. Different NLM types have different file extensions. **Figure 8-1** lists the NLM extensions and their purpose.

The Host Adapter Module (HAM) provides support for storage adapters, and the Custom Device Module (CMD) provides support for storage devices. HAM and CMD storage drivers are referred to by Novell as NWPA drivers.

NLMs with the LAN extension are referred to as LAN drivers. These drivers support network interface cards.

Name space support allows a NetWare volume to store files from other operating systems. These NLMs have the NAM extension. Examples of name space support NLMs are long.nam, which supports Windows long file names, and nfs.nam, which supports UNIX/Linux file names.

There are many types of NLMs. Some NLMs enhance the kernel by providing hardware and software support. Other NLMs are programs and utilities. NLMs that provide additional features to enhance the kernel include those listed in **Figure 8-2.**

NLMs that are programs and utilities are often loaded by an administrator. **Figure 8-3** lists of some of those NLMs.

This is just a small representation of common NetWare NLMs. A complete list of modules that are loaded on a NetWare server can be viewed by typing **modules** at the console prompt. NLMs can be loaded and unloaded as needed. In other words, if the network administrator needs a certain service, they can load it from the console prompt by using the **load** command followed by the name of the NLM. The administrator can also enter just the name of the NLM as long as it is located in the SYS:system directory or in a search path. To unload an NLM, the administrator uses the **unload** command followed by the name of the NLM.

Figure 8-1.
NLM extensions and their purpose.

NLM Extension	NLM Purpose
HAM and CMD	Storage drivers
LAN	Network interface card drivers
NAM	Name space support
NLM	Utilities, programs, hardware support, and software support

NLM	NLM Name
cdbe.nlm	Netware Configuration Database engine
mm.nlm	NetWare Media Manager
filesys.nlm	NetWare File System NLM
nwpa.nlm	NetWare Peripheral Architecture NLM program
connmgr.nlm	NetWare Connection Manager NLM program
ncp.nlm	NetWare Core Protocol™ (NCP™) Engine
dsloader	eDirectory module
ipxspx.nlm	NetWare IPX/SPX Protocol Stack NLM program
ethersm.nlm	Novell Ethernet Topology Specific Module

Figure 8-2.
Some NLMs that provide additional features to enhance the kernel.

Third party software companies are free to develop their own set of NLMs and programs to support NetWare. NetWare 5 and NetWare 6 NLMs can be developed using JAVA programming. JAVA was selected because of its programming power, worldwide acceptance, platform independence, and ability to provide a GUI. By using JAVA programmed NLMs, it is now possible for a network administrator to control the entire network from any Windows-based machine using a graphical user interface. One such interface is called ConsoleOne. ConsoleOne will be discussed later in the chapter.

NetWare Boot Process

The NetWare operating system boots from a DOS partition on the server's primary hard drive. During installation, the nwserver directory is created in the root of drive C. The C:\nwserver directory contains files needed to boot the server to the NetWare operating system. It also contains the server.exe file.

Server.exe boots the NetWare operating system, loads itself into memory, and loads several bound-in modules. These modules are NLMs that are a part of the server.exe file. They provide the functionality for NetWare to take control of the server.

Server.exe can be added to the autoexec.bat file so that it can be invoked automatically at startup.

Tech Tip

NLM	NLM Name
dsrepair.nlm	Repair Utility for Novell eDirectory
dstrace.nlm	Trace Utility for Novell eDirectory
edit.nlm	NetWare Text Editor
hdetect.nlm	NetWare Hardware Installation Utility
monitor.nlm	NetWare Console Monitor
nwconfig.nlm	NetWare Configuration Utility
nwping.nlm	NetWare Ping Command
nwtrace.nlm	Traceroute: TCPIP Diagnosis Command

Figure 8-3.
Some NLMs that are programs and utilities.

Server.exe executes the startup.ncf file that is also located in C:\nwserver. Startup.ncf loads the drivers needed by NetWare to access the hard disk drive or drives and sets some memory management parameters. With the disk drivers loaded, server.exe mounts the SYS: volume. The SYS: volume is NetWare's system volume. It contains the directories listed in **Figure 8-4.**

Server.exe executes the autoexec.ncf file located in the SYS:system directory. The autoexec.ncf file contains commands to set the time synchronization, load communication protocols, load LAN drivers, bind protocols to LAN drivers, execute server parameters, and load programs and utilities. The autoexcec.ncf file also contains the name of the server and an IPX internal network number. The *IPX internal network number* is a unique hexadecimal number that serves to identify a NetWare server on the network.

IPX internal network number
a unique hexadecimal number that serves to identify a NetWare server on the network.

Tech Tip

To rename a server when eDirectory is installed, use the ConsoleOne utility. Changing the server name in the autoexec.ncf file does not affect server name settings in eDirectory.

The following is a summary of the NetWare boot process performed by server.exe:

1. Boots the NetWare operating system.
2. Loads itself into memory.
3. Loads all in-bound modules.
4. Executes C:\nwserver\startup.ncf.
5. Mounts the SYS: volume.
6. Executes SYS:system\autoexec.ncf.

The NetWare Console

console
the keyboard and monitor located at the NetWare server.

The *console* is the keyboard and monitor located at the NetWare server. When the NetWare operating system is fully booted, the console may display a console prompt or the servertop depending on how the server is configured.

The **startx** command loads the servertop. It is placed by default in the autoexec.ncf file. To boot to the console prompt, the administrator can either delete this command from the autoexec.ncf file or comment it out by placing the pound sign (#) in front of the command line. This is similar to placing the **rem** command in front of a command line in the config.sys and autoexec.bat files in DOS.

Figure 8-4.
Directories in the SYS: volume and their description.

SYS: Volume Directory	Description
Deleted.sav	Stores deleted files so that they can be salvaged.
Etc	Contains sample configuration files.
Login	Contains the files necessary, such as login.exe, for a user to log on to the system.
System	Contains NetWare operating system files and programs and utilities for the administrator.
Public	Contains programs and utilities for users.
JAVA	Contains JAVA support files.

Many programs, utilities, and commands can run on a NetWare server at the same time. Each instance of a running program, utility, or command is referred to as a *screen.* To display a list of screens, press [Ctrl] [Esc]. Screens can be accessed by entering the screen menu number or by pressing [Alt] [Esc] to page through the screens.

screen
an instance of a running program, utility, or command on a NetWare server.

Some screens you may see loaded are System Console, Logger Screen, and X Server—Graphical Console. The System Console is the screen in which commands are entered. This screen displays the console prompt. System Console is always listed as number one because it is the first screen loaded.

The Logger Screen displays system messages and alerts. This includes messages and alerts that occur as the server boots and while it is running. The information listed in the Logger Screen can aid an administrator in troubleshooting the server and the network system.

The X Server—Graphical Console is the NetWare servertop. It is called X Server in the menu because it is based on the X Windows system. The X Windows system is a graphical user environment that originated with the UNIX operating system.

Console Commands

Although NetWare provides for GUI administration at the console, the network administrator must be completely versed in the various text-based commands to properly administer the NetWare server and the NetWare network.

The commands listed in **Figure 8-5** are commonly entered at the console. The commands are not case-sensitive. They can be typed using lowercase or uppercase letters. To display a complete list of console commands, type . To display help on a specific console command, type the command **help** followed by the command. For example, **help *console_command*.**

Servertop

The *servertop* is NetWare's GUI. From the servertop you can run JAVA-based utilities such as ConsoleOne, Install, File Browser, Editor, and RConsoleJ. You can also run console commands and add programs to the menu or edit the menu. Through the RConsoleJ utility, you can access and control the consoles of other NetWare servers on the network.

servertop
the GUI on a NetWare server.

The servertop should load by default when the server is booted unless the **startx** command has been deleted or commented out of the autoxec.ncf file. If the servertop does not load automatically when the server boots or if it has been unloaded, you can start it by entering **startx** at the console prompt. To close the servertop, you can do one of the following:

- From the servertop, select **Novell | Close GUI**, and then click **Yes**.
- From the servertop, press [Ctrl] [Alt] [Backspace].
- From the console, enter **stopx**.

ConsoleOne

ConsoleOne is an administrative program that can be run on a server or a workstation. It is used to manage objects in NetWare's eDirectory. It is similar in design and presentation to the NWAdmin utility. See **Figure 8-6** and **Figure 8-7** to compare the ConsoleOne and NWAdmin utilities.

Figure 8-5.
Commands commonly
used at the console.

Console Command	Description
Broadcast	Followed by a message enclosed in quotes, **broadcast** displays a message to all or to specified users on the network. **Broadcast** is often used to warn users of upcoming maintenance as in the following example: **broadcast "Save all of your data and shut down your computer before leaving work today. We will be performing routine maintenance on the server tonight at 6:30 pm. Normal operations will be back in service at 7:00 am Tuesday."**
Config	Displays information about the server configuration such as the node address, IRQ, I/O port, slot information, network broadcast frame type, and much more. The **config** command is useful when troubleshooting server or network problems.
Dismount *volume_name*	Dismounts the specified volume. Dismounting a volume makes it unavailable.
Down	Properly shuts down the NetWare server by closing all open files and applications and dismounting all volumes. Returns the server to the DOS prompt. To properly shut down the NetWare server from the GUI, simply press the key combination [Ctrl] [Alt] [Backspace] and then enter **down** at the console prompt.
Help	Displays a listing of all commands available. Typing **help** followed by a command provides details about the command. For example, typing **help set time** displays information about the **set time** command similar to the following: set time [month/day/year] [hour:minute:second] Sets the file server data and time. Example: set time 9 October 2002 5:25:00 pm
List devices	Lists all the storage devices present in the server.
Memory	Displays the total amount of RAM installed on the server.
Mount *volume_name*	Mounts the specified volume. Mounting a volume makes it available on the network.
Restart server	Brings down the server and restarts the server. Similar to selecting **Restart** in the Windows **Shut Down** menu.
Volume	Displays details about the volumes mounted on the file server.

ConsoleOne is written entirely in JAVA. ConsoleOne allows for network administration from anywhere in the world via the Internet. The ConsoleOne utility uses snap-in modules, like NWAdmin, that expand its capabilities. The main difference between ConsoleOne and NWAdmin is that ConsoleOne uses snap-in modules written in JAVA, while NWAdmin uses Windows DLL files. JAVA snap-in modules can be written by any third-party software vender to enhance the administration of the NDS. Windows DLL files are limited to the Microsoft family of products, while JAVA-based applications are universal and can be run on any operating system platform.

Monitor

The Monitor utility provides a great deal of comprehensive information about the server such as CPU and memory utilization, the services that are running, and volume information. It allows an administrator to see who is logged on to the server and from where, disconnect users, and view the files that are in use. Monitor also allows an administrator to set server parameters.

Figure 8-6.
The ConsoleOne utility.

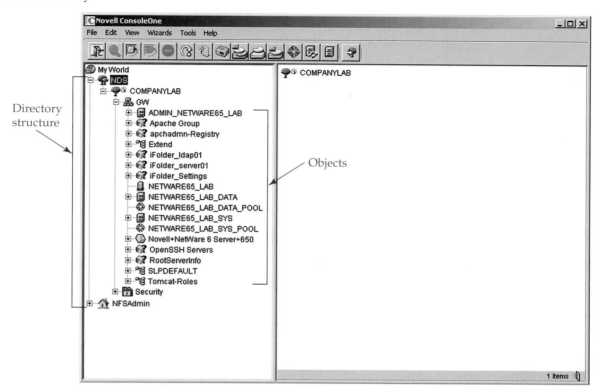

Figure 8-7.
The NWAdmin utility.

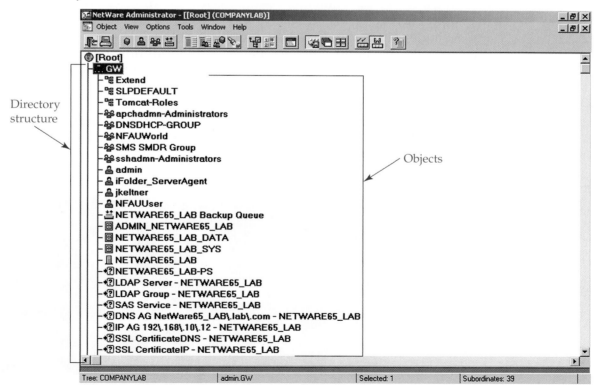

The Monitor utility is the chief administrative tool to monitor, diagnose, and configure NetWare servers. Monitor is accessed by entering **monitor** at the console prompt. Monitor is not a JAVA-based utility with a GUI, but rather a text-based program similar to DOS-based programs. See **Figure 8-8.** Many of the features of Monitor are available through iManager. The iManager utility is covered later in this chapter.

Network+ Note:

For the Network+ Certification exam, only the knowledge of the console's existence is important. You will not be tested on console commands, programs, or navigation.

NetWare File Systems

NetWare has two file systems: NetWare File System (NWFS) and Novell Storage Services (NSS). The *NetWare File System (NWFS)* is NetWare's traditional file system. The partitioning methods of NWFS are similar to DOS or Windows. Any attempt to change a partition that contains data or system files renders the files useless.

NetWare File System (NWFS)
NetWare's traditional file system.

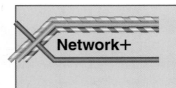

Before NetWare 5, NetWare only had one file system—NWFS.

An NWFS partition can be divided into a maximum of eight volumes. Dividing a NWFS partition into volumes is like dividing an extended partition into logical drives.

A single volume can also span multiple partitions. This is useful if a volume is reaching its maximum storage capacity and the administrator wants to quickly alleviate the problem. An additional drive can be added to the server and an NWFS partition created on the drive. An existing volume can then be expanded to use the space on the new partition.

To gain a better understanding of how NWFS partitions and volumes are created, look at **Figure 8-9.** A DOS partition that will hold the C:\nwserver

Figure 8-8.
The Monitor utility.

General information about CPU, Cache, and memory usage

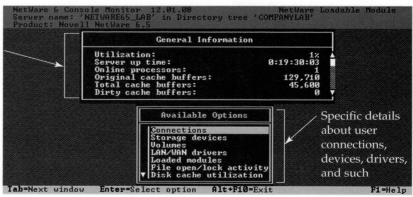

Specific details about user connections, devices, drivers, and such

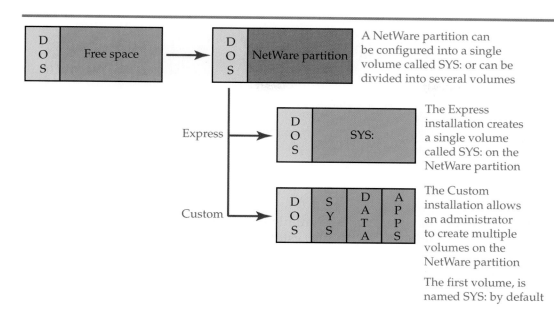

Figure 8-9.
A NetWare partition
is created before a
NetWare volume. The
first NetWare volume is
always named **SYS:**.

directory and serve as a platform from which the NetWare operating system boots is created on a hard disk drive during the NetWare installation. The rest of the hard drive space is left unpartitioned. If the administrator chooses the express installation method, NetWare automatically makes the rest of the drive a NetWare partition with a single volume called SYS:. If the administrator chooses the custom installation method, the administrator can create a NetWare partition and divide the partition into multiple volumes. The first volume is automatically named SYS:. The administrator cannot rename this volume. The administrator can name additional volumes anything they like. Some common names are DATA for data storage and APPS for applications.

System files (those installed by the NetWare installation) should be kept on the SYS: volume, and data and applications should be stored on other volumes.

Partitions and volumes can be modified and created at any time using ConsoleOne or iManager. iManager is discussed later in this chapter.

NWFS has many limitations. A single partition can have only a maximum of 8 volumes, and a maximum of 64 volumes can be mounted at one time. Also, a NetWare server can have only a maximum of 10,000 open files. This is minimal and impractical for a database server or for a file server in a large corporation. *Novell Storage Services (NSS),* NetWare's enhanced file system, overcomes these limitations. Look at the chart in **Figure 8-10** for a comparison of NWFS and NSS.

NSS volumes mount faster than NWFS volumes and provide for better recovery after a crash. This is due to the journaling file system. The *journaling file system* originated with UNIX systems. It has been recently introduced to Linux systems and adopted by Novell. NSS loads faster than traditional NetWare volumes because it does not need to scan the complete file system before mounting the volume. It requires only 1MB of RAM to mount.

When an NWFS volume is mounted, the whole file system is scanned and a table is created and stored in RAM. The greater the file system, the greater the

Novell Storage Services (NSS)
NetWare's enhanced files system.

journaling file system
a file system that relies on a journaling file to attain a faster boot time, use less system memory, and prevent file corruption.

Figure 8-10.
NetWare File System
(NWFS) and Novell
Storage Services (NSS)
comparison chart.

	NWFS	NSS
Volumes per partition	8	Limited only to the physical space available on the hard disk drive.
Volume size	4GB	8 TB
Files per volume	2 million	8 trillion
Open files	10,000	1,000,000
Mounted volumes	64	Unlimited

amount of RAM that is needed to store file system data. NSS stores only a small amount of information about the file system in RAM and stores it there until the volume is mounted. Once the volume is mounted, the file system data is removed from RAM. When a file is accessed on the volume, only information about that file is stored in memory. The system obtains this information from a file called the journaling file rather than from a file allocation table (FAT). The system also writes all transactions to the journaling file. If the NetWare server crashes, the system scans the journaling file for errors and incomplete transactions. When an error is found, it is fixed. Incomplete transactions are restored to a previous state.

storage pools
collections of storage devices with logical volumes.

The NSS file system manages volumes in storage pools. **Storage pools** are collections of storage devices with logical volumes. The logical volumes are called NSS volumes. One advantage of using storage pools is that they allow for growth. An NSS volume can be configured to grow dynamically, or a storage pool can be increased by adding more NetWare partitions. However, no more than four partitions can exist in a storage pool.

To better understand the structure of NSS, look at **Figure 8-11.** The storage pool in the example consists of three 2GB NetWare partitions. NetWare partitions are always created before storage pools.

Next, volumes are created. In the example, two NSS volumes are created: DATA: and APPS:. A storage pool can have an unlimited number of volumes. However, in reality, this is limited to the amount of physical disk space available.

When creating an NSS volume, the administrator can allocate a set amount of storage space to a volume or allow the volume to grow dynamically. In the example, both NSS volumes have been set to use 5GB of storage space. Notice that the total sum of storage space equals 10GB, which is greater than the set size of the storage pool, 6GB. Setting the total amount storage space to be greater than the size of the storage pool does not mean that storage space will be allowed to be greater than the storage pool size. It simply means that each volume can be allowed to grow to 5GB; however, not at the same time. The combined amount of data on all NSS volumes in a storage pool cannot exceed the size of the storage pool. In the example, both or one of the drives could have been set to grow dynamically. However, when set to grow dynamically, the same rule applies. Therefore, if the DATA volume grows to 5GB, the APPS volume cannot exceed 1GB.

overbooking
when the combined amount of declared NSS volume sizes is greater than the size of the storage pool.

When the combined amount of declared NSS volume sizes is greater than the size of the storage pool, it is called **overbooking.** You can think of a storage pool as a resource from which NSS volumes can draw hard disk drive space. This allows for flexibility.

NetWare supports communications and data exchange with major operating systems such as Windows, Macintosh, and UNIX. However, namespace support

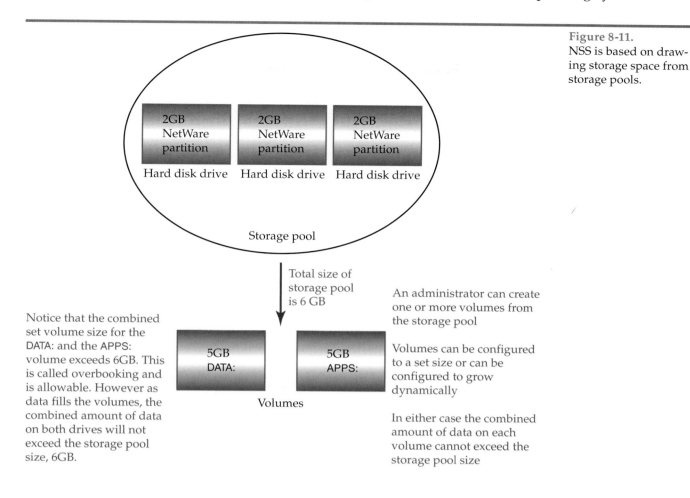

Figure 8-11.
NSS is based on drawing storage space from storage pools.

for the major operating system must be added to the NWFS volume. For example, to allow a Macintosh client to access and store files on an NWFS volume, the Macintosh name space must be loaded and then added to the volume:

load mac.nam

add name space *mac* to volume *volume_name*

The chart in **Figure 8-12** shows the name spaces available and a description of each. NWFS recognizes DOS file names by default. NSS provides support for DOS, Windows, Macintosh, UNIX systems by default.

eDirectory

eDirectory or *Novell Directory Services (NDS)* is a database of network resources similar to Microsoft's Active Directory. The ConsoleOne utility is used to display the contents of the NDS. Look at **Figure 8-13.** Notice that the eDirectory

eDirectory
a database of network resources similar to Microsoft's Active Directory.

Figure 8-12.
Name spaces available for NWFS volumes.

Name Space	Description
long.nam	Used by IBM OS/2 and Windows systems.
mac.nam	Used by Macintosh systems.
nfs.nam	Used by UNIX and Linux systems.

Figure 8-13.
The eDirectory structure as displayed through the ConsoleOne utility.

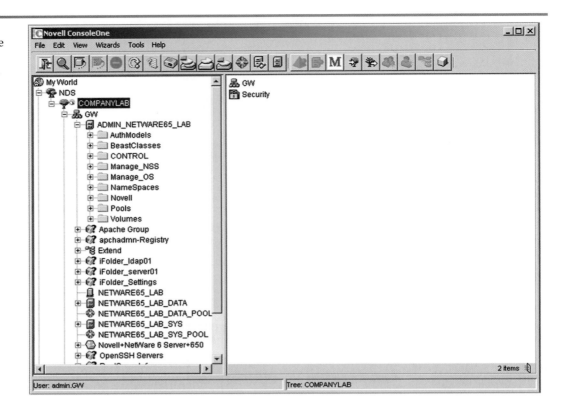

structure looks similar to Microsoft's Active Directory. This is because they are both based on the same Lightweight Directory Access Protocol (LDAP) technology.

Tech Tip

Originally, NDS was the acronym for NetWare Directory Services. It was changed to Novell Directory services in 1996 because NDS could be installed on platforms other than NetWare. When NetWare 6 was introduced, it was renamed eDirectory.

server-centric database
a database that is used for a single server.

network-centric database
a single database that serves an entire network.

NetWare operating systems before NetWare 4 used the term *bindery* for the files that contained security information for a single server. Each NetWare 4 server had its own bindery. Novell described the bindery as a server-centric database. A *server-centric database* means that the database is used for a single server. A single log on to multiple NetWare servers is not possible with a server-centric database. Later, NDS, now known as eDirectory, was developed to replace the bindery. NDS is network-centric. A *network-centric database* serves the entire network, making a single log on possible. See **Figure 8-14.**

eDirectory Organization

Like Microsoft's Active Directory, eDirectory is arranged in a hierarchical structure. However, eDirectory has a different naming scheme and system of organizing the structure. The tree container is always at the top or start of the directory. See **Figure 8-15.** (Earlier versions of NDS called the top of the directory the root.) You can think of the tree as the first object in the overall directory structure. All other objects are relative to the tree container. A tree container usually contains organization (O) container objects, country (C) container objects, and alias container objects. Organization container objects and country container

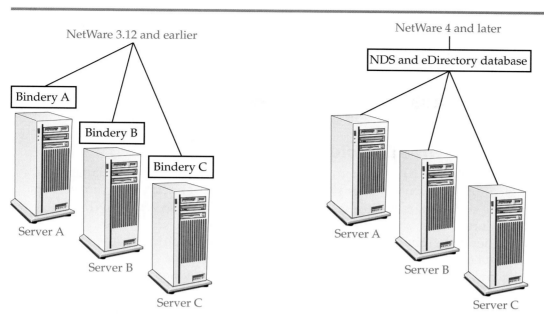

NetWare 3.12 and earlier

NetWare 4 and later

NDS and eDirectory database

Bindery A

Bindery B

Bindery C

Server A

Server B

Server C

Server A

Server B

Server C

Figure 8-14.
NetWare 3.12 and earlier used server-centric database called the bindery. NetWare 4 and later used a network-centric database. Versions 4 through 5 call this database NDS. NetWare 6 refers to it as eDirectory.

objects help an administrator arrange the tree to reflect the company structure or geographic layout. An alias object is a pointer to an object in a tree and serves as a shortcut.

Organization and country container objects usually contain organizational unit (OU) objects. Organizational units are also container objects. They help to

Figure 8-15.
The eDirectory structure is composed of containers and leaf objects arranged under the tree, which serves as the root of the structure.

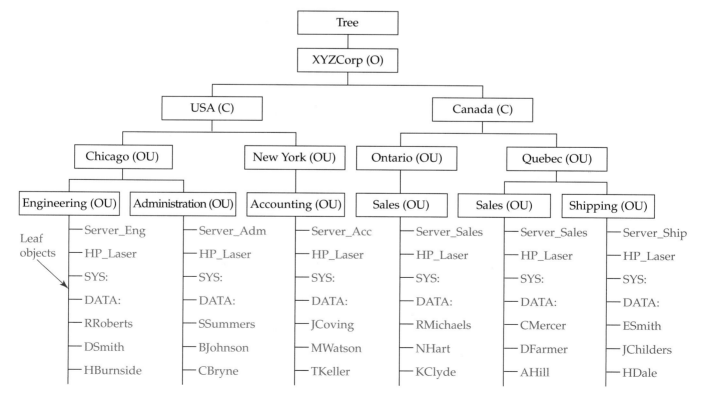

Tree

XYZCorp (O)

USA (C)

Canada (C)

Chicago (OU)

New York (OU)

Ontario (OU)

Quebec (OU)

Engineering (OU)

Administration (OU)

Accounting (OU)

Sales (OU)

Sales (OU)

Shipping (OU)

Leaf objects

Engineering	Administration	Accounting	Sales	Sales	Shipping
Server_Eng	Server_Adm	Server_Acc	Server_Sales	Server_Sales	Server_Ship
HP_Laser	HP_Laser	HP_Laser	HP_Laser	HP_Laser	HP_Laser
SYS:	SYS:	SYS:	SYS:	SYS:	SYS:
DATA:	DATA:	DATA:	DATA:	DATA:	DATA:
RRoberts	SSummers	JCoving	RMichaels	CMercer	ESmith
DSmith	BJohnson	MWatson	NHart	DFarmer	JChilders
HBurnside	CBryne	TKeller	KClyde	AHill	HDale

further organize the directory by matching it to the company's organizational structure. Organizational units usually contain organizational units or leaf objects.

Leaf objects are the lowest level of the eDirectory structure. They are associated with the CN abbreviation. CN stand for common name. The leaf objects represent users, groups, printers, servers, volumes, and such. Each leaf object is considered a network resource. They can be stored under any of the container objects. **Figure 8-16** shows the many types of leaf objects.

eDirectory Tree Structure

The eDirectory tree structure allows the entire network to be managed as a single unit. It does not matter if the network consists of many different servers running different operating systems such as Windows Server or UNIX. All network resources can be managed through eDirectory.

When managing objects, NetWare often needs the official name of an object as it appears in its context. The location of an object in the NDS tree is called its *context*. The context can be displayed as a *typeless name* and a *typeful name*. A *typeless name* is the name of the object followed by the path to the top of the directory. Each object in the path is separated by a dot (.). See **Figure 8-17.**

A *typeful name* also contains the name of the object and the path to the top of the directory. However, each object is preceded by the object's abbreviation, such as CN, O, and OU, and an equal sign (=). Each object is followed by a dot (.) as they are in a typeless name. Look at the example of a typeful name in Figure 8-17.

To log on to eDirectory, a user needs to input four pieces of information into a user logon box. The four pieces of information are username, password, directory context, and the name of the eDirectory tree.

context
the location of an object in the NDS tree.

typeless name
the name of an object followed by the path to the top of the directory. Each object in the path is separated by a dot (.).

typeful name
the name of an object preceded by the object's abbreviation, such as CN, O, and OU, and an equal sign (=).

Figure 8-16.
Notice the many leaf objects that can exist and their related icons.

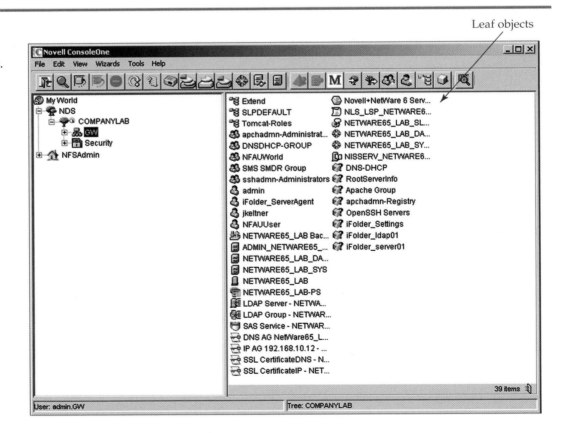

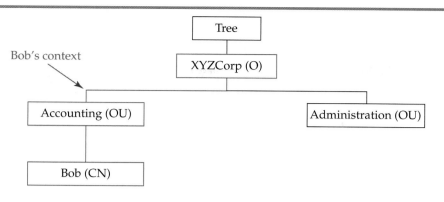

Bob's context

Typeless name: bob.accounting.XYZCorp

Typefull name: CN=bob.OU=accounting.O=XYZCorp

Figure 8-17.
Example of a typeless name and typefull name.

Network+ Note:

You will most likely be asked on the Network+ Certification exam what information is needed to log on to the eDirectory.

NetWare Security

NetWare security consists of user authentication, object rights, and directory and file rights. You should recall that for versions of NetWare that relied on a bindery for security, a user had to log on to each server they wished to use. Versions of NetWare that use NDS or eDirectory allow for a single log on. In these versions, a user logs on to the directory tree, not on to a single server. A user is then able to access resources within the tree to which they are granted rights.

Two types of rights can be granted: object rights and directory and file rights. Object rights relate to objects within the NDS directory or eDirectory. A user or group can be granted certain types of access to objects within the directory, such as browse, create, and rename. These rights are called *object rights*. **Figure 8-18** shows a complete list of object rights and their meaning. A user or group that is granted rights to an object is called a *trustee.* Each object in the directory has an Access Control List (ACL) just like objects do in the Microsoft Active Directory. The ACL contains a list of trustees and their related rights.

object rights
rights that allow a trustee certain types of access to an object.

trustee
a user or group that is granted rights to an object. Also, a user, group, or container object that is granted rights to directories and files.

Object Rights	Description
Browse	Allows a user to view an object within the directory tree.
Create	This right applies only to container objects. It allows a trustee to create an object and view objects within the container object to which they are a trustee.
Delete	Allows a trustee to delete an object.
Rename	Allows a trustee to rename an object.
Supervisor	Grants a trustee all rights to the object.

Figure 8-18.
NDS and eDirectory object rights.

inherited rights
rights that are granted to a trustee through inheritance.

explicit rights
rights that are specifically granted to a trustee, and are not granted through inheritance.

Web-based management utility
a management utility based on existing open source standards such as HTTP and XML.

JAVA-based management utility
a management utility based on the Sun Microsystems JAVA programming language.

Directory and file rights secure access to directories and files. User, groups, and container objects can be granted rights to directories and files. A user, group, or container object that is granted rights to directories and files is also called a trustee. **Figure 8-19** shows a complete list of directory and file rights and their meaning.

Object rights and directory and file rights are inheritable. This means that rights applied at one level of the directory tree or file system directory flow down to lower levels unless they are filtered, or blocked. Rights that are granted to a trustee through inheritance are called *inherited rights.* Inherited rights can be overruled by assigning explicit rights to trustees. *Explicit rights* are rights that are specifically granted to a trustee, and are not granted through inheritance.

NetWare 6 Administration

NetWare 6 provides an administrator with JAVA-based, browser-based, and Web-based management utilities. A *Web-based management utility* is based on existing open source standards such as HTTP and XML. It allows a client to access a server or network device and perform functions within the limit of the programming language. A *JAVA-based management utility* is based on the Sun Microsystems JAVA programming language and does not necessarily follow the open standards as defined by the W3C. A *browser-based management utility* is typically associated with, but not limited to, a thin client. The thin client connects to the network device and then runs scripts hosted on the network device, not on the thin client. Browser-based management utilities combine the ability to view local directory structures and to interact or modify the configuration using additional utilities.

Tech Tip

All three management types use a basic Web browser and then modify the browser features by adding additional programming. The browser connects to the Internet to access remote devices or through an intranet to connect to a local network device. Often the term *browser-based* is used in conjunction with Web-based or JAVA-based management utilities. Other times the term *browser-based* is used to represent all three types of utilities. The exact definition depends on the manufacturer when describing their software products. At the time of this writing, there is not a widely accepted definition for the three types of Internet management utilities.

Figure 8-19.
NetWare directory and file rights.

Directory and File Rights	Description
Read	Allows a trustee to open and read files or to execute a program.
Write	Allows a trustee to open a file and to add to or modify its contents.
Modify	Allows a trustee to rename a file or a directory and to change their attributes.
Create	Allows a user to create a file or a subdirectory.
Erase	Allows a user to delete a file or directory.
File scan	Allows a user to view files and directories.
Access Control	Allows the trustee to modify the ACL such as add or delete trustees or to change a trustee's rights to an object.
Supervisor	Grants a trustee all rights to a directory or file.

You are already familiar with ConsoleOne. ConsoleOne is a JAVA-based utility that can run on Windows, NetWare, Linux, Solaris, and UNIX Tru64 platforms. It has come to replace NWAdmin, which was the main management utility starting with NetWare 4. Although NWAdmin can be used to manage objects in a NetWare 6 network, it is not recommended. ConsoleOne provides more functionality than NWAdmin. It supports new Novell products and enhancements such as Single Sign-On and Certificate Server, allows faster browsing of large eDirectory trees, and more.

Another JAVA-based utility is Remote Server Management also known as RConsoleJ. The Remote Server Management utility provides a remote connection to a NetWare server so that you can enter console commands just as if you were physically at the server's console. You can also load and use NLM programs such as Monitor. **Figure 8-20** shows RConsoleJ used to access the server's console. The **Server Screens** list at the top of the RConsoleJ screen allows an administrator to change the view to a different screen. The administrator can also scroll through the screens by clicking the right or left arrows that are next to the **Server Screens** list.

The Remote Server Management utility can be configured to provide a secure connection and is backward compatible NetWare 5. It consists of three components: RConsoleJ Client, RConsoleJ Agent, and the RConsoleJ Proxy Agent. The RConsoleJ Client is the JAVA-based utility that runs on a workstation or a NetWare server. The RConsoleJ Agent (rconag6.nlm) is an NLM that is loaded on the target server. It handles all RConsoleJ requests to that server. The RConsoleJ Proxy Agent (rconprxy.nlm) provides translation between the IPX and IP protocols. It is only needed if a server or client it wishes to communicate with is using the IPX protocol. It can be loaded on any NetWare server that is running both IPX and IP.

Remote Manager is a browser-based utility that provides the same functionality and more as Monitor. Unlike Monitor, it displays information about the server's health through a graphical user interface. It requires two NLM programs to be loaded on the server: portal.nlm and httpstk.nlm.

iManager is a Web-based utility that allows an administrator to manage eDirectory objects, DNS, DHCP, and Novell Licensing Services (NLS). It does not require any special software to be installed on the client. All the client needs is a supported Web browser.

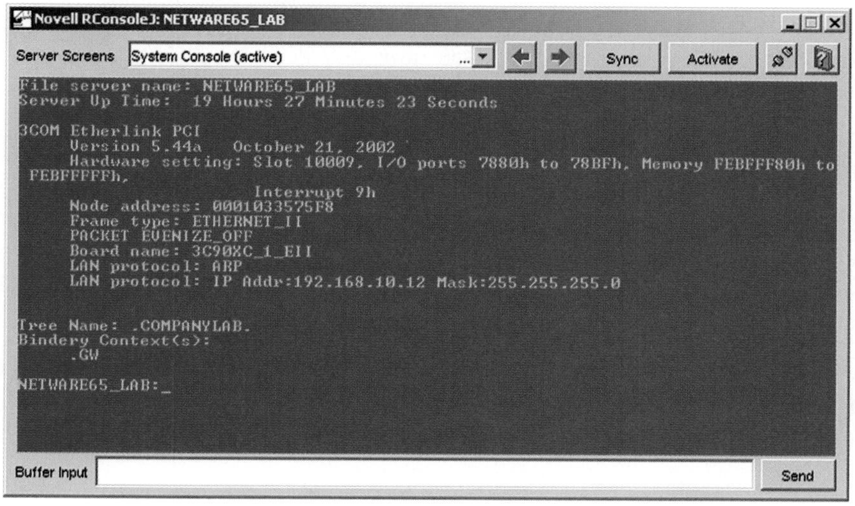

browser-based management utility a management utility typically associated with, but not limited to, a thin client. The thin client connects to the network device and then runs scripts hosted on the network device, not on the thin client.

Figure 8-20.
The RConsoleJ utility is used here to access the server's console.

NetWare Connectivity and Interoperability

Prior to NetWare 5, the IPX/SPX protocol was Novell's native protocol. IPX protocol addressing was the standard way to identify and communicate on a Novell network. Microsoft clients running NetBEUI could communicate with a NetWare server by installing the IPX/SPX compatible protocol (NWLink) and Client Services for NetWare. Microsoft clients could also opt to install the Novell Client, which included Novell's IPX/SPX protocol. Installing Gateway Services for Netware (GSNW) on a Microsoft server was also an option. GSNW enabled Microsoft clients to use their native protocols, NetBEUI and the Server Message Block (SMB), to communicate with a NetWare server. They did not need to install the Novell Client.

Beginning with NetWare 5, Novell provided direct support for IP addressing and the TCP/IP protocol suite. You should recall from earlier chapters the primary protocol for communication on the Internet is TCP/IP. Microsoft networking systems, with the introduction of Active Directory and its emphasis on Internet connectivity, also moved to using TCC/IP as its standard protocol. The evolution of protocols for most every network operating system has come to accept TCP/IP as its primary protocol.

Network+ Note:

When a question is presented about network protocols for NetWare, look closely to see if the question references NetWare 5 or a later version. Remember, NetWare 5 and later can be installed with TCP/IP. Prior to NetWare 5, the default protocol was SPX/IPX.

redirector
a software package
that determines
where a client
request for a resource
should be routed.

With the TCP/IP protocol as the standard protocol for most every network operating system, the only concern for connecting to a NetWare server is redirector support. A redirector is not unique to Novell NetWare. A *redirector* is a software package that determines where a client request for a resource should be routed. For example, if a client requests to access a local resource, such as a file on local drive C, then the redirector directs the request to the local computer's operating system. If a client requests to access a resource on a networked computer, such as a shared printer, the redirector sends the request to the networked computer with the shared printer.

Network+ Note:

Remember that the redirector operates at the presentation layer of the OSI model. You may be asked such a question on the Network+ Certification exam.

NetWare 4 and NetWare 5 only provide the Novell Client as a means of accessing a NetWare server. NetWare 6 provides many ways to access a NetWare server—through the Novell Client, the Native File Access Pack, and the Internet.

Novell Client

The Novell Client is NetWare's redirector. It provides full support for all of eDirectory's features and must be installed on a workstation that will run ConsoleOne. Microsoft's CSNW cannot be used because it does not provide support for the IP protocol. NetWare provides two versions of the Novell Client—one for Windows 95/98 and the other for Windows NT/2000/XP. The Novell Client can be downloaded for free from Novell's Support Web site.

Native File Access Pack (NFAP)

The Native File Access Pack allows Microsoft, UNIX/Linux, and Macintosh clients to access a NetWare server using their own native file access protocols. A *file access protocol* provides the means for a client to access networked files and print services. **Figure 8-21** lists the native file access protocol of each operating system.

file access protocol provides the means for a client to access networked files and print services.

Microsoft uses the Common Internet File System (CIFS) for file and print services. It is part of the Microsoft client. Macintosh uses the AppleTalk File Protocol (AFP). It installs with the AppleShare client. UNIX and Linux computers use the Network File System (NFS).

When NetWare 6 is installed, the CIFS and AFP file access protocols are loaded and configured by default. Loading the CIFS protocol on a NetWare server enables Windows clients to see and access NetWare servers through Network Neighborhood. To do this, however, the Microsoft client software must be installed so that the CIFS protocol is available. Loading the AFP on a NetWare server allows Macintosh clients to use the Chooser and Go menu to access files on a NetWare server. Macintosh computers are by default network ready. That is, the AppleShare client is installed by default. No further client installation is needed.

To make NFS services available on a NetWare server, the NFS Server software (xnfs.nlm) is loaded by default. Loading the NFS Server NLM provides support for NFS and allows UNIX/Linux clients to access files on a NetWare server. First, however, the NFS requires a file system that is to be shared, be exported. Exporting a file system makes it available to NFS clients on the network. NFS clients can then mount the shared file system. Mounting a file system makes the file system appear as if it was a part of the client's file system. Exporting and mounting files systems is covered in Chapter 9—UNIX/Linux Network Operating Systems.

Web-Based Access

Web-based access is made possible by using open standard protocols such as HTTP, FTP, and IPP. Web-based access provides a way for users to access files and printers without the need for the Novell Client. This allows for mobility and flexibility. The four features that make this type of access possible are iFolder, iPrint, NetStorage, and NetDrive.

Client	File Access Protocol
Windows	Common Internet File System (CIFS)
UNIX/Linux	Network File System (NFS)
Macintosh	AppleTalk File Protocol (AFP)

Figure 8-21.
Native file access protocols for the major operating systems.

iFolder

The iFolder feature allows users to access their local files from anywhere in the world. This is made possible through synchronization with an iFolder server. When a computer with the iFolder client connects to the network or the Internet, files in the local computer's iFolder directory and those in the server's iFolder directory are synchronized. This means that an updated copy of the local file is always stored and available on an iFolder server. If the user moves to another computer and connects to the network or to the Internet, iFolders on both the server and the local computer are again synchronized so that the updated files are copied to the local computer. A user can access an iFolder server without an iFolder client; however, the user must manually copy updated files to and from the iFolder on the server.

The components that make up iFolder are the iFolder server software, iFolder JAVA applet, and the iFolder client. The iFolder server software creates a Web site and makes the user's iFolders available on the Internet. The JAVA applet allows access to iFolder files. The iFolder client handles the file synchronization.

iPrint

The iPrint feature allows a user to send a print job to a remote location through the Internet. iPrint relies on the Internet Printing Protocol (IPP). Print jobs are sent in an encrypted form. Printing can be secured by requiring users to log on to eDirectory. Unsecured connections can be established for users who do not have an eDirectory account.

To print to an iPrint printer, a user needs an iPrint client and a JAVA-enabled Web browser installed on their computer. The iPrint client is installed automatically when a user selects a printer from an available iPrint Web page. At that time, the appropriate print driver is installed on the user's computer and the printer is made available through the Windows **Printers** folder.

NetStorage

NetStorage serves as an interface between the Internet and the files on a NetWare server. Internet protocols, such as HTTP, HTTPS, WebDAV, and XML, are used to access a NetStorage server. The NetStorage server accesses the requested files by using the NetWare Core Protocol (NCP), NetWare's native file access protocol. Unlike iFolder, NetStorage is used to access files on a server volume, not on the server's Web site. Therefore, NCP must be used.

NetDrive

NetDrive allows users to map network drives without the need for the Novell Client. Mapped drives appear as they normally do in Microsoft Explorer. To access files on the mapped drive, NetDrive can be coupled with NetStorage. NetDrive can also work with iFolder.

Summary

- NWAdmin is a windows-based administration tool introduced with NetWare 4 to replace the Syscon utility.
- NetWare 5 introduced a GUI for the NetWare console called the servertop and a JAVA-based administration program called ConsoleOne.

- A more robust version of the servertop and ConsoleOne was introduced with NetWare 6.

- ConsoleOne is similar in function to NWAdmin. In NetWare 6, ConsoleOne has come to replace NWAdmin.

- NetWare 6 uses the LDAP protocol to communicate directory information to other operating systems.

- NetWare 6 no longer requires the Novell Client to be installed on a workstation to support communications with a NetWare server.

- Server.exe is the core component of the NetWare operating system.

- Server.exe is enhanced by loading individual programs referred to as NetWare Loadable Modules (NLMs).

- The list of Novell NLMs can be viewed by entering **modules** at the console prompt.

- Server.exe boots the NetWare operating system, loads itself into memory, loads all in-bound modules, executes C:\nwserver\startup.ncf, mounts the SYS: volume, and executes SYS:system\autoexec.ncf.

- The term *console* refers to the keyboard and monitor at the NetWare server.

- NetWare's two file systems are called NetWare File System (NWFS) and Novell Storage Services (NSS).

- Before NDS, Novell used the bindery as a container for all security information for a single server.

- eDirectory is a database of network resources and allows for a single log on.

- The main administration programs of NetWare 6 are ConsoleOne, RConsoleJ, Remote Server Manager, and iMonitor.

- NetWare 5 first presented the option to select either IPX/SPX or TCP/IP as the default network protocol.

- A redirector is a program that redirects a client's request to the appropriate server and volume on that server.

- NetWare 6 provides three ways for a client to access a NetWare server—through the Novell Client, the Native File Access Pack (NFAP), and the Internet.

- The Native File Access Pack (NFAP) allows a client to use its native file access protocol to access files on a NetWare server.

- Web-based access uses open standard protocols such as HTTP, FTP, and IPP to communicate with a NetWare server.

Review Questions

Answer the following questions on a separate sheet of paper. Please do not write in this book.

1. Which version of NetWare incorporated some features from the UNIX/Linux open standard software programs?

2. What is the core component of the NetWare operating system?

3. What is an NLM?

4. What command is entered at the console prompt to view loaded NLMs?

5. Describe the NetWare operating system boot process.

6. How do you properly shut down a NetWare server from the console prompt?

7. How can you display information about a command?

8. What command is used to display information about a NetWare server's configuration?

9. What is the main purpose of ConsoleOne?

10. With which programming language is ConsoleOne written?

11. What is the name of NetWare's traditional file system?

12. Which NetWare file system manages volumes in storage pools?

13. Which NetWare file system has a quicker mount time and provides for better recovery after a crash?

14. Which volume does NetWare create by default during an installation?

15. What is the eDirectory?

16. Why does NDS resemble Microsoft's active directory?

17. What does OU represent in a typeful name?

18. What does CN represent in a typeful name?

19. What does the O represent in a typeful name?

20. What does the term *context* mean in relation to NetWare?

21. What are two ways the context can be displayed?

22. What is another name for a typeless name?

23. What four pieces of information are needed to log on to eDirectory?

24. What types of objects can exist in an eDirectory?

25. What are some examples of leaf objects?

26. Name four common NetWare administration utilities and give a brief description of each.

27. Which version of NetWare first allowed the choice of using the IPX/SPX protocol or TCP/IP?

28. What three ways can a client use to access and communicate with a NetWare 6 server?

29. Briefly describe how NFAP works.

Sample Network+ Exam Questions

Answer the following questions on a separate sheet of paper. Please do not write in this book.

Network+

1. What utility is especially designed to provide access to a Novell NetWare 5 server from a Windows 98 computer?
   - A.   Winsock
   - B.   Novell Client
   - C.   FTP
   - D.   Novell Access

2. What is the typical protocol used by the Novell NetWare 6.5 networking system to communicate across a LAN or WAN?
   - A.   TCP/IP
   - B.   NetBEUI
   - C.   IPX/SPX
   - D.   POP3

3. What text-based program is used at the console to provide system information and modify a NetWare server?
   - A.   ConsoleOne
   - B.   NWAdmin
   - C.   Monitor
   - D.   Console

4. What type of partition is required to support the boot operation and installation process of a NetWare 5 server?
   - A.   DOS
   - B.   NetWare
   - C.   NTFS4.0
   - D.   NTFS5.0

5. Which of the following commands starts a NetWare file server that was downed?
   - A.   **install**
   - B.   **setup**
   - C.   **server**
   - D.   **startup**

6. Which network operating system uses an autoexec.bat file?
   - A.   Windows Server 2003
   - B.   Microsoft NT 4.0
   - C.   Windows 2000 Server
   - D.   NetWare 5

Network+

7. What are the two main file systems used by NetWare servers?
    A.   FAT16 and FAT32
    B.   NTFS4.0 and FAT32
    C.   NFS and NSS
    D.   NFFS5.0 and FAT32

8. What does the acronym GSNW represent?
    A.   Global Services Net Watch
    B.   Gateway Services for NetWare
    C.   Global Search for NetWare
    D.   Gateway Search for NetWare

9. What is the name of a software program used to enhance the function of server.exe?
    A.   API
    B.   NLM
    C.   Daemon
    D.   FTP

10. What is the name of the Novell NetWare redirector?
    A.   Novell Client
    B.   JAVA1
    C.   iManager
    D.   Syscon

Suggested Laboratory Activities

1. Visit the Novell Web site www.novell.com to see all of Novell's services and its technical support.

2. Acquire a beta version of NetWare by accessing the Novell Web site and downloading the beta version to a PC. Follow Novell's instructions that are presented on the download site for burning a copy of the software to a CD. Practice a NetWare installation.

3. Use the **help** command to access the command list on a NetWare server. Practice using the commands listed.

4. Add a user to a NetWare network.

5. Install the Novell Client on a Windows PC and then access a NetWare server.

Interesting Web Sites for More Information

www.novell.com

Chapter 8
Laboratory Activity

Installing the Novell Client

After completing this laboratory activity, you will be able to:

- Describe the proper sequence for installing the Novell Client.
- Identify the four protocol options available in the Novel Client.
- Identify the common reasons a user might have a problem logging on to a NetWare network.
- Explain the cause of common logon error messages.
- Explain the purpose of the **ipxroute config** command.

Introduction

In this laboratory activity, you will install the Novell Client for Windows NT/2000/XP. The Novell Client enables successful communications with a Novell server and makes available a full selection of Novell services. Once the Novell Client is properly installed and configured and the appropriate user rights have been designated, the Windows workstation will be able to access the full range of NetWare services, such as file and print sharing.

Note:
Novell has two main clients for Microsoft operating systems: Novell Client for Windows 95/98 and Novell Client for Windows NT/2000/XP. You must use the appropriate client for a given workstation.

When performing a custom installation, the Novell Client presents four protocol options. These include the following:

- IP only.
- IP with IPX compatibility mode.
- IP and IPX.
- IPX only.

Note:
The selected option depends on the protocol or protocols installed on the NetWare network. Starting with NetWare 5.x, the IP protocol became the default choice of network protocol. Prior to NetWare 5, the default choice was IPX/SPX. To communicate with an IPX/SPX NetWare server, you must use the IPX/SPX protocol.

The first protocol option is **IP only**. You would select this option for a network that contains only NetWare 5.x or later and is configured for the TCP/IP protocol. The second option is **IP with IPX compatibility mode** for NetWare 5.x or later. This selection allows IPX applications to run by converting the IPX packet into compatible IP packets. The third option is **IP and IPX**. This option allows the workstation to operate with both an IP network and an IPX network. This would be used, for example, for a network that contained a mixture of Microsoft servers running TCP/IP and NetWare servers running IPX. The fourth option is **IPX only** and is used for IPX only networks. If you choose to do a typical installation, instead of a custom installation, the Novell Client will automatically select the best protocol option based on the protocols already installed on the system.

When first using the **Novell Login** dialog box, you will need to identify the tree, context, and server associated with the NetWare network. The term *tree* is the name given to the location at the top of the hierarchal structure of eDirectory or Novell Directory Services (NDS) that contains organizational objects. The term *context* means the exact location of an object in eDirectory or NDS. When logging on to a NetWare network, a user must indicate where their associated user object is located in the directory tree. The server is the name of the NetWare server the user wishes to log on to. Entering a server name is not required. Look at the following screen capture of the **Advanced** screen in the **Novell Login** dialog box. Note the three buttons labeled **Trees**, **Contexts**, and **Servers**. If you do not know the name of the tree, context, or server, you can click the **Trees**, **Contexts**, **Servers** buttons to see what is available in the network.

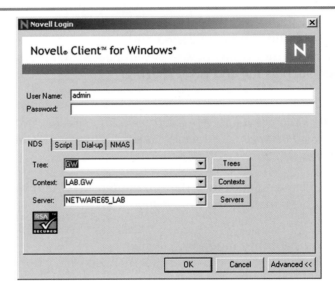

As a network administrator, you may be called on to solve logon problems. Some of the most common problems a user might experience when connecting to a NetWare network include the following:

- The user may not have a user account established. Remember, a user must have a user name and an assigned password to log on to the NetWare network.

- A physical connection may not exist. Check the network cable to be sure it is attached. Try to ping the NetWare server.

- The user has accidentally pressed the [Caps Lock] key, causing the password to be entered in the wrong case. Have the user check the [Caps Lock] key.

- The NetWare server may be down. Check with the network administrator or check the server room to verify.

There are some things you should be aware of when working with the Novell Client. These include the following:

- When upgrading an operating system to Windows XP, you should also upgrade the Novell Client.

- Even with the Novell Client installed, to access NetWare files, printers, and directory services, the user must have a user account on the NetWare server and appropriate user rights to access resources.

- To view the frame type and network number of a Windows XP workstation running IPX/SPX, use the **ipxroute config** command from the command prompt. (Note: The IPX protocol must be bound to the network adapter for this command to work. This does not happen automatically when using the Novell Client.)

- To reveal the frame type and network address configured on a NetWare server running IPX/SPX, use the RConsoleJ utility from a workstation or go to the NetWare console and run the **config** command.

Note:
Before you begin installing the Novell Client, you must create a Restore Point. Creating a Restore Point will enable you to return the Windows XP computer to the condition it was in before you installed the Novell Client.

Equipment and Materials

- Workstation with Windows XP.

- A network connection to a NetWare 6.5 server. (The lab activity can be modified for a Novell 5.x server.)

- User account set up on the NetWare server. (The lab activity can still be performed without a user account, but you will not be able to log on to the server.)

- An Internet connection. (Required to download the client program.)

Record the following information on a separate sheet of paper:

NetWare user name: _____

NetWare password: _____

Tree: _____

Context: _____

Server: _____

Procedures

1. _____ Report to your assigned workstation.

2. _____ Boot the workstation and verify it is in working order.

3. _____ Create a Restore Point with the System Restore utility (**Start | All Programs | Accessories | System Tools | System Restore**) before going further into the lab activity.

4. _____ If your instructor has not provided you with the Novell Client program on a CD, create a directory to contain the download version of the client software. Label the directory Novell Client Download. Connect to the Internet and download the latest Novell Client for Windows NT/2000/XP. Look in the following URL: www.novell.com/products/clients/ or conduct a Google search using the following terms: *Novell NetWare 6.5 Windows XP Client products*. Download the product after reading the download and installation procedures located at the Web site. The Novell Client will download as a self-extracting file. Extract the file in the directory you created.

5. _____ Start the Novell Client for Windows NT/2000/XP installation.

6. _____ Accept the license agreement by clicking **Yes**.

7. _____ The next screen presents you with two installation options: **Typical** and **Custom**. Select the **Typical** option and proceed to the next step. You will see a progress window detailing the transfer of files required for the installation. Also, the status of the installation process will be displayed on the screen.

8. _____ When the installation is complete, reboot the workstation so that the changes can take effect. If you have installed the Novell Client by CD, remove the CD before rebooting.

9. _____ When the workstation reboots and the Novell Client **Begin Login** dialog box appears, press the [Ctrl] [Alt] [Del] key combination. This will invoke the **Novell Login** dialog box.

10. _____ Take note of the **Workstation only** checkbox located in the lower-left corner of the **Novell Login** dialog box. When selected, the user will log on to the workstation, not the NetWare server. When unselected, the user will log on to the NetWare server. Leave the checkbox unselected at this time.

11. _____ Enter your user name, password, tree, context, and the name of the server to which you want to log on. Click **OK**.

12. _____   To verify a successful log on to the NetWare network, check if a **Novell** folder with Novell programs is listed in the **Start** menu. Your Novell program entry should look similar to the following:

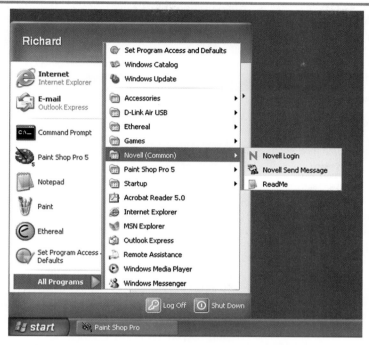

Another way to verify the client installation was successful is to look for a red *N* in the notification area (lower-right corner of the desktop) as shown in the following screen capture.

13. _____   Open the **Local Area Connection Properties** dialog box for your network adapter. The Novell Client for Windows entry should be listed.

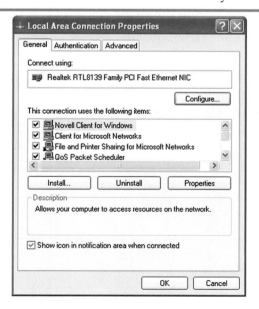

14. _____ Now, open the **My Network Places** to view the connection to the NetWare network. To access **My Network Places,** right-click **My Computer** (located off of the **Start** menu or on the desktop) and select **Explore**. In the **Folders** pane, click **My Network Places**. After opening **My Network Places**, you should see images similar to those in the right-hand pane of the following screen capture.

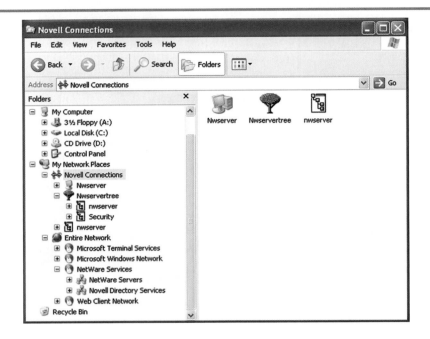

By double-clicking the **Nwserver** icon, you can reveal the server's directory structure and available files, if you have the proper rights. If you have failed to establish a connection to the NetWare network, call your instructor.

15. _____ After you have established a connection to the NetWare network, you may explore the NetWare server's directory structure. Do not make any changes to the structure. Simply look, and then proceed to the next step.

16. _____ Reboot the workstation.

17. _____ When the workstation reboots, attempt to log on under a user name that does not have an account on the server and observe the message that is generated. You can try any combination of letters that you know is not a user name, such as *ABCD*. Do the same for the password. On a separate sheet of paper, record your answers to the following questions:

What error message was displayed?

_____

_____

What is the name or title of the dialog box in which the message is displayed?

_____

_____

18. _____ Have the instructor disconnect the network cable from the server or shut down the NetWare server and then attempt to log on again to see the message generated. On a separate sheet of paper, record your answers to the following questions:

What error message was displayed?

_____

_____

How long did it take for the message to appear as compared to when an incorrect user name and password was entered?

_____

_____

19. _____ Have the instructor reconnect or start the NetWare server.

20. _____ Log on once more to verify the connection has been established.

21. _____ Now, uninstall the Novell Client by using the **Add/Remove Program** utility located in **Control Panel**. After uninstalling the Novell Client, open **My Network Places** to see if you can access the NetWare network.

22. _____ Answer the review questions.

Review Questions

Answer the following questions on a separate sheet of paper. Please do not write in this book.

1. What should you do before installing the Novell Client?

2. What command is issued from the Windows XP command prompt to view the NetWare frame type and network number of a workstation running the IPX/SPX protocol?

3. What are the four protocol options for the Novell Client?

4. What is the name of the Novell Client to be used on a Windows 98 workstation?

5. What is the name of the Novell Client to be used on a Windows XP workstation?

6. Name two ways you can verify that the Novell Client has been installed.

7. Explain how to verify that a connection has been established with a NetWare server.

8. What are some of the reasons a user might have trouble connecting to a NetWare network?

9. What error message was generated when you entered the wrong user name and password?

10. What error message was generated when the server was disconnected from the network?

9

UNIX/Linux Operating Systems

After studying this chapter, you will be able to:

❏ Describe the major features in the UNIX/Linux operating system.

❏ Define *daemon* and explain its purpose.

❏ Explain the purpose of the Grub and LILO bootloaders.

❏ Identify the common shell programs associated with Linux.

❏ Describe the file systems associated with Linux.

❏ Compare and contrast the file system structure of UNIX with other common file system structures.

❏ Define the file and directory permissions used with the Linux file system.

❏ Explain how UNIX/Linux can establish communications with a Microsoft operating system.

Network+ Exam—Key Points

Typical Network+ Certification exam questions address issues such as TCP/IP settings for UNIX/Linux systems. TCP/IP is covered in later chapters and includes information about UNIX/Linux systems at that time. Other questions concern basic Linux system issues such as logging on, logging off, and Linux file systems. You should also be aware of the role SAMBA plays in system interoperability.

You should use a Linux operating system for approximately eight hours of guided instruction, or forty or more hours of unguided instruction. Unguided instruction means using the available help files and tutorials to become familiar with Linux. A guided instruction means that the program is instructor-led. Guided instruction can save you a lot of time when becoming familiar with Linux.

While it is possible to pass the Network+ Certification exam without ever using a Linux operating system, it is highly advisable you spend some time using any of the numerous Linux systems available. It is suggested that Red Hat or SuSE versions of Linux be used because of their extensive documentation and support.

Key Words and Terms

The following words and terms will become important pieces of your networking vocabulary. Be sure you can define them.

bootloader	mount point
Common UNIX Printing System (CUPS)	Samba
daemon	session
desktop environment	shell
device file	tar ball
inode	windows manager
journal file	X client
kernel	X server
module	X Windows

UNIX is the fountainhead of modern network operating systems. After its inception, it served as a model of how network operating systems should operate. There have been many variations of UNIX. None of the variations produced a large-scale change in the networking industry until Linux was released. Linux has become a robust network and desktop operating system. Many software companies, such as IBM, Novell, and Apple Computer, have absorbed the open-source technology of Linux into their networking operating systems.

While Microsoft Windows has dominated the desktop computer world in the United States, Linux is a preferred desktop for most C and C++ programmers and computer science majors. Linux allows programmers complete control over the operating system and the kernel. In fact, a programmer can write a new kernel if they have sufficient skills. This is why Linux is popular with serious programmers.

This chapter presents a basic foundation of UNIX and Linux operating systems. The UNIX and Linux information in this chapter is combined because talking about one operating system is like talking about the other.

UNIX

UNIX was developed at the AT&T Bell Laboratories in the 1960s. It was the first major operating system for networking. Scientific, academic communities, and governmental agencies readily accepted UNIX. The original version of UNIX has spawned countless variations, such as Berkeley Software Distribution (BSD), Sun Solaris, HP, and IBM.

Many networking concepts originated from the early versions of UNIX. While UNIX was still in its early development, the TCP/IP protocol was developed to support communications between universities, government sites, and scientific research companies. UNIX adopted TCP/IP as its native communications protocol.

What makes UNIX and Linux unique when compared to NetWare and Windows network operating systems is the source code used for the components is readily available to anyone. Novell and Microsoft have guarded their source code by keeping it proprietary. The availability of the source code has made the UNIX/Linux system a favorite among programmers and programming enthusiasts. If a program needs to be modified or to have a new procedure developed, the administrator can simply write the code for the change. The most common programming language used for writing UNIX/Linux code is the C programming language. If you cannot program, you can usually find the code you need through an Internet search. There are many organizations devoted to developing and sharing program code.

Beginning with NetWare 6, Novell incorporated Linux open-source programs into the NetWare operating system.

UNIX/Linux consists of the system kernel surrounded by modules, **Figure 9-1.** A *kernel* is the core of an operating system. A *module* is a small program, such as hardware driver or a kernel enhancement. Modules are similar to Novell Netware NLMs. They are used to expand the kernel and to allow for flexibility. An administrator can load and unload modules as needed or configure the kernel to load them automatically during the system boot.

The UNIX/Linux system also consists of daemons. A *daemon* is a program that runs in the background and waits for a client to request its services. It is much like a service in the Windows and NetWare operating systems. An example of a daemon is the Simple Mail Transfer Protocol (SMTP). SMTP waits for e-mail to arrive. When e-mail does arrive, SMTP either transfers the e-mail to another server or delivers the e-mail to the recipient's mailbox. All of this is handled in the background without the user's knowledge—until a message pops up on the screen notifying the user that they have mail. **Figure 9-2** shows a list of currently running programs and daemons on a Linux computer. Note that the file names of all daemons end in the letter *d*.

kernel
the core of an operating system.

module
a small program, such as a hardware driver or kernel enhancement.

daemon
a program that runs in the background and waits for a client to request its services.

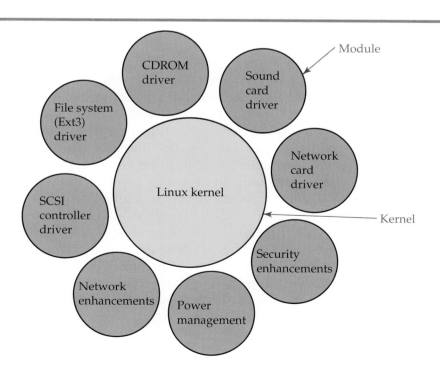

Figure 9-1.
The UNIX/Linux kernel is surrounded by modules. A module is typically a hardware driver, file system driver, or other kernel enhancement.

Figure 9-2.
In this example, a utility called **top** has been run to reveal currently running programs and daemons. Files on a UNIX/Linux system that are daemons, typically end with the letter *d*. Some of the daemons listed in this example have been pointed out. Can you identify other daemons in this example?

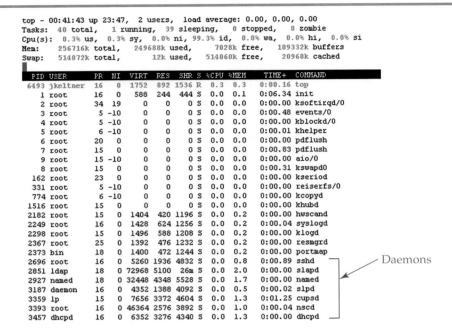

```
top - 00:41:43 up 23:47,  2 users,  load average: 0.00, 0.00, 0.00
Tasks:  40 total,   1 running,  39 sleeping,   0 stopped,   0 zombie
Cpu(s):  0.3% us,   0.3% sy,   0.0% ni, 99.3% id,   0.0% wa,   0.0% hi,   0.0% si
Mem:    256716k total,   249688k used,     7028k free,   109332k buffers
Swap:   514072k total,       12k used,   514060k free,    20968k cached

  PID USER       PR  NI  VIRT  RES  SHR S %CPU %MEM    TIME+  COMMAND
 6493 jkeltner   16   0  1752  892 1536 R  0.3  0.3  0:00.16 top
    1 root       16   0   588  244  444 S  0.0  0.1  0:06.34 init
    2 root       34  19     0    0    0 S  0.0  0.0  0:00.00 ksoftirqd/0
    3 root        5 -10     0    0    0 S  0.0  0.0  0:00.48 events/0
    4 root        5 -10     0    0    0 S  0.0  0.0  0:00.00 kblockd/0
    5 root        6 -10     0    0    0 S  0.0  0.0  0:00.01 khelper
    6 root       20   0     0    0    0 S  0.0  0.0  0:00.00 pdflush
    7 root       15   0     0    0    0 S  0.0  0.0  0:00.83 pdflush
    9 root       15 -10     0    0    0 S  0.0  0.0  0:00.00 aio/0
    8 root       15   0     0    0    0 S  0.0  0.0  0:00.31 kswapd0
  162 root       23   0     0    0    0 S  0.0  0.0  0:00.00 kseriod
  331 root        5 -10     0    0    0 S  0.0  0.0  0:00.00 reiserfs/0
  774 root        6 -10     0    0    0 S  0.0  0.0  0:00.00 kcopyd
 1516 root       15   0     0    0    0 S  0.0  0.0  0:00.00 khubd
 2182 root       15   0  1404  420 1196 S  0.0  0.2  0:00.00 hwscand
 2249 root       16   0  1428  624 1256 S  0.0  0.2  0:00.04 syslogd
 2298 root       15   0  1496  588 1208 S  0.0  0.2  0:00.00 klogd
 2367 root       25   0  1392  476 1232 S  0.0  0.2  0:00.00 resmgrd
 2373 bin        18   0  1400  472 1244 S  0.0  0.2  0:00.00 portmap
 2696 root       16   0  5260 1936 4832 S  0.0  0.8  0:00.89 sshd
 2851 ldap       18   0 72968 5100  26m S  0.0  2.0  0:00.00 slapd
 2927 named      18   0 32448 4348 5528 S  0.0  1.7  0:00.00 named
 3187 daemon     16   0  4352 1388 4092 S  0.0  0.5  0:00.02 slpd
 3359 lp         15   0  7656 3372 4604 S  0.0  1.3  0:01.25 cupsd
 3393 root       16   0 46364 2576 3892 S  0.0  1.0  0:00.04 nscd
 3457 dhcpd      16   0  6352 3276 4340 S  0.0  1.3  0:00.00 dhcpd
```

Daemons

The kernel combined with modules and daemons defines the total operating system and its capabilities, **Figure 9-3.** An administrator can customize the operating system and load only the modules and services they desire. By building a custom operating system, overall performance is improved. The operating system is more efficient than when loading an operating system with all of its possible components. The administrator can also unload modules and services no longer needed. The ability to load or unload separate modules to support the kernel results in an efficient use of system resources, such as RAM. As each module is loaded, valuable RAM is used. There are also more demands on the CPU. Processor time is required for each running module.

Linux

In 1991, the University of Helsinki student, Linus Torvalds, developed the Linux (pronounced Lih-nucks) kernel and freely distributed the source code rather than retaining proprietary rights. What makes this operating system unique is that the code is a collaboration through the years of countless numbers of developers, programmers, students, and engineers. Linux is distributed for free or for a relatively small cost. It is licensed under the GNU (pronounced guh-noo) General Public License (GPL), which allows users to freely modify, copy, and distribute the Linux software system. We will discuss more about this subject later.

While Linux has been presented as an open source operating system, this can be misleading. The kernel and many other Linux components are free to download, but many versions are a mix of open source and proprietary programs and components. For example, Corel Linux includes a proprietary file management system, proprietary applications, and proprietary utilities. It also contains several third-party applications such as Netscape Communicator and Adobe Acrobat Reader.

Since Linux was designed to emulate UNIX, most of its features are the same as UNIX. You can think of Linux as another version in the evolution of

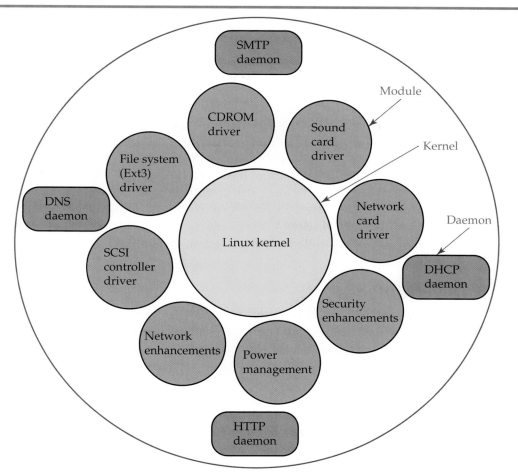

UNIX/Linux Operating System

Figure 9-3.
The kernel, modules, and daemons make up the UNIX/Linux operating system.

UNIX. There are many variations of Linux available such as Red Hat, Caldera, Mandrake, SuSE, Turbo, and Corel. You will see many similarities between the different versions of Linux. These similarities are due to all of the Linux operating systems using the most common, available programs to enhance the kernel, such as Gimp, KDE, GNOME, Mozilla, Xfree86, Apache Web Server, Evolution, and Samba.

Linux has evolved rapidly and has been incorporated into other operating systems. Linux was once difficult to install, mainly because of the lack of hardware driver support by manufacturers. Since Novell and IBM have endorsed Linux, hardware manufacturers have increased their driver support.

There are also various Linux certifications available. They range from vendor-specific, such as Red Hat certification, to vender-neutral, such as Linux certification offered by CompTIA and IBM. IBM uses a vender-neutral approach because they endorse Red Hat, SuSE Linux, and TurboLinux versions.

Linux will continue to grow because it has incorporated all of the common protocols associated with other operating systems. Protocols such as SMB, LDAP, and SNMP make the operating system more compatible with Microsoft and Novell operating systems. The expansion of operating system interoperability has helped to develop the popularity of Linux.

Linux Advantages

The features that have made the versions of Linux popular are its flexibility and price. Linux is flexible because it can be easily modified or customized. Since the program code is open source, an experienced programmer can customize the operating system and security to meet their needs. Generally, Linux is priced reasonably when compared to other network operating systems. In fact, an entire system can be downloaded for free. Prices for network operating systems that allow only a limited number of users to access the server typically start at several thousands of dollars. As the number of users increases, so does the cost of the network operating system.

Linux Disadvantages

Commercial versions of Linux are tested on various systems and have somewhat of a guarantee they will not harm your computer. Downloaded versions of Linux and associated files are not guaranteed. If the software destroys, causes loss of data, or crashes the system, the user has no recourse, no one to blame, or cost recovery source.

Many Linux companies do not have a highly structured support system. Commercial versions of Linux, however, do provide support, either for free or for a small cost. Commercial versions Linux also have the broadest range of hardware compatibility when compared to the free download versions. Remember the old saying, "You get what you pay for." If the operating system is free, the support is typically less than desirable.

Also, since Linux is an open-source operating system, the code is not secret. It can be hacked or cracked more easily than some other operating systems.

Another concern is independent-vendor support. There are not as many venders that provide driver support for UNIX/Linux in the same way as they do for Microsoft and Novell products. Also, not all versions of UNIX/Linux services or programs are interchangeable and compatible. You must be very careful when adding a free service or utility. It may not be compatible with the version of Linux you are running. Other operating systems may have problems too, but not to the extent as the Linux operating system. It is best to stick with a well-established Linux operating system such as Red Hat or SuSE.

Copyright and Copyleft

One of the most interesting and often misunderstood aspects of Linux is the copyright. Linux is copyrighted under the GNU General Public License (GPL), often referred to as copyleft. See **Figure 9-4.** Many people are under the impression that you cannot sell Linux software. This is not true. The intent of the GNU GPL is to make the Linux kernel source code freely available. This allows anyone to modify the code to suit his or her needs. In short, the GNU GPL allows anyone the right to copy, modify, and distribute software without the permission of the original copyright holder, as long as they pass the same rights on to the next user. In other words, all users who receive a copy of the source code may freely modify the programs and freely distribute the programs.

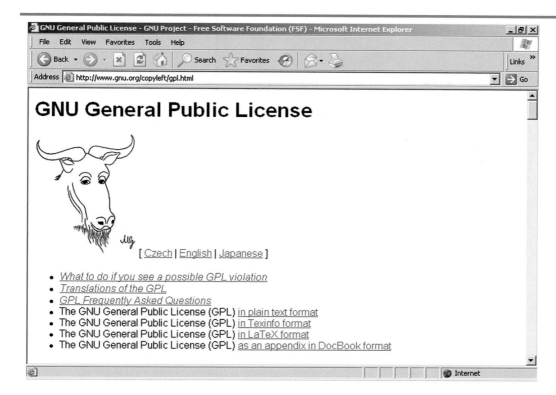

Figure 9-4.
The GNU General
Public License (GPL)
can be viewed in
various forms and
languages on the
www.gnu.org/
copyleft/gpl.htm
Web page.

Therefore, this may lead to the question of how do companies have the right to sell Linux? Commercial companies extend all the rights associated with the GNU GPL. The commercial companies distribute the source code and allow copies to be made and distributed under the same guidelines as stated above. What the companies are charging for is a neat package with manuals and support via the Internet and telephone for the consumer. The consumer actually pays for the enhancements while not being charged for the software. In fact, you can go to the commercial sites and download an image of the software to use on a computer. The fact that you can download and install the software from the commercial site and access the source code is what keeps the commercial sites in compliance with the GNU general public license. For more information on the GNU GPL visit www.gnu.org/copyleft/gpl.html.

The Free Software Foundation (FSF), a nonprofit organization founded in 1983, develops many different software packages that complement Linux systems. The FSF promotes the free distribution of software and source code similar to the GNU GPL. See **Figure 9-5.** The FSF develops utilities and programs that interface with GNU/Linux operating systems. For more information on the FSF visit www.gnu.org/fsf/fsf.html.

You should note that not all versions of Linux are under the free distribution plan. Many commercial versions are independent of the GNU GPL license and are very expensive.

UNIX/Linux Basics

This section introduces the basics of the UNIX/Linux operating system. It is not comprehensive in its coverage. It would take a full textbook to cover the basics in detail. The intention of this section is to introduce you to some UNIX/Linux terminology and to the basic differences and likenesses between UNIX/Linux and other operating systems.

Figure 9-5.
The www.gnu.org/
fsf/fsf.html Web page
provides information
about the Free Software
Foundation (FSF) and
its purpose.

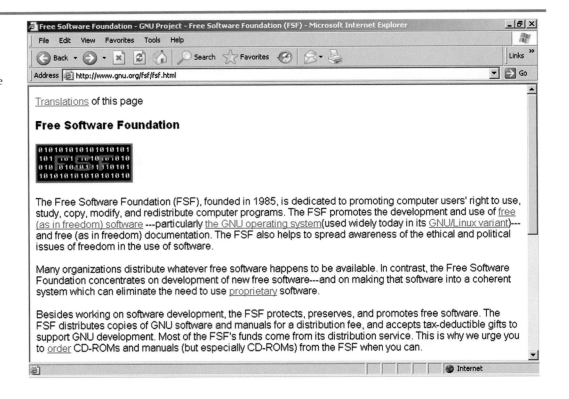

LILO and GRUB

The acronym LILO represents Linux Loader. GRUB represents Grand Unified Bootloader. LILO has been around for many years, but GRUB is relatively new.

LILO and GRUB are bootloaders. A **bootloader** is a program that starts the operating system load process. Linux bootloaders are used in much the same way as ntldr is used in Windows NT–based operating systems. LILO and GRUB install in the master boot record (MBR) of a Linux partition. After the Power On Self-Test (POST), the system looks for and transfers control to the bootloader. The bootloader loads the kernel. When the kernel loads, the kernel loads the operating system.

LILO and GRUB can be used to load other operating systems. You will most commonly use LILO in a multiboot situation. When LILO or GRUB is configured on a computer with multiple operating systems, the LILO or GRUB bootloader takes control after the system POST and before the operating system is selected. It allows the user to choose between two or more operating systems and then boots directly into the selected operating system.

bootloader
a program that starts
the operating system
load process.

Shells

A *shell* is a user interface that interprets and carries out commands of the user similar to the way the DOS command interpreter (command.com) interprets and carries out commands. The shell directly interfaces with the operating system kernel. There are numerous shells available: some are freeware and others are proprietary. Some of the more common shells are the Bourne Again Shell, Korn Shell, and Turbo C Shell. See **Figure 9-6** for a list of common UNIX/Linux shells.

The Bourne Again Shell is the default shell for Red Hat Linux. **Figure 9-7** shows the **set** command issued at the Bourne Again Shell. The UNIX/Linux

shell
a user interface that
interprets and carries
out commands of
the user similar to
the way the DOS
command interpreter
(command.com)
interprets and carries
out commands.

Shell	Full Name
bash	Bourne Again Shell
bs	Bourne Shell
csh	C Shell
ksh	Korn Shell
tcsh	Turbo C Shell

Figure 9-6.
Common UNIX/Linux shells.

set command is similar to the DOS **set** command. Entered without options, the **set** command displays a list of a user's shell variables, including the name of the current shell. In the example, you can see that the user's shell is set to bash, which is the Bourne Again Shell.

A user can change to a different shell by typing the desired shell name at the prompt. For example, to activate and use the Turbo C Shell, you would simply type **tcsh** at the prompt. In **Figure 9-8,** the user has issued the **tcsh** command and then entered **set** at the shell prompt. Since each shell has its own set of variables, the SHELL variable now reads /bin/tcsh. A user can also run another session of the current shell by entering the shell name. For example, to run another session of the bash shell, the user would enter bash at the command prompt.

Note that each time a user enters the name of a new shell or the name of the shell they are currently using, a new session is invoked. One of the greatest advantages of the UNIX/Linux operating system is the ability to run multiple sessions on a single computer. A *session* is a logical connection with the Linux computer. Each session provides its own set of variables as shown in Figure 9-7 and Figure 9-8. A session allows for remote users to log on to a Linux computer and use the computer at the same time as the user sitting at the computer. It also allows a user to run a lengthy batch process in one session and perform other tasks in another session concurrently.

session
a logical connection with the Linux computer.

```
jkeltner@linux:/> set
ACLOCAL_PATH=/opt/gnome/share/aclocal
BASH=/bin/bash
BASH_VERSINFO=([0]="2" [1]="05b" [2]="0" [3]="1" [4]="release" [5]="i586-suse-linux")
BASH_VERSION='2.05b.0(1)-release'
COLORTERM=1
COLUMNS=142
CPU=i686
CVS_RSH=ssh
DEFAULT_WM=kde
DIRSTACK=()
EUID=1000
GNOME2_PATH=/usr/local:/opt/gnome:/usr
GNOMEDIR=/opt/gnome
GNOME_PATH=:/opt/gnome:/usr
GROUPS=()
GTK_PATH=/usr/local/lib/gtk-2.0:/opt/gnome/lib/gtk-2.0:/usr/lib/gtk-2.0
G_BROKEN_FILENAMES=1
HISTCONTROL=ignoreboth
HISTFILE=/home/jkeltner/.bash_history
HISTFILESIZE=1000
HISTSIZE=1000
HOME=/home/jkeltner
HOST=linux
SHELL=/bin/bash
SHELLOPTS=braceexpand:emacs:hashall:histexpand:history:interactive-comments:monitor
SHLVL=1
SSH_CLIENT='::ffff:192.168.10.40 3032 22'
SSH_CONNECTION='::ffff:192.168.10.40 3032 ::ffff:192.168.10.20 22'
SSH_TTY=/dev/pts/0
TERM=xterm
TEXINPUTS=:/home/jkeltner/.TeX:/usr/share/doc/.TeX:/usr/doc/.TeX
UID=1000
USER=jkeltner
```

SHELL variable

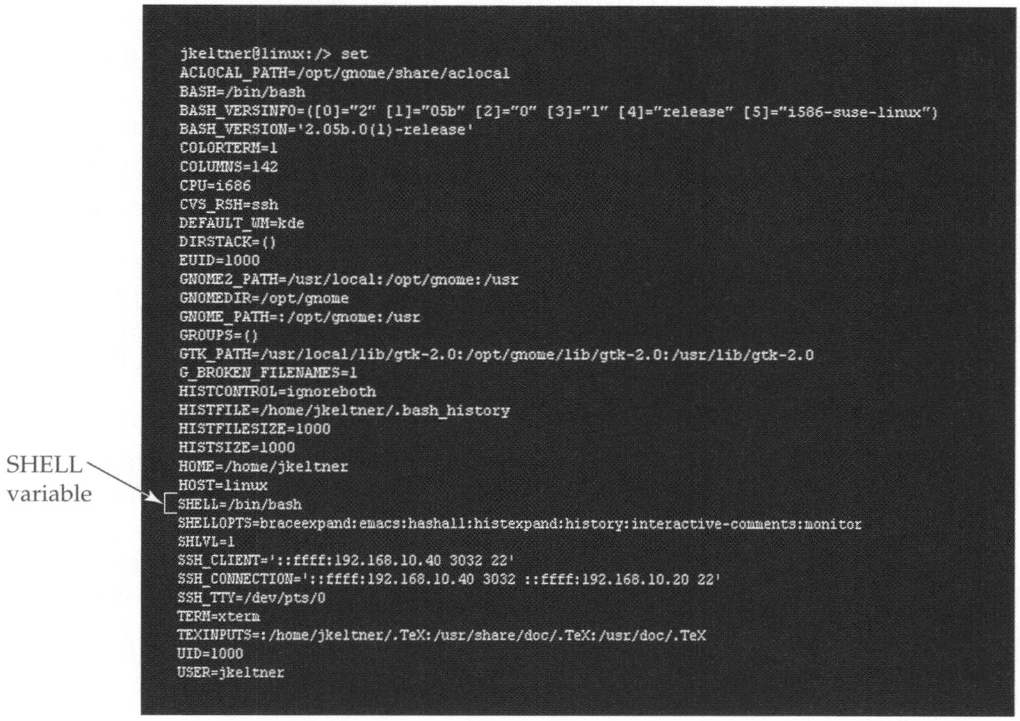

Figure 9-7.
The **set** command lists the current shell's variables. Note that the value of the SHELL variable, /bin/bash. This means that the shell is currently set to use the bash shell.

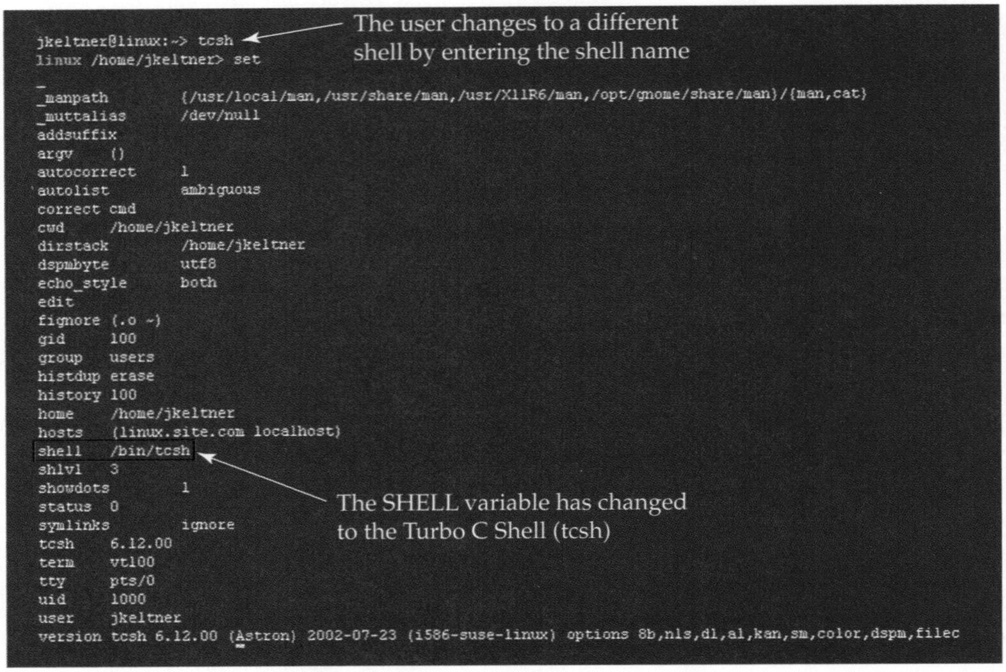

Figure 9-8.
The user has entered
tsch at the shell prompt
to change to the Turbo
C Shell. The SHELL
variable, as indicated
by the **set** command,
now reads /bin/tsch.

```
jkeltner@linux:~> tcsh          The user changes to a different
linux /home/jkeltner> set       shell by entering the shell name

_
_manpath        {/usr/local/man,/usr/share/man,/usr/X11R6/man,/opt/gnome/share/man}/{man,cat}
_muttalias      /dev/null
addsuffix
argv    {}
autocorrect     1
autolist        ambiguous
correct cmd
cwd     /home/jkeltner
dirstack        /home/jkeltner
dspmbyte        utf8
echo_style      both
edit
fignore {.o ~}
gid     100
group   users
histdup erase
history 100
home    /home/jkeltner
hosts   {linux.site.com localhost}
shell   /bin/tcsh              The SHELL variable has changed
shlvl   3                      to the Turbo C Shell (tcsh)
showdots        1
status  0
symlinks        ignore
tcsh    6.12.00
term    vt100
tty     pts/0
uid     1000
user    jkeltner
version tcsh 6.12.00 {Astron} 2002-07-23 (i586-suse-linux) options 8b,nls,dl,al,kan,sm,color,dspm,filec
```

The user can end a session by entering the **exit** command. This will return a user to a previous session. If there is not a previous session, the **exit** command will log the user off the system.

Commands

Linux commands are similar among the various versions of Linux, but they do not match perfectly. This lack of standardization can be frustrating. You can usually type the command followed by **--help** (two dashes before the word *help*) at the shell prompt to obtain help about a particular command. You can also access the manual pages for a particular command by typing **man** followed by the command. However, even these commands are not universal.

The command syntax is similar to that of DOS commands. The command may stand alone or be followed by options, an argument, or both. The following shows the complete syntax of a Linux command:

command *<options> <argument>*

An *option* alters the output of a command. Some options must be preceded with a dash (-) or with two dashes (--). Some options do not need a dash at all. In any case, a space must exist between the command and the option. Options, like commands and file names, are case-sensitive.

An *argument* is what the command is to act on. It can be a file name, a directory name, wild card characters, such as the asterisk (*) or question mark (?), or other piece of information designated as an argument for a particular command.

Figure 9-9 shows an example of using a command without options and an argument. The **ls** command is used to list the contents of the current directory. Now look at **Figure 9-10.** In this example, entering **ls** with the **l** option displays the directory listing in a long form. The long form lists details about each file and directory, such as permissions and file size. The **ls** command can also be used with an argument. The argument can be a path name to list the contents of a directory

```
jkeltner@linux:/> ls
bin  boot  dev  etc  home  lib  media  mnt  opt  proc
root  sbin  srv  sys  tmp  usr  var  windows
jkeltner@linux:/> █
```

Figure 9-9.
Example of issuing the **ls** command.

other than the working directory. The argument can also be a file name or part of a file name with wild card characters. Look at **Figure 9-11.** In this example, the **ls** command is used with an argument that contains a wildcard character. This argument tells the **ls** command to list only files and directories that begin with the letter *S*. The **-l** option tells the **ls** command to list the results in long form. See **Figure 9-12** for a list of other common Linux commands and their function.

File Systems

Several choices of Linux file systems are available: Ext2, Ext3, ReiserFS, and JFS. Ext2 is the original Linux file system. It supports long file names (255 characters) and stores information in blocks much like the DOS and Windows file systems. Information about each file is stored in an inode. An *inode* is a table entry that contains information such as permissions, file size, name of the owner, the time stamps of the file's creation, modification, and last access, and a pointer to where the file is stored.

The Ext3 file system is basically the Ext2 file system with the journaling feature added. In fact, each of the Linux file systems, except for Ext2, is a journaling file system. You should recall that Novell's NSS is also a journaling file system. A journaling file system ensures file integrity whenever an unexpected system shutdown occurs, such as a power outage. The journaling file system maintains a log of all file activity. This log is called the *journal file.* Three time stamps that contain information about the creation of the file, the last date the file was modified, and the last date the file was accessed are added to the journal file. The journaling file system uses the journal file to restore a file to its most usable state. This can be accomplished even if the file is corrupted or incomplete. A journaling file system also allows for faster boot time and consumes less memory than other file systems. For more information on journaling file systems, visit www.linux-mag.com/2002-10/jfs_01.html.

inode
a table entry that contains information such as permissions, file size, name of the owner, the time stamps of the file's creation, modification, and last access, and a pointer to where the file is stored.

journal file
a log of all file activity on a journaling file system.

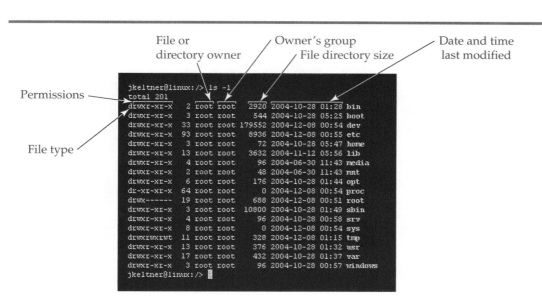

Permissions

File type

File or directory owner

Owner's group
File directory size

Date and time last modified

```
jkeltner@linux:/> ls -l
total 201
drwxr-xr-x   2 root root    2920 2004-10-28 01:28 bin
drwxr-xr-x   3 root root     544 2004-10-28 05:25 boot
drwxr-xr-x  33 root root  179552 2004-12-08 00:54 dev
drwxr-xr-x  93 root root    8936 2004-12-08 00:55 etc
drwxr-xr-x   3 root root      72 2004-10-28 05:47 home
drwxr-xr-x  13 root root    3632 2004-11-12 05:56 lib
drwxr-xr-x   4 root root      96 2004-06-30 11:43 media
drwxr-xr-x   2 root root      48 2004-06-30 11:43 mnt
drwxr-xr-x   6 root root     176 2004-10-28 01:44 opt
dr-xr-xr-x  64 root root       0 2004-12-08 00:54 proc
drwx------  19 root root     688 2004-12-08 00:51 root
drwxr-xr-x   3 root root   10800 2004-10-28 01:49 sbin
drwxr-xr-x   4 root root      96 2004-10-28 00:58 srv
drwxr-xr-x   8 root root       0 2004-12-08 00:54 sys
drwxrwxrwt  11 root root     328 2004-12-08 01:15 tmp
drwxr-xr-x  13 root root     376 2004-10-28 01:32 usr
drwxr-xr-x  17 root root     432 2004-10-28 01:37 var
drwxr-xr-x   3 root root      96 2004-10-28 00:57 windows
jkeltner@linux:/> █
```

Figure 9-10.
Issuing the **ls -l** command displays a long directory listing. Items listed include permissions, file or directory owner, file size, and date and time last modified.

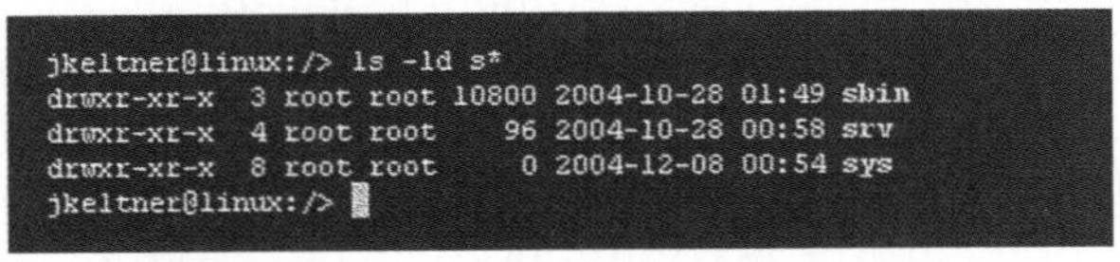

Figure 9-11.
The **ls** command has been listed here with an argument. The argument in this example includes a wildcard character, the asterisk (*). Together with the letter *S*, it tells the **ls** command to list only those files and directories that begin with the letter *S*. The **-l** option tells the **ls** command to output the result of the argument in a long listing format.

```
jkeltner@linux:/> ls -ld s*
drwxr-xr-x  3 root root 10800 2004-10-28 01:49 sbin
drwxr-xr-x  4 root root    96 2004-10-28 00:58 srv
drwxr-xr-x  8 root root     0 2004-12-08 00:54 sys
jkeltner@linux:/>
```

The ReiserFS file system is SuSE's default file system. It allows for better storage capacity by efficiently writing files, especially small files, to the hard disk drive. It does this by storing files back-to-back rather than by storing them in blocks. With DOS and Windows file systems, a hard drive is divided into blocks, and only one file can be stored in a block. If the block size is 4k and the file is 1k, then 3k of hard drive space is wasted. The ReiserFS file system treats the hard disk drive as a database. Information about the file, such as permissions, are stored next to the file rather than in an index. This helps to speed file access.

The Journaled File System (JFS) was originally developed by IBM for the OS/2 Warp operating system. JFS is similar to Ext3 in that information is stored in blocks and metadata is stored in inodes. However, JFS is a 64-bit file system, which allows for a larger file system and larger file sizes.

Figure 9-12.
Common UNIX/Linux commands.

Linux command	Description
cd	Changes the directory to the user's home directory.
cd ..	Changes to the parent directory of the current working directory.
cd <directory>	Changes the directory to the specified directory.
cp <filename> <directory>	Copies a file to a specified directory.
ls	Lists the contents of the current working directory.
ls <directory>	Lists the contents of the specified directory.
ls l ls l <directory>	Lists the contents of a directory, including file and subdirectory details.
mkdir <new directory>	Makes (creates) a directory.
mv <filename> <directory>	Moves a file to a new directory.
mv <filename> <new filename>	Renames a file.
pwd	Displays the path of the current working directory.
rm <filename>	Removes (deletes) a file.
rmdir <directory>	Removes (deletes) a directory.

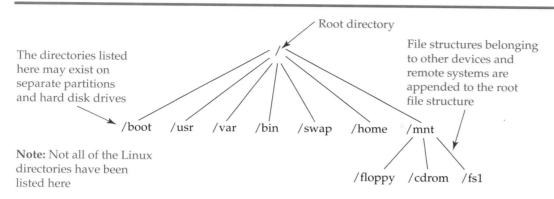

Figure 9-13.
The root directory is the top of the directory structure. File structures belonging to other devices, such as hard disk drives, CD-ROMs, and remote systems, are appended to the root directory when they are mounted.

File Structure

The UNIX/Linux file structure looks similar to the DOS file structure. The most noticeable difference, however, between the UNIX/Linux file structure and the DOS file structure is the lack of drive letters. This means that your system does not have a storage limit based on the number of drive letters.

The root (/) directory is the top of the directory structure, **Figure 9-13.** All files and directories are under the root directory. The root directory can be subdivided into additional directories and files like it can with the DOS file structure. Unlike DOS, UNIX/Linux has the ability to append file structures from devices such as floppy disk drives, hard disk drives, CD-ROMs, and ZIP drives and to append the file structures from remote computers. This ability makes the file structures appear to be part of the root file structure. To append a file structure to the root, the file structure or device that holds the file structure must be mounted.

Do not confuse the root of the directory (/) with the /root directory. The /root directory belongs to the root user, also known as the superuser. It is their home directory.

Tech Tip

Mounting a device makes it appear as part of the directory structure. **Figure 9-14** shows the contents of the /etc/fstab file. This file indicates the drives and partitions that are to be mounted during the system boot. The first entry represents the device file associated with the device. In UNIX/Linux, each device on the local system has an associated device file in the /dev directory. The *device file* serves as a means of communication between the driver and the device. Devices are referred to by the device file name. For example, a primary drive designated as the master drive is called hda. The device file for this drive is

device file
a file that serves as a means of communication between a driver and a device.

Device file   Mount point   File system   Options

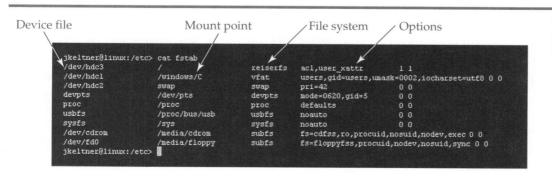

Figure 9-14.
Contents of the /etc/fstab file.

Figure 9-15.
Storage device naming
scheme.

Hard Drives	Type	Description
hda	IDE	Master or single hard drive on primary controller
hdb	IDE	Slave hard drive on primary controller
hdc	IDE	Master or single hard drive on secondary controller
hdd	IDE	Slave hard drive on secondary controller
sda	SCSI	First SCSI drive in chain
sdb	SCSI	Second SCSI drive in chain, and so on

/dev/hda. **Figure 9-15** shows the storage device naming scheme and related device files. Partitions are indicated by numbers, starting with the number one for the first partition. For example, the first partition on /dev/hda would be indicated as hda1.

Look again at Figure 9-14. The second entry in the /etc/fstab file is the mount point. The **mount point** is the location in the directory structure where a device is mounted or inserted. For example, a mount point of /mnt/cdrom mounts the file structure of the /dev/hdc device in the /mnt/cdrom directory.

mount point
the location in the
directory structure
where a device is
mounted or inserted.

The third entry indicates the type of file system on the device. Note that various file systems can exist on the devices and still be a part of the complete file structure.

The fourth entry lists the options that should be used by the mount command to configure the device. For example, the **ro** option indicates that the device is to be mounted as read only. The **users** option indicates that all users should be allowed to mount and unmount the device. Finally, the fifth and sixth entries set administrative parameters. These will not be discussed because the details about these parameters are beyond the scope of this chapter.

Typical PC devices, such as the keyboard, mouse, and most common file systems are mounted during the system boot process. Before you can save a file to a device such as hard disk drive, CD, or floppy disk, the device must be mounted. Before removing a CD disc or floppy disk, the device should be unmounted. This will ensure that the file can be read by other systems. If the device is not unmounted, the file may be unreadable.

Tech Tip   Some Linux operating systems perform the unmount function automatically.

Common Files and Directories

There are many directories commonly associated with UNIX/Linux systems. This is not to say that all UNIX/Linux file system structures are identical. They do vary among operating systems and versions. **Figure 9-16** lists the most universal directories associated with UNIX/Linux systems.

While it is not necessary for the common user to navigate the directory structure, it is required for the system administrator who commonly modifies software packages designed for Linux. Typically, a common user has access to only part of the directory structure and can only modify their directories and files.

UNIX/Linux file and directory names are case-sensitive. This means that two files can be spelled the same but be recognized by the operating system as two separate files. For example, Myfile and myfile are two separate files. Also, be aware that UNIX/Linux files typically do not use file extensions. For example, when looking for an executable file, you will not see the familiar exe

Directory	Description
/dev	Contains system device files.
/bin	Contains system commands.
/etc	Contains system configuration files.
/tmp	Used as a temporary storage area for files created by utilities and executables. It is emptied when the system is rebooted.
/var	Contains variable data associated with administrative, logging, and temporary files.
/home	Contains user directories and files. The startup files are located here by user.

Figure 9-16.
Common UNIX/Linux directories.

extension appended to the file name. An executable file is identifiable by the file's permissions accessed through one of the utilities that display the file name, directory, and permissions or by using the **ls -l** command.

File and Directory Security

Like other network operating systems, the UNIX/Linux operating system has the ability to assign permissions to files and directories. Look at the long directory listing in **Figure 9-17.** Note the first entry for each file and directory. This entry lists the file or directory type followed by permissions. There are ten spaces available for this information. The file or directory type occupies the first space. Some of the characters that may fill this space are *d*, *l*, or a dash (-) The following list shows the meaning of each:

$$d = directory$$

$$l = link$$

$$- = normal file$$

The file and directory permissions occupy the other nine spaces. The permissions represent the file and directory permissions assigned to user, group, and other. The user is typically the owner of the file or directory. The group is typically the user's default group. Other is anyone else who is not the file's owner or a member of the group.

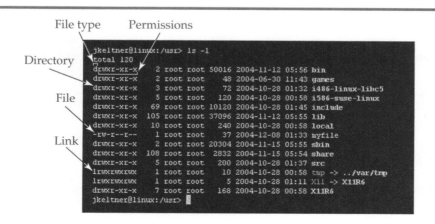

Figure 9-17.
Note the permissions for the files and directories in this long directory listing.

Figure 9-18.
File and directory permissions and their meaning.

File Permissions	Description
read	Open a file and view its contents.
write	Change a file's contents.
execute	Run a file. This attribute is assigned to executable files.

Directory Permissions	Description
read	Display the contents of a directory.
write	Create, modify, or delete files in the directory that you own.
execute	Change to that directory.

Note the letters *r, w,* and *x* as well as the dash (-) symbol. A dash within any of the file and directory permissions spaces indicates that no permission for that entry has been set. The permissions associated with the letters are as follows:

r = read

w= write

x = execute

These permissions vary in meaning depending on whether they are set for a file or for a directory. See **Figure 9-18.**

Each of the nine permission spaces is a placeholder for the read, write, and execute permissions for user, group, and other. If any of these values are not set, then a dash displays in its place. To better understand how to interpret permissions, look at **Figure 9-19.** Note that user, group, and other is assigned three placeholders. Each of these placeholders can be filled with the read, write, or execute permission. In the example, the user and group have permission to read and write to the file. All other users are only able to read the file.

The administrator or user with equal administrative powers can set file and directory permissions. They can even change the owner of the file or of the default group. The owner of a file or a directory can also change permissions or the owner and group.

Network Authentication

Procedures for logging on and off UNIX/Linux systems are similar to other network operating systems. Logging on to the system can take place through a command-line prompt or a GUI-based prompt, depending on how Linux is configured. In either case, the user must supply a user name followed by a password. The user name and password is compared to that in the /etc/passwd file. The /etc/passwd file stores user account information. After the user is authenticated, they may use the services for which they are authorized.

Figure 9-19.
In this example, the user and default group are granted the read and write permissions to a file. All other users can only read the file name.

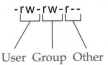

User Group Other

The highest level of administration is the superuser and has the default user account name of *root*. The initial password for *root* is set during the installation of the operating system. The system administrator uses the root account to make changes in the system configuration. Changes in the system configuration or file structure can render the entire system inaccessible. When the administrator is using the system for normal tasks, the administrator logs on under a common user name and password similar to any typical user. Mistakes made by a common user can be fixed, and typically, the common user cannot access critical areas of the system.

File and Print Sharing

File sharing is made possible through the Network File System (NFS) protocol. NFS was originally developed by Sun Microsystems and later released to the public as an open standard. NFS allows a UNIX/Linux system to share its files with users from remote locations. The system that shares its files is called the NFS server. It runs the NFS service. The NFS service consists of several daemons. These daemons handle services such as authentication and client mount requests. NFS also relies on support from the kernel.

The system that accesses files on the NFS server is called the client. It relies on support from the kernel and from several NFS client daemons.

An administrator makes an NFS server's files available to remote users by listing in the /etc/exports file the directories to be shared, associated permissions, and the systems that are allowed to access each directory. The administrator then runs the **exportfs** command to export the directories. To view the exported directories, a user at the client computer must mount the directory. Mounting the directory makes the directory appear as if it were part of the client's file structure.

The line printer daemon (lpd) handles remote and local printing services. It listens for incoming print jobs and sends the jobs to the printer. Security is set through the /etc/host.equiv or /etc/host.lpd file. Remote access can be denied through the /ect/printcap file. Denying remote access allows only local users to send print jobs. The **lpr** command is used to send print jobs to the line printer daemon (lpd). For example, to print a file named chapter9 located in the /home/rroberts directory, the user would enter lpr /home/rroberts/chapter9 at the command prompt.

A relatively new printer daemon called *Common UNIX Printing System (CUPS)* has been designed to support network printing using the Internet Printing Protocol (IPP). CUPS is used in many Linux operating systems as well as MAC OS X. Since CUPS is based on IPP, it is designed to locate and configure printers located on the LAN and WAN through the use of a browser. Linux systems typically provide the user with a choice of using the traditional lpd printing system or CUPS.

Common UNIX Printing System (CUPS)
a printing system that supports network printing using the Internet Printing Protocol (IPP).

Remote Access Features

UNIX/Linux is designed to share resources but not in the same way that other networking systems share resources. For example, when you create and set up a share on a Windows network, the share is run on the client's computer. In UNIX, when a share is set up on the UNIX computer and is then accessed by the client, the program runs on the UNIX computer. This means that the client only needs minimal resources, but the resources of the UNIX computer must be increased as the number of clients increases.

Any operating system can access a UNIX/Linux system by using remote access protocols and services. There are several remote access protocols and services that you have probably heard of. They are FTP, TFTP, Telnet, and Apache. These various protocols and services provide remote access to clients anywhere in the world. Linux also provides the SFTP service as a way to secure the remote connection. The SFTP protocol differs from FTP and TFTP in that it encrypts the user password and data.

FTP

The File Transfer Protocol (FTP) is designed to transfer files between a server and a client. Using FTP, you cannot run a program or manipulate a file on the server. FTP transfers files rather than runs or manipulates them. In other words, using FTP, you can only download or upload files. If you need to edit a file on the server while using FTP, you can download the file, modify it, and then upload it to the server.

FTP can require a user name and password, or it can permit anonymous access. Most often, however, anonymous access requires some form of authentication, such as the client's e-mail address.

TFTP

Like FTP, the Trivial File Transfer Protocol (TFTP) is designed to transfer or copy files to and from a remote system. The main difference between FTP and TFTP is the protocol it uses to transfer the information. FTP uses the Transmission Control Protocol (TCP) to transfer data, while TFTP uses the User Datagram Protocol (UDP) to transfer data. TFTP does not require a user name or password and does not guarantee data delivery. It is a connectionless type of protocol.

Telnet

Telnet is designed to remotely access any computer running as a Telnet server. The Telnet service allows a user to log on using a user name and password. After the connection is made, the client communicates with the Telnet server using text-based commands. The Telnet session allows a user to manipulate a file on the remote computer. However, the amount of manipulation is limited to the set of Telnet commands. Telnet is often referred to as a terminal emulation program because it emulates a terminal connected to a mainframe. Telnet is commonly used to remotely control WAN servers and routers. See **Figure 9-20** for a comparison of FTP, TFTP, and Telnet.

Apache

Apache is Web server software developed in the public domain to run on UNIX. Many Web servers use the Apache Web server software. It is popular because not only is it a robust and stable system, it is free. The Apache Web server software was designed as a free distribution by a group of approximately 20 developers. Apache derives its name from the fact that it consists of a series of software patches for the original NCSA Web server software. Hence, the name Apache, which is a play on the words *"a patchy."* Apache is the default Web server package in most commercial Linux operating systems.

X Windows
a system that
provides a GUI for
UNIX and Linux
systems.

X Windows System

The *X Windows* system provides a GUI for UNIX and Linux systems. Other operating systems integrate the GUI into the operating system. The operating system cannot be separated from the GUI to function as single unit. The X Windows

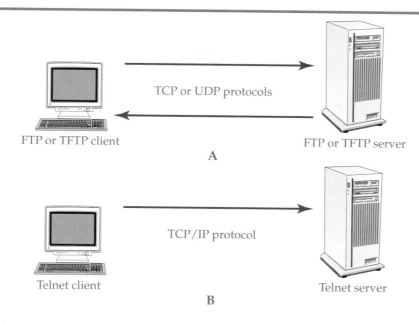

Figure 9-20.
FTP, TFTP, and Telnet comparison. A—FTP and TFTP allow a client to transfer files to and from an FTP or TFTP server. FTP uses the TCP protocol and TFTP uses the UDP protocol. B—Telnet allows a client to manipulate files while they reside on the server.

system is separate from the operating system. A user can operate a UNIX/Linux computer without ever installing the GUI.

The X Windows system consists of the X server and the X client. *X server* is a program that communicates with the computer hardware, such as the keyboard, mouse, and monitor. The *X client* is an application, such as a word processing program. The X client can also be a windows manager. A *windows manager* controls the display. In other words, it provides the GUI such as the icons, boxes, and buttons. **Figure 9-21** shows the relationship between the hardware, X server, and X client applications and windows manager.

X server
a program that communicates with the computer hardware, such as the keyboard, mouse, and monitor.

X client
an application or a windows manager.

X Windows is also referred to as X, X11, or XFree86.

Tech Tip

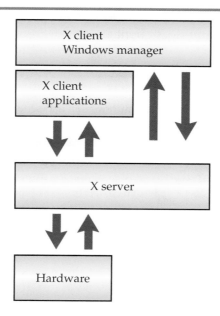

Figure 9-21.
Relationship between hardware, X server, and X client applications and windows manager.

Figure 9-22.
The X server and X client are typically installed on the same computer.

X server

X client

windows manager an X client that controls the display. In other words, it provides the GUI such as the icons, boxes, and buttons.

The X Windows system has as a client/server architecture. The X server and the X client can run on the same computer, **Figure 9-22,** or on separate computers, **Figure 9-23.** However, in contrast to a typical client/server architecture in which the client receives resources from the server, the X Windows system works in the opposite manner. The X client provides the processing power and display and the X server communicates with the hardware. The X server component serves as an interface between the system hardware and X clients. It allows the windows manager and X clients to operate independently of the hardware.

When the X client and the X server are installed on separate computers, the computer that runs the X clients is typically the more powerful of the two computers.

When X server is installed, it is configured for the system hardware. Typical configuration settings include mouse, keyboard, and video card type and monitor settings, such as the refresh rate and resolution.

An incorrect configuration of the monitor identification or frequency can permanently damage your monitor.

desktop environment a software package that includes many applications (X clients) and a windows manager. The desktop environment provides a common look among X clients.

The X Windows system is an open source program for which many X clients or applications have been written. The various applications differ in appearance. To provide a common look among the various X clients, a desktop environment is often installed. A *desktop environment* is a software package that includes many applications (X clients) and a windows manager. The desktop environment provides a common look among X clients.

Figure 9-23.
The client/server architecture of the X Windows system allows the X server and X client to be installed on different computers on the network. Unlike the typical client/server design, the X client acts as the server in that it provides the processing power. The X client also sends display information to the X server.

The X server sends display information to the hardware

X server

An X client processes information and sends the display to the X server

X server sends user input received from the hardware to the X client

X client

The hardware displays the information on the monitor

The hardware also sends user input to the X server

The term *X client* can refer to an application, windows manager, or the computer running the X client software. The term *X server* can refer to the X server program or to the computer running the X server program.

Tech Tip

Two of the most commonly used desktop environments are K Desktop Environment (KDE) and GNU Network Object Model Environment (GNOME). **Figure 9-24** shows the KDE GUI. Notice that it is similar to the Windows GUI in that it contains a desktop area, icons, and a menu.

Desktop environments offer a large assortment of software packages included as part of the installation or available for download. Many popular software packages are designed for both KDE and GNOME desktop environments. Software packages that are not included in the installation can be downloaded from the Internet, typically in a file format called a tar ball.

A *tar ball* is a compressed file containing one or more software programs. Once downloaded, it must be decompressed. Additional configuration may be required before the programs are installed. You should always read the documentation that is included in the tar ball. You will typically find a readme file, which contains an overview of the package, and an install file, which includes instructions for installing the software.

tar ball
a compressed file containing one or more software programs.

The X Windows system is started by entering **startx** at the command prompt. Most Linux systems can be configured to load and run the X Windows system automatically. When loaded automatically, a logon dialog box prompts for a user name and password. Many of the tasks an administrator performs at the command line, such as adding users or configuring services, can be performed in the GUI. **Figure 9-25** shows the YaST Control Center, which is included in SuSe Linux Enterprise Server 9.

Interoperability

The Common Internet File System (CIFS) protocol and the Network File System (NFS) protocol can be used to support file access between Microsoft Windows and UNIX/Linux network systems. This section looks at the connectivity software that incorporate these protocols.

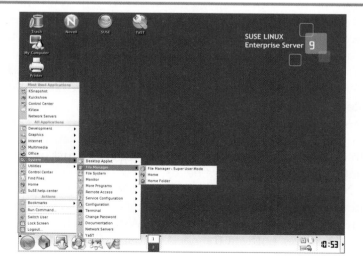

Figure 9-24.
The K Desktop Environment (KDE).

Figure 9-25.
The YaST Control Center provides a GUI alternative to typical administrative tasks. A—**Security and Users** display.
B—**Network Services** display.

A   B

Samba

Samba
a free software package that allows UNIX and Linux systems to share files and printers with Windows-based clients.

Samba is a free, software package included in most UNIX/Linux versions. It allows UNIX and Linux systems to share files and printers with Windows clients. Windows clients are able to view shared UNIX/Linux resources through **Network Neighborhood** or **My Network Places**, **Figure 9-26.**

The protocol that Samba uses to support communications between UNIX/Linux systems and Microsoft systems is the Common Internet File System (CIFS). Microsoft has implemented CIFS as a replacement for SMB, Microsoft's original, file-sharing protocol. CIFS is an improvement of SMB in that it is able to run over TCP/IP.

Samba is installed on a UNIX or Linux server. The Client for Microsoft Networks, however, is required for the Windows workstation.

Windows and NFS

Microsoft Windows can communicate seamlessly with Linux and UNIX systems through the Network File System (NFS) protocol. NFS allows files to be exchanged between Windows NT, 2000, XP, and any UNIX or Linux operating system. NFS works in conjunction with the TCP/IP protocol when transferring data between networks. When an NFS server shares (exports) its file system and a client mounts the shared file system, it is transparent to the user. The user is typically not aware that the files exist on a hard disk drive controlled by another operating system. Look at **Figure 9-27.**

As you can see, through the NFS protocol, Windows clients can access a UNIX server and UNIX clients can access a Windows server. When a Windows client needs to access a UNIX server, the Client for NFS must be installed on the Windows client. When a UNIX client needs to access files on a Windows server, the Server for NFS software must be installed on the Windows server.

In the gateway arrangement, the Gateway for NFS software on the Windows server performs the needed translation for the Windows clients. Windows clients can access a UNIX server without the need to install Client for NFS.

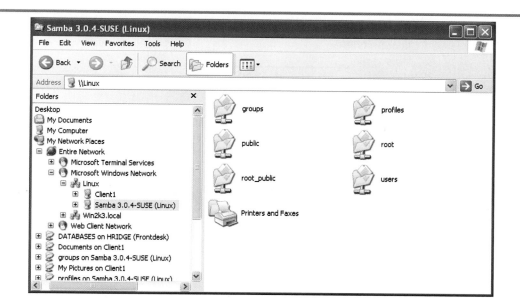

Figure 9-26.
When Samba is installed on a UNIX/Linux system, UNIX/Linux resources can be viewed through **Network Neighborhood** or **My Network Places** on a Windows workstation.

It is important to note that NFS was designed in the early days of UNIX. At that time UNIX was strictly a command-line system. As the operating system evolved, GUIs became popular. Many of the file sharing programs are based on the original NFS structure and then expanded by graphic capabilities to minimize the need to use text-based commands. Also, remember that companies often develop their own software to perform the same function as NFS. For example, Microsoft Services for UNIX is developed using the NFS structure.

You can also set up communications through other means between a Linux system and a Windows system, but it requires significant expertise and may have some communication limitations. For example, you may not be able to browse the Linux system from Windows using **Network Neighborhood** or **My Network Places**. You can, however, access files in the Windows system from Linux. A common set of protocols must be installed such as Simple Management Protocol (SMP), TCP/IP, and NFS. The amount of compatibility between the two operating systems depends on the version of Linux and of the associated programs.

Network+ Note:

When taking the Network+ Certification exam and you see a question referring to NFS, you can generally bet the right answer involves UNIX.

MAC OS X Server

MAC OS X Server is the latest in the series of Apple server network operating systems. The MAC OS X Server is included in this chapter because it is basically a Linux system designed from open source software. The MAC OS X Server operating system contains all the features found in a typical Linux system, such as SAMBA, Apache, X Windows, CUPS, and LDAP.

The MAC OS X Server can be configured as a primary domain controller to control user authentication. When configured as a primary domain controller, Windows workstations can log directly on to the network using the MAC OS X Server

Figure 9-27.
The NFS protocol allows for file exchange between Windows and UNIX/Linux computers. A—Using the Client for NFS software, a Windows client can access files on a UNIX/Linux server. B—The Server for NFS software, when installed on a Windows server, allows a UNIX/Linux client to access files on a Windows server. C—When the Gateway for NFS software is installed on a Windows server, Windows clients can access files on a UNIX/Linux server without the need to install the Client for NFS software.

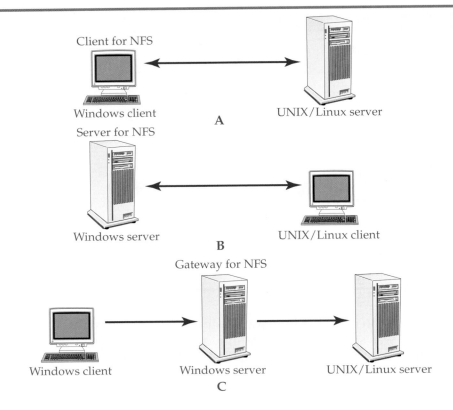

for authentication. You can connect workstations running MAC OS X, Microsoft Windows, or Linux. A MAC OS X Server can also be integrated into a network that is managed by Microsoft or Novell servers.

Summary

- Linux uses a General Public License (GPL), known as a copyleft, to allow Linux users to copy, modify, and distribute Linux software.
- UNIX/Linux consists of a system kernel surrounded by modules.
- A *daemon* is a program that runs in the background and waits for a client to request its services.
- The Linux kernel is further enhanced by many software packages, such as GNOME, KDE, Apache, and Samba, which have become the de facto standard for distribution with Linux.
- Linux is a variation of the UNIX operating system.
- The term *Linux* refers to the operating system kernel.
- LILO and GRUB are bootloader programs that start the Linux operating system load process and can be used to load other operating systems.
- A *shell* is a user interface that interprets and carries out commands.
- The most common Linux file systems are Ext2, Ext3, ReiserFS, and JFS.
- Ext3, ReiserFS, and JFS are journaling file systems.
- A journaling file system ensures file integrity by maintaining a log of all file activity whenever an unexpected system shutdown occurs.

■ The UNIX/Linux file structure looks similar to the DOS file structure; however, UNIX/Linux does not use drive letters.

■ The top of the UNIX/Linux directory structure is called the root directory and is indicated by the forward slash (/) symbol.

■ All file structures from storage devices, such as floppy drives and CD-ROM drives and shared resources from remote systems, are appended to the root through the mounting process.

■ Hardware such as the floppy drive and CD-ROM must be mounted before they can be used.

■ The read, write, and execute permission can be set for a file or directory's owner, default group, and other users.

■ The highest level of administration is the superuser and has the default user account name of *root*.

■ The Network File System (NFS) was developed by Sun Microsystems for exchanging files on a network.

■ NFS works in conjunction with TCP/IP to transfer files across a network system.

■ Local and remote printing services are handled by the line printer daemon (lpd).

■ UNIX/Linux file and directory names are case-sensitive.

■ Various protocols and services, such as FTP, TFTP, Telnet, and Apache provide remote access to clients anywhere in the world.

■ Apache is the default Web server software package used with UNIX/Linux.

■ The two main desktop environments used with Linux are Gnome and KDE.

■ Samba is a free software package that allows UNIX/Linux systems to share files and printers with Windows-based clients.

■ A gateway can be used to provide communications between multiple Windows workstations and a UNIX/Linux network without the need to set up each workstation with a compatible UNIX protocol.

■ The three products used for allowing NFS communications between a UNIX/Linux system and a Windows system are Client for NFS, Server for NFS, and Gateway for NFS.

■ MAC OS X Server is designed from open source software and contains all the features found in a typical Linux system, such as SAMBA, Apache, X Windows, CUPS, and LDAP.

Review Questions

Answer the following questions on a separate sheet of paper. Please do not write in this book.

1. What is the native network protocol used for UNIX?

2. Why are UNIX and Linux referred to as a programmer's operating system?

3. What are daemons?

4. What is the purpose of LILO and GRUB?

5. What is the function of a shell?

6. List four common shells used with UNIX/Linux.

7. What is Ext2?

8. What is Ext3?

9. What is ReiserFS?

10. What is JFS?

11. What is a journaling file system?

12. Using the DOS file system as a comparison, describe the UNIX/Linux file system.

13. All files and all directories on a UNIX/Linux system are located under the _____ directory.

14. What does mounting a device do?

15. What is a mount point?

16. What is the /bin directory used for?

17. What are the three Linux file and directory permissions and how are they indicated?

18. What is the difference between FTP and TFTP?

19. What is Telnet used for?

20. What is X Windows?

21. What are the two most common desktop environments associated with X Windows?

22. What is the purpose of the Samba program?

23. What does the acronym NFS represent?

24. What is the purpose of NFS?

25. An office running Windows XP must connect to the home office that uses a Linux system. What are some possible methods of connecting the Windows XP clients to the Linux server at the home office?

Sample Network+ Exam Questions

Network+

Answer the following questions on a separate sheet of paper. Please do not write in this book.

1. What program is most commonly used on UNIX/Linux systems to provide Web services?

   A. Samba

   B. TFTP

   C. Telnet

   D. Apache

2. In a Linux system, how is the second partition on the first hard disk drive identified?

 A. /dev/hda1

 B. /dev/hda2

 C. /hda/dev2

 D. /dev0/part2

3. What program is used to allow Windows workstations to access files on a Linux server?

 A. Samba

 B. Network File System

 C. TCP/IP

 D. Microsoft Client

4. Which file system is not typically associated with a Linux system?

 A. Ext2

 B. Ext3

 C. ReiserFS

 D. NTFS

5. What protocol is used for communication between a UNIX and a Microsoft domain?

 A. SMB

 B. NetBEUI

 C. RIP

 D. DHCP

6. Which of the following server operating systems is based on UNIX design and open source software?

 A. MAC OS X

 B. Microsoft Enterprise Server 2003

 C. Microsoft NT 4.0

 D. Novell Netware 5.0

7. What is the default name of the server administrator in UNIX?

 A. Admin

 B. Administrator

 C. Root

 D. Man

8. Which of the following protocols is used by Windows XP to support communication with a UNIX system?

 A. IPX/SPX

 B. CIFS

 C. NetBEUI

 D. AppleShare

Network+

9. Where is the C: partition located in a Linux directory structure?
   A. Directly under the root.
   B. Between the B: and D: partition.
   C. Each user has their own C: partition created when their user account is created.
   D. There is never a C: partition in the Linux directory structure.

10. What is required to run a graphical user interface on a Linux server?
    A. Samba
    B. Apache
    C. X Windows
    D. Inode

Suggested Laboratory Activities

1. Download and install a version of Linux. After the Linux system is installed, explore the basic file structure.

2. Visit the CompTIA Web site and look up the requirements for the Linux+ Certification.

3. Add a user to the Linux system.

4. Create a file and set up permissions such as read, write, and execute.

5. Create a group and set group permissions.

6. Create a user with limited permissions.

7. Set up a Linux system to communicate with a Windows XP or Windows 2000 system.

8. Set up a printer share on a Windows XP or Windows 2000 system to support a UNIX client.

Interesting Web Sites for More Information

www.debian.org

www.fsf.org

www.gnu.org

www.gnu.org/copyleft/gpl.html

www.gnu.org/fsf/fsf.html

www.isu.edu/departments/comcom/unix/workshop/unixindex.html

www.kde.org

www.Linux.org/index.html

www.redhat.com

www.samba.org

www.SuSE.com

Chapter 9
Laboratory Activity

Adding a New User in SuSE Linux

After completing this laboratory activity, you will be able to:
- Add a new user account using the YaST utility.
- Log on to the system using the newly created account.
- Describe how to edit a user account.
- Describe how to remove a user account.

Introduction

In this laboratory activity, you will create and edit a new user account on a SuSE Linux computer or SuSE Linux Open Enterprise Server. The utility you will use to create and edit the user account is called YaST Control Panel, or YaST for short. The Acronym YaST represents Yet Another Setup Tool. YaST is similar to Microsoft's Control Panel in that it is used to modify the system configuration. The following screen capture shows the YaST display.

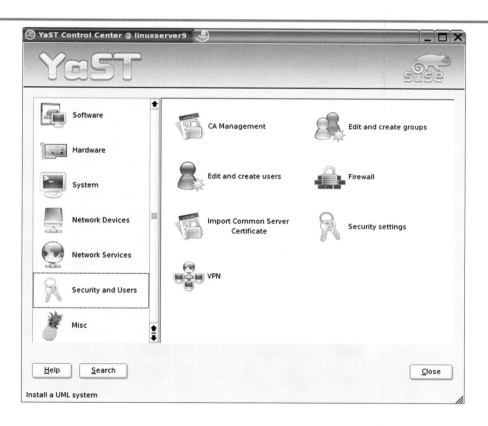

In the left frame, notice the major configuration categories, such as **Software**, **Hardware**, **System**, **Network Devices**, **Network Services**, **Security and Users**, and **Misc**. For this laboratory activity, you will select the **Security and Users** category and then select **Edit and create users** located in the right frame.

Only the root user has permission to create a new user account. If you are logged on to the system and are not the root user, you will be prompted for the root user password. This is only true for X Windows utilities such as YaST. When working at the command line, the command to create a new user will only be recognized if you are logged on as the root user.

Equipment and Materials

■ Workstation running SuSE Linux 8.0 or later or SuSE Linux Open Enterprise Server 8.0 or later.

■ The following information recorded on a separate sheet of paper:

Root user and LDAP password

Assigned user name: _____

Root password: _____

LDAP server password (if required): _____

New user account information

First name (Example: student): _____

Last name (Example: one): _____

Login name (Example: student1): _____

Password: _____

Procedures

1. _____ Report to your assigned workstation.

2. _____ Boot the workstation and verify it is in working order.

3. _____ Log on to the KDE desktop using your assigned user name or as root.

4. _____ At the KDE desktop, open the YaST utility through **N** (main menu icon) **I System | YaST** or by clicking the **YaST** icon located in the panel or on the desktop. If you logged on to the system using an account other than root, the **Run as root** dialog box will display prompting you for the root password. Enter the root password. The YaST utility will display.

5. _____ Click **Security and Users** in the left frame and then click **Edit and create users** in the right frame. You may be prompted to enter the root password for the LDAP server if an LDAP server is configured. Next, the **User and Group Administration** dialog will display.

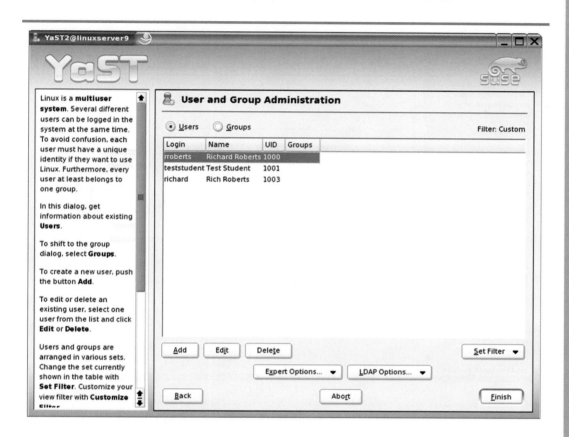

You will see a list of users already configured for the server. Notice that each user has a User ID (UID). The UID is a unique number that identifies each user on the system. Also, any groups the user belongs to are listed. In the example, the users do not belong to any groups. Look at the **Add**, **Edit**, and **Delete** buttons at the bottom of the dialog. As indicated on the button, this is where you make the selection to add, edit, or delete a user account. Also, notice the helpful information provided in the left frame. The information displayed always matches the task at hand.

6. _____ Click **Add** to add a new user. A dialog similar to the following will appear.

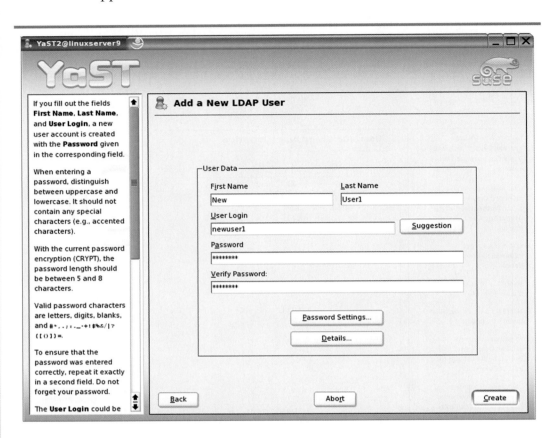

This dialog is similar to other operating systems new user dialogs. You are asked for the user's first and last name, login name, and password. Notice the **Suggestion** button to the right of the **User Login** text box. SuSE Linux will automatically create a user login name if you desire. SuSE Linux and many other versions of Linux and UNIX typically request that the user's login name be in lowercase letters. Uppercase letters may create problems for some e-mail systems.

7. _____ Enter the user login name using an uppercase letter for the first letter in the name. This action should generate a warning message. Write the warning message on a separate sheet of paper, and then respond to the warning message by clicking **No**.

8. _____ Change the login name to all lowercase letters and then enter the password. Click **Create**. The new user account should now appear in the list of users. If not, call your instructor.

9. _____ Click **Finish** to save the new user's account settings and to return to the YaST utility window.

10. _____ Click **Close** to exit the YaST utility.

11. _____ Logout of the Linux session by selecting **Logout** from the main menu.

12. _____ Test the account by logging on to the system using the user login name and password of the user account you created. If the logon is successful, you have completed the laboratory activity correctly. If not, call your instructor.

13. _____ Access the **User and Group Administration** dialog (**YaST | Security and Users | Edit and create users**).

14. _____ Select the new account you just created and then click **Edit**. The **Edit an Existing LDAP User** dialog will display. You can change the user login name or password by entering a new login name or password.

15. _____ Click the **Password Settings** button. A dialog similar to the following will display.

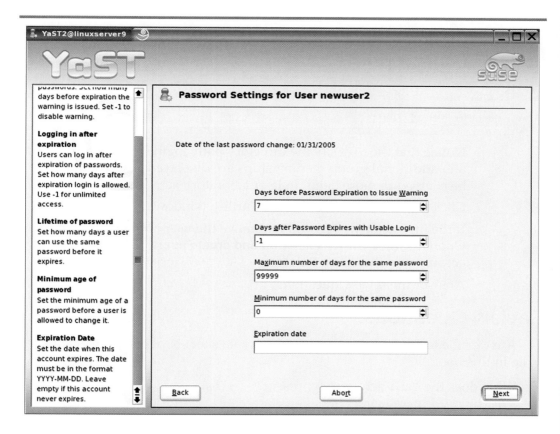

This dialog displays the default values for the user account password. Any of the default values can be changed. For now, leave the default values as they are and click the **Back** button to return to the **Edit an Existing LDAP User** dialog.

16. _____   Click the **Details** button. A dialog similar to the following should appear.

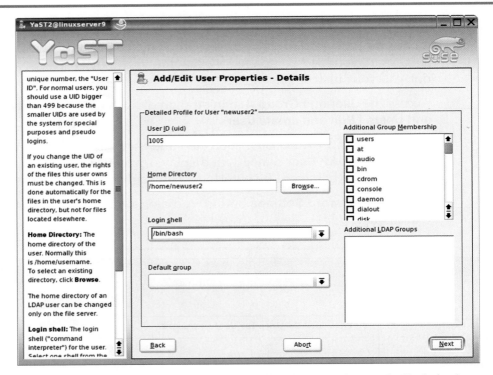

Notice that the user's home directory path, login shell, default group, and additional group memberships are displayed. These items will be covered in more detail in later laboratory activities.

17. _____   Click **Abort** to return to the YaST utility window.

18. _____   If the instructor wishes you to remove the user account, do so now. Open the YaST utility, click **Edit and create users**. Select the new account and then click **Delete**.

19. _____   Answer the review questions.

Review Questions

Answer the following questions on a separate sheet of paper. Please do not write in this book.

1. What does the YaST acronym represent?

2. Who can create a new user account?

3. Why did the message appear warning that the user login name contained an uppercase letter?

4. What is the UID?

5. What button is used to delete a user account?

6. What is the name of the button used to modify an existing user account?

7. What is the default value of the number of days a user may keep the same password?

8. What is the default number of days before a password expires that a warning will be issued?

10

Introduction to the Server

After studying this chapter, you will be able to:

- ❏ Describe various roles of servers in a network environment.
- ❏ Explain the function of a thin server.
- ❏ Explain the function of a thin client server.
- ❏ Identify and explain major components that distinguish servers from PCs.
- ❏ Identify and explain Small Computer Systems Interface (SCSI).
- ❏ List the four types of system resources and explain their function.
- ❏ Identify and describe common RAID systems.
- ❏ Explain network attached storage (NAS) technology.
- ❏ Explain storage area network (SAN) technology.
- ❏ Describe common Fibre Channel technology applications.
- ❏ Identify and describe the Fibre Channel topologies.

Network+ Exam—Key Points

The Network+ Certification exam specifically references fault tolerance and network storage. Pay particular attention to the characteristics of RAID 0, RAID 1, RAID 5, and network storage, such as storage area network (SAN) and network attached storage (NAS).

Key Words and Terms

The following words and terms will become important pieces of your networking vocabulary. Be sure you can define them.

backplane
blade server
Direct Memory Access (DMA) channel
disk mirroring
disk striping
duplexing
error correction
fabric switch
fault tolerance
Fibre Channel
firmware
hot swapping
hot-swap technology
Input/Output (I/O) port

interrupt request (IRQ)
logical unit number (LUN)
memory address assignment
network attached storage (NAS)
parallel processing
power on self-test (POST)
Redundant Array of Independent
    Disks (RAID)
Small Computer Systems Interface
    (SCSI)
storage area network (SAN)
thin client
thin client server
thin server

A complete understanding of server hardware can take a great deal of study. This chapter provides an overview of the server and identifies different server types and their roles. You will be introduced to some of the hardware that makes the server unique from the ordinary PC, and you will learn about RAID and other storage systems.

Server Types and Services

Servers provide a variety of services. Some of the services a server can provide are authentication and security, Web, mail, and print. A server can be called by many names. For example, it can be called an authentication and security database server, Web server, mail server, and print server, **Figure 10-1.**

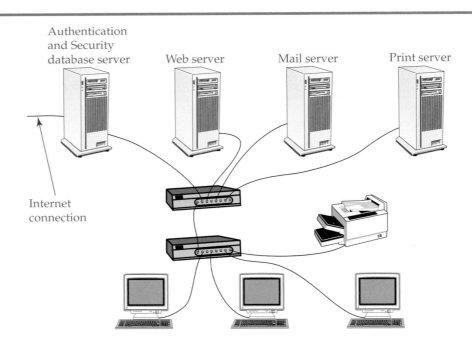

Figure 10-1.
Servers provide a variety of services, such as authentication and security, Web, mail, and print.

Authentication and Security database server

Web server

Mail server

Print server

Internet connection

A network may have a single server that provides a variety of services, or it may have a group of servers, each providing a specific service.

A small network usually has one server set up to handle many different services. A large network usually has several servers, each providing a different service or set of services. For example, a large corporation may use one server to handle e-mail requests and Web hosting, another server to serve as a domain controller to provide security for the entire network, and another server to provide application software, a database for its clients, and support for print operations.

The more services provided and the numbers of clients create a demand on the server. A single server with a limited amount of services may be fine for a relatively small number of users, typically less than 25. A commercial enterprise spanning across the country or world may require hundreds of servers. The network administrator or network designer decides the number and capacity of individual servers needed by taking into account the number of users and the system's predicted network traffic.

Each network system is uniquely designed, even though each network has many similarities. Some network equipment providers have software programs that help you design a network. You simply enter information, such as the number of clients, offices, cities, and countries and the type of software and services to be provided. After all the information is collected, the software program provides an estimate of the size and number of servers required. We will discuss this topic later in Chapter 20—Designing and Installing a Network.

Thin Servers

A *thin server* is a server that has only the hardware and software needed to support and run a specific function, such as Web services, print services, and file services. It is more economical to use a thin server as a print server than to tie up a more expensive server simply to handle printing on a network. IBM markets a thin server than consists of a sealed box rather than the traditional server. The sealed box contains only the essential hardware and software required for supporting the server's dedicated function.

Thin Client Servers

A *thin client server* is a server that provides applications and processing power to a thin client. Thin client servers run terminal server software and may have more RAM and hard disk drive storage than needed. A *thin client* is a computer that typically has a minimal amount of processing power and memory. It relies on the server's processor and memory for processing data and running software applications. Thin clients are becoming common in industry for such applications as hotel bookings, airline ticketing, and medical record access.

thin server
a server that has only the hardware and software needed to support and run a specific function, such as Web services, print services, and file services.

thin client server
a server that provides applications and processing power to a thin client.

thin client
a computer that relies on a thin client server's processing power and memory.

When budgets are tight, you can install what is normally considered an obsolete PC on a thin client network. The obsolete PC is obsolete because of its processing speed, lack of storage, and inability to run new software. The thin client server can provide all the services that are required by the obsolete PC, thus making the obsolete PC a useful workstation.

Tech Tip

Do not confuse a thin client with a dumb terminal. A dumb terminal sends user input to a mainframe. Dumb terminals have absolutely no computing power, operating system, hard disk drive, BIOS, and RAM. A thin client may not need a hard disk drive or an operating system; however, they are still a full-fledged computer because they have a CPU and processing power.

Windows 2000 Server and Windows Server 2003 come with Terminal Services software. Once Terminal Services is set up on the server, you have the option to generate a disk that will assist you in automatically setting up thin client workstations.

Server Classification by Number of CPUs

Servers vary a great deal by size and power. The number of processors they contain usually classifies them. For example, Sun Microsystems has three classifications of servers: entry-level, midrange, and high-end. Sun Microsystems defines these classifications by the number of processors they contain. A Sun Microsystems entry-level server has up to 8 processors. A midrange server has up to 30 processors, and a high-end server has up to 106 processors. Other vendor's definitions vary somewhat, but this gives you a general idea. In this chapter, we focus on the HP ProLiant DL740 server. The HP ProLiant DL740 is one of the most powerful entry-level servers manufactured.

Major Server Components

It is assumed you have some PC hardware background from completing a course on PC service and repair or have a CompTIA A+ Certification. If not, it is strongly advised that you take a course in PC repair or, at the very least, complete a home study program. A good background in PC hardware proves to be beneficial to anyone working with network servers. As a matter of fact, IBM and other companies require the CompTIA A+ Certification as part of their progression toward certification as server technicians.

Warning

When handling electronic components, it is standard practice to use electrostatic discharge (ESD) safety practices. Electronic components based on complimentary metal oxide semiconductor (CMOS) technology can be easily damaged by static electricity. The human body and clothing can build up a tremendous static electricity charge. When touching a static-sensitive electronic component, the static electricity can discharge and cause the component to be destroyed. To avoid damaging electronic components, use an antistatic wrist strap, which is designed to drain static charges safely.

The major components of a server are similar to a typical PC. In fact, you can use a typical PC as a server for a small network. A network server can be as simple as a typical PC or as complex as a piece of equipment designed exclusively for networks. Some server models with multiple CPUs, large amounts of RAM, and vast amounts of storage space can be thought of as a "small" mainframe rather than as a PC. There are many server designs available. As part of your study of network servers, it is highly advisable to check the IBM, Sun Microsystems, HP, and Dell Web sites to see the available designs. This section looks at the following components: case, hot-swap components, power supply, motherboard, BIOS, CPU, and SCSI.

Figure 10-2.
The ProLiant DL740 can
be mounted into a rack.
(Hewlett-Packard
Company)

Case

Server case styles vary greatly. Some are similar to desktop models. Some are designed to be mounted into a rack like the server in **Figure 10-2.**

Large enterprise servers are actually a group of individual servers mounted in a rack system or cabinet, **Figure 10-3.** The group of servers acts as one unit that has multiple CPUs and a large amount of system resources, such as hard disk drive storage and RAM.

A *blade server* derives its name from its size and shape. It is extremely thin compared to other servers and fits into a rack, **Figure 10-4.** Blade servers are

blade server
a powerful server
that is extremely
thin. It is designed
to be mounted in
a small space with
other blade servers.

Figure 10-3.
This cabinet can store
many servers, allowing
them to act as a single
unit. (Hewlett-Packard
Company)

Figure 10-4.
A blade server is
thin, yet powerful. It
is designed to allow
many blade servers to
be installed in a small
area. A—A single blade
server. B—Many blade
servers installed in a
single rack. (Hewlett-
Packard Company)

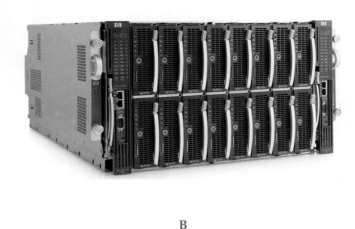

A   B

Figure 10-4.
A blade server is thin, yet powerful. It is designed to allow many blade servers to be installed in a small area. A—A single blade server. B—Many blade servers installed in a single rack. (Hewlett-Packard Company)

especially designed to allow a large number of servers to be mounted in a small space. Blade servers are not to be confused with thin servers. A blade server can contain more than one processor and be quite powerful. They are often selected for applications where there is a requirement for many servers to perform the same or similar function. For example, they may be used in a large Web service facility or a file server farm.

Hot-Swap Components

hot-swap technology allows a component to be removed or installed while the system is running.

Servers are generally designed to provide continuous service with minimal interruptions and data loss. The main feature that most servers use to provide this continuous service is hot-swap technology. *Hot-swap technology* allows a component to be removed or installed while the system is running. There is no need to power down the system while replacing or adding major components. Some common hot-swap components are hard disk drives, memory modules (RAM), CPUs, and power supplies.

Never assume a component is hot-swappable. Always check the system manual before removing any component while the system is powered on.

Do *not* use an antistatic wrist strap when servicing any component with live voltage present. Do *not* open a CRT (monitor) while wearing an antistatic wrist strap, even if the CRT is unplugged. CRT screens can hold a high voltage charge for long periods of time after they have been disconnected from electrical power.

Power Supply

The power supply converts standard 120-volt or 240-volt AC power into lower DC voltage levels that can be used by the motherboard and other devices inside the computer case. A typical power supply provides an assortment of output plugs. The plug style matches the intended piece of equipment. For example, a standard ATA drive uses a Molex plug, which supplies 5 volts DC and 12 volts DC to the ATA drive. An ATX-style motherboard plug supplies 3.3 volts, 5 volts, and 12 volts as well as a ground connection to the motherboard. It is also designed to plug into the motherboard correctly. Older style connections could be incorrectly plugged into the motherboard if you did not pay attention to the color-coding.

Entry-level servers are typically equipped with two power supplies, **Figure 10-5.** The two power supplies balance the load. If one power supply fails, the other power supply carries the full load. Some servers are designed with "hot-swappable" power supplies, which permits the defective power supply to be changed while the server is running.

To prevent the loss of data, most server manufacturer's recommend disconnecting the server from the network while replacing hot-swappable power supplies.

Tech Tip

Motherboard

Servers generally use a backplane. A *backplane* is a simple motherboard designed with minimal components. It typically serves as the interface of all the major components. It is designed to allow major components to be added or removed without powering down the system. Removing components without shutting down the system is call *hot swapping.*

Figure 10-6 shows the general backplane layout of the HP ProLiant DL740 server with and without the added modules. An I/O board attaches to the backplane. It contains six PCI hot-swappable slots. The processor boards each contain 4 CPUs, for a total of 8 CPUs. Each memory module in the system is hot-swappable and contains 8 DIMMS. The HP ProLiant DL740 can contain up to 40GB of RAM. The multiple CPUs and vast amount of RAM are required to provide the many different services to numerous network clients with minimal

backplane
a simple motherboard designed with minimal components. It typically serves as an interface of all major components in the system.

hot swapping
the process of removing system components without shutting down the system.

Power supplies

Figure 10-5.
Back view of the ProLiant ML350. This system comes with two power supplies that evenly distribute power throughout the system. When one power supply fails, the other carries the full load. (Hewlett-Packard Company)

Figure 10-6.
The backplane layout of the ProLiant DL740.

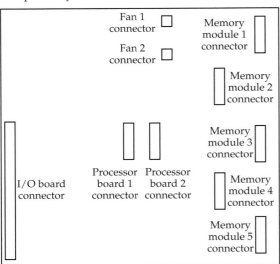

Backplane

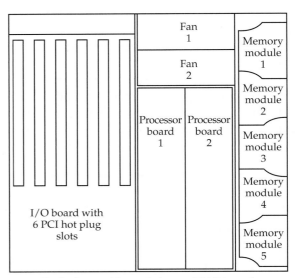

Backplane with modules and I/O board

delay. The CPUs need not be the latest, high-speed processors available because there are numerous processors in the server.

BIOS

The BIOS contains a small software program that starts the server boot operation when power is applied to the server. The combination BIOS chip and the software program are usually referred to as *firmware*. The BIOS is responsible for the *power on self-test (POST)*, which is performed at startup. The POST does a quick, initial check of the major components, such as memory, disk drives, keyboard, mouse, and monitor, to be sure that a minimum working system is available. After a general check is made of the major hardware components, the BIOS turns control over to the operating system software. The operating system completes the boot process by loading more advanced hardware drivers than the BIOS did. It also performs a more sophisticated check of hardware and software systems. If all the software and hardware components appear to be in working order, the final screen, which serves as a user interface, appears.

Central Processing Unit (CPU)

One major difference between a server and a typical PC is that many servers have multiple CPUs installed rather than just one. When multiple CPUs are installed, the server can perform *parallel processing*. This means that a program can be processed through more than one CPU simultaneously. Another advantage of multiple CPUs is several clients may be serviced at the same time rather than waiting their turn to access a single CPU. Super computers and enterprise servers may contain over 1,000 CPUs.

A small office server can get by with one processor, especially if demand on the server is low. A large enterprise system requires multiple processors to meet the demands of the server's clients. Some processor modules are hot-swappable. When a single CPU fails, it can easily be removed and replaced while the server is running and providing services to clients.

firmware
the combination of the BIOS chip and the software program within the chip.

power on self-test (POST)
a quick initial check of the major components in a computer, such as memory, disk drives, keyboard, mouse, and monitor, to be sure that a minimum working system is available.

parallel processing
processing a program through more than one CPU simultaneously.

Common Name	Class	Devices	Bus Width in bits	Speed	MBps
SCSI-1	SCSI-1	8	8	5 MHz	4-5 MBps
Wide SCSI	SCSI-2	16	8	5 MHz	10 MBps
Fast SCSI	SCSI-2	8	8	10 MHz	10 MBps
Fast/Wide SCSI	SCSI-2	16	16	10 MHz	20 MBps
Ultra SCSI	SCSI-3	8	8	20 MHz	20 MBps
Ultra/Wide	SCSI-3	8	16	20 MHz	20 MBps
Ultra2	SCSI-3	8	8	40 MHz	40 MBps
Ultra2/Wide SCSI	SCSI-3	16	16	40 MHz	80 MBps
Ultra3 SCSI	SCSI-3	16	16	40 MHz	160 MBps

Figure 10-7.
SCSI technology specifications.

Small Computer Systems Interface (SCSI)

Small Computer Systems Interface (SCSI), pronounced *skuzzy,* is a computer bus technology that allows you to connect multiple devices to a single controller. The benefits of SCSI technology can be readily observed in the chart in **Figure 10-7.** SCSI technology not only allows multiple devices to connect to a single controller, it also supports high-data transfer rates. This is quite a performance improvement when compared to the traditional IDE or ATA attachment. SCSI is ideal for servers that must hold large amounts of data that is accessed by numerous clients.

Most SCSI drives are hot-swap devices. Often, SCSI drives are arranged in the server with hot-swap bays open to the outside of the case to provide easy access. This allows you to connect or disconnect a drive without opening the server case. See **Figure 10-8.**

SCSI technology has evolved over the years. With this evolution, new names have emerged to describe the improved technology. The term *Wide* is used to indicate 16-bit data transfers in place of 8-bit data transfers. To reflect the increase in frequency, the term *Fast* was used and then the term *Ultra*. Combination of the words, such as *Ultra/Wide,* are also used to express the newer technologies. Ultra/Wide means the SCSI device is faster and supports 16-bit transfers. A close study of the table in Figure 10-7 will help you understand the evolution of the SCSI technology. To learn more about SCSI technology visit the Adaptec Web

Small Computer Systems Interface (SCSI) a computer bus technology that allows you to connect multiple devices to a single controller.

Hot-swap SCSI drive

Figure 10-8.
Hot-swap drives are typically accessible from the front of the server for easy access. (Hewlett-Packard Company)

page at www.adaptec.com. There you will find in-depth knowledge about SCSI systems and other technologies.

A SCSI hardware system consists of a host adapter or controller card, a flat ribbon cable to connect SCSI devices, and SCSI devices such as disk drives, CD-ROMs, and tape drives. SCSI cables come in a variety to match the many different classifications of SCSI. The cables come as DB-25, 50-pin, 68-pin, and 80-pin styles.

Look at **Figure 10-9** to see the way a typical SCSI host adapter, cable, and devices might appear. The host adapter and devices can connect to any part of the chain. The chain of SCSI devices must be terminated. Termination is typically completed at the last device with a termination block.

SCSI ID numbers

Each device, including the host adapter card, must have a unique ID number. The numbers start at zero and end at seven for an eight-device SCSI chain and end at fifteen for a sixteen-device SCSI chain. The host adapter typically is assigned the highest number.

SCSI ID numbers are assigned through pins and jumpers on the SCSI drive. The pins are arranged in pairs. When a jumper is applied across the pins, an electrical connection is made. The pairs of pins represent the binary number system. The jumpers are applied in a binary pattern that represents the SCSI ID number. **Figure 10-10** illustrates each of the binary patterns and the related SCSI ID number.

Figure 10-9.
Typical SCSI drive arrangement.

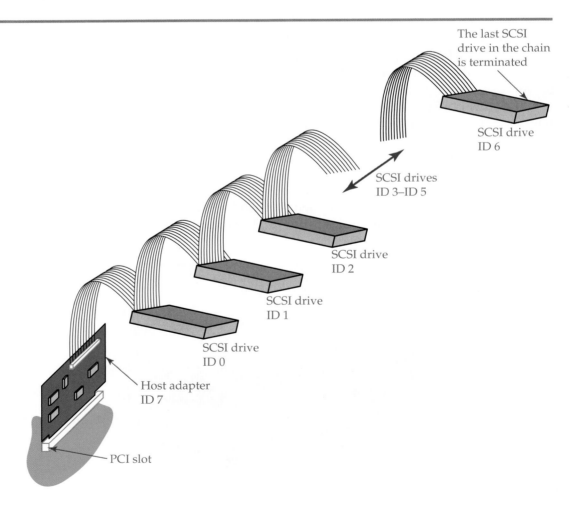

The last SCSI drive in the chain is terminated

SCSI drive ID 6

SCSI drives ID 3–ID 5

SCSI drive ID 2

SCSI drive ID 1

SCSI drive ID 0

Host adapter ID 7

PCI slot

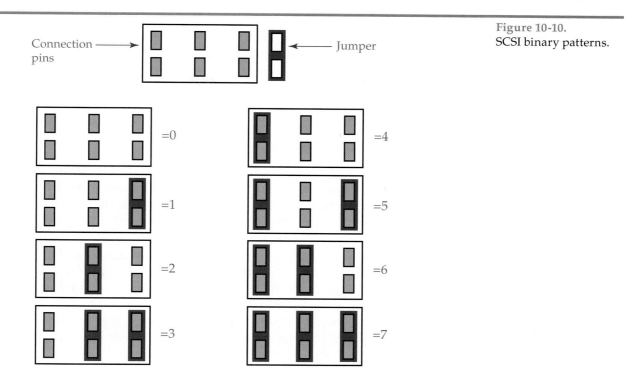

Figure 10-10.
SCSI binary patterns.

Logical unit number (LUN)

SCSI devices are not limited to internal devices. The SCSI chain can extend outside the case by adding an extender card to the SCSI chain. The extender card allows more devices to be attached to existing SCSI systems. The extender card is an integrated circuit card connected to the SCSI chain as a SCSI device. See **Figure 10-11.**

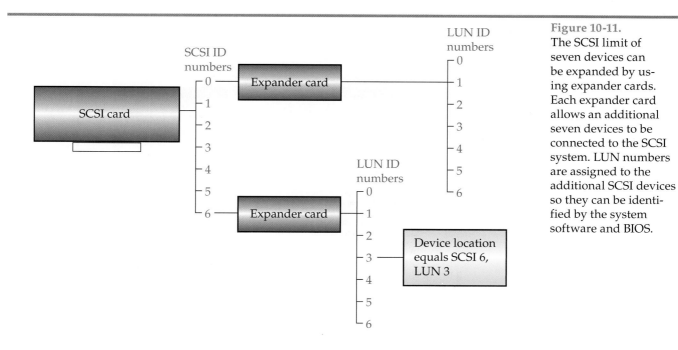

Figure 10-11.
The SCSI limit of seven devices can be expanded by using expander cards. Each expander card allows an additional seven devices to be connected to the SCSI system. LUN numbers are assigned to the additional SCSI devices so they can be identified by the system software and BIOS.

When additional devices are attached, they are identified separately from the original chain of devices. Each additional device connected to the SCSI extender is identified with a *logical unit number (LUN)*. SCSI bus extenders are also referred to as SCSI expanders, repeaters, and regenerators.

logical unit number (LUN)
a numbering scheme to identify SCSI devices attached to an extender card.

External SCSI device chains are commonly used in RAID systems. RAID systems are covered later in this chapter.

System Resources

System resources refer to resources such as interrupt requests, Direct Memory Access channels, input/output ports, and memory. System resources are assigned to components installed in the computer system such as hard disk drives, keyboards, and mice. In a Microsoft-based server, you can view system resource assignments in Device Manager.

Not all devices have all four types of resources assigned to them. For example, some devices do not require a DMA channel assignment. System resources typically cannot be shared between two devices. Sharing the same system resource causes a system resource conflict, referred to simply as a conflict. Conflicts must be resolved before the system can operate properly. An exception to this rule is motherboard chips used to bridge different bus systems. A motherboard chip can share an IRQ with a hardware device.

Interrupt Request (IRQ)

interrupt request (IRQ)
a circuit that communicates with the CPU.

An *interrupt request (IRQ)* is a circuit that communicates with the CPU. Hardware devices send an electrical signal to the CPU using an assigned IRQ circuit. There are 16 IRQ assignments numbered from 0 to 15, **Figure 10-12.** Many of the assignments cannot be changed, but some can. When Plug and Play devices are used, the IRQ is automatically assigned. Typically, each hardware device must use a separate IRQ to communicate with the CPU. If two hardware devices are assigned the same IRQ, an IRQ conflict occurs. Once the first device contacts the CPU using the IRQ assignment, the other device cannot communicate with the CPU. An IRQ conflict can lead to problems such as an inoperable device, a system crash, and a system lockup. An IRQ can typically be

Figure 10-12.
IRQ assignments.

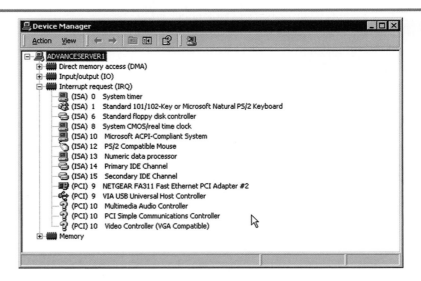

assigned manually by first entering the BIOS setup program and disabling Plug and Play detection. Changes to the IRQ can then be made through Device Manager. After the problem device has been assigned the proper IRQ, the Plug and Play detection can be activated without a problem.

Yellow question marks next to devices in Device Manager indicate that there is a problem with the device or assignment.

Direct Memory Access (DMA) Channel

A *Direct Memory Access (DMA) channel* is a circuit that allows devices to communicate and transfer data to and from RAM without the need of CPU intervention. Computer systems were first designed with the CPU handling all communications. Every bit of data had to travel through the CPU. DMA technology was introduced to save valuable processor time.

DMA works in conjunction with the chipsets on the motherboard, BIOS software, and the CPU. Large blocks of data that need to be transferred between hardware devices and memory are transferred through a DMA channel that is assigned to the device. **Figure 10-13** shows DMA assignments for the standard floppy disk controller, ECP printer port, and the Direct Memory Access controller.

Direct Memory Access (DMA) channel
a circuit that allows devices to communicate and transfer data to and from RAM without the need of CPU intervention.

Input/Output (I/O) Port

The *Input/Output (I/O) port* is a small amount of memory assigned to a device that temporarily holds small amounts of data. It is used to transfer data between two locations. The data remains in the I/O port assignment until it can be moved.

The system has many hardware devices and software programs that depend on transferring data between hardware and memory locations. When a device or software program wants to transfer data, it may need to wait until the CPU is finished with its current process. In fact, it may need to wait until several

Input/Output (I/O) port
a small amount of memory assigned to a device that temporarily holds small amounts of data and is used to transfer data between two locations.

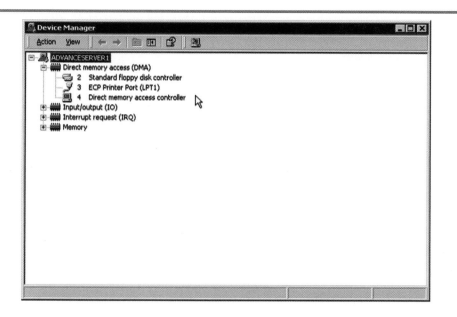

Figure 10-13.
DMA assignments.

Figure 10-14.
I/O port assignments.

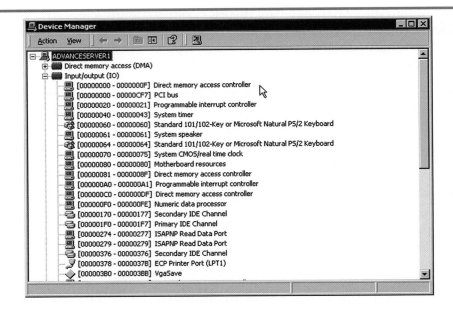

processes are completed before the CPU can handle the data. The I/O port stores this data until the CPU is free. **Figure 10-14** shows the I/O port assignments associated with a system's hardware devices.

memory address assignment
a large block of memory assigned to a device and used to transfer data between two locations.

Memory Address Assignment

A *memory address assignment* is a large block of memory assigned to a device and is used to transfer data between two locations. A device is assigned a range of memory addresses. Some devices such as video and sound cards require a great deal of memory. The blocks of RAM assigned to a device cannot be used by any other device. **Figure 10-15** shows the memory assignments associated with a system's hardware devices.

Figure 10-15.
Memory assignments.

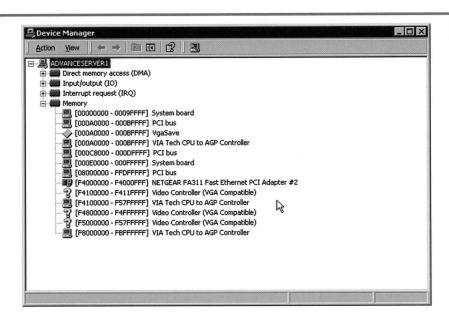

RAID Systems

Redundant Array of Independent Disks (RAID) is a system of disks arranged for speed or fault tolerance, or both. **Fault tolerance** as applied to RAID systems means the ability to recover from a hard disk or hard disk controller failure without the loss of stored data. Fault tolerance as applied to a network infrastructure means the ability to continue operation during a system hardware or software error. To achieve a high-data transfer rate, a technique known as disk striping is used. **Disk striping** involves dividing the data into separate sections and writing the data across several hard disk drives at the same time. This reduces the total amount of time it takes to store large amounts of data.

Redundant Array of Independent Disks (RAID)
a system that can provide fault tolerance, increase disk speed when large blocks of data need to be constantly accessed, or both.

Another accepted representation of the RAID acronym is Redundant Array of Inexpensive Disks.

Tech Tip

Another RAID technique is **error correction,** which can use traditional error checking code (ECC) or parity. ECC is obsolete as a method for RAID because most hard disk drive systems use ECC as a standard way to protect data. Incorporating ECC into a RAID system would be redundant.

Parity is a technique that allows data to be recovered if one of the hard disk drives fails in a multiple disk drive system. A minimum of three hard disk drives must be used for this technique. If three hard disk drives are used, one hard disk drive is used to store parity and the other two are used to store data. Data is spread evenly between the two data storage drives, and the parity code for the sum of the two data storage drives is stored on the parity drive.

Parity is a binary code that represents the total data pattern shared between the data storage drives. If any data storage drive fails, the system can use the parity bit to rebuild the missing data.

RAID systems are referred to as RAID levels. There are many RAID levels, each incorporating a different technique to increase speed or to provide data redundancy, or both.

Things to consider when selecting a RAID level are cost, reliability, and speed. There is no single, best RAID level. The appropriate level depends on the value of the data being protected and the amount of budget available. If a bank were protecting financial data, cost would most likely be of little concern. A RAID 0/5 with a tape backup would be quite appropriate. If you were protecting a personal computer, you would not likely need a RAID system.

Repairing a failed RAID system means the drive must be physically replaced and then the data reconstructed. In a Microsoft Windows XP, 2000, or 2003 system, the Disk Management utility is used to reconstruct data. It is important to note that when implementing a RAID system in a Microsoft Windows XP, 2000, or 2003 computer, all drives must be dynamic disk. If using proprietary SCSI drives, data is reconstructed by the data array manufacturer software package.

See **Figure 10-16** for a list of RAID levels and a short description of each. The following section looks at each RAID level in detail.

fault tolerance
the ability to recover from a hard disk or hard disk controller failure without the loss of stored data. When applied to a network infrastructure, fault tolerance means the ability to continue operation during a system hardware or software error.

disk striping
a storage technique that divides data into sections and writes the data sections across several hard disk drives at the same time.

error correction
a RAID technique that uses traditional error checking code (ECC) or parity.

RAID 0

RAID 0 uses disk striping across a group of independent hard disk drives, **Figure 10-17.** This technique increases data flow but provides no fault tolerance. If one hard disk drive fails, all data is lost and cannot be reconstructed. This RAID level is most useful when speed is important, not fault tolerance.

Figure 10-16.
RAID levels.

RAID Level	Description
RAID 0	Striping.
RAID 1	Mirroring or duplexing.
RAID 2	Error checking code (ECC).
RAID 3	Byte-level striping with parity.
RAID 4	Block-level striping with parity.
RAID 5	Block-level striping with distributed parity.
RAID 0/1 or 10	Disk mirroring with striping.
RAID 0/5 or 50	Block striping with parity and striping.

RAID 1

disk mirroring
the act of writing the same information to two hard disk drives at the same time.

RAID 1 uses the technique of disk mirroring. *Disk mirroring* is the act of writing the same information to two hard disk drives at the same time, **Figure 10-18.** Each of the two hard disk drives contains the same data. If one hard disk drive fails, a copy exists on the other hard disk drive.

duplexing
a technique of placing each mirrored hard drive on a separate hard disk drive controller.

Another form of RAID 1 is duplexing. *Duplexing* is the technique of placing each mirrored hard drive on a separate hard disk drive controller. Duplexing adds another level of fault tolerance. If one hard disk drive controller fails, the other is still operable. Mirroring contains the risk of the hard disk drive controller failing and causing both hard disk drives to fail. RAID 1 provides data protection at the cost of speed when compared to RAID 0. When there is a small demand for disk read/writes, RAID 1 is an appropriate technique.

RAID 2

RAID 2 uses common ECC error correction code to provide fault tolerance. All drives provide some sort of ECC, so this form of RAID is basically obsolete.

RAID 3

RAID 3 stripes data across multiple drives and stores parity on a separate drive. Data striping is performed at the byte level. See **Figure 10-19.**

RAID 4

RAID 4 stripes data across several drives and stores parity on a separate drive. The main difference between RAID 3 and RAID 4 is RAID 4 stores data in blocks. The size of the block can vary.

RAID 5

RAID 5 is also called block striping with distributed parity. RAID 5 distributes parity across all drives rather than writing parity to one drive. It provides fault tolerance and some increase in read-write data transfer. See **Figure 10-20.**

Figure 10-17.
RAID 0 divides data and writes it to multiple drives. It provides excellent data transfer rates but no data loss protection.

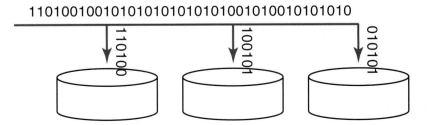

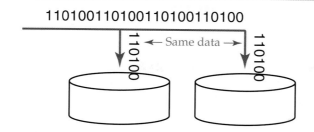

110100110100110100110100

←— Same data —→

110100   110100

Figure 10-18.
RAID 1, or disk mirroring, provides a copy of data on two disk drives.

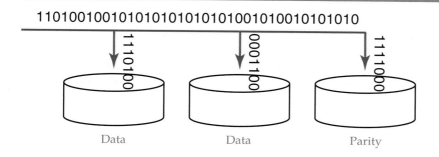

110100100101010101010101001010010101010

1110100   0001100   1111000

Data   Data   Parity

Figure 10-19.
RAID 3 writes across multiple drives with parity stored on separate drives.

RAID 0/1

RAID 0/1, or RAID 10, uses multiple RAID 1 (mirroring) disks systems. It also incorporates disk striping across each of the mirrored disk sets. When compared to RAID 1, this system provides better fault tolerance because data is duplicated on multiple mirrored disk sets. It also provides relatively fast data-transfer speeds by striping data across the mirrored disk sets.

RAID 0/5

RAID 0/5, or RAID 50, uses several RAID 5 (block striping with parity) sets and combines them with RAID 0 (disk striping). It provides increased fault tolerance when compared to a RAID 5 system.

Network+ Note:

Pay particular attention to RAID 0, 1, and 5. These systems are the most commonly used and are the ones most likely to be covered on the Network+ Certification exam.

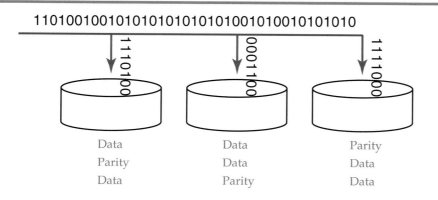

110100100101010101010101001010010101010

1110100   0001100   1111000

Data   Data   Parity

Parity   Data   Data

Data   Parity   Data

Figure 10-20.
RAID 5 provides both disk striping and parity. Parity and data is distributed across all drives.

External Storage Systems

Data is often stored separately from a server. There are network devices that serve specifically as storage containers for network data. These devices are typically called network attached storage (NAS). There are also facilities external to the local area network that provide data storage. These devices are typically called storage area network (SAN). This section discusses both types of storage systems.

Network-Attached Storage (NAS)

*network attached
storage (NAS)*
a device or collection
of devices used to
provide storage for
network data.

Network attached storage (NAS) is a device or collection of devices that provide storage for network data. Network attached storage units are typically composed of disk arrays or tape arrays. The file systems used on the disk array and file array do not typically match the file system used by network clients or servers. The file systems are proprietary and are made to facilitate access speed and data integrity.

This is how data on a NAS device is accessed. A client requesting data makes a request to the file server. The file server in turn makes a request to the NAS device. The NAS device retrieves the data and sends it to the file server. The file server, in turn, sends the data to the client. See **Figure 10-21.**

Storage Area Network (SAN)

*storage area
network (SAN)*
a separate, high-
speed network used
to provide a storage
facility for one or
more networks.

A *storage area network (SAN)* is a separate, high-speed network that provides a storage facility for one or more networks. Typically, a SAN uses a high-speed access media such as Fibre Channel. Fibre Channel is discussed in the next section of the chapter. See **Figure 10-22.**

Tech Tip

Exact definitions of storage area network vary according to different manufacturers. Often the differences are blurred.

Figure 10-21.
In a network attached storage (NAS) design, clients make requests for data through the network file server. The file server then makes a request to the NAS device. The NAS device retrieves the data and sends it to the file server. The file server forwards the data to the client.

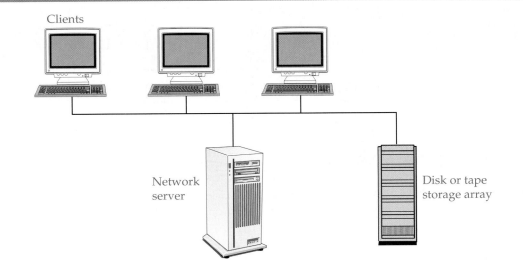

Clients

Network
server

Disk or tape
storage array

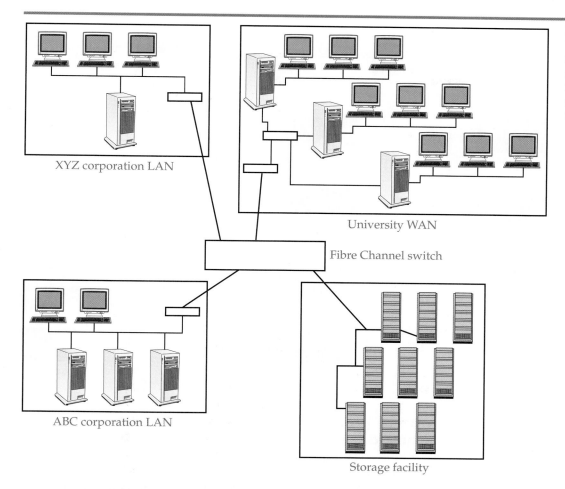

Figure 10-22.
The storage area network (SAN) is a separate, high-speed network that provides a storage facility for other networks.

XYZ corporation LAN

University WAN

Fibre Channel switch

ABC corporation LAN

Storage facility

Fibre Channel

Fibre Channel is a high-speed access method that typically uses fiber-optic cable as network media. It may also use copper core cable and wireless. The term *Fibre Channel* also refers to a set of standards and a protocol. The Fibre Channel standard provides a relatively high-data transmission rate between supercomputers, mainframes, servers, and desktops. Fibre Channel is often used for storage area network (SAN) access. There are three common Fibre Channel topologies: point-to-point, arbitrated loop, and fabric, **Figure 10-23.**

Fibre Channel
a high-speed access method that typically uses fiber-optic cable as media.

Fibre Channel Point-to-Point Topology

A Fibre Channel point-to-point topology is simply a straight connection or channel between two points, such as a mainframe and a server. The channel is isolated from other channels or media, and total bandwidth is dedicated to the channel between the two devices.

Fibre Channel Arbitrated Loop Topology

The Fibre Channel arbitrated loop is a common Fibre Channel topology. It looks similar to the Token Ring topology. The arbitrated loop differs from Token Ring in that two devices in the loop set up a direct communication link, or channel, for the duration of the data transfer. In Token Ring network, a token is passed around the ring while one device on the ring controls communication.

Figure 10-23.
The three major topologies associated with Fibre Channel systems are point-to-point, arbitrated loop, and fabric switched.

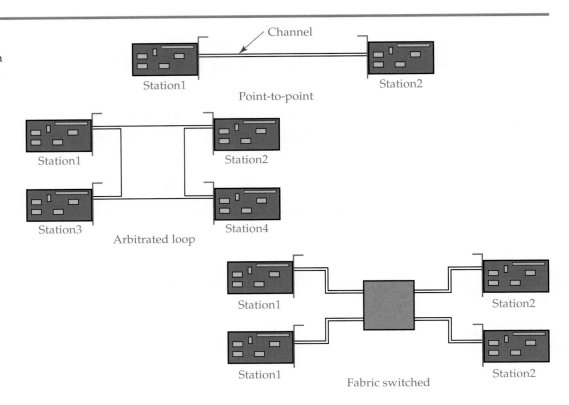

Token Ring has a maximum control time of approximately 8 milliseconds in contrast to the arbitrated loop standard, which allows two devices to communicate until all the data requested has been transferred.

The arbitrated loop is limited to 127 connections. It is used to move large volumes of data between two devices such as a disk storage array and a server. It is not used for sporadic communications between multiple nodes as it is on a Token Ring network.

Fibre Channel Fabric Switched Topology

The fabric switched topology uses a device known as a fabric switch. A *fabric switch* is a switch designed especially for Fibre Channel networking. The fabric switch provides a direct or switched connection between two points. When two points wish to communicate, a private link, or channel, is set up between the two devices. This method is best used when numerous nodes wish to access the same disk storage array. For example, a university consisting of many buildings can connect each building's server to a common disk storage array using the fabric switched topology. This way, departments in each building can have access to research documents stored in the disk storage array. They can also access servers in other buildings through a high-speed connection.

fabric switch
a switch designed especially for Fibre Channel networking.

You may be interested in further study and possible certification as a server technician. The CompTIA organization has a certification called Server+ available to test your knowledge of server hardware. The Server+ Certification is designed to verify your competence for installing, repairing, and maintaining servers. You can download a complete description of the exam objectives from the CompTIA Web site.

Many of the areas in the Server+ Certification exam overlap those on the A+ Certification exam. However, even if you have scored high on the A+ Certification exams, there are sufficient differences between the certifications that will prevent you from passing the Server+ Certification exam without specialized study.

For additional study materials and more detailed information about server hardware, access the IBM and the Sun Microsystems Web sites. For detailed information on SCSI technology, access the Adaptec Web site.

Summary

- Servers play many roles in a network, such as print server, file server, database server, application server, backup server, Web server, and mail server.

- A thin server has only the hardware and software needed to support and run a specific function, or role.

- A thin client server provides applications and processing power to a thin client.

- The thin client relies on a thin client server's processor and memory for processing data and running software applications.

- Typically, entry-level servers contain 1 to 8 processors; mid-range servers contain 9 to 30 processors; and high-end servers contain 31 to 106 processors.

- Servers typically incorporate hot-swap technology to allow components to be removed or installed while the system is running.

- System resources associated with hardware are Direct Memory Access (DMA), interrupt request (IRQ), I/O port address, and memory address assignment.

- When two hardware devices share the same resource assignment, a system conflict occurs.

- The most common RAID levels are RAID 0, RAID 1, and RAID 5.

- RAID 0 is known as striping. Striping increases data read/write speed but does not provide fault tolerance.

- RAID 1 is mirroring or duplexing two hard disk drives.

- RAID 5 is known as striping with parity, which combines data read/write speed with fault tolerance.

- Network Attached Storage (NAS) is a device or collection of devices used to provide storage for a local area network. The NAS shares the bandwidth of the local network.

- Storage Area Network (SAN) is a separate network dedicated to data storage. It does not affect the bandwidth of the other connected networks.

Review Questions

Answer the following questions on a separate sheet of paper. Please do not write in this book.

1. What are some typical services that servers provide?

2. What is a thin server?

3. What is a thin client server?

4. What does the term *hot-swap* mean?

5. What type of motherboard is commonly used for servers?

6. Which server component is responsible for the POST?

7. How many devices can be connected to a Fast/Wide SCSI-2 chain?

8. What is the bus width of Fast SCSI-2?

9. What is the bus width of Fast/Wide SCSI-2?

10. What must be done to the last device on a SCSI chain?

11. What does the acronym LUN represent?

12. Name the four system resources that may be assigned to a hardware device?

13. How many IRQ settings are there in a typical computer?

14. What does the acronym RAID represent?

15. How is parity used to replace data lost on a damaged disk?

16. Which RAID level(s) provide no fault tolerance?

17. Which RAID levels provide disk mirroring?

18. What is the difference between disk mirroring and duplexing?

19. Which RAID level provides both striping and parity?

20. Your system storage is accessed many times throughout the day by many users. The data is not critical. Rather, it is a collection of application software. A tape backup system is used to restore data from a system failure. Which RAID level would *most likely* be appropriate and why?

21. Which RAID level is denoted by the use of two hard disk drives each containing identical data?

22. What does the acronym NAS represent?

23. What does the acronym SAN represent?

24. What is the difference between NAS and SAN?

25. What type of media is used for Fibre Channel?

26. What are the three common Fibre Channel topologies?

Sample Network+ Exam Questions

Answer the following questions on a separate sheet of paper. Please do not write in this book.

Network+

1. Which RAID level provides the fastest reading and writing data transfer time?
   A. RAID 0
   B. RAID 1
   C. RAID 5
   D. RAID 0 and RAID 1

2. Which network RAID level provides a duplicate copy of one hard disk drive on another hard disk drive?
   A. RAID 0
   B. RAID 1
   C. RAID 5
   D. RAID 4

3. How does SAN differ from NAS?
   A. SANS provides access to a SCSI disk exclusively while NAS provides access to SCSI, ATA, and EIDE drives.
   B. SANS offers encryption and NAS does not.
   C. SANS does not increase local network traffic and NAS does.
   D. SANS is connected to the local network system while NAS is accessed remotely via a WAN link.

4. What is the minimum number of drives required for implementing a RAID 5 installation?
   A. 1
   B. 2
   C. 3
   D. 4

5. What does the term *fault tolerance* mean?
   A. The ability to find faults in data before it is backed up.
   B. The ability of the network administrator to tolerate system policy breeches by users.
   C. The ability to recover from a system failure.
   D. The ability to continue normal operation despite a system failure.

6. Which of the following describes the ability to change a module such as hard disk drive or power supply on a server without the need to shut down the server?
   A. System substitution
   B. On the fly changes
   C. Hot-swap
   D. Infinite run ability

Network+

7. What is the function of disk striping?
   A. Reduces the total amount of time it takes to store large amounts of data.
   B. Stores duplicate data across two or more drives.
   C. Ensure data integrity.
   D. Aids in disaster recovery.

8. How does a client access files on a NAS device?
   A. Through NAS client software.
   B. Through a server that is attached to the NAS device.
   C. Directly through Category 5e cable.
   D. Directly through high-speed fiber-optic cable.

9. An investment broker is installing a network system for his employees and to provide customer access. Money transactions will occur on a continuous basis. Which RAID level would you recommend for this business?
   A. RAID 0
   B. RAID 1
   C. RAID 1 (with duplexed drives)
   D. RAID 5

10. Multiple CPUs in a server can perform which of the following?
    A. Unilateral processing
    B. Multiplexed processing
    C. Synchronous processing
    D. Parallel processing

Suggested Laboratory Activities

Do not attempt any suggested laboratory activities without your instructor's permission. Certain activities can render the PC operating system inoperable.

1. Create several partitions on a hard disk drive and then format the partitions using FAT32.
2. Convert a basic FAT32 partition to an NTFS partition and then convert the NTFS partition to dynamic disk.
3. Use the **diskpart** command to view the contents of a hard disk drive. Familiarize yourself with the **diskpart** command line options.
4. Install a RAID 0, RAID 1, and RAID 5 system. Remove a hard disk drive to simulate a disk failure and then replace the hard disk drive with a new hard disk drive and rebuild the missing data.

Interesting Web Sites for More Information

http://computer.howstuffworks.com/scsi.htm

http://serverwatch.internet.com

www.acnc.com/raid.html

www.molex.com

www.pcguide.com/topic.html

www.raid-advisory.com

Chapter 10
Laboratory Activity

Using the DiskPart Command Interpreter

After completing this laboratory activity, you will be able to:

- Access the DiskPart command interpreter.
- Use common DiskPart commands.
- Access DiskPart **help** command to reveal a list of DiskPart commands.
- Identify the number of disks installed in a system.
- Retrieve information about a system of disks.
- Identify and retrieve information about disk partitions or volumes.

Introduction

Disk management is usually performed in the Microsoft Management Console (MMC). The MMC is a GUI utility that manages many administrative functions. The DiskPart command interpreter is an alternative way to manage disks and partitions. At times the DiskPart command interpreter may be the only alternative for inspecting and manipulating the master boot record (MBR) and partition table should the GUI fail. Also, certain tasks can be performed using the DiskPart command interpreter that cannot be performed using the MMC. For example, you must use the DiskPart command interpreter to convert a dynamic disk to a basic disk and to delete or replace the MBR in the active partition. The MMC does not allow the performance of these tasks because it is designed to protect the integrity of the disk structure from accidental deletion.

You can think of the DiskPart command interpreter as an updated version of the **fdisk** command. Windows XP no longer supports the **fdisk** command. The DiskPart utility can be used in place of **fdisk** to inspect, create, and delete partitions or volumes on a disk drive.

The DiskPart command interpreter will not make changes to removal media such as a USB drive and an IEEE-1394 drive. These types of media are identified as a "super floppy" by the DiskPart utility rather than as a hard disk drive.

In this laboratory activity, you will perform some basic DiskPart operations to become familiar with this utility. While you most likely will never need to use the full capability of this utility, more information can be obtained at the Microsoft Tech Support Web site.

Look at the following screen capture. Notice that the **detail disk** command has been entered at the DiskPart command line. The **detail disk** command reveals details about the selected disk, such as the hard disk drive manufacturer and type. It also shows that the hard disk drive has two partitions or volumes. Volume 1 is assigned letter *C* and volume 2 is assigned letter *D*. Notice that both volumes are indicated as healthy.

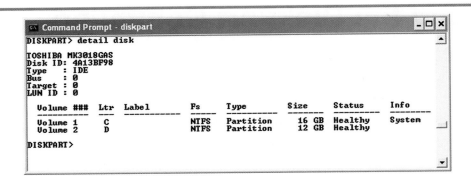

The **list** command can be used with the disk, volume, and partition object. When using the DiskPart command interpreter, disks, partitions, and volumes are referred to as *objects*. Look at the following screen capture. It shows the **list disk**, **list partition**, and **list volume** commands. The **list disk** command lists the disk(s) installed starting with disk 0. The **list volume** command lists the volumes contained in the system. The **list partition** command lists the partitions located on a selected disk. Notice how the DVD-ROM drive is identified after issuing the **list volume** command. Also, notice how the removable drive is identified even though it is no longer present.

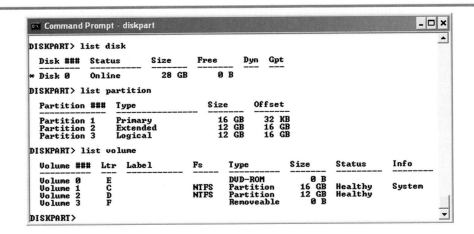

Before issuing the **list partition** command, you must select the disk you wish to view. For example, to view the partition information for a single hard disk drive, you would issue the **select disk** command. Then you would issue the **list partition** command. Otherwise, you will generate an error message stating, "There is no disk selected to list partitions."

Selecting the disk is also known as setting focus. When you set focus, you focus the **list partition** command on a selected disk. You must also set the focus on the partition you wish to list. If you do not, the default partition you are presently running the DiskPart command interpreter from is automatically selected.

For more information on the DiskPart command interpreter, read the following Microsoft Technical Article: Q300415—A Description of the DiskPart Command Line Utility.

Equipment and Materials

■ A typical PC running Windows XP Professional. (A server is not required for this laboratory activity.)

Procedures

1. _____ Report to your assigned workstation.

2. _____ Boot the workstation and verify it is in working order.

3. _____ Open the command prompt (**Start | All Programs | Accessories | Command Prompt**, or **Start | Run** and enter **cmd** in the **Run** dialog box).

4. _____ At the command line type and enter the **DiskPart** command. If successful, the prompt will display > DISKPART to let you know that you are now using the DiskPart command interpreter.

5. _____ Enter the **Help** or **/?** command to reveal a list of commands.

6. _____ Enter the **list** command to display the objects that can be specified with the **list** command. Disk, partition, and volume are listed.

7. _____ Try each of the following commands to reveal information about the disk drive(s): **list disk, list partition, list volume.** Remember, the **list partition** command requires focus. You must use the **select** command to select the disk drive you wish to list partition information about. For example, you would enter **select disk 0** to set the focus on the first disk in the system. The **select** command is only required for the **list partition** command. After running each command, record on a separate sheet of paper the information revealed.

8. _____ The **detail** command reveals similar content but with more detailed information about an object. Enter each of the following commands and then record on a separate sheet of paper the information revealed: **detail disk, detail volume, detail partition**.

9. _____ Practice using the commands covered thus far until you are very familiar with them. They are difficult to learn at first because some involve setting focus and using two words for a valid command.

10. _____   Use the **help** command to reveal what the following commands are used for. Make a list of the commands and the purpose of each on a separate sheet of paper. Do not attempt to use any of the following commands without explicit permission from the instructor.

Command	Purpose
add	
active	
assign	
break	
clean	
create	
delete	
detail	
exit	
extend	
help	
import	
inactive	
list	
online	
rem	
remove	
repair	
rescan	
retain	
select	

11. _____   Enter the **exit** command to return to the command prompt and then again to return to the Windows desktop. Leave your workstation in the condition as specified by your instructor.

12. _____   Answer the review questions.

Review Questions

Answer the following questions on a separate sheet of paper. Please do not write in this book.

1. What are objects when using the Diskpart command interpreter?

2. What command is used to reveal information about a particular partition?

3. Which command is used to set focus on a particular drive?

4. What does the command prompt look like when you are running the DiskPart command interpreter?

5. How can you see a list of the various DiskPart commands?

6. A particular computer has only one hard disk drive. What command is used to set focus on that one hard disk drive?

7. What command will stop the DiskPart command interpreter and return you to the default prompt?

An enterprise business typically depends on hundreds of servers to meet user demands.

11

TCP/IP Fundamentals

After studying this chapter, you will be able to:

❏ Explain the differences between IPv4 and IPv6.

❏ Explain the purpose and operation of the Network Address Translation (NAT) protocol.

❏ Determine the IP address and subnet mask on a workstation.

❏ Explain the purpose and operation of the Domain Name System (DNS).

❏ Describe how UDP, TCP, and IP relate to the OSI model.

❏ Explain the purpose and operation of the Windows Internet Naming Service (WINS).

❏ Explain the purpose and operation of the Dynamic Host Configuration Protocol (DHCP).

❏ Identify an Automatic Private IP Addressing (APIPA) address.

❏ Interpret the displays of TCP/IP troubleshooting utilities.

Network+ Exam—Key Points

The Network+ Certification exam requires extensive knowledge of how IP addresses function in a network system. You need to have a solid grasp of the relationship of IP addresses and subnet mask as well as the common command line utilities, **ping**, **tracert**, **nbstat**, **ipconfig** and **winipcfg**. Be able to explain how WINS, DNS, DHCP, and APIPA services function in a network. You cannot study this chapter too much! Be sure to perform laboratory activities relating to the topics in this chapter.

Key Words and Terms

The following words and terms will become important pieces of your networking vocabulary. Be sure you can define them.

Address Resolution Protocol (ARP)
Automatic Private IP Addressing (APIPA)
Bootstrap Protocol (BOOTP)
Class A network
Class B network
Class C network
default gateway address
diskless station
Domain Name System (DNS)
dynamic addressing
Dynamic Host Configuration Protocol (DHCP)
dynamic IP assignment
Fully Qualified Domain Name (FQDN)
host
Integrated Network Information Center (InterNIC)
Internet Corporation for Assigned Names and Numbers (ICANN)
IPv6
loopback address
Network Address Translation (NAT)
octet
port number
registrar
resolver
Reverse Address Resolution Protocol (RARP)
socket
static addressing
static IP assignment
subdomain
subnet mask
subnetwork
Windows Internet Naming Service (WINS)

The TCP/IP protocol is used for network communications by all major network operating systems. It was especially designed for the Internet. Entire textbooks have been written on the subject of TCP/IP, and it accounts for a large portion of the Network+ Certification exam and the i-Net+ Certification exam. Other certifications from Novell, Nortel, Microsoft, Linux, and IBM also rely heavily on TCP/IP principles for their certification exams.

As a network administrator, you need to be able to configure a host to use the TCP/IP protocol and to troubleshoot network connection problems in a TCP/IP environment. To be able to do this, you need to understand the Domain Name Service (DNS) structure, TCP/IP addressing, and how services such as Dynamic Internet Connection Protocol (DHCP) work. You also need to be able to use TCP/IP troubleshooting utilities and interpret their output to solve communication problems.

This chapter introduces TCP/IP addressing and the Domain Name Service (DNS), which provide functionality and structure to the Internet. It takes a detailed look at how TCP/IP hosts are identified and relates the OSI model to TCP/IP communications. Lastly, a brief overview of some of the many TCP/IP troubleshooting utilities is presented.

IP Addressing

The TCP/IP protocol was developed by the Department of Defense. It was specifically designed for communication over the Internet. As the Internet evolved, it became evident that a means of identification of network nodes was needed. The IP address was the solution and was approved in 1980.

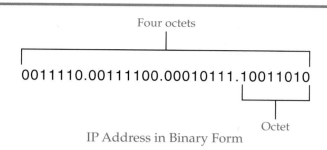

Four octets

0011110.00111100.00010111.10011010

Octet

IP Address in Binary Form

Figure 11-1.
An IPv4 address
consists of four octets
(8-bit units).

TCP/IP can also be used on a LAN in place of NetBEUI. In fact, Windows XP no longer installs NetBEUI as one of the default protocols. Instead, it uses TCP/IP for LAN communications.

Tech Tip

There are two types of IP addressing schemes: IPv4 and IPv6. IPv6 addressing is covered later in this chapter. An IPv4 address consists of four octets, **Figure 11-1.** An *octet* is an 8-bit (or one byte) value. Each octet in an IPv4 address is composed of eight digits of 0s and 1s. Periods separate each octet from neighboring octets.

Although IP numbers are computed in the binary form, they are interpreted and used in the decimal form. When an octet is interpreted as a decimal number, the values range from 0 through 255. This range is based on the maximum possibilities of an eight-position binary number. **Figure 11-2** shows an IP address specified in binary and decimal. As you can see, the decimal expression of an IP address is much shorter than the binary expression of the same address. It is also much easier to read and interpret.

IP addressing is a method of identifying every node or host on a network. The terms *host* and *node* are used interchangeably to identify individual PCs, printers, and network equipment that may require an address. *TCP/IP addressing* and *IP addressing* are also interchangeable terms.

A near perfect analogy of IP addressing is the telephone system. The telephone system is a hierarchical system that uses area codes to identify areas outside your local area. The first three digits identify the area code and the last seven digits identify the telephone inside the area code. See **Figure 11-3.**

IP addresses work in much the same way. An IP address has two parts: the network number and the host number. The network number identifies the network you are trying to connect to and the host number identifies the host or

octet
an 8-bit (or one byte) binary value.

host
a name for a computer, printer, or network device associated with an IP address on a TCP/IP network.

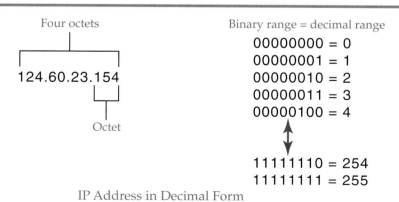

Four octets

124.60.23.154

Octet

IP Address in Decimal Form

Binary range = decimal range
00000000 = 0
00000001 = 1
00000010 = 2
00000011 = 3
00000100 = 4

↕

11111110 = 254
11111111 = 255

Figure 11-2.
An IPv4 address is typically written in decimal form. Each octet in decimal form can range from 0 through 255. This figure is based on the maximum number of binary positions in a binary octet.

Figure 11-3.
The IP addressing scheme can be compared to the telephone system. In a telephone system, a three-digit number indicates the area the telephone is in and the telephone number indicates the telephone. In an IP address, part of the IP address indicates the network the host is in and the other part indicates the host.

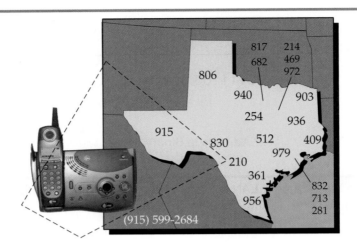

network node. The relationship of IP addresses to the network system is the same as telephone numbers to the telephone system. Each telephone is identified with a unique telephone number, and each node on a network is identified by a unique IP number.

The term *IP address* could mean either IPv4 or IPv6.

Network Class

For the purpose of assigning IP addresses, networks are divided into three major classifications: Class A, Class B, and Class C. Large networks are assigned a Class A classification. A *Class A network* can support up to 16 million hosts on each of 127 networks. Medium-sized networks are assigned a Class B classification. A *Class B network* supports up to 65,000 hosts on each of 16,000 networks. Small networks are assigned a Class C classification. A *Class C network* supports 254 hosts on each of 2 million networks.

Networks are assigned an IP address based on their network classification. Look at **Figure 11-4.** In the table, you can see that the class of the network determines the numeric value of the first octet in its IP address. The range for a Class A network is from 1 to 127, for a Class B network it is from 128 to 191, and for a Class C network it is from 192 to 223.

IP addresses for Class A networks use only the first octet as the network address. The remaining three octets define nodes on the network. The first two octets of a Class B network identify the network. The remaining two octets identify nodes on the network. A Class C network uses the first three octets to identify the network and the last octet to identify the nodes. For example, a typical Class C network might have a TCP/IP address of 201.100.100.12. The network is identified by the 201.100.100, and the node is identified as 12.

Class A network
a TCP/IP network classification that supports up to 16 million hosts on each of 127 networks.

Class B network
a TCP/IP network classification that supports up to 65,000 hosts on each of 16,000 networks.

Class C network
a TCP/IP network classification that supports 254 hosts on each of 2 million networks.

Network+ Note:
For the Network+ Certification exam, be sure you can identify IP addresses by their class.

Class	Range	Number of Networks	Number of Hosts
Class A	1–127	127	16,000,000
Class B	128–191	16,000	65,000
Class C	192–223	2,000,000	254

Figure 11-4.
IPv4 address
classifications.

Subnet Mask

A company may divide their network into several, smaller networks. A network within a network is known as a *subnetwork*. A *subnet mask* is a number similar to an IP address. It is used to determine in which subnetwork a particular IP address belongs. The subnet mask can be used to identify the class of network, but it is really intended to allow the network address to be broken down into subnetworks.

When the subnet mask is encountered, it is usually viewed in the decimal form—a series of four three-digit numbers. At first glance, a subnet mask may appear identical to an IP address. However, a subnet mask is distinguishable from an IP address because it begins with one or more octets of 255. An IP address cannot begin with 255.

The octets of a subnet mask correspond to octets in the IP address. The numbers found in a subnet mask depend on the class of the network and the number of subnetworks into which the network is divided. The subnet mask is combined with the IP address using the bitwise AND operation. The details of this operation are beyond the scope of this textbook. The resulting address is the *subnet address*. For now, just remember that the subnet mask is used to identify any subnetwork at a network address. Chapter 12—Subnetting covers the subnet mask in more depth.

subnetwork
a network within a network.

subnet mask
a number similar to an IP address that is used to determine in which subnet a particular IP address belongs.

Reserved IP Addresses

A number of IP addresses are reserved for private networks. They are often used for offices sharing an Internet address or for experimentation. They are not valid for use as a direct connection to the Internet. Look at the following chart of private IP addresses.

Class	Range	Subnet Mask
A	10.0.0.0 – 10.255.255.255	255.0.0.0
B	172.016.0.0 – 172.31.255.255	255.255.0.0
C	192.168.0.0 – 192.168.255.255	255.255.255.0

Network+ Note:

For the Network+ Certification exam, be sure you can identify a private IP address.

Microsoft Windows uses the Class C IP range for implementing a SOHO network. Typically, one computer is connected to the Internet through an ISP and other computers share the connection by creating a small TCP/IP network. The host is assigned the 192.168.0.1 IP address while the clients are assigned IP addresses

Network Address Translation (NAT) a protocol that translates private network addresses into an assigned Internet address, and vice versa. In other words, it allows an unregistered private network address to communicate with a legally registered IP address.

192.168.0.2 through 192.168.0.255. While the term *host* is used typically to describe any device on a network that has an IP address, the host in this scenario is the computer that hosts the Internet connection. Computers that connect to the Internet through the host are called clients.

Microsoft refers to this type of connection sharing as Internet Connection Sharing (ICS). ICS uses the **Network Address Translation (NAT)** protocol to translate private network addresses into an assigned Internet address, and vice versa. NAT was especially designed for implementing private network configurations. It allows an unregistered private network address to communicate with a legally registered IP address. For example, a client that wishes to access the Internet sends a request to the host. The host, in turn, uses the IP address assigned by the ISP to connect to the Internet. See **Figure 11-5.** Three main advantages of using NAT are as follows:

- Provides a firewall type of service by hiding internal IP addresses.
- Allows computers on a network to share one recognized IP address to access the Internet without the need of multiple IP addresses to be assigned to the subnetwork.
- Allows multiple ISDN connections to be combined into one Internet connection.

Network+ Note:

For the Network+ Certification exam, be sure to know the purpose of the Network Address Translation (NAT) protocol.

It is important to note that a private network such as an Intranet may use any possible IP address as long as it does not connect to the Internet. Using a number that is assigned to a URL would not only confuse the network system, it is illegal. The reserved private addresses can be used even when the network connects to the Internet. The Internet does not recognize the private network addresses, so they do not affect the system. This is why private network addresses are ideal for creating a subnetwork.

Figure 11-5.
Internet Connection Sharing (ICS) uses the Network Address Translation (NAT) protocol to translate unregistered IP addresses to a registered IP address. In this example, an ICS client sends an Internet request to the IP address of the ICS host, 192.168.0.1. The ICS host uses the ISP assigned address to access the Internet.

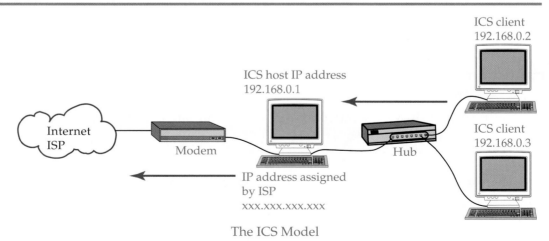

The ICS Model

The NAT protocol is commonly installed automatically in Microsoft Windows 2000 and later versions when the Network Setup Wizard is used. For example, when installing Windows 2000 or Windows XP through the Network Setup Wizard, the user is asked a series of questions. Based on the responses that are entered by the user, the operating system automatically sets up a connection share for a single Internet connection.

Viewing IP Configuration Settings

The IP configuration is a group of settings made on a host that allows it to communicate in a TCP/IP environment. The IP configuration consists of the host's IP address, the subnet mask, and the default gateway. The **default gateway address** is the address of the computer that provides a connection to the Internet. The Internet service provider's software or the network DHCP server usually determines these settings automatically. Occasionally, you may have to adjust the settings manually or verify the information while working with an ISP to troubleshoot a connection. On a Windows-based computer, you can use the IP Configuration utility to verify the settings.

Windows IP Configuration is accessed in Windows 95, 98, and Me with the **winipcfg** command. This command can be entered in the **Run** dialog box located in the **Start** menu. Windows NT, 2000, and XP use the **ipconfig** command issued from the command prompt to reveal similar information. See **Figure 11-6.** The assigned IP address, the subnet mask, and the default gateway address are displayed. To reveal more information about the connection, use the **ipconfig /all** command. See **Figure 11-7.** Notice that the MAC address (physical address), DHCP server address, and DNS server are revealed as well as other information. The additional information provided, such as the IP address of the DHCP server and the DNS server, will become more relevant to you later in the chapter.

To reveal more information about a connection using the **winipcfg** command, click the **More Info** button in the **IP Configuration** dialog box. This reveals a second dialog box.

default gateway address
the address of the computer that provides a connection to the Internet.

Domain Name System (DNS)
a system that associates a host or domain name with an IP address, making it easy to identify and find hosts and networks.

The terms *physical* and *logical* are often used to describe network systems and devices. It is important to be able to determine the difference between a physical and a logical component. For example, the IP address is a logical address assigned to a network host. It is logical because it is assigned and can be changed at any time. In contrast, the MAC address is a physical address electrically etched into the network interface card associated with each host.

Tech Tip

Domain Name System (DNS)

Since IP addresses are difficult to remember, the Internet uses a *Domain Name System (DNS)* or *Domain Name Service* to make it easier for people to

Figure 11-6.
IP Configuration display when the **ipconfig** command is issued.

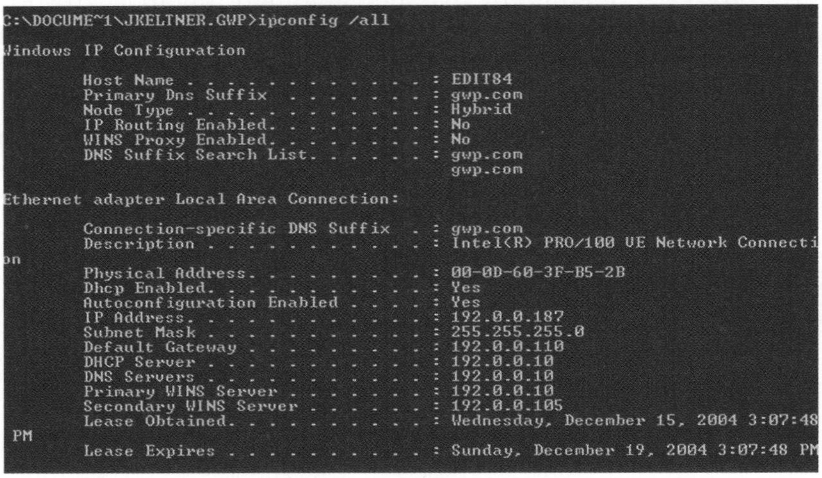

Figure 11-7.
The **IP Configuration**
display when the
ipconfig /all command
is issued.

```
C:\DOCUME~1\JKELTNER.GWP>ipconfig /all

Windows IP Configuration

        Host Name . . . . . . . . . . . . : EDIT84
        Primary Dns Suffix  . . . . . . . : gwp.com
        Node Type . . . . . . . . . . . . : Hybrid
        IP Routing Enabled. . . . . . . . : No
        WINS Proxy Enabled. . . . . . . . : No
        DNS Suffix Search List. . . . . . : gwp.com
                                            gwp.com

Ethernet adapter Local Area Connection:

        Connection-specific DNS Suffix  . : gwp.com
        Description . . . . . . . . . . . : Intel(R) PRO/100 VE Network Connecti
on
        Physical Address. . . . . . . . . : 00-0D-60-3F-B5-2B
        Dhcp Enabled. . . . . . . . . . . : Yes
        Autoconfiguration Enabled . . . . : Yes
        IP Address. . . . . . . . . . . . : 192.0.0.187
        Subnet Mask . . . . . . . . . . . : 255.255.255.0
        Default Gateway . . . . . . . . . : 192.0.0.110
        DHCP Server . . . . . . . . . . . : 192.0.0.10
        DNS Servers . . . . . . . . . . . : 192.0.0.10
        Primary WINS Server . . . . . . . : 192.0.0.10
        Secondary WINS Server . . . . . . : 192.0.0.105
        Lease Obtained. . . . . . . . . . : Wednesday, December 15, 2004 3:07:48
PM
        Lease Expires . . . . . . . . . . : Sunday, December 19, 2004 3:07:48 PM
```

Integrated Network Information Center (InterNIC)
a branch of the United States government that was responsible for regulating the Internet, overseeing the issue of domain names, and assigning IP addresses to them.

Internet Corporation for Assigned Names and Numbers (ICANN)
a company that manages domain name registrations by allocating domain name registration to select, private companies.

registrar
a select, private company that is assigned a pool of IP addresses from ICANN and handles domain registration.

identify and find networks on the Internet. It does this by allowing a name, called a domain name, to be associated with a network address. You can communicate on the Internet by using a domain name or an IP address. For example, when you use a Web browser you can enter the domain name for the URL or an IP address. The IP address connects faster to a distant location than a domain name. When a domain name is entered, it must be converted into an IP address before communication can begin. The amount of time it takes varies according to conditions. Many of these conditions are discussed throughout this chapter.

Internet Corporation for Assigned Names and Numbers (ICANN)

IP addresses were regulated and assigned through the government organization known as the *Integrated Network Information Center (InterNIC)*. InterNIC was a branch of the United States government under the direction of the Department of Commerce. It was responsible for regulating the Internet, overseeing the issue of domain names, and assigning IP addresses to them. InterNIC managed the domain root for .com, .net, .org, .edu, and .gov top-level domains. Management of the root is contracted to ICANN.

In 1999, the *Internet Corporation for Assigned Names and Numbers (ICANN)* took responsibility to manage domain name registrations. It does so by allocating domain name registration to select, private companies. Each of the select, private companies is called a *registrar.*

To obtain a domain name and an IP address, a business or organization places an application through a registrar. Once the registrar processes the application, the business or organization receives an official second-level domain name and an IP address that is registered to the domain name they have been issued.

The Internet Assigned Numbers Authority (IANA) was the original authority that the US contracted with in the early 1990s. IANA established the Internet Network Information Center (InterNIC) funded by the National Science Foundation. Late in the 1990s the government felt that the assigned numbers and names should be somewhat privatized, so they subcontracted the responsibilities to ICANN, a not-for-profit organization. ICANN assumed many of the responsibilities of IANA, but the ultimate authority for assigned numbers and names rests with the US government.

To find out more about ICANN, visit their Web site at www.icann.org.

Fully Qualified Domain Name (FQDN)

To locate a host on a large network, you need to use the complete name of the host, which includes a combination of the host name and domain name. The complete name of the host is referred to as the *Fully Qualified Domain Name (FQDN).* For example, a host called *Station12* in the domain xyzcorp.com would have a Fully Qualified Domain Name of station12.xyzcorp.com. *Station12* is the host name, and xyzcorp.com is the domain name.

As you have seen in previous chapters, network operating systems follow the domain name structure for naming servers, workstations, and other network devices. This was not always the case. Earlier systems, such as NetBIOS and NetBEUI, used their own naming schemes. These naming schemes are not consistent with naming rules used for Fully Qualified Domain Names. Network operating systems such as Linux, Windows 2000, and NetWare 4.0 and later follow the rules of Fully Qualified Domain Names when creating their network structure. However, these operating systems can still communicate with older networking technologies.

Realizing the difference between a host name and a NetBIOS name can help you understand how communication can occur between systems that use Fully Qualified Domain Names and those that use NetBIOS names. A NetBIOS name is a computer name limited to a maximum of 15 characters. A host name can be a Fully Qualified Domain Name or the first part of the Fully Qualified Domain Name. A host name is limited to 63 characters. Many of the symbols allowed in NetBIOS names cannot be used in Fully Qualified Domain Names. These symbols include the following:

$$; : " < > * + = \backslash \ / \ ? \ ,$$

To allow for communications between a system that uses NetBIOS names and one that uses host names, computer names are converted automatically for compatibility. For example, consider a network that consists of Windows 2000 and Windows NT 4.0 computers. Windows 2000 uses host names or Fully Qualified Domain Names to identify hosts on a Microsoft network. Windows NT 4.0 uses NetBIOS names to identify computers on a Microsoft network. The Windows 2000 computers can use the Fully Qualified Domain Name to communicate with each other. However, to communicate with the NT 4.0 computers, the Fully Qualified Domain Names are shortened automatically to match the NetBIOS maximum length of 15 characters. Windows NT 4.0 computers can use NetBIOS names to communicate with other Windows NT 4.0 computers on the network. However, when communicating with Windows 2000 computers, any symbols in the NetBIOS name need to be replaced with hyphens.

DNS Structure and Operation

DNS is hierarchical, similar to a typical file system structure comprised of directories and files, **Figure 11-8.** At the top of DNS is the root. Directly under the root are the top-level domains. The top-level domains start the process of dividing the organizational structure into specialized areas. The most common top-level domains are .com, .edu, .gov, .int, .mil, .net, and .org.

Fully Qualified Domain Name (FQDN) a combination of a host name and a domain name.

Figure 11-8.
The Domain Name
System structure.

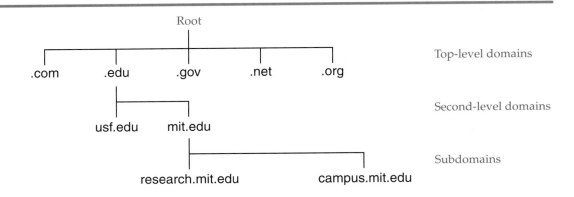

The next level of the structure is composed of second-level domains. Second-level domains represent the domain names as assigned by ICANN. They identify various companies and organizations. For example, the Massachusetts Institute of Technology domain name is known as mit.edu. The structure is further divided into subdomains controlled by the second-level domains. A *subdomain* is any level domain located beneath the secondary domain. For example, the mit.edu, second-level domain can be divided into subdomains such as faculty.mit.edu, research.mit.edu, and students.mit.edu. A subdomain can also be called a lower-level domain.

subdomain
any level domain located beneath the secondary domain. Also called a lower-level domain.

The Internet Corporation for Assigned Names and Numbers (ICANN) is a private, not-for-profit organization that coordinates the assignment of domain names and IP addresses. InterNIC is a Web site that provides services and information about Internet assigned names and numbers and is a registered service mark of the US Department of Commerce. The Department of Commerce licensed InterNIC to ICANN, which operates the InterNIC Web site. The US Department of Commerce is responsible for the overall assignment of the assigned numbers and names in the United States.

DNS root servers are located at the top of the domain name hierarchy structure. The root domain server stores the ultimate database for resolving an Internet domain name to a specific IP address. Currently there are thirteen DNS root servers. A top-level domain server stores DNS information of all top-level domains. A second-level domain server stores DNS information of all second-level domains.

DNS operation begins when a resolver requests that a domain name be resolved to an IP address. The *resolver* is a software program located on a host that queries a DNS server to resolve a host name to an IP address. The first DNS server the resolver contacts is a subdomain server, **Figure 11-9.** For example, when the resolver attempts to connect to *research.mit.edu*, it queries the first subdomain server it encounters. This could be the domain server the host is connected to as part of a workgroup or domain, or it could be the Internet service provider (ISP) for a home PC.

A subdomain server may be queried for the same information many times. Because the request is likely to occur again from a different host, domain names are placed in a cache on the domain server for future requests. This method speeds up the process of supplying IP address information when the request is

resolver
a software program located on a host that queries a DNS server to resolve a host name to an IP address.

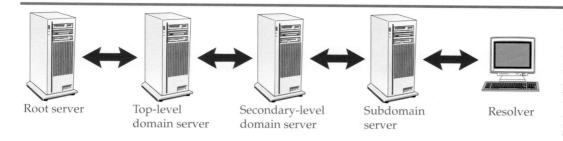

Root server   Top-level   Secondary-level   Subdomain   Resolver
                domain server   domain server   server

Figure 11-9.
The resolver queries the closest DNS server for IP information. If the DNS server does not have this information, the DNS server passes the request to a higher-level DNS server.

repeated. If the IP address for the domain name is not in the cache, the domain server queries higher-level domain servers until reaching the root. When the information is found, it is returned to the resolver. The resolver then connects to the intended host.

The domain name service is dynamically changing all the time. It is like a live entity that continues to grow and update with new information and changes. During the update period, errors can be encountered while trying to resolve a domain name to an IP address. Most servers will cache domain name information for approximately three days. The network administrator controls the time period. If a domain changes its IP address during the three-day period, an error can be generated attempting to connect to the domain site. This is because the old information about the IP address is still in the server cache and has not been updated.

Let's take a closer look at how DNS is used to acquire information about a domain name and its associated IP address. The following are the possible steps that occur when a resolver requests the IP address of a domain name. Refer to **Figure 11-10.**

1. First, the user types in www.g-w.com in the browser.
2. The browser checks if the domain name g-w.com exists in its own browser cache.
3. If not, the browser sends a DNS query to the network's domain name server. The domain server checks its own cache for the requested domain IP address. If this information is in the cache, it connects the host to the destination IP address.
4. If the information is not stored in the domain server cache, it is then requested from a higher-level domain server. The highest level is the top-level domain server located beneath the root.
5. If the domain name is an authorized domain name, the top-level domain server returns the IP address information. This information is added to the domain server and workstation cache.
6. The computer that originally attempted to connect to www.g-w.com now uses the IP address to connect to the Web server located at www.g-w.com.

Hosts and Lmhosts Text Files

Originally, all computer names were resolved using a text database. The text database is simply a text file that contains IP addresses and related information. The two text files used to resolve host and NetBIOS names to IP addresses are hosts and lmhosts. These files can still be used as a backup in case other means of resolving computer names to IP addresses fail.

Figure 11-10.
This example shows the possible steps that can occur while resolving a domain name to an IP address.

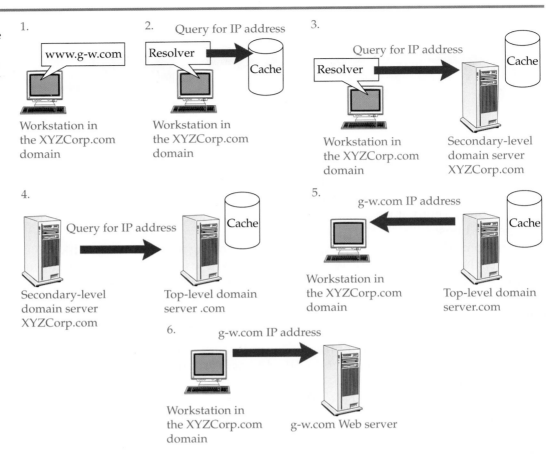

The text file hosts was the original method used by UNIX systems to match computer names to corresponding IP addresses. The following is an example of the contents of a hosts file.

127.0.0.1	Localhost
192.168.0.25	PcsAreUs.com
10.100.50.25	Station23.XYZcorp.com
172.63.10.24	USF.edu

Note that Fully Qualified Domain Names are considered host names and can appear in a hosts file. A list of IP addresses and corresponding host names are created using a simple, plain-text editor and saved as a hosts file. The hosts file is referenced automatically to assist in locating computers, printers, and servers on a network. The hosts file is used when the automatic functions provided by a DNS server fails.

The lmhosts file is similar to the hosts file but is designed for use with NetBIOS name resolution to IP addresses. Look at the following example of the contents of an lmhosts file.

192.168.23.104	Station12
10.23.105.21.1	Sales4
173.76.22.231	Computer25

Note that the file contains NetBIOS names of computers matched to IP addresses. The lmhosts file is read when the Windows Internet Naming Service (WINS) server fails. A ***Windows Internet Naming Service (WINS)*** server resolves NetBIOS names to IP addresses.

The hosts and lmhosts file must be edited with a plain-text editor such as Notepad. A more sophisticated editor uses many different fonts that make the file unreadable by the network client software. The hosts and lmhosts file must be in plain, ASCII text to be read. The key point to remember is that the hosts file resolves host names to IP addresses related to DNS, and the lmhosts file resolves NetBIOS names to IP addresses related to WINS.

Windows Internet Naming Service (WINS)
a service that resolves NetBIOS names to IP addresses.

The IP, TCP, and UDP Protocols

TCP/IP is actually a suite of protocols. The three most commonly used protocols in the TCP/IP suite are Internet Protocol (IP), Transmission Control Protocol (TCP), and User Datagram Protocol (UDP). IP is designed to control data transmission between two nodes and to ensure that a link has been established between source and destination. IP does not care if the packet is delivered correctly. IP simply transmits the data. IP is a connectionless protocol. UDP functions similar to IP. It establishes a link, but does not ensure that data is delivered correctly. It is also a connectionless protocol.

TCP ensures packets arrive intact and in correct order. TCP is a connection-oriented protocol. The main difference between TCP and UDP is that TCP can break large amounts of data into smaller packets and UDP cannot. UDP sends a single packet to transmit control information and data. This section looks at how the IP, TCP, and UDP move data from source to destination.

Relationship to the OSI Model

Figure 11-11 shows the relationship of some TCP/IP protocols to the OSI model and typical data formats. In the illustration, you can see that the three upper layers—application, presentation, and session—take raw data and send it down to the transport layer. At the transport layer, large blocks of raw data are broken into smaller units referred to as segments. The segments are placed into

OSI Layer	Data Format	Protocol
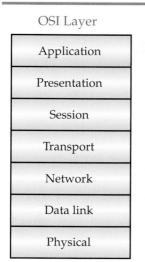	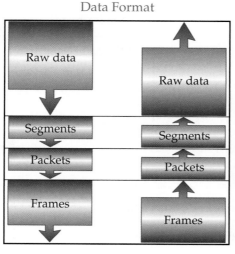	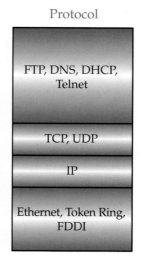

Figure 11-11.
The relationship of TCP/IP protocols and data formats to the OSI model.

the data portion of the TCP or UDP frame. The selection of TCP or UDP frame depends on three factors:

- ■ The software used in the upper layers of the OSI model.
- ■ The amount of data to be sent.
- ■ If a connectionless or connection-oriented transport of data is desired.

Both the TCP and UDP frame formats provide the port number of the source and destination. For UDP, the port number is optional, but for TCP, the port number is required. There will be more about port numbers later in the chapter.

The content of the TCP or UDP protocol is then moved to the network layer. At the network layer, the TCP or UDP frame is enclosed in an IP protocol frame. The entire unit is referred to as a packet. The IP packet contains a source and destination IP address and is sent to the data link layer.

The data link layer performs the final encapsulation of the combined frame formats (TCP or UDP and IP). The data link layer supplies the physical address of the destination and source. The physical address is the MAC address associated with the network card of the source and destination computer. The collective unit of frame formats is referred to as a frame.

The frame is placed on the physical media as a series of digital pulses. If the network is Ethernet, the frame is broadcast to all nodes in the segment of the network. When the frame arrives at the destination, the process is reversed. As the data moves up the OSI model, each of the protocol frame formats is removed until the raw data is received at the highest level of the OSI model, the application layer.

Frame Formats

Now lets take a closer look at the frame formats of IP, TCP, UDP, and Ethernet to get a better understanding of the mechanics of the process. Look at **Figure 11-12.** In the illustration, a UDP frame is inserted into the data portion of an IP frame.

The UDP frame format is simple in construction. The header information contains only four significant items: source port, destination port, message length, and checksum. The frame's structure reveals that only port addresses are used to identify the source and destination. There is no way to identify the source or destination by IP address or MAC address. UDP and TCP rely on the IP frame

Figure 11-12.
UDP is framed in an IP packet, and then the IP packet is framed in an Ethernet frame.

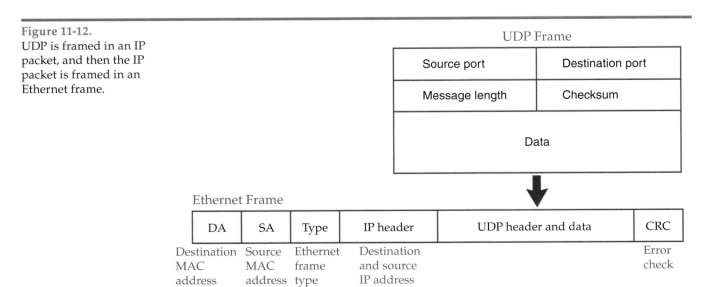

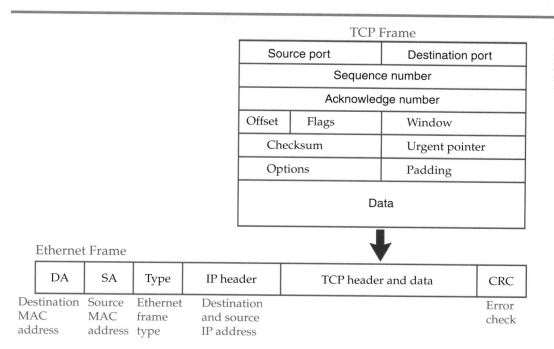

Figure 11-13.
TCP is framed in an IP
packet, and then the IP
packet is framed in an
Ethernet frame.

format for the IP address of the destination and source. The IP frame relies on the Ethernet frame for the source and destination MAC address.

Now look at **Figure 11-13.** Notice that the TCP header contains more information than the UDP header. The TCP header contains additional information such as a sequence number, which is used to reassemble data in a correct order. UDP does not need a sequence number. It is used to transfer small blocks of data or commands.

As the protocol frames reach the data link layer of the OSI model, the final step is to encapsulate the entire set of protocol frames into a frame matching the network standard used, such as Ethernet, Token Ring, and FDDI. Look at **Figure 11-14.** In the

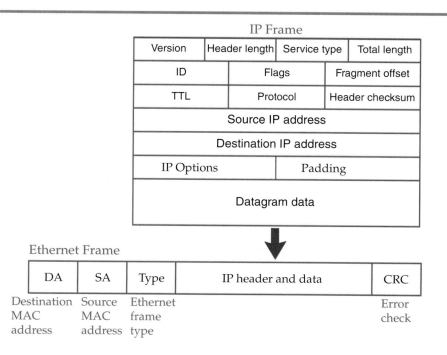

Figure 11-14.
IP is framed in an
Ethernet frame. The
Ethernet frame is then
placed on the network
media.

example, Ethernet has been chosen. The Ethernet frame relies on the MAC addresses for transmitting between workstations. Remember that a unique MAC address is embedded into each network interface card. The Ethernet frame is broadcast to all the network interface cards on the LAN. When the destination network interface card receives the entire frame of protocol formats and data, it removes the Ethernet frame and passes the remaining protocol frame formats to the OSI model at the destination. If the destination is on the same LAN as the source, then the remaining formats are removed until the data reaches the application layer.

If the destination is at a distant location that requires navigation across the Internet or a WAN link, a network device such as a router removes the Ethernet frame. The IP address is used to deliver the entire contents to the distant location. At the distant location, the remaining protocol formats are removed until the raw data is revealed.

This is a simplified explanation of the role of the various protocol frame formats. You will rely on these basic concepts to understand how bridges, switches, routers, and gateways operate. Other protocols classified as application protocols, such as FTP, TFTP, Telnet, and SNMP, fit inside a TCP or UDP frame.

To learn more about various protocol formats associated with TCP/IP visit www.protocols.com. This Web site has an extensive listing and a complete explanation of each field within the frame format. There are also hot links, acronyms, and technical papers available, which will assist you in furthering your network expertise.

Assigning IP Addresses

static IP assignment an IP address that is entered manually for each host on a network.

There are several methods for assigning IP addresses. The assignment of IP addresses falls into two broad categories: static and dynamic. A *static IP assignment* means that an IP address is entered manually for each host on the network. This is typically done when the network interface card is configured. A *dynamic IP assignment* means that the IP address is issued automatically, typically when the computer boots and joins the network.

dynamic IP assignment an IP address that is issued automatically, typically when the computer boots and joins the network.

Windows Internet Naming Service (WINS)

Windows Internet Naming Service (WINS) is used to resolve computer NetBIOS names to IP addresses. WINS is unique to Microsoft networks. WINS maintains a dynamic database of computer names and IP addresses. The computer names and IP addresses are automatically updated as computers log on and off the network, **Figure 11-15.**

Starting with Windows 2000, the WINS service is no longer required. In the Windows 2000 and later, a new service called Dynamic DNS incorporates the features of WINS with DNS.

Network+ Note:

WINS and DNS are often confused. For the Network+ Certification exam, be sure you can distinguish between DNS and WINS.

Network+

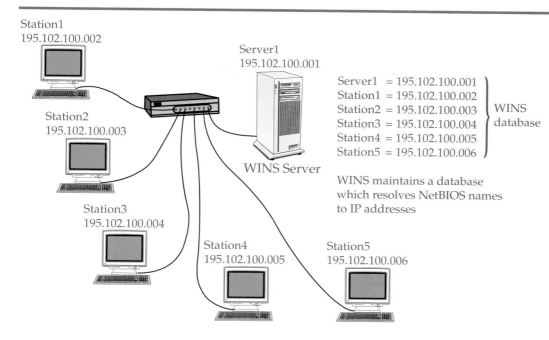

Figure 11-15.
The Windows Internet
Naming Service
(WINS) resolves
NetBIOS computer
names to IP addresses.

Station1
195.102.100.002

Server1
195.102.100.001

Station2
195.102.100.003

Server1 = 195.102.100.001
Station1 = 195.102.100.002
Station2 = 195.102.100.003
Station3 = 195.102.100.004
Station4 = 195.102.100.005
Station5 = 195.102.100.006

WINS
database

WINS Server

WINS maintains a database
which resolves NetBIOS names
to IP addresses

Station3
195.102.100.004

Station4
195.102.100.005

Station5
195.102.100.006

Dynamic Host Configuration Protocol (DHCP)

Originally, computers on a network had to have their IP addresses assigned manually. This method was called *static addressing*. Static addressing was a time-consuming operation if a large number of PCs were on a network. A log of computer names, locations, MAC addresses, and the assigned IP addresses had to be recorded manually. An administrator or network technician had to be careful not to use the same IP address on another computer. This is because each IP address has to be unique. Using the same IP address for two different computers causes communication conflicts, resulting in erratic behavior.

Dynamic Host Configuration Protocol (DHCP) is designed to replace the manual setup of IP addresses on a network. When a server runs DHCP, the IP addresses are assigned automatically. The act of automatically assigning IP addresses is known as *dynamic addressing.*

The DHCP server has a pool, or list, of IP addresses to draw from, **Figure 11-16.** Each computer that logs on to the network is assigned an address from the pool. The IP address assignment is temporary. The address is released after a period of time and may be reissued to another computer.

This is how a client is assigned an IP address from a DHCP server. First, when *Station3* is booted, it automatically requests an IP address assignment from the DHCP server. *Station3* does not know where the DHCP server is, so it sends out the request for an IP address as a broadcast to all nodes on the network segment. The protocol used for the request is UDP, which is especially designed for broadcast communication.

Second, the DHCP server receives the request and returns an unassigned IP address from its pool of IP addresses. This is also done in the form of a broadcast. The server returns the information as a UDP broadcast so that any other device connected to the same network as *Station3* will have the IP address information.

Third, *Station3* accepts the assigned IP address and acknowledges this with a UDP broadcast to all nodes on the network. This allows any other device, like a backup DHCP server, to update their database of IP address assignments.

static addressing
assigning an IP
address manually.

*Dynamic Host
Configuration
Protocol (DHCP)*
a service that assigns
IP addresses auto-
matically to the hosts
on a network.

dynamic addressing
automatically assign-
ing IP addresses.

Figure 11-16.
The Dynamic Host Configuration Protocol (DHCP) on a DHCP server automatically assigns a temporary IP address to a host. The IP address is randomly selected from a pool of addresses. The temporary IP address is returned to the pool when the lease expires. Certain types of equipment, such as servers, must maintain the same address all of the time. They are assigned an IP address manually. When an IP address is manually assigned, it is said to be a static IP address.

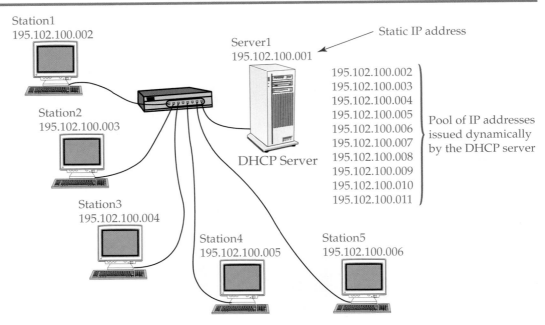

Each workstation receives a temporary IP address from the DHCP server

Fourth, the DHCP server acknowledges that *Station3* has accepted the assigned IP address. This completes the four-step process.

The temporary assignment of the IP address to *Station3* requires four broadcast packets. This is not a lot of network traffic compared to several hundred or even thousands of computers logging on to the network at the beginning of the workday. Each computer generates four broadcasts to acquire an IP address from a DHCP server. In addition, each computer generates traffic when downloading e-mail. The traffic on a network in this scenario is heavy, and the number of collisions can greatly increase.

During the IP address negotiation between the DHCP server and *Station3,* other information is transmitted, such as the lease period for the assigned IP address, the subnet mask, and the location of the default gateway. The IP address assignment can be easily confirmed by using **winipcfg** or **ipconfig**.

It is important to note that when you are setting up a network using a DHCP server, you must assign static IP addresses to all servers and printers and any other equipment required to have a static IP address. When the DHCP server assigns an IP address to a host, it does not necessarily assign the same IP address to the same piece of equipment each time. The IP address is randomly assigned. For computers to be able to consistently find devices such as servers and printers, the devices must have static IP addresses assigned to them.

DHCP Lease

The DHCP server sets a lease period for IP address assignments. The default lease period is 72 hours, but the network administrator can modify the lease period. The length of the lease period depends on factors such as the type of network, number of users, typical user profiles, and available number of IP addresses assigned to the network. A large corporate network would need a longer lease period. It is up to the professional judgment of the network administrator to determine the lease period. If the network belongs to an ISP, the ISP administrator would issue a shorter lease period.

If the lease period spans 30 days, for example, the client would not need to request an IP address from the DHCP server the next time they log on as long they log on during the 30-day time period. If the lease period expires, the computer must request a new temporary IP address from the DHCP server.

However, when a lease has expired, it is important that a user properly log off or shut down their computer when ending a session. The IP address is only released when a user logs off or properly shuts down the computer. If a person simply shuts down the computer using the power switch, the computer cannot properly release the IP address. The IP address will still be in use and will not be available to other clients.

Automatic Private IP Addressing (APIPA)

A special set of IP addresses have been set aside by IANA to use when a DHCP server cannot be reached. The range is 169.254.0.0 through 169.254.255.255. If a network interface card fails to connect to a DHCP server and if the card has not been set up with a static IP address, the workstation automatically generates an IP address in the range from 169.254.0.1 to 169.254.255.255. It uses this IP address to communicate with other workstations on the same segment. Once the DHCP server comes back on line and can resume issuing IP addresses from its pool of IP addresses, the workstation releases the 169.254.xxx.xxx IP address and takes one from the DHCP server.

This type of addressing is called *Automatic Private IP Addressing (APIPA)* and is compatible with all Microsoft products starting with Windows 98. This process was not necessary in prior versions of Windows because the NetBEUI protocol was automatically configured when a network interface card was installed. If the TCP/IP protocol failed because the DHCP server did not issue an IP address to the workstation, the workstation used the NetBEUI protocol. The NetBEUI protocol uses the MAC address to communicate with workstations on the local segment. On Windows 98 and later workstations, NetBEUI is not installed by default. When the DHCP server fails, there is no way for the computers to communicate on the local segment.

APIPA is configured in the **Internet Protocol (TCP/IP) Properties** dialog box, **Figure 11-17.** At the top of the dialog box is the **Automatic private IP address** option. Notice that the other option, **User configured**, allows the workstation to be installed on more than one network by assigning an alternate set of IP settings and preferred DNS and WINS server settings. The **User configured** option is an excellent feature for laptop computers that may be moved to more than one location.

Bootstrap Protocol (BOOTP)

The *Bootstrap Protocol (BOOTP)* was developed in the mid-1980s. It uses a centralized database of the MAC addresses and IP assignments of all devices on the network. The network administrator is required to enter every MAC address and IP address into the database. When a computer or other network device is booted, it automatically queries the BOOTP server and requests an IP assignment. The BOOTP process is similar to the DHCP IP assignment process.

BOOTP is ideal for workstations that do not have a hard disk drive. A workstation that does not have hard disk drives installed is referred to as a *diskless station.* The term *diskless station* is also used to refer to workstations that do not contain floppy disk drives.

automatic private IP addressing (APIPA)
a feature that assigns an IP address from a special set of IP addresses that have been set aside by IANA when a DHCP server cannot be reached.

Bootstrap Protocol (BOOTP)
a service that uses a centralized database of the MAC addresses and IP assignments of all devices on the network and assigns the appropriate IP address to a host when it boots.

diskless station
a workstation that does not have hard disk drives. It can also mean a workstation that does not have a floppy disk drive.

Figure 11-17.
The **Internet Protocol
(TCP/IP) Properties**
dialog box allows a
workstation to be
assigned alternate
TCP/IP settings.

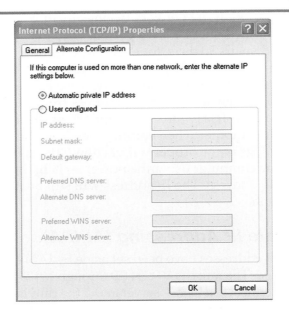

Figure 11-17.
The **Internet Protocol
(TCP/IP) Properties**
dialog box allows a
workstation to be
assigned alternate
TCP/IP settings.

TCP/IP Ports and Sockets

port number
a number that is
associated with the
TCP/IP protocol and
is used to create a
virtual connection
between two
computers running
TCP/IP.

socket
a port number com-
bined with an IP
address.

A *port number* is associated with the TCP/IP protocol and is used to create a virtual connection between two computers running TCP/IP. A computer, especially a server, runs many services at the same time. For example, a server could coordinate many e-mail requests, directory lookups, file transfers, Internet search requests, and Web page requests. If it were not for port numbers, communications would be confused and errors would occur.

Port numbers are assigned to network services. When the port number is combined with an IP address, it is referred to as a *socket.* The following is an example of a socket: 192.168.20.45:80. Notice that the IP address 192.168.20.45 is connected with a colon (:) to port 80.

To understand how port numbers work, think of an old-time telephone switchboard that you see in the old movies. A human operator plugs a connection cord into specific, physical sockets to complete a connection between two telephones. The connection is maintained for the duration of the call. The switchboard has many simultaneous conversations going on at the same time. Like the old-time telephone switchboard, socket technology allows multiple, simultaneous services to run between a server and a workstation.

Look at **Figure 11-18.** You can see how the server uses port numbers combined with an IP address to sort the various communications taking place. Note that two sockets are created between the workstation and the server for file transfer and another socket is created for Web page browsing. Port number assignments are not limited to client/server communications. They are used to communicate between any two devices on a network.

Figure 11-19 lists some typical port number assignments. Looking at the chart you can see some familiar port assignments such as port 80, which is associated with a Web server; port 20, which is assigned to data transfers using FTP; and port 21, which is used for FTP control commands. For a complete list of assigned ports, visit www.iana.org/assignments/port-numbers.

The port number is derived from a sixteen-bit binary field, which means there are over 65,000 different possible port numbers available for use. Ports 0

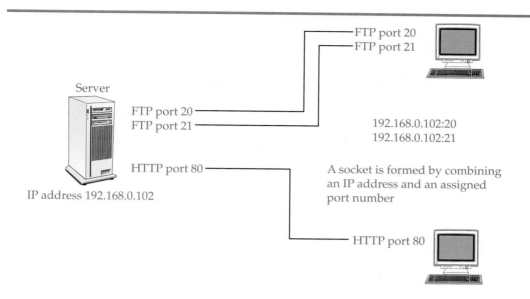

Figure 11-18.
Communications can be sorted by the creation of a virtual circuit between a server and a workstation or between a workstation and a workstation. The virtual circuit is a combination of an IP address and a port number. This combination is called a socket. The socket creates and maintains a logical communication channel between two points on a network.

through 1023 are assigned by IANA and are called *well-known port numbers* because they have typical assignments. The port numbers higher than 1023 are referred to as *upper-level port numbers*. They do not have typical assignments. Private companies commonly use the numbers from 1023 through 49151 to support their software utilities. The remaining numbers, 49152 through 65535, can be used by anyone.

Network+ Note:

To be prepared for the Network+ Certification exam, memorize the common port numbers associated with TCP/IP and their function. The most common port numbers encountered on exam are 20, 21, 23, 80, and 110.

TCP/IP Troubleshooting Utilities

There can be many types of problems associated with TCP/IP technology. Fortunately, there are a number of useful utilities that can identify the problem. These utilities are **netstat**, **nbtstat**, **ping**, **tracert**, **arp**, and **nslookup**. This section only introduces these tools. Further experience can be gained by using these tools in the lab and in the field.

Netstat

The **netstat** command displays current TCP/IP and port statistics. It can be used to determine network problems such as excessive broadcasts on the network. It also allows the user to monitor network connections.

Figure 11-20 shows the results of the **netstat -a** command issued at the command prompt. Notice that the **netstat -a** command has returned a list of services, port addresses, and type of packet sent (TCP or UDP). The **-a** switch is just one of the many switches available to use with the **netstat** command.

Figure 11-19.
Typical port
assignments.

Port #	Service	Description
7	ECHO	Echo a reply
20	FTP	File Transfer Protocol data
21	FTP	File Transfer Protocol control commands
22	SSH	Secure Shell
23	TELNET	Terminal emulation connection
25	SMTP	Simple Mail Transfer Protocol
43	NICNAME	Who Is
49	LOGIN	Login Host Protocol
53	DNS	Domain Name Server
69	TFTP	Trivia File Transfer Protocol
80	HTTP	Hypertext Transfer Protocol
110	POP	Post Office Protocol
119	NNTP	Network News Transfer Protocol
123	NTP	Network Time Protocol
137	NETBIOS	NetBIOS name service
143	IMAP4	Internet Message Access Protocol version 4
161	SNMP	Simple Network Management Protocol
389	LDAP	Lightweight Directory Access Protocol
443	HTTPS	HTTP Security
500	IPSEC	IP Security
1723	PPTP	Point-to-Point Tunneling Protocol
5631	pcAnywhere data	pcAnywhere data
5632	pcAnywhere status	pcAnywhere status
3389	Windows Remote Desktop	Remote access to desktops

Nbstat

The **nbstat** command is used to display NetBIOS over TCP statistics. In **Figure 11-21,** you can see a list of NetBIOS names of equipment connected to the local area network and workgroup names.

Remember, NetBIOS does not use IP addresses to identify computers on a network. Windows-based networks use a protocol called Server Message Block (SMB) and NetBIOS names to identify computers on a network. For TCP/IP to resolve NetBIOS names, the WINS utility must be running or the lmhosts file must be manually configured.

Tech Tip

SMB has been upgraded, or replaced, by the Common Internet File System (CIFS) protocol. SMB and CIFS are the Microsoft network protocols that allow print and file sharing on a Microsoft network.

The **nbstat** command can be issued from the command prompt with various switches that affect the outcome of the command. Remember, switches for commands issued from the command prompt can be revealed by using the **/?** or **/help** switch after typing the command at the command prompt. Be sure to leave a space between the switch and the end of the command.

Ping

The **ping** command is used to send a packet from one host to another on a
network and then echo a return reply. It is commonly used to quickly check the
connection state of network media between two hosts on a network. The **ping**
command can be used on LANs and WANs. To run the **ping** utility, you simply
access the command prompt and then issue the **ping** command followed by the
IP address or the host name with which you wish to test communications.

In **Figure 11-22,** you see two ways to issue the **ping** command. In the top
portion of the screen, the command is issued using the destination host name
multiboot. When a host name is used, the reply displays the destination IP address.
In this example, it is 192.168.0.3. In the middle portion of the screen, the **ping**
command has been issued using the destination IP address. The same results as
in the previous command are displayed. Note that each reply from the
destination indicates the number or bytes, lapse time in microseconds, and the
time to live (TTL) of the packet sent.

The **ping** command uses the Internet Control Message Protocol (ICMP) to
carry troubleshooting data across the network. The Internet Control Message

Figure 11-22.
The **ping** command sends a packet from one host to another on a network and then echoes a return reply.

```
Command Prompt                                          _□×
Pinging multiboot [192.168.0.3] with 32 bytes of data:

Reply from 192.168.0.3: bytes=32 time<10ms TTL=128
Reply from 192.168.0.3: bytes=32 time<10ms TTL=128
Reply from 192.168.0.3: bytes=32 time<10ms TTL=128
Reply from 192.168.0.3: bytes=32 time<10ms TTL=128

Ping statistics for 192.168.0.3:
    Packets: Sent = 4, Received = 4, Lost = 0 (0% loss),
Approximate round trip times in milli-seconds:
    Minimum = 0ms, Maximum =  0ms, Average =  0ms

F:\>ping 192.168.0.3

Pinging 192.168.0.3 with 32 bytes of data:

Reply from 192.168.0.3: bytes=32 time<10ms TTL=128
Reply from 192.168.0.3: bytes=32 time<10ms TTL=128
Reply from 192.168.0.3: bytes=32 time<10ms TTL=128
Reply from 192.168.0.3: bytes=32 time<10ms TTL=128

Ping statistics for 192.168.0.3:
    Packets: Sent = 4, Received = 4, Lost = 0 (0% loss),
Approximate round trip times in milli-seconds:
    Minimum = 0ms, Maximum =  0ms, Average =  0ms

F:\>
```

Protocol (ICMP) is a part of the TCP/IP suite of protocols and provides the ability to remotely troubleshoot and monitor devices on network systems.

Tracert or Traceroute

You can think of the trace route utility, **tracert** or **traceroute**, as an advanced **ping** utility. Unix/Linux systems use the **traceroute** command, while Microsoft systems use the **tracert** command. In this section, examples of the trace route utility use the **tracert** command.

The trace route utility sends a packet to a destination host, gathers statistics and information along the way, and displays the information on the monitor of the originator. The final destination point echoes information such as who owns the site's domain name, the IP address, and geographic location. Also displayed is each connection point along the route of the trace with information about its location. Most of the connection points displayed are routers.

The trace route utility also displays the approximate hop lapse times between points along the route. The amount of time delay can help analyze network failure or problems caused by excessive time delays.

Tech Tip

There are much more advanced utilities that can be used to trace a route to a destination host. One such program is NeoTrace developed by NeoWorx Inc. Their Web site is located at www.neoworx.com.

Figure 11-23 shows the information revealed when a **tracert** command has been issued from the local host to the www.novell.com Web site. A series of hops, or router connections, are revealed along with the amount of time between each hop. Trace route is a good utility for troubleshooting a path to a distant destination.

ARP

Address Resolution Protocol (ARP)
a service that maps a MAC address to an IP address.

Address Resolution Protocol (ARP) is used to map the host MAC address to the logical host IP address. For a computer to be able to run ARP, ARP must know the computer's IP address. ARP is a protocol in the TCP/IP suite of protocols. Since it is dependent on TCP/IP, it needs to have an IP address assigned before it can reveal any information. ARP also caches information about contacts on the network.

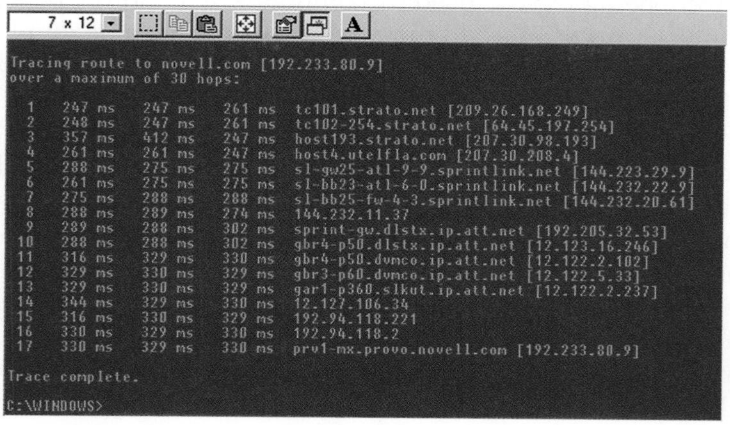

Figure 11-23.
The trace route utility
is used to gather and
display statistics and
information about the
route to a destination
host.

Reverse Address Resolution Protocol (RARP) is used to find the MAC address of the host when the IP address is known. It serves the opposite function of ARP. RARP was originally developed to support the BOOTP protocol. Remember that BOOTP is used to obtain an IP address automatically from the BOOTP server.

In **Figure 11-24,** the **arp -a** command lists IP addresses resolved to MAC addresses. The information displayed comes from the ARP cache on the 192.168.0.3 host. The ARP cache contains the IP and MAC addresses the 192.168.0.3 host has communicated with in the last two minutes. The information presented is displayed in a database format. The **arp** command can be used to verify IP address and MAC address assignments.

Reverse Address Resolution Protocol (RARP) a service that finds the MAC address of a host when the IP address is known.

Nslookup

The **nslookup** command is a UNIX/Linux utility used to query domain servers when seeking information about domain names and IP addresses. The **nslookup** command maps, or resolves, domain names to IP addresses. This is a convenient tool when looking for information about a particular domain or IP addresses. **Nslookup** is very similar to the **whois** command found at many Web server providers.

The IPv6 Standard

The newest Internet addressing scheme is called *IPv6.* It is also referred to as a classless IP addressing scheme because there is no need for a subnet mask based on network class. This scheme was developed because the number of

IPv6 an Internet addressing scheme that uses 128 bits to represent an IP address. The 128 bits are divided into 8 units of 16 bits. These units can be represented as a 4-digit hexadecimal number separated by colons.

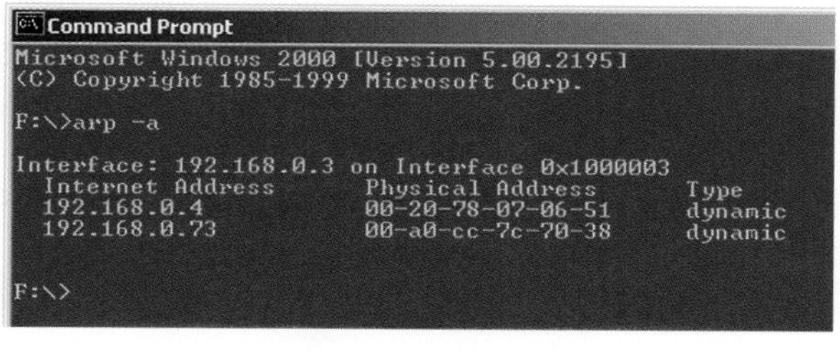

Figure 11-24.
The **arp -a** command
reveals information
found in the ARP
cache. The ARP cache
stores the IP and MAC
addresses the host has
communicated with in
the last two minutes.

IP addresses were being rapidly used by computers, telephones, pagers, fax machines, printers, televisions, and more. The pool of available numbers is rapidly dwindling and a need for a new address system that will not run out of numbers in the near future is evolving. To solve the problem, the IPv6 standard was created, which increased the pool of IP addresses available.

Modern computer equipment and software readily accepts IPv6 addresses and uses them to communicate across a network. The IPv6 standard presents an address of 128 bits. The 128 bits are divided into 8 units of 16 bits. These units can be represented as a 4-digit hexadecimal number separated by colons. See **Figure 11-25**. The IPv6 address is assigned to network interface cards and equipment in a similar fashion as the IPv4 standard.

It is interesting to note that with the development of the NAT protocol, the numbers of IP addresses are lasting longer than expected. By using the NAT protocol to translate private IP addresses on a home or office network, thousands, if not millions, of IP addresses are saved.

Loopback Address

loopback address
the reserved IP address
of the network inter-
face card.

The *loopback address* is the reserved IP address of the network interface card. It is used to test if an IP address is configured for the network interface card and if the network interface card is functioning normally. The loopback address for IPv4 is 127.0.0.1. The loopback address for IPv6 is 0:0:0:0:0:0:0:1 or ::1. Note that the double colon eliminates fields containing only zeros.

IPv6 MAC Address

The original IEEE 802 MAC address consists of a 24-bit manufacturer's ID and a 24-bit unique ID associated with each network interface card. The IPv6 standard uses a 24-bit manufacturer's ID and a 40-bit unique ID to identify each network interface card. See **Figure 11-26**.

IPv6 has been slowly introduced into networking systems and will probably not be fully implemented until the current set of IPv4 addresses are depleted. Traditional IPv4 utilities will not operate on IPv6 hosts. The IPv4 utilities were designed for 32-bit IP addresses and cannot resolve 128-bit IP addresses associated with IPv6. The **ping** command associated with IPv4 is not valid with IPv6. You must use the **ping6** command. For example, type **ping6 ::1** at the command prompt to verify the localhost connection.

Figure 11-25.
The IPv6 addressing
scheme.

128 bits

0011001101010011000011110101001111111100010100011001100101011001000111
0010101010101111000101000100010010000110001101010010 1010

8 units of 16 bits

| 0011001101010011 | 0000111101010011 | 1111110001010001 | 1100110010101110 |
| 0100011100101010 | 1010111100010100 | 0100010010000110 | 0011010100101010 |

8 units of hexadecimal numbers

3353 : F53 : FC51 : CCAE : 472A : AF14 : 4486 : 352A

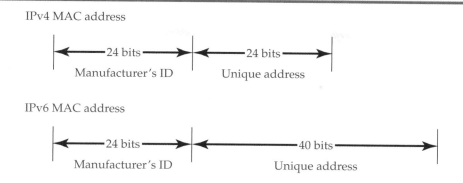

Figure 11-26.
IPv4 and IPv6 MAC
address comparison.

Summary

- An IP address is used to identify a host on a TCP/IP network.
- There are two types of IP addressing schemes: IPv4 and IPv6.
- IPv4 uses four octets separated by periods to uniquely identify each host.
- The decimal number located in the first octet of an IP number can identify its class. For example, a Class A network is identified by the 0–127 range, a Class B network by the 128–191 range, and a Class C network by the 192–223 range.
- The IPv4 assignment uses a subnet mask that identifies which portion of the IP address is the network address and which portion is the host address.
- The three reserved, or private, IP addresses are: 10.0.0.0, 172.16.0.0, and 192.168.0.0.
- The Network Address Translation (NAT) protocol provides the technology to allow multiple workstations to share one common Internet connection.
- A Fully Qualified Domain Name (FQDN) consists of a host name and a domain name.
- The Domain Name System (DNS) matches host and domain names to IP addresses.
- Once a domain name is matched to an IP address, a subdomain DNS server places the domain name and correlating IP address information in its cache for future reference.
- DNS is a hierarchical system composed of top-level domains, second-level domains, and subdomains.
- Servers are typically assigned a static IP address.
- A hosts file is a text file that contains a listing of computer names and corresponding IP addresses.
- A hosts file is used to resolve computer names to IP addresses when a DNS server is unavailable.
- WINS matches IP addresses to NetBIOS names.
- WINS dynamically updates a database of NetBIOS names and IP addresses.
- WINS works solely inside a LAN.
- The Dynamic Host Configuration Protocol (DHCP) is used to dynamically, or automatically, assign IP addresses to workstations.
- Automatic Private IP Addressing (APIPA) is a technique that temporarily issues an IP address in the range of 169.254.0.1 to 169.254.255.255 in case of a DHCP server failure.

■ IPv6 uses eight sets of 16-bit hexadecimal numbers to uniquely identify each host.

■ IPv6 does not require a subnet mask.

■ IPv6 is downward compatible with IPv4.

■ The loopback address for an IPv4 address is 127.0.0.1.

■ The loopback address for an IPv6 address is ::1.

Review Questions

Answer the following questions on a separate sheet of paper. Please do not write in this book.

1. What is the default protocol for communication across the Internet?

2. How many bytes are in an IPv4 address?

3. How many bits are in an IPv4 address?

4. How many octets are in an IPv4 address?

5. What is the maximum and minimum range of numbers in an octet?

6. List the IP addresses commonly assigned as private networks.

7. The _____ protocol allows an unregistered private network address to communicate with a legally registered IP address.

8. Briefly explain how the NAT protocol works.

9. A technician using Windows 2000 wishes to inspect the subnet mask used for a workstation at which they are sitting. How might they find the subnet mask information?

10. What two parts make up an FQDN?

11. What two methods are used to assign an IP address to a host?

12. Compare and contrast DNS and WINS.

13. What is contained in a typical lmhosts file?

14. What is contained in a typical hosts file?

15. Identify each of the following protocols as either connectionless or connection-oriented.

    A. TCP

    B. IP

    C. UDP

16. Briefly describe how TCP, UDP, and IP relate to the OSI model.

17. On what does the TCP frame rely to identify source and destination IP addresses?

18. On what does the UDP frame rely to identify source and destination sIP addresses?

19. On what does the IP frame rely to identify the source and destination MAC addresses?

20. What is the function of DHCP?

21. Briefly describe how DHCP works.

22. What type of addressing automatically generates an IP address in the range of 169.254.0.1 to 169.254.255.255 if a network interface card fails to connect to a DHCP server?

23. What port is indicated by the following address: 192.168.23.45:80?

24. What utility is most likely used to quickly verify a connection between two network points?

25. What TCP/IP utility can reveal the number of hops between the source and destination host?

26. What TCP/IP utility displays current network protocol statistics, such as the number of packets sent?

27. What TCP/IP utility displays a current listing of NetBIOS names and their associated IP addresses?

28. How many bits are in an IPv6 address?

Sample Network+ Exam Questions

Answer the following questions on a separate sheet of paper. Please do not write in this book.

1. Which network operating system(s) use TCP/IP as a default protocol? (Select all that apply.)
   A. Windows 2000
   B. Linux
   C. NetWare 4.0
   D. UNIX

2. Which technology supports the automatic assignment of IP addresses to network hosts?
   A. WINS
   B. DHCP
   C. ARP
   D. DNS

3. Which command reveals the assigned IP address and MAC address on a Windows XP computer?
   A. **macip**
   B. **winmac**
   C. **ipconfig**
   D. **winipcfg**

4. Which IP address is commonly assigned to private networks?
   A. 123.244.12.0
   B. 192.168.0.0
   C. 255.255.255.255
   D. 1.1.1.0

5. Identify the subnet mask of a Class B network.
   A. 255.0.0.0
   B. 255.255.0.0
   C. 0.0.255.255
   D. 255.255.255.0

6. What is the loopback address for an IPv4 protocol?
   A. 127.0.0.1
   B. 255.255.255.255
   C. 000.000.000.000
   D. 255.000.000.000

7. What function does a WINS server perform?
   A. Automatically assigns a Mac address to a network adapter.
   B. Resolves IP addresses to MAC addresses.
   C. Resolves NetBIOS names to IP addresses.
   D. Resolves domain names to IP addresses.

8. Which is an example of a Fully Qualified Domain Name?
   A. 3F 2B 12 CC D2 1F
   B. //workstation1/server
   C. //server1/client23
   D. workstation1.auditing.abc.com

9. A technician types **ipconfig** at the command prompt of a Windows XP workstation and sees the assigned IP address for the workstation is 169.254.1.12. What can the technician surmise from the assigned address?
   A. The workstation is using a static IP address.
   B. The DHCP server is down.
   C. The WINS server is not on line.
   D. The network is functioning normally.

10. What type of information is stored in the ARP cache?
    A. IP and MAC addresses.
    B. IP addresses only.
    C. IP addresses and domain names of workstations.
    D. MAC addresses only.

Interesting Web Sites for More Information

www.iana.org/assignments/port-numbers

www.icann.org

www.internic.org

www.neoworx.com

Suggested Laboratory Activities

1. Locate the sample lmhosts file called lmhosts.sam on a Windows computer by using the **Search** option located on the **Start** menu. Once you've located the lmhosts.sam file, open it in Notepad. A sample file should be displayed with information about how to create and modify the file.

2. Run **ipconfig** or **winipcfg** from the command prompt to reveal information about a computer such as its IP address assignment and MAC address. Run the utility when the computer is not connected to the Internet and then again after connected to the Internet to reveal changes in the IP address information. Remember, run **winipcfg** for Windows-based operating systems and use **ipconfig** for NT-based operating systems.

3. Run **netstat** from the command prompt and use the **/?** option to reveal information about the command. Experiment with the command options while the computer is connected as part of a peer-to-peer network and when connected to the Internet.

4. Run **nbtstat** from the DOS prompt and use the **/?** option to reveal information about the command. Experiment with the command options while the computer is connected as part of a peer-to-peer network and when connected to the Internet.

5. Set up a DHCP server and then configure a workstation to connect to it.

6. Create identical hosts files on several computers to resolve PC names to IP addresses without a DNS server.

7. Use a protocol analyzer to capture some typical TCP/IP frames on a network. Identify the parts of the frame, such as the IP address of the source and destination, MAC address of the source and destination, and contents of the data section.

8. Using Windows XP Professional, install the IPv6 protocol. Use Windows' Help and Support for assistance.

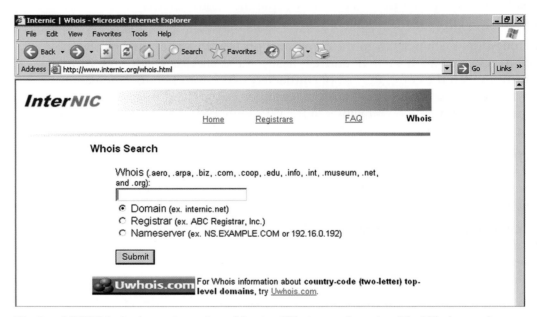

The InterNIC Web site (www.internic.org) hosts a Whois search engine. The Whois search engine provides information about registered domains, registrars, and nameservers.

Chapter 11
Laboratory Activity

Configuring a DHCP Server

After completing this laboratory activity, you will be able to:
- ■ Explain how IP addresses are assigned from an address pool.
- ■ Describe the purpose of reserved addresses.
- ■ Describe the purpose of a lease period.
- ■ Configure a DHCP scope on a DHCP server.

Introduction

In this laboratory activity, you will configure a Windows 2003 DHCP server. A DHCP server is responsible for automatically issuing IP addresses to DHCP clients. The DHCP server draws IP addresses from a pool of addresses indicated in the scope. The scope is an administrative grouping of DHCP clients. A Windows XP workstation is configured by default to receive an IP address, called a dynamic address, from a DHCP server. A workstation that is configured to receive a dynamic address is called a DHCP client.

An IP address dynamically assigned to a DHCP client has a maximum lease period. The default lease period for Windows Server 2003 is eight days. After the lease period expires, the IP address is released and is made available to the IP address pool. Before the lease expires, the client will attempt to contact the DHCP server and renew the lease. The lease period prevents a DHCP client from using an IP address from the pool when it is no longer needed. Some networks set the lease period quite short, such as an hour or less. This is especially true when there are an insufficient number of IP addresses for the number of DHCP clients on the network.

Not all network devices receive dynamic addresses. Certain devices, such as servers and printers, must use static addresses so that clients who require their services can locate them. When included in a DHCP scope, these types of static addresses are considered reserved addresses. Reserved addresses are matched to the MAC address of the network device requiring a constant IP address.

Equipment and Materials

- ■ Windows XP workstation.
- ■ Windows 2000 or 2003 server.

Your instructor will provide the following information. You will need this information to complete the lab. Record this information on a separate sheet of paper.

Scope name: _____

DHCP address pool: _____ to

_____

Reserved address block: _____ to

_____

MAC addresses of reserved address devices: _____

_____

Lease duration:

_____ days

_____ hours

_____ minutes

Procedures

1. _____ Report to your assigned workstation.

2. _____ Boot the workstation and verify it is in working order.

3. _____ At the server, open the DHCP utility (**Start | All Programs | Administrative Tools | DHCP**). You should see a screen similar to the following.

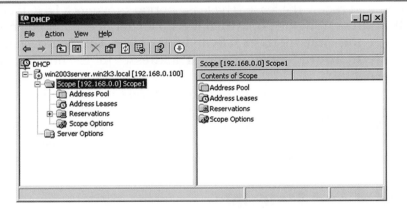

4. _____ Right-click the **Scope** folder and then select **Properties** from the short-cut menu. A dialog box similar to the following should appear. This dialog box lists the scope name, starting and the ending IP addresses, and the lease duration for the DHCP clients.

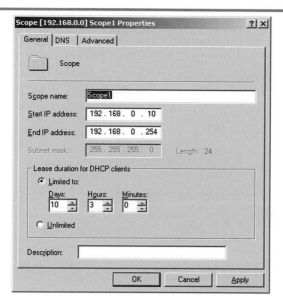

5. _____ Set up the DHCP server according to the information provided by your instructor. After the DHCP scope is configured with the given information, call your instructor to check your work.

6. _____ After your instructor has checked your work, you will need to inspect the workstation that will serve as a DHCP client. See if it is configured for a static or a dynamic IP address. To check this setting, open the **Properties** dialog box for the network adapter by accessing **Start | Control Panel | Network Connections**. Right-click **Local Area Connection** and select **Properties** from the shortcut menu. The **Local Area Connection Properties** dialog box will display. Select the **Internet Protocol (TCP/IP)** and then click **Properties**. The **Internet Protocol (TCP/IP) Properties** dialog box will display. Make sure the **Obtain an IP address automatically** option is selected. Also make sure the network adapter is configured to automatically look for a DNS server. This is the default condition of a newly installed network adapter in a Windows XP client.

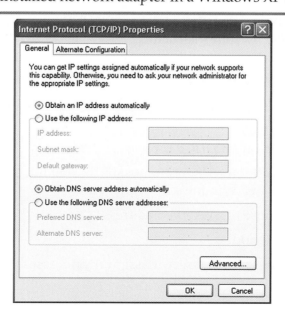

7. _____ Reboot the workstation.

8. _____ After the workstation reboots, open the command prompt and enter the **ipconfig** command. Note the IP address assigned to the workstation. The IP address should be the first address available from the DHCP address pool. All leased IP addresses are issued in sequential order starting from the first available address in the pool.

9. _____ Go to the server and open the **Address Leases** folder in the **Scope** directory to view all IP address assignments. Your display should look similar to the following.

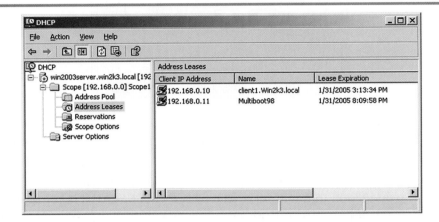

In the example, you can see two IP addresses leased to two different DHCP clients. If the workstation you are using as a DHCP client is not listed, call your instructor.

10. _____ Practice configuring a DHCP scope by creating a different pool of IP addresses. Review this laboratory activity until you feel comfortable configuring the DHCP scope.

11. _____ Answer the review questions.

12. _____ Leave the workstation and server DHCP scope in the condition specified by your instructor.

Review Questions

Answer the following questions on a separate sheet of paper. Please do not write in this book.

1. What does the acronym DHCP represent?

2. What is the default IP configuration (static or dynamic) of an installed network adapter in a Windows XP workstation?

3. What is the purpose of a reserved address?

4. What is a lease period?

5. In what order does a DHCP server assign IP addresses?

12 Subnetting

After studying this chapter, you will be able to:

❏ Count using the binary number system.
❏ Calculate a specific subnet mask needed for a set of conditions.
❏ Identify subnet network characteristics by inspecting the subnet mask.
❏ Explain the purpose, advantages, and disadvantages of subnetting.
❏ Explain the characteristics and purpose of a Virtual LAN (VLAN).

Network+ Exam—Key Points

The Network+ Certification exam requires only a basic understanding of TCP/IP addressing and its relationship to the subnet mask. You should know how to access TCP/IP properties for all major operating systems. You should also know how to identify a network class by looking at the first octet of an IPv4 IP address. The numeric value range of the first octet is how the network is classified. This can be a very difficult question to answer correctly on an exam unless you memorize the decimal values associated with each network class.

Key Words and Terms

The following words and terms will become important pieces of your networking vocabulary. Be sure you can define them.

dotted decimal notation

Fixed Length Subnet Mask (FLSM)

routing table

subnetting

Variable Length Subnet Mask (VLSM)

Virtual LAN (VLAN)

This chapter covers the most difficult concept to master concerning networking fundamentals: subnetting. Subnetting means taking a single IP address issued from a registrar or an ISP and creating two or more subnetworks from it. For example, if a small business is issued only one IP address, they can increase security or subdivide the network by creating two different networks from the issued IP address. They can do this by using the subnet mask technique explained in this chapter.

To learn how to subnet, pay close attention to the information presented in this chapter, and practice the calculations. Before going any further in this unit, you must master binary to decimal conversion and decimal to binary conversion. The term *mastery* means a complete understanding and ability to use and apply a concept, not simply becoming familiar with it. A simple familiarity is not sufficient. Without fully understanding binary counting and conversion to decimal, the following information is difficult to understand and impossible to apply in the field.

For this unit, you may be able to use a calculator for making binary to decimal conversions. One such calculator is available in Microsoft Windows 98 or Windows XP under **Accessories**. However, some certification exams require subnet masks to be calculated manually. If you plan to take one of the other various certification exams, please check if subnet mask calculations by hand are required. If they are, make sure you get plenty of practice making binary to decimal conversions without a calculator or reference table.

The Binary Number System

The binary number system is a perfect match for digital electronic systems, such as computer systems. Digital electronics consists of circuits that have only two electrical states: on and off. The binary number system consists entirely of ones (1s) and zeros (0s). The ones typically represent electrical energy, and the zero represents the absence of electrical energy.

Digital circuit patterns can be represented by binary bit patterns. The major problem with the binary number system is reading and interpreting the numeric values of the binary bit patterns. We are used to the decimal number system and our minds are trained to interpret and do calculations based on decimal numbers. In addition, decimal numbers require much less space than binary numbers when written out.

The fact that large numeric expressions written in binary can be quite lengthy adds to the difficulty of using the binary number system and quickly interpreting their numeric value. For this unit, you will need to be able to convert binary numbers to their decimal equivalents. You may use the tables provided as you progress through this chapter, but you should be able to convert an eight digit binary number to its decimal equivalent without a table to fully understand the technology. Think of the exercises in this chapter as learning an entirely new numbering system, such as Roman numerals.

Following this paragraph is a simple table that can help you convert binary numbers to their decimal equivalents. The table is limited to eight positions. This is equal to one byte, the most common unit you will encounter. It is also equal to one octet of an IPv4 IP address.

1	1	1	1	1	1	1	1
128	64	32	16	8	4	2	1

To convert a binary bit pattern to a decimal value, simply insert the decimal value for each individual bit position represented by a one, and then add the decimal values together. Study the following example.

1	0	0	1	0	1	0	1
128	0	0	16	0	4	0	1
128 + 16 + 4 + 1 = 149							

Dotted Decimal Notation

Typically, the IP address numeric values encountered in computer technology are displayed in decimal form, not binary. This form is referred to as *dotted decimal notation.* Look at **Figure 12-1** to see an example of a Class B IP address and subnet mask expressed as both binary and dotted decimal notation.

The figure shows a network IPv4 address of 130.50.125.25 with a subnet mask of 255.255.000.000. The subnet mask, like the IP address, is divided into two parts: the network address and the host address. In the example, the subnet mask network address is 130.50.0.0 and the host address is 0.0.125.25, or simply 125.25. All IPv4 addresses indicate a host address and a network address.

dotted decimal notation
an IP address displayed in decimal form, not binary.

Subnetting

You should recall from Chapter 11—TCP/IP Fundamentals that a subnet mask serves to identify the class of network and to allow the network address to be divided into subnetworks. Dividing a network into subnetworks, or subnets, is called *subnetting.* Router technology also relies on subnetting for creating large sections of network communication paths.

subnetting
dividing a network into subnetworks, or subnets.

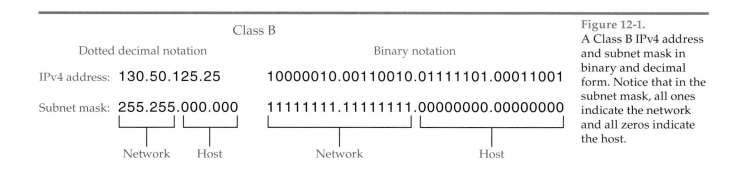

Figure 12-1.
A Class B IPv4 address and subnet mask in binary and decimal form. Notice that in the subnet mask, all ones indicate the network and all zeros indicate the host.

A subnet is a network created by borrowing bits from the host portion of an assigned network IP address. Look at **Figure 12-2.** In the example, a Class B network subnet mask has been extended into the first host octet. The first three bits borrowed from the host portion of the address are used to identify the subnets created from the original IP address. By extending the network portion of the address by borrowing the first three bits from the host portion, six subnets may be formed. Take particular notice of the new dotted decimal notation of the subnet mask. It is now 255.255.248.000. The original subnet mask was 255.255.000.000.

The three bits borrowed from the host can create a total of eight binary patterns. Refer to Figure 12-2. Notice that only six of the eight possible binary patterns can be used to identify the subnets. The binary pattern containing all zeros and the binary pattern of all ones are reserved values. The fact that they are reserved means they cannot be used to identify a particular network. When messages are broadcast to all network nodes, a pattern of all ones is used in the destination address. This indicates that the message is intended for all network addresses. A bit pattern of all zeros is normally used for special functions such as a temporary address of a network card when it is set up for an automatic IP address assignment from a DHCP server. The all zero pattern indicates to the DHCP server that an IP address needs to be issued to the network card.

So when you are looking at binary bit patterns for subnet masks, be sure to subtract all zeros and all ones from the total possible number of patterns. When a binary bit pattern is converted directly to a decimal equivalent, subtract one from the total. For example, an extended portion of a subnet mask consisting of three binary ones equals 4 + 2 + 1 = 7. Subtract one from the total to derive the total number of subnets. See **Figure 12-3.**

A subnet reduces broadcasts and provides additional security. Broadcasts on Ethernet networks are typically limited to their immediate network, which is identified by the network portion of an IPv4 address. When traffic becomes too heavy, creating subnets reduces the number of broadcasts. Users with similar duties can be placed into the subnet areas to reduce the number of broadcasts affecting the immediate network. Subnets also increase security because they do not normally appear when viewed by intruders from outside the network area.

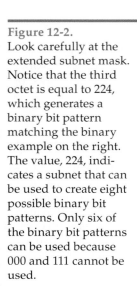

Figure 12-2.
Look carefully at the extended subnet mask. Notice that the third octet is equal to 224, which generates a binary bit pattern matching the binary example on the right. The value, 224, indicates a subnet that can be used to create eight possible binary bit patterns. Only six of the binary bit patterns can be used because 000 and 111 cannot be used.

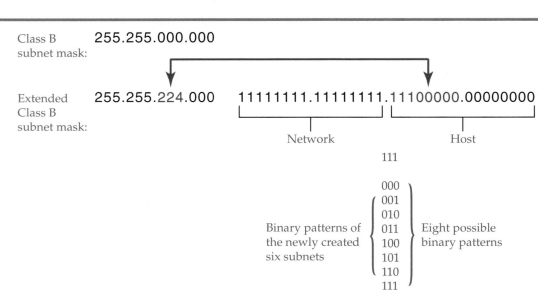

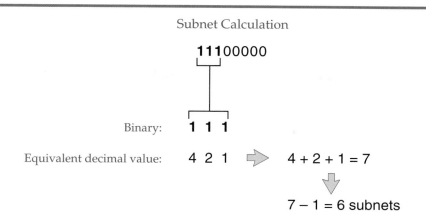

Figure 12-3.
To calculate the total number of subnets, simply add the value of the three bits together and then subtract one from the total.

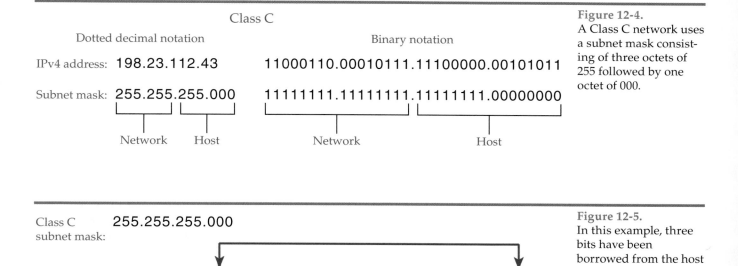

Figure 12-4.
A Class C network uses a subnet mask consisting of three octets of 255 followed by one octet of 000.

Figure 12-5.
In this example, three bits have been borrowed from the host of a Class C network.

Now let's look at a Class C IP address and subnet mask, **Figure 12-4.** (Class B and Class C are the most common networks to be used to create subnets.) When a Class C network is used to create subnets, bits are borrowed from the fourth octet, which is the host portion of the address.

In addition to determining the number of subnetworks that can be created, the number of actual hosts on each subnetwork needs to be determined. This time, the host portion is used to determine the number of hosts for each subnet. Look at **Figure 12-5.**

The number of zeros in the address determines the number of hosts. To calculate the number of hosts available for each subnet, simply convert the bit pattern to a decimal number and then add them together. Subtract one from the total to find the number of hosts per subnet, **Figure 12-6.**

Figure 12-6.
To calculate the total number of hosts per subnetwork, convert the binary value to decimal and then subtract one from the total.

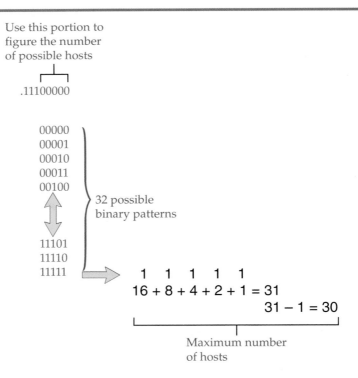

A point of normal confusion for students is the difference between total possible binary patterns and total decimal equivalent value. The total number of binary bit patterns always includes a bit pattern consisting of all zeros. When completing a conversion of binary to decimal, a pattern of all zeros is not considered. Just remember to subtract one from the total number of hosts when converting from binary to decimal. Also, you can use the tables in **Figure 12-7, Figure 12-8,** and **Figure 12-9** to assist you. The first table, Figure 12-7, is designed for determining the dotted decimal value based on the subnet mask bit pattern. In the last column of the table are the decimal values that correspond to the binary bit patterns displayed in the table. This is valuable for determining the subnet mask in dotted decimal format.

The second table, Figure 12-8, will help you determine the number of subnets and the number of hosts based on the subnet mask of a Class B network. The table has fourteen entries. This is because a Class B network has two host octets.

Figure 12-7.
Binary bit pattern to decimal conversion table. This table can be used to convert decimal values to subnet mask bit patterns.

128	64	32	16	8	4	2	1	
1	0	0	0	0	0	0	0	128
1	1	0	0	0	0	0	0	192
1	1	1	0	0	0	0	0	224
1	1	1	1	0	0	0	0	240
1	1	1	1	1	0	0	0	248
1	1	1	1	1	1	0	0	252
1	1	1	1	1	1	1	0	254
1	1	1	1	1	1	1	1	255

Number of Bits	Subnet Mask	Number of Subnets	Number of Hosts
2	255.255.192.0	2	16,382
3	255.255.224.0	6	8,190
4	255.255.240.0	14	4,094
5	255.255.248.0	30	2,046
6	255.255.252.0	62	1,022
7	255.255.254.0	126	510
8	255.255.255.0	254	254
9	255.255.255.128	510	126
10	255.255.255.192	1,022	62
11	255.255.255.224	2,046	30
12	255.255.255.240	4,094	14
13	255.255.255.248	8,190	6
14	255.255.255.252	16,382	2

Note: Values for all binary patterns of all ones and all zeros in the network portion of the subnet mask have been eliminated from the table.

Figure 12-8.
Common values used for Class B subnetting. Notice that a Class B subnet mask can span across two octets.

When creating subnets, it is possible to borrow bits from both host octets to create the subnet mask.

The third table, Figure 12-9, is used to determine the number of subnets and hosts corresponding to a Class C network. Notice that there are only six possibilities for a Class C subnet mask. Also, pay attention to the fact that a Class C and a Class B network have several matching subnet masks in dotted decimal form, but the quantity of subnets differ.

A Closer Look at Subnets

In **Figure 12-10**, a subnet mask of 255.255.255.192 has been applied to a Class C IP address of 202.130.46.125. When applied, two subnets containing 62 hosts each are created. Note that the only two valid network addresses are 202.130.046.64 and 202.130.046.128. The other two calculated network addresses 202.130.046.000 and 202.130.046.192 cannot be used. The first address is based on a network binary bit pattern of all zeros and the other is based on a binary pattern of all ones. Remember that you cannot use those two binary bit patterns while subnetting.

Number of Bits	Subnet Mask	Number of Subnets	Number of Hosts
2	255.255.255.192	2	62
3	255.255.255.224	6	30
4	255.255.255.240	14	14
5	255.255.255.248	30	6
6	255.255.255.252	62	2

Note: Values for all binary patterns of all ones and all zeros in the network portion of the subnet mask have been eliminated from the table.

Figure 12-9.
Common values used for Class C subnetting. Notice that the Class C subnet mask values match many of the Class B subnet mask values and host values. However, for the same network subnet masks, the quantities of subnets generated do not match.

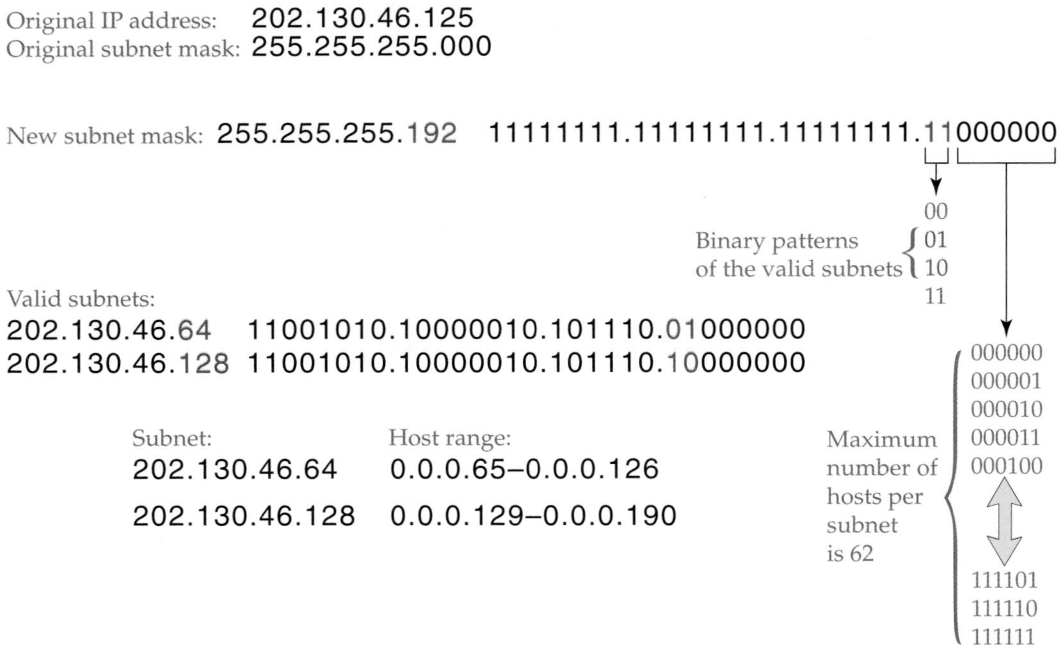

Figure 12-10.
Subnetting four subnets.

Original IP address: 202.130.46.125
Original subnet mask: 255.255.255.000

New subnet mask: 255.255.255.192   11111111.11111111.11111111.11000000

Binary patterns
of the valid subnets { 00 01 10 11

Valid subnets:
202.130.46.64   11001010.10000010.101110.01000000
202.130.46.128   11001010.10000010.101110.10000000

Subnet: Host range:
202.130.46.64   0.0.0.65–0.0.0.126

202.130.46.128   0.0.0.129–0.0.0.190

Maximum
number of
hosts per
subnet
is 62

000000
000001
000010
000011
000100

⇕

111101
111110
111111

Variable Length Subnet Mask (VLSM)
a subnet mask that is not expressed in standard eight-bit or one-byte values. A VLSM occurs when a subnet is further divided into smaller subnets, which are not equal to the original subnet in length or in the number of hosts.

Fixed Length Subnet Mask (FLSM)
a subnet mask that is expressed in standard eight-bit or one-byte values. An FLSM has subnets that are equal in range and have an equal number of hosts.

Some software/hardware combination equipment is able to use a network address using all zeros and all ones that increases the total possible number of subnets and hosts. You will encounter both systems until one becomes the dominant system. For now, consider the subnet mask system that does not allow the use of consecutive ones or zeros in the network portion of the subnet mask as the legacy system. The system that does allow the use of ones and zeros is the preferred system because it allows the use of more possible combinations of IP addresses. RFC 1874 allows the use of all ones and all zeros in the extended portion of the subnet mask.

A subnet mask that is not expressed in standard, eight-bit or one-byte values is often referred to as a **Variable Length Subnet Mask (VLSM)**. The term *Variable Length Subnet Mask* is derived from the fact that by subnetting, the total length of the binary one-bit pattern can vary in length rather than be held to the traditional series of eight-bit or one-byte patterns. Look at **Figure 12-11.** Notice how two additional subnets have been created out of two subnets. This makes a total of four subnets—two original subnets plus two smaller subnets created from the originals. The original subnets are referred to as *fixed length* because each subnet is of equal length. A **Fixed Length Subnet Mask (FLSM),** therefore, has subnets that are equal in length and have an equal number of hosts. With the additional subnets, however, there is more than one length of subnet mask. The term *Variable Length Subnet Mask* now applies because multiple subnets with more than one length of subnet mask have been created. The length refers to the binary ones used to identify the network and the host.

Subnetting can cause confusion, which has led to IP addresses being presented in a different written manner at times. For example, you may see an IP address that contains a Variable Length Subnet Mask written as 202.130.046.125/26. The /26 indicates that a mask with a total length of 26 bits is used for the mask. Do not confuse the slash (/) symbol with the colon (:) symbol, which follows an IP address to identify a port number.

Figure 12-11.

A Variable Length Subnet Mask (VLSM) is formed when a subnet is used to create an additional subnet, which is not equal to the original subnet in length or in the number of hosts.

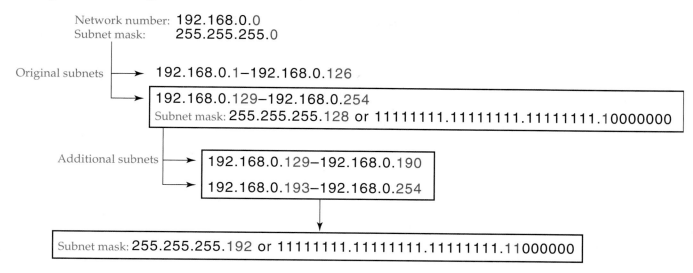

Network number: **192.168.0.0**
Subnet mask: **255.255.255.0**

Original subnets → **192.168.0.1–192.168.0.126**

192.168.0.129–192.168.0.254
Subnet mask: **255.255.255.128** or **11111111.11111111.11111111.10000000**

Additional subnets → **192.168.0.129–192.168.0.190**
→ **192.168.0.193–192.168.0.254**

Subnet mask: **255.255.255.192** or **11111111.11111111.11111111.11000000**

There are many software programs available that can automatically calculate the number of subnets, hosts, subnet mask, and all addresses for a given IP address. A free network subnet calculator is available from www.wildpackets.com.

Tech Tip

Unfortunately, you cannot use software or a calculator on any actual certification exams at the time of this writing. This procedure may change in the future. For now, you must learn to calculate these values manually to pass most certification exams that involve IP addressing and the TCP/IP protocol. Many certification authorities feel that when students can successfully perform the necessary calculations, they prove they have mastered the basic concept of subnetting networks.

Advantages of Subnetting

As stated earlier, subnetting creates a more secure network by placing hosts on separate networks. Network users cannot readily access resources on a host of a different subnet unless the system administrator creates the necessary user profile to allow it. Subnetting creates two or more segments out of the original network. When the original network is divided into two or more segments, broadcast packets from one network are not forwarded to the other network. If broadcast packets are not forwarded, the amount of data traffic is reduced. When the amount of data traffic is reduced, fewer collisions occur. As the number of collisions is reduced, the network appears faster to users. Subnetting, therefore, also reduces the amount of collisions on a network segment by dividing the original segment into two or more segments.

Disadvantages of Subnetting

Subnetting by using calculated subnet mask assignments could be difficult to manage because of the inherent confusion of the IP address assignments. There is also confusion because some equipment is designed to use subnets based on all one and all zero bit patterns to increase the total number of possible subnets and hosts.

Virtual LAN (VLAN)

Virtual LAN (VLAN)
a LAN that is created using software and hardware techniques to connect work-stations on separate network segments as though they were on the same segment.

A *Virtual LAN (VLAN)* is a LAN that is created using software and hardware techniques to connect workstations on separate network segments as though they were on the same segment. The VLAN acts similar to a subnet. All computers on the VLAN communicate with each other seamlessly as though they are on the same segment. VLANs can be created from MAC address or IP addresses. When network switches (layer 2 devices) are used, the MAC address of the workstation is used to identify the VLAN. When routers are used, the IP address of the workstation is used to create the VLAN. Some manufacturers have produced a layer 3 switch. A layer 3 switch uses either a MAC address or an IP address to communicate across network segments. Layer 3 switches are designed for use on LAN and MAN topologies. The layer 3 switch is not used for WAN at the time of this writing. Routers are still the dominant devices used for WANs.

Figure 12-12 shows how two LAN segments have been created by the use of a router. The router passes packets based on IP addresses, not MAC addresses. This is why inserting a router into a LAN can divide the LAN into two separate segments, thus reducing the number of collisions caused by broadcasts based on MAC addresses. The router also allows a VLAN to be created by using a routing table. A *routing table* is a database of IP addresses and computer names used to direct packet flow to each side of the router. By controlling the flow of packets between designated workstations on the two network segments, a VLAN is created.

routing table
a database of IP addresses and com-puter names used to direct packet flow to each side of a router.

Summary

- The subnet mask uses a series of ones to identify the network portion of an IP address and a series of zeros to identify the host portion.

- To create a subnet, bits are borrowed from the host portion of an IP address to expand the network portion of the IP address.

- An IP address with all zeros in the host portion cannot be assigned to any network device.

- An IP address with all ones in the host portion cannot be assigned to any network device.

Figure 12-12.
A VLAN occurs when two or more network connections communi-cate as though they were on the same network segment. Typi-cally, a table of IP addresses identifies connections in the VLAN. Even if the devices are moved to a new location in the network system, the VLAN remains intact.

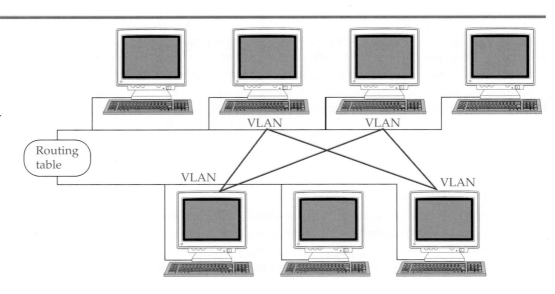

- A Fixed Length Subnet Mask (FLSM) has two or more equal-sized subnets.
- A Variable Length Subnet Mask (VLSM) occurs when you create additional subnets from a subnet.
- Subnetting provides security because subnetworks cannot be seen from outside the subnetwork.
- Subnetting reduces network traffic by isolating broadcasts to each subnet.
- A Virtual LAN (VLAN) is a communication path created between workstations that are on separate segments of a LAN.

Review Questions

Answer the following questions on a separate sheet of paper. Please do not write in this book.

1. Convert the following binary bit patterns to decimal number values. Do *not* use a calculator for this activity. You may use the following tables for support.

1	1	1	1	1	1	1	1
128	64	32	16	8	4	2	1

1	0	0	1	0	1	0	1
128	0	0	16	0	4	0	1

A.

0	0	0	0	1	1	0	0

B.

0	0	0	0	1	1	1	1

C.

1	1	0	0	0	0	0	0

D.

0	0	0	1	0	1	0	1

E.

1	1	0	1	1	1	1	1

2. A subnet mask can be divided into two descriptive sections. What are the two sections called?

3. Why would a network administrator create subnets?

4. What do the binary ones in a subnet mask indicate?

5. What do the binary zeros in a subnet mask indicate?

6. Using the information provided, identify which part of the IP address 190.134.124.112 with a subnet mask of 255.255.255.0 is the network address and which part is the host address?

7. How many bits are borrowed from a Class B host address to create 30 new subnets?

8. How many bits are borrowed from a Class C host address to create 6 new subnets?

9. How many subnets can be formed using a subnet mask of 255.255.255.224 for a Class C network and using hardware that allows the use of all zeros and all ones in the network portion of the subnet mask?

10. How many hosts are created from the following Class C subnet mask: 255.255.255.240?

11. A network administrator decides to partition an existing Class C network into six equal parts. Each part of the network must hold at least 22 computers. What subnet mask must be used to satisfy the requirement?

13. How many subnets can be created from the following Class C subnet mask: 255.255.255.224?

14. How many subnets can be created from the following Class C subnet mask: 255.255.255.248?

15. What is a VLAN?

16. What types of addresses are required to create a VLAN using a router?

17. VLANs that are created using a router use which layer of the OSI model?

18. At what layer of the OSI model are MAC addresses used?

Sample Network+ Exam Questions

Network+

Answer the following questions on a separate sheet of paper. Please do not write in this book.

1. What subnet mask is used to identify a typical Class A network and host?
   A. 255.255.255.255
   B. 255.255.255.000
   C. 255.255.000.000
   D. 255.000.000.000

2. How do you access the **TCP/IP Properties** dialog box on a Windows 98 computer?
   A. Right-click the **Network Neighborhood** icon, click **Properties**, highlight the TCP/IP protocol, and then click the **Properties** button.
   B. Open **Control Panel**, double-click the **System** icon, open **Device Manager**, select the network adapter, and then click on the **Properties** button.
   C. Open **Control Panel** and then double-click the **TCP/IP** icon.
   D. TCP/IP settings cannot be changed. They are unique for each network adapter card.

3. What is the network class *most likely* associated with the IP address 150.150.23.34? (Hint: The network class can be identified by the first octet.)

A. Class A

B. Class B

C. Class C

D. Class D

4. Which of the following IP addresses is a reserved address and cannot be used when connecting to the Internet?

A. 123.001.001.21

B. 192.168.0.23

C. 154.23.168.32

D. 200.200.200.200

5. What is the host address for an IP address of 199.200.12.45 with a subnet mask of 255.255.255.000?

A. 199.200.12

B. 45

C. 255.255.255

D. .000

6. What subnet mask is used to identify a typical Class B network and host?

A. 255.255.255.255

B. 255.255.255.000

C. 255.255.000.000

D. 255.000.000.000

7. What is the network class *most likely* associated with the IP address 198.150.25.10? (Hint: The network class can be identified by the first octet.)

A. Class A

B. Class B

C. Class C

D. Class D

8. What is the host address for an IP address of 128.30.45.15 with a subnet mask of 255.255.000.000?

A. 45.15

B. 128.30

C. 15

D. 000.000

9. What is the network address for an IP address of 50.100.30.3 with a subnet mask of 255.000.000.000?

A. 50.100.30.000

B. 50.100.000.000

C. 3

D. 50.000.000.000

Network+

10. What is the network class *most likely* associated with the IP address 120.168.10.15? (Hint: The network class can be identified by the first octet.)

   A. Class A

   B. Class B

   C. Class C

   D. Class D

Interesting Web Sites for More Information

www.3com.com

www.cisco.com/warp/public/701/8.html

www.faqs.org/rfcs/rfc1878.html

www.solarwinds.net

www.telusplanet.net/public/sparkman/netcalc.htm

www.wildpackets.com/products/ipsubnetcalculator

www.windowsitlibrary.com/Content/386/02/2.html

Suggested Laboratory Activities

1. Create two subnets using the reserved IP address 192.168.0.0. This should be a team project. You will need one PC acting as a server and two or more PCs to serve as workstations in different subnets.

2. Install Internet Connection Sharing (ICS). Use at least two computers: one for the host and the other for the client.

Chapter 12
Laboratory Activity

Subnet Mask Calculator

After completing this laboratory activity, you will be able to:

■ Download and install the WildPackets IP Subnet Calculator.

■ Use the WildPackets IP Subnet Calculator to determine the number of hosts and subnetworks associated with a given subnet mask.

■ Identify the binary pattern associated with a given subnet mask or IP address.

Introduction

In this laboratory activity, you will download the WildPackets IP Subnet Calculator from the WildPackets Web site. The calculator is free. If the WildPackets IP Subnet Calculator is not available, you can conduct an Internet search to locate another subnet calculator. The WildPackets IP Subnet Calculator is a very handy utility.

Note:

While the Network+ Certification exam does not require network subnet mask calculations at this time, some other network certifications do require subnet mask calculations by hand. You cannot use a calculator during these exams. Always check for the latest test specifications concerning this area.

The calculations for subnets can be very difficult for students and also for many technicians. It is one of the most difficult aspects of networking fundamentals. The concept of creating a subnet out of an assigned IP address is very simple. The subnet mask is divided into two portions: the network portion and the host portion. For example, the subnet mask of a typical Class C network is 255.255.255.000. Each of the first three octets has the number *255* representing the network portion of the subnet mask and the number *000*, or *0*, representing the host portion of the subnet mask. The value 255 is equal to eight binary ones, or an entire octet filled with ones, such as 11111111. The series of ones represents the network portion of the subnet mask.

To create a subnet from a Class C network assignment, the host assignment of 000 is changed to some other value, such as 160, 192, or 224. The new value extends the network portion into the host portion of an assigned address. Remember, when subnetting, bits are borrowed from the host portion of an IP address. The borrowed bits are the extended portion of the network address. In a Class C subnet mask, this means that the fourth octet will begin (starting at the left of the octet) with a series of ones.

To calculate the decimal number that represents the series of binary ones is very difficult for most people—especially if they do not have a background in digital electronics. This is where the WildPackets IP Subnet Calculator comes in handy. The calculator allows you to choose a class of network, such as Class A, Class B, or Class C, and select the number of subnets you desire. The WildPackets IP Subnet Calculator instantly gives you all the information you need. Let's see how this is done.

When provided with an IP address, the calculator immediately displays the network class, the bit-map pattern of the host and network, and the hexadecimal equivalent of the address. The following screen capture shows the information given for the IP address 128.15.10.5.

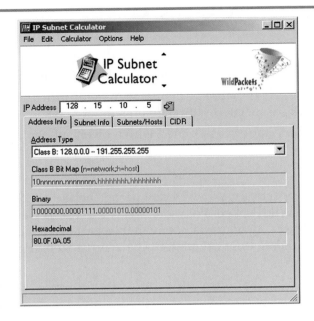

Now look at the following screen capture. Notice the tabs beneath the IP Address textbox. The **Subnet Info** tab allows you to enter values for the calculation based on the number of subnets required or the number of hosts required for each subnet. These two values are the main values you will typically work with.

Notice the **Allow 1 Subnet Bit** option. This option allows all zeros and all ones to be used in the extended portion of a network address. Look at the following table. Notice the first two binary positions of the host portion of the address.

Network	Network	Network	Host	
NA	NA	NA	0000000000	**All zeros**
NA	NA	NA	0100000000	
NA	NA	NA	1000000000	
NA	NA	NA	1100000000	**All ones**

Note: NA = Not Applicable

Originally, all zeros and all ones were not allowed in the lead position of the host address used for subnetting. The binary pattern containing all zeros and all ones were reserved values. A bit pattern of all zeros is typically used for special functions, such as a temporary address of a network card when it is configured for an automatic IP address assignment from a DHCP server. Some routers recognize the use of the binary one in the first part network address as reserved for broadcasts. The generally accepted convention today, however, is to allow all zeros and ones. Notice that in the previous table, when all zeros and ones are allowed, four subnets can be created.

Note:
While running the laboratory activity, select and unselect the **Allow 1 Subnet Bit** *option to see the effect on the number of hosts and subnetworks.*

Now look at the following screen capture. Notice that a complete list of all possible host IP addresses is automatically listed under the **Subnets/Hosts** tab. As you can see, the WildPackets IP Subnet Calculator is a handy tool. The best way to learn how to use this calculator is by downloading it and experimenting with different scenarios.

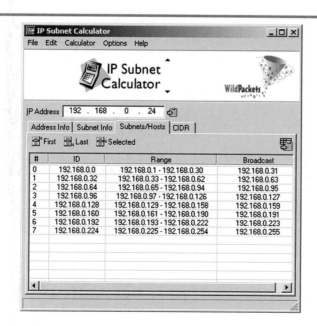

Equipment and Materials

■   PC with Internet access.

■   E-mail address. (An e-mail address is typically required by WildPackets. The e-mail address is used to supply a user name and password for completing the download and the installation of the software package.)

Procedures

1. _____ Report to your assigned workstation.

2. _____ Boot the workstation and verify it is in working order.

3. _____ Use the following URL to locate the WildPackets IP Subnet Calculator: www.wildpackets.com. The WildPackets IP Subnet Calculator is found under the **Downloads** link.

4. _____ Install the WildPackets IP Subnet Calculator.

5. _____ Open the WildPackets IP Subnet Calculator program.

6. _____ In the **IP Address** textbox, enter the following IP addresses and identify the class of network. Record your answers on a separate sheet of paper.

IP Address	Network Class
192.168.000.000	
202.111.0.0	
68.0.0.0	
234.0.0.0	
254.0.0.0	
127.1.1.1	
126.0.0.0	
128.0.0.0	
192.0.0.0	
191.0.0.0	

7. _____ Now, enter the IP address 192.168.0.0 and then select the **Subnet Info** tab. Check if the **Allow 1 Subnet Bit** option is selected. If it is, unselect it.

8. _____ Select the **Subnets/Hosts** tab to reveal the number of subnets and hosts generated.

9. _____ Select the **Subnet Info** tab and then select the **Allow 1 Subnet Bit** option.

10. _____ Return to the **Subnets/Hosts** tab and observe the effect on the total number of hosts and subnets.

11. _____ Experiment with the calculator.

12. _____ When you have finished experimenting with the calculator, answer the review questions.

13. _____ Return your workstation to its original condition.

Review Questions

Answer the following questions on a separate sheet of paper. Please do not write in this book.

1. What is the generally accepted convention of having all zeros and ones in the extended portion of a network address?

2. What is the number of hosts for any Class C network?

3. What is the maximum number of hosts for one subnet on a Class C network that has been divided into two subnets?

4. What is the maximum number of hosts for one subnet on a Class C network that has been divided into six subnets?

5. What is the maximum number of hosts for a Class C network that has been divided into 14 subnets?

6. A Class C network is divided into six equal subnets. What is the range of IP addresses of host IP assignments for the first subnet when the **Allow 1 Subnet Bit** option is selected?

7. A Class C network is divided into six equal subnets. What is the range of IP addresses of host IP assignments for the first subnet when the **Allow 1 Subnet Bit** option is not selected?

8. What is the broadcast IP address of the first IP range for a Class C network that is divided into six equal subnets with the **Allow 1 Subnet Bit** option selected?

9. What decimal number is represented by 11110000?

10. What decimal number is represented by 01000000?

Schools and large businesses often rely on subnetting to create two or more subnetworks from an assigned IP address.

13

ATM and VoIP

After studying this chapter, you will be able to:

- ❏ Describe how signals are converted between analog an digital.
- ❏ Define and explain latency as it affects the transfer of audio and video signals.
- ❏ Explain how data is compressed and decompressed.
- ❏ Describe the X.25 protocol.
- ❏ Describe the Frame Relay protocol.
- ❏ Describe how ATM is used to transfer multimedia data across networks.
- ❏ Explain how the TCP/IP protocol (VoIP) can be used as a transport media for telephone conversation.
- ❏ Describe the purpose of the H.323 protocol.

Network+ Exam—Key Points

There will most likely be at least one question concerning the transfer of multimedia across a network system. Since ATM is used extensively to transport multimedia, be prepared to answer questions about the ATM protocol and multiplexing.

Key Words and Terms

The following words and terms will become important pieces of your networking vocabulary. Be sure you can define them.

Analog to Digital Converter (ADC)
Asynchronous Transfer Mode (ATM)
Available Bit Rate (ABR)
bit rate
codec (compressor/decompressor)
Committed Information Rate (CIR)
Constant Bit Rate (CBR)
Digital to Analog Converter (DAC)
Frame Relay
H.323 standard
jitter
MPEG (Moving Picture Experts Group)
permanent virtual circuit (PVC)

Plain Old Telephone Service (POTS)
Public Switched Telephone Network (PSTN)
Quality of Service (QoS)
sampling frequency
sampling rate
Unspecified Bit Rate (UBR)
Variable Bit Rate (VBR)
Variable Bit Rate-Non Real Time (VBR-NRT)
Variable Bit Rate-Real Time (VBR-RT)
Voice over IP (VoIP)
X.25

This chapter introduces network technologies that support the transfer of audio, video, and multimedia (a mixture of audio, video, text, and images). When transferring these types of data, there are some concerns that are not found in typical file transfer situations consisting of text only documents. The beginning of the chapter introduces some basic concepts about audio and video that will help you better understand the technologies used to transport multimedia. Later in the chapter, protocols such as X.25, Frame Relay, ATM, and VoIP and the technologies that support them are introduced.

This area of networking can be quite complex and challenging. It is the fastest growing area of network technology. Soon, all industries, such as telephone, television, music, and movie, will use network technology to deliver their data all over the world. Individuals are already using personal Web cams to transmit video anywhere in the world. Telephone calls can be made to any location a computer can reach without necessarily using the standard telephone systems. Data can also be sent across existing telephone lines or a mixture of network, telephone, and wireless systems. As you will see, the transfer of audio, video, and multimedia requires special techniques.

Voice and Audio Signals

Anyone reading this sentence has most likely transmitted their voice over a section of network owned by a long distance telephone company. Digitizing a voice and transporting it over existing telephone lines began in the early 1970s. It was brought on in part by the rapid development of digital electronics.

The human voice and other audio signals are analog signals. Remember from earlier chapters: an analog signal fluctuates in amplitude (voltage level). For an audio signal (analog signal) to be transported across a network system, it must be converted into a series of digital pulses that represent the analog waveform. After the digital pulses reach their destination, the digital pulses must be converted into an analog signal so that the human ear can detect and understand it.

Converting digital pulses to an analog waveform is called digital to analog conversion. Converting an analog waveform to digital pulses is called analog to digital conversion. The following section takes a closer look at both of these processes.

Signal Conversion

Audio signals must be converted into digital signals and placed in packets before transmitted across a network. When the packets of digital signals reach their destination, they are reassembled and converted into an analog signal. The quality of the analog signal depends on the detail of the digital coding.

The analog signal is converted into a digital code by taking samples of the analog signal's amplitude at specific times. The number of times the sample is taken is referred to as the *sampling rate* or sampling frequency. The number of bits used to represent the amplitude of the analog signal is referred to as *bit rate* or *bit resolution.* The quality of the sound conversion is directly related to the sampling rate and the bit resolution. In other words, the higher the sampling rate and the greater the bit resolution, the better the sound.

How often amplitude is measured per second is referred to as the *sampling frequency.* Sampling frequency varies significantly. A typical voice sampling frequency is approximately 11 kHz, or 11,000 times per second. The sampling frequency of quality sound applications, such as music, is approximately 44 kHz. Look at **Figure 13-1.**

sampling rate
the number of times a signal is sampled during a specific period of time.

bit rate
the number of bits used to represent the amplitude of an analog signal.

sampling frequency
how often the amplitude of an analog signal is measured per second.

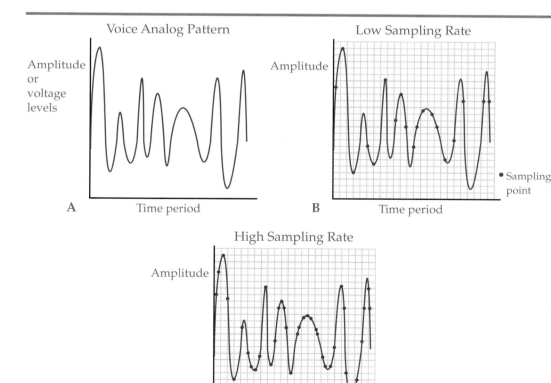

Figure 13-1.
The sampling rate of the human voice. A—In this graph, the amplitude of the human voice is plotted on the vertical axis. The time period represents the frequency of the human voice and the point and is plotted on the horizontal axis. B—This graph depicts a low sampling rate. Each point on the graph represents the point at which a sample is taken. C—This graph depicts a high sampling rate. In a high sampling rate, more data about the original signal is collected.

A typical human voice or any other audio signal fluctuates over a period of time. Figure 13-1A shows how an audio signal's amplitude is plotted against time. Amplitude is plotted vertically on the graph, and the time period is plotted horizontally on the graph. The quality of the analog to digital conversion depends on the sampling frequency, which is based on the number of bits available to store the sampling.

Compare the two sampling rates in Figure 13-1B and Figure 13-1C. Notice that the lower sampling rate has a much smaller number of sampling points to represent the original analog signal. The higher sampling rate contains more sampling points to represent the original analog signal. As the sampling rate increases, a better representation of the original signal is made.

The maximum number of sampling points represented in digital form depends on the number of bits used to represent the analog signal at any given point of time. The maximum number of possible bits per sample represents the bit rate or bit resolution. The higher the number of bits used, the better the digital description of the analog signal. For example, if only 4 bits are used to represent the voltage level, a total of 16 distinct voltage levels are represented. Four bits can have a maximum of 16 binary patterns. Look at **Figure 13-2.** Notice that 16 voltage levels (0 volts through 15 volts) are represented by 16 four-bit binary patterns. A distinct binary pattern represents each voltage level. Now look at **Figure 13-3.** In this example, 8 bits are used to represent the voltage levels 0 volts through 15 volts. A maximum of 256 distinct voltage levels are represented. Notice the intervals at which the voltage increases. In the first example, the voltage increases by 1 volt with each increment. In this example, the voltage increases by 0.05859 volts with each increment. If 16 bits are used, a total of 65,536 levels can be used to represent voltage levels 0 volts through 15 volts. With each increment, the voltage level would increase by 0.000229 volts. Therefore, the greater the number of bits used, the more levels of voltage can be sampled resulting in a better representation of the analog signal.

Figure 13-2.
Voltage levels, 0 volts through 15 volts, represented by 4-bit binary patterns.

Voltage Level	Binary Code
0 volts	0000
1 volts	0001
2 volts	0010
3 volts	0011
4 volts	0100
5 volts	0101
6 volts	0110
7 volts	0111
8 volts	1000
9 volts	1001
10 volts	1010
11 volts	1011
12 volts	1100
13 volts	1101
14 volts	1110
15 volts	1111

Voltage Level	Binary Code
0 volts	0000 0000
0.05859 volts	0000 0001
0.11718 volts	0000 0010
0.17577 volts	0000 0011
0.23436 volts	0000 0100
0.29295 volts	0000 0101
0.35154 volts	0000 0110
0.41013 volts	0000 0111
0.46872 volts	0000 1000
0.52731 volts	0000 1001
0.58594 volts	0000 1010
0.64449 volts	0000 1011
0.70308 volts	0000 1100
0.05859 volts	0000 1101
0.82026 volts	0000 1110
0.87885 volts	0000 1111
0.93744 volts	0001 0000
0.99603 volts	0001 0001
1.05462 volts	0001 0010
⬇	⬇
14.94141 volts	1111 1110
15 volts	1111 1111

Figure 13-3.
Voltage levels, 0 volts through 15 volts, represented by 8-bit binary patterns.

The human voice can be converted to digital using a relatively low sampling rate. Speech samples can be achieved using an 8-bit, or 1-byte, voltage level sample at 11 kHz time increments. See **Figure 13-4.** An 8-bit sampling rate of an analog signal's amplitude results in 256 binary codes. The 8-bit sampling is taken at a sampling rate, or frequency rate, of 11.025 kHz.

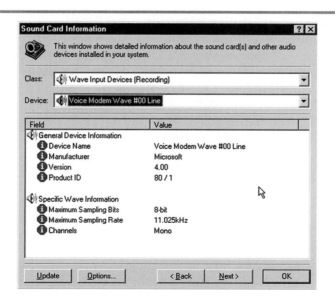

Figure 13-4.
Notice the details listed under **Specific Wave Information**. These details indicate the typical sampling rate of the human voice.

Figure 13-5.
The details listed under **Specific Wave Information** indicate the typical sampling rate of audio, such as music and singing voices.

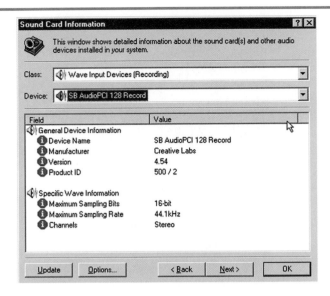

When working with sounds that require a higher degree of quality, such as music and singing voices, a higher sampling rate is needed. Musical instruments have a much wider range of fluctuating sound patterns compared to human speech. Look at **Figure 13-5.** To capture the wide range of sound patterns, the number of sampling bits is set to 16 and the sampling rate is set to 44 kHz. This provides a high-quality recording of the analog sound patterns. The 16-bit sampling rate of the signal's amplitude equals 65,536 possible levels of amplitude based on a sampling rate of 44.1 kHz, or 44,100 times per second. As you can see, the higher the sampling rate, the larger the amount of digital data needed to represent the analog signal.

A high sampling rate requires a high bandwidth, or throughput. A practical compromise between the maximum obtainable bandwidth of the network media and the sampling rate of the audio signal must be reached. If the sampling rate is too high, there will be too much data to transmit across the network, and some of the packets will be dropped to accommodate the inefficient bandwidth. This results in jitter. *Jitter* is the small staggers or hesitations in the delivery sequence of audio or video data caused by latency or missing packets. A good example of jitter can be seen at times during a television news broadcast. A reporter located in some distant country is talking in real time to a news correspondent in the United States over a videophone or a similar device. You will often see a disrupted picture and time delay of responses to questions while the broadcast takes place. If the sampling rate is too low, the quality of the sound suffers, resulting in an artificial sound produced at the destination point.

Multimedia data transmitted across a large network requires network media such as T1 or fiber-optic. Lower quality data, such as voice, can be transmitted over traditional copper cabling, such as Cat 5. The distance traveled by the data and the type of data (voice, music, and video, multimedia) determines the cabling requirements.

jitter
small staggers or hesitations in the delivery sequence of audio or video data caused by latency or missing packets.

Latency

The term *latency* means the delay of data as it travels to its destination. Latency may not be a problem for text file transfers or e-mail exchange. Even the common practice of downloading music has an acceptable latency period. Audio files contain a tremendous amount of data. When music is downloaded, it is buffered before it is played. *Buffering* means data is stored in memory until a sufficient amount of data is stored. When a sufficient amount is stored, the music begins to play. If buffering did not occur, there would be many momentary interruptions in the music.

A music file contains a large amount of data. This data must be processed at a very high data rate. If something acts as a bottleneck, the flow of data could slow. For example, if a music file required a data rate of 1 Mb per minute to produce a steady stream of music and there was a restriction between the origin and the destination, such as a modem, the music would start and stop every time the data-transfer rate fell lower than the required data rate to produce a steady sound. When buffering takes place, a large volume of data is stored in memory before the music begins to play. This reserve ensures a steady flow of music data is provided while the music is playing. This same technique is used for video.

When a real-time application such as a telephone conversation occurs, only a minimal latency period can be tolerated. The human ear can tolerate small breaks in the flow of audio data. If the breaks are small, they will go completely unnoticed. For example, a break in the data flow of less than 150 milliseconds (150/1000 or .15) can be tolerated. The human ear will not detect any break in the conversation. When a latency period of approximately 250 milliseconds is present, the listener will notice it but not tolerate it. A period of 500 milliseconds, however, is unacceptable to the listener.

Latency occurs because of several factors. One factor can be equipment such as gateways that change the data frame from one format to another while connecting two dissimilar network systems. Satellite transmission can generate delays from 50 milliseconds to 500 milliseconds. The time it takes a software compression program to compress data can increase latency. Lost or destroyed frames can add to the latency.

The total amount of latency between data packets produces jitter. Remember, jitter is the effect on the quality of sound or images when they arrive at their destination in other than regularly spaced, minimum intervals. See **Figure 13-6.** As the data packets enter the Internet system, latency periods between the packets may increase to an undesirable level. In the illustration, notice that each of the packets leave the source in regularly spaced intervals, but after navigating long distances, irregular spacing develops. If the spaces are too large (250 milliseconds), jitter is produced at the destination.

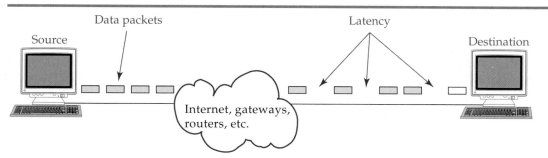

Figure 13-6.
Packets leaving the source are evenly spaced with minimal delay. After they traverse the Internet, latency increases causing packets to arrive at their destination in other than regularly spaced, minimum intervals. The total amount of latency between data packets produces jitter.

Video

Video is a result of flashing still images in front of a viewer at a rate faster than the human eye can detect. For example, a typical movie consists of a series of still images running through a projector at a rate of 24 frames per second. In other words, 24 images flash in front of the viewer every second. A frequency lower than this would result in a choppy picture and a jitter affect would be detected.

Look at **Figure 13-7.** This figure shows a set of still images taken from a digital movie of a dog named Daisy who is returning a ball in a game of catch. Look closely at Daisy's front legs to see the gradual change in position. When the series of frames are displayed on a monitor at 24 frames per second, a smooth video motion is displayed. Full motion video for Internet and Intranet require the same number of frames per second. However, there are more variables than frame rate that need to be considered, such as picture resolution, color depth, the processing speed of the computer, the amount video RAM available, and the type of network medium used to transfer the video.

Video resolution is described in pixels and is measured horizontally and vertically across a display. For example, a typical VGA resolution is 640 × 480, or 640 pixels on each of the 480 lines of the display. See **Figure 13-8.**

Assuming that an image was at a VGA resolution in full color (32-bit), the amount of RAM required to contain the image would be approximately 1 MB (480 × 640 × 32) per second. If 24 frames (images) were required for full-motion video, approximately 24 MB of data would need to be delivered across the network every second. A three-minute video sequence would require approximately a 4320 MB (180 sec × 24 MB) data stream.

As you can see, video requires many computer resources and a lot of network media bandwidth. While most computers in use today will support video as well as audio, many network systems will not. Full motion video with audio requires a high bandwidth network medium, or the color resolution and frame rate will need to be significantly reduced. Think of the actual bandwidth of a telephone modem rated for 56 kbps. It would be impossible to produce a live videoconference using a 56 kbps modem.

Compression

All video systems use a form of compression and decompression to reduce the total number of bytes required for each image. The term *codec (compressor/ decompressor)* is used to represent software, hardware, or combination of software and hardware that compress and decompress video and audio information. Some common codecs are MPEG, Indeo, and Cinepak.

Let's take a closer look at one of the most common codecs in use—MPEG. *MPEG (Moving Picture Experts Group)* is an industry standard that ensures compatibility between different cameras, displays, and other multimedia equipment.

The compression technique that MPEG uses is simple to understand. Look again at Figure 13-6 of the dog playing ball. Notice how much of the image is redundant. Most of the background image does not change. MPEG capitalizes on this typical video feature. The MPEG compression software is written to predict which areas of the next frame will change and which areas will not. The prediction is based on the likelihood that if an area has not changed from one frame to the next, it is likely not to change in the next frame. The areas that do not

codec (compressor/ decompressor) software, hardware, or a combination of software and hardware that compress and decompress video and audio information.

MPEG (Moving Picture Experts Group) an industry standard developed to ensure compatibility between different cameras, displays, and other multimedia equipment.

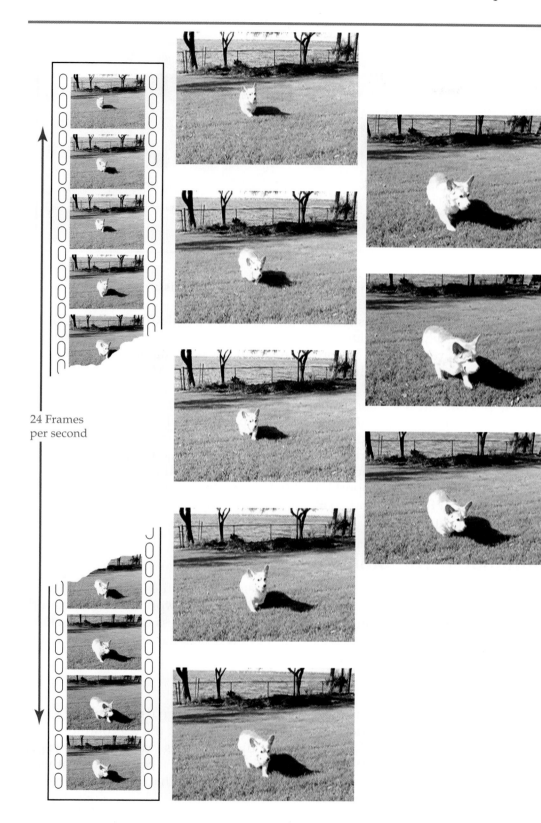

24 Frames per second

Figure 13-7.
A typical strip of movie film must change at a rate of 24 frames per second. The film consists of a string of still prints. Each frame contains a slight change of position of the object in motion.

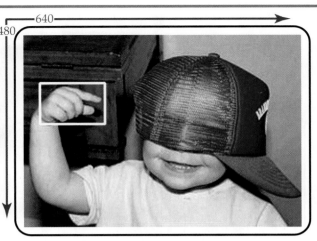

A closeup of the hand area reveals the pixel pattern

Figure 13-8.
The image in the display represents a VGA resolution of 640 × 480 pixels, or a total of 307,200 pixels. Each pixel location may require as many as 3 bytes to represent a color depth of 256 possible colors. This means that one frame of an image would require approximately 1MB of memory or storage.

change do not need to be transmitted or stored. The result is a reduction of the total amount of data that needs to be stored or broadcast.

There are various MPEG standards such as MPEG-1 and MPEG-2. MPEG-1 produces a relatively low-resolution image of approximately 320 × 240 pixels with a data rate of approximately 1.5 Megabits. This is an excellent standard for applications such as e-mail, where high resolution is not required. MPEG is also used to compress audio files. MPEG-2 produces a resolution of 720 × 480 pixels with a data rate of approximately 6 Mb and quality CD audio. The development of MPEG-3 was canceled and never fully developed. It was to support HDTV (High Definition Television). The MPEG-2 standard, however, is adequate to support HDTV.

MPEG-4 is a new standard that is not yet complete. When complete, MPEG-4 will be used on multimedia applications for the Internet. MPEG-4 is designed to meet the requirements to support broadcast quality audio and video. For full motion video, MPEG-4 supports data rates higher than 1 Gb per second. This is a vast improvement over MPEG-2 data compression.

Multimedia Transmission Protocols

Early network systems were used for transferring text files and still images. Today, multimedia is incorporated into many different network systems. Network videoconferencing is becoming commonplace in the business world. This section looks at some of the protocols designed to support high data rate audio and video streaming.

X.25

X.25
a protocol that uses analog signals to transmit data across long distances.

The *X.25* protocol is unusual because it uses analog signals to transmit data across long distances. It was developed in the early years of the telecommunications industry, prior to the widespread use of digital signals. An analog transport design for data was natural. What makes X.25 unusual when compared to other analog technologies is that it uses packet switching rather than permanent, electrical connections. Analog signals are routed around the world in packets that many times follow different routes to reach their destinations. X.25 laid the foundation for packet switching technology. X.25 is not commonly used in the United States, but many locations overseas are still committed to X.25. The X.25 protocol is limited to a maximum data rate of 56 kbps.

Frame Relay

Frame Relay was designed to replace X.25. Frame Relay is a packet switching protocol that typically uses leased lines such as T1 to carry data over long distances. You can think of Frame Relay as an upgraded digital version of the analog X.25 protocol. Frame Relay allows for a data rate as high as 1.544 Mbps compared to X.25, which has a limited data rate of 56 kbps.

Both X.25 and Frame Relay can be configured as a *permanent virtual circuit (PVC)*. A PVC behaves like a hard-wired connection between the destination and source. It can follow many different paths while transmitting data between the destination and source. A PVC is not committed to following a single path such as a T1 line.

This may sound confusing because Frame Relay is generally transmitted over T1 lines to ensure a minimum data rate known as a Committed Information Rate (CIR). A *Committed Information Rate (CIR)* is a guaranteed bandwidth a commercial carrier will provide a subscriber. Many times, a T1 line is shared with other users. When the T1 is shared, the data rate can fluctuate because of the variations in traffic, which includes the number of users and the type of data on the line. Frame Relay data transfer speeds are limited to the media used. A T1 line has a bandwidth of 1.544 bps, and T-3 is limited to 45 Mbps.

ATM

Asynchronous Transfer Mode (ATM) is a widely used protocol that is especially designed for carrying audio, video, and multimedia. It can support a bandwidth of 622 Mbps. ATM technology was designed in the 1970s at Bell Labs for use as a telephone technology. At that time, digital electronics was rapidly expanding and replacing electronic analog devices. Bell Labs was experimenting with a way to transmit analog voice signals as digital signals across telephone lines and networks. ATM technology was developed very early during the evolution of network communications and really did not come into common use for WAN applications until the 1990s.

ATM is designed specifically to transfer audio, video, and multimedia. ATM maintains a constant stream of voice or video from the source to the destination and avoids jitter. Traditionally, audio and video was transferred over analog media, such as telephone lines and radio and television frequencies. An analog signal can be converted into a digital signal through the use of specialized electronic chips referred to as *Analog to Digital Converter (ADC)* and *Digital to Analog Converter (DAC)* chips. As the names imply, the chips are designed to change an analog signal into a digital signal and vice versa.

Figure 13-9 is a simplification of ATM technology. The sound of a person's voice is converted into electrical energy. The electrical energy is a series of voltage fluctuation patterns known as an analog signal. The voltage fluctuation patterns are converted into a series of digital signals. At the destination, the digital signal is converted into an analog signal. The analog signal is amplified and applied to a speaker, which converts the analog signal into the sound of the voice.

ATM data transmission model

ATM is designed to divide text and audio/video into cells of 53 bytes each. **Figure 13-10** illustrates how ATM technology works. Notice that the cells are placed in sequence giving higher priority to the audio/video cells. Remember that the audio/video packets cannot tolerate excessive latency. Any delay in

---

Frame Relay
a packet switching protocol that typically uses leased lines such as T1 to carry data over long distances.

permanent virtual circuit (PVC)
a type of connection that behaves like a hard-wired connection between the destination and source.

Committed Information Rate (CIR)
a guaranteed bandwidth a commercial carrier will provide a subscriber.

Asynchronous transfer mode (ATM)
a protocol especially designed for carrying audio, video, and multimedia. It can support a bandwidth of 622 Mbps.

Analog to Digital Converter (ADC)
a computer chip designed to change an analog signal to a digital signal.

Digital to Analog Converter (DAC)
a computer chip designed to change a digital signal to an analog signal.

Figure 13-9.
An example of analog-to-digital and digital-to-analog conversion.

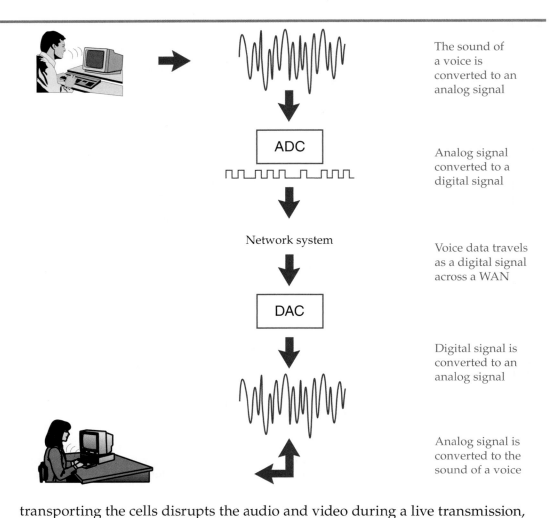

The sound of a voice is converted to an analog signal

ADC

Analog signal converted to a digital signal

Network system

Voice data travels as a digital signal across a WAN

DAC

Digital signal is converted to an analog signal

Analog signal is converted to the sound of a voice

transporting the cells disrupts the audio and video during a live transmission, resulting in jitter.

ATM can be transmitted at a Constant Bit Rate (CBR) or as a Variable Bit Rate (VBR). A *Constant Bit Rate (CBR)*, as the name implies, is a steady stream of ATM cells moving at a predictable rate. CBR is best used for video conferencing, telephone conversations, television broadcasts, or anywhere real-time data transfer is required. When a CBR is desired, a dedicated circuit or a virtual circuit must be established between two points. By establishing a constant connection between the two points, the data rate can be guaranteed. A guaranteed data rate is required for a live or a real time exchange of data.

A *Variable Bit Rate (VBR)* is a cell rate that automatically adjusts to support time sensitive data. It uses multiplexing techniques to provide a minimum CBR for time sensitive audio and video transmissions while controlling the data rate of not time sensitive data, such as text or plain e-mail. A VBR may be used when cost prohibits a CBR system. For example, a high bandwidth system equipped with a multiplexer at one end and a demultiplexer at the other end can transfer telephone conversations, e-mail messages, and files across a single carrier. The trick is to maintain a minimum data transfer rate for the telephone conversation and video, both of which are time-sensitive, while allowing delay to occur in the e-mail and file transfer. The ATM cells carrying audio and video are given the highest priority so that a constant flow of the time-sensitive data can be maintained. The cells carrying e-mail and file transfers can be delayed.

Constant Bit Rate (CBR)
a steady stream of ATM cells moving at a predictable rate. CBR is best used for video conferencing, telephone conversations, television broadcast, or anywhere real-time data transfer is required.

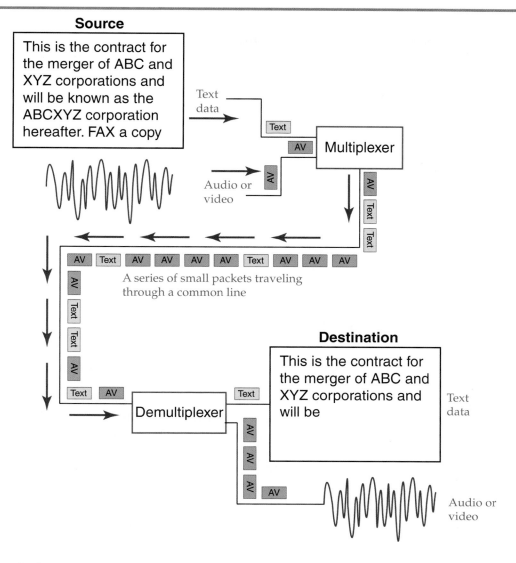

Source

This is the contract for the merger of ABC and XYZ corporations and will be known as the ABCXYZ corporation hereafter. FAX a copy

Text data

Text

AV

Multiplexer

Audio or video

AV

AV Text Text

AV Text AV AV AV AV Text AV AV AV

A series of small packets traveling through a common line

AV
Text
Text
AV
Text AV

Destination

This is the contract for the merger of ABC and XYZ corporations and will be

Text data

Demultiplexer

Text

AV
AV
AV AV

Audio or video

Figure 13-10. In this example, two forms of data, text and audio/video are separated into cells and multiplexed. All cells are placed on a single line as a series of packets. Priority is given to the time-sensitive audio/video cells. At the destination, a demultiplexer separates the two forms of data.

Before ATM, separate systems were required to accommodate the various types of data. ATM allows text, voice, and video to be carried on the same network system. This can prove to be cost effective in certain instances. For example, a corporation spanning the globe generates a large number of long distant telephone calls. It is less expensive to use voice transmission over the corporate WAN than using separate telephone lines. ATM communication may also prove to be an effective way of holding corporate meetings, which would otherwise require employees from various locations to converge and meet at a central location. Conferencing over a network is more economical than traveling, and, in many instances, it is a more effective use of employee time.

ATM switch

The ATM switch provides virtual circuits between various points on the network system and is used with the ATM protocol. An ATM switch can provide a separate virtual circuit between several computers at the same time.

Look at **Figure 13-11.** In the illustration, the ATM switch provides virtual circuit connections between *Computer A*, *Computer B*, and *Computer C* and at the same time provides a connection between *Computer D* and *Computer E*. The computers are

Variable Bit Rate (VBR)
a cell rate that automatically adjusts to support time sensitive data. It uses multiplexing techniques to provide a minimum CBR for time sensitive audio and video transmissions while controlling the data rate of not time sensitive data, such as text or plain e-mail.

Figure 13-11.
Computer A provides
video to *Computer B*
and *Computer C* while
Computer D exchanges
a file with *Computer E.*
The ATM switch creates
a separate virtual
circuit between these
computers, which
behave as though they
were wired directly to
each other.

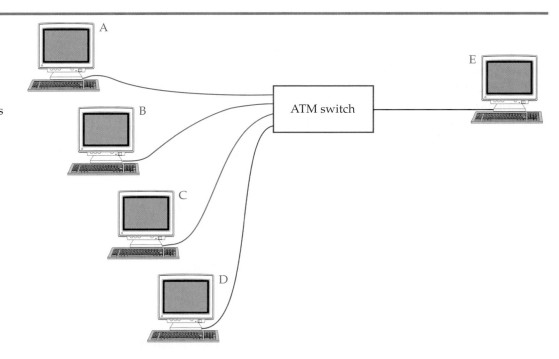

connected as virtual circuits, not permanent circuits. During the communication
period between two or more computers, the ATM switch makes the connection appear
to be a permanent connection. When the transfer of text, audio, or video is complete,
the circuits disconnect. A computer can then connect to a different computer.

In the illustration, the ATM switch is used to provide a temporary connection
between designated computers. The switch can also be set up to provide a
permanent, virtual connection that provides a guaranteed bandwidth between
two or more computers.

ATM data transfer classifications

There are five ATM classifications: CBR, VBR-NRT, VBR-RT, ABR, and UBR.
Each classification describes the degree to which the virtual circuit is dedicated
to ATM cell transmission.

You have already learned that CBR, Constant Bit Rate, is a steady stream of
ATM cells moving at a predictable rate. CBR is used for applications such as
video conferencing and telephone communications. Variable Bit Rate (VBR) has
the ability to change the rate of flow in contrast to CBR. VBR is divided into two
major classes: VBR-Real Time (VBR-RT) and VBR-Non Real Time (VBR-NRT).

Variable Bit Rate-Non Real Time (VBR-NRT) allows cells to move at a
variable rate. The rate of movement depends on the type of data contained in
each cell. For example, an e-mail that contains a text and multimedia attachment
contains two types of data. Each type of data needs to be handled with a
different priority. The multimedia data would have the highest priority. However,
there is no need for the cells that contain multimedia data to be transferred in
real time because the communication is strictly in one direction, not two-way like
in a telephone conversation.

Variable Bit Rate-Real Time (VBR-RT) is similar to VBR-NRT in that it
moves cells at a variable rate depending on the cell's contents. The main
difference between the two bit rates is VBR-RT adjusts the bit rate to support real-
time audio and video transfers. For example, when transferring text files and

*Variable Bit Rate-
Non Real Time
(VBR-NRT)*
an ATM data transfer
classification that
allows cells to move
at a variable rate. The
rate of movement
depends on the type
of data contained in
each cell.

*Variable Bit Rate-
Real Time (VBR-RT)*
an ATM data transfer
classification that
allows cells to move
at a variable rate to
support real-time
audio and video
transfers.

telephone conversation at the same time, the telephone conversation is transferred in real time while the text file is sent at a lower rate.

Available Bit Rate (ABR) is most appropriate for file transfer. It uses the available bit rate associated with the networking medium. The speed of a file transfer using the ABR classification is affected by the amount of traffic on the network system.

Unspecified Bit Rate (UBR) is as implied. It does not guarantee any speed or meet requirements of any special application such as multimedia or telephony. This classification is typically applied inside TCP/IP frames. A good example is a Web cam that updates the image every thirty seconds or so, rather than providing full-motion video.

The ATM data transfer classifications reflect a matching cost factor. The highest quality is CBR, and it is the most expensive to use. The least expensive is UBR, which is minimal to no cost when used across the Internet as part of a TCP/IP frame. The minimal cost is the cost of the ISP.

ATM cell design

The ATM cell is simple in design. The maximum size of the cell is 53 bytes, consisting of a 5-byte header and a 48-byte payload. See **Figure 13-12.**

Available Bit Rate (ABR) an ATM data transfer classification most appropriate for file transfer. It uses the available bit rate associated with the networking medium. The speed of a file transfer using the ABR classification is affected by the amount of traffic on the network system.

The ATM cell is at times referred to as a packet or frame. This is incorrect according to the ATM forum organization.

Tech Tip

The most significant characteristic of the ATM cell is it carries only a 48-byte payload. While this may seem small, you must consider that the idea behind the design is to be able to mix many different cells on the same network medium. Remember, it is important not to cause too much latency or destroy a multimedia cell. Any cells that are destroyed or do not reach the destination are not retransmitted over the media. There is no time to retransmit a cell or to reorder cells if they become out of order when traveling across a network.

The 4 bits in the Generic Flow Control (GFC) were originally designed to contain information about the function of the cell. It is not used at the time of this writing. The 8 bits in the Virtual Path Identifier (VPI) and the 16 bits in the Virtual Channel Identifier (VCI) identify the path from the source to the destination using a series of ATM switches. The 3 bits in the Payload Type (PT) field indicate the type of payload carried in the cell, such as data or control instructions. The 1 bit in the Cell Loss Priority (CLP) field indicates if the cell should be dropped if it encounters severe congestion on the network. The CLP

Unspecified Bit Rate (UBR) an ATM data transfer classification that does not guarantee any speed or meet requirements of any special application such as multimedia or telephony. This classification is typically applied inside TCP/IP frames.

Generic Flow Control (GFC)
Virtual Path Identifier (VPID)
Virtual Channel Identifier (VCID)
Payload Type (PT)
Cell Loss Priority (CLP)
Header Error Control (HEC)

Figure 13-12.
The ATM protocol is designed for efficiency. Compared to other protocols, such as IP, it is quite small.

field keeps time-sensitive transmissions, such as live video and audio, up-to-date and prevents latency issues. The 8-bit Header Error Control (HEC) field detects errors in the header portion of the cell, not in the payload.

ATM is a connection-oriented protocol. This means that it establishes a logical connection between the source and destination. Logical connections are also referred to as virtual connections. Remember that a virtual connection is not a permanent, physical connection but rather a connection that behaves as a physical connection. A virtual connection is set up and controlled by a software program. There are two types of virtual connections associated with ATM, and they are identified in the header of the ATM cell. The two types of connections are Virtual Path Connections (VPC) and Virtual Channel Connections (VCC). See **Figure 13-13.**

The illustration shows a backbone running between two ATM switches. The backbone is considered a virtual path. The virtual path contains the individual virtual circuits, or virtual channels, between specific computers or network equipment. In the illustration, there are two virtual channels, A and B. The virtual channels merge into the virtual path when traveling between the switches and then split into separate virtual channels when serving the computers identified as A and B.

To better understand virtual path connections and virtual channel connections, a complete highway system can be used as an analogy. A complete highway system includes state highways and an interstate highway. The state highways can be compared to virtual channels and an interstate highway can be compared to a virtual path. State highways connect at different points along the interstate highway. A car travels from its home (source) across a state highway, merges onto the interstate highway where it travels with other cars from various state highways, and then exits onto a state highway to reach its destination.

Figure 13-13.
The ATM backbone contains paths called virtual paths, which consist of two or more virtual channels. The backbone is referred to as the virtual path.

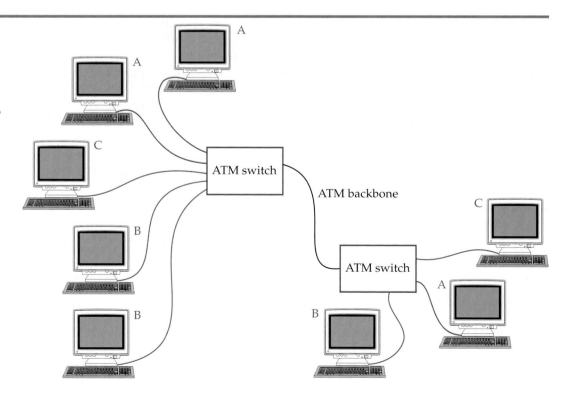

In reality, there can be many virtual path connections between ATM switches. The virtual path connections can be set up as Permanent Virtual Circuits (PVC) or Switched Virtual Circuits (SVC). For example, a corporation may have a PVC established between its Chicago and St. Louis offices to maintain a constant update of a critical database. Other connections can be set up as an SVC. The SVC would be used as needed. It would typically be used by individuals who only connect to resources at the other locations on a temporary basis. Remember that connections across long distance lines are very expensive. Even virtual connections are expensive. A dedicated, physical connection between two locations is more expensive than a switched circuit that is only connected temporarily. The ATM and other cell switching technologies are designed with flexibility in mind. However, because of the flexibility, much confusion can arise.

ATM switches are designed to support ATM protocol functions. The ATM switch processes millions of cells per second and provides multiplexing, which is required to deliver each cell in a timely fashion while sharing a common media. Each ATM switch maintains routing tables based on information contained in the ATM cell header. A routing table for the ATM switch is similar in design and purpose to the routing table used for a router. Remember that the virtual path identifier and the virtual channel identifier are used to locate and maintain the connection between two devices. Based on the information contained in the header, the ATM switch sets up the virtual channels and paths between end users.

The ATM switch controls the bandwidth allocated to each virtual circuit and controls the cells by priorities set in the Cell Loss Priority (CLP) field in the ATM cell header. The switch also drops some of the cells if cell traffic exceeds the capacity of the switch, media attached to the switch, or both.

The typical ATM switch is designed with multiple ports. The switch is used to make a physical connection to various network media such as fiber-optic cable, twisted pair cable, and coaxial cable. Because of costs, ATM switches are mostly found on long-haul communication portions of a WAN rather than on a LAN.

VoIP

Voice over IP (VoIP), also known as Internet telephony, is similar in concept to ATM in that it is designed for high performance data delivery and quality of service. VoIP, however, relies on existing TCP/IP technology and existing TCP/IP networking equipment. ATM is a protocol separate from the TCP/IP stack and uses special equipment designed for the protocol, such as ATM switches and dedicated communication lines. ATM requires expensive equipment for support, while support for VoIP is inexpensive.

Voice over IP (VoIP) an Internet telephony protocol that relies on existing TCP/IP technology and existing TCP/IP networking equipment.

At the time of this writing, VoIP is still evolving and is considered an inexpensive alternative to ATM. VoIP is less expensive for long distance calls when compared to existing telephone technologies. This is why it is appealing to so many people, especially to corporations communicating through long distance with overseas corporations.

VoIP typically uses a series of UDP packets to send voice data across the network. The construction of the UDP packet is similar to that of ATM. Remember that ATM designed its packet for efficiency. The ATM packet is a short packet with little overhead like the UDP packet. UDP provides a best effort delivery of data with no mechanism to check if the data arrived in sequence or intact. The UDP packet does not generate high latency because it does not attempt to resend lost packets.

VoIP uses the TCP protocol for applications such as video or audio streaming. Streaming is the continuous downloading of multimedia files without interruption. The TCP protocol has a sequence number in the header, which ensures packets are arranged in proper sequence when they arrive at the final destination. Remember, TCP uses a sequence number as part of its header information so packets can be reassembled at the destination. UDP does not have a sequence number, so it cannot guarantee assembly of data packets in their proper sequence.

For TCP/IP to be an excellent means of carrying multimedia, a media carrier from source to destination with a high bandwidth is required, such as Gigabyte Ethernet or fiber-optic cable.

One of the major technical problems with Internet communications has been the location of the last few hundred feet of cabling to the home computer, known as the local loop. The local loop is typically designed for low-bandwidth analog voice communications. Since the bandwidth is low, it is the bottleneck for the entire system, even if the rest of the communication link uses the very latest technology. The last few hundred feet of the system can affect the overall latency of packet transmission, which negates the latest technology. Traditional telephone lines cannot adequately carry audio and video in real-time. For VoIP to work successfully, there must be adequate bandwidth. Without adequate bandwidth, latency becomes a problem. DSL and cable for Internet access provide the adequate bandwidth for VoIP to run without latency. DSL and cable are not connected to the local loop.

There are several ways VoIP can be used. These include the following:

■ Communicating from PC to PC using TCP/IP.

■ Mixing TCP/IP telephone technology with existing and modern telephone technologies.

■ Communicating from a PC to a cell phone via traditional LAN cabling network, a Public Switched Telephone Network (PSTN), and then wireless technology.

VoIP can communicate from PC to PC using only the TCP/IP protocol. This is easy to set up and is similar to the technology known as NetMeeting. With NetMeeting, two or more PCs are connected via a LAN to support conferencing. NetMeeting allows the transfer of text, voice, and video depending on the bandwidth of the media.

Public Switched Telephone Network (PSTN)
an older telephone technology that uses twisted pair cabling and analog signals rather than digital.

Plain Old Telephone Service (POTS)
an older telephone technology that uses twisted pair cabling and analog signals rather than digital.

VoIP can be used by mixing TCP/IP with modern telephone technologies. For example, a PC can connect to a standard telephone via a TCP/IP network and Public Switched Telephone Network (PSTN) and the more modern ISDN and FDDI technologies. *Public Switched Telephone Network (PSTN)* usually refers to the older telephone technology that uses twisted pair cable and analog signals rather than digital. The older telephone system is also referred to as *Plain Old Telephone Service (POTS)*. Mixing TCP/IP with telephone technologies is very complex. It requires translating TCP/IP into traditional telephone protocols and vice versa. There can be substantial delays or latency using a hybrid, or mixed, protocol system required for communications over the variety of media encountered.

VoIP can also be used to communicate from a PC to a cell phone via traditional LAN cabling, a Public Switched Telephone Network (PSTN), and then wireless technology. This type of communication is even more complicated than the last scenario and can create even longer delays. Look at **Figure 13-14.**

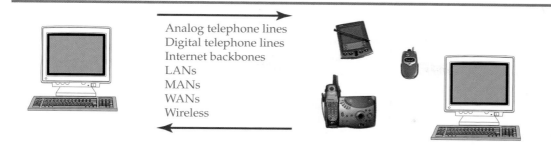

Figure 13-14.
Computers can communicate with a wide variety of communication devices such as cell phones, pagers, palmtops, personal digital assistants, and landline telephones. The wide variety of equipment and devices require a wide variety of communication lines and equipment.

A PC can communicate with a wide variety of communication devices. There are many communication scenarios that mix PC communication with conventional telephone, wireless cell phones, personal digital assistants, and pagers. Each of these communication technologies requires special protocols and protocol conversions. Each set of protocols varies according to the communication path. The path can be over public analog telephone lines, public digital telephone lines, private company lines, fiber-optic cable, infrared, satellite, and wireless.

A telephone gateway is a specialized piece of equipment that connects a packet-style network communication system to a telephone system using the H.323 protocol. The **H.323 standard** is the telecommunications standard for audio, video, and data communications using IP or packet-type networks defined by the International Telecommunication Union (ITU). On one side of the telephone gateway is the normal telephone system and on the other side is the network system. The telephone system side is the H.323 side. You can think of the gateway as the H.323 server and the computers on the network side as the H.323 clients. Look at **Figure 13-15.**

H.323 is not one specific protocol but rather an entire suite of protocols similar to the TCP/IP suite of protocols. There are over 20 different protocols designed to work with H.323. The protocols are systems that control processes such as name to telephone number conversion, call forwarding, caller ID, call blocking, conversion from TCP/IP to wireless, and conversion to European telephone systems.

H.323 standard
a telecommunications standard for audio, video, and data communications using IP or packet-type networks defined by the International Telecommunication Union (ITU).

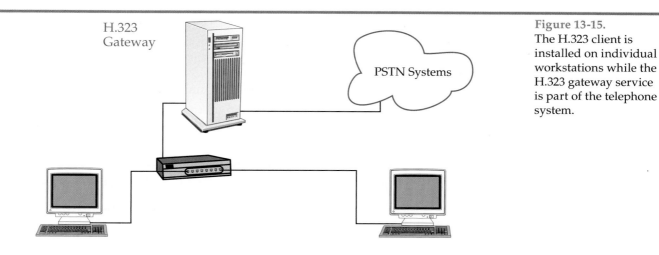

Figure 13-15.
The H.323 client is installed on individual workstations while the H.323 gateway service is part of the telephone system.

Figure 13-16.
The QoS Packet
Scheduler is an optional
protocol that is avail-
able in Windows XP.
The QoS protocol gives
time-sensitive packets a
higher priority than
data packets.

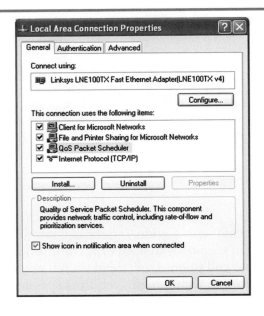

*Quality of Service
(QoS)*
a protocol that gives
time-sensitive
packets, such as
those carrying tele-
phone conversations,
a higher priority than
data packets.

The biggest problem in using packet-based networking systems is latency. A protocol, called *Quality of Service (QoS),* was developed to minimize latency. The QoS protocol gives time-sensitive packets, such as those carrying telephone conversations, a higher priority than data packets. The QoS Packet Scheduler is available in Windows XP, **Figure 13-16.**

Frame Relay and ATM are sometimes carried over typical Internet backbones packaged inside the IP protocol. These technologies are referred to as ATM over IP (ATMoIP) and Frame Relay over IP (FRoIP). We take a closer look at how these systems work in Chapter 15—Remote Access and Long Distance Communications.

Nortel Networks specializes in integrating telecommunications and networking. You may want to visit the Nortel Web site at www.nortelnetworks.com to learn more about the telecommunications industry as a possible source of employment. Nortel also offers a variety of certifications. Many of these certifications involve a detailed understanding of networking.

Summary

- An analog signal is converted into a digital code by taking samples of the analog amplitude pattern at specific times.
- A sampling rate is the number of times a signal is sampled during a specific period of time.
- Jitter is the small staggers or hesitations in the delivery sequence of audio or video data caused by latency or missing packets.
- Video images typically occur at a minimum rate of 24 frames per second.
- The term *latency* means the delay of data as it travels to its destination.
- While latency with a delay of approximately 250 milliseconds will be noticed by a user, it is acceptable.

■ The term *codec* represents software, hardware, or combination of software and hardware that compresses and decompresses video and audio information.

■ X.25 is one of the early protocols used to transfer data over telephone lines.

■ Frame Relay is a packet switching protocol that typically uses leased lines such as T1 to carry data over long distances.

■ The Asynchronous Transfer Mode (ATM) protocol is designed to carry audio, video, and multimedia.

■ The five ATM classifications are CBR, VBR-NRT, VBR-RT, ABR, and UBR.

■ The maximum size of an ATM cell is 53 bytes, consisting of a 5-byte header and a 48-byte payload.

■ Voice over IP (VoIP) is a technique that relies on the TCP/IP protocol suite to carry audio and video data.

■ VoIP uses the UDP protocol when sending time-sensitive data, such as a telephone conversation.

■ VoIP uses the TCP protocol when sending audio and video streaming data.

■ The H.323 protocol is used by telecommunications and telephone equipment.

■ A gateway is used to convert the TCP/IP protocol to the H.323 protocol.

Review Questions

Answer the following questions on a separate sheet of paper. Please do not write in this book.

1. Converting digital pulses to an analog waveform is called _____ conversion.

2. Converting an analog waveform to digital pulses is called _____ conversion.

3. What is jitter?

4. The delay of data from its origin to the final destination is called _____.

5. What is the principle used for MPEG compression?

6. What communication protocol is characterized by analog packets rather than digital packets for communications?

7. What is the maximum data rate of X.25?

8. What protocol could be called an upgrade of the X.25 protocol?

9. What is the maximum data rate of Frame Relay?

10. Which scenario requires the Constant Bit Rate (CBR) classification of ATM communications?

    A. Live telephone conversation.
    B. FAX.
    C. Real-time broadcast of a movie.
    D. Business conference call.
    E. Cell phone conversation.
    F. Transfer of an image from an archive.

11. Which ATM classification provides the best support for video conferencing?

12. What does the term *Variable Bit Rate* mean in reference to ATM classifications?

13. What is the total length of a typical ATM cell including the header?

14. What happens when one of the ATM cells is destroyed or does not reach the destination?

15. Why is the ATM cell payload so small (48 bytes)?

16. An ATM virtual path may contain two or more _____.

17. A Switched Virtual Circuit (SVC) is a (temporary or permanent) _____ connection between two devices on a network.

18. Based on what two factors will an ATM switch drop a cell?

19. When using VoIP technology, what protocol is used for time-sensitive data such as telephone conversations?

20. When using VoIP technology, which protocol is used for streaming data?

21. Which protocol is designed for multimedia traffic on packet-type networks?

22. What piece of equipment is responsible for converting IP protocols to H.323 protocols?

Sample Network+ Exam Questions

Network+

Answer the following questions on a separate sheet of paper. Please do not write in this book.

1. The delay of video/audio data packets is referred to as _____ in networking terminology.
   A. Packet Loss Ratio
   B. latency
   C. Variable Rate Overflow
   D. heading

2. Which ATM data transfer classification provides the highest quality of ATM service?
   A. VBR
   B. CBR
   C. UBR
   D. ABR

3. Which upper layer protocol is typically used to package voice data that is to be transferred using Voice over IP (VoIP) technology?
   A. PPP
   B. SLIP
   C. UDP
   D. CHAP

4. Which definition best describes multiplexing as it applies to ATM?

   A. Combining multiple signals or sequencing different data packet streams over a single medium.

   B. Combining several different protocols to use as a single protocol.

   C. Combining several different types of networking media to form one cohesive media.

   D. Combining separate data packets into one large data packet before transmitting across a single network line.

5. Which of the following statements are true concerning ATM? (Select all that apply.)

   A. The ATM protocol allows voice, video, and text data to be carried on the same media.

   B. The maximum cell size for ATM is 1500 bytes.

   C. ATM is a connection-oriented protocol.

   D. ATM is a connectionless protocol.

6. Which protocol transmits data in the form of an analog signal?

   A. IP

   B. IPX

   C. X.25

   D. ATM

7. Which compression method is used to achieve the highest data rates for multimedia over a network?

   A. MPEG-1

   B. X.25

   C. SLIP

   D. MPEG-4

8. Which protocol is used to support Internet telephony?

   A. CHAP

   B. VoIP

   C. SLIP

   D. AppleTalk

9. Which protocol is designed specifically to ensure the delivery of time-sensitive packets and reduce latency?

   A. QoS

   B. TCP

   C. X.25

   D. MPEG

10. Which telephone protocol suite is designed to support audio, video, and data communications?
   A. SLIP
   B. RADIUS
   C. RAS
   D. H.323

Interesting Web Sites for More Information

www.ectf.org

www.imtc.org

www.nortelnetworks.com/corporate/technology/voip/index.html

www.webtutorials.com

Suggested Laboratory Activities

1. Open **Control Panel** in Windows 2000 and see if there is an ATM protocol selection for setting up a NIC card?

2. Install two Web cams in the lab area and monitor the performance between the two cameras. You may wish to use a protocol analyzer to measure the amount of traffic generated between the two.

3. Access a Web cam on the Internet and analyze, using a protocol analyzer, the amount of network traffic required to support the download of images from the Web cam.

Chapter 13
Laboratory Activity

NetMeeting

After completing this laboratory activity, you will be able to:
- Set up and configure Microsoft NetMeeting.
- Modify the NetMeeting configuration.
- Explain how a firewall can affect NetMeeting communication.
- Host a NetMeeting and select the tools to be used in the meeting.

Introduction

NetMeeting is designed to support online collaboration, live video, Chat, Whiteboard technology, and file sharing and exchange. In this laboratory activity, you will set up and configure Microsoft NetMeeting and then participate in a live conference call with a person at another workstation. NetMeeting can be used to support a conference between two or more clients on the same local network or over the Internet and can be used with all of the Microsoft operating systems. You can even use it to communicate with operating systems other than Microsoft that use a third-party video conferencing software package.

NetMeeting can be configured to automatically look for a directory server at startup. A directory server keeps a list of NetMeeting users. It is an optional service and is not required for NetMeeting. Microsoft provides an Internet directory server. There are also many third-party servers used to keep the same type of information. An Intranet server can also be configured to serve as a directory server for NetMeeting.

The first screen to appear when NetMeeting opens is the main NetMeeting user interface. It is used to set up or join a meeting from the host. The following figure shows the main NetMeeting user interface.

The main user interface can be configured in several different views depending on your needs and the equipment used, such as a video cam. The main user interface can be configured as not compact, like in the previous figure, compact, data only, or dial pad. The following figure shows the compact, data only, and dial pad views.

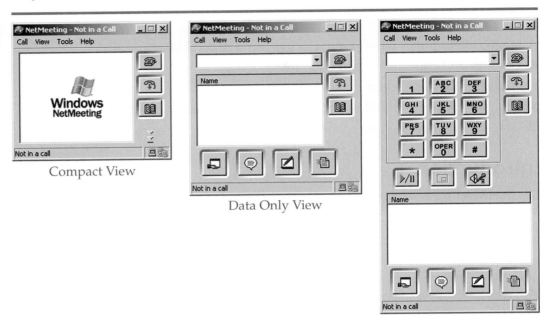

Compact View   Data Only View   Dial Pad View

Other screens will appear depending on the features used. For example, if the Chat feature is used, the Chat screen will display. Features include Sharing, Chat, Whiteboard, and File Transfer.

A NetMeeting connection is established between two of more workstations by e-mail address, computer name, IP address, or telephone number. Network security often incorporates a firewall for protecting network clients. Firewalls typically block certain ports and leave others unblocked to support required services. The ports that most likely are required for a successful NetMeeting are TCP ports 389, 522, 1503, 1720, 1731, and several additional ports located between 1024 and 65535. Network and workstation firewalls can interfere with the correct operation of NetMeeting. You should make sure that the ports you will use for NetMeeting communications are not blocked by a firewall.

NetMeeting will communicate with most other third-party network meeting utilities because it is based on two common International Telecommunications Union (ITU) communications standards: H.323 and T.120. By adhering to these protocols, the two end points can use different proprietary software designed for network conferencing and be able to communicate effectively with each other.

You may be wondering why two different standards are used and not just one. Audio and video typically use a "best effort" means of delivering data from source to destination. This is because live video and audio cannot rebroadcast lost packets, yet it still must achieve a quality transmission. The H.323 is a best effort specification to ensure the data (video and audio) is delivered quickly. The T.120 standard ensures that all data is delivered completely with no missing packets. Thus, the two protocols work together to achieve a quality transmission.

Note:
Microsoft has released LiveMeeting, which is an enhanced version of NetMeeting. It is expected to replace NetMeeting.

Equipment and Materials

■   (2) Workstation running Windows XP connected to a peer-to-peer network.

■   (2) Web cam (optional).

■   (2) Microphone (optional).

■   (2) Set of speakers (optional).

Note:
Two students can perform this laboratory activity, one at each workstation.

Procedures

1. _____   Gather all required materials and report to your assigned workstation.

2. _____   Boot the workstation and verify it is in working order.

3. _____   To automatically install NetMeeting, open the **Start** menu, select **Run**, and then type **conf** into the **Run** dialog box. The NetMeeting wizard will automatically start. The first screen presents an overview of NetMeeting features. Click **Next**.

4. _____   You will be asked if you want to log on to a directory server when NetMeeting starts and if you want your name listed in the directory. Deselect **Log on to a directory server when NetMeeting starts** and select **Do not list my name in the directory**. Click **Next**.

5. _____   Now you are presented with four connection type choices: **14400 bps modem; 28800 bps or faster modem; Cable, xDSL or ISDN;** and **Local Area Network**. For this laboratory activity, select **Local Area Network**. Click **Next**.

6. _____   You will be prompted with the following choices: **Put a shortcut to NetMeeting on my desktop** and **Put a shortcut to NetMeeting on my Quick Launch bar**. Select both, and then click **Next**.

7. _____   You will see a series of windows prompting you to adjust and test the video and audio settings. Respond to the simple prompts accordingly. When prompted, click **Finish** to complete the wizard. NetMeeting will automatically start.

8. _____   Install NetMeeting on the other workstation.

9. _____   After NetMeeting is installed on both workstations, use one of the workstations to call the other. Use the computer name to establish the connection.

Note:
NetMeeting must be running on each workstation to establish a call.

10. _____ After establishing a connection between the two workstations, open the **Chat** feature located under the **Tools** menu. Send several text messages until you are comfortable using the text message system.

11. _____ Open the **Whiteboard** feature located under the **Tools** menu. Experiment with the Whiteboard feature until you feel confident about using it.

12. _____ Complete this step only if a Web cam is installed on the workstation. If not, proceed to the next step. Open the **My Video** option located under the **View** menu. After the **My Video** window is filled with the Web cam image, click **Options** located under **Tools** menu. The **Options** dialog box will appear.

13. _____ In the **Options** dialog box select the **Video** tab. The **Video** options page should appear similar to the one in the following figure. Experiment with the video settings. Change the image size and the video quality and then observe the effect on the video image. You may need to move your hand across the front of the Web cam to observe changes in the video quality adjustment. The changes will be easiest to observe when a moving image is captured. After you are familiar with the video adjustments, proceed to the next step.

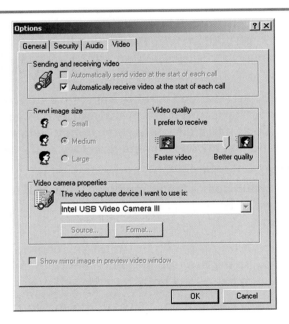

14. _____ Select the **Audio** tab and then the **Security** tab to see what options are available. After familiarizing yourself with these options, close NetMeeting on each workstation.

15. _____   Now, you will host a meeting. When you host a meeting you are in control of the NetMeeting environment. To host a meeting, simply start the NetMeeting program and then select **Host Meeting** from the **Call** menu. The **Host a Meeting** dialog box should look similar to the one in the following figure.

Notice that there are two major categories of options in the dialog box: **Meeting Settings** and **Meeting Tools**. Under the Meeting Settings options, you can provide a name for the meeting and password. The password can be used to control who can access the meeting. Under the Meeting Tools options, notice that the host selects which tools can be used for the meeting, such as the Whiteboard, Chat, File Transfer, and Sharing. As you can see, the main difference between hosting a meeting and simply calling another workstation is controlling the meeting environment. Now go on and set up a meeting, including a password. Have the other workstation access the meeting after you have configured it.

16. _____   Answer the review questions. You may use **Help Topics** located under the **Help** menu to assist you.

17. _____   Leave the workstation in the condition specified by your instructor and return all materials to their proper storage area.

Review Questions

Answer the following questions on a separate sheet of paper. Please do not write in this book.

1. What information can be used to establish a connection with another computer using NetMeeting?

2. How can a firewall affect NetMeeting?

3. What ports are commonly associated with NetMeeting?

4. What standard is used for audio and video conferencing across telecommunication lines?

5. What standard is used to support Whiteboard technology across a telecommunications line?

6. What connection types does NetMeeting support?

7. What environmental factors do you control when you host a meeting?

14

Web Servers and Services

After studying this chapter, you will be able to:

- ❑ Explain the difference between Internet, intranet, and extranet.
- ❑ Explain how the Domain Name System (DNS) relates to Web servers.
- ❑ Describe the purpose of a Web browser.
- ❑ Explain how markup languages work.
- ❑ Describe how e-mail messages are transported.
- ❑ Explain the difference between common e-mail protocols.

Network+ Exam—Key Points

The Network+ Certification exam presents questions about Web server basics. It does not contain in-depth questions in this area. Rather, it tests your knowledge of protocols such as HTTP, HTML, POP3, IMAP4, SMTP, FTP, and SOAP, commonly used in conjunction with Web servers and e-mail servers. You need a basic understanding of how these protocols are used. Also, be sure you can configure a workstation for e-mail.

Network+

For in-depth questions on Web servers, the CompTIA organization has developed the iNet+ Certification, which covers Web service support. The iNet+ Certification exam concentrates on basic networking technology, Web services, and the basics of Web programming languages, such as HTML, XML, and JAVA. It also covers an introductory level of common Web page authoring programs, such as Adobe GoLive and Microsoft FrontPage. A candidate must be very familiar with graphics and animation to successfully pass the exam.

Tech Tip

Key Words and Terms

The following words and terms will become important pieces of your networking vocabulary. Be sure you can define them.

Extensible Markup Language (XML)

extranet

HTML tag

hyperlink

Hypertext Markup Language (HTML)

Hypertext Transport Protocol (HTTP)

Internet

Internet Message Access Protocol (IMAP)

intranet

mail filter

mail gateway

Multipurpose Internet Mail Extensions (MIME)

Network News Transfer Protocol (NNTP)

newsgroup

Post Office Protocol (POP)

Secure File Transfer Protocol (SFTP)

Simple Mail Transfer Protocol (SMTP)

Simple Object Access Protocol (SOAP)

spam

spammer

spamming

Trivial File Transfer Protocol (TFTP)

Uniform Resource Locator (URL)

Web browser

Web server

Web site

The need for Web servers has greatly increased over the years. Web servers provide a means to display informational pages to viewers across the World Wide Web or a local area network. E-mail has rapidly grown to become one of the most common forms of communication.

In this chapter you will learn the differences between the three types of networks that provide Web page distribution. You will also be introduced to two of the most popular Web server software packages and to the basic operation of an e-mail system.

Internet, Intranet, and Extranet

There are three types of networks that provide Web page distribution: Internet, intranet, and extranet. The type of access allowed to a network's Web pages (Web site) determines each classification.

Internet

Internet
a collection of interconnected networks from all around the world.

An *Internet* is a collection of interconnected networks from all around the world. Therefore, when you install a Web server for the Internet, you are setting up a computer to be accessed by anyone in the world. The Internet relies on the TCP/IP protocol and its suite of protocols to transport Web pages. The Internet is also referred to as the World Wide Web because when the various network cable connections that make up the Internet are diagrammed, the diagram resembles a spider's web.

There are other types of networks that function similar to the Internet but on a limited or restricted basis. These networks are called intranet and extranet and are commonly found in business environments.

Intranet

A Web server need not be connected to the Internet to serve as a way for companies to communicate and to distribute information. A Web server can be connected to an intranet. An *intranet* is a private network that serves a specific group of users within a LAN. For example, a corporation or an educational institution may set up a private Web server to be accessed only by a designated group of users from within its local area network. In the corporation, an intranet is designed for only the employees. In the educational setting, it may be accessed only by school employees and students. Normally, an intranet cannot be freely accessed by anyone. If the intranet is accessible from the World Wide Web, the intranet uses a firewall to isolate it from the outside world. The main reason for access limitations is security.

An intranet can be used to post forms, explanations and examples of forms, policy books, maps, address books, and such. The advantage of using an intranet is users already know how to surf the Internet and can start using the system immediately. In a traditional network, a new user often needs training in how to access network shares and navigate directory structures.

intranet
a private network that provides Web page distribution to a specific group of users within a LAN.

Extranet

An *extranet* allows internal access to Web pages and allows authorized personnel from outside the network to access the network's Web pages. An extranet is often designed to allow employees, business partners (other businesses), and customers access to the intranet from outside the network, through the Internet. Access to the extranet is available only if the user has a valid user name and password. The user name and password determines what areas of the extranet are accessible. For example, members of partner companies may have only limited access to the Web site while employees may have full access to the Web site.

Look at **Figure 14-1** for a comparison of an intranet and an extranet. Remember that an extranet does not let the general public access the Web site—only partner companies, employees, and authorized customers.

extranet
a network that allows internal and external access to Web pages by authorized personnel.

Domain Name and URL Resolution

Since the TCP/IP protocols use IP addresses to transfer data, the Internet must use the Domain Name Service (DNS) to translate domain names and Uniform Resource Locators (URLs) to an IP address. A *Uniform Resource Locator (URL)* is a user-friendly name such as http://support.microsoft.com/. When a URL is typed into a Web browser, the Web browser uses a DNS server to translate or resolve the URL name to an IP address.

A user can directly access a Web page by typing in the IP address of the Web server that hosts the Web page. This eliminates the need to use a DNS server to resolve the URL name to an IP address and may expedite access. However, people do not generally communicate by IP addresses. They use domain names.

Uniform Resource Locator (URL)
a user-friendly name, such as http://support. microsoft.com/, which allows users to access Web-based resources.

Network proxy servers store IP and URL addresses in their cache to expedite the connection process. A proxy server intercepts all requests for Web sites and then provides the requested Web page without the need to connect to the Web site. If the proxy server does not contain a copy of the requested page, it forwards the request to the Web site.

Tech Tip

Figure 14-1.
An intranet allows employees limited access to a company Web server. The Web server is accessible only through the company LAN. It is not accessible through the Internet. An extranet provides limited access to the company Web server to both employees within the LAN and to customers, partner companies, and employees through the Internet.

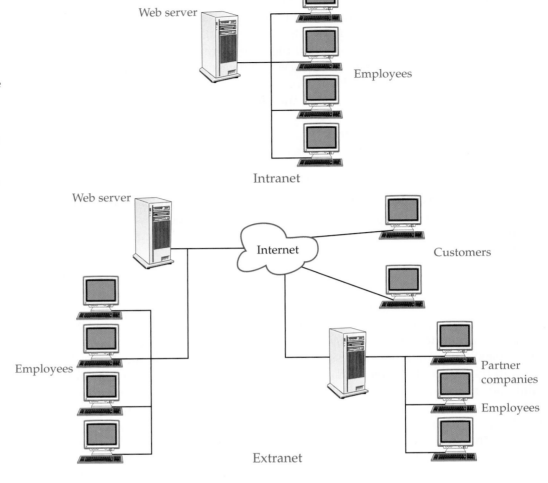

A complete URL can be seen in **Figure 14-2.** Note the way the URL is expressed. The first item in the URL is the protocol. Typical choices are HTTP and FTP. The next part of the URL is the domain name of the Web server hosting the site followed closely by the path to the resource or page desired. The path is actually a directory or directory and subdirectories on the Web server that lead to the Web page file. In the example, the index.html Web page is located in the /books directory. The port number at the end of the URL is optional and is not generally used. The default port number of a Web page TCP/IP connection is 80.

DNS is organized as a hierarchical structure with the root at the top. As you have seen in previous chapters, Microsoft Active Directory (AD) and the Novell Directory Services (NDS) have a similar structure to DNS. NetBIOS domain names of earlier systems have come to look like a URL. Lightweight Directory

Figure 14-2.
Structure of a URL.

URL

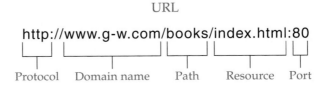

Access Protocol (LDAP) has made it possible for NDS, Microsoft AD, and DNS to look and function similarly. In each structure, the entire directory appears as though it is one structure on one computer; when in reality, it can comprise many computers spread across the world. You may wish to review the DNS structure in Chapter 11—TCP/IP Fundamentals.

Web Servers

A *Web server* is, as the name implies, a server configured to provide Web services, such as Web pages, file transfer, and HTML-based e-mail. To create a Web server, a software package, such as Apache HTTP Server or Microsoft Internet Information Services (IIS), is installed on a computer. The computer can be a server or a PC. Typically, Web server software is installed on a server that is running a network operating system, such as Windows 2000 Server. You can also install a light version of IIS on a PC running Microsoft Windows 98, Windows 2000 Professional, or Windows XP Professional. The only requirement is the computer should provide sufficient resources such as memory, CPU speed, and hard disk drive space to accommodate the software package.

A *Web site* is traditionally defined as a location on the World Wide Web. The Web site contains a collection of Web pages and files, which can be accessed through the Internet. A Web site is owned and managed by an individual, company, or organization.

Web server
a server configured to provide Web services, such as Web pages, file transfer, and HTML-based e-mail.

Web site
a location on the World Wide Web, which contains a collection of Web pages and files. A Web site is owned and managed by an individual, company, or organization.

An intranet has almost all of the characteristics of a Web site, but it is not a true Web site by strict definition.

 **Tech Tip**

For a Web server to operate while connected to the Internet, it must be assigned a domain name and IP address. A domain name can be acquired through a domain name provider or a Web hosting service. A domain name must be registered before an IP address can be assigned to a Web server. To obtain an IP address and domain name combination, the user submits an application to a domain name provider or Web hosting service for a nominal fee. Once the information is processed, an IP address is assigned to the domain name.

A domain name must be unique. In other words, the domain name you choose cannot be in use by another person or company. You can check if the domain name you've selected is unique by using the **whois** utility on the InterNIC Web site (www.internic.net). If the domain name you enter is in use, the **whois** utility displays information such as the domain name's registrar, the names of the name servers on which the domain name is listed, the date the domain name was assigned, and the date the domain name expires.

Figure 14-3 shows the **whois** utility screen prior to a request for information. **Figure 14-4** shows the detailed results of a search for the g-w.com Web site. The **whois** utility can conduct searches by the domain name, registrar name, or nameserver. The nameserver can be listed as a domain name or an IP address.

There are several ways to set up a Web server. A Web server can be set up on the same server as the network server, a server dedicated to only Web services, and a Web hosting company's Web site.

A Web server can run on the same server as the network server, especially if the LAN has only one server. In fact, you can run Web server services from a

Figure 14-3.
The **whois** utility can be accessed through the InterNIC Web site.

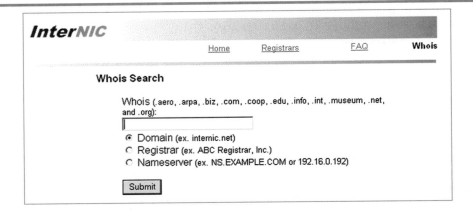

workstation. For example, installing Internet Information Service (IIS) on a Windows 2000 Professional workstation allows the workstation to run a single Web site that can be accessed by a maximum of ten simultaneous users.

A dedicated server can be used to host Web services. A dedicated server serves only one function. However, more than one Web site can be hosted from a single server.

A company or organization may choose to have a Web hosting company host their Web site. A Web hosting company sells space on its Web server, allowing a company or organization to set up a Web site without the need of owning and supporting a Web server.

Figure 14-5 and **Figure 14-6** illustrate the two most common scenarios of Web hosting. Each has advantages and disadvantages. For a large Web site consisting of many pages and a large volume of traffic, it is probably less expensive to install a Web server ivn-house. For a smaller Web site with a low volume of traffic, using a Web hosting company is more appropriate. Using a company that provides most of your Web hosting needs is called outsourcing.

Figure 14-4.
The results of a **whois** search.

Whois Search Results

Search again (.aero, .arpa, .biz, .com, .coop, .edu, .info, .int, .museum, .name, .net, or .org):

`g-w.com`

⦿ Domain (ex. internic.net)
○ Registrar (ex. ABC Registrar, Inc.)
○ Nameserver (ex. ns.example.com or 192.16.0.192)

[Submit]

```
Whois Server Version 1.3

Domain names in the .com and .net domains can now be registered
with many different competing registrars. Go to http://www.internic.net
for detailed information.

   Domain Name: G-W.COM
   Registrar: NETWORK SOLUTIONS, LLC.
   Whois Server: whois.networksolutions.com
   Referral URL: http://www.networksolutions.com
   Name Server: NSF.ALGX.NET
   Name Server: NSE.ALGX.NET
   Status: REGISTRAR-LOCK
   Updated Date: 10-oct-2003
   Creation Date: 01-jun-1998
   Expiration Date: 31-may-2005
```

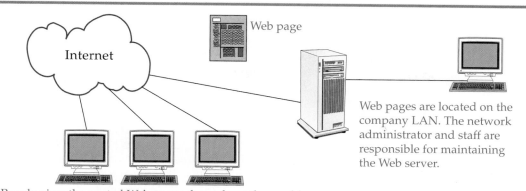

People view the posted Web pages throughout the world

Figure 14-5.
When a Web server is part of a company's LAN, the network administrator and staff are responsible for maintaining the server.

Look at the chart in **Figure 14-7** to see how some of the factors of Web site operation compare between hosting your own Web site and outsourcing your Web site.

The CompTIA organization offers a certification called iNet+ that is designed to measure the skills of a Web support technician. Many of the skills learned in networking fundamentals can be applied toward earning this certification. You may want to pursue this additional certification.

Tech Tip

Apache HTTP Server

The most widely used Web server software is Apache HTTP Server. The main reason for its popularity is its price. Programmers with a common interest developed the Apache HTTP Server software for free. The Apache HTTP Server software can be downloaded from the Apache Software Foundation Web site at www.apache.org, or it can be bought as a package with documentation and extra utilities for a modest price. Documentation and additional references are available at the Apache Software Foundation Web site. Once Apache is installed and started on a server, a manual can be accessed through http://localhost/manual/index.html. The manual is a complete user guide with information about all aspects of installing and using Apache HTTP Server.

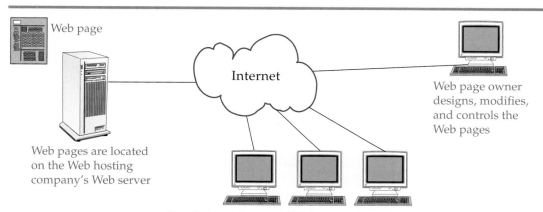

People view the posted Web pages throughout the world

Figure 14-6.
Web hosting companies can provide Web services. In this scenario, the Web page owner is responsible for the content and design of Web pages. The Web hosting company is responsible for supporting and maintaining the server.

Figure 14-7.
Comparison of out-
sourcing a Web site to
hosting your own
Web site.

Outsourcing Your Web Site	Hosting Your Own Web Site
Server space used is prorated.	Owner must purchase and install server hardware and software.
Technical support is provided.	An expert technician is required.
Connectivity is provided. Fee varies according to traffic volume.	Owner must obtain and pay for connectivity such as DSL, cable modem, T1 line, or other high bandwidth connection.
Most services, such as administration and maintenance, are provided.	Frequent administration, maintenance, and security checks are needed.
Web site design is not provided. Some companies offer a one-time design service for free on a limited number of pages.	Need Web page design software and personnel competent to use it.

Apache HTTP Server software is generally associated with Linux and UNIX systems. However, versions of Apache HTTP Server software are available for all the major network operating systems, such as Microsoft and Novell.

Internet Information Service (IIS)

The Internet Information Service (IIS) is the Microsoft Windows Web server default service. IIS supports Web site creation and management. Some service protocols that are supported by IIS are the Network News Transfer Protocol (NNTP), File Transfer Protocol (FTP), and Simple Mail Transfer Protocol (SMTP). These protocols are covered in detail later in this chapter.

IIS is not installed by default on a Microsoft server. Services such as IIS, FTP, SMTP, and NNTP must be installed as additional services. **Figure 14-8** shows a common dialog box that lists additional services (Windows components). The listing for IIS has a checkmark in its box, indicating that IIS should be installed.

Figure 14-9 shows a listing of the IIS subcomponents. The subcomponents add extra functionality to IIS. For example, the FrontPage 2000 Server Extensions subcomponent is used to support the Microsoft FrontPage Web page creation tool.

Figure 14-8.
Internet Information
Services (IIS) can be
installed through the
Windows Components
Wizard.

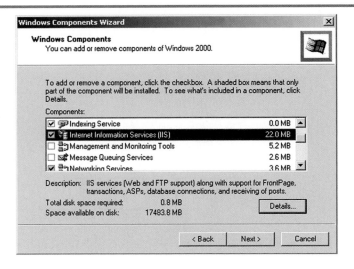

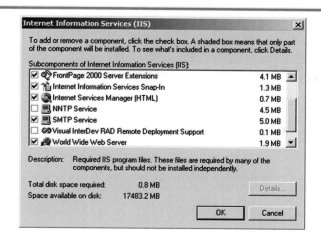

Figure 14-9.
Internet Information
Services (IIS) subcom-
ponent list.

Figure 14-10 shows the results of using Microsoft Internet Explorer to access
the default Web page after IIS has been installed on a Windows XP workstation.
This page contains information about IIS applications and management. The page
is displayed on the Windows XP workstation. Clients connecting to Windows XP
workstation's Web site will see the familiar and often annoying "Under
Construction" message displayed similar to the screen capture in **Figure 14-11.**

You can install IIS on a Windows XP Professional computer and support up to ten simultaneous connections. This will allow you to experiment with Web page and Web server
technologies.

Tech Tip

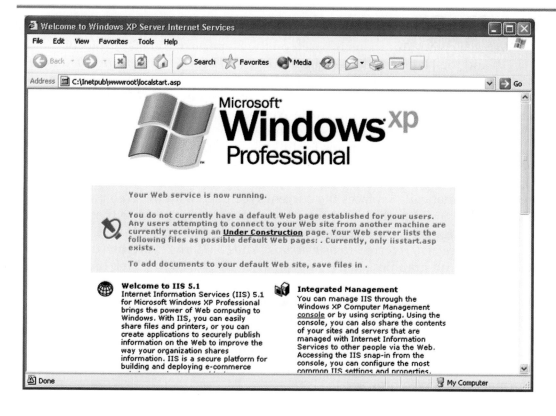

Figure 14-10.
Default IIS Web page
displayed on the Web
server.

Figure 14-11.
Default IIS Web page displayed on a client.

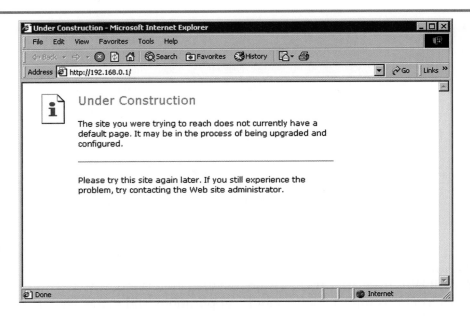

Web Browsers

Web browser
a software program that permits a user to navigate the World Wide Web and display and interpret Web pages.

A *Web browser* is a software program that permits a user to navigate the World Wide Web and display and interpret Web pages. Most Web browsers incorporate an e-mail client. Some of the most common Web browsers are Internet Explorer (IE), Netscape Navigator, Mozilla, and Konqueror. Internet Explorer is the most commonly used browser because it is the default Web browser installed with the most predominant desktop operating system in the world, Microsoft Windows.

Web Site Communications

Hypertext Transfer Protocol (HTTP)
a protocol designed for communications between a Web browser and a Web Server.

Web site communications is based on the Hypertext Transfer Protocol (HTTP) and the Hypertext Markup Language (HTML). The *Hypertext Transfer Protocol (HTTP)* is a protocol designed for communications between a Web browser and a Web Server. *Hypertext Markup Language (HTML)* is a programming language used to create Web pages. It can be interpreted by any type of computer with any type of Web browser.

Hypertext Markup Language (HTML)
a programming language used to create Web pages.

HTML also provides a means for linking to other Web pages or to an area on the same Web page. A link to another Web page or to an area on the same Web page is called a *hyperlink.* A user simply clicks the hyperlink and the Web browser automatically replaces the current Web page with the Web page designated in the hyperlink or "jumps" to a specified area within the Web page.

hyperlink
a link to another Web page or to an area on the same Web page.

A Web page contains text and special symbols defined by HTML tags. HTML tags are also referred to as *markups.* An *HTML tag* is an instruction for how the text and graphics should appear when displayed in a Web browser. For example, HTML tags determine the size, style, and color of the text font and identify the placement of graphics. A Web browser interprets the HTML tags to make the text and graphics appear on the display as specified.

HTML tag
an instruction for how text and graphics should appear when displayed in a Web browser.

Figure 14-12 shows the Goodheart-Willcox Publisher Web page as it appears in Internet Explorer. **Figure 14-13** shows a section of coding used to represent the page. You can see how the HTML tags and information combine to present the information in a Web browser.

Figure 14-12.
A Web browser interprets the HTML tags in the Web page coding to display the text and graphics.

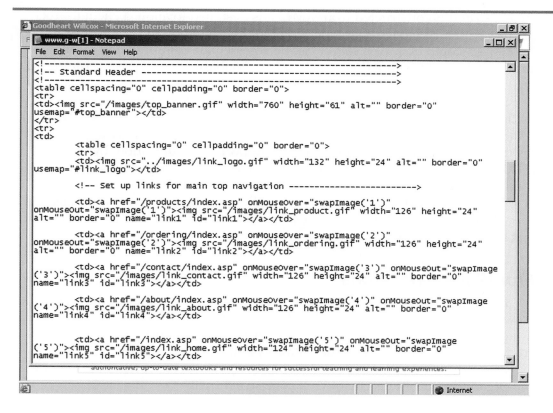

Figure 14-13.
Example of coding used to create the Goodheart-Willcox Publisher home page.

The HTML standard was developed by the World Wide Web Consortium (W3C). As you have previously learned in Chapter 1—Introduction to Networking, W3C is a non-profit organization dedicated to the development of voluntary standards for the Web. The consortium has spent many years developing the standards for SGML, HTML, and XML.

Standard Generalized Markup Language (SGML) is the original standard for both HTML and XML. HTML is a more compact version of the original SGML Web standard. XML is the latest in the evolution of Web standards.

Extensible Markup Language (XML)

Extensible Markup Language (XML) a markup language based on the same principles as HTML, with the added ability to create custom tags.

Extensible Markup Language (XML) is greatly enhanced compared to HTML. XML was designed especially for large enterprise business applications running on Web servers. While XML is based on the same principles as HTML, it has one very important difference: XML allows the creation of custom tags. XML also identifies the data contained inside the tags, while HTML only identifies how to display the data identified by the tag. For example, HTML uses tags to identify the size, font, or color of a block of text. XML can do the same, but it can also identify the contents, such as a numerical calculation for a database or spreadsheet. This allows the contents to be easily accessed and manipulated by a database or spreadsheet program. XML was designed with business applications in mind. XML can be easily converted to HTML and displayed through a browser. This ensures downward compatibility for systems.

SOAP

Simple Object Access Protocol (SOAP) an XML message format that allows a client to freely interact with a Web page on the Web server, rather than download it.

Simple Object Access Protocol (SOAP) is a set of rules for Web-based messages and is constructed from XML. It is designed to allow a client to freely interact with a Web page on a Web server, rather than download it. For example, a database stored on a Web server can be accessed and manipulated by an authorized client without downloading any database records. Also, a program can be run on the Web server while viewed by the client.

SOAP combined with XML and HTML technologies solves the problem of exchanging data and information between two incompatible systems such as Windows and UNIX. Normally, the two operating systems require a gateway or the installation of additional protocols to support communication. By using HTML, XML, and SOAP technologies, the two incompatible systems are able to communicate without additional modifications.

The use of SOAP to access and run applications on a Web server has lead to a new phrase coined by Microsoft—*Web Services*. The phrase *Web Services* means that services can be provided through Web pages.

FrontPage Extensions

FrontPage extensions are a set of programs that support the authoring, browsing, and administration of the Microsoft FrontPage program. If you are using Microsoft FrontPage to design your Web site, you must install IIS and the FrontPage extensions. This is true for a corporate intranet and Internet service. Extensions provide software tools for managing site security, checking site traffic, and organizing the hierarchy of Web page structures.

File Transfer Protocol (FTP)

The File Transfer Protocol (FTP) can be incorporated into a Web server to support file transfers between a client and server. FTP uses TCP packets and establishes a connection between the client and server. FTP is accessed using an FTP client that is either text-based or GUI-based. When using a text-based FTP client, a series of commands are issued at an FTP prompt similar to the way commands are issued at a DOS prompt. The following is a list of some common FTP commands and their function.

bye	Exit the FTP program.
cd/*directory*	Change the directory on the FTP site.
get	Transfer a file from the FTP site to the client.
help	Display FTP commands.
lcd	Change the directory on the client.
open	Open a connection to an FTP site.
put	Transfer a file from the client to the FTP site.
pwd	Display the current directory of the FTP site.
quit	Close the FTP session.

To start the FTP utility, go to the DOS prompt and enter the **ftp** command. Use the **help** command to display a list of FTP commands, **Figure 14-14.** To get help with a specific command, type **help** followed by the command. To exit the FTP program, enter the **bye** command.

Most Web page software has its own utility for uploading a file to the Web server FTP directory. There are also many third-party utilities available.

You can also use most Web browsers to access an FTP site by simply changing the protocol in Web browser's address bar from http: to ftp:. One important difference between most Web sites and an FTP site is you can upload files to an FTP site but not to a Web site. You can, however, download files from a Web site or an FTP site.

Figure 14-15 shows the Microsoft FTP site accessed through Internet Explorer using the ftp: protocol at the beginning of the URL. Look closely at how the FTP site is indicated in the DNS host name portion of the URL. Also notice that in the bottom, right side of the Web browser the user is identified as

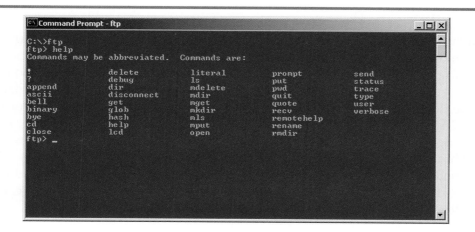

Figure 14-14.
A list of FTP commands can be displayed by entering **help** at the FTP command prompt.

Figure 14-15.
Example of an FTP site
displayed through the
Internet Explorer
browser.

anonymous. An anonymous user does not require a password or user name. Anyone can access the FTP site to download files.

One of the main disadvantages of an FTP site is the lack of security. FTP does not encrypt passwords or the contents of transferred files. If security is a concern, another utility and protocol should be used to ensure secure data transfers. One such tool is Microsoft File Transfer Manager. This tool provides security and guaranteed file delivery over the Internet. It is available as a free download from Microsoft.

Two other variations of the FTP protocol are Trivial File Transfer Protocol and Secure File Transfer Protocol. *Trivial File Transfer Protocol (TFTP)* is a lightweight version of FTP. TFTP never requires the use of a user name and password because it uses UDP packets for transferring data. Since UDP packets are used, a connection is never established between the client and server. The client is allowed to transfer files, but not to view the directory listing at the FTP site. TFTP uses fewer commands than FTP.

Secure File Transfer Protocol (SFTP) is a secure version of FTP. SFTP uses TCP packets and therefore establishes a connection using a user name and password. It encrypts the user name, password, and data to provide the highest level of security compared to FTP and TFTP. SFTP should be used when transferring sensitive data and when security is required.

Anonymous FTP

One of the easiest FTP sites to administer is an anonymous FTP site. An anonymous FTP site allows anyone to access the site and download or upload files. No password or any other form of authentication is required. Since no real form of authentication is required, the site is always at risk of being vandalized. It is better to require some form of authentication such as an e-mail address or to have a list of authorized user names and passwords. With proper authentication required, you can determine who accessed the site, when they accessed the site, and what files they downloaded or uploaded.

Another major problem with anonymous FTP sites is the unauthorized use of the site. If anyone can access the site using the *anonymous* user name, they can store illegal files on your site. It is a common practice of crackers to use

---

Trivial File Transfer Protocol (TFTP) a lightweight version of FTP, which does not require a user name and password because it uses UDP packets for transferring data. It allows a client to transfer files, but not to view the directory listing at the FTP site.

Secure File Transfer Protocol (SFTP) a secure version of FTP, which encrypts the user name, password, and data to provide the highest level of security compared to FTP and TFTP.

anonymous FTP sites to store their stolen files. They do not run the risk of being caught with stolen files on their own computer. They can gain access to their files on the anonymous site at any time in the future.

A *cracker* is an unauthorized user that infiltrates the network to create problems or to disrupt or steal information. A *hacker* simply breaks in to see if they can but does no real damage.

Tech Tip

If you set up the FTP site as a secure FTP site, always test the security. Try accessing the FTP site from another computer using some of the commands mentioned earlier.

IIS can activate a log of activities at the Web site or FTP site. Common information contained in the IIS log file is client address, user name, date and time of access, server name, IP address, and number of bytes downloaded or uploaded.

Network News Transfer Protocol (NNTP)

Network News Transfer Protocol (NNTP) is part of the TCP/IP suite of protocols. It is designed to distribute news messages to NNTP clients and NNTP servers across the Internet. News articles are stored in a central NNTP server and distributed automatically to other NNTP servers and clients.

News articles are arranged in groups or categories referred to as newsgroups. A *newsgroup* is also referred to as a discussion group. There are many different newsgroups available. **Figure 14-16** shows a list of available newsgroups run by Microsoft. Each newsgroup is dedicated to a particular Microsoft product and contains a collection of messages posted by individuals. The messages consist of

Network News Transfer Protocol (NNTP) a TCP/IP protocol that is designed to distribute news messages to NNTP clients and NNTP servers across the Internet.

newsgroup news articles that are arranged in a group or category on an NNTP server. Newsgroups are also referred to as discussion groups.

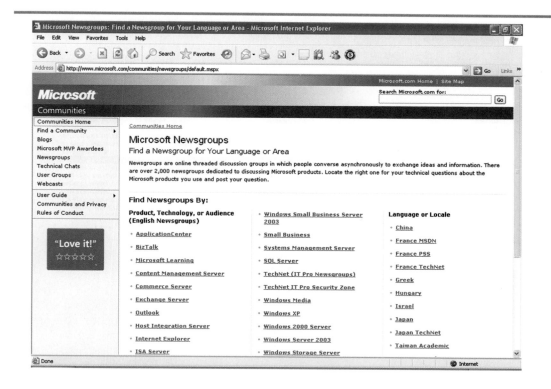

Figure 14-16. A partial list of Microsoft newsgroups.

questions and answers. You can join a discussion or post a question to a newsgroup. Members of the group will post responses to your question. A complete listing of all responses is available for reading.

A client is required to access, download, and read the contents of an NNTP server. There are various newsreader clients designed for accessing and reading the messages posted on an NNTP server. You can use Microsoft Outlook Express or Outlook to access a network news server and read and post messages.

Figure 14-17 shows an example of a newsgroup message posted on one of Microsoft's newsgroups. The message concerns a problem with a virus. The message is displayed in Outlook Express. Pay particular attention to the **New Post** and **Reply Group** buttons in the upper-left corner of the toolbar. These buttons are available when connected to a newsgroup. Also, notice in the left pane of the screen how the Microsoft newsgroup is displayed as a separate directory with the Internet address indicted rather than a folder name.

E-Mail

You are probably familiar with e-mail as a user, but not with its underling protocols and standards. Early in the history of networking, e-mail was viewed as a novelty rather than a necessity. Early e-mail messages contained text information only and not the rich formats viewed today that include various fonts, pictures, sounds, and multimedia.

In the simplest terms, an e-mail system consists of a mail server and an e-mail client. E-mail clients are typically installed automatically when an operating system is installed. For example, when you install Windows XP, you automatically install the e-mail client Outlook Express. Other e-mail client programs available from other companies are Eudora and Pegasus.

Figure 14-17.
Example of a newsgroup message displayed in Microsoft Express.

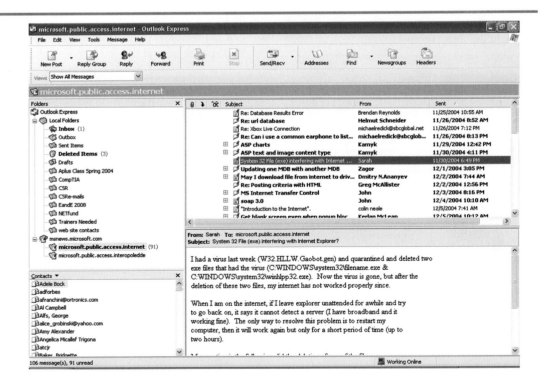

The mail server is responsible for forwarding e-mail to other servers and storing e-mail until a client can retrieve them. The user must use an e-mail client, also referred to as an e-mail agent, to communicate with the mail server and to retrieve the messages. An ISP or a private network, such as one run by a corporation, can provide a mail server. In other words, the e-mail service could be limited to a private network or could span the world using the Internet.

In the following sections, the protocols most commonly associated with e-mail are covered. They are SMTP, POP, IMAP, and the newest HTTP e-mail also referred to as HTML e-mail or Web e-mail.

SMTP

Simple Mail Transfer Protocol (SMTP) is part of the TCP/IP protocol suite and is designed to transfer plain text e-mail from an e-mail client to a mail server and from a mail server to a mail server. Look at **Figure 14-18.**

In the illustration, you can see that a mail server is connected to the Internet where it forwards messages to other mail servers. The PC uses an e-mail client, with the POP or IMAP protocol, to access messages on a mail server. The mail server uses the POP or IMAP protocol to communicate with and to download e-mail to the client.

When setting up an e-mail account, the mail server is designated as an SMTP server for sending e-mail and as a POP3 or IMAP server for retrieving e-mail. Two separate mail servers may be used—one for sending e-mail and the other for receiving e-mail—or one server may be used for sending and receiving e-mail. In either case, two server names typically need to be supplied when setting up an e-mail account.

Simple Mail Transfer Protocol (SMTP)

a protocol that is part of the TCP/IP protocol suite and is designed to transfer plain text e-mail messages from an e-mail client to a mail server and from a mail server to a mail server.

An e-mail server that hosts the POP or IMAP protocol can be called a POP server or an IMAP server. A server that hosts the SMTP protocol can be called an SMTP server.

Figure 14-19 shows a typical Windows XP **Internet Connection Wizard** dialog box associated with setting up an e-mail account. Notice the requirements, such as selecting the type of incoming mail server, the name of the incoming mail server, and the name of the outgoing mail server. If the same server is used for incoming and outgoing e-mail, the same server name will appear in each text box. Note the protocol choices for the incoming mail server: POP3, IMAP, and HTTP. The outgoing mail server is limited to the SMTP protocol.

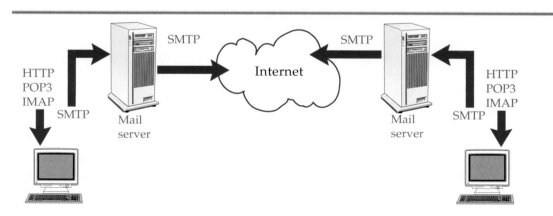

Figure 14-18.
The SMTP protocol is used to send mail from an e-mail client to a mail server and from a mail server to a mail server. The POP3, IMAP, and HTTP protocols are used to retrieve e-mail from mail servers.

Figure 14-19.
Setting up an e-mail account with the Windows XP **Internet Connection Wizard.** Notice the choices for the type of incoming mail server. Also note that an incoming mail server and an outgoing mail server must be specified.

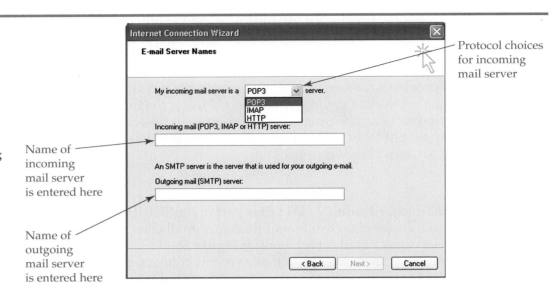

Protocol choices for incoming mail server

Name of incoming mail server is entered here

Name of outgoing mail server is entered here

SMTP is the most, well-known e-mail transfer agent. Two other agents are Unix-to-Unix Copy Protocol (UUCP) and X.400. You will most likely not encounter the other mail transfer agents unless working internationally or with UNIX systems.

Post Office Protocol (POP)

Post Office Protocol (POP)
a mail access protocol designed to access a mail server and download messages to the e-mail client.

The *Post Office Protocol (POP)* is a simple protocol designed to access a mail server and download e-mail to the e-mail client. In contrast to IMAP, POP does not allow you to store messages on the mail server. Stored e-mail resides on the user's local hard drive, not the mail server's hard drive. Typically, as soon as you open a POP mail account, the server begins downloading e-mail to the PC. There are various versions of POP: POP, POP2, and POP3. Today, these protocols are simply referred to as POP or POP3.

Internet Message Access Protocol (IMAP)

Internet Message Access Protocol (IMAP)
a sophisticated mail access protocol that can manipulate e-mail while it is on the server. It also allows a user to access their e-mail and then leave the e-mail on the server.

Internet Message Access Protocol (IMAP) is used to access messages stored on a mail server in similar fashion to POP. There are various releases of IMAP such as IMAP, IMAP2, IMAP3, and IMAP4. When compared to POP, IMAP is a more sophisticated protocol because it can manipulate the e-mail while it is on the server. You can view the e-mail headers and content before downloading the e-mail. IMAP allows you to select the e-mail that you want to download rather than automatically downloading all of the e-mail for you. You can also store the e-mail on the mail server rather than on the client. One advantage of storing e-mail on the server is the server may be automatically backed up each day. Important e-mail can be retained even if your PC's hard drive fails.

The IMAP protocol also allows the user to access their e-mail from more than one PC. A user can access their e-mail and then leave the e-mail on the server. This means that a user can access their e-mail from home, work, or while traveling. Copies of the e-mails can be stored on more than one computer such as your home computer or laptop. With the POP protocol, all e-mail is downloaded to a PC and is deleted from the e-mail server. If a user accesses the mail server from another PC, the previously downloaded e-mail is no longer available. Any new e-mail on the mail server is downloaded instead. E-mail is spread across the total number of computers you use to access your e-mail rather than stored in one central location.

The downside to the way IMAP handles e-mail storage is the mail server must provide sufficient space for all e-mail users. Storage space could be limited, which forces the e-mail user to remove some of their e-mail from the server to provide space for new e-mail. Also, a large e-mail, such as one with a graphic or sound attachment, might very well use all the storage space provided for the e-mail account, causing other e-mail delivery attempts to be rejected.

It may seem that all e-mail clients should use IMAP rather than POP. This is not always the case. For example, a corporate organization may choose to use POP rather than IMAP because POP is less resource intensive than IMAP. POP e-mail would not burden the network server the way IMAP would. Also, IMAP would create more network traffic than POP because IMAP allows a greater amount of port numbers.

HTTP E-Mail

The newest version of an Internet e-mail client is HTTP e-mail. It is also referred to as Web-based e-mail. To use the HTTP e-mail service, you need only a Web browser for the e-mail client. You do not need a specific e-mail client designed strictly for e-mail. There are numerous HTTP e-mail clients based on the HTML protocol.

HTTP e-mail is usually advertised as free e-mail. The HTTP e-mail client is designed similar to Web pages. You can transfer messages that contain animation, graphics, and sound. It also allows users to access their e-mail from anywhere the Internet is accessed. All the user needs to do is access the Internet to connect to their mail server and then enter their user name and password.

There are a few disadvantages of using HTTP e-mail. One disadvantage associated with HTTP-based e-mail is security. HTTP e-mail uses HTML coding that can easily hide viruses and worms. A Trojan horse can easily be hidden in the contents of an HTTP e-mail. When the client opens the e-mail, the computer is immediately infected. The only way to use HTTP securely is with certificates and a secure Web site. This ensures a secure connection. When a secure connection is used, the protocol in the Web address is identified with *https* rather than *http*. The *s* at the end of *http* indicates a secure site.

Another disadvantage to HTTP e-mail is it is typically associated with free e-mail. With free e-mail, you have a small amount of designated storage space. Once that set amount is reached, the provider of the free e-mail service will charge you for additional space. They may also attempt to sell you many add-ons to enhance your e-mail, like fancy paper, animated characters, and sounds attached to your e-mail message. While these items are fun, they can cause you to use up valuable storage space.

For example, you send a cute e-mail to a friend that has an animated character and plays a familiar tune when they open it. The free e-mail company will probably solicit your friend to download a free version of the e-mail client. After they do, they compose an e-mail and return the message to you. Soon, everyone you know is using the free e-mail client and your inbox is exceeding the maximum allowed capacity. All those fancy additions to the e-mail, like music and animated characters, take up a lot of storage space because they are coded into the e-mail message. You eventually will be prompted to buy space at a nominal monthly fee or suffer the consequences of losing e-mail messages.

The main advantages of Web mail is it can be obtained free of charge, has many graphic display enhancements, and can typically be accessed from anywhere on the Internet, which makes it especially appealing to travelers. There are numerous free e-mail services available: some reputable and some not so reputable.

Multipurpose Internet Mail Extensions (MIME)

Multipurpose Internet Mail Extensions (MIME)
a protocol that encodes additional information known as mail attachments to e-mail protocols that normally could not transfer attachments such as graphics.

Multipurpose Internet Mail Extensions (MIME) is a protocol that encodes additional information known as mail attachments to e-mail protocols that normally could not transfer attachments such as graphics. SMTP is designed only to transfer text material between servers and from clients. MIME allows e-mail attachments to be transferred as separate files using SMTP as the transport protocol. Not all mail servers support MIME, but most do. Some do not support MIME by design to reduce the amount of traffic on a network. Remember that mail attachments such as music and animation require a lot of bandwidth and can cause congestion on a network. Simple text messages do not cause severe congestion.

E-Mail Address Format

The e-mail address is based on the domain name system of addresses. The following are typical e-mail address formats:

someone@Somewhere.com
someone@NewAccounts.Somewhere.com
someone@Somewhere.com.uk

The user name or mailbox name in the example is *Someone*. The @ (at) symbol is called the *axon*. The host name follows the @ symbol. Then the type of domain is identified after the period. The type of domain in the example is .com for commercial. There are many domain types and names, such as .gov for government or .org for organization. For addresses outside the United States, there is an additional country code such as .uk for United Kingdom.

E-Mail Structure

Figure 14-20 shows an example of an e-mail structure. In the example, you can see how HTML commands, or tags, are used to form e-mail. The HTTP **<message>** tag is used with the attributes **to =**, **from =**, and **subject =**. These attributes form the e-mail header. The HTML **<text>** tag identifies the text portion of the e-mail.

Figure 14-20.
Example HTML commands, or tags, used to form an e-mail

```
<message to="KeltnerJ@g-w.com" from="RRoberts@strato.net"
    subject="Here is the latest chapter revision with XML info">
    <text>
    Both XML and HTML can be used to create e-mail. HTML and XML tags
    simply combine with HTTP commands to create an e-mail message.
    </text>
/message>
```

Of course, e-mail clients are much more sophisticated than the previous example. The example is used to illustrate how HTTP commands are combined with HTML tags to form a basic e-mail structure. Typical e-mail incorporates icons and visual elements from other programming languages to hide HTML coding. The same message would appear in an e-mail client as in the screen capture shown in **Figure 14-21.**

Mail Filter

A *mail filter* filters unwanted e-mail messages such as *spam. Spamming* is the distribution of unsolicited e-mail. E-mails are sent as broadcasts to numerous mailboxes using a list of known mailboxes or programs that randomly generate mailbox addresses. A *spammer* often sends e-mail with some sort of advertisement as a probe. They often include a line that says something like, "If you received this e-mail by mistake or want to be removed from our mailing list, simply reply to this e-mail address." If the recipient responds to the e-mail, the spammer receives an e-mail reply with a legitimate e-mail address to add to their list of e-mail addresses. The spammer can use this e-mail address in the future or sell it. Many legitimate companies or organizations do offer a genuine removal from an e-mail list service. If you receive unsolicited e-mail from an unknown source, the best advice is do not reply!

Mail filters can be placed on servers or desktops. The filtering technique varies between the different mail filter systems available. Most e-mail clients that come with an operating system contain some sort of filtering feature. For example, Microsoft Outlook Express allows you to identify undesirable sources of e-mail. Look at **Figure 14-22.**

Outlook Express has two features available called "block sender" and "message rule." Block sender blocks messages based on the mailbox address of the sender. Message rule filters mail contents based on header or letter body contents, or both. The filter can be based on one or more words. For example, you can filter all incoming mail that contains the phrase "act now" or "immediate response required" in the message header or body. By guessing the most common sales terminology used, you can block unwanted spam, but you must be careful. The filter acts solely on the words or phrases present in the message. If a friend sends e-mail to you with the same phrase, their e-mail will also be filtered.

mail filter
filters, or blocks, unwanted e-mail messages.

spam
an unwanted e-mail, such as advertisement.

spamming
the distribution of unsolicited e-mail.

spammer
a person who engages in distributing unsolicited e-mail or sending e-mail with some sort of advertisement as a probe.

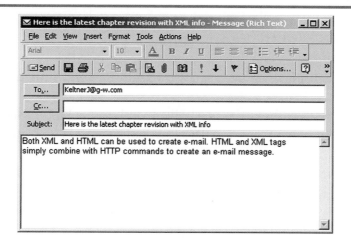

Figure 14-21.
E-mail client software hides the HTML coding from the user.

Figure 14-22.
The **New Mail Rule** dialog box in Outlook Express allows a user to filter e-mail based on the e-mail's header or contents, or both.

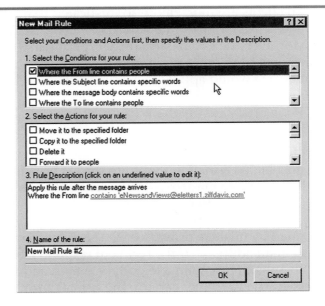

Mail Gateway

mail gateway
a special software and computer used to connect two normally incompatible e-mail systems.

A *mail gateway* is a special software and computer used to connect two normally incompatible e-mail systems. For example, two e-mail systems that cannot be directly translated require a gateway to act as an intermediate. The gateway translates e-mail protocols so they can be read by the other e-mail system. Sometimes the contents can be translated, but not the attachment. In other words, the contents of the text message can be transferred between the two systems, but the attached graphic or audio file cannot.

When troubleshooting e-mail systems, *always* attempt to send an e-mail that contains a short, plain text message and no fancy paper backdrops or attachment. Many times an e-mail problem between two different systems is a result of the enhanced e-mail or a lesser-known e-mail client. If the plain text e-mail is successfully delivered, it will ensure that the network infrastructure is intact and that the SMTP system is working correctly. You can then turn your attention to e-mail client compatibility.

Summary

- The Internet is a worldwide network that anyone can use for communications.

- An intranet is typically a private, Web-based network limited to a select set of clients.

- An extranet is a combination of a private and public Web-based network.

- A Uniform Resource Locator (URL) is a user-friendly name that resolves to an IP address.

- A Domain Name System (DNS) server translates URLs to IP addresses.

- Microsoft Active Directory (AD) and Novell Directory Services (NDS) rely on the Lightweight Directory Access Protocol (LDAP) to communicate directory information between clients and servers.

- Web servers provide services, such as Web page, file transfer, and e-mail, to clients.
- The **whois** utility provides information about domain name owners.
- Apache is the most widely used Web server software package.
- Web browsers are designed to navigate the World Wide Web and to display Web page contents.
- The foundation of Web communications is based on HTTP and HTML standards.
- The Hypertext Transport Protocol (HTTP) is a protocol designed for communications between a Web client and a Web Server.
- The Hypertext Markup Language (HTML) is a programming language.
- XML is an enhanced markup language that supports client interaction with server applications.
- SOAP is a protocol designed to support XML and run applications on a Web server.
- The File Transfer Protocol (FTP) supports file transfers between a client and an FTP server.
- TFTP is a simpler version of FTP that uses fewer commands.
- SFTP provides a more secure transfer of data by encrypting user names, passwords, and data.
- An anonymous FTP site permits file transfers without the need of a required password or user name.
- Incoming mail server choices are typically limited to POP3, IMAP, and HTTP.
- An outgoing mail server is typically an SMTP server.
- One server may function as both the incoming and outgoing mail server.

Review Questions

Answer the following questions on a separate sheet of paper. Please do not write in this book.

1. What are the three types of networks that provide Web page distribution?

2. A(n) _____ is a network in which employees can only access a company Web server from within the company.

3. A(n) _____ allows internal access to Web pages and allows authorized personnel from outside the network to access the network's Web pages.

4. What is the purpose of a DNS server?

5. What protocol is common to Microsoft Active Directory (AD) and the Novell Directory Services (NDS)?

6. What is the purpose of a Web server?

7. What does the **whois** utility do?

8. What are some things to consider when choosing to set up your own Web server rather than choosing to use a Web hosting service?

9. Name three ways to conduct a **whois** search?

10. Name two software packages that are designed to turn a server into a Web server?

11. What does the acronym IIS represent?

12. What is the name of the default Web server software installed on a Microsoft server?

13. Name four common Web browsers?

14. What is the difference between HTML and HTTP?

15. What is a markup?

16. What is the purpose of HTML tags?

17. What does the acronym W3C represent?

18. What is the difference between HTML and XML?

19. SOAP is closely associated with which markup language?

20. What is an anonymous FTP site?

21. What protocol is used to send e-mail to a mail server or to transfer e-mail from one mail server to another mail server?

22. What protocols download e-mail from a server?

23. Which e-mail protocol is used to download e-mail automatically without the option to store it on the mail server?

24. How does MIME enhance e-mail?

Sample Network+ Exam Questions

Network+

Answer the following questions on a separate sheet of paper. Please do not write in this book.

1. What protocol is used to transport Web page information from a Web server to a client?
   A. HTML
   B. HTTP
   C. IPX/SPX
   D. SMTP

2. Which protocol is used to download files from a Web site?
   A. ATM
   B. FTP
   C. SNMP
   D. ICMP

3. A(n) _____ is a network in which only employees from within a company can access the company Web server.

   A. Internet

   B. Intranet

   C. Extranet

   D. LAN Net

4. The typical port number used for making a Web server connection is _____.

   A. 33

   B. 80

   C. 125

   D. 440

5. Which protocol is specifically designed to store e-mail messages on a mail server?

   A. FTP

   B. POP

   C. IMAP

   D. HTML

6. Which type of server is designated to receive e-mail messages from an e-mail?

   A. DNS

   B. DHCP

   C. SMTP

   D. POP

7. Your company is setting up an e-mail system for the sales department. The sales department wants to be able to access and read their e-mail while traveling or from any location outside the office. Which type of e-mail server will best meet their needs?

   A. POP3

   B. FTP

   C. TFTP

   D. IMAP

8. Which is the correct e-mail format for a salesman from the United Kingdom with a user name of Bjones?

   A. BJones@Sales.Homeoffice.com.uk

   B. BJones@uk.com/homeoffice.sales

   C. BJones@Sales/HomeOffice/com/uk

   D. BJones.HomeOffice.Sales.com@uk

9. What type of server translates URLs to IP addresses?

   A. DNS

   B. DHCP

   C. IMAP

   D. IIS

Network+

10. Which command would you use to determine who owns and operates a specific Web site?

A. **netstat**

B. **tracert**

C. **whois**

D. **ping**

Interesting Web Sites for More Information

www.apache.org

www.arin.net

www.internic.net

www.linux.org

www.microsoft.com

www.novell.com

www.redhat.com

Suggested Laboratory Activities

1. Set up an intranet for your network class. Install Microsoft IIS and then use a Web-authoring tool, such as FrontPage, to design and install a Web site. The Web site should permit students to access information. Each student can design a page for the Web site.

2. Download and explore anyone of several free HTML authoring tools available and design a simple Web page. Use the authoring utility to expose the hidden HTML code.

3. Set up an FTP site as part of the Web site you created in the first laboratory activity. Load a file on the FTP site that can be distributed to all the students using anonymous access.

4. Set up restrictions on the FTP Web site so that only users with a user name and password can access the site.

5. Explore the InterNIC Web site or conduct a Google search for the **whois** utility. Look up some familiar URL names and see what information is revealed.

6. Search for three different Web-hosting services and compare prices and services.

7. Configure a workstation for a connection to an e-mail server. Obtain step-by-step instructions from your ISP's Web site.

**Chapter 14
Laboratory Activity**

Installing Internet Information Services (IIS)

After completing this laboratory activity, you will be able to:

■ Explain the limitations of Internet Information Service (IIS) on a Windows XP Professional PC.

■ Explain the purpose of Internet Information Service (IIS).

■ Install Internet Information Service (IIS).

■ Access and display the default Internet Information Service (IIS) Web page.

■ Use the Microsoft Management Console (MMC) to explore the IIS directory structure.

Introduction

In this laboratory activity, you will install Internet Information Service (IIS) on a Windows XP Professional computer to provide Web site and FTP services to a maximum of ten, simultaneous computer connections. You can install IIS on a workstation that is part of a peer-to-peer network to form a company intranet. IIS can also be installed on a standalone computer to test Web page development before posting the Web pages to the company Web site.

IIS is not installed during a typical Windows XP installation. To install IIS, you must access **Control Panel | Add/Remove Programs** and then select **Add Windows Components**. In the following screen capture, you can see **Internet Information Service (IIS)** is highlighted. Select this component by clicking the checkbox. Notice that the description of the service is displayed below the **Components** list box. Also, notice the **Details** button at the bottom, right side of the window. The **Detail** button reveals a detailed list of subcomponents. For this laboratory activity, you will select all of the IIS subcomponents. You will need the Windows XP installation CD to complete the installation.

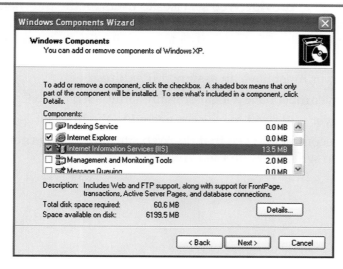

Setting up a Windows XP Professional computer as a Web server limited to LAN access is equal to an intranet. Compared to the Internet, an intranet is limited in scope. The Internet is a World Wide Web service while an intranet is limited to a single corporation, business, school, or such. The intranet is intended for a specific audience rather than for users on the World Wide Web. The intranet can be used as a means of accessing company information and provide a mechanism for downloading files. Many employees find it easier to navigate a Web site than a network environment. The Web site can be set up to provide to all employees a newsletter, a business calendar, company forms, a company policy book, employee benefits, and files for download. The use of the intranet is limited only to the designer's imagination.

After IIS has been installed, you can test the service by typing http://localhost/ in the Internet Explorer address bar. The first part of the URL is the type of protocol used: typically, HTTP or FTP. The second part of the URL after the double slash (//) is the location of the Web server. The location will be the Web server's domain name or IP address. To resolve a domain name, a DNS server is required. A direct connection using the IP address does not require a DNS server.

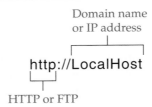

You may use Internet Explorer in the offline mode because there is no need to use a dial-up or any type of Internet connection. If Internet Explorer has been configured for automatic dial-up, you will not be able to directly access the localhost. You will need to reconfigure the Internet connection to connect through the local area network. This will allow you to use Internet Explorer in the offline mode. To allow LAN clients to access the IIS default Web page, you will need to share the wwwroot folder.

IIS can be used to provide Web services to users on the Internet; however, there are several requirements with which you need to be familiar. Your local Internet Service Provider (ISP) can provide you with the first three items on the following list:

- The Web server must have an assigned IP address.
- The Web server must have a legal domain name, such as MyCorporation.com.
- The Web server will need to stay connected to the Internet 24/7.
- A router is recommended so that the Web site and the Internet connection can be shared with users on the LAN.

Equipment and Materials

■ (2) Typical PCs connected as peer-to-peer. (You will install IIS on one PC and access the Web server from the other. If this is a two-student project, simply take turns, switching the responsibility for the client and the Web host.)

■ Windows XP installation CD.

Do not install IIS on a workstation connected to a client/server network. The IIS installation can create unexpected results, such as system lockups.

Note:

You must use Windows XP Professional for this project. Windows XP Home Edition will not support this laboratory activity.

Procedures

1. _____ Report to your assigned workstation.

2. _____ Boot the workstation and verify it is in working order.

3. _____ Create a Restore Point to uninstall IIS if a problem arises.

4. _____ Open **Control Panel** and access the **Add/Remove Windows Components**.

5. _____ Select **Internet Information Services (IIS)** and then click **Details**. Select all of the available subcomponents, such as **File Transfer Protocol** and **FrontPage Service Extensions**, and then click **OK**.

6. _____ Click **Next**. The Windows Components Wizard will automatically configure the Internet Information Services (IIS) component. You will be prompted for the Windows XP installation CD. Insert the CD into the CD-ROM drive and then click **Next**. Installation may take several minutes.

7. _____ After Internet Information Services (IIS) and its subcomponents have been installed, the wizard will display a window informing you the components have been successfully installed. Click **Finish** to exit the wizard and then remove the CD.

8. _____   To verify IIS has been installed, open the **Microsoft Management Console (MMC)**. The MMC can be accessed by right-clicking **My Computer** and selecting **Manage** from the shortcut menu. You should see the Internet Information Services folder listed under the Services and Applications folder. If you do not see the Internet Information Services folder, call your instructor.

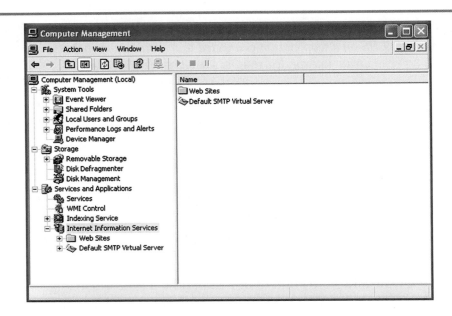

9. _____   To access the Web site over the LAN, you must create a share for the wwwroot folder. The location is C:\Inetpub\wwwroot\. Simply right-click the wwwroot folder and select the **Sharing** option. You will know that the share has been established when the wwwroot folder appears with an opened hand under it. If you cannot share the wwwroot folder, you will need to turn off the Simple File Sharing feature. To turn off the Simple File Sharing feature, right-click **My Computer** and select **Explore** from the shortcut menu. Select **Tools |Folder Options | View**. Scroll to the bottom of the **Advanced settings** list and deselect **Use Simple File Sharing**.

10. _____   From the other workstation (the one without IIS), enter the IP address of the Web server and the default Web site folder—for example: http://192.168.000.075/wwwroot/. A Web page should appear stating that the Web site is under construction. If this Web page does not appear, call your instructor.

11. _____   Enter the following URLs and observe the results. First, test the URL address from the Web server and then from the workstation. Use the IP address of the Web server in place of the 192.168.0.75 used in the examples.

Web Server:

- Use http://localhost at the Web server to display the default Web page.

- Use http://192.168.0.75/iishelp or http://localhost/iishelp at the Web server to display the IIS 5.1 Documentation.

- Use http://192.168.0.75/localstart.asp at the Web server to display the Web server start page. This request requires the administrator's user name and password.

Workstation:

- Use http://192.168.0.75 at the workstation to display the under construction message.

- Use ftp://192.168.0.75 at the workstation to expose any files available for file transfer. The Web server must have the File Transfer Protocol (FTP) Service subcomponent installed.

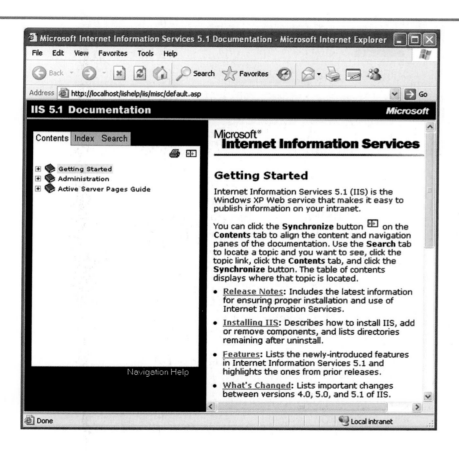

The IIS 5.1 Documentation page contains a wealth of basic information about setting up and managing a Web site and a Web server. You should briefly look at the available information during this laboratory activity. If you are interested in Web page development and Web site administration, you can install IIS on your home computer and look at the information in more detail.

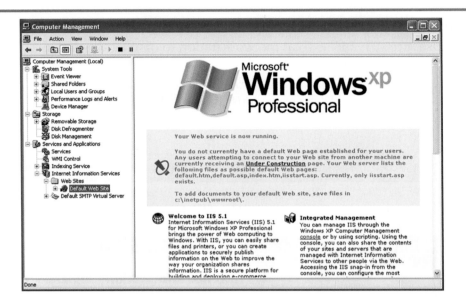

12. _____ If you have a Web page authoring utility, make a Web page to replace the default IIS Web page. Simply create a Web page and save it as default.htm in the C:\inetpub\wwwroot directory. You could use Microsoft Word to create a Web page. To create a Web page using Microsoft Word, simply open Word and create a document. When you save the document, save it in the HTML format. Use the **Save As** command from the **File** menu. Locate the C:\inetpub\wwwroot directory in the **Save in** list box and then select **Web Page** from the **Save as type** list box. Enter the file name default in the **File name** text box and then click **Save**.

13. _____ Answer the review questions.

14. _____ Leave the workstation and IIS server in the condition specified by your instructor.

Review Questions

Answer the following questions on a separate sheet of paper. Please do not write in this book.

1. What does the acronym IIS represent?

2. On what version of Windows XP can IIS be installed?

3. What is the maximum number of simultaneous connections IIS allows when installed on Windows XP?

4. Is IIS installed by default when a typical installation of Windows XP is performed?

5. What is the difference between the URL of a Web page and the URL of a FTP site?

6. What are the minimal requirements for a small-office/home-office (SOHO) Web server that will serve Internet users? (Select all that apply.)

   A. The Web server must have an assigned IP address.

   B. The Web server must have a legal domain name, such as MyCorporation.com.

   C. The Web server will need to stay connected to the Internet 24/7.

   D. A router is recommended so that the Web site and the Internet connection can be shared with users on the LAN.

7. How do you know that the wwwroot folder has been set up as a shared folder?

8. What addresses can be used on the Web server to display the IIS default Web page and the IIS 5.1 Documentation?

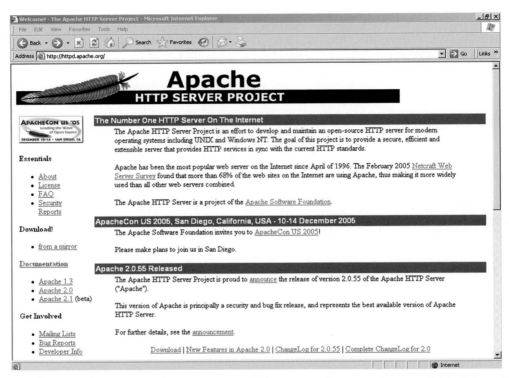

Apache HTTP Server is the most popular open-source Web server software package available. It is continuously developed and maintained by the Apache Software Foundation (www.apache.org).

15

Remote Access and Long Distance Communications

After studying this chapter, you will be able to:

❏ Explain the functions of the major parts of the public telephone system.

❏ Describe the various media and technologies used to support long distance communication.

❏ Explain how remote access to a network is achieved.

❏ Give a brief overview of common remote access protocols.

❏ Explain the purpose of a Virtual Private Network (VPN).

Network+ Exam—Key Points

The Network+ Certification exam covers the various types of media associated with the telecommunications industry. Typical questions are based on descriptions of the media and media performance limits. You will need to be able to distinguish between remote access protocols, such as SLIP, PPP, PPPoE, and PPTP. Be careful of questions involving SLIP and accessing a UNIX system. While not often encountered, SLIP can still be a valid correct response for a question concerning remote access to a UNIX server.

Network+

Key Words and Terms

The following words and terms will become important pieces of your networking vocabulary. Be sure you can define them.

Bandwidth Allocation Protocol (BAP)

Basic Rate ISDN (BRI-ISDN)

Broadband ISDN (B-ISDN)

Channel Service Unit/Data Service Unit (CSU/DSU)

dial-up networking

Digital Subscriber Line (DSL)

head-end

Integrated Services Digital Network (ISDN)

International Telecommunications Union (ITU)

Layer 2 Forwarding (L2F)

Layer 2 Tunneling Protocol (L2TP)

loading coil

Local Central Office

Local Exchange Carrier (LEC)

local loop

Multilink Point-to-Point Protocol (MLPPP)

Point of Presence (PoP)

Point-to-Point Protocol (PPP)

Point-to-Point Protocol over Ethernet (PPPoE)

Point-to-Point Tunneling Protocol (PPTP)

Primary Rate ISDN (PRI-ISDN)

remote access server

Remote Desktop Protocol (RDP)

Serial Line Internet Protocol (SLIP)

sunspots

T-carrier

trunk lines

upconverter

Virtual Private Network (VPN)

This chapter explains the requirements for communicating remotely with a network and for communicating between LANs and to and from a MAN or a WAN. Remote communications rely on either public communication systems or private communication systems. The communication system we are most familiar with is the public telephone system.

Remote connection technology is at the very heart of a MAN and WAN. To understand a MAN and WAN, you must have an excellent understanding of remote connection technologies, such as T1, ISDN, and SONET. This chapter explores remote connection technologies and explores the different types of media used to transfer data across vast distances. We will begin the chapter with an introduction to the telecommunications system.

Introduction to Telecommunications Systems

The telecommunications industry has existed for almost a hundred years. During most of this period, it existed as an analog system and has been steadily converting to a complete digital system. The typical telephone system with which you are familiar consists mainly of two systems: local carriers and long distance carriers.

Originally, local and long distance carriers were the same entity. Rates were controlled to an extent by the government. The breakup, or deregulation, of telephone companies began as a series of legal disputes beginning in the late 1940s and ending in 1984 with the complete deregulation of the telephone industry. Large companies such as AT&T and Bell Telephone Company were separated into individual companies.

The two major service parts, local service and long distance service, were also separated. This meant the customer would use one company for local telephone service, or local calls, and another for long distance calls. Deregulation eliminated the monopolistic nature of the telecommunications industry. The concept was to make the market more competitive so that the consumer could benefit from the competition. To read more about this subject, visit www.iec.org.

Deregulation created many local service and long distance carriers. A local carrier is often referred to as a *Local Exchange Carrier (LEC)* and is made of one or more Local Central Offices. The *Local Central Office* is where the customer's telephone lines connect to the switchgear. The Local Central Office connects to long distance carriers to provide long distance access to individual residencies. Look at **Figure 15-1.** Notice that the *local loop* is the section of wiring between customer premises and the Local Central Office. The Local Central Office can be tied to other Local Central Offices via trunk lines. *Trunk lines* consist of hundreds of pairs of twisted pair cable or fiber-optic cable.

The *Point of Presence (PoP)* is the point where the telephone company line connects to the subscriber line. The subscriber line begins at the customer's premises. Before deregulation, the telephone company was responsible for all wiring and equipment provided to the customer. The telephone company provided, installed, and repaired telephones and telephone lines inside the customer's home or business. Today, as you know, you must purchase a telephone and are responsible for repairing any telephone lines and equipment inside your home or business.

Local Exchange Carrier (LEC)
a local carrier. It is often made of one or more Local Central Offices.

Local Central Office
the location where the customer's telephone lines connect to the switchgear. The Local Central Office connects to long distance carriers to provide long distance access to individual residencies.

local loop
the section of wiring between customer premises and the Local Central Office.

trunk lines
the hundreds of pairs of twisted pair cable or fiber-optic cable that connect Local Central Offices.

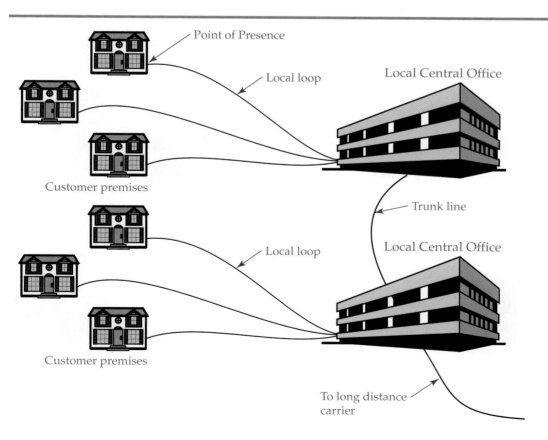

Point of Presence

Local loop

Local Central Office

Customer premises

Trunk line

Local Central Office

Local loop

Customer premises

To long distance carrier

Figure 15-1.
A local loop is typically a low-bandwidth, twisted pair line between the telephone company's Local Central Office and the customer premises.

Point of Presence (PoP)
the point where the telephone company line connects to the subscriber line.

Remote Connection Technologies and Media

The worldwide communication system is quite complex, offering a variety of media to use and many different long distance providers. While a LAN usually has a limited scope of communications media, a MAN and a WAN have a large variety of communications media and technologies from which to choose.

Both public and private communication companies often control the media. Communication companies are referred to as carriers or providers. These companies provide and maintain the media for data transmission.

Small businesses and private residencies typically connect to their ISP via the telephone Local Central Office, **Figure 15-2.** Therefore, they usually use the low-bandwidth local loop between their site and the Local Central Office. Enterprises or large businesses require high bandwidth for telecommunications and Internet access. They typically avoid the low-bandwidth local loop by using high-speed leased lines, such as a T1 line, to access other business locations and their ISP.

The geographic area of a residence or business often affects the choice or availability of remote connection media. Connecting to a remote network from an office in a metropolitan area does not typically present the same bandwidth limitations as connecting to a remote network in a rural or suburban area. Access to the Local Exchange Carrier from a metropolitan area is usually accessible using media that provides a much higher bandwidth than is typically found in suburban residential or rural areas.

In a metropolitan area, there is the option of using leased lines, which provide private and high-bandwidth communications. Telecommunication companies make leased lines available between major metropolitan areas all over the world. In general, private communications are typically available in densely populated, metropolitan areas that contain numerous businesses with a need for high-speed communications. In contrast, public communications are typically located everywhere but do not always offer the same high bandwidth.

Figure 15-2.
While typical residencies use a low-bandwidth, twisted pair line for telecommunications and Internet access, Internet Service Providers (ISPs) and enterprise businesses use a high-bandwidth line, such as T1.

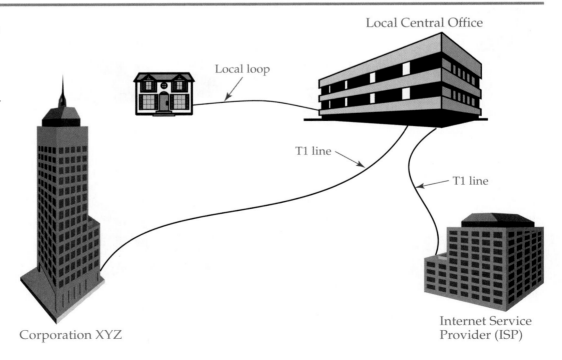

Local Central Office

Local loop

T1 line

T1 line

Corporation XYZ

Internet Service Provider (ISP)

As you will see in the following sections, there are many different types of media, such as PSTN, ISDN, Cable, DSL, satellite, and T-carrier lines, available for communicating over long distances. Each of the media covered is offered at various speeds and costs.

Public Switched Telephone Network (PSTN)

As you have previously learned in Chapter 13—ATM and VoIP, a Public Switched Telephone Network (PSTN) is an older telephone technology that uses twisted pair cabling and analog signals rather than digital. Another name for a Public Switched Telephone Network (PSTN) is Plain Old Telephone Service (POTS).

The public telephone system was originally analog. After digital electronics emerged, the telephone industry began switching the public telephone system equipment and lines to that which supports digital technology. Today, the public telephone system is a mixture of digital and analog systems. While long distance connections are usually digital, the short run between a residence or small business to the local loop is usually analog. Connections between the major telephone equipment is digital. It is possible, however, to have an entirely digital system that includes local loop. Several options to achieve a totally digital connection are discussed later in the chapter.

Typically, the PSTN is used when making a dial-up connection. A dial-up connection is a type of connection made using a traditional telephone line to reach a distant computer or network system. For example, a businessman using a laptop will typically connect the laptop modem to a telephone jack and dial a telephone number. This will connect the laptop modem to the modem at the distant network. Home users use a dial-up connection to reach their ISP when accessing their ISP through the telephone line.

It is important to note that a remote connection to a distant network or other computer has a total bandwidth equal to its weakest link. The weakest link is typically the local loop. This is especially true when attempting to connect to a remote computer from a residence. The existing residential telecommunications technology is largely influenced by the original telecommunications technology.

The original telephone system was an analog system requiring 48 volts to communicate across telephone lines. The telephone lines were designed to carry voice, not digital data. The original line did not require the twist in the conductor pairs, which is required for a high data rate. The telephone lines were designed to carry a maximum frequency of 4 kHz. This frequency is more than adequate for voice communications. Historically, the first network systems were designed to run at these slow transmission rates to take advantage of the millions of miles of existing telephone cables. Although this speed is usable, it is unacceptable by today's standards.

ISDN

Integrated Services Digital Network (ISDN) was developed in the 1980s. It provides the means for a fully digital transmission. ISDN lines are commonly found in small businesses, but there are some homes that are wired with ISDN. ISDN requires an ISDN modem. The ISDN modem can be installed either internally in the PC as an expansion card or externally, connecting through a serial port.

There are three categories of ISDN from which to choose: basic rate, primary rate, and broadband. *Basic Rate ISDN (BRI-ISDN)* consists of three conductors: two B channels, referred to as *bearer channels*, and one D channel, referred to as the *delta channel*. See **Figure 15-3.** The B channels carry data, voice, video, or a

Integrated Services Digital Network (ISDN) a long distance connection technology that provides a means of a fully digital transmission over channels that are capable of speeds of up to 64 kbps each.

Basic Rate ISDN (BRI-ISDN) a category of ISDN that consists of three conductors: two B channels, referred to as bearer channels, and one D channel, referred to as the delta channel.

Figure 15-3.
A Basic Rate ISDN (BRI-ISDN) cable consists of two 64 kbps B channels and one 16 kbps D channel. The two B channels can be combined for a total of 128 kbps bandwidth.

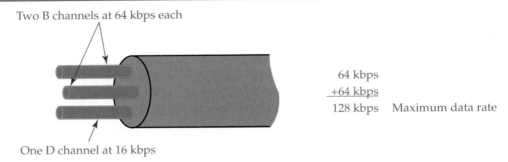

Two B channels at 64 kbps each

One D channel at 16 kbps

64 kbps
+64 kbps
128 kbps   Maximum data rate

Basic Rate ISDN (BRI-ISDN) Cable

combination of voice and data. Each B channel carries a maximum of 64 kbps or a combined maximum data rate of 128 kbps. This data rate is approximately two to five times faster than the traditional telephone modem, which is 56 kbps. The D channel has a maximum bandwidth of 16 kbps. The D channel carries control signals. The controls signals are used to set up, maintain, and terminate transmission on the B channels.

Primary Rate ISDN (PRI-ISDN) consists of twenty-three B channels and one D channel. It has a total data rate of 1.544 Mbps. *Broadband ISDN (B-ISDN)* is designed to carry multiple frequencies. It has a total data rate of 1.5 Mbps. Remember that Broadband refers to media that can carry multiple frequencies. Baseband refers to media that can carry only one frequency. Basic Rate ISDN and Primary Rate ISDN are digital technologies and can carry only one frequency; hence, they are both Baseband.

Primary Rate ISDN (PRI-ISDN)
a category of ISDN that consists of twenty-three B channels and one D channel. It has a total data rate of 1.544 Mbps.

Broadband ISDN (B-ISDN)
a category of ISDN that is designed to carry multiple frequencies.

Cable Internet Service

Cable Internet service uses the Cable television distribution system to provide Internet access. Typically, Cable Internet service requires a Cable modem connection and a twisted pair or USB connection. The Cable modem connection connects to the F-Type connector on the Cable service coaxial cable. The RJ-45 or USB connection on the Cable modem connects via twisted pair or USB cable to the network interface card on the computer. The RJ-45 or USB connection on the Cable modem may also connect to a hub to provide a shared Internet connection.

Tech Tip

A Cable modem is also called a transceiver. The word *transceiver* is a combination of the words *transmit* and *receive*.

Cable Internet service is an asymmetrical form of communication. The term *asymmetrical* is used to describe Cable Internet service communication because the uplink and downlink have two different transfer speeds. Upstream connections vary between 320 kbps and 10 Mbps, while downstream data rates vary between 27 Mbps and 36 Mbps. The term *downstream* is used to describe the data flow direction from the carrier or provider site to the customer. The term *upstream* is used for the data flow direction from the customer to the carrier. These speeds are advertised by the Data Over Cable System Interface Specification (DOCSIS), which is the industry standard.

The DOCSIS standard was developed by a group of Cable industry representatives before the IEEE 802.14 standard was developed. The IEEE 802.14 standard was an early attempt to standardize cable modems and high-speed access. It failed because of a lack of support by the manufacturers who opted to follow the DOCSIS standard.

Tech Tip

The speeds advertised by DOCSIS are theoretical speeds and are far from the actual speeds you can expect. In fact, it is difficult to obtain the actual speed because speed is affected by various factors, such as time of day, type of data downloaded, and overall distance from the Cable provider.

The amount of bandwidth available is mostly influenced by how many other subscribers are using the Cable service in the local area. The local area distribution point is referred to as the *head-end*. See **Figure 15-4.** There can be hundreds of subscribers connected to the cable before accessing the head-end. As the volume of subscribers increases, the bandwidth decreases.

head-end
the local area distribution point of a Cable television service provider.

The actual bandwidth fluctuates according to the number of users on line at the same time and the type of data downloaded. For example, in the middle of the night, a typical user should have remarkable download speeds because there are very few users on line. In contrast, the late afternoon hours offer slower speeds due to the large number of users, typically just home from school. At that time, students are on the system doing homework and downloading music and games. All this traffic puts a stress on the bandwidth, thus decreasing it substantially.

As stated earlier, Cable Internet service requires a special piece of equipment called a Cable modem. The Cable modem provides a connection to the Cable television service media and to the network interface card in the PC. **Figure 15-5** shows how a splitter can be used to share the Cable television coaxial cable with the Cable modem.

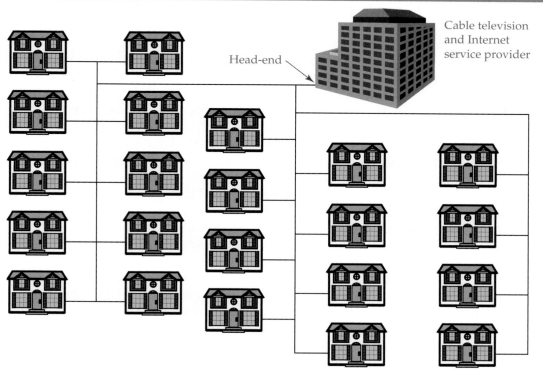

Head-end

Cable television and Internet service provider

Figure 15-4.
While Cable Internet access offers a high-bandwidth Internet connection to residencies, the total number of simultaneous Internet connections through the Cable system affects the amount of bandwidth used by each residence.

Figure 15-5.
The television Cable service can provide Internet access. A Cable modem is needed to connect the PC to the Cable service.

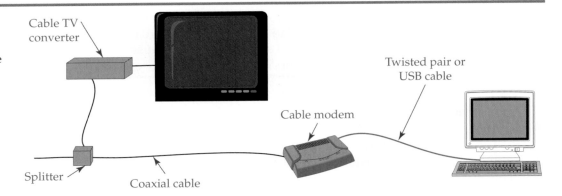

Not all Cable companies can provide Internet service because Internet access requires special equipment to distribute Internet access through the Cable company's distribution system. Original Cable distribution equipment was designed for the downstream distribution of video and music and not designed to accept upstream communications. The cabling and the equipment at the Cable company's distribution center usually have to be upgraded. The original coaxial cable is replaced with hybrid fiber-coax cable. The fiber-optic portion supports the high bandwidth while the coaxial portion supports the control.

Digital Subscriber Line (DSL)

Digital Subscriber Line (DSL) is a high-speed Internet access technology that uses existing local loop telephone lines. It is much faster than ISDN and is designed to replace it. DSL comes in a variety of standards that vary in the techniques used to make data transmission faster.

DSL uses modulation techniques similar to those discussed in Chapter 4—Wireless Technology. Multiple frequencies are used as separate channels on the existing telephone local loop. The multiple channels combine to carry more data than the original telephone modem design. The original telephone modem design does not use multiple frequency signals. It uses only one.

Maximum distance for DSL is limited due to the high frequencies transmitted. The typical maximum distance for DSL is 1,000 feet to 18,000 feet and is measured from the DSL modem to the telephone company's Local Central Office. The exact limit depends on the variation of DSL used and any special equipment that might exist on the telephone line, such as loading coils, or a change in the media. A *loading coil* is used to amplify voice signals, which are analog, and will not amplify DSL signals, which are digital. In fact, the loading coil reduces or blocks the higher frequency DSL signal.

Another factor is the length of the cable. Signal strength is affected in direct proportion to the length of the cable. A short cable can pass a much higher frequency than a long cable. This means that the longer the cable, the lower the applied DSL frequency, resulting in lower data transfer rates.

The media may also change between the subscriber location and the Local Central Office. For example, the copper conductor may be changed to a fiber-optic cable at some point, which will prevent the application of DSL. The DSL technique is applied only to copper core cable, not fiber-optic cable.

DSL varies in upstream and downstream bandwidth based on the transmission technique used. Some varieties of DSL are listed in **Figure 15-6.** For more information see www.xdsl.com or www.paradyne.com. The main points to

Digital Subscriber Line (DSL)
a leased line dedicated to networking that uses multiple frequencies as separate channels on the existing telephone local loop. The multiple channels combine to carry more data than the original telephone modem design.

loading coil
a device used to amplify voice signals.

DSL Type	Description	Upstream Data Rate	Downstream Data Rate	Maximum Distance between DSL Modem and Central Office
ADSL	Asymmetrical DSL	1.544 Mbps	1.5 Mbps-8 Mbps	12,000 ft.-18,000 ft.
SDSL	Symmetric DSL	1.544 Mbps	1.544 Mbps	10,000 ft.
HDSL	High bit-rate DSL	1.544 Mbps	1.544 Mbps	14,000 ft.
VDSL	Very high bit-rate DSL	1.5 Mbps-2.3 Mbps	13 Mbps-52 Mbps	1000 ft.-4500 ft.

Figure 15-6.
DSL technologies.

remember about DSL is it is a high-speed Internet access technology that uses existing local loop telephone lines, and the actual download rate is affected by the length of the cable.

Satellite

A satellite system can be used for Internet access and for data communication, **Figure 15-7.** A typical satellite system consists of a satellite dish at the satellite service provider location, a satellite, and a satellite dish at the consumer location. Satellite system installations are available in consumer-grade and business-grade. The typical consumer-grade installation consists of a small dish for downloading and a landline, such as a traditional telephone line, for uploading. Before you can connect to the Internet, you need to make an initial connection to the satellite Internet service provider through a modem. Thereafter, you must use the telephone line connection via the modem to communicate (upload) Web page requests and to send e-mail.

Typical download speeds for consumer satellite communications are 400 kbps to 500 kbps. Upload speeds are limited to the particular land-based technology used for upload. For example, a telephone modem offers speeds of 56 kbps and an ISDN modem offers speeds of 1.54 Mbps and 128 kbps. Often, the only real advantage for consumer satellite technology is access from remote locations.

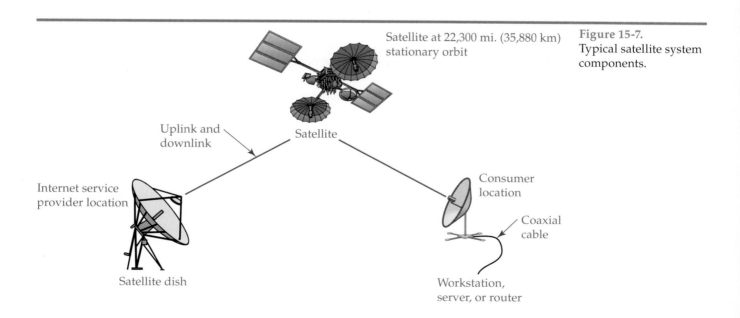

Satellite at 22,300 mi. (35,880 km) stationary orbit

Uplink and downlink

Satellite

Internet service provider location

Satellite dish

Consumer location

Coaxial cable

Workstation, server, or router

Figure 15-7.
Typical satellite system components.

Business-grade satellite communications offer higher speeds, typically up to 1.5 Mbps in the US and up to 2 Mbps in Europe and other countries. The speed difference between the US and other countries is due to different wireless regulations. Business-grade satellite communications is designed with more expensive equipment while consumer-grade satellite communications is designed with less expensive equipment. Using less expensive equipment for consumer-grade satellite communications allows satellite communications to be competitively priced among other Internet access technologies. Business-grade satellite communications, however, offer uplink and downlink capabilities. Uplink requires a radio transmitter to be integrated into the satellite dish. The radio transmitter is referred to as an ***upconverter.***

upconverter
a radio transmitter integrated into a satellite dish.

The main advantages of using satellite communications is it is ideal for temporary connections such as sporting events or remote locations where landlines are not readily available. The disadvantages of satellite communications are latency, security, and interference. As you have already learned in Chapter 4—Wireless Technology, the satellite is located 22,300 miles from the surface of the earth in a geosynchronous orbit (synchronized with the earth's rotational speed). The great distance the radio wave must travel causes the transmitted signal to experience propagation delay. The propagation delay for a signal transmitted from a satellite to a satellite dish is approximately 250 milliseconds. The propagation delay does not noticeably affect the downloading of Web pages with text-based contents, but it is unacceptable for telephone support or full-duplex video conferencing.

Security is also a problem because all transmissions are based on radio transmissions, which can be easily intercepted. An encryption program is needed to secure sensitive data. Radio wave transmissions to and from the satellite can be adversely affected by certain conditions.

Atmospheric conditions due to stormy weather (rain, lightning, etc.) affect satellite communications, causing data loss. Another phenomenon known as sunspots also affects communications. *Sunspots* are magnetic energy storms that occur at the surface of the sun. The magnetic fields produce charged particles that are released from the surface of the sun and projected into outer space. The charged particles can enter the earth's atmosphere and affect electronic communications. A sunspot transmits high levels of electromagnetic energy across the universe and affects to some degree all radio transmissions.

sunspots
magnetic energy storms that occur at the surface of the sun.

The radio frequency used for satellite transmission is classified as microwave. Microwaves have a high frequency pattern with short wavelengths. The high frequency, short wavelength radio waves are easily disrupted by trees, buildings, bridges, and other such obstacles. A clear line of sight is required between the receiving satellite dish and the orbiting satellite. This is in contrast to lower frequency radio waves like the one used for wireless networks. Wireless network radio waves can penetrate some objects that are not very dense. In summary, satellite communications require a clear line of sight between the destination and source. See **Figure 15-8.**

The same principle used to transmit data between the satellite and satellite dish can be used between a series of microwave satellite dishes. You can see microwave dishes located on towers all over the country. The microwave dishes are commonly used by telephone communications, especially in areas where laying cables is difficult or impossible.

Another location where microwave dish transmission is often used is across bodies of water, such as bays and wide rivers, or over large tracts of flat land.

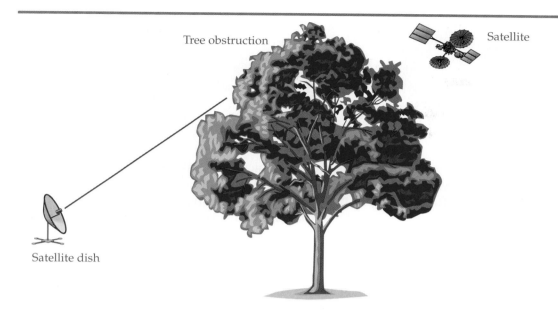

Tree obstruction

Satellite

Satellite dish

Figure 15-8.
There must be a clear line of sight between the satellite dish and the satellite. Trees, buildings, bridges, and such interfere with satellite radio transmissions.

Again, the line of sight principle must be applied, and the series of dishes must be precisely aimed at one another.

T-Carrier

A **T-carrier** is a leased line that follows one of the standards known as T1, fractional T1, T2, or T3. The T-carrier is a dedicated, permanent connection that is capable of providing a high bandwidth. A T1 line consists of 24 channels, each with a data rate of 64 kbps or a total bandwidth of 1.544 Mbps. A fractional T1 only uses part of a T1 line's full capability, usually in increments of 64 kbps. See **Figure 15-9** for a summary of T-carrier standards.

The European counterpart to a T1 leased line is E1, which has a maximum rate of 2.048 Mbps. The Japanese counterpart is J1. Notice in **Figure 15-10** that data rates for T1, E1, and J1 differ.

T-carriers were developed to carry data and voice signals. You will often see T1 lines as a mixture of voice and data lines. To accomplish the task of sharing T1 media with different signals, a pair of multiplexers is used. One multiplexer is placed on each end of the T-carrier line. The multiplexer combines the different signal types (voice and data) into an organized manner and places the different signal types onto a single T1 line for transmission. Look at **Figure 15-11.**

When mixing voice and data, voice packets must be placed onto the T-carrier line in close proximity to one another. If voice packets are spaced too far apart

T-carrier
a leased line that follows one of the standards known as T1, fractional T1, T2, or T3. The T-carrier is a dedicated, permanent connection that is capable of providing a high bandwidth.

T-Carrier Standard	Number of T1 Lines	Number of Channels	Maximum Data Rate
T1	1	24	1.544 Mbps
T2	4	96	6.312 Mbps
T3	28	672	44.736 Mbps
T4	168	4032	274.176 Mbps

Figure 15-9.
T-carrier technologies.

Figure 15-10.
Comparison of T1, E1, and J1 lines.

	USA		Europe		Japan
T1	1.544 Mbps	E1	2.048 Mbps	J1	1.544 Mbps
T2	6.312 Mbps	E2	8.448 Mbps	J2	6.312 Mbps
T3	44.736 Mbps	E3	34.368 Mbps	J3	32.064 Mbps
T4	274.176 Mbps	E4	139.264 Mbps	J4	97.728 Mbps
NA	NA	E5	564.992 Mbps	J5	397.200 Mbps

(interspersed between data), the received voice transmission will have undesirable pauses. It is the multiplexer's responsibility to place the voice packets on the T-carrier line often enough so that there will not be any detectable sound gaps or pauses when reassembled at the destination multiplexer. It is not as critical for data packets to be placed in close proximately to one another.

Often, the multiplexer and a Channel Service Unit/Data Service Unit (CSU/DSU) are incorporated into a single unit. A *Channel Service Unit/Data Service Unit (CSU/DSU)* converts signals from the LAN to signals that can be carried by the T1 line, and vice versa.

Channel Service Unit/Data Service Unit (CSU/DSU) electronic circuitry that converts signals from the LAN to signals that can be carried by the T1 line, and vice versa.

FDDI

The Fiber Distribution Data Interface (FDDI) is a dual ring, fiber-optic arrangement. FDDI was covered earlier in Chapter 3—Fiber-Optic Cable. FDDI is mentioned here because it is often used in a MAN distribution system because of its reliability and its high bandwidth. FDDI can be used for distances of up to 62 miles. This distance is more than adequate for MAN installations.

SONET

Synchronous Optical Network (SONET) is similar in design to T-carrier technology except SONET bases its technology on fiber-optic cable. Because SONET is based on fiber-optic cable, it can provide a much higher bandwidth—as high as 9.9 Gbps. SONET is the standard choice for connecting global-sized networks spanning across the nation and oceans. See **Figure 15-12** for a summary of SONET levels.

SONET is similar to T-carrier in that special termination equipment, such as a multiplexer, is needed at the customer location. SONET often uses multiplexer technology to carry a mix of data, voice, and video on the same channel.

SONET is designed as a ring topology similar to the dual-ring structure of FDDI. One ring is the primary communication ring and the other ring serves as a backup. The backup ring ensures reliability. If the primary ring fails, data transmission is automatically switched to the backup ring.

Figure 15-11.
A multiplexer typically combines multiple lines of incoming data into a single series of data packets on a T1 line. A demultiplexer at the destination reverses the process.

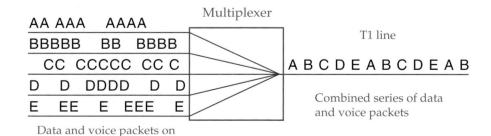

SONET Level	Maximum Data Rate
OC-1	51.84 Mbps
OC-3	145.52 Mbps
OC-12	622 Mbps
OC-24	1244 Mbps
OC-48	2488 Mbps
OC-192	9953 Mbps

Figure 15-12.
Optical Carrier (OCx) levels.

Enterprise companies spanning long distances, such as across the nation or the globe, most often use SONET. Small or mid-size companies do not typically use SONET because it is quite expensive to lease when compared to other technologies. It has, however, the best reliability design.

X.25

X.25 is a packet switching analog network technology developed in the 1970s. X.25 can support a maximum bandwidth of 56 kbps. For over twenty years, X.25 was used to transmit data over long distances. It was used almost exclusively for transmissions around the world because the *International Telecommunications Union (ITU)* recognized it as a standard. Remember that the ITU is responsible for standardizing communications on an international level. The ITU regulates radio, television, and satellite frequencies and standards and telephony standards. Although X.25 dominated worldwide communications, Frame Relay has been the first choice since the early 1990s.

X.25 is not usually available in the USA. However, knowledge about X.25 is required for working with WAN systems that connect all over the world. This protocol is often included in the list of protocols available in many different operating systems.

International Telecommunications Union (ITU) an organization responsible for standardizing communications on an international level. The ITU regulates radio, television, and satellite frequencies and standards and telephony standards.

Frame Relay

Frame Relay uses digital media rather than analog like X.25. Frame Relay is a packet switching technology similar to X.25. Both X.25 and Frame Relay are Permanent Virtual Circuits. A Permanent Virtual Circuit (PVC) typically uses numerous paths between two points. A packet switching technology reduces data into smaller units (packets) and uses a number of different routes (switching) to transport the packets. Remember, a packet contains a destination and source address and a packet sequence number. The sequence number allows the packets to be arranged in their original order when they reach the destination.

One main advantage of Frame Relay technology is, because it is a shared, leased line, it can be leased at a reduced rate. This can mean a significant cost reduction when compared to T-carrier, especially for expensive overseas communications. The fact that Frame Relay is a shared leased line can also be viewed as a disadvantage. The available bandwidth could fluctuate from 56 kbps to 1.544 kbps. For example, the bandwidth available to a user in the morning can be significantly reduced because of the traffic generated by personnel checking for data transactions from the night before. At another time of day, the full 1.544 Mbps bandwidth could be available. See **Figure 15-13** for a summary of WAN connection technologies.

Figure 15-13.
Summary of WAN connection technologies.

WAN Connection Technology	Access Method	Data Rate	Comments
ATM	Direct connection	25 Mbps-622 Mbps	Virtually private.
Cable Internet Service	Direct connection	Downstream: 27 Mbps-37 Mbps (theoretical). 300 kbps-500 kbps (actual) Upstream: 320 kbps-10 Mbps	TV Cable service. Requires a cable modem. Cable service uses MPEG-2 for data compression.
FDDI	Direct connection	100 Mbps	High reliability and bandwidth.
Fractional T1	Direct connection	64 kbps	T-carrier technology. Can use multiple channels to increase speed by increments of 64 kbps.
Frame Relay	Virtual connection	56 kbps-45 Mbps	Virtually private.
ISDN	Dial-up	64 kbps and 128 kbps	Requires a leased line and an ISDN modem. Can carry only one frequency.
PRI-ISDN	Direct connection	1.544 Mbps	Requires a leased line and an ISDN modem. Can carry only one frequency.
Public Switched Telephone Network (PSTN)	Dial-up	56 kbps	This is also known as Plain Old Telephone Service (POTS). Requires a telephone modem.
Satellite Internet Service	Direct connection	400 kbps-2Mbps	May use a dial-up service to connect to the satellite Internet provider. Experiences propagation delay.
SONET	Direct connection	51 kbps-9.9 Mbps	Commonly used to span long distances such as across the nation and overseas.
T1	Direct connection	1.544 Mbps	T-carrier technology.
T3	Direct connection	44.736 Mbps	T-carrier technology.
X.25	Virtual connection	56 kbps	Packet switching, analog technology.
xDSL	Direct connection	1.544 Mbps-52 Mbps	Data rate depends on the version of DSL. Requires an xDSL modem. Limited distance from the DSL modem to the telephone company's central office is between 1,000 ft.-18,000 ft.

dial-up networking
a network system in which a dial-up connection is used to access a remote server.

Dial-Up Networking

Remote connections can be made to distant networks using existing telephone lines. The term *dial-up networking* describes a network system in which a dial-up connection is used to access a remote access server. A *remote access server,* therefore, is a server that is accessed through dial-up networking. To set up a dial-up network, a modem must be installed at the client location and on the remote access server. Remote access software must also be installed. The remote access software specific to Microsoft operating systems are Dial-Up Networking (DUN) and Remote Access Service (RAS). DUN is installed on the client side of the remote access network and RAS is installed on the remote access server.

remote access server a server that is accessed through dial-up networking.

Windows 2000 Server and Windows 2003 Server use an enhanced version of RAS called Routing and Remote Access Service (RRAS).

Tech Tip

Look at **Figure 15-14.** A laptop at a distant location is connected to a telephone line. The laptop connects to the home-office network through the remote access server. Once a connection is established, the remote access client can check their office e-mail, print out customer orders, access files on their office desktop computer, and access the company database. There are three common ways to set up remote dial-in access on a server.

- Allow dial-in to the remote access server only.
- Allow dial-out from the remote access server only.
- Allow full-service dial-in and dial-out on the remote access server.

Dial-In Only

The dial-in only configuration allows remote access clients to connect to the network server, but does not allow hosts on the network to dial out. This prevents personnel from surfing the World Wide Web, but allows the hosts on the network to accept incoming calls. This is a good business scenario that prevents workers from wasting time when they should be providing service to customers.

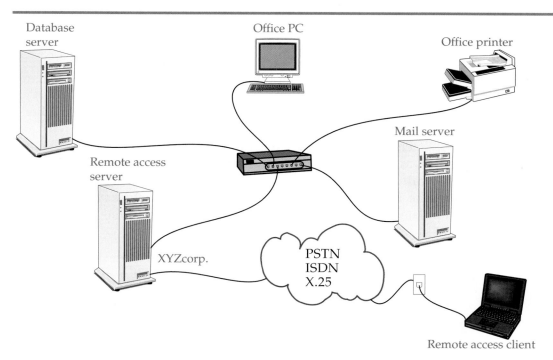

Figure 15-14. A laptop computer can connect to the corporate office by using dial-up networking software. A server must be set up with remote access server software. Once this is accomplished, the remote client can access items such as their office computer, printer, company database, and e-mail.

Dial-Out Only

Dial-out only access adds security to a network system by allowing calls to be placed from the network but not into the network. This prevents unauthorized persons from accessing the network through a telephone modem connection.

Full Service

Full service access allows the remote client to dial-in to the network and also allows network hosts to dial-out to remote locations. This type of access is perfect for a small-office environment or a limited number of users in an enterprise business environment. Security policies can control individual user access to telephone features in much the same way file-sharing access can be controlled. Specific users can be identified and granted various levels of modem use and specific hours or times of modem access.

Remote access can be configured to allow the business to bear the cost of a long distance call. For example, a remote access client calls into the office server. The office server detects the incoming call but does not accept it. The call is terminated and the server immediately returns the call to the remote access client so that the server side bears the cost of the long distance call rather than the remote access client.

A network may also be accessed remotely through an Internet connection if the network has a network server with a modem and the remote access service running. Not all offices have a network server. In this scenario, remote access must take place from telephone modem to telephone modem between a network PC and a remote access PC.

Remote Desktop Protocol (RDP)

Remote Desktop Protocol (RDP) a presentation protocol that allows Windows XP Professional computers to communicate directly with Windows-based clients.

The *Remote Desktop Protocol (RDP)* is a presentation protocol that allows Windows XP Professional computers to communicate directly with Windows-based clients. Client software is required for Windows 2000, Windows Me, Windows 98, Windows 95, and Windows NT. The client can be downloaded from the Microsoft support Web site.

The Remote Desktop Protocol is transmitted across any TCP/IP connection. RDP provides security by encrypting the contents of packets sent across the TCP/IP network. Port 3389 is the default port used for RDP and must be opened when a firewall is used.

RDP is an extension of the original Terminal Services technology, an earlier version of the presentation type protocol. Look at **Figure 15-15.** The original Terminal Service technology allowed a Terminal Service Client to send keyboard and mouse control to the Terminal Server and the Terminal Server to send its display to the Terminal Service Client. For example, after the client establishes a connection with the Terminal Server, it sends commands via the keyboard or mouse to manipulate the directory and file structure. The Terminal Server sends

Figure 15-15.
Remote Desktop and Remote Assistance is based on Terminal Services technology.

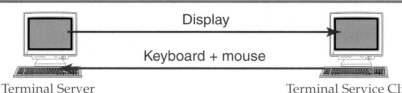

Display

Keyboard + mouse

Terminal Server   Terminal Service Client

the screen image to the client as the files are manipulated. The client experiences the session as though they were sitting at the Terminal Server.

When using RDP with a Windows XP Professional system, the roles of Terminal Service Client and Terminal Server are simply called the client and host, respectively. The client uses the Remote Desktop Connection software to connect to the host.

RDP is based on the International Telecommunications Union (ITU) T.120 protocol.

Tech Tip

Remote Access Protocols

Dial-up access uses special protocols for negotiating a connection between a PC and a network, two networks, or two PCs. Some of the most common remote access protocols are briefly introduced in this section. You will see that many remote access protocols were developed to better address the issues of security. See **Figure 15-16** for a summary of remote access protocols.

PPP and SLIP

Serial Line Internet Protocol (SLIP) and Point-to-Point Protocol (PPP) enable a PC to connect to a remote network using a serial line connection, typically through a telephone line. The *Serial Line Internet Protocol (SLIP)* protocol was

Serial Line Internet Protocol (SLIP) a remote access protocol that enables a PC to connect to a remote network using a serial line connection, typically through a telephone line. SLIP is an asynchronous protocol that supports only IP.

Remote Access Protocol	Full Name	Description
L2F	Layer 2 Forwarding	Enhances security and makes use of a VPN using the public Internet. L2F requires special equipment on the host side to support data transfers.
L2TP	Layer 2 Tunneling Protocol	Combines the features of L2F and PPTP.
MLPPP	Multilink Point-to-Point Protocol	Combines two or more physical links in such a way that they act as one, thus increasing the supported bandwidth. Works with ISDN lines, PSTN lines, and X.25 technology.
PPP	Point-to-Point Protocol	A synchronous protocol that supports multiple protocols such as IPX and AppleTalk. Enables a PC to connect to a remote network using a serial line connection, typically through a telephone line.
PPPoE	Point-to-Point Protocol over Ethernet	Supports multiple clients connected to an ISP over Ethernet.
PPTP	Point-to-Point Tunneling Protocol	Enhanced version of PPP that makes use of a VPN using the public Internet. It encapsulates the existing network protocol (IP, IPX, AppleTalk, and such) into the PPTP protocol.
SLIP	Serial Line Internet Protocol	An asynchronous protocol that supports only IP. Enables a PC to connect to a remote network using a serial line connection, typically through a telephone line.

Figure 15-16.
Summary of remote access protocols.

Point-to-Point Protocol (PPP)
a remote access protocol that enables a PC to connect to a remote network using a serial line connection, typically through a telephone line. PPP is a synchronous protocol that supports multiple protocols such as IPX and AppleTalk.

introduced with UNIX and supports only TCP/IP. **Point-to-Point Protocol (PPP)** is a vast improvement over SLIP and can support multiple protocols such as IPX, AppleTalk, and TCP/IP.

SLIP is strictly an asynchronous protocol while PPP is both asynchronous and synchronous. Remember from earlier chapters that synchronous transmission relies on a separate timing signal so that both the transmitting and receiving equipment are in step during the exchange of data. Asynchronous means that the data is transferred without a timing signal. Data is transferred from one point to another without being in step.

PPP is much easier to set up than SLIP because most of the required settings can be automatically negotiated. SLIP requires information such as the IP address to be entered manually. You will seldom encounter SLIP because it doesn't support authentication, encryption, or VPN connections.

PPTP

Point-to-Point Tunneling Protocol (PPTP)
a remote access protocol that is an enhanced version of PPP. It is designed to enhance security and to make use of a virtually private network using the public Internet.

Point-to-Point Tunneling Protocol (PPTP) is an enhanced version of PPP. It is designed to enhance security and to make use of a virtually private network using the public Internet. The Internet is used by millions of people. While it is an excellent system of communication, it has security issues. The term *virtually private* means that the network connection appears to be a private network, similar to a dedicated network media such as a T1 line. A T1 line offers a much higher level of security than an Internet connection because it is a privately leased line in contrast to a publicly accessed connection.

When using PPTP, you can gain access to a corporate network by using a dial-up connection through an ISP and then negotiating data transfers over the public Internet. The term *tunneling* refers to encapsulating the existing network protocol (IP, IPX, AppleTalk, and such) into the PPTP protocol.

PPTP incorporates methods to encrypt data. The PPTP protocol also supports the necessary remote access service and network access authentication.

PPPoE

Point-to-Point Protocol over Ethernet (PPPoE)
a remote access protocol that provides one or more host on an Ethernet network the ability to establish an individual PPP connection with an ISP. PPPoE frames the PPP protocol so that the PPP frame can travel over an Ethernet network.

Point-to-Point Protocol over Ethernet (PPPoE) provides one or more hosts on an Ethernet network the ability to establish an individual PPP connection with an ISP. PPPoE frames the PPP protocol so that the PPP frame can travel over an Ethernet network.

Look at the **Figure 15-17.** Notice that each computer using the PPPoE protocol has the ability to establish and maintain an individual connection with an ISP. In ICS, one computer establishes a connection to the ISP and must remain on for the other computers to have the ability to connect to the Internet. With the PPPoE protocol, each PPPoE client has direct access to the ISP. The PPPoE clients are arranged as a typical Ethernet network and use a DSL or Cable modem to connect to the ISP. The DSL or Cable modem may be incorporated into a router or serve as an individual piece of equipment. A PPPoE access server is located at the ISP. The server provides individual connections as needed to each of the PPPoE clients.

The PPPoE connection has two stages: discovery and session. In the discovery stage, the PPPoE client attempts to discover the location of the PPPoE server. There may be more than one PPPoE server. The client must locate and establish a connection with only one PPPoE server. Typically, the PPPoE client and the PPPoE server exchange network identification consisting of an IP address and MAC address. After addressees are exchanged, a connection is established

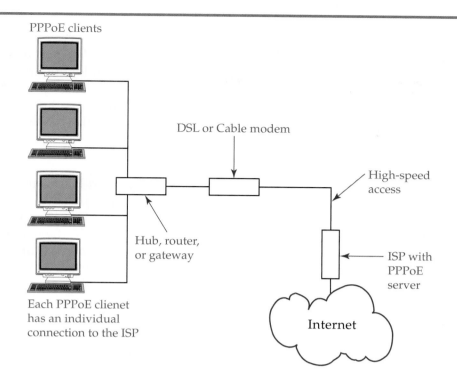

PPPoE clients

DSL or Cable modem

High-speed access

Hub, router, or gateway

ISP with PPPoE server

Internet

Each PPPoE clienet has an individual connection to the ISP

Figure 15-17.
PPPoE is an effective means of creating individual high-speed connections for Internet access.

and maintained indefinitely. The established connection is referred to as a session. Each session is independent of other connections to the PPPoE.

In Figure 15-17, all computers are on the same network. In another scenario, the computers connecting to the ISP may reside in an office building consisting of many different businesses. Individual users may need to be billed rather than the entire site. The MAC address provides a way of billing the user and establishing a user account for authentication purposes. The ISP can then provide billing based on a per-user, rather than per site basis.

PPPoE establishes a virtual serial connection to the ISP in the same way a single PPP connection creates a true serial connection. In other words, when using PPPoE, each computer establishes a virtually private PPP connection to the Internet.

MLPPP

Multilink Point-to-Point Protocol (MLPPP), also known as Multilink dialing, can combine two or more physical links in such a way that they act as one, thus increasing the supported bandwidth. For MLPPP to operate, the access server and the client must have MLPPP enabled. MLPPP works with ISDN lines, PSTN lines, and X.25 technology.

MLPPP can use the *Bandwidth Allocation Protocol (BAP).* BAP allows the number of lines used by MLPPP to change on demand. For example, if simple text is downloaded, a single line will suffice for the low bandwidth data. If multimedia is downloaded to multiple workstations, such as for a videoconferencing, BAP will automatically increase the number of lines to support the increase in bandwidth. Once the high bandwidth data transmission has ended, the number of lines is reduced to minimum. Remember, there is a high cost associated with telephone and leased lines. Controlling the number of lines used based on demand is cost effective.

Multilink Point-to-Point Protocol (MLPPP)
a remote access protocol that can combine two or more physical links in such a way that they act as one, thus increasing the supported bandwidth. This protocol works with ISDN lines, PSTN lines, and X.25 technology.

Bandwidth Allocation Protocol (BAP)
a protocol that can change the number of lines or channels used in communication according to current bandwidth.

Layer 2 Forwarding (L2F)
a remote access protocol, similar to PPTP, designed to enhance security and to make use of a virtually private network using the public Internet. It was developed by Cisco Systems and then released for use by other venders. L2F requires special equipment on the host side to support data transfer whereas PPTP does not.

Layer 2 Tunneling Protocol (L2TP)
a tunneling protocol that has the best features of both L2F and PPTP protocols. It uses IPSec to encrypt the contents of the encapsulated PPP protocol.

Virtual Private Network (VPN)
a simulated, independent network created by software over a public network.

L2F

Layer 2 Forwarding (L2F) is a protocol similar to PPTP. It was developed by Cisco Systems and then released for use by other venders. One big advantage of L2F is that it can work directly with ATM and Frame Relay, and it does not require TCP/IP. L2F also supports more than one connection. PPTP does not. However, L2F requires special equipment on the host side to support data transfer. The only requirement for PPTP is that the host is installed with a Microsoft server class of operating system.

L2TP

Layer 2 Tunneling Protocol (L2TP) combines the L2F and PPTP protocols. Microsoft and several other companies developed PPTP, and Cisco Systems Inc. developed L2F. The Internet Engineering Task Force wanted to create a protocol that had the best features of both L2F and PPTP protocols. Jointly, the IETF, Cisco Systems, Microsoft, and many others designed L2TP. L2TP has become an industry standard tunneling protocol that uses IPSec to encrypt the contents of the encapsulated PPP protocol.

Virtual Private Networks (VPN)

A *Virtual Private Network (VPN)* is a simulated, independent network created by software over a public network. An example of a public network is the Internet. Since anyone can access and use a public network, there is a possibility of unauthorized access to any network connected to it. A VPN makes it possible to create a private connection over a public network with no additional cost. For example, a company may wish to use the Internet to connect to their many branch offices spread throughout the country. The main reason for using the Internet is it is inexpensive when compared to leased lines and other physical private connections. However, the company may wish to keep their data and transmissions secure while using the Internet for data transfers and communications. The answer to this dilemma would be to implement a VPN.

A VPN is created through a software package that provides security. Adding special equipment, such as a firewall, can further increase the security of the VPN. VPN capability is typically provided by the latest operating systems such as Windows XP and many versions of Linux.

Tech Tip

Not all firewalls are equipment-based technologies. Some firewalls are strictly software packages.

VPN software encapsulates transport protocol packets, such as IP, IPX, and AppleTalk, inside a secure protocol packet and encrypts the contents. Security was not considered a real threat at the time protocols such as IP, IPX, and Apple-Talk were designed. In fact, information about the designs of these protocols was made public to encourage the development of the Internet by private sectors.

Four of the most common protocols used in a VPN are PPTP, L2F, L2TP, and IPSec. These protocols allow a VPN to provide four basic features common to all VPN connections: authentication, access control, confidentiality, and data integrity. *Authentication* (in relation to a VPN) is the process of assuring the person who is accessing the remote system is really the person who is authorized to use the system. *Access control* is the act of allowing only authorized users access to the VPN. *Confidentiality* is the act of preventing anyone else from reading the data that flows through the VPN. *Data integrity* is the process of ensuring the data that leaves the source and arrives at the destination has not been tampered with. These topics are covered in detail in Chapter 16—Network Security.

Summary

- The deregulation of the telecommunications industry allows for separate local telephone systems and long distant telephone systems, which increases competition and reduces overall costs.

- The Point of Presence (PoP) is the point where the public telephone line ends and the customer premises telephone line and equipment begin.

- Since the deregulation of the telecommunications industry, owners of residences and businesses are responsible for the telephone system on their side of the Point of Presence (PoP).

- There are many types of media available for communicating over long distances, such as Cable, DSL, ISDN, PSTN, satellite, SONET, and T1. Each is offered at various speeds and costs.

- Basic Rate ISDN (BRI-ISDN) consists of two B channels and one D channel with a total bandwidth of 128 kbps.

- Primary Rate ISDN (PRI-ISDN) consists of twenty-three B channels and one D channel with a total data rate of 1.544 Mbps.

- Broadband ISDN (B-ISDN) is a new category of ISDN and is designed to carry multiple frequencies; hence, the term *Broadband*.

- Cable Internet service offers data rates as high as 36 Mbps downstream and 10 Mbps upstream. However, 10 Mbps is a theoretical speed.

- Cable Internet service is defined by the Data Over Cable System Interface Specification (DOCSIS) standard developed independently of IEEE 802.14.

- DSL is offered in many varieties, such as ADSL, SDSL, HDSL, and VDSL.

- The maximum allowable distance between a DSL modem and the Local Central Office is typically 1,000 feet to 18,000 feet. This distance varies according to the type of DSL used and any special equipment that might exist on the telephone line, such as loading coils or a change in the media.

- DSL bandwidth varies from 1.5 Mbps to 52 Mbps depending on the DSL variation used and the distance from the central office.

- Satellite communications is ideal for remote locations and offers various speeds based on cost of equipment and service.
- Consumer satellite service bandwidth is typically 400 kbps to 500 kbps.
- Commercial satellite service bandwidth is a maximum 1.5 Mbps in the USA and 2 Mbps in Europe.
- Satellite communications use microwave technology and must have a clear line of site between the satellite and the satellite dish.
- T-carriers are leased lines and range from fractional T1 to T4. The most common T-carrier lines encountered in networking are fractional T1, T2, and T3.
- A T1 line consists of 24 channels with 64 kbps of bandwidth available per channel.
- T1 offers a maximum bandwidth of 1.544 Mbps, and T3 offers 44.736.
- Synchronous Optical Network (SONET) offers the highest bandwidth and the greatest distances.
- X.25 is an older telephone line technology that is rarely encountered in the USA but still exists in other parts of the world.
- Frame Relay is the replacement for X.25 and employs a Permanent Virtual Circuit (PVC).
- Frame Relay and X.25 are packet switching technologies.
- Dial-up networking allows a user to access a network remotely using telephone lines.
- Dial-up networking requires a modem and remote access software to be installed on the client side of a remote connection and on a remote access server.
- Serial Line Interface Protocol (SLIP) is an older protocol used to support remote connections over telephone lines. It has been replaced by PPP and PPTP.
- Point-to-Point Tunneling Protocol (PPTP) encapsulates other protocols inside the PPTP packet, resulting in a secure connection.
- A Virtual Private Network (VPN) is a secure connection over a public network system based on software, not hardware.

Review Questions

Answer the following questions on a separate sheet of paper. Please do not write in this book.

1. What is the name used for the telephone line running between customer premises and the Local Central Office?

2. Where is the Point of Presence (PoP) located?

3. What does the acronym ISDN represent?

4. How many channels are there in a BRI-ISDN cable assembly?

5. What is the maximum bandwidth of a BRI-ISDN service?

6. What is the maximum bandwidth for a telephone modem?

7. What is the purpose of the D channel in an ISDN cable assembly?

8. What is the maximum theoretical bandwidth for Cable Internet access?

9. What factors influence the speed of a Cable Internet service connection?

10. What is the difference between a T1 and a fractional T1 line?

11. How many T1 lines make up a T3 line?

12. Which has the highest bandwidth: a T1 connection or a telephone modem connection?

13. Which has the highest bandwidth: SONET or ISDN?

14. Which has the highest bandwidth: T1 or X.25?

15. What high-speed technology is used for long distance communications especially overseas via underwater?

16. Which has a highest bandwidth: ISDN or consumer satellite service?

17. What does the acronym CSU/DSU represent?

18. What type of media is a CSU/DSU associated with?

19. Match the following:

    1. ISDN   A. Fiber optic system used for overseas connections.

    2. DSL   B. Typically the slowest connection technology.

    3. SONET   C. Two B channels and one D channel.

    4. T3   D. Typically 45 Mbps bandwidth.

    5. Telephone modem.   E. Digital telephone line.

20. What is the main difference between SLIP and PPP?

21. In what situation is the PPPoE protocol used?

22. What is the main difference between PPTP and PPP?

23. What is the difference between a Virtual Private Network (VPN) and a network that uses a T1 line to connect two branch offices?

Sample Network+ Exam Questions

Answer the following questions on a separate sheet of paper. Please do not write in this book.

1. Which technology provides the most bandwidth when connecting two networks located approximately 18 miles apart?
   A. PSTN
   B. DSL
   C. ISDN
   D. T1

2. What is the typical bandwidth of one ISDN B channel?
   A. 56 kbps
   B. 64 kbps
   C. 128 kbps
   D. 1.5 Mbps

3. What is the typical data rate of a fractional T1 line?
   A. 56 kbps
   B. 64 kbps
   C. 128 kbps
   D. 1.544 Mbps

4. What would *most likely* affect the bandwidth of a DSL line?
   A. The number of subscribers sharing the DSL line.
   B. The distance from the DSL modem to the DSL provider.
   C. The distance to the ISP.
   D. The amplitude of the signal modulation.

5. What network media does DSL use?
   A. Shielded twisted pair
   B. Coaxial cable
   C. Fiber-optic cable
   D. Telephone line

6. Which of the following protocols is commonly used to create a VPN?
   A. FTP
   B. HTTP
   C. PPTP
   D. SLIP

7. Which of the following protocols would be most appropriate for creating multiple, individual dial-up connections from an Ethernet network through a common DSL or Cable modem to an ISP.

   A. PPP

   B. PPPoE

   C. SLIP

   D. TFTP

8. Which medium provides the best security when transmitting data between two points?

   A. Satellite

   B. Internet

   C. T1 line

   D. IEEE 802.11a

9. Which of the following statements is true concerning X.25 and Frame Relay technologies?

   A. Frame Relay uses analog technology and X.25 does not.

   B. X.25 uses analog technology and Frame Relay does not.

   C. Frame Relay and X.25 both use analog technology.

   D. Frame Relay and X.25 both use digital technology.

10. Which communications medium is used for SONET?

   A. Wireless 2.5 GHz

   B. UTP

   C. RG-58

   D. Fiber-optic

Suggested Laboratory Activities

1. Install and configure a variety of modems such as a Cable modem, DSL modem, and an ISDN modem.

2. Using the **tracert** command from the DOS prompt, trace the route taken by the packet to a remote location. Some test locations are www.novell.com and www.microsoft.com. Identify the amount of time it takes to reach each router along the route. (Each router along the route to the destination is called a *hop*.)

3. Go to the Microsoft support Web page and research Remote Desktop and VPN.

4. Using the knowledge gained from the previous suggested laboratory activity, set up a remote connection to a PC across the Internet.

5. Research the average cost for the following services if available to your school or geographic area. Do *not* write in this book. Use a separate sheet of paper to copy and fill in the chart. Access the information by using the Internet.

Technology	Availability Yes/No	Cost	Comments
Telephone line			
ISDN			
DSL			
Cable			
Fractional T-1			
T-1			
Satellite			

6. Set up Microsoft NetMeeting on a LAN. You will first need to download a copy of NetMeeting from the Microsoft Web site.

Interesting Web Sites for More Information

www.cabledatacomnews.com

www.cablemodemhelp.com

www.howstuffworks.com

www.iec.org

www.protocols.com

Chapter 15
Laboratory Activity

Routing and Remote Access Service (RRAS)

After completing this laboratory activity, you will be able to:

- Explain the various Routing and Remote Access Service (RRAS) options.
- Install and configure the Routing and Remote Access Service.
- Explain the Callback feature.

Introduction

In this laboratory activity, you will configure the Routing and Remote Access Service (RRAS), which is available in Windows Server 2003. Microsoft provides a RRAS Setup Wizard, which automatically walks a technician through RRAS configuration. RRAS allows a server to pose as a remote access server or as a network router.

When configured as a router, the server is used to connect various network segments or subnets. You will not configure the server as a router during this laboratory activity. You will only configure the server as a remote access server accessible by telephone.

The following figure shows one way a client might access a company server. A sales person that is traveling across the United States might use a dial-up connection from a laptop to the home office to place an order, check e-mail, or access files on the server. The server has a telephone modem installed and configured to receive incoming calls. This modem is dedicated to remote access and is not used for Internet access.

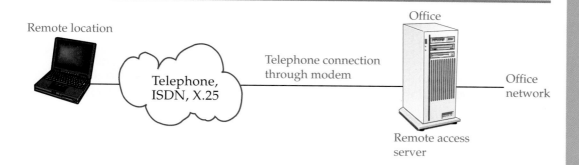

Remote location

Telephone, ISDN, X.25

Telephone connection through modem

Office

Office network

Remote access server

More than one telephone modem can be installed in a server to provide multiple access points to the server. Each modem is assigned a unique telephone number. Several manufacturers make a dial-up telephone modem bank. A dial-up telephone modem bank is a separate group of telephone modems that provide multiple access points to the server. Large, corporate network servers that have a high demand for remote access commonly use a modem bank.

A small office server might have two modems installed in a single server: one for dial out and the other for dialing in. The server may also be configured as a multiple connection location for Virtual Private Network (VPN) connections or may be configured as a Network Address Translation (NAT) server providing access to the Internet for multiple workstations through a modem or high bandwidth connection.

After a server is configured for dial-in remote access, user profiles must be configured so that users can access the server using a dial-up connection. Look at the following figure of the properties for user Richard Roberts.

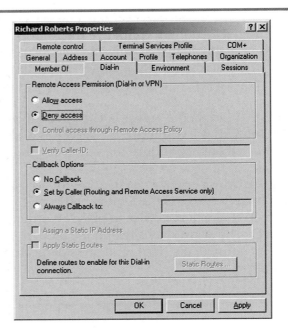

Take note of the **Allow access** option in the **Remote Access Permission (Dial-in or VPN)** section. By default, all users are denied access. By selecting the **Allow access** option, the user can dial in to the server from a remote location. Also, take note of the **Callback Options** section. A Callback number can be entered next to the **Always Callback to:** option. This enables a call back from a specified telephone number and thus enhances security because the user must call from the specified telephone number.

At the server, properties for the remote configuration can be set up to limit telephone calls to specific days and hours and for the maximum length of a connection. For example, all dial-in calls can be limited to a maximum of five minutes. This way, no one single user can monopolize the dial-in service.

Extensive information is available through **Help and Support** accessed from the **Start** menu. Also, the RRAS Setup Wizard provides a link to more information about RRAS.

At the end of this laboratory activity, your instructor will most likely want you to uninstall the RRAS service. Be sure to check with your instructor before you begin the laboratory activity.

Equipment and Materials

- Windows 2003 server with a modem installed. (All features of the RRAS Setup Wizard may not be available for viewing without a modem installed.)

Procedures

1. _____ Report to your assigned workstation.

2. _____ Boot the workstation and verify it is in working order.

3. _____ To access the Routing and Remote Access Setup Wizard, open **Start | Administrative Tools | Routing and Remote Access**. A screen similar to the following will appear.

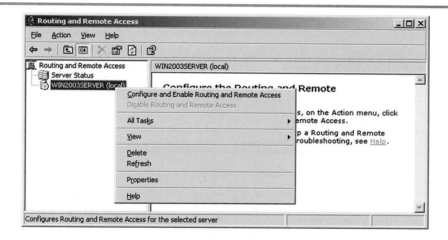

4. _____ Notice the Microsoft Management Console with the **Routing and Remote Access** snap-in. Right-click the server located beneath **Routing and Remote Access** and then select **Configure and Enable Routing and Remote Access** from the shortcut menu.

Note:
*If your workstation does not appear, select **Add Server** from the **Action** menu and then select **This computer**.*

The RRAS Setup Wizard will display and look similar to that in the following figure. The RRAS Setup Wizard provides an easy to follow step-by-step process for setting up the server for remote access and routing.

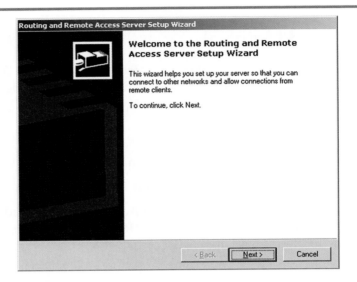

5. _____   Click **Next**. The RRAS Setup Wizard will display various configuration options, such as **Remote access (dial-up or VPN)** and **Network address translation (NAT)**. Record these configuration options on a separate sheet of paper.

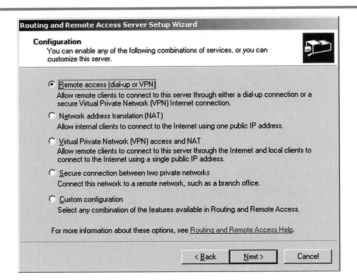

6. _____ Select the first option, **Remote access (dial-up or VPN)**. At the bottom of the screen is a link to **Routing and Remote Access Help** where you can find more information about RRAS. Take a few minutes and explore this option. After exploring this option, return to the current RRAS Setup Wizard screen.

7. _____ Click **Next**. You will be presented with two options: **VPN** and **Dial-up**. Select **Dial-up**.

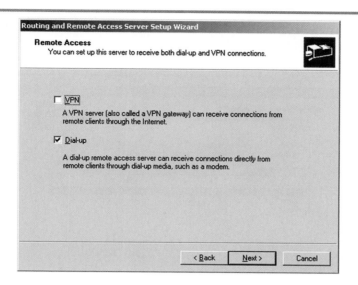

8. _____ Click **Next**. If you have more than one connection available from the server, the **Network Selection** screen will display and prompt you to select which connection to use; otherwise, the **IP Address Assignment** screen will display. The following figure shows **Network Selection** screen. Notice that there are two connections listed: modem and wireless. If you have more than one connection available from the server, and the **Network Selection** screen displays, select only the modem connection and then click **Next**.

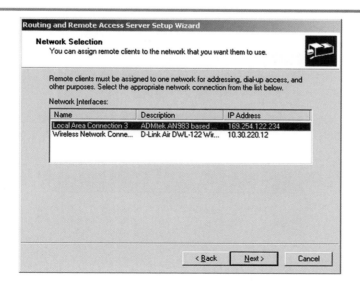

9. _____ The **IP Address Assignment** screen allows you to select the IP address assignments method. You can select either **Automatically** or **From a specified range of addresses**. To be able to use the **Automatically** option, you must have DHCP configured on this server or another server. If you are not using a DHCP server, you must statically assign IP addresses by selecting the **From a specified range of addresses** option. When using a specified range, you only need to reserve one or more IP addresses to be used by remote access clients. For this laboratory activity, select **From a specified range of addresses** option and then click **Next**. The **Address Range Assignment** screen will display.

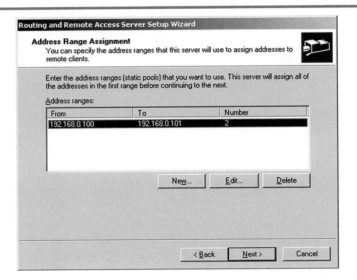

10. _____ To enter a range of IP addresses, click **New** and then enter the starting IP address and ending IP address. Alternatively, you can enter the starting IP address and then the number of IP addresses desired. In the following figure, notice that two IP addresses are reserved: 192.168.000.100 and 192.168.000.101.

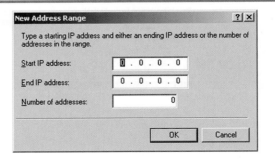

11. _____ Click **OK** and then click **Next**. The **Managing Multiple Remote Access Servers** screen will display. You will be asked if you want to set up this server to work with a RADIUS server. RADIUS is typically used on large networks that have more than one Wireless Access Point (WAP) for accessing the network system. For this laboratory activity, select **No, use Routing and Remote Access to authenticate connection requests**.

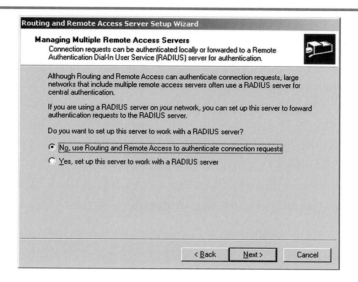

12. _____ Click **Next**. A summary screen will display and will look similar to the following figure. Click **Finish** to complete the RRAS configuration. If you have another telephone line available in the lab, you can test RRAS by dialing in on this line from a separate computer.

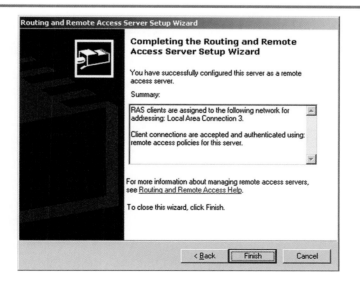

13. _____ In this step, you will delete the Routing and Remote Access Service from the server by deleting the server from the Routing and Remote Access snap-in. To remove RRAS from the server, open the Routing and Remote Access utility (**Start | Administrative Tools| Routing and Remote Access**). When the RRAS snap-in appears, right-click the server icon and then select **Delete** from the shortcut menu.

14. _____ After you have removed the Routing and Remote Access Service, answer the review questions.

Review Questions

Answer the following questions on a separate sheet of paper. Please do not write in this book.

1. How do you access the Routing and Remote Access snap-in?
2. A technician has just successfully configured remote dial-in access for a server, but no one can access the server. What might be the problem?
3. Where is the **Callback** option configured for a client?
4. What are the five configuration options available in the RRAS Setup Wizard?
5. What are the two options for assigning an IP address to the dial-in clients?

16

Network Security

After studying this chapter, you will be able to:

❏ Identify common network security breaches and vulnerabilities.

❏ Explain the difference between symmetrical and asymmetrical encryption.

❏ Explain the role of a Certificate Authority (CA).

❏ Explain the security process associated with the Challenge Handshake Access Protocol (CHAP).

❏ Describe the characteristics of a secure password.

❏ Describe how a firewall and proxy server are used to secure network access.

❏ Describe how to monitor network activities.

Network+ Exam—Key Points

You will most likely encounter questions about the topic of security on the Network+ Certification exam. Pay particular attention to the subject of passwords and password best practices. Port numbers are also a source of questions, especially common port numbers like the ones listed in this chapter and in the Chapter 11—TCP/IP Fundamentals. There could also be a question about software upgrades or patches and how they affect security.

Network+

Key Words and Terms

The following words and terms will become important pieces of your networking vocabulary. Be sure you can define them.

application gateway
asymmetric-key encryption
authentication
Authentication Header (AH)
backdoor
biometrics
Certificate Authority (CA)
Challenge Handshake Authentication
    Protocol (CHAP)
ciphertext
circuit-level gateway
cracker
cryptology
Denial of Service (DoS)
digital certificate
Encapsulation Security Payload (ESP)
encryption
firewall
hashing
IP Security (IPSec)
Kerberos
key
man in the middle (MITM)
Media Access Control (MAC) filter
Microsoft Challenge Handshake
    Authentication Protocol (MS-CHAP)

packet filter
packet sniffer
Password Authentication Protocol
    (PAP)
phishing
protocol analyzer
proxy server
Remote Authentication Dial-In User
    Service (RADIUS)
replay attack
Secure Copy Protocol (SCP)
Secure HTTP (S-HTTP)
Secure Shell (SSH)
Secure Sockets Layer (SSL)
service pack
smart card
social engineering
spoofing
symmetric-key encryption
transport mode
Trojan horse
tunnel mode
worm
Zero Configuration (Zeroconf)

authentication
the process of identifying a user and ensuring the user is who they claim to be.

This chapter presents the basics of network security. Network security is a vital part of network operation. It ensures data integrity and privacy. Network security is comprised of two main elements: authentication and encryption. *Authentication* is the process used to identify a user and ensure the user is who they claim to be. Typically, authentication is done with a combination of a user name and password. This method is probably the one you are most familiar with. There are other forms of authentication such as digital certificates, smart cards, and biometrics. We will look at these and other methods of authentication as well as security methods, such as encryption, and security protocols, such as IPSec and SSL. The dos and don'ts of network security as recommended by industry leaders and the basic tools to monitor system security are also covered in this chapter.

The field of network security is vast. This chapter only covers the basics of network security. A complete presentation of network security would require an entire course with a textbook completely dedicated to security and an intensive set of laboratory activities.

This chapter deals with basic security. Students are often tempted to use the knowledge gained in this chapter to experiment with hacker techniques. This is not only unethical, it is illegal. If you are using this textbook, you are undoubtedly at an introductory level of networking technology. Do not ruin your chances of ever securing a good job in the IT industry by being accused or convicted of a computer crime-related incident. You are just beginning your career, and the people defending the Web site or network have many years of experience. The odds are overwhelmingly in the favor of the defenders. It is not the intent that you use the skills taught in this chapter to be applied in any attempt to crack any computer system. The skills taught here are to be used for your basic understanding of best security practices.

Hackers, Crackers, and Intruders

The term *hacker* originally described any computer enthusiast who lacked formal training. The term has now been redefined to identify anyone who gains unauthorized access to a computer system. A **cracker** is defined as anyone who gains access to a computer system without authorization and with the intent to do harm or play pranks. They are said to "crack" a password or security system. The terms are debatable. The exact meaning of the term *hacker* depends on the content in which it is used and by whom. A self-proclaimed hacker does not think of himself or herself as a cracker. For the sake of clarity, in this textbook, anyone who gains access to a computer system they are not authorized to access is called an intruder.

cracker
anyone who gains access to a computer system without authorization and with the intent to do harm or play pranks.

Common Network Security Breaches

The topic of network security is very complex and would easily fill thousands of pages. In this next section, the most commonly encountered security breaches are covered to provide you with a basic understanding of network vulnerability. As you will see, people, not equipment, generate most of the security breaches.

Unprotected Network Shares

Many times employees will set up a network share to allow other employees to access or copy their files. They do not realize that setting up a network share with minimal to no security sets up a possible entry point for an attack from the outside. Many hacker tools can probe and access available shares on a network. This is a common way networks are compromised. Employees also create network shares with minimal to no security so that they might be able to access their own files from home. Network shares with minimal to no security and remote access enabled combine for a security breach waiting to happen. See **Figure 16-1**.

Social Engineering

Social engineering is a term used to describe the manipulation of personnel by the use of deceit to gain security information. One of the weakest security areas is the gullibility of typical users and their respect for assumed authority. For example, consider the following scenario. A telephone rings on a user desk and the following conversation takes place:

Caller: Hello. This is Bob down at IMS operations conducting a security check. We believe we may have an intruder in our system.

social engineering
the manipulation of personnel by the use of deceit to gain security information.

Figure 16-1.
In this example, an
employee has given the
group *Everyone* full
access to their C drive.

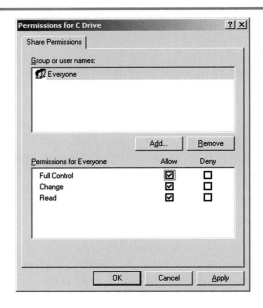

Gene Gullible: Yes. What can I do to help?

Caller: Well Gene, I need to look at your PC files to see if there have been any possible intrusions. I need your user name and password to inspect your system. It will only take a minute and will save me a lot of time rather than coming down there.

Gene Gullible: Sure. My user name is ggullible and the password is toocool.

You see, that is all there is to it. Gene's system has been compromised by the social engineering technique.

Open Ports

Ports, such as HTTP port 80, are used to connect to computers during communication sessions. Legitimate users normally use them for browsing a Web page and performing file transfers, downloads, and network meetings. Open ports can also be a way for intruders to gain access to the network system. Open ports are one of the most common security problems for any site. All unused ports should be closed, and all ports should be monitored for activity. A third-party utility or the **netstat** utility can be used to check for open ports. The **netstat** utility is covered later in this chapter.

Zero Configuration (Zeroconf)
a standard developed by IEEE which recommends how to design a device that automatically detects other devices on the same network or on a nearby network segment without the need of intervention by an administrator or a DHCP or DNS server.

Zero Configuration (Zeroconf)

Zero Configuration (Zeroconf) is a standard developed by IEEE which recommends how to design a device that automatically detects other devices on the same network or on a nearby network segment without the need of intervention by an administrator or a DHCP or DNS server. Identification is based on MAC addresses, which are unique for each device. Network devices are any devices that can be identified by a MAC address and can include communication devices and electronic appliances.

Zeroconf devices must have a unique IP address, subnet mask, and gateway address issued automatically. The Zeroconf protocol must also be able to resolve duplicate IP addresses automatically and translate between IP addresses and host names.

The main reason for the development of Zeroconf protocol is to enable an easy method of networking devices without the use of an administrator. Microsoft operating systems as well as MAC OS 9 already incorporate the Zeroconf standard. For example, the Microsoft wireless device protocol automatically configures a wireless network card when it is physically inserted into the computer.

The downside of Zeroconf is it makes a network less secure. A person could simply purchase a similar device and connect it to a network system and then gain access. An example is a wireless device, such as a WAP, that uses a default SSID and automatic IP addressing. If the device is installed with the default settings and it uses Zeroconf to configure the clients, anyone with a similar device installed can access the same network.

Denial of Service (DoS)

Denial of Service (DoS) is one of the most common attacks on servers. Denial of Service is, as the name implies, denying access to a server by overloading the server with bogus requests. The Denial of Service attack overloads the server to the point that the server will crash or will not be able to complete legitimate user requests.

Man in the Middle (MITM)

Man in the middle (MITM) is a method of intercepting a network transmission, reading it, and then placing it back on route to its intended destination, **Figure 16-2.** It is a serious security breach. The contents may or may not be modified. This method can also be used for a replay attack. A *replay attack* occurs when the data in the network transmission is copied and stored. Later, information such as an IP address and MAC can be used to establish an unauthorized connection to the destination by impersonating the original source. Replay attacks are generally avoided by using a time stamp as part of the communication structure. The time stamp is used to verify the time of each transmission. If the time that is recorded in the time stamp exceeds a specific value, the entire packet is dropped because it is assumed to be bogus.

Denial of Service (DoS)
denying access to a server by overloading the server with bogus requests.

man in the middle (MITM)
a method of intercepting a network transmission, reading it, and then placing it back on route to its intended destination.

replay attack
a method of using information, such as an IP and MAC address, that had previously been copied and stored during a network transmission to establish an unauthorized connection with the destination.

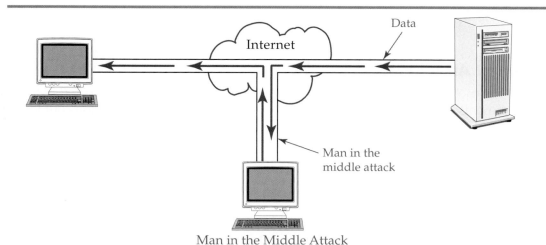

Man in the Middle Attack

Figure 16-2.
A man in the middle attack occurs when a third party intercepts a data transmission with the intent to use vital information it contains for a later attack. Once the third party reads and stores the data transmission, it is placed back on route to its destination.

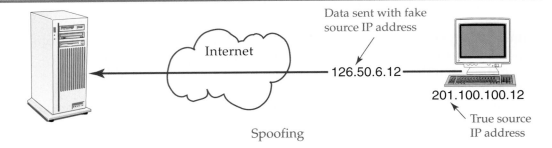

Figure 16-3.
Spoofing is the act of using a fake IP address to gain access to a network.

Spoofing

spoofing
a method of fooling the destination by using an IP address other than the true IP address of a source to create a fake identity.

Spoofing is fooling the destination by using an IP address other than the true IP address of a source to create a fake identity, **Figure 16-3.** For example, to gain access to a network, an unauthorized person may try to use what they think is a valid IP address to fool a server. Another example is to use a bogus IP address and ID when sending an unsolicited e-mail to someone. The e-mail can be used to solicit information or release a worm.

It is a good policy never to respond to an unsolicited e-mail. A common ploy is to send unsolicited e-mails to millions of possible legitimate e-mail addresses using a software program that generates possible e-mail addresses. In the e-mail, a line is included that states: "If this e-mail was sent by mistake and you wish to be removed from our e-mail list, please click the link below." The link is intended to generate a list of everyone who replies. The list is valuable because it contains a list of real e-mail addresses that can be sold or distributed to interested parties, such as advertisers. By responding to the removal link, you verify the e-mail address used was a genuine e-mail address.

Trojan Horse

Trojan horse
a program designed to gain access to a computer while pretending to be something else.

backdoor
a software access port to a computer a Trojan horse has infected.

The *Trojan horse* is a program designed to gain access to a computer while pretending to be something else. For example, a user downloads what they think is a free game. A real game may indeed be inside the download, and for a good Trojan horse design, it should be. However, in addition to the game program code is code of a malicious nature. The malicious code could contain a virus, worm, or what is known as a backdoor. A *backdoor* is a software access port to the computer the Trojan horse has infected. Another common purpose for a Trojan horse is to download a program that imitates the logon screen of a legitimate operating system. When the user logs on using the Trojan horse fake logon screen, the user's name and password is sent to an unauthorized user for later use in gaining access to the computer.

E-mail Attachments

The most commonly encountered viruses come as e-mail attachments. E-mail attachments can be written from a variety of software programs. A person could use Visual Basic, C, C++, Visual C++, Java, and other programming tools, such as a common word processing macro tool. Once the program is written, it is attached to an e-mail. When the recipient opens the e-mail attachment, the program is activated.

Macro Virus

One of the most common and most easily written virus programs is a macro virus. Many legitimate software programs, such as word processing packages, contain a special tool called a macro writer. The macro writer records and saves certain keystroke sequences as a convenience for the user. For example, a user can make a macro that inserts their return address into a document when they press a specific key sequence. The macro program is handy for reducing the amount of work involved for repetitive tasks. The same principle can be used to create viruses. A series of commonly used keystrokes can be linked to a virus. The macro is sent as an e-mail attachment and is launched when the e-mail recipient opens the attachment. The macro virus may infect a template file like normal.dot and then execute when the user presses a certain combination of keys.

Worm

A *worm* is a software program that can spread easily and rapidly to many different computers, **Figure 16-4.** A typical virus cannot spread automatically. There are many famous worm programs, and more are released each day. The most common worm programs use e-mail to replicate and spread. The most sophisticated worm programs use the list of e-mail contacts on the target computer to spread automatically.

The only way to keep ahead of a new worm is to check one of the many centers dedicated to worm and virus detection. Check with sans.org, cern.org, and commercial antivirus Web sites. A common practice of securing a network system against worms is to install a dedicated e-mail server to limit the infection. The e-mail server can be interrupted without disrupting service from the database server, Web server, or any other server. Placing the e-mail service on a server that performs other services could cause the other services to fail.

worm
a software program that can spread easily and rapidly to many different computers.

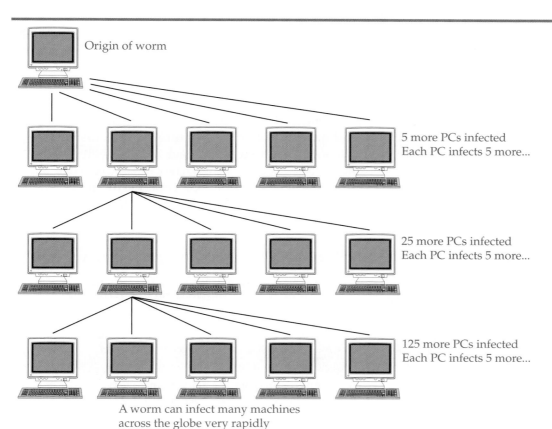

Origin of worm

5 more PCs infected
Each PC infects 5 more...

25 more PCs infected
Each PC infects 5 more...

125 more PCs infected
Each PC infects 5 more...

A worm can infect many machines
across the globe very rapidly

Figure 16-4.
E-mail is the most common method of transmitting a worm. Using this method, the worm multiplies at an exponential rate.

Phishing

phishing
a method of Internet fraud, which involves using e-mail to steal a person's identity as well as other sensitive information, such as financial.

Phishing (pronounced *fishing*) is the latest method of Internet fraud that involves using e-mail to steal a person's identity and other sensitive information, such as financial. In a phishing attack, e-mail is sent that appears to be from a legitimate enterprise in an attempt to solicit personal information. For example, a person may receive an e-mail that appears to be from the legitimate company eBay. The e-mail may even use the eBay artwork and logo. The contents of the e-mail requests the person's personal information, such as a bank account PIN number, credit card number, password, social security number, and anything else that can be used for identity theft.

Another example is an e-mail that poses as a known and often accessed company, such as a credit card company. The e-mail states that a virus has compromised all user accounts and that the company must reconstruct all user account information. The e-mail contains a Web site link. The e-mail recipient is directed to go to the Web site and enter the requested information. The e-mail recipient is directed to an illegal site created to look like the Web site of the legitimate company. However, the domain name of the illegal site is slightly different than the legitimate domain name of the company. The bogus domain name goes undetected by the e-mail recipient because of slight changes in the characters. Look at **Figure 16-5.** Notice how similarly shaped characters and numbers have been substituted.

You can see how easy it is to create a bogus Web site that could fool a trusting user. Also be aware that rolling the mouse over a Web site link embedded in the text of an e-mail created with HTTP can automatically take you to a site that does not resemble the Web site indicated in the link. The Web site link can be easily programmed to respond to a different Web site altogether. Never send personnel information in response to e-mail, even if the e-mail looks legitimate. Call the company first to confirm the legitimacy of the e-mail.

Administrator Laziness

Administrator laziness, rather than a lack of knowledge, causes the most common security breaches. Common security breaches are a result of quick fixes, lack of user training, and naivety about being a target. Administrators and users often think, "Why would anyone want to hack into this site?" The truth is, many intruders simply want to access any network. The network's contents are not the important issue. The challenge of a successful, unauthorized access is sufficient reward for many intruders. Never assume that any network or PC is safe from an intruder.

Figure 16-5.
A domain name can be disguised by making slight changes in the characters. Notice how a lowercase *l* looks similar to the number *1* and an uppercase *O* looks similar to the number *0*.

Legitimate Site	Bogus Site	Look at the following in the bogus Web site:
www.paypal.com	www.paypa1.com	The number *1* used in place of the letter *l*.
www.firstfederal.com	www.firstfedera1.com	The letter *l* again.
www.payonline.com	www.pay0nline.com	The number *0* for the letter *O*.

Security Methods and Protocols

The two basic building blocks of secure network communications are authentication and encryption. These security methods are supported by security protocols. The concepts presented in this section are critical for a basic understanding of the sophisticated world of network security.

Encryption

Encryption is the method of using an algorithm to encode data. The algorithm converts the data into a form that cannot be easily decoded. The science of encoding data is called *cryptology.* Encrypted data is referred to as *ciphertext.* The number of bits used in the encryption code is often used to express the level of security provided. The maximum number of code variations depends on the maximum number of bits used to encode the message. For example, a 48-bit code can produce fewer combinations compared to a 128-bit encoding scheme.

This is how the system of encryption works. A sender writes a message and then encrypts the message using an algorithm. An algorithm used to encode data is called a *key.* The key is unique for each transmitted message. It is randomly generated and then applied to the message. The key can scramble the original message and add additional misleading characters to the text. The message is transmitted and then received by the destination. At the destination, the message is decrypted using a key. There are two main types of key encryption methods: symmetric-key encryption and asymmetric-key encryption.

Symmetric-key encryption, or *secret-key cryptography,* uses a key that only the sender and the receiver know. Both parties use the same key to perform encryption and decryption, **Figure 16-6.** The term *symmetric* is used because the same key is used for both coding and decoding the message.

Symmetric keys are typically used when large amounts of data are to be encrypted. The symmetric key encodes and decodes faster than more complex methods and can be just as secure as more complex methods of encoding data.

Asymmetric-key encryption, or *public-key cryptography,* uses two keys: a private key and a public key, **Figure 16-7.** Typically, the originator of the encryption system owns the private key. For example, a teacher who wishes to communicate with his or her students across the Internet but ensure privacy would

encryption
the method of using an algorithm to encode data.

cryptology
the science of encoding data.

ciphertext
encrypted data.

key
an algorithm used to encode data.

symmetric-key encryption
a key encryption method in which the sender and receiver use the same key.

asymmetric-key encryption
a key encryption method in which the sender and the receiver use two different keys: a private key and a public key.

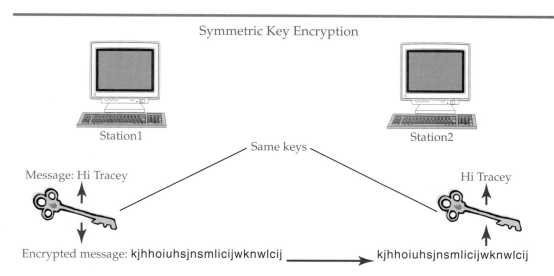

Symmetric Key Encryption

Station1

Station2

Same keys

Message: Hi Tracey

Hi Tracey

Encrypted message: kjhhoiuhsjnsmlicijwknwlcij ⟶ kjhhoiuhsjnsmlicijwknwlcij

Figure 16-6.
The symmetric key encryption system uses the same key to encrypt and decrypt data.

Figure 16-7.
The asymmetric key encryption system uses two different keys—one key for encrypting the data and the other for de-crypting the data.

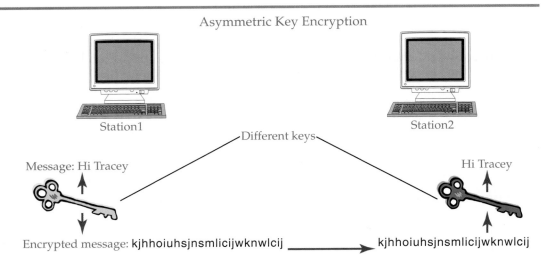

Asymmetric Key Encryption

Certificate Authority (CA)
a service that contains a security list of users authorized to access the private key owner's messages, using a public key.

digital certificate
a file that commonly contains data such as a user's name and e-mail address, the public key value assigned to the user, the validity period of the public key, and issuing authority identifier information.

retain a private key. A public key is issued to all students. A message is encoded using the private key and can only be decoded using the public key. The owner of the public key can send a message in return that can only be decoded by the private key.

In this method, a message created with a private key can only be decoded with a public key, and a message encrypted with a public key can only be decoded with a private key. A message created with a public key cannot be decoded with another public key. This means that all the encryption and decryption is centered on the owner of the private key. It takes both the public and private keys to code and decode an encrypted message.

Typically, a service referred to as the *Certificate Authority (CA)* contains the security list of users authorized to access the private key owner's messages, using a public key. See **Figure 16-8.** The private key owner consults the CA for this information before sending a message. The CA sends the requested information in the form of a digital certificate. A *digital certificate* is a file that commonly contains data such as the user's name and e-mail address, the public key value assigned to the user, the validity period of the public key, and issuing authority identifier information.

Figure 16-8.
The CA issues a private key to the private key owner and then issues public keys to public key users approved by the private key owner.

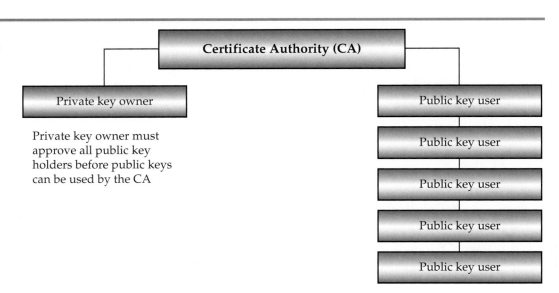

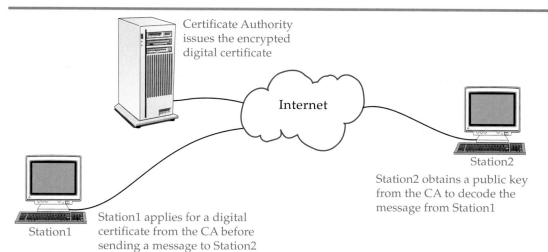

Figure 16-9.
Station1 applies for and receives a digital certificate from the CA. *Station2* uses the public key provided by the CA to decode the message and to verify the sender is really *Station1*. After verifying and receiving the message from *Station1*, *Station2* can send an encrypted reply to *Station1*.

This is how the encryption/decryption process works, **Figure 16-9.** *Station1* wants to send an encrypted message to *Station2*. *Station1* applies for a digital certificate from a CA. The CA issues a digital certificate, which contains the user's public key/private key and security identification information known only to *Station1* and the CA. *Station1* uses the private key to encrypt the message. *Station1* can now send a message to *Station2*. When *Station2* receives the message, it uses the public key to decode the encrypted message. If *Station2* sends an encrypted reply to *Station1* using the public key, *Station1* will decode the reply using the private key.

VeriSign is one of the largest commercial digital certificate companies in the world. A certificate from VeriSign can be obtained by applying for one at their Web site. **Figure 16-10** shows a VeriSign security certificate. You can see the list of protection services offered by this particular certificate. In **Figure 16-11,** you can see some of the details about this certificate. Looking closely, you can see the validity period of the certificate, the particular algorithm used, and the length of the public key, which is 1024 bits.

Figure 16-10.
A digital certificate issued from VeriSign.

Figure 16-11.
Details of a digital certificate from VeriSign.

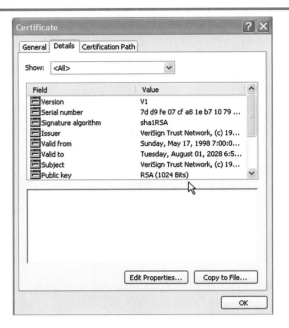

Secure Sockets Layer (SSL)

Secure Sockets Layer (SSL)
an Internet security protocol developed by Netscape to ensure secure transmissions between Web servers and individuals using the Internet for such purposes as credit card transactions.

Secure Sockets Layer (SSL) developed by Netscape is an Internet security protocol. Both Netscape Navigator and Microsoft Internet Explorer use SSL. SSL was primarily developed to ensure secure transmissions between Web servers and individuals using the Internet for such purposes as credit card transactions. SSL is based on the principle of public key encryption. To identify a Web server that requires SSL, the Web server's default Web page starts with https: instead of http:. The *s* indicates *secure*. When the Web site is accessed, a padlock icon appears in the browser's status bar, indicating a secure connection has been made available.

Both Netscape Navigator and Microsoft Internet Explorer are based on the X.509 security protocol. X.509 is the most widely used standard for digital certificates. However, the X.509 security protocol is an ITU recommendation, not an adopted standard. Because the X.509 protocol is not an adopted standard, the security protocols in Netscape Navigator and Microsoft Internet Explorer do not exactly match. You may need to have both browsers available on your system to ensure communication.

Secure HTTP

Secure HTTP (S-HTTP)
an Internet security protocol that secures individual messages between the client and server rather than the connection.

Another protocol used to securely send messages across the Internet is *Secure HTTP (S-HTTP)*. S-HTTP and SSL are used for similar purposes. S-HTTP was developed by Enterprise Integration Technologies (EIT), which was taken over by Verifone, Inc. Not all Web browsers support S-HTTP. S-HTTP uses symmetric keys not public keys like SSL. SSL is the most widely used method of secure Web connectivity. The main difference between S-HTTP and SSL is SSL creates a secure connection between the client and server. S-HTTP secures individual messages between the client and server rather than the connection.

Tech Tip   Both S-HTTP and HTTPS are often referred to as Secure Hypertext Transfer Protocol. This is incorrect. S-HTTP and HTTPS are two different protocols.

IP Security (IPSec)

IP Security (IPSec) is a protocol that secures IP packets across the Internet. It works with IPv4 and IPv6. IPSec supports encryption in two modes: transport and tunnel mode. *Transport mode* encrypts only the payload (data) portion of the packet, while *tunnel mode* encrypts the payload and the header information. IPSec works hand in hand with L2TP when used in tunnel mode. The L2TP protocol establishes a tunnel between the destination and the source. IPSec secures the transmitted data.

Look at **Figure 16-12.** Notice the difference between an IP packet that is sent in transport mode and one that is sent in tunnel mode. In transport mode, only the payload is encrypted. The IP header is left in its original form. In tunnel mode, the IP header and payload is encrypted. A new IP header is added to the packet. The new IP header enables the packet to be delivered to the destination's gateway or router.

The IPSec protocol can be used with one of two protocols: Authentication Header (AH) or Encapsulation Security Payload (ESP). *Authentication Header (AH)* provides authentication and anti-replay but does not encrypt the data payload inside the packet. *Encapsulation Security Payload (ESP)* provides authentication and anti-replay and encrypts the payload data inside the packet. ESP provides greater security but at the cost of more protocol overhead because it encrypts the data.

Sometimes, it is only necessary to secure the connection, not the data. At other times, data needs to be secured. For example, when transmitting credit card information, it is imperative the data inside the packet is protected. If you are simply exchanging e-mail without sensitive contents, only the connection between the destination and source needs to be secured. Securing the connection protects identities so that they cannot be intercepted and used by unauthorized personnel.

SSH

Secure Shell (SSH) is a protocol that provides secure network services over an insecure network medium such as the Internet. SSH was originally designed for UNIX systems to replace Remote Login (**rlogin**), Remote Shell (**rsh**), and Remote Copy (**rcp**). These utilities are known as r commands by UNIX and Linux users. The r commands do not directly support encryption, which means files and commands are sent in plain text and can be intercepted and read by a protocol sniffer or analyzer. SSH is associated with TCP/IP port 22. It requires the

---

IP Security (IPSec)
a protocol that secures IP packets across the Internet. It works with IPv4 and IPv6 and supports two modes of encryption: transport and tunnel mode.

transport mode
an IPSec mode that encrypts only the payload portion of a packet.

tunnel mode
an IPSec mode that encrypts the payload and the header information of a packet.

Authentication Header (AH)
an IPSec protocol that provides authentication and anti-replay but does not encrypt the data payload inside the packet.

Encapsulation Security Payload (ESP)
an IPSec protocol that provides authentication and anti-replay and encrypts the payload data inside the packet.

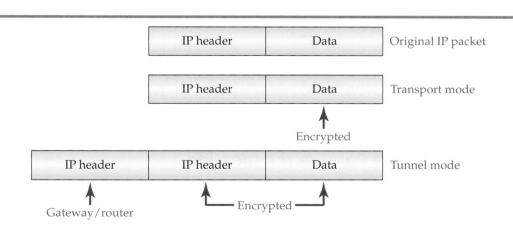

Figure 16-12.
Transport mode encrypts only the packet data. Tunnel mode encrypts both the IP header information and the data. Tunnel mode requires a gateway or router IP header address, which accounts for the additional IP header at the front of the packet.

Secure Shell (SSH)
a protocol that pro-
vides secure network
services over an
insecure network
medium such as the
Internet.

use of a private and a public key as well as a password. The r commands do not require a password, but do require to be issued by the root user. While the r commands require root privileges, which provide a level of system security, many security experts do not consider this a sophisticated means of authentication.

Today, the open source SSH protocol can be used on any operating system that supports the TCP/IP suite. This means SSH can be used on Microsoft and MAC OS. Microsoft does not directly support SSH because it uses its own security mechanisms. However, a third-party software program that supports SSH can be installed.

SCP

**Secure Copy
Protocol (SCP)**
a protocol that pro-
vides a secure way of
transferring files
between computers.
It is the replacement
for **rcp**.

Secure Copy Protocol (SCP) provides a secure way of transferring files between computers. It is the replacement for **rcp**. The **rcp** command does not require a password similar to anonyms FTP.

SSH, SCP, and SFTP are all issued as commands in the UNIX / Linux environment. Linux commands are case sensitive, so they must be issued in lowercase form, such as **ssh**, **scp**, **sftp**. There are third-party GUI programs that incorporate the open source SSH, SCP, and SFTP protocols so that the user does not need to issue text commands at the command prompt.

Tech Tip

The letters SFTP can represent two different acronyms in networking. Secure File Transfer Protocol and Simple File Transfer Protocol. Secure File Transfer Protocol usually uses the following acronym S/FTP but not always. S/FTP is actually a software program rather than a true protocol. When asked a question about SFTP, look at the context of the question for clues.

Wireless Security

Wireless devices are inherently insecure because the wireless medium, the radio wave, is an unbound medium. Wired networks can be installed inside walls, ceilings, floors, and metal conduits, which provide a degree of physical security. Someone attempting to tap into a wired network would need to at least access the inside of the building. The wireless network, because of the nature of the radio wave, can extend beyond the physical limits of a building. This means that anyone can automatically have physical access to the network without entering the building if certain security measures are not set in place. For example, a small wireless network may have no security enabled or have only limited security provided by the Wireless Access Point (WAP).

The original solution to wireless network security involves three mechanisms: Service Set Identifier (SSID), Media Access Control (MAC) filtering, and Wired Equivalent Privacy (WEP). Wireless network security has evolved at a remarkable rate. In less than seven years, four major security implementations have been developed: WPA, WPA2, 802.1x, and 802.11i. This section covers the original solution to wireless network security, SSID, MAC filtering, and WEP, plus the most recent security implementations, WPA, WPA2, 802.1x, and 802.11i.

Wireless Access Point Authentication

Wireless device authentication can be achieved in more than one way. A WAP is the first authentication mechanism. However, not all Wireless Access

Points implement authentication based on user name and password. Typically, a WAP is set by default to allow anyone with the same brand of wireless network device to connect automatically to the WAP, and thus to the wired network.

If a WAP has more sophisticated options, such as a user name and password feature, it is referred to as a wireless gateway or router.

Tech Tip

Wireless networks use a Service Set Identifier (SSID) to identify the wireless network. The SSID is similar to a workgroup name. Multiple wireless networks can coexist within range of each other and operate independently by using different SSIDs. All wireless devices have a default SSID. To increase security, the SSID should be changed when the WAP is installed.

Media Access Control (MAC) Filter

A *Media Access Control (MAC) filter* is a feature that allows or restricts WAP access based on the MAC address of a wireless network card. To set up a MAC filter, an administrator creates an Access Control List (ACL). The ACL contains a list of MAC addresses belonging to authorized wireless network devices. The ACL is stored in the WAP. When a wireless network device attempts to access the network through the WAP, the WAP checks the ACL to see if the wireless network device is authorized to access the network.

Media Access Control (MAC) filter a feature that allows or restricts WAP access based on the MAC address of a wireless network card.

Wired Equivalency Privacy (WEP)

Wired Equivalent Privacy (WEP) was the first attempt to secure by encryption the data transferred across a wireless network. It was part of the original IEEE 802.11 wireless standard. Not long after WEP was implemented, it was discovered that there were flaws in the encryption method. The WEP algorithm was not as complex as it was first thought to be. A determined hacker could crack the encryption in several hours. In fact, several tools are available on the Internet that can be used to crack WEP encryption keys.

While WEP might be adequate for a low-risk network, such as a home network that does not participate in financial transactions, it is inadequate for high-risk networks, such as a business where financial transactions are commonplace. Creating a VPN, however, can compensate for WEP vulnerability. A VPN can incorporate an authentication and an encryption method, adding to the security set in place by WEP.

Wi-Fi Protected Access (WPA)

Wi-Fi Protected Access (WPA) was developed by the Wi-Fi organization and is not an IEEE standard. When vulnerabilities were discovered in the algorithm used for WEP, a more restrictive encryption was needed to protect data transferred across a wireless network. The Wi-Fi organization sponsored the development of Wi-Fi Protect Access (WPA) as a solution to the vulnerabilities discovered in WEP. WPA uses a more complex encryption technique to protect data. It has become the replacement for WEP. WPA is designed to be compatible with 802.11 standard devices. WPA encryption is selected from the **Associations** page of the **WLAN Properties** dialog box, **Figure 16-13.**

Figure 16-13.
WPA encryption is set in the **Network Authentication** box on the **Associations** page of the **WLAN Properties** dialog box.

Wi-Fi Protected Access 2 (WPA2)

Wi-Fi Protected Access 2 (WPA2) was developed by the Wi-Fi organization as an enhanced version of its original WPA. It is designed to be compatible with the IEEE 802.11i standard.

802.11i

The IEEE ratified the IEEE 802.11i standard in June of 2004 to remedy the original security flaws in 802.11. The 802.11i standard specifies the use of a 128-bit Advanced Encryption Standard (AES) for data encryption. It also incorporates a mechanism for generating a fresh set of keys for each new connection session. This means enhanced security because the keys are constantly changed rather than reused. 802.11i is downward compatible with existing 802.11 devices. However, this does not mean that the security standards of the 802.11 devices are improved. It simply means that an 802.11i device will use WEP for security when communicating with an 802.11 device.

802.1x Authentication

802.1x provides port-based, network access control. Port-based, network access control supports authentication for Ethernet network access. The term *port-based* refers to any location point represented as a point of access. Do not confuse the term *port-based* with the term *port*, which is used in conjunction with an IP address to identify a service, such as port 80 for HTTP. 802.1x is primarily used for client/server-based networks. It allows the network server to authenticate a wireless network device when the wireless network device attempts to connect to the wired network through a WAP. Older wireless network hardware does not support AES and therefore cannot fully support the 802.1x enhancements.

802.1x requires three components: supplicant, authenticator, and authentication server, **Figure 16-14.** The *supplicant* is the wireless network device that is requesting network access. The WAP functions as the *authenticator* and does not allow any type of access to the network without proper authentication. A server running Remote Authentication Dial-In User Service (RADIUS) acts as the

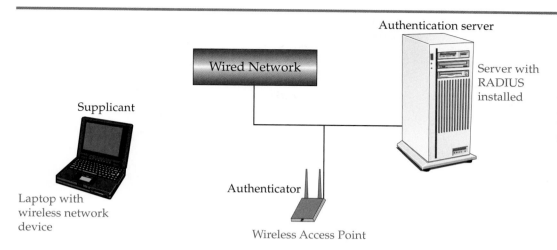

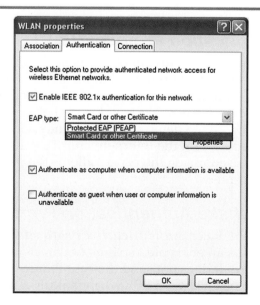

Authentication server

Wired Network

Server with
RADIUS
installed

Supplicant

Laptop with
wireless network
device

Authenticator

Wireless Access Point

Figure 16-14.
The 802.1x standard
provides centralized
authentication for wire-
less LANs. It incorpo-
rates three components:
supplicant, authentica-
tor, and authentication
server.

*authentication server. **Remote Authentication Dial-In User Service (RADIUS)** is*
a service that allows remote access servers to authenticate to a central server.

802.1x provides a much easier way to manage numerous Wireless Access
Points. By using a server to provide centralized authentication, there is no need to
maintain an ACL at each WAP. For example, a university campus might incorporate hundreds of Wireless Access Points, which permit students and faculty to
access the system from anywhere on campus. Every semester, new students and
faculty need to be added to the ACL. 802.1x allows for a much easier security model
by allowing an administrator to manage security from a centralized location.

To set up IEEE 802.1x on a wireless network client, IEEE 802.1x must be
enabled. Look at **Figure 16-15.** Notice the option to enable IEEE 802.1x authentication is located in the **WLAN Properties** dialog box. When IEEE 802.1x is enabled,
the Extensible Authentication Protocol (EAP) type must also be selected. EAP is
an extension of the Point-to-Point Protocol (PPP) and is used for authentication
for LAN segments. EAP allows for different security mechanisms, such as smart
cards and certificates, to be used as part of the authentication process using PPP.

*Remote Authentica-
tion Dial-In User
Service (RADIUS)*
a service that allows
remote access servers
to authenticate to a
central server.

Figure 16-15.
IEEE 802.1x authenti-
cation is enabled in the
WLAN Properties dialog
box.

Figure 16-16.
IEEE 802.1x cannot be used for authentication if encryption is disabled for the wireless client. It also cannot be used on a peer-to-peer network because a peer-to-peer network does not have a server to perform authentication.

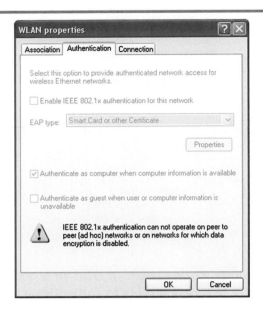

If encryption is disabled for the wireless client or the wireless client is part of a peer-to-peer network, the message as shown in **Figure 16-16** will display. The message says that if encryption is disabled for the wireless client, IEEE 802.1x cannot be used for authentication—nor can 802.1x be used on a peer-to-peer network. Remember, a peer-to-peer network does not have a server to verify authentication, which is one of the requirements of 802.1x.

Authentication Protocols

Chapter 15—Remote Access and Long Distance Communications introduced SLIP, PPP, and many variations of PPP such as PPTP, PPPoE, and MLPPP. While PPP and SLIP are primarily concerned with remote connection security over dial-up telephone lines and ISDN, there are many protocols designed to provide secure connections over Ethernet networks and the Internet, such as PAP, CHAP, and Kerberos. This section takes a look at these protocols.

Password Authentication Protocol (PAP) an authentication protocol that sends a user name and password in plain text format.

Password Authentication Protocol (PAP)

Password Authentication Protocol (PAP) is the basic password authentication technique used for HTTP and remote dial-up access. PAP sends the user name and password in plain text format, also referred to as clear text. The user name and password are sent over the network and then compared to a database of user names and passwords to determine if they may access the server. PAP was developed when security on the Internet using TCP/IP was not a real problem. The clear text used inside the packet allows the password and user name to be easily intercepted.

Challenge Handshake Authentication Protocol (CHAP) an authentication protocol that sends an encrypted string of characters that represent the user name and password. It does not send the actual user name and password.

Challenge Handshake Authentication Protocol (CHAP)

The *Challenge Handshake Authentication Protocol (CHAP)* is an authentication protocol that sends an encrypted string of characters that represent the user name and password. It does not send the actual user name and password. CHAP was designed to be used with PPP when making a remote connection to a server. **Microsoft Challenge Handshake Authentication Protocol (MS-CHAP)** is an enhanced version of CHAP that encrypts not only the user name and password but also the data package. MS-CHAP must be used with Microsoft operating systems. It is not compatible with other operating systems.

This is how CHAP works. Using the PPP protocol, a computer connects to a remote system. After a connection is established, the server, also known as the authentication agent, sends a challenge to the client, or peer. The authentication agent sends a key to the client so that it can encrypt its user name and password. The client responds with an encrypted key representing the password and user name. The server either accepts or rejects the client user name and password based on a matching encryption key. The actual user name and password are not sent. Only a key generated from the characters used in the user name and password is sent.

The authenticator randomly generates challenges to verify it is still connected to an authorized peer and is not an impostor that has intercepted packets. CHAP prevents the replay attack by repeating the challenge at random intervals to detect an unauthorized connection. The technical names *authentication agent* and *peer* are used because CHAP can be used for more than server access, such as to authenticate two routers when using a tunneling protocol or VPN connection.

Kerberos

The Massachusetts Institute of Technology (MIT) developed a security authentication system called Kerberos. The name Kerberos (also spelled *Cerberus*) comes from Greek mythology. It is the name of the three-headed dog that guarded the entrance to Hades. **Kerberos** allows two computers to communicate securely over a network that is not typically secure, such as the Internet. While Kerberos was developed and distributed as an open protocol, it has been incorporated into many proprietary software systems.

Kerberos provides both authentication and encryption services. It uses a two-way method of authentication. In this method, the client is authenticated to the server and the server is authenticated to the client. When encrypting and decrypting data, Kerberos uses a key to ensure privacy. Two different keys are used similar to the public and private key concept—one key to encode the data (encryption key) and the other key to decode the data (decryption key). Possession of only one key is of no use to the intruder because the same key cannot be used to encode and decode the data.

As you can see in **Figure 16-17**, a key encodes the text message. The message is transmitted from the source to the destination. At the destination, another key is applied to the ciphered text to decode it into plain text. Remember, two different keys are used in the Kerberos security system. If the data is intercepted along the way, it is unreadable because it cannot be deciphered without the destination key.

Security Implementations

This section covers the various ways an administrator can implement network security. Several recommended security practices for new network installations are installing the latest software updates and patches, setting up an account for daily administrative tasks, and changing the default administrator's name. A network administrator needs to constantly educate system users in

Microsoft Challenge Handshake Authentication Protocol (MS-CHAP) an enhanced version of CHAP that encrypts the user name, password, and data. MS-CHAP must be used with Microsoft operating systems.

Kerberos a security authentication system that provides both authentication and encryption services. It uses a two-way method of authentication.

Figure 16-17.
Kerberos uses two dif-
ferent key codes. The
first key code is used to
encrypt the original
data and the second
key code is used to
decrypt the data.

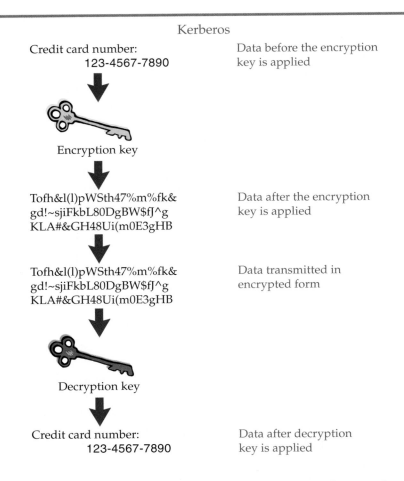

Figure 16-17.
Kerberos uses two different key codes. The first key code is used to encrypt the original data and the second key code is used to decrypt the data.

routine security practices. Yearly seminars, e-mail alerts, and reminders at department meetings are all good efforts.

An administrator should add software or hardware devices that block open ports or filter incoming and outgoing traffic to secure the network. Physically securing the server and other vulnerable points of the network, such as wiring closets, hubs, and router, should also be seriously considered.

Software Installation Patches

Security begins immediately after the initial installation of the network operating system and associated software programs. There are hundreds of known vulnerabilities to operating systems and software packages. For example, when a network operating system such as Windows 2000 is installed and configured, it has many known vulnerabilities. Before the installation can be considered complete, the latest patch must be installed. Software patches contain many of software fixes that close security holes and fix software bugs. Microsoft releases collections of patches and fixes referred to as a *service pack.*

service pack
a collection of
patches and fixes.

Administrator Account

Many network operating systems are installed with a default administrator account. During installation, you are given the opportunity to enter a password of your choice for this account. The default user name, such as *Administrator,* provides one half of authenticity to a potential intruder. To secure this potential breach, choose a new administrator name to use in place of the default administrator name, assign the account full administrative privileges, and then delete the default account.

For example, Windows 2000 has the default system administrator name of *Administrator*. A new account should be made with a new name such as *Operat0r1$*. It should be created with full administrative privileges. After the new account has been created, the original account, *Administrator*, should be deleted from the system.

The ability to delete or rename the administrator account varies according to operating system and version.

It is also recommended an account be set up that the administrator can use to perform daily duties and that does not require a complete set of administrative powers. The idea behind using a limited administrator account is to protect the system in the event an intruder compromises the account. If an intruder were to gain access of the system using a Trojan horse, the intruder could use a password-stealing program to acquire the administrator's password when they log on to maintain the system. The intruder could use the administrator account to set up their own account that has administrative powers.

If an administrator uses an account with less than full administrative powers, they limit the powers of an intruder if the system is cracked. A user account cannot create another user account that is more powerful than itself. It can only create an account that has equal or lesser powers. By using a less powerful account for daily business, you leave the most powerful account in reserve for when it is really needed. Remember, if the intruder has equal power to the administrator, they can delete everything the system administrator account has created!

Some network administrators set up a fake administrator account and limit the fake administrator's abilities to control or modify the network resources. This is intended to mislead unauthorized intruders.

User Account Passwords

After installing the network operating system and patches, the next major item of concern is password protection. The network administrator can do much to ensure proper passwords and techniques are used. An administrator can educate system users on choosing a proper password, and they could set up password policies. Educating users includes teaching them about poor and secure passwords. Setting up password policies includes setting defaults for password histories, age, and length.

Poor passwords contain common names, words, or sequential numbers or letters. The following are some examples of poor passwords:

Jamie	TopSecret	AbCdE
Reds	love	A1B2C3
password	12345	1q2w3e4r
secret	abcde	

Poor passwords match words commonly found in the dictionary or contain names familiar to the password's owner. Poor passwords also include keyboard combinations that are easy for the password owner to remember. Look at the password *1q2w3e4r* in the previous list. The *1q2w3e4r* password may look like a secure password. However, by locating the letters and numbers on your keyboard to reveal the pattern, you will see that it is not. Keyboard patterns are not secure, but they are better than most typical poor passwords.

Another common password that should be avoided is a social security number. A social security number is easily identified by its nine number sequence. When a social security password is cracked, the intruder can gain access to other personal information. Never use your social security number as a password under any circumstance.

Secure passwords do not match words commonly found in a dictionary. Instead, they incorporate numbers and special characters, which makes them much more difficult to crack. The following are some examples of secure passwords:

AceHat$_301   Bob_$ecure4PC   Open_Lock231!

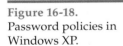

hashing
a technique that relies on an algorithm or an encryption device based on mathematical algorithms for guessing a password.

No password is 100% secure; however, there are passwords that are difficult to guess or hash. *Hashing* is a technique that relies on an algorithm or encryption device based on mathematical algorithms for guessing a password. Passwords are so critical to authentication that network operating systems can control many of the important characteristics required of good passwords.

An administrator can set policies or default settings that can aid in password security. **Figure 16-18** shows the Domain Security Policy dialog box associated with Windows 2000. You can readily see the list of password policies available such as history, age, length, and complexity.

Password history

As passwords are changed, the old passwords can be stored and used for comparison against the most recent password. For example, a user may be

Figure 16-18.
Password policies in Windows XP.

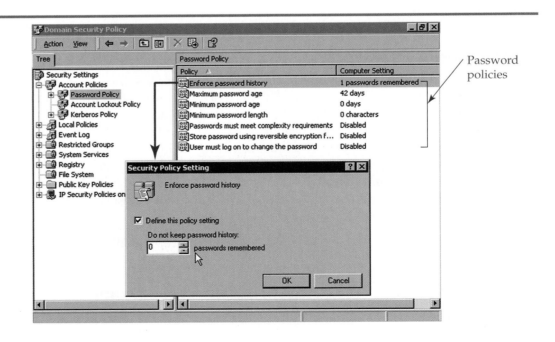

required to change their password every 90 days. This practice is compromised if the user constantly switches between two passwords such as *MySecret* and *Secure*. Not only are these poor passwords, by constantly switching between the two, there is little protection offered by changing the password. Some network operating systems, therefore, allow the administrator to set a minimum password history, which forces the user to use a new password that does not match any of their old passwords. For example the history password policy can be set to store the last 24 passwords. By enforcing the password history, users are forced to continually use new passwords. This can also be easily compromised by ingenious techniques, such as adding a number to the end of a password and then simply incrementing the number by one each time a new password is required—for example, *MySecret1*, *MySecret2*, and *MySecret3*.

Password age

Passwords should be changed frequently, but not so often that it becomes a real annoyance to the users. A good rule of thumb is to require passwords to be changed every 90 days.

Minimum password length

The exact password length depends on the company and network administrator's perception of the need for security. Passwords that are too long are not practical for most applications. A password should be of sufficient length to allow a variety of characters and symbols to be used but not so long that the support desk is constantly reassigning forgotten passwords.

A good rule of thumb is eight characters minimum. Administrators and special operators, such as department heads with administrative privileges to assign passwords to their individual workgroups, should be required to use a password of at least 12 characters.

Other Password Security Measures

For further password security, it is a recommended practice to move the location of the password storage file. Depending on the network operating system, it is also a good practice to relocate other security files from the default location. When security files are left in the default location, intruders can easily locate them. Also, when a database with user name and password information is saved on a computer, the database should be encrypted and placed under an unassuming file name such as Tax Report Summaries rather than Personnel Security Passwords.

Any password can be hacked, but not all intruders can hack all password methods. For example, an intruder can hack a highly secure password, but it can be very difficult and take even a month or more. A non-secure password can be compromised very easily and in a very short time.

You can audit the network security passwords by using special software packages designed for this purpose. Many operating system Web sites provide such tools at no cost.

Firewall

A *firewall* is designed to monitor and pass or block data packets as they enter or exit the network system, **Figure 16-19.** A firewall may consist of either hardware or software or a combination of both. Servers and routers may be used

firewall
a hardware device or software package that passes or blocks data packets as they enter and exit the network system.

Figure 16-19.
A firewall inspects data packets and either allows the packet to pass or blocks the packet.

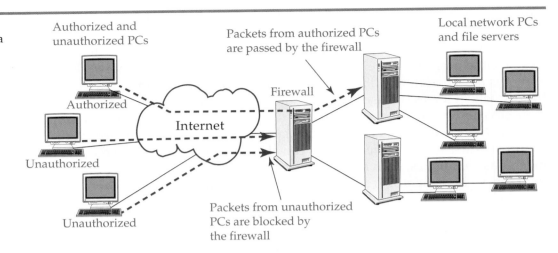

as firewalls. Routers are often used because the basic function of the router is based on IP addresses. Since the IP address is the basis of the Internet addressing scheme, the router can easily be used to screen out certain IP addresses.

There are several classifications of firewalls such as packet filter, application gateway, and circuit level gateway. A typical firewall consists of two or more of classifications. Also, gateways are often classified as firewalls because firewall technologies are incorporated into the gateway. In addition, a firewall may use an encryption technique to secure the data transferred or rely on the encryption designed into the networking operating system or protocol.

Any PC can be turned into a firewall by installing two network interface cards and placing the PC at the threshold of the Internet connection. In this scenario, the PC also functions as a gateway. See **Figure 16-20.** Most operating systems offer a firewall proxy server that can be set up automatically on a server or a PC. Windows XP offers a firewall.

Figure 16-21 shows the Windows XP Firewall **Advance Settings** dialog box. It lists the various services running on the network that Internet users can access. Placing a checkmark in the appropriate box permits the access of a specific service to Internet users. Limiting services limits possible areas of attack by an intruder.

Figure 16-22 shows the various Internet Control Message Protocol (ICMP) properties that can be controlled. ICMP is used to send networking error messages and messages generated by such programs as **ping** and **tracert**. The items listed in the dialog box can be disabled to limit information disclosed to an

Figure 16-20.
Two NICs can be installed in a PC to serve as a gateway or firewall.

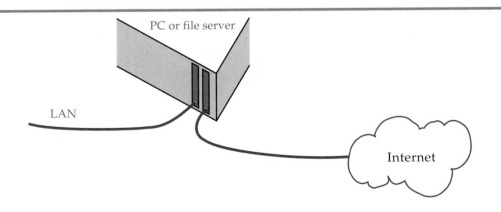

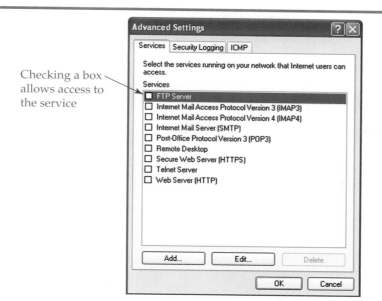

Checking a box allows access to the service

Figure 16-21.
Windows Firewall
Advanced Settings
dialog box. On the
Services page, services
can be selected to allow
access to by Internet
users.

intruder's probe. The same condition applies to a technician troubleshooting the system. If the ability to **ping** the server has been disabled, then the technician will not get the desired reply from the server when testing connections. The firewall will need to be disabled or reconfigured for troubleshooting the network.

When the **Logging Options** on the **Security Logging** page are enabled, events regarding dropped packets and successful log on connections are recorded in a log file, **Figure 16-23**. A log file can be viewed later to reveal information about a network attack from an intruder. As you can see from the examples, a firewall is a powerful security tool and a necessity for networking.

Packet Filter

A *packet filter* inspects each packet entering or leaving the network. Typically, the packet filter is transparent to users on the network. Packets are either accepted or rejected based on a set of rules set up by the network administrator.

packet filter
a hardware device or
software package
that accepts or rejects
packets based on a
set of rules set up by
the network
administrator.

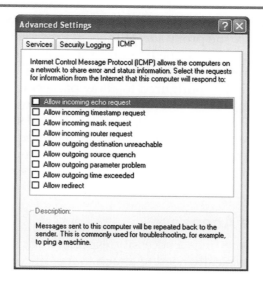

Figure 16-22.
The **ICMP** page of the
Windows Firewall
Advanced Settings
dialog box allows an
administrator to select
which types of request
for information the com-
puter will respond to.

Figure 16-23.
The **Security Logging** page of the Windows Firewall **Advanced Settings** dialog box allows an administrator to select the type of events that will be recorded in a log file. The administrator can also set the location and file name of the log and set a size limit.

Application Gateway

application gateway
a hardware device or software package that provides security for specific applications by accepting traffic based on the exact match of the application permitted.

An *application gateway* provides security for specific applications such as FTP and Telnet servers. The gateway is configured to accept traffic based on the exact match of the application permitted.

Circuit-Level Gateway

circuit-level gateway
a hardware device or software package that monitors a connection until it is successfully established between the destination and source host.

A *circuit-level gateway* monitors a connection until it is successfully established between the destination and source host. After the connection is established, packets can flow freely between the two hosts. Because the packet sequence is encoded, it is usually quite difficult for an intruder to access the stream of data moving between the hosts.

Proxy Server

proxy server
a firewall component that is typically installed on a server and resides between Internet servers and the LAN hosts.

A *proxy server* is a firewall component that is typically installed on a server and resides between Internet servers and the LAN hosts. It appears as a destination host while hiding the address of the true host inside the LAN. To anyone outside the network, only the proxy server is seen. Proxy servers replace the IP address of outgoing packets with the IP address assigned to the proxy server. For incoming packets that are allowed to flow into the network, the reverse is true.

The proxy server can be configured to allow packets to flow into and out of the network if they meet certain conditions. The conditions configured can be items such as specific IP addresses, certain protocols, and server names or URLs.

Proxy servers may also cache information such as frequently visited Web sites. By caching the Web sites and their IP addresses, connections can be made faster than when searching for the Web site. Proxy servers are sometimes referred to as gateways.

Securing Remote Access

Remote access is sometimes a necessity, especially for sales personnel who travel. Sales personnel typically need to attach to the office file server to check their e-mail, place customer orders, and check an order's status. While remote access is necessary, it can also pose as a vulnerable access point.

It is interesting to note that remote access can occur without the direct knowledge of network administrators. An employee can secretly attach a modem to their workstation so that they may access their files from home. Software such as pcAnywhere and Windows XP Remote Desktop Connection are designed for such capabilities. A hacker can gain access the exact way the employee did.

Most intruders are aware that telephone numbers assigned to an office run in successive numerical order. For example, if the business telephone number is 333-1234, it is probably a safe bet that other telephones in the business are assigned telephone numbers such as 333-1235, 333-1236, 333-1237, and 333-1238. The pattern is easily revealed. This is also true for telephone number extensions.

When an employee attaches a telephone modem to their PC at work, they leave a backdoor open to an intruder. As stated earlier, they most likely have created a file share on their workstation so they can access the file share from home or while traveling.

Physical Security

Physical security is an important aspect of network security. When we discuss physical security, we are talking about the location of the physical system. File servers should be placed in a secure room. A secure room means that it is physically locked and can only be accessed by authorized personnel. This policy should apply to the file server room, wiring closets, point of presence location, and anywhere along data lines where someone can gain access. Devices, such as hubs and routers, which connect the workstations to the network, must also be secured but generally not at such a high level as the file server and wiring closets.

Workstations should also be physically secured. Many users leave their workstation connected to the network when they go home for the day. When they leave, the workstation is still logged on to the network with their user account. This is an open invitation to unauthorized personnel. This is especially true in a large open office environment when 50 or more workers have open access to every computer in the office area. What if a sensitive document such as a salary schedule for the entire corporation was downloaded, reproduced, and posted on the company lounge bulletin board. Management would most likely be upset. The network administrator could trace the event to your workstation using a standard event monitoring utility. You would then have to explain this incident and hope you can save your job in the process.

Biometrics

Biometrics is the science of using the unique physical features of a person to confirm their identification for authentication purposes. Some examples of unique physical features are fingerprints, speech, eye color patterns, and facial features. One or more of these physical traits can be scanned and encoded as data to be used for comparison when a person attempts to enter a secure area. See **Figure 16-24.** Biometrics can also be used in conjunction with traditional authentication methods, such as user name and password. It is assumed that because biometrics is unique it cannot be compromised. It is a well-known fact that even foolproof security models can be compromised.

A group of college students once cracked a biometrics fingerprint scanner by getting a sample of a network user's fingerprint from a glass. Next, they made a mold of the fingerprint and poured a plastic substance similar to rubber into the

biometrics
the science of using the unique physical features of a person to confirm their identification for authentication purposes.

Figure 16-24.
A biometric device that uses the fingerprint for authentication. (Precise Biometrics)

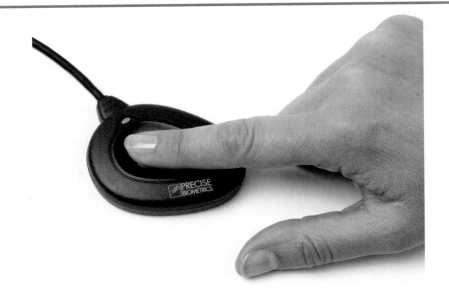

mold. They used the rubber form of the fingerprint to crack a biometrics fingerprint scanner. As this story exemplifies, any security model can be broken. However, certain models are much more difficult to crack than others.

Smart card

smart card
a technology that incorporates a special card into the security system, which is used in conjunction with a personal identification number (PIN).

A *smart card* incorporates a special card into the security system, which is used in conjunction with a personal identification number (PIN), **Figure 16-25.** A card reader is attached to the computer system via any standard port such as a serial or USB port. Once the smart card is inserted into the card reader, the user is prompted for a PIN number. After the correct pin number is entered, access to the computer system is granted. Smart card technology is not only designed for a single PC, it can also be used to access a network workgroup or domain. Access to the network is limited to the permissions assigned to the user account by the system administrator.

Figure 16-25.
A smart card security system typically uses a smart card and a PIN for authentication. (Photo courtesy of Gemplus)

Security Tools

There are numerous security tools available on the market, which are designed to identify common network security weaknesses by probing the network and searching for vulnerabilities. The tools are similar to tools used by hackers and crackers to probe a network. Security tools can help you determine if a potential problem exists. One such tool is LANguard Network Scanner by GFI.

In **Figure 16-26,** the LANguard Network Scanner utility has revealed some of the potential problems that exist on a network. Since this information is taken from an actual site, the IP address has been blocked out for protection. As you can see, numerous possible security problems exist on this network. A person with some expertise could easily penetrate this site and cause a tremendous amount of damage. The LANguard Network Scanner utility is designed to check security problems on the local area network, but can be used by a hacker or cracker as well.

Netstat Utility

The **netstat** utility can help determine which ports are open on a computer. To check for open ports using the **netstat** utility, enter **netstat -a** at the command prompt. A display will appear similar to the one in **Figure 16-27,** which shows the TCP and UDP protocols, the port number of each port currently opened, and the name of the computer associated with the protocol. Any port listed that is not being used should be closed. A utility that periodically detects open ports and alerts you to unauthorized intrusions should also be used.

Audit Tools

User authentication and encryption may not be a sufficient measure of security. A system of auditing user activities should be set up. Not all network attacks or probes come from outside the network. Many network attacks come from inside the network by employees. The activities of users or intruders can be recorded in a log, or the network can be set up to generate messages to alert the

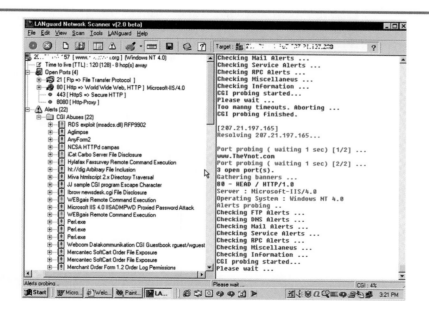

Figure 16-26.
The LANguard Network Scanner utility can be used to check for security problems on a local area network.

Figure 16-27.
An administrator can use the **netstat** utility to detect open ports on a computer.

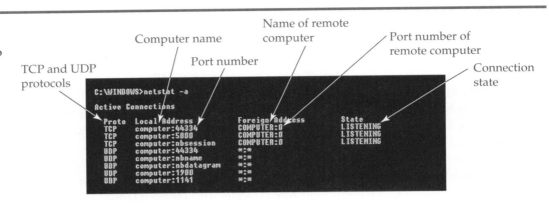

administrator of a possible attack. Activities such as repeated logon failures are a good sign of intruder activity.

Look at the Event Viewer security log events of a typical Windows XP system in **Figure 16-28.** The security log lists logon events, successes and failures, dates, and times. By clicking a failed event in the security log, specifics of the event are revealed in more detail. In the screen capture, the **Event Properties** dialog box reveals a failed logon attempt. In the failed logon event, a user, MrX, in the domain GOODHEAR-XML01L attempted to log on to the computer system and was denied access because of either a bad user name or bad password. The Event Viewer allows network activities to be monitored that may indicate an attempted or successful system intrusion.

Self-Hack Tools

Several companies have self-hack tools available. These tools are typically built into a security package and offered as an additional feature. For example, the LANguard Network Scanner utility is not only designed to provide security measures for a network, it can also test network security using common methods

Figure 16-28.
The Event Viewer can be used to log security events. This utility can aid an administrator in detecting an attempted or successful system intrusion.

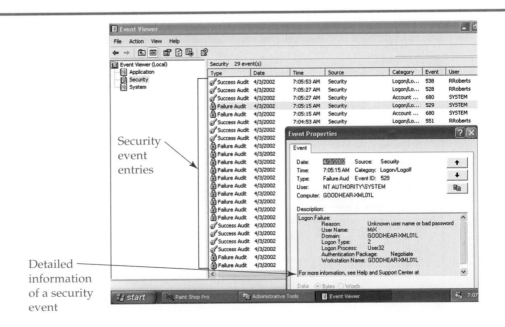

such as password cracking. The LANguard Network Scanner utility probes the LAN looking for open communication ports and general security weaknesses. It also provides a password-cracking tool.

Protocol Analyzer

A *protocol analyzer* is a special tool used to inspect protocol activity and contents. It can reveal information about protocols such as the source and destination IP address, MAC address, port address, time of transmission, and the contents of unencrypted packets.

protocol analyzer
a special tool used to inspect protocol activity and contents.

Figure 16-29 shows an e-mail transmission. Take a close look at the contents of the e-mail, which are translated on the right side of the screen. In the translation, you will see the password and the user name issued to the recipient because the e-mail was not encrypted. The contents of this e-mail are revealed to anyone using a protocol analyzer or protocol sniffer on this network.

Another excellent tool is the Fluke Protocol Analyzer. In fact, Fluke has an entire series of network analyzing tools to help you inspect a network system.

Packet Sniffer

A *packet sniffer* is a network monitoring utility that captures data packets as they travel across a network. Look at the packet sniffer utility in **Figure 16-30.** As you can see, a packet sniffer provides a vast amount of information such as packet size, protocol, and the source and destination address expressed as an IP address and a MAC address. On the right side of the screen, you can see some of the information contained inside the decoded packet. This particular packet sniffer can capture and do a complete analysis of TCP, IP, ARP, and UDP protocols. This utility can be very useful in analyzing network problems. However, in the wrong hands, it can be a security threat.

packet sniffer
a network monitoring utility that captures data packets as they travel across a network.

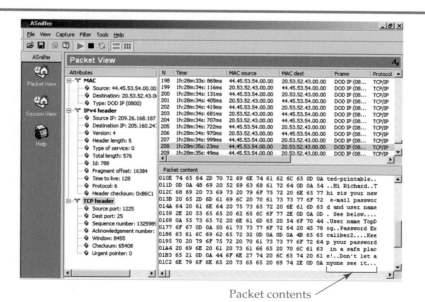

Figure 16-29.
Notice how the contents of an unencrypted e-mail can be revealed to anyone using a protocol analyzer or protocol sniffer.

Packet contents translation

Figure 16-30.
A packet sniffer provides a vast amount of information about a packet.

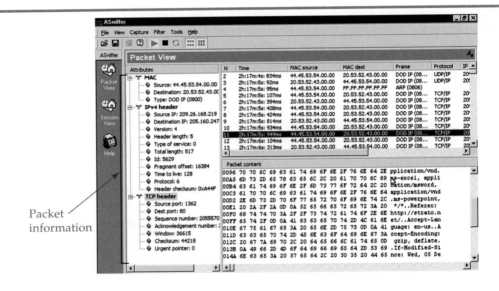

Packet
information

The terms *packet sniffer* and *protocol analyzer* are commonly used interchangeably. A packet sniffer is mainly designed as a tool to capture packet contents and header information and to provide limited information. The protocol analyzer provides these same functions and much more, such as analyzing network traffic patterns, producing graphical representations of protocol characteristics and the network infrastructure, and analyzing network problems.

System Backups

Backups are necessary for recovering lost data. However, the common system backup will restore most data lost from an attack, but will not restore all the data. Any data saved during routine operation, which has not been backed up since the last incremental backup, will be lost.

For example, a system is backed up every night at 7:00 p.m. The next day, the office opens at 8:00 a.m. and begins business for the day. At 2:00 p.m., an intruder enters the system and formats the hard drive of the file server and then releases a virus on the workstations. All data stored since 8:00 a.m. is lost and cannot be recovered.

Network security is a rapidly growing field. It has its own special certification. The certification requires an intensive study of security far above the limited scope of this textbook. If you think you would like to become an expert in network security, visit the SANS organization Web site to find out more about the requirements. This certification is well respected and much in demand.

Summary

- Authentication is the procedure of verifying a user's identity.
- Social engineering is an activity that uses personal skills rather than technical knowledge to gain access to a network or secure area.
- Spoofing is fooling the destination by using an IP address other than the true IP address of a source to gain access to a system.
- A Trojan horse often contains a program that can reveal user names and passwords.
- The most common way worms and other undesirable content are spread is through e-mail attachments.
- A Macro virus is written using readily available macro writers that come as part of a word processing program.
- Encryption is a method of using an algorithm to encode data.
- A key is a software code used to encode or decode data.
- The two types of keys associated with encryption are public and private.
- A public key is shared and a private key is retained by the original key holder.
- A symmetric key is a key classification that uses the same key to encrypt and decrypt data.
- An asymmetric key is a key classification that uses two different keys to encrypt and decrypt data.
- Digital certificates are issued by Certificate Authorities and are used to verify identities.
- Digital certificate keys are typically asymmetric.
- The Secure Socket Layer (SSL) protocol was developed by Netscape to secure HTTP messages or transactions.
- IPSec is a protocol designed to secure IP packets on an unsecured network medium.
- The original solution to wireless device security includes a Service Set Identifier (SSID), Media Access Control (MAC) filtering, and Wired Equivalent Privacy (WEP).
- IEEE 802.11i is a standard that was developed to remedy the original security flaws in 802.11.
- 802.1x provides port-based, network access control, which supports authentication for Ethernet network access.
- 802.1x is composed of three components: supplicant, authenticator, and authentication server.
- Password Authentication Protocol (PAP) was an early development of user name and password authentication but transmits the user name and password as clear text.

- The Challenge Handshake Authentication Protocol (CHAP) encrypts the user name and password.
- Microsoft Challenge Handshake Authentication Protocol (MS-CHAP) is a Microsoft version of CHAP that can only be used with Microsoft operating systems.
- Kerberos was designed at the Massachusetts Institute of Technology and is used to authenticate the client and the server.
- The Kerberos system uses two different keys similar to a public and private key for security.
- After installing a network operating system, you should immediately apply any available software patches to the system.
- The system administrator default user name should be changed to lessen the chance of the system being cracked.
- A strong password is composed of letters, numbers, and special symbols.
- Network operating systems typically incorporate password policy features such as password history, minimum length, age, and complexity.
- A firewall monitors data packets as they enter or exit the network system and either blocks or passes them.
- A proxy server is a special firewall designed to hide clients inside the network from unauthorized personnel outside the network.
- Biometrics is the use of physical characteristics for the basis of user authentication.
- A smart card integrates electronics into a card for authentication.
- A protocol analyzer is a tool that inspects protocol activity and contents.
- A packet sniffer is a type of network monitoring utility that inspects data packets.

Review Questions

Answer the following questions on a separate sheet of paper. Please do not write in this book.

1. What is using a false IP address or identity called?
2. What is the purpose behind using a Trojan horse?
3. Where are Macro writers commonly found?
4. Through what method are worms typically spread?
5. What is the difference between symmetrical key encryption and asymmetrical key encryption?
6. Another name for symmetric key encryption is _____.
7. What is the purpose of a Certificate Authority (CA)?

8. Who developed SSL?

9. What is the purpose of SSL?

10. What is IPSec?

11. What is the difference between IPSec tunnel mode and transport mode?

12. What two protocols are common to IPSec?

13. What are the three original wireless device security mechanisms?

14. What encryption mechanism was developed as a replacement for WEP?

15. What is 802.11i?

16. What is 802.1x?

17. What are the three components of an 802.1x configuration?

18. Why is PAP considered an unsecured system of authorization?

19. What does the acronym CHAP represent?

20. Why is CHAP considered an improvement over PAP?

21. What operating system is MS-CHAP designed exclusively for?

22. Of the three different protocols, PAP, CHAP, and MS-CHAP, which encrypts the contents of the packet?

23. What is Kerberos?

24. The initial network operating system should not be considered complete until the latest software _____ is installed.

25. Describe the characteristics of a poor password.

26. Describe the characteristics of a good password.

27. How is a firewall used to secure network access?

28. How is a proxy server used to secure network access?

29. What is biometrics?

30. What technology uses a plastic card with embedded electronics for identification?

31. What TCP utility displays open ports?

32. What is a protocol analyzer used for?

33. What is a packet sniffer?

Sample Network+ Exam Questions

Network+

Answer the following questions on a separate sheet of paper. Please do not write in this book.

1. What common security measure should be performed immediately after the new network operating system has been installed?
   A. Change the default administrator user name and give it a difficult password to crack.
   B. Immediately write down the administrator user name and password, and then delete the administrator account to prevent it from being penetrated by unauthorized persons.
   C. Access the BIOS settings and change the network administrator's password.
   D. Install a protocol analyzer to see if a cracker detected the installation.

2. Which of the following software packages or utilities is used to analyze the contents of individual packets on an Ethernet network?
   A. **arp**
   B. Packet sniffer
   C. **tracert**
   D. **ping**

3. Which of the following statements is true concerning CHAP?
   A. CHAP off nds an encrypted string of characters representing the user name and password.
   B. CHAP can be used to encrypt packet content.
   C. CHAP can provide the same features as Kerberos.
   D. CHAP sends an encrypted string of characters representing the user name and password.

4. Which utility can record security events as they take place on a Windows 2000 network?
   A. Event Viewer
   B. Network Monitor
   C. Directory Service
   D. Device Manager

5. Which is the best example of a secure password?
   A. BigDog
   B. Star$Read1345
   C. NtsysFive
   D. Pass123456789

6. Which security term is used to describe the act of obtaining a person's identity through fraudulent means?
   A. Trojan
   B. Phishing
   C. Man in the middle
   D. DoS

7. Which is of the following is the most recent version of wireless security based?

A. WEP

B. WPA

C. IEEE 802.11i

D. IEEE 802.11b

Network+

8. Which of the following is used to identify a wireless network?

A. MAC filter

B. SSID

C. EAP

D. Subnet mask

9. Which of the following is most responsible for the spread of computer viruses and worms?

A. Uploading data from a disc.

B. Newly installed hardware devices.

C. Opening e-mail attachments.

D. Newly installed software utilities.

10. Which protocol was designed to provide secure communication between a Web server and a Web browser?

A. SSL

B. FTP

C. PPP

D. RADIUS

Suggested Laboratory Activities

1. Install and configure Windows XP Internet Connection Sharing (ICS) and the Internet Connection Firewall (ICF).

2. Using Windows 2000 or a similar network operating system, set up and experiment with password policy settings.

3. Search the Internet and download trial versions of security software.

4. Download a trial version of a firewall or other security software.

5. Install or activate Windows Firewall on a small network.

6. Install two network interface cards in one PC and set up a proxy server/gateway.

7. Apply for a free, personal digital certificate from a CA such as VeriSign. After obtaining the digital certificate, try it out with a classmate.

8. Open Windows 2000 Help and Support and look up the powers of the administrators, power users, users, backup operators, and guest account. Compare and contrast the various users allocated to the various Windows 2000 default accounts.

9. Use the Event Viewer on a Windows XP, NT, or 2000 workstation to generate a log of activities, such as failed logon attempts.

10. Go to various US government Web sites and see if there are any computer security articles available.

11. Visit the SANS Web site for the latest security information.

Interesting Web Sites for More Information

http://web.mit.edu/kerberos/www/

www.cert.org/security-improvement/

www.cert.org/tech_tips/home_networks.html#II-L

www.hackingexposed.com/tools/tools.html

www.incidents.org

www.insecure.org

www.microsoft.com

www.microsoft.com/technet/security/tooles/tools.asp

www.sans.org

**Chapter 16
Laboratory Activity**

Security Event Monitoring

After completing this laboratory activity, you will be able to:
- Access the security log.
- Interpret a failed security event.
- Identify the type of events that are recorded in the security log.
- Modify the security log configuration.

Introduction

In this laboratory activity, you will access and modify the security log configuration. The security log can be accessed by entering **eventvwr.msc** in the **Run** dialog box off the **Start** menu. The msc file extension is optional. You can also access the security log by selecting **Start**, right-clicking **My Computer**, and selecting **Manage** from the shortcut menu. When **Microsoft Computer Management** opens, you will see an **Event Viewer** folder listed in the left-side pane of the window. In the right-side pane, you will see the **Security** log.

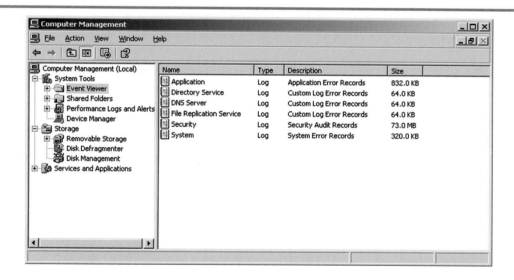

The security log contains security audit records. The security log is a list of recorded events, such as successful and unsuccessful logon attempts and access to shares that require authentication. When the **Security** log is double-clicked, the entire list of security audit events is displayed similar to that in the following figure.

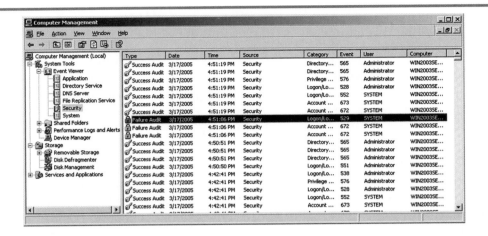

Pay particular attention to the key and lock icons to the left of the security event listings. The key represents a successful event and the lock represents a failure. Security events are listed by Type, Date, Time, Source, Category, Event, User, and Computer. The security events are listed in chronological order by default. For convenience, security events can be sorted by each of the individual columns. For example, to sort all events by user name, click the top of the User column.

When a particular event in the event log is double-clicked, a box opens displaying detailed information about that particular event. To open an event, you may also right-click the event and then select **Properties** from the shortcut menu. The following figure shows the properties of a failed security event.

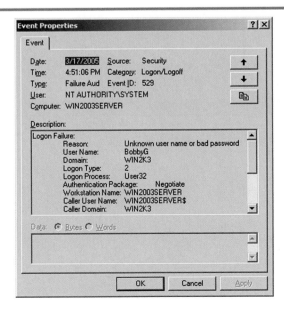

As you can see, the information gathered is about an attempted logon failure that took place at 3/17/2005 at 4:51:06 p.m. The user name entered for authentication was *BobbyG* and it occurred in the Windows 2000 domain on the WIN2003SERVER. This record can be copied to the clipboard as a single event and then printed so the administrator has a hard copy to work from or to add to a security report. As you can see, Event Viewer is an important utility that can monitor user authentication. If a series of failed events occur using the same name, an administrator will know that either the user *BobbyG* is having a problem with his password or that someone is attempting to access the server using the *BobbyG* account name and guessing the password.

Certain security log properties can be modified, such as the size of the security log file. The default size is 16MB. It can be increased in increments of 64kB until a maximum of 4GB is reached. The entire file can be saved to a separate folder for storage or review.

Look at the following figure and note the three overwrite options listed in the Log Size section of the Security Properties dialog box: **Overwrite events as needed, Overwrite events older than seven days**, and **Do not overwrite events (clear log manually)**. The default setting is **Overwrite events as needed**. This means when the allotted amount of file space is used, the oldest event will be erased as new ones are generated. **Overwrite events older than seven days** is as indicated. **Do not overwrite events (clear log manually)** is not often used because if the log reaches its limit, the computer will generate an error message or a blue screen and then lock the computer when a new event needs to be recorded. The server must then be restarted and the log manually cleared.

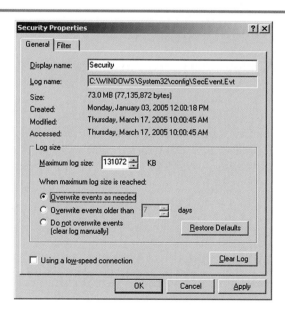

Events may also be filtered. You can select to filter events by the event types listed at the top of the **Filter** page of the **Security Properties** dialog box as shown in the following figure. For example, you may wish to uncheck successful audits to save log space. To find out more about Event Viewer and the security log, check the **Help and Support** located off the **Start** menu.

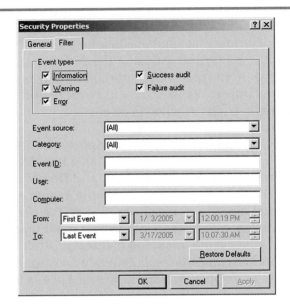

Equipment and Materials

■   Computer with Windows 2003 Server installed. You will need administrative privileges.

Note:
If a Windows 2003 server is not available, you can perform this laboratory activity on a Windows XP Professional workstation with your instructor's permission. The features are very similar to those in Windows 2003 Server.

Procedures

1. _____   Report to your assigned workstation.

2. _____   Boot the workstation and make sure it is in working order.

3. _____   Access Event Viewer by entering **eventvwr** in the **Run** dialog box located off the **Start** menu.

4. _____   After Event Viewer opens, exit Event Viewer by selecting **Exit** from the **File** menu.

5. _____ Now access Event Viewer by selecting **Start**, right-clicking **My Computer**, and selecting **Manage** from the shortcut menu.

6. _____ Expand the **Event Viewer** folder so that you can see the three event logs: **Application**, **Security**, and **System**.

7. _____ Double-click **Security**. Look at the security events listed in the right-side pane of the window. If your system is new, there may be few events listed. Take note of what the latest event is by double-clicking the event. The **Event Properties** dialog box will display. When you are finished viewing the event details, click **OK** to close the **Event Properties** dialog box.

8. _____ Sort the events by type, date, user, and computer by clicking on the top of each column.

9. _____ Log off the server.

10. _____ Generate a failed logon event by attempting to log on to the server using a bogus user name and password.

11. _____ Now log on to the server using your assigned user name and password.

12. _____ Open the security log and see if the failed event was recorded. If the failed logon attempt is not listed, call your instructor for assistance.

13. _____ View the details of the failed event by double-clicking the event or by right-clicking the event and selecting **Properties** from the shortcut menu. Answer the following questions on a separate sheet of paper.

Is the bogus name listed?

What other information is listed in the **Event Properties** window?

14. _____ Close the **Event Properties** window by clicking **OK**.

15. _____ Right-click the **Security** folder listed under **Event Viewer** and then select **Properties**. Look at the options listed on the **General** page. Practice changing the log size and the type of overwrite events. When you are finished experimenting, click **Restore Defaults**.

16. _____ Answer the review questions and then return the Windows 2003 server to the condition specified by your instructor.

Review Questions

Answer the following questions on a separate sheet of paper. Please do not write in this book.

1. How can you access the security log?

2. What does the key icon represent?

3. What does the lock icon represent?

4. What are the names of the security event column headings in Event Viewer?

5. What is the maximum size for the security log file?

6. What is the default size of the security log file?

7. In what kB increments can the size of the security log file be increased or decreased?

8. What are the three write options for the security log?

9. What is the default write option?

17

A Closer Look at the OSI Model

After studying this chapter, you will be able to:

❑ Compare the OSI model to the DoD model.

❑ Describe the function of the IEEE logical link control (LLC) and the media access control (MAC) sublayers.

❑ Compare various network hardware to the OSI model.

❑ Explain the function of each layer of the OSI model.

❑ Compare various IEEE standards to the OSI model.

❑ Describe the encapsulation process.

❑ Compare the TCP/IP protocol suite to the OSI model.

❑ Compare the IPX/SPX protocol suite to the OSI model.

❑ Compare the AppleTalk protocol suite to the OSI model.

Network+ Exam—Key Points

The Network+ Certification exam relies heavily on general knowledge of the most encountered protocols in the TCP/IP suite. Only a minimal knowledge of Novell and Apple protocols is required. Since there are hundreds of protocols used in networking, you need to check the CompTIA Web site and be sure you are familiar with all protocols mentioned in the Network+ Certification exam objectives. Memorize the OSI model and be familiar with the major protocols, their characteristics, and where they align with the OSI model.

Key Words and Terms

The following words and terms will become important pieces of your networking vocabulary. Be sure you can define them.

buffer	Routing Information Protocol (RIP)
dynamic IP address table	Service Access Point (SAP)
Internet Group Management Protocol (IGMP)	static IP address table
	token
multicasting	windowing
Open Shortest Path First (OSPF)	

An in-depth understanding of the OSI model has been one of the most difficult concepts for students to master. Simply explaining the OSI model once in any text-book is insufficient. Only after the student masters the concepts of the various protocols, hardware, and software can they truly begin to grasp the concepts related to the OSI model. In this chapter, a more detailed study of the OSI model is presented with special emphasis on the physical, data, network, and transport layers.

This chapter presents the OSI model as it relates to the TCP/IP protocol stack, how it compares to network equipment, and how it relates to various network operating systems. It is the intent of this chapter to clear much of the fog surrounding the functions of the OSI model and its relationship to hardware and software. A detailed study and review of this chapter will help you prepare for the Network+ Certification exam and provide you with an excellent refresher before going on to the following chapters, which cover maintaining, troubleshooting, designing, and installing networks.

History and Purpose of the OSI Model

The original model, or design, of TCP/IP communications over the Internet was first developed by the Department of Defense (DoD) in the 1970s. The DoD model is simpler in design than the OSI model. See **Figure 17-1.** Notice that the DoD model has four layers and the OSI model has seven.

Figure 17-1.
The original DoD model was much simpler in design than the OSI model. Note that there are two versions of the DoD model.

OSI Model

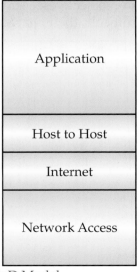

Variations of the DoD Model

When the DoD model was designed, the Internet structure was new and concepts were much simpler. The communication protocol was TCP/IP. There were not an overwhelming number of applications used in the early days of the Internet. Security was not considered a real problem, and encryption was not used. There was only one dominant operating system—UNIX. As long as all hardware models adapted to the DoD model and used the TCP/IP protocol, compatibility was not an issue.

The original network media was existing telephone lines. Only plain ASCII text was exchanged, not graphics, sound, or animation. There was no exchange of e-mail, as we know it today. The DoD model was based on the simple concept of exchanging text materials using TCP/IP over the Internet. As networking systems became more robust with full multimedia features, many more types of networking media and security features evolved along with a more complex model.

The requirements of such advances led to the development of the OSI model, as we know it today. Software (protocol) developers and network equipment manufacturers use the OSI model as an outline when developing software and hardware devices.

The Institute of Electrical and Electronic Engineers (IEEE) is concerned with the electronic aspects of the OSI model. This organization is composed of engineers, students, scientists, and manufacturing representatives who form standards that promote interoperability of hardware devices. In other words, they set standards that ensure hardware, such as network adapter cards, cables, cable connectors, switches, and routers, will easily interconnect and support each other rather than be proprietary.

The IEEE organization is nonprofit and many members donate their time toward the development of the standards. Because the IEEE is so concerned about the electrical aspects of network communication, they redesigned the OSI model to reflect their interests. The IEEE version of the OSI model splits the data link layer into two sublayers: logical link control (LLC) and media access control (MAC). This version became known as the IEEE 802 model. See **Figure 17-2.**

Notice how the data link layer of the OSI model has been subdivided into the LLC sublayer and the MAC sublayer. The LLC sublayer communicates with the

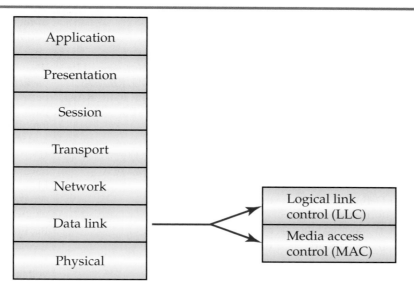

Figure 17-2.
The revised OSI model referred to as the IEEE 802 model divides the data link layer into sublayers: logical link control (LLC) and media access control (MAC).

IEEE 802 Model

upper-level protocols and is software in nature. It is responsible for framing the contents of the upper levels and includes the node address of the source as identified by the network interface card's MAC address.

The MAC sublayer is concerned with accessing the network medium and converting the entire content of the frame into a serial stream of bits that will be carried by copper core cable, fiber-optic cable, or by wireless means. There will be more about the data link layer later in the chapter.

The OSI Layers

This section takes a closer look at the layers of the OSI model. Compare the characteristics of the various OSI layers in **Figure 17-3.** The top four layers are

Figure 17-3.
OSI layer comparison table.

OSI Layer	Function	Hardware	Protocols	Keywords
Application	User interface	Gateways	HTTP, FTP, WWW, SNAP, SMB, SMTP, Telnet	Browser, e-mail, network applications
Presentation	Convert to common format such as ASCII, data encryption, and compression	Gateways	MPEG, WAV, MIDI, QuickTime	ASCII, Unicode, EBCDIC, CODEC, and bit order
Session	Establish and close communication between two nodes Coordinates communication	Gateways	NFS, DNS, SQL, RPC, NetBIOS, X.25, SMB	Establish and terminate a communication session, log on, user name, password, authentication, assign services through port numbers
Transport	Sequence packets Ensure error free delivery Takes over after the session has been established	Gateways (layer 4 switches)	TCP, UDP, SPX	Segments, windowing, flow control, transport packets, error checking (if required), port numbers
Network	Navigates outside of the LAN	Routers (layer 3 switches)	IP, IPX, AppleTalk, ICMP, RIP, ARP, OSPF, IGRP, RARP, EGRP, BGP, NLSP	IP address, routing, packets, datagrams, network address, packet switching, logical address, best and shortest route
Data Link	Prepares data for media access Defines frame format	Bridges, switches, Wireless Access Points, network interface cards	CSMA/CD, CSMA/CA	MAC address, hardware address, LLC, CRC, frame types, frames, topologies, contention
Physical	The physical aspect of the network	Copper core cable, fiber-optic cable, wireless, hubs, repeaters, transceivers, amplifiers, transducers	NA	Bit, byte, cable, media, topology, transmission, voltage, digital signals

primarily descriptions of software functions and are hardware independent. The software in these upper layers are Web browsers like Internet Explorer, Netscape, and Mozilla, e-mail clients, instant messaging, FTP, and such. The lower three levels are directly related to specific hardware types.

The physical and data link layers are dominated by hardware, especially the physical layer. Hardware consists of such devices as routers, switches, hubs, network interface cards, and cabling. Notice that the gateway is assigned to the upper layers of the OSI model. This is because the gateway acts as a translator when two different operating systems attempt to communicate. Remember that an operating system is software. The term *gateway* is often confused to mean a specific piece of equipment when in reality it is a software service typically installed on a router or server. The gateway translates the different file formats used to form packets. By now, you know that different network operating systems use different formats to construct packets and identify network equipment. The gateway interprets the packets and then translates the contents.

The data link layer, is where the data in the packets are linked to or converted to the network medium in the form of electrical digital pulses. The network layer is where hardware and software combine to allow packets to traverse great distances using the IP addressing scheme. It is also where the router is located. The router is responsible for routing packets based on the IP address contained within the packets. The network layer is designed with the WAN in mind. It is typically concerned with using IP for routing packets to and from distant hosts. IP can be used for a LAN as well.

There are many exceptions to the OSI model. No one networking system follows the model perfectly. Always remember that the OSI model is a suggested, voluntary model for network communications and is not mandatory. It is an attempt to help unify network communications by serving as a design guideline. The following sections cover each OSI layer in more detail. Refer to Figure 17-3 as you read through them.

Application Layer

The application layer is where the networking application interfaces to the OSI model. The term *application* should not be confused with software applications such as spreadsheet, word processing, and database programs. You can think of the application layer as the location of the networking specific application. For example, FTP, HTTP, SMTP, SMB and Telnet reside at the top of the OSI model. Each of these high-level protocols typically represents a network application. The protocols at the application layer perform services such as Web browsing, exchanging e-mail messages, and file transfer, **Figure 17-4.**

For example, when an e-mail client (application) receives e-mail, it uses one of the various protocols designed for e-mail retrieval, such as POP3 or IMAP. The e-mail client interface is a GUI. The user is completely unaware of the complexities behind the scenes, such as the data flowing up and down the OSI layers to receive the e-mail.

Presentation Layer

The presentation layer is concerned with negotiating a common set of symbols that the source and destination hosts can interpret. For example, the content of a message can be in ASCII format or coded as EBCDIC. You should recall that EBCDIC is the coding format IBM mainframes use to represent data, and ASCII

Figure 17-4.
The application layer is the OSI layer in which networking specific applications, such as SMTP, HTTP, SMB, and FTP, operate.

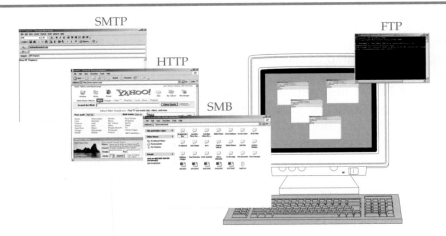

is the format PCs use to represent data. The two coding schemes do not directly convert but rather must be interpreted to exchange data, **Figure 17-5.**

Both the destination and source must agree on the format used for coding information and compression. Compression techniques are generally referred to as CODECs. When security is an issue, the contents should be encrypted. The encryption method used will be one that can be interpreted by both ends of the communications link.

The exact sequence of bits is also of major concern of the presentation layer. The bit order of data can be read from left to right or from right to left. For example, the Motorola 68000 processor and the IBM 8086 processor read the stream of bits in the opposite order. In other words, the binary stream 11110000 from one computer is represented as 00001111 on the other computer. The data binary pattern is reversed. This must be corrected before transmitting data. It is the responsibility of the presentation layer and associated protocols to correct the problem of bit order. In short, the presentation layer is concerned with packaging the raw data into a form that both the source and the destination can interpret.

Session Layer

The session layer establishes, maintains, and terminates the connection with the destination. It determines what the correct virtual connection is by matching the port number to the service required to support the layers above it. For example, when connecting to a distant location, not only does the IP address need to be encapsulated into the packet, the port number that corresponds with the provided service must also be encapsulated. If the requested port number is busy,

Figure 17-5.
The presentation layer negotiates a common set of symbols that the source and destination hosts can interpret. In this example, an ASCII text file is interpreted as EBCDIC.

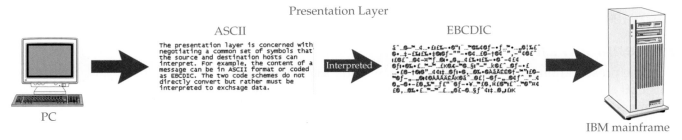

the session layer determines if the connection can wait or if another port number can be assigned. System logon activities and passwords can take place during the session layer.

The communications between the two hosts must be synchronized. Synchronized means that each host must recognize the other is ready to communicate or end communications. Properly ending a communications session between two hosts is important. The connection must be released at both ends so the port addresses are freed for future communications. Therefore, there must be a way of letting both computers know when the session is over. To do this, the session layer resynchronizes the connection by regularly checking if both hosts desire to continue the communication session.

The session layer is where differences in the programming mechanics or code are worked out for communicating between two nodes in the same network or distant networks. The session layer solves problems, such as how to establish a connection between nodes on the same system architecture and diverse architecture. For example, the session layer solves communication problems between a Windows 2000 client on an Ethernet network and an IBM Linux server on a Token Ring network.

Negotiations between the destination and source are accomplished using a token. A *token* is the name given to the packet that exchanges session information between the source and destination. Do not confuse the term *token* as used here with the term *token* used in Token Ring network technology.

The decisions determined at the session layer concerning the connection are carried out at the next layer, the data layer. For example, in the data layer, the port number is inserted into each segment that reflects the service requested or provided.

Some of the protocols closely associated with the session layer are DNS, X.25, NetBIOS, RPC, SMB, and NFS. When you think of the session layer, think of establishing, maintaining, and terminating a connection. See **Figure 17-6.**

token
the name given to the packet that exchanges session information between the source and destination.

Transport Layer

The transport layer provides reliable end-to-end data transmissions and error-checking techniques based on sequence packet numbers and software programs. For a TCP/IP connection, the transport layer uses either UDP or TCP to encapsulate the upper layer data. If a connectionless protocol is required, UDP is used. If a connection-oriented protocol is required, TCP is used.

The integrity of the data is checked at the transport layer. The transport layer checks the contents of a frame and for the proper sequence of frames to ensure

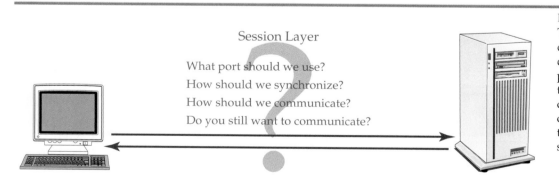

Session Layer

What port should we use?
How should we synchronize?
How should we communicate?
Do you still want to communicate?

Figure 17-6.
The session layer is concerned with deciding on a communications port and synchronization, establishing a common form of communication, and terminating the session.

the data will be reassembled in the correct order at the destination. The transport layer is also where port numbers are inserted into the segments. Remember that the port number is assigned in the session layer, and the IP address is assigned in the network layer. Together, the port number and IP address form what is known as a *socket*.

The transport layer divides large blocks of data into smaller blocks of data. Dividing large blocks of data into smaller blocks of data is called segmenting and results in a segment. The segment is associated with the transport layer. The size of a typical segment is negotiated with the destination host. Together, the destination host and the source host decide the maximum size of each segment and the amount of segments that will be sent before requiring an acknowledgment, **Figure 17-7.** The process is often referred to as ***windowing,*** or *flow control.*

windowing
the process of deciding the maximum size of each segment and the amount of segments that will be sent before requiring an acknowledgment.

Controlling the amount of data between two hosts is essential. A window size must be established between the destination and the source before large volumes of data can be exchanged in the most efficient manner. The large blocks of data can be moved between two nodes using several different techniques. For example, a live telephone conversation needs to be moved as rapidly as possible without any form of acknowledgment between the two nodes. A data transmission consisting of a large, detailed graphic may require that each block of data delivered be confirmed to ensure integrity.

In the last scenario, the size of the packet must be determined. Both the source and destination must agree on the maximum size that can be transmitted between the two nodes. The size is based on physical factors such as the buffer size at each location. A *buffer* is a small amount of memory used to temporarily store data. If the packets sent exceed the buffer size at the destination, only the packets that fill the buffer are left intact. The rest are discarded.

buffer
a small amount of memory used to temporarily store data.

The next major issue of the windowing technique concerns the efficiency of delivering a large quantity of data packets. For instance, is it necessary to send a confirmation or acknowledgment after each packet has reached its destination? It may be more efficient to send an acknowledgment after several packets are received.

Part of the process of windowing is determining the error rate associated with data packet delivery. If the number of errors in delivery is low or nonexistent, then a large number of packets can be delivered before returning an acknowledgment to the sender. If the error rate is high, then an acknowledgment after every packet may be required. The size of the packets and how many are delivered before an acknowledgment from the destination is how windowing affects the efficiency of data exchange.

Figure 17-7.
The transport layer is concerned with determining the segment and window size and how often an acknowledgement should be sent.

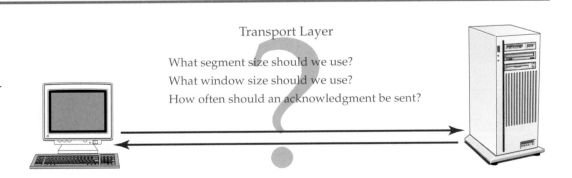

Transport Layer

What segment size should we use?
What window size should we use?
How often should an acknowledgment be sent?

Network Layer

As the name of this layer implies, the network layer is concerned with navigating the network by using IP addresses, **Figure 17-8.** The network layer is where the IP address is added to the segment. The segment is turned into a packet by including the IP address of the destination and source with the port numbers that were inserted in the transport layer.

Packets are also referred to as data grams by some reference materials.

Tech Tip

The network layer is associated with routers. You should recall from Chapter 1—Introduction to Networking that routers are devices that route packets to various networks based on the IP address inside the packet. This is the way packets are able to traverse the complexity of the World Wide Web.

The entire World Wide Web structure is composed of thousands of routers that are capable of directing the flow of packets to and from destination and source hosts based on IP addresses. The way each packet is forwarded depends on the router protocol used. The most common router protocols are BGP, IGP, IGRP/EIGRP, NLSP, OSPF, and RIP.

Routers can be programmed to operate from static IP address tables or dynamic IP address tables. A *static IP address table* is a table of IP addresses entered manually. A *dynamic IP address table* is a table of IP addresses generated by software programs that communicate with nearby routers. The software program allows routers to exchange IP address information automatically and build reference tables based on information that is exchanged with other routers.

The information contained in the router tables determines which network path is used to deliver the packet. The router protocols are programs designed with specific variables in mind for forwarding packets. The variables include things to be considered such as the shortest path based on physical distance, time, and the cost of the connection. The cost is very important because there are numerous private-leased lines that can be used to transmit a packet. The cost may vary considerably between different vendors.

Routing Information Protocol (RIP) is one of the first protocols developed for routing. RIP periodically exchanges an entire table of routing information

static IP address table
a table of IP addresses entered manually.

dynamic IP address table
a table of IP addresses generated by software programs that communicate with nearby routers.

Routing Information Protocol (RIP)
a routing protocol that periodically exchanges an entire table of routing information with nearby routers.

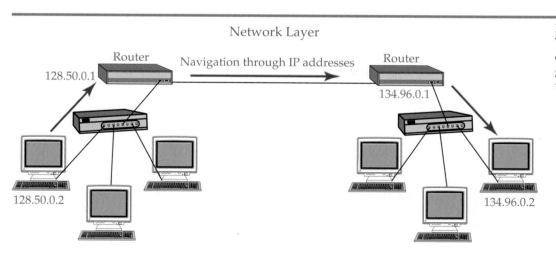

Network Layer

Router   Navigation through IP addresses   Router

128.50.0.1

134.96.0.1

128.50.0.2

134.96.0.2

Figure 17-8.
The network layer is concerned with navigating the network using IP addresses.

Open Shortest Path First (OSPF)
a routing protocol that only exchanges the most recently changed information in the routing table.

with nearby routers. This seemed to be the best way to keep the most accurate information until OSPF was developed. **Open Shortest Path First (OSPF)** is a routing protocol that only exchanges the most recently changed information in the table. OSPF saves bandwidth because it does not transfer the large amounts of data associated with RIP.

Many of the other router protocols are designed to provide special functions and enhancements of previous router protocols. A detailed study of routers and brouter protocols is beyond the scope of this textbook. To learn more about routers and routing protocols, visit the CISCO Systems Web site. CISCO Systems sponsors training academies throughout the United States dedicated to training CISCO router specialist. You may want to do a search to find the nearest CISCO academy to pursue additional training.

Some other protocols closely associated with the network layer are Internet Control Message Protocol (ICMP), Address Resolution Protocol (ARP), Reverse Address Resolution Protocol (RARP), and IGMP. You should recall that ICMP is used to generate messages and carry commands inside an IP packet. Some of the more common messages deal with attempts to reach a destination. ICMP carries error messages regarding unreachable hosts and message timeouts based on the TTL value.

Internet Group Management Protocol (IGMP)
a TCP/IP protocol designed to support multicasting. IGMP informs a multicast router the names of the multicast group to which a host belongs.

Internet Group Management Protocol (IGMP) is a TCP/IP protocol designed to support multicasting. **Multicasting** is sending the same data packet to a group of hosts identified by one IP address. Multicasting can conserve bandwidth on a network because packets are identified for delivery to hosts that belong to a particular multicast group. When a message is sent to a multicast group, the single set of packets for that group reach a router and are then forwarded only to members of the multicast group. Without multicasting and the IGMP protocol, the same message would need to be sent to each member of the multicast group as individual sets of packets. For example, if a message contained 5 packets and it was to be sent to 20 different hosts, it would result in sending 100 packets. When using multicast, only one set of 5 packs are sent across the local network, thus limiting local area network traffic.

multicasting
a process of sending the same data packet to a group of clients identified by one IP address.

IGMP informs a multicast router the names of the multicast group to which a host belongs. A host broadcasts its group membership on the network segment. This information is recorded in a database on a multicast router, a server, or a client acting as a database container. Membership information includes the IP address of each client and the IP address of the identified group. For example, a local multicast group might be identified as 224.0.3.1.

Data Link Layer

As mentioned in the beginning of this chapter, the data link layer was first defined by the ISO when they developed the OSI model. Later, the IEEE redefined the data link layer by dividing it into two sublayers called the LLC and the MAC. See **Figure 17-9** for a summary of LLC and MAC sublayer functions.

The IEEE published extensive details concerning the LLC, known as the IEEE 802.2 standard. The LLC sublayer is concerned with framing the contents of the upper levels and including the MAC addresses of the source and destination. The LLC sublayer deals with physical addresses rather than logical addresses. Remember that the MAC address that is programmed into every network interface card is used to uniquely identify the network nodes along the network media.

Data Link Layer

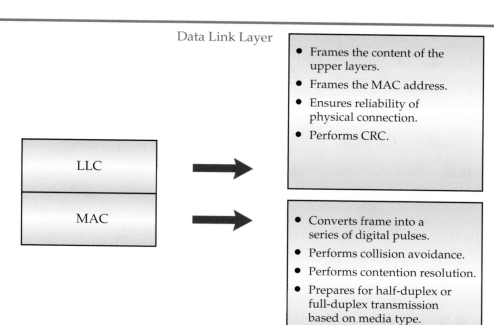

- Frames the content of the upper layers.
- Frames the MAC address.
- Ensures reliability of physical connection.
- Performs CRC.

- Converts frame into a series of digital pulses.
- Performs collision avoidance.
- Performs contention resolution.
- Prepares for half-duplex or full-duplex transmission based on media type.
- Negotiates transmission speed.

Figure 17-9.
LLC and MAC sublayer functions.

The LLC sublayer also ensures the reliability of the physical connection rather than of the data contained within the frame. The error check performed at the LLC sublayer is much simpler in design than that performed at the transport layer. It consists of a Cyclic Redundancy Check (CRC). The CRC is incorporated into the Frame Check Sequence (FCS) field of the frame and represents the overall size of the frame, which includes the size of the header and the data. The CRC check uses a simple mathematical formula that checks the size of the entire frame when it is delivered at the destination data link layer or while in route to the destination.

Network+ Note:

It is very easy to confuse the transport layer and the data link layer when asked a question about error checking in the OSI model. You must carefully read any questions concerning the OSI model and error checks and concentrate on what type of error is being checked. The transport layer is concerned about the content of the protocol data unit (PDU) while the data link layer is concerned about the overall content of the frame, which includes the data content. Think of the transport layer as ensuring error-free data or contents of the PDU. The data link layer looks at the entire frame based on the CRC and then decides to either pass the frame along or drop it entirely. The corruption may be in the header rather than the data content.

The MAC sublayer converts the frame and its contents into a series of digital pulses to be carried on the media (physical layer). For example, the binary electrical pulse sequence for Token Ring is different than wireless or Ethernet framing. Novell has used four different Ethernet framing techniques for their NetWare network communications. Each of the four is slightly different. If two different frame types are used on two different computers in a Novell NetWare network, the computers will not be able to communicate properly.

Figure 17-10.
The majority of IEEE standards deal with the physical layer and the MAC sublayer.

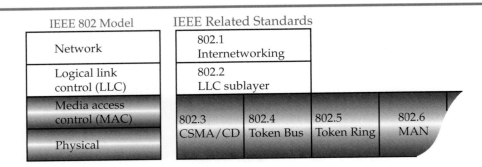

The IEEE standards such as Ethernet 802.3, Token Bus 802.4, and Token Ring 802.5, define the data link layer and the physical layer. **Figure 17-10** shows the relationship of many different standards that define various network media architecture and physical attributes. The standards are very detailed so that programmers and hardware designers can use it as a model for designing software and hardware devices. In other words, the standards address the type of materials to be used.

The data link layer uses Service Access Points (SAPs) to communicate with protocols at various layers of the OSI model. The MAC and LLC also each have a SAP and use it to communicate with one another. The SAP is used in similar fashion as a port address, which is used to exchange information between computers. The *Service Access Point (SAP)* consists of a combination MAC address and a number that identifies a protocol and its service. SAP addresses exist for the source and destination computer. The SAP provides a means of direct communication between protocols at different layers of the OSI model and between the source and destination computers.

Service Access Point (SAP)

a combination of a MAC address and a number that identifies a protocol and its service. It provides a means of direct communication between protocols at different layers of the OSI model and between the source and destination.

Physical Layer

The physical layer provides the path for the raw digital pulses that are moved along cables and connectors. The various network topologies and IEEE standards for equipment are identified at this layer. For example, the type of connectors used for various mediums such as BNC, RJ-45, D-shell, RS-232, ST, and SC connectors. Exact dimensions of connectors, materials, voltage level, frequency, data rates, Baseband and Broadband signaling techniques, and any other concerns about the physical communications circuit between the destination and source is described at this layer.

The Encapsulation Process

Encapsulation is the process of adding information to the segment that identifies such things as the source address, the destination address, the end of the segment, and the size of the segment. The size of a segment is determined by the protocol selected to communicate between two points. Segments have additional information added which is necessary to complete the transmission from one point to another. For example, if the data is transferred across a vast distance using the Internet, information such as the destination IP address and the source IP address must be added. Also, a sequence number must be added to ensure the segments are rearranged in the proper order and that no segments are missing.

The following is a summary of the encapsulation process:

- User data is converted into standard data format such as ASCII and EBCDIC.
- Data is broken into segments.
- Segments are placed into packets and header information is added.
- Packets are placed into frames and more header information is added.
- Frames are converted into a serial stream of data.
- The serial stream of data is converted to digital pulses.

Refer to **Figure 17-11** as the process of encapsulating raw data into a frame at the source and then disassembling the frame at the destination is reviewed. To transfer data between two locations, the source and destination must first establish a connection with each other. We will assume that the connection between the two points has been established and the data format, communication rules, encryption (if necessary), security, and any other communications details have been agreed on. All these concerns are accomplished mainly at the top three layers: application, presentation, and session.

Next, the data is divided into small units. This process is referred to as data segmentation or simply segmenting. A segment of data is simply a portion of the total amount of raw data broken into sizes that can be handled. The size is decided when the source and destination negotiate their connection and set up their rules for communication.

Each segment is placed into a packet. A segment is typically enclosed in either a UDP or TCP envelope. If a large amount of data is to be transferred, TCP is used for constructing the envelope. Since TCP is used to carry the data segment, a port number must be assigned. The port number is necessary to set up a constant connection with the other workstation and to acknowledge the data has arrived intact or not. If only a small amount of data such as a command were sent between the source and destination, UDP could be used. UDP does not require a port number because it does not require a verification that the data was delivered. The expected action of the command sent is the verification.

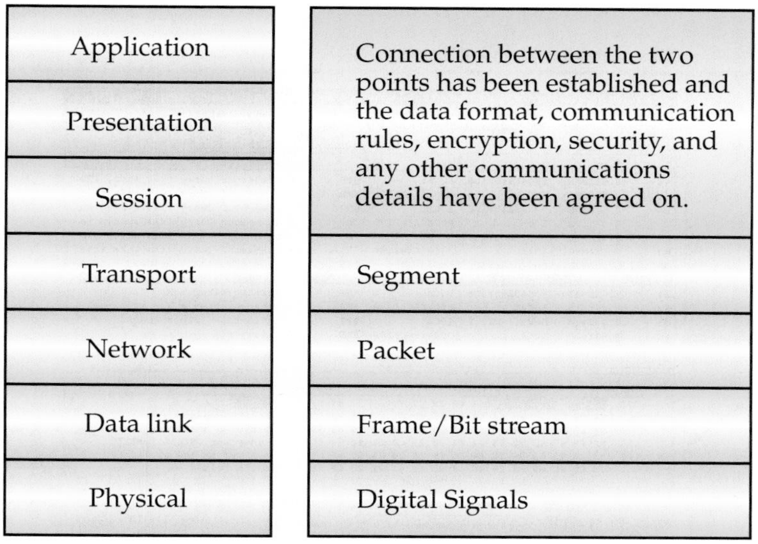

Application	Connection between the two points has been established and the data format, communication rules, encryption, security, and any other communications details have been agreed on.
Presentation	
Session	
Transport	Segment
Network	Packet
Data link	Frame/Bit stream
Physical	Digital Signals

Figure 17-11.
Data flow and packaging in relation to the OSI model.

After the segment has been enclosed in a TCP format, it is enclosed in a packet. The packet adds the IP address to the envelope carrying the data. An IP address for each of the communication locations, destination, and source are included in every packet.

Next, each packet is enclosed in a frame. A frame adds the MAC address for the destination and source to the packet. The frame is converted into a serial stream of data and then converted to digital pulses and placed on the network media. The type of media, such as Ethernet, Token Ring, or wireless has been identified in the frame. The MAC address is used to move the frame through the LAN. The MAC address identifies the workstation and equipment along the LAN path until the frame reaches the outer edge of the LAN. When the frame encounters the first piece of Internet equipment, typically a router, the MAC address is stripped away, leaving the IP address of the source and destination. Remember that the Internet uses IP addresses to locate nodes, not MAC addresses. The frame can travel over thousands of miles using only an IP address to find the destination network.

When the frame reaches the destination LAN, the network operating system uses ARP to resolve the IP address to a MAC address. Inside the destination LAN, the MAC address identifies the final destination node or workstation. At the destination, the envelopes around the data are stripped away until the raw data is revealed. The segments of raw data are put back together in proper sequence to display the data at the destination. The data could be plain text or multimedia. Any missing segments invoke a request for retransmission.

This is the classic explanation of how data is transported across the Internet. At this point, it might seem that each level of the OSI model only takes one or two frames for communication to take place, but that is far from reality. Even a simple request takes many steps to perform. The OSI model is at best an abstract model of how network communications should work. In reality, an actual network communication sequence is more complex.

Figure 17-12 shows a session taken from the Network Monitor utility on a Windows 2000 server. Notice the series of frames captured on the network. There are over fifty frames captured in the session. The frames are a series of communications between two computers on the same LAN. To set the stage, the two computers were set up to share a file. Then the Network Monitor utility was set up to capture frames as they are generated between the two nodes. A simple right-click

Figure 17-12.
The Network Monitor utility can be used to capture and analyze frames.

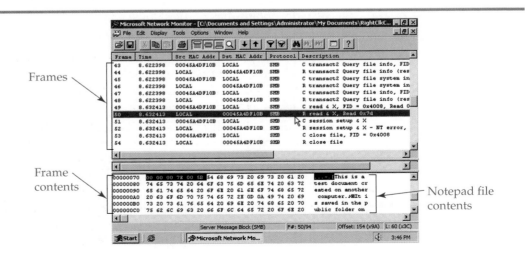

Frames

Frame contents

Notepad file contents

to open the content of a Notepad file on the other workstation was the only request made. The content then opened onto the screen.

Only a request to open the file was made, but over fifty frames were generated. At exactly frame fifty, the contents are revealed as you can see in the lower, right side of the Network Monitor screen. Releasing the connection took an additional 43 frames. The entire session took only .48 seconds. As you can see, performing a simple task such as opening a shared file for viewing takes many frames and causes a great deal of network traffic. Most of the steps were accomplished by using the Server Message Block (SMB) protocol.

For a detailed look at protocol encapsulation, check out the protocol encapsulation chart at www.wildpackets.com/support/resources. This chart includes a brief description of each OSI layer and a visual representation of protocols belonging to various protocol suites, such as TCP/IP, AppleTalk, and Novell NetWare.

The OSI Model and the Major Protocol Suites

The OSI model was developed after the majority of network protocols were already in use. As a result, not all network protocols match neatly to the OSI model design. For example, many protocols must span more than one layer to be properly represented when compared to the OSI model. Typically, higher-level protocols can perform multiple functions that span several layers of the OSI model. This section takes a closer look at how the dominate protocols of the major network operating systems compare to the OSI model.

TCP/IP Protocols

Look at **Figure 17-13.** The TCP/IP protocol suite is collected into three main areas: the top layers, which consist of the application, presentation, and session layer, the transport layer, and the network layer. The top layers are concerned with application protocols, such as HTTP, FTP, and Telnet. The transport layer packages the upper-level protocols into either TCP or UDP segments. The network layer encapsulates the TCP or UDP segments into packets. The protocols used at the network layer are IP, ICMP, and ARP.

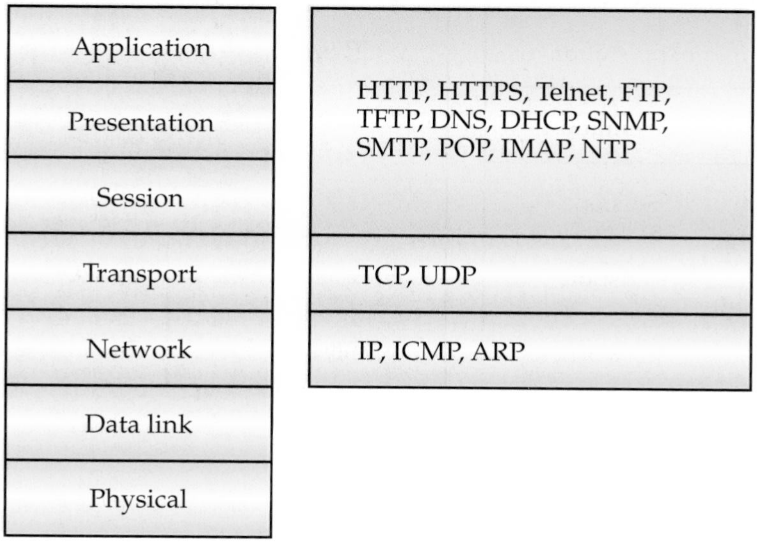

Application	HTTP, HTTPS, Telnet, FTP, TFTP, DNS, DHCP, SNMP, SMTP, POP, IMAP, NTP
Presentation	
Session	
Transport	TCP, UDP
Network	IP, ICMP, ARP
Data link	
Physical	

Figure 17-13.
The TCP/IP protocol suite is collected into three main areas. The top layers are concerned with application protocols, such as HTTP, FTP, and Telnet. The transport layer packages the upper-level protocols into either TCP or UDP protocol units. The network layer encapsulates the TCP or UDP into IP packets.

Novell IPX/SPX Protocols

Novell Network developed a proprietary set of protocols referred to as the Internetwork Packet Exchange/Sequenced Packet Exchange (IPX/SPX) suite. The IPX/SPX protocol suite functions in a similar fashion as TCP/IP. **Figure 17-14** shows how the NetWare protocols align with the OSI model. The NetWare Core Protocol (NCP) handles the upper layer functions associated with application, presentation, and session layers. NCP provides services between the server and clients and provides support for file services, Novell Directory Services (NDS), printing, messages, security, network synchronization, auditing, and more. The Service Advertising Protocol (SAP) is used to locate servers and services on the network. Tables of information, such as the server hosting the service and its location, are typically located on routers to assist in locating the desired service and server.

Internetwork Packet Exchange (IPX) is the original network layer protocol used to route packets. IPX is a connectionless protocol responsible for identifying destination and source network, node, and socket. Sequenced Packet Exchange (SPX) is a connection-oriented protocol designed to enhance IPX by guaranteeing packet delivery. SPX is similar in function to TCP, while IPX can be compared to IP.

RIP and NLSP, as mentioned earlier in this chapter, are routing protocols. NetWare Link Services Protocol (NLSP) was developed by Novell to overcome some of the limitations of RIP and SAP. Both RIP and SAP work on small networks, but for large networks a better protocol, such as NLSP, is needed. For example, RIP is limited to 15 hops and NLSP is limited to 127 hops. A *hop* is a term used to represent network distances between routers. NLSP uses smaller packet sizes than RIP and SAP and also supports multicasting.

The lower layers are designed to be physical and data link independent, which means that Novell NetWare can communicate over various vendor equipment.

Tech Tip

As TCP/IP has become the de facto standard for Internet communications, Novell has offered TCP/IP as an option to IPX/SPX since NetWare 5. While IPX/SPX is still offered as a choice of protocols mainly to ensure downward compatibility within existing Novell systems, the trend is to eventually convert to TCP/IP as the default protocol. Additionally, as the Lightweight Directory Access Protocol (LDAP) gained popularity, it has allowed modern network systems to communicate directly without the need for additional protocol translation.

Figure 17-14.
The IPX/SPX protocol suite.

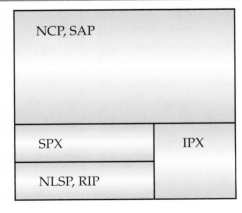

AppleTalk

The AppleTalk Filing Protocol (AFP) handles functions associated with the application and presentation layers. As the name implies, AFP provides services for file sharing, printing, messages, and other upper layer services. The session layer aligns with the Zone Information Protocol (ZIP), AppleTalk Session Protocol (ASP), and AppleTalk Data Stream Protocol (ADSP). ADSP provides functions for full-duplex communications. Note that ADSP spans both the session layer and the transport layer. Not all protocols align perfectly with the OSI model. See **Figure 17-15.**

ASP provides general support for communications between the server and clients. ZIP provides support for zone information and name resolution. AppleTalk divides the network into zones, which is the equivalent to groups or workgroups.

The transport layer aligns with AppleTalk Transaction Protocol (ATP), AppleTalk Echo Protocol (AEP), Name Binding Protocol (NBP), and Routing Table Maintenance Protocol (RTMP). These various protocol functions combine collectively to ensure reliable data communications.

At the network layer is the Datagram Delivery Protocol (DDP). It is a best effort, connectionless protocol similar in design to IP and IPX. There is no one matching protocol to ensure data delivery at this level. DDP relies on upper-level protocols to ensure accurate delivery of data.

The lower level of the AppleTalk suite is denoted by the LocalTalk protocol. LocalTalk supports communications on an Apple LAN by providing support for Apple networking devices. Apple can also communicate with other standard interfaces such as Ethernet, Token Ring, and FDDI.

Both Apple and Novell protocols can be placed inside IP and be transported across the Internet.

Tech Tip

AppleShare IP

AppleShare is a networking suite developed by and for Apple computers based on the AppleTalk protocol. AppleShare IP is an improvement over Apple-Share and is a protocol suite based on AppleShare and TCP/IP. AppleShare IP

OSI Layer	AppleTalk Protocol	
Application	AFP	
Presentation		
Session	ASP, ZIP	ADSP
Transport	AEP, ATP, NBP, RTMP	
Network	DDP	
Data link	LocalTalk	
Physical		

Figure 17-15. The AppleTalk protocol suite is similar in function to TCP/IP and IPX/SPX protocol suites. Both IPX/SPX and AppleTalk are designed to be data link independent, which allows both protocol suites to be used with a variety of technologies.

provides services over the Internet or LAN using IP addresses instead of node identification as in AppleTalk. By using the IP address as a means of location identification, AppleShare IP can provide remote printing and file sharing using the TCP/IP suite of protocols. AppleShare IP provides three types of file sharing technology: Apple File Sharing (AFP) over TCP/IP, SMB/CIFS over TCP/IP, and File Transfer Protocol (FTP). In addition, it provides common network services such as Web hosting, e-mail, and DNS.

Request For Comments (RFC)

The Request For Comments (RFC) is a series of technical reports and proposals concerning protocols and network standards. The Internet Engineering Task Force and other closely related organizations publish RFCs. RFCs continuously evolve as new technologies are implemented. Each RFC is assigned a unique identification number and name such as RFC 2131 Dynamic Host Configuration Protocol. There are several classifications of RFC documents such as the approved Internet standard, proposed Internet standard, and Internet best practices.

When a proposal is first submitted, it is viewed as a proposed Internet standard and is circulated in draft form. The draft form tells all concerned that the proposal is open for discussion and comment but has not yet been formally adopted.

Approved Internet standards have been examined by a special committee and formally adopted as a standard. Standards are provided to help manufacturers, computer programmers, scientists, and such to develop resources conforming to standard lines of communications. You have already seen how protocols have a set sequence of fields that provide certain types of information. Once the protocol standard has been accepted, the standard can be used as a basis for designing other software and hardware to interface with the protocol.

Anyone can write a proposal for an RFC. It is an open forum, but there are some guidelines that an RFC must follow before it can be accepted as a proposal. The guidelines spell out the written format of the proposal. For example, it may only be written in standard ASCII text with a maximum of 72 characters per line; a maximum height of 58 lines followed by an end of page character, such as a form feed (FF); a suggested table of contents; and rules about the size of the page and margins. There can be no special additional features added to the ASCII characters, such as underlining or double strike. The text must be kept as plain as possible because it will be converted into other forms for storage and transmission, such as FTP. You can visit www.rfc-editor.org to see the typical layout of an RFC.

Summary

- The OSI model is an abstract representation of data communications developed by the ISO.
- Network systems do not necessarily exactly match all layers of the OSI model.
- The application layer is where the user interfaces with the network system using network programs such as a Web browser, FTP, Telnet, and an e-mail client.
- The presentation layer is concerned with bit sequence and with using an acceptable data format such as ASCII, EBCDIC, and CODEC.

- The session layer is concerned with establishing, maintaining, and terminating a connection between two nodes.

- The transport layer is concerned with providing reliable, accurate data in the most efficient manner.

- The network layer is where IP addressing is inserted into a packet and where decisions based on IP addresses are made.

- The data link layer is where frames are converted into a serial stream of data and decisions based on MAC addresses are made.

- The physical layer provides the path for digital signals and represents the media, connectors, and passive components.

- While the OSI model tends to make encapsulation appear simple and direct, actual communication is complex and generates hundreds or even thousands of frames while carrying out the simplest of networking tasks.

- Novell NetWare IPX is a connectionless protocol, and SPX is a connection-oriented protocol.

- AppleTalk Datagram Delivery Protocol (DDP) is a connectionless protocol used to route packets across a network.

- AppleTalk Datagram Delivery Protocol (DDP) relies on upper-level protocols to form connection-oriented communication.

- The Apple LocalTalk protocol is used to access network media in an Apple network environment.

- AppleShare IP is a protocol suite based on AppleShare and TCP/IP.

Review Questions

Answer the following questions on a separate sheet of paper. Please do not write in this book.

1. How does the ISO OSI model compare to the DoD model?

2. Why is the DoD model so simple when compared to the OSI model?

3. What is the major difference between the ISO version of the OSI model and the IEEE 802 version of the OSI model?

4. The IEEE divided the data link layer into two sublayers. What are the two sublayers called?

5. Briefly describe the function of each IEEE data link sublayer.

6. At which layer of the OSI model does a network interface card operate?

7. At which layer of the OSI model does a hub operate?

8. At which layers of the OSI model does a gateway operate?

9. At which later of the OSI model does a router operate?

10. Match the following terms to their definitions.
   10.1.   Application
   10.2.   Presentation
   10.3.   Session
   10.4.   Transport
   10.5.   Network
   10.6.   Data link
   10.7.   Physical
      a.   Cabling and connectors.
      b.   Converts frames codes to a serial stream of data.
      c.   Responsible for connecting across WAN systems.
      d.   Repackages long messages into smaller units.
      e.   Establishes rules for communication between two computers.
      f.   Translates data into a common compatible format.
      g.   File transfer service, e-mail, and Web browser programs.

11. At which layer does windowing take place?

12. At which layer is the MAC address inserted into the frame?

13. To which two layers of the OSI model do the IEEE 802.3 standards relate?

14. To which two layers of the OSI model do the Token Ring standards relate?

15. Briefly describe the encapsulation process.

16. Which Novell protocols align with the upper layers of the OSI model?

17. Which Novell protocol is similar in function to IP?

18. Which Novell protocol is similar in function to TCP?

19. Which Apple protocol is similar in function to IP?

20. What Apple protocol is used to access the network media?

21. Which Apple protocol aligns with the upper two layers of the OSI model?

22. How does the Apple protocol suite ensure delivery of data if the Datagram Delivery Protocol (DDP) is a connectionless protocol?

23. What does the acronym RFC mean?

24. Who may write an RFC proposal?

Sample Network+ Exam Questions

Network+

Answer the following questions on a separate sheet of paper. Please do not write in this book.

1. What is the name of the two sublayers of the data link layer as defined by the IEEE 802 standard?
   A. Network layer, session access control
   B. Logical link control, media access control
   C. Protocol control, data control
   D. Physical layer, network layer

2. Which is the correct order of encapsulation moving from the application layer to the data link layer?

   A. Packet, segment, frame, raw data.

   B. Raw data, segment, frame, packet.

   C. Frame, packet, segment, raw data.

   D. Raw data, segment, packet, frame.

3. Which of the following protocols are closely or directly related to the network layer? (Select all that apply.)

   A. HTTP

   B. ARP

   C. RIP

   D. TCP

   E. UDP

   F. IP

   G. FTP

4. Which of the following protocols are connectionless? (Select all that apply.)

   A. UDP

   B. TCP

   C. IP

   D. SPX

5. Which of the following statements best describe the transport layer?

   A. Provides an end-to-end connection with reliable and efficient data transfer.

   B. Provides routing information using IP addresses when sending packets across a WAN.

   C. Converts the data into an acceptable form such as ASCII.

   D. Provides services to the client such as Web browsing.

6. Which IEEE standard is responsible for detailed information concerning the LLC?

   A. 802.2

   B. 802.4

   C. 802.6

   D. 802.16

7. Which two items are used to create a socket? (Select all that apply.)

   A. IP address

   B. MAC address

   C. Port number

   D. NetBIOS name

Network+

8. At which layer of the OSI model do routers operate?
   A. Application
   B. Session
   C. Network
   D. Data link

9. Which OSI layer is responsible for checking frame errors based on Cyclic Redundancy Checks (CRC)?
   A. Application
   B. Session
   C. Network
   D. Data link

10. With which layer of the OSI model would an RJ-45 connector be associated?
   A. Application
   B. Transport
   C. Physical
   D. Network

Suggested Laboratory Activities

1. Install Network Monitor. Generate frames by using common functions such as **ping, tracert,** and opening a share. Capture the series of frames with Network Monitor. An excellent capture is to set up Network Monitor with no activity on a peer-to-peer network. Turn on one other PC while Network Monitor is running. Capture and observe the activity created when a workstation boots and joins the network.

2. Go to the CompTIA Web site and look at the Network+ Certification exam objectives. List all of the protocols that are listed in the objectives. Write out the complete name for each acronym. Write a brief description of each.

3. Visit the www.iso.org Web site and see what standards they are responsible for.

Interesting Web Sites for More Information

http://developer.apple.com/techpubs/mac/Networking/Networking-21.html

http://split.org/storage/osireference.pdf

http://whatis.techtarget.com/definition/0,289893,sid9_gci212725,00.html

www.certyourself.com

www.cisco.com/univercd/cc/td/doc/cisintwk/ito_doc/introint.htm#xtocid5

www.decodes.com

www.lex-con.com/osimodel.htm

www.protocol.com

www.wildpackets.com/support/resources

Chapter 17
Laboratory Activity

Ethereal OSI Model Exploration

After completing this laboratory activity, you will be able to:

- Use the Ethereal protocol analyzer to compare protocols to the OSI model.
- Compare OSI model layers to the major TCP/IP protocols.
- Identify the protocols responsible for MAC address, IP address, and port numbers.
- Explain how protocols are used to encapsulate data and other protocols.

Introduction

In this laboratory activity, you will use the Ethereal protocol analyzer to compare the relationship of specific protocols to the layers of the OSI model. You can typically determine the OSI model layer a protocol aligns with by looking at its relative position in the hierarchy of protocols in the Ethereal program. Look at the following screen capture. Notice that the hierarchy of protocols is displayed in the middle pane of the Ethereal screen. Each frame that is selected displays its own hierarchy. The typical hierarchy lists at the top of the pane the protocol related to the bottom-most OSI layer.

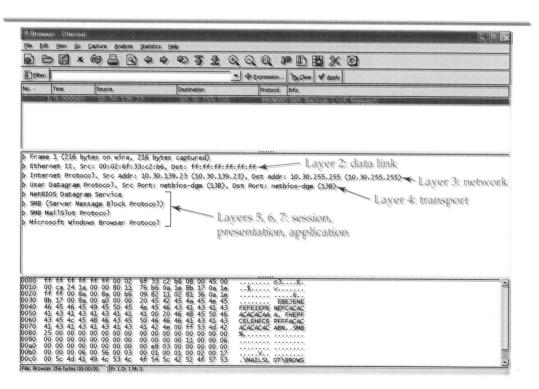

Clicking the box with a plus sign in it can reveal more information about each protocol listed in the hierarchy. This information may reveal items such as source, destination, MAC, and port addresses.

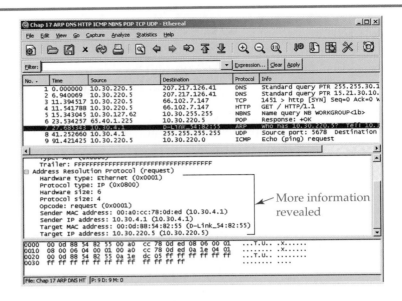

More information revealed

According to the OSI model, the Ethernet protocol is a layer 2 protocol. It contains the MAC address of the destination and source. The Internet Protocol (IP) is a layer 3 protocol. It contains the IP address of the destination and source. The User Data Protocol (UDP) is a layer 4 protocol. It contains the source and destination port numbers. The remaining protocols are upper-level protocols (layers 5, 6, and 7). Unless you know the function of these upper-level protocols, it is not easy to determine their location in the OSI model. Many protocols span more than one layer because they perform more than one function.

In this laboratory activity, you will open and inspect specific Ethernet frame captures (provided by your instructor) and identify the OSI model layer the protocol to which it should be assigned. Layers 5, 6, and 7 are combined for this activity. Layer 1 does not exist for this activity. To open a frame from disk, select **Open** from the **File** menu.

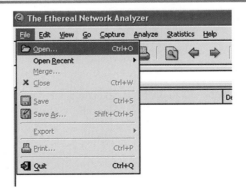

A dialog box similar to the following will appear. The **Ethereal: Open Capture File** dialog box will allow you to locate the files required for this laboratory activity. As you can see, it is a user-friendly graphical interface. The drives are listed on the left side, and the contents of each drive are listed on the right side.

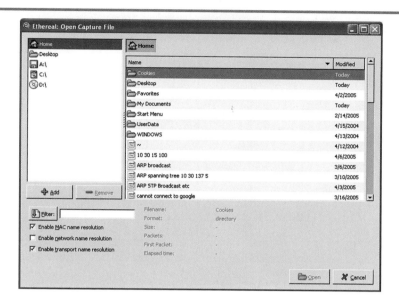

You will look for files with a name starting with "Chap 17 OSI" followed by a protocol name or names, such as Chap 17 OSI DNS. After opening each file, you will examine its contents and will identify which layer(s) of the OSI model closely match the protocols listed. You will record your answers in a chart that is provided later in this laboratory activity. Draw the chart on a separate sheet of paper. Do not write in this book.

Equipment and Materials

- Workstation configured with Windows XP. Ethereal must be installed on the workstation.
- Ethereal frame capture files. (These files will be provided by your instructor either on a floppy disk, local drive, or networked drive.)

Procedures

1. _____ Report to your assigned workstation.

2. _____ Boot the workstation and verify it is in working order.

3. _____ Start the Ethereal utility.

4. _____ Use the **Open** command from the **File** menu to open one of the frame capture files provided by your instructor.

5. _____ Copy the following chart onto a separate sheet of paper. For each protocol displayed in the frame capture file, place an X in the corresponding OSI model layer of the chart and write out the protocol name.

Protocol	Layer 2	Layer 3	Layer 4	Layer 5, 6, and 7	Protocol Name
Ethernet					Ethernet
ARP					Address Resolution Protocol
IP					
TCP					
UDP					
SMTP					
POP					
NBNS					
HTTP					
SMB					
DNS					

6. _____ Repeat steps 4 and 5 for each frame capture file.

7. _____ After completing the chart, go on to answer the review questions.

8. _____ Return the workstation to its original condition.

Review Questions

Answer the following questions on a separate sheet of paper. Please do not write in this book.

1. Which protocol was found at layer 2 throughout this laboratory activity?

2. Which protocols were found at layer 3 throughout this laboratory activity?

3. Which protocols were found at layer 4 throughout this laboratory activity?

4. Which protocols were not specifically identified, but typically found at layers 5, 6, and 7?

5. What protocols are used to encapsulate the upper-level (5, 6, and 7) protocols?

6. Which protocols are used to identify the location of the destination and source MAC address?

7. Which protocol is used to identify the source and destination IP address?

8. Which protocols are used to identify the destination and source port address?

9. Explain how protocol encapsulation provides all the required information for data packet delivery.

18

Maintaining the Network

After studying this chapter, you will be able to:

❏ Explain why a baseline is established.

❏ Describe how to perform a baseline.

❏ Explain the purpose and proper procedure for installing patches, upgrades, and service packs.

❏ Describe the commonly accepted practices for protecting data.

❏ Explain fault tolerance.

❏ Describe server data backup strategies.

❏ Explain the purpose of an Uninterruptible Power Supply (UPS).

❏ List commonly accepted antivirus procedures and policies.

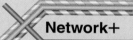

Network+ Exam—Key Points

Questions concerning maintaining the network are drawn from such topics as patches, upgrades, and fixes; data back up strategies; antivirus procedures; disaster recovery; and fault tolerance. Maintaining the network requires knowledge about all aspects of networking and of some basic concepts associated with PC support.

The CompTIA organization does not list malicious software in the Network+ Certification exam objectives, but it is implied because the test objectives state that the candidate is responsible for A+ type certification questions, which may appear on the Network+ Certification exam.

Key Words and Terms

The following words and terms will become important pieces of your networking vocabulary. Be sure you can define them.

antivirus suite
archive bit
average utilization
backdoor virus
baseline
blackout
brownout
bug
cluster
cold spare
continuous UPS
differential backup
disaster recovery
electrical spike
electrical surge
frame size average
frame size peak
full backup
generator
hoax
hot spare
incremental backup
isolation transformer
joke program

lightning arrestor
logic bomb
macro
macro virus
malware
MBR virus
password virus
peak utilization
polymorphic virus
power conditioning
service pack
scan engine
Simple Network Management Protocol (SNMP)
software upgrade
standby UPS
stealth virus
swap file
system migration
Trojan horse
Uninterruptible Power Supply (UPS)
virus
virus pattern file
worm

Network+

Network+ Note:
Be aware that you may see A+ type questions on the Network+ Certification exam.

A network only comes to a user's attention after it fails or is slow to respond to a user's requests. In reality, after a network is installed, it requires constant maintenance to be in near perfect condition and to ensure data integrity. Only with constant maintenance will the network be in good condition to ensure user satisfaction and smooth company operation.

This chapter introduces some new concepts and reintroduces some concepts that have been taught much earlier. Earlier concepts are applied directly to network maintenance. Most of the maintenance of a network system is the result of constant monitoring by the network administrator. While the network technician may perform routine daily maintenance under the direction of the network administrator, the administrator is ultimately responsible for the type, extent, and frequency of the maintenance to be performed.

Monitoring the Server and Network

The network system needs to be constantly monitored to determine when failures occur and to predict possible or imminent failure. Activity can be monitored, recorded, and expressed statistically to predict failure. For example, as a company grows, so will the amount of network traffic generated by the number of employees and customers. Statistical data based on network media bandwidth, total CPU activity, data storage space, and memory utilization can be monitored and a predicted failure rate can be forecasted through data analysis.

The predictions are based on system history. For example, if a company consumes storage space at a rate of 6% per month, network storage will be compromised in less than 18 months. A server's CPU activity can be monitored, such as when and how often CPU utilization reaches 100%. Another example is memory (RAM) activity. When exceeded, it typically causes more storage space activity. For example, operating systems use hard disk drive space to assist RAM. A block of hard disk drive space is used to augment RAM. When RAM is full, the operating system uses the block of hard disk drive space to supplement the RAM.

Hard drive disk space that supplements the RAM is referred to as a *swap file*, or *page file*. The swap file on the hard disk drive serves the same purpose as RAM. When the swap file is used, data processing slows because the hard disk drive cannot perform at the same high speed as the RAM. The slower swap file causes data processing to slow and affects the time it takes to retrieve data across the network. While the network is not the cause for the slow down, the network appears to be responding slowly to the user.

swap file
hard drive disk space that supplements the RAM.

Establishing a Baseline

The only way to objectively determine the performance of a network or a server is to establish a baseline after the network or server is first installed. A *baseline* is a measurement of performance characteristics that can be used at a later date to objectively determine if the network or server is performing satisfactorily. Typically, a baseline can be established by software utilities especially designed for such tasks. Many network operating systems have their own set of utilities that can perform routine baseline assessments.

One of the most important aspects of establishing a baseline is testing the network or server under normal and stressed conditions. Some software packages are designed to simulate network traffic so that the network can be stress-tested.

One of the best times to monitor a network or server is early in the morning. Start the monitoring process before the first shift of personnel arrives. At this time, activity should be at a minimum. As personnel report to their office and start the workday, network traffic increases dramatically. For example, the mail server will experience the most activity when the greatest number of personnel check their e-mail and send return e-mails. The times that typically generate the most e-mail traffic are first thing in the morning, after the lunch break, and at the end of the workday.

It is important to understand that a baseline is not a collection of data at a particular point in time, but rather a collection of data over a period of time. A baseline of the system should be established soon after a new network or server is installed and is operational. After that, the baseline information should be routinely collected and reviewed to predict events that could affect network or server performance. By predicting events, such as reaching an unacceptable level of collisions or running out of data storage space, preventive maintenance can be scheduled.

baseline
a measurement of performance characteristics that can be used at a later date to objectively determine if the network or server is performing satisfactorily.

Figure 18-1.
A simple form like the
one in this example or a
more detailed form can
be created to record
baseline information.

| Network Segment ID _____ | | |
| Date and Time Period _____ | | |

Segment	Value	Comments
Peak Utilization		
Average Utilization		
Frame Size Peak		
Frame Size Average		
Number of Protocols		
Number of Nodes		
Most Active 10 Nodes		
Collisions		
Packets Dropped		

Network baseline

Look at a sample form in **Figure 18-1** for collecting and recording baseline information for a network. Baselines can be very detailed, but the one in the example is brief. It is meant only to present you with an idea of what type of data is gathered and why.

Peak utilization is the highest level of utilization experienced by the network. Peak utilization occurs when there is the most traffic on the network segment. This can happen several times during a day—for example, first thing in the morning, lunchtime, and at the end of the day. This is normal and is mainly generated by users logging on to the network, opening files (especially large ones), and reading and sending e-mail. Peak utilization should not exceed 80% for long periods of time.

Average utilization is as the name implies—the average amount of traffic on the network in the time period monitored. The average utilization for the network segment should not exceed 40%. Most new network segments will operate at 10% to 25%. The average and peak utilization is of major concern. When high rates occur, the network should be evaluated as to the cause. After the cause is determined, a plan for corrective action should be proposed.

Frame size peak is a record of the largest frame size recorded during the time period monitored. *Frame size average* is the average of all frame sizes during the period monitored. Frame size varies, but the factors that affect frame size are the type of protocol and equipment used, the amount of traffic on the segment, and the type of data transmitted. An average TCP frame will be approximately 500 bytes. Very small TCP frames can indicate a problem.

Number of protocols is the number of different protocols transmitted in the segment. A large number of protocols can indicate an improperly configured network interface card. For example, a card that has been configured for the Microsoft Client, Novell Client, and more may include more protocols than what

peak utilization
the highest level of
utilization experienced by the
network.

average utilization
the average amount
of traffic throughout
the network in the
time period
monitored.

frame size peak
a record of the largest
frame size recorded
during the time
period monitored.

frame size average
the average of all
frame sizes during
the time period
monitored.

is needed and may generate an unusual amount of network traffic. This is especially true if the network interface card has been configured for frame type auto detection. Novell recognizes four different Ethernet frame types. When communicating across the network, it is possible to generate the same frame in all four frame formats before successfully delivering the data. This automatically generates four times the amount of needed network traffic. Configuring the network interface card for the correct frame type can reduce traffic.

Number of nodes is the number of nodes communicating on the network segment. The number of nodes normally increases as time goes by. This is because economically healthy organizations grow. When comparing segment statistics over long periods of time (years), the number of nodes generating network traffic needs to be taken into account.

The *Most active 10 nodes* entry is an indicator of possible network problem locations. For example, a workstation with a bad cable connection may be dropping an unusually large number of frames, which in turn causes new frames to be generated and transmitted. A node that must retransmit many frames will most likely be in the top ten of this list.

The *Collisions* entry represents frame or packet collisions on the network segment. Collisions are the primary indicator of a problem. There should be very few collisions on a healthy Ethernet segment. Less than 1% is not unusual. *Packets dropped* also indicate a problem. Packets dropped are a good indication that there is a cable or connection problem.

While a form was used in the example, many systems can typically collect data and export the data to spreadsheet format. The spreadsheet format can be incorporated into a word processor application or simply printed out as a direct report with graphs. Most protocol analyzers can also record system baseline information.

Be aware that a server, not just network devices, can cause network bottlenecks. The three main causes of server bottlenecks are insufficient resources, unbalanced client loads, and incorrectly configured service(s). Insufficient server resources include too many users for the amount of server resources available. This is especially true of server memory and hard disk drive activity. To alleviate this type of problem, add more memory or configure more hard disk drives and map users to the additional drives.

When resource sharing is unbalanced, too many users are logging on to the same server when other servers are available. In this case, the default server logon for a portion of the users should be changed to balance the load.

A service may be incorrectly configured to handle the given client load. For example, a DHCP server may be configured with a limited IP address pool or with a long lease period duration. In this case, the administrator may need to increase the number of available IP addresses or shorten the lease period.

Server baseline

Some of the most commonly monitored categories for establishing a server baseline are memory utilization, hard disk activity, CPU utilization, and server access activity. Memory utilization can help detect if more memory should be installed. Baselines include determining the amount of RAM available and paging file (swap file) activity. The amount of RAM available should not be less than 4 MB. Paging file activity should not reach higher than 70%. Higher activity indicates that more memory (RAM) should be added. The recommended page file size is approximately 1.5 times the amount of RAM. In other words, if the server contains 1 GB of RAM, the hard disk space reserved for the page file should be set to approximately 1.5 GB. The exact size required varies according to the network operating system installed on the server.

Hard disk drive activity, or disk I/O activity, is monitored to determine when additional disks need to be added to the system. When a hard disk drive reaches approximately 80% capacity, storage space should be added.

CPU utilization determines if the CPU is overloaded with requests. CPU utilization should not exceed 85%. When the 85% threshold is reached, a faster CPU or CPUs should be added. Additional servers can also be added. Servers can be configured in clusters and configured to balance the resource load.

Server access can be monitored by network connection activity. In addition to network connection activity, telephone connection activity will need to be monitored if the server is configured for remote access using a telephone modem.

It is important to note that running a performance monitoring utility adversely affects server performance. Server performance monitoring utilities use resources such as memory and processor time. A healthy server should be able to handle some limited performance monitoring. However, if the server is already overburdened, the use of a performance monitoring utility can cause a severe increase in response time to client requests.

Monitoring Tools

Network events need to be monitored to detect and predict problems. The exact problems can vary a great deal, from simple user problems to detecting an intruder's attempts to break into the system. All major operating systems have monitoring utilities. There are also many good third-party tools available. Monitoring tools not only monitor events but also record the events into a log, which is then saved as a file. Event logs can be used to assist in analyzing system problems involving hardware, software, and security issues.

Performance monitors

Most major operating systems come with some kind of performance monitoring utility. Microsoft includes the System Monitor utility and Network Monitor utility. The System Monitor utility displays selected information about hardware activities of disks, memory, CPU(s), printing, and networking. The results of system monitoring are displayed as a graph. The System Monitor utility can be used to plot server performance and send alerts to the system administrator. See **Figure 18-2**.

A utility like System Monitor provides the system administrator with objective facts on which to base decisions about server hardware. It is an excellent tool for predicting system failures and determining the cause of poor performance. When used properly, an administrator can schedule routine hardware replacement and upgrades that will minimally affect user access. For example, if the CPU is at maximum performance most of the time, a better CPU or additional CPUs can be added during a period of low usage, such as on the weekend when server down time will affect a minimal number of users. Scheduling routine repairs and upgrades prevents unexpected and predictable network system failures. Many failures can be avoided by routine monitoring of system activities.

Microsoft Network Monitor is a network protocol analyzer that is used to capture network traffic in much the same way as other protocol analyzers. It can be used to establish baseline information for the network server or a particular network segment. The Network Monitor (**netmon.exe**) utility is not

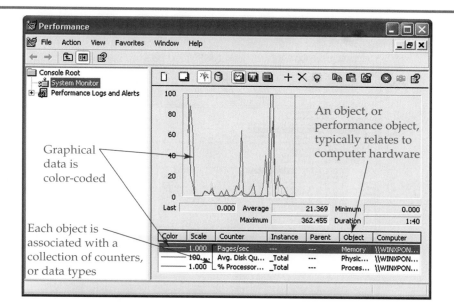

Figure 18-2.
The System Monitor utility can be used to monitor server performance.

installed by default. It must be installed using **Control Panel | Add Remove Programs | Add/Remove Windows Components**.

Figure 18-3 shows the various types of data gathered by Network Monitor. Network statistics such as network utilization, frames per second, and bytes per second, can help determine network performance. **Figure 18-4** shows a list of frames captured during one session. Any frame can be examined in more detail simply by clicking its entry. Details, such as the IP address and MAC address of the destination and source, the contents of a packet contained inside the frame, and a brief explanation of the frame's purpose, can be revealed. A filter can be configured to inspect specific frame types or addresses.

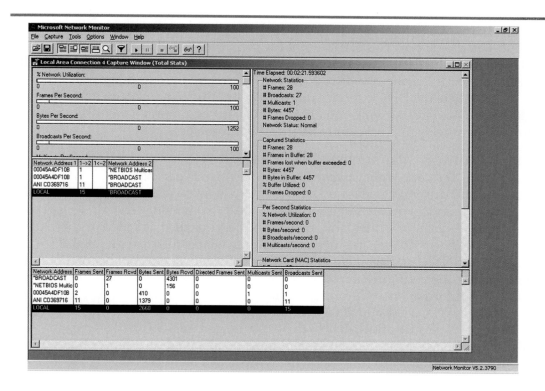

Figure 18-3.
The Network Monitor utility is a basic protocol analyzer that can be used to monitor network performance.

Figure 18-4.
Details of a captured frame, such as the IP address and MAC address of the destination and source, and the contents of a packet contained inside the frame can be examined by clicking the frame's entry.

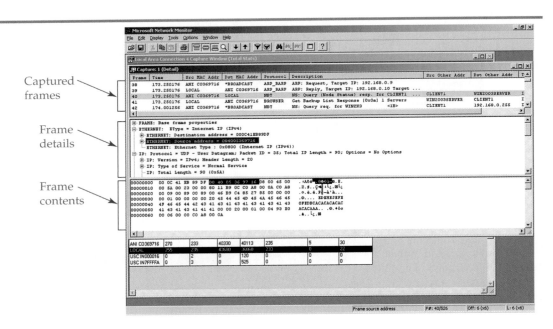

Captured frames

Frame details

Frame contents

Event viewers

All major operating systems used on servers and many desktops include some form of an event monitoring utility. Microsoft provides in many of their operating systems a utility called Event Viewer. It is designed to record events that occur during the operation of the server or desktop system. Some typical events are the starting or stopping of a service, user logon activities, share access, file access, and hardware and software information. The Event Viewer is an excellent utility for troubleshooting system failures.

Figure 18-5 shows some of the various events recorded by Event Viewer. Pay particular attention to the DHCP warning in the Event Viewer screen in the illustration. Event details for a specific event can be viewed by double-clicking that event. This will open the **Event Properties** dialog box, **Figure 18-6.** In the **Event Properties** dialog box, you can see that the computer failed to receive an IP address from the DHCP server and was issued IP address 169.254.166.252, which is an Automatic Private IP Address (APIPA) issued by the computer.

Not only is the DHCP service monitored, but the WINS service and other networking services as well. You can see the value of a monitoring utility such as this one to quickly and efficiently locate system problems.

Simple Network Management Protocol (SNMP)

Simple Network Management Protocol (SNMP) is part of the TCP/IP suite of protocols and designed to allow an administrator to manage and monitor network devices and services from a single location. SNMP can monitor network devices such as servers, workstations, hubs, and routers, and services such as DHCP and WINS.

An SNMP service consists of an SNMP management system and SNMP agents. The SNMP management system queries SNMP agents for information, such as the status and configuration of network devices and services. The SNMP agents gather this information and store it in a management information base (MIB).

Windows XP, 2000, and 2003 operating systems include the SNMP Service. However, this service allows a host to be configured only as an SNMP agent. A third-party SNMP management system must be installed on a server that will act as the SNMP manager to fully implement the SNMP service.

Simple Network Management Protocol (SNMP)
a protocol that is part of the TCP/IP suite and is designed to allow an administrator to manage and monitor network devices and services from a single location.

DHCP warning

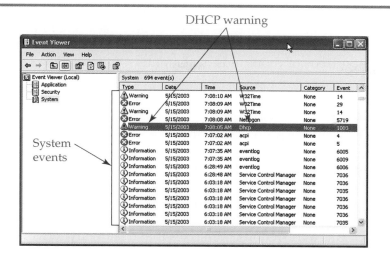

System events

Figure 18-5.
System events listed in
Event Viewer.

Maintaining System Software

Software packages typically need routine maintenance. When software developers design their programs, they do the very best job they can, but problems do occur. A software program error is referred to as a **bug**. Software patches, or fixes, are developed after the discovery of a bug or security problem. Software is usually updated through patches, service packs, and system upgrades.

bug
a software program error.

Patches

Software patches for operating systems are released frequently. Patches should be tested on a limited number of computers or servers before installing them across the entire network. Patches can cause problems with your existing system because of unknown or untested hardware software applications. After carefully and thoroughly testing the effects of a patch, you may then proceed to install the patch on other computers in the network. Another reason for the

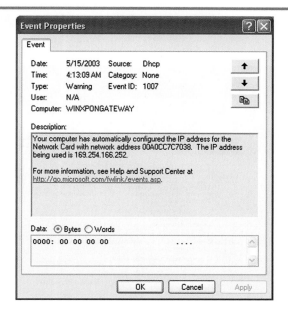

Figure 18-6.
Event Properties dialog
box displaying the
details of a DHCP
warning event.

caution is most patches cannot be uninstalled. Once installed and an unforeseen critical problem develops, you may be forced to do a back up of important data, wipe the system clean, and then do a new operating system installation. When talking about hundreds or thousands of computers, this could be very time consuming.

Service Packs

service pack
a collection of software patches, or fixes.

Microsoft uses the term *service pack* to describe a collection of software patches, or fixes. After every release of a major operating system, a series of service packs is provided free of charge to users to update the system. A typical service pack consists of program modifications that provide security fixes and updated drivers for new hardware technologies. For example, Windows NT 4.0 went through a series of six major service packs. The first service pack released for Windows XP was a collection of 324 fixes that dealt with network, hardware, security, and basic operating system issues. You can download a detailed list of each fix included in the service pack. Details about the nature of each problem are also provided.

Figure 18-7 shows the beginning of a list of individual fixes that were released with Service Pack 1. There are a total of 324. To the left of each fix is a technical article reference number. Clicking the technical article reference number displays an article with complete details about the fix.

Keeping the operating system up-to-date with service packs and individual patches as they are released is good insurance against program security flaws and other problems that are discovered. Service packs should be tested before installing them on the network.

Upgrades

software upgrade
a major improvement or enhancement to existing software programs.

A *software upgrade* is a major improvement or enhancement to existing software programs. For example, when Microsoft developed major improvements and enhancements for Windows 98, they re-released Windows 98 as Windows 98 Second Edition. When the total amount of service packs and individual patches warrant it, a system upgrade is often released. The term *upgrade* also refers to changing from one operating system to another, such as upgrading from Windows 2000 Server to Windows Server 2003.

Figure 18-7.
The beginning of a list of individual fixes that were released with Windows XP Service Pack 1.

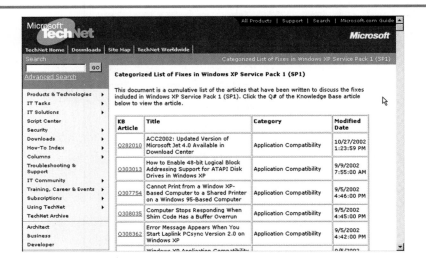

Software Installations

New software programs should be tested before they are installed on a network server or distributed to network clients. Software developers do a good job of testing their software on numerous types of computers with various types of hardware and software configurations. Often, a beta version of the software package is released so the developer can gain more knowledge about the software compatibility of the software package.

Microsoft and other software developers routinely allow users to have access to beta versions of an upcoming release of a software package. For example, Microsoft Windows Server 2003 was released as a beta version months before it was distributed through retail outlets and installed on distributed computers. By releasing beta versions, thousands of copies of the software can be tested by many different users running various software packages and using many different types of hardware. Numerous flaws can be detected and remedied before releasing the software package.

Hardware manufacturers work hand in hand with software developers to write driver programs to support their hardware. If the hardware drivers are developed in time, they are included in the beta version of the operating system so they can also be tested. After the release of new operating system hardware, developers must provide drivers for the system by packaging a disk or disc with the hardware device or offering free download drivers from their Web site. As you can see, the development of a new operating system can be quite involved.

Before you install a new software package on a server or client, it should be fully tested. A test network consisting of a few workstations and a server can be set up using the same software and hardware that exists in the real network. The test network is separate from the real network and is used to test software package compatibility with hardware and software applications. A corporation or institution should have a lab or training room set aside that can be used for testing purposes.

Installing a new software system on a network can cause unpredictable problems. Some problems are not reversible. Any administrator or technician would want to avoid having to reformat the server hard drive, reinstall the operating system and software applications, and restore the data that was last backed up. During this time period, all personnel who use the network for work purposes would be idle. Production loss can be very expensive. Think about the hundreds or even thousands of employees that are being paid while waiting for the network to be functional. Think about how many customers a business could lose.

Before installing any new hardware or software on a network system, you should check the hardware or software compatibility list at both the hardware manufacture Web site and the operating system Web site. **Figure 18-8** shows a Microsoft Hardware Compatibility List (HCL) for Windows Server 2003. It lists the server and cluster systems that support Windows Server 2003.

Often, information on one Web site will conflict with another. For example, the operating system Web site might indicate that the hardware device is not compatible with the operating system. By going to the hardware manufacturer's Web site, you may find that a driver has been released that will allow the hardware to work on that operating system. Conflicting information about hardware compatibility is a common occurrence.

Operating systems often offer patches or fixes for security problems as they are discovered. Security is an on going issue. Breaches in network security

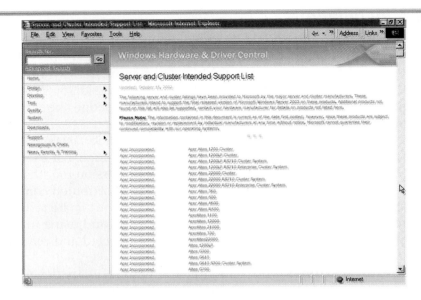

Figure 18-8.
A Microsoft Hardware Compatibility List (HCL) for Windows Server 2003, listing the server and cluster systems that support Windows Server 2003.

happen constantly. E-mail newsletters are offered free of charge at all leading operating system Web sites. You should subscribe to these newsletters and read them often to become aware of the latest developments. You must stay ahead of intruders. Leading antivirus developers also provide newsletters. Not only do antivirus developers provide you with timely information, they typically release fixes in 24 hours or less of initial virus detection.

Maintaining System Hardware

The proper maintenance of system hardware is cost effective. A system failure during business hours can be very costly to the company. Whenever possible, all system hardware maintenance should be performed when business and clients will be least affected.

Scheduling Downtime

Scheduling downtime is an important aspect of maintenance. A small network with a single server should have maintenance performed after business hours to avoid the possibility of a system crash while workers are present. You never want to perform maintenance that may possibly jeopardize the use of the network system. Most maintenance such as backups, upgrades, and virus updates should be performed when the network is not being used. A notice should be sent to all users so that they know that the network may be down for a period of time. The notice should be sent even if the maintenance is going to be performed after regular business hours. Workers may be planning to work late or to work at home by accessing the network remotely.

Some businesses operate on a 24/7/365 schedule. In this case, two or more servers are routinely installed. One may be taken off-line to perform maintenance while the other carries the network. A special type of server installation is referred to as cluster. A *cluster* is a group of servers that share the network demand. One of the servers can easily be taken off-line for maintenance without disrupting network activities.

cluster
a group of servers that share the network demand.

Major Network Hardware Upgrades

Backbones, additional servers, routers, switches, and such may be added to the network system. As a company grows, it will undoubtedly require more networking capacity. This means that additional throughput capacity and additional workstations will be required. As the network system grows, other additional hardware may be required, such as routers and switches. Routers and switches are sometimes installed on existing systems as the number of workstations increase. The additional hardware is used to create Virtual LANs (VLANs). A VLAN increases the available bandwidth. The bandwidth is increased because each VLAN acts as its own separate network, thus reducing the amount of collision zones.

Each VLAN has the same throughput as the original LAN. For example, it is possible to divide an existing 10BaseT network consisting of 100 workstations into two separate VLANs by using switch technology. The network is divided according to workgroups. Workgroups often communicate within their own workgroup but seldom outside of the same workgroup. By dividing the network into two separate LANs, the bandwidth is technically doubled by creating two networks, each with 10 Mbps bandwidth. However, this is not always practical. Bandwidth is based on the actual working conditions of two or more workgroups within the network.

Maintaining System Integrity

The completeness and accuracy of data is referred to as system, or data, integrity. Maintaining system integrity requires protecting the data while the server is running. To protect data, fault tolerance and disaster recovery must be designed into the network. You should recall that fault tolerance is a system's ability to continue operation during a system hardware or software error. Fault tolerance strategies include some RAID systems and electrical power systems. Electrical power systems are covered in detail later in this chapter. **Disaster recovery** is the restoration of a system to normal operation after a disaster has occurred. It includes data backups, hot and cold spares, and hot, warm, and cold sites.

*disaster recovery
the restoration of a
system to normal
operation after a
disaster has occurred.*

Network+ Note:

For the Network+ Certification exam, be sure you know the difference between fault tolerance and disaster recovery and can identify devices, systems, or strategies related to each. Remember, fault tolerance includes RAID systems, server clustering, and UPS units. Disaster recovery includes data backups and data backup media storage. RAID 0, however, does not provide fault tolerance.

Fault Tolerant RAID Systems

RAID will only be quickly reviewed in this chapter. For a detailed explanation of each RAID system, review Chapter 10—Introduction to the Server. While there are many different RAID styles, the predominant RAID systems that provide fault tolerance are RAID 1 and RAID 5.

RAID 0 does not offer fault tolerance. It is designed strictly for data storage speed.

Tech Tip

RAID 1 consists of two disk drives with matching data on each drive. If one drive fails, the other has a complete copy of data. RAID 1 is also known as *disk mirroring*. RAID 5 is parity and striping combined to provide an economical way to protect data should a hard disk drive fail. At least three hard disk drives are required for a RAID 5 configuration. Data blocks are striped across two hard disk drives, and parity for those two blocks is stored on a third hard disk drive. Parity and data blocks alternate across the drive system so that each hard disk drive consists of approximately two part data blocks and one part parity. This method ensures equal restoration time if any of the three hard disk drives fail.

Backup Data Methods

Even with fault tolerant RAID systems, data can still be corrupted or damaged beyond repair. For example, a virus could infect each disk drive in the RAID system. A natural disaster, storm, hurricane, tornado, flood, or a fire could destroy the computer system. A backup data method must be employed to restore valuable data in case any of these events occur. Backing up data is part of a disaster recovery system. Backups should be made at regular intervals. This way the data can be restored up to the last, good backup.

The term *backup* refers to making copies of the data on a storage system. Backups can be made manually or automatically through the use of software applications designed for such purposes. Copies of data can be made to most any type of data storage medium such as floppy disks, CD-RW drives, and tapes. The most common storage device is a tape backup drive, but CD-RW drives and DVD-W drives are becoming more appropriate for small storage systems. Three types of backups can be performed: full, incremental, and differential.

Full backup

full backup
a backup method that backs up all designated data.

As the name implies, a *full backup* is a complete backup of all designated data. You do not need to backup everything on a server during a full backup. You need only copy the data. If the hard disk drive is replaced because it failed, the operating system and applications can be reinstalled from discs. The data is the real concern. A full backup operation copies all identified data during a single backup period.

Incremental backup

incremental backup
a backup method that backs up only data that has changed since the previous backup.

An *incremental backup* copies only data that has changed since the previous backup. Compared to full backups, incremental backups are time savers.

A typical scenario of using the incremental backup method is to perform a full backup once a week and then incremental backups daily. In a week's time, this would equal seven backup sessions: one full backup and six incremental backups. This process saves time when compared to doing full backups daily. The only disadvantage is that when data is restored, the last full backup must be restored first and then each incremental backup in sequential order. If one incremental backup is restored out of sequence, the data restoration is flawed. See **Figure 18-9**. Keep in mind that large data systems may require more than one tape each day for differential as well as incremental backups. One tape was used in the example to keep the concept simple.

Differential backup

differential backup
a backup method that backs up all data that has been changed since the last full backup.

A *differential backup* backs up all data that has been changed since the last full backup. There is no need to use a series of tapes. One tape for the differential

Figure 18-9.
An incremental backup requires separate media for each incremental backup.

backup will do because all data changes since the last full backup are recorded. See **Figure 18-10.**

A typical scenario of using the differential backup method is to perform a full backup once a week and a differential daily on the other days of the week. To restore data, the full backup is restored and then the last differential backup.

You may wonder how the technology knows if a full backup has been performed or an incremental or a differential change has occurred. A file attribute, called archive, provides this information. Remember that files have file attributes such as read only and system. The archive attribute is actually a bit that is set when changes are made in the file's contents. The **archive bit** identifies if a file has changed since the last full or incremental backup. By indicating which files have changed since the last full backup, the backup system can copy only the changed files. The archive bit is reset every time a full backup or an incremental backup is performed. The archive bit is not reset when a differential backup is performed or when a file is copied using commands such as **copy** and **xcopy**.

Remember, an incremental backup copies only files that have changed since the last full backup or last incremental backup. Both incremental and full backups reset the archive bit. When the contents of a particular file changes by being opened and then closed during a business transaction, the archive bit is set, indicating that it should be backed up during the next backup operation.

A differential backup does not reset the archive bit. It simply copies all files that have changed since the last full backup. Note that differential and incremental backups are never mixed. Only one type of backup, incremental or differential, can be performed after a full backup.

archive bit
a file attribute that identifies if a file's contents has changed since the last full or incremental backup.

Figure 18-10.
A differential backup can use the same media for each backup.

Mon	Tue	Wed	Thr	Fri

Last full backup   +   +   =   Data recovered

Data backup and restoration methods

Backup data consists of a collection of many different files. Not all files are accessed on any given day. For example, if a bank maintains one file per customer, then that single file's content remains unchanged until the customer performs a bank transaction. A bank may have thousands or even millions of customers, which equates to millions of files. There is no need to back up every file every day, especially if the vast majority of files have not changed. In this scenario, performing a full backup once a week with incremental backups during the week is sufficient and will speed the backup process. Small data systems, on the other hand, can perform a full backup everyday or a full backup once a week followed by a differential for every other day.

Remember, when restoring data from an incremental backup, each incremental tape is needed as well as the last full backup. The incremental backups must be restored in the exact order they were created since the last full backup. When restoring data from a differential backup, only the last full backup and the last differential backup created are needed.

Storing backup data

Backups of critical information are a must, but what if the facility experiences a disaster such as a fire, flood, severe storms, or vandalism? The tape backups must be stored in a secure, climate-controlled, off-site environment. You may have additional copies of the same backups on-site for easy access if needed, but critical data must be stored away from the site. Key points concerning data backup storage include the following:

- Store backups off-site.
- Store backups in a secure area with limited access.
- Store backups in a climate-controlled storage area.
- Avoid storing backups in direct sunlight.
- Avoid storing backups near electrical panels or in equipment rooms.
- Avoid exposing backups to magnetic fields generated by electrical equipment, such as motors and speakers.

Never store magnetic tape, disc media, or any computer equipment in direct sunlight. Direct sunlight can damage sensitive magnetic media. The magnetic tracks on backup media can lose part or all of their magnetic qualities due to exposure to direct sunlight and the heat generated by the exposure. Backup media should never be stored in equipment rooms of buildings that contain electrical equipment, HVAC (Heat Ventilation and Air Conditioning), or other forms of electrical equipment and electrical motors. Storage areas need to be climate-controlled. Backup media, especially tapes, are easily damage by excessive heat and humidity. Always store backup media in a secure area that has limited access by personnel. Someone can easily throw away a collection of backup media while on a routine cleanup assignment of a dusty storage area. Also, remember that backup media will prove to be of little value if the disaster that destroys the original data such as fire also destroys the backup copies.

Hot and Cold Spares

A *hot spare* is a backup component that can automatically replace a failed system component without the intervention of a technician. A hot spare is typically a spare disk drive that is part of a RAID system but not in regular use.

hot spare
a backup component that can automatically replace the failed system component without the intervention of a technician.

It is a standby unit that activates when one of the disk drives in the RAID unit fail. On failure of any disk drive, the hot spare automatically replaces the failed disk drive. The disk controller will automatically begin rebuilding the lost data on the new disk drive. The process of rebuilding the data onto the new drive is only possible with a fault tolerant type of RAID system, such as RAID 1 or RAID 5. The hot spare provides excellent fault tolerance and zero down time for the server.

A failed disk drive can also be replaced by a cold spare. A *cold spare* is any compatible disk drive that is in storage and is used to replace a failed disk drive. If the server supports hot swap drives, the cold disk drive can be hot swapped with the failed disk drive. As soon as the failed drive is hot swapped with the cold spare, the disk controller will automatically begin to rebuild the data onto the new disk drive. The advantage of a hot spare over the cold spare is obvious. There is a minimum amount of server down time when hot spares are used.

cold spare
any compatible drive that is in storage and used to replace a failed drive.

Hot, Cold, and Warm Sites

The terms hot, cold, and warm are used to describe the readiness of an off-site data storage facility to recover from a failure at the primary site. A cold site refers to a data storage facility where backup data is stored. A warm site consists of hardware and data, but the data has not yet been loaded on to the hardware. A hot site is a data storage facility where a backup of data is stored as well as a running system containing the most up-to-date data. The system is ready to serve users in case of a complete system failure at the original site. The only thing missing from the system is the last incremental or full backup.

As you can obviously see, the choice of off-site storage is directly related to cost and the value placed on the time it takes to recover from a data loss scenario. The cold site is the cheapest but takes the longest to set up and recover. The hot site offers the quickest recovery but is the most expensive to maintain.

Maintaining Stable Electrical Power

A good supply of electrical power is essential for keeping the server running. Most people assume the electrical power coming from the 120-volt outlet is a steady stream of 120-volt electricity. The fact is, the electrical power coming from the outlet is not a steady stream and, most times, is not exactly 120 volts. Commercial electrical power suffers from conditions such as spikes, surges, brownouts, and blackouts. This section contains a detailed discussion of electrical power conditions and the electrical power devices used to overcome the ill effects of these conditions.

Electrical Surges and Spikes

Electrical surges and spikes are a very common occurrence in commercial electrical power. An *electrical surge* is a higher-than-normal voltage level, typically caused by lightning. An *electrical spike* is a very short burst of abnormally high voltage, typically caused by electrical equipment. Electrical surges and spikes happen many times throughout the day and typically go unnoticed. The magnitude of the voltage level varies. Normally, electrical spikes are not of sufficient magnitude to harm most electrical systems, even delicate electronic equipment. Over the years, electronic designers have incorporated techniques that limit the effect of most electrical surges and spikes. Occasionally though, they are of sufficient strength to damage equipment.

electrical surge
a higher-than-normal voltage level, typically caused by lightning.

electrical spike
a very short burst of abnormally high voltage, typically caused by electrical equipment.

lightning arrestor
a special piece of electrical equipment designed to dampen the effects of an electrical surge caused by lightning.

The most severe electrical surge is caused by lightning strikes. A thunderstorm need not be in close proximity to equipment to damage it. After the lightning strikes a power line, the surge can travel over twenty miles before reaching its final destination. Equipment along the pathway of the electrical surge can be damaged. A *lightning arrestor* is a special piece of electrical equipment designed to dampen the effects of the electrical surge caused by the lightning. While they work most times, they do not always prevent the destruction caused by the lightning. Most network systems are adequately protected from lightning damage by a good quality Uninterruptible Power Supply (UPS). UPS devices are discussed in detail later in this section.

Another cause of electrical surges through electrical systems is automobile accidents. Many times an automobile will strike a utility pole. When the automobile strikes the pole. The power lines swing together causing two lines to touch. This causes a momentary electrical explosion, which is followed by an electrical surge and occasionally a brownout or blackout condition.

Electrical equipment, such as motors, welders, and even electrical switches, cause electrical spikes. The spike is the result of an electrical phenomena referred to as *inductive reactance*. Inductive reactance is very common and is associated with the windings (coiled wire) inside motors, transformers, appliances, tools, and various other pieces of electrical equipment. When an electrical circuit that uses alternating current (AC) is opened or closed and the circuit has inductive properties (windings), an electrical spike is produced. The exact voltage level of the spike depends on the electrical characteristics of the circuit. In short, there are many electrical spikes during a normal 24-hour period. Most electrical spikes go unnoticed because of the way electrical systems are designed. The amount of information required to adequately understand inductive reactance is far beyond the scope of this book. In fact, it is one of the most difficult principles to master in the study of electronics. For now, just remember that electrical equipment, such as the items mentioned, can generate electrical spikes during normal usage.

Standards and best practices are used to reduce the effects of electrical spikes. For example, computer system electrical circuits are typically isolated from electrical circuits that provide power to lighting and electrical equipment. By isolating the electrical circuits serving computer equipment, the possibility of electrical spike produced by equipment in the building are reduced. For example, commercial buildings, hospitals, and schools use a bright orange color to identify electrical outlets to be used strictly for computer or sensitive electronic equipment, such as hospital monitoring or life support equipment. These outlets should never be used for equipment that can produce electrical noise or spikes, such as vacuum cleaners, drills, microwaves, and other common equipment found in the workplace. Many electrical appliances and power tools produce unwanted distortions in the AC signal, which may damage sensitive electronic equipment or corrupt data. A worker using a drill plugged into one of the designated orange outlets will produce thousands of small spikes that will affect equipment on the entire dedicated circuit.

Brownouts and Blackouts

brownout
a partial loss of electrical energy, or voltage, in which the voltage level is lower than normal or average levels.

blackout
a total loss of electrical energy, or voltage.

Brownouts and blackouts are a classification of electrical energy loss dependent on the amount of voltage lost. A *brownout* is a partial loss while a *blackout* is a total loss of electrical energy. For example, during a blackout, there is no voltage available. During a brownout, there is a measurable, lower-than-normal voltage level. A brownout sometimes appears to the naked eye as a dimming effect.

Blackouts are routinely caused by equipment failure, automobiles striking utility poles, fires, and natural disasters such as earthquakes, storms, tornadoes, and hurricanes. While a blackout can cause a loss of data, a brownout can actually damage electrical equipment. When a brownout occurs, the voltage level at the electrical equipment is lower than the equipment normally expects. Most electrical equipment will attempt to compensate for the lower voltage level by allowing the amperage to increase. The increased amperage will be in excess of what the system can safely handle. Normally, a fuse will blow or a breaker will trip, thus protecting the equipment. If not, the high electrical current (amperage) will damage components in the electrical system. The effects of blackouts and brownouts on the network server and other related equipment could be prevented by the use of an Uninterruptible Power Supply (UPS).

Brownouts may occur on any power system. The main cause of brownouts is excessive consumption of electricity. Excessive consumption happens most often when there are severe weather conditions, such as extremely high or low temperature. Electrical heating and cooling equipment requires large amounts of electrical energy during extreme temperature conditions. Most electrical devices specify that the ideal voltage for equipment should be within 10% of the indicated voltage level of the appliance or machine. For example, an electrical device rated for 120 volts AC requires a safe operating voltage level plus or minus 10% of 120 volts. That would be a range of 108 volts to 132 volts. Any voltage level outside this range could have serious side effects.

Uninterruptible Power Supply (UPS)

An *Uninterruptible Power Supply (UPS)* is a device that ensures constant and consistent network performance by supplying electrical energy in case of a power failure or blackout. UPS systems are also designed to provide protection from electrical surges, spikes, and brownouts. A UPS provides a steady voltage level to equipment even when the input voltage level is higher or lower than needed. The exact length of time the UPS supplies power during a blackout depends on the amount of electrical load it must support and the size or amp-hour capacity of the batteries in the UPS unit. A UPS also provides power conditioning. *Power conditioning* is the process of eliminating spikes as well as any type of variation in the desired AC signal pattern.

There are two major categories of UPS units: standby and continuous. Many other names are used to describe UPS units, such as off-line, on-line, and line-interactive. Off-line and line-interactive are standby UPS units. On-line is a continuous UPS unit. Manufacturers often coin these names. What you need to understand is the characteristics of each. This will help you know which one is the correct style for your computer system's protection. For this chapter, we will use the terms *standby* and *continuous* for the two major categories.

A *standby UPS* waits until there is a disruption in commercial electricity before it takes over the responsibility of supplying electrical energy. A *continuous UPS* provides a steady supply of electrical energy at all times, even when there is no electrical problem. The reason for using a continuous UPS is to eliminate the possibility of data corruption. A standby UPS quickly takes over to supply electrical energy when commercial power is lost. The standby UPS can switch from standby mode to full operation in less than 5 milliseconds, or 5/1000 of a second. While that may seem fast, it is not fast enough to ensure error-free data during computer data processing. The best solution is to use a continuous UPS. Look at **Figure 18-11** to see how the two compare.

Uninterruptible Power Supply (UPS) a device that ensures constant and consistent network performance, by supplying electrical energy in case of a power failure or blackout.

power conditioning the process of eliminating spikes and any type of variation in the desired AC signal pattern.

standby UPS a UPS unit that waits until there is a disruption in commercial electricity before it takes over the responsibility of supplying electrical energy.

continuous UPS a UPS unit that provides a steady supply of electrical energy at all times, even when there is no electrical problem.

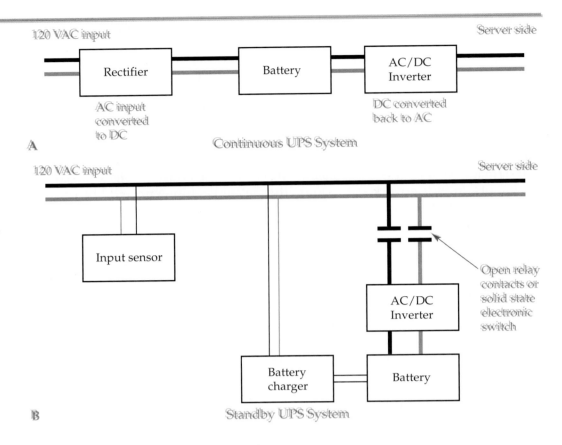

The two major categories of UPS unit are continuous and standby. A—A continuous UPS supplies a constant flow of electrical power to the server. When the commercial power goes off, there is no momentary interruption of electrical power to the server because power is continuously supplied. B—A standby UPS does not provide a steady continuous supply of electrical power. There is always a momentary power interruption while switching to backup power.

Both UPS systems use an inverter and a battery. The main difference is the continuous UPS provides a constant flow of electrical energy to equipment without the slightest interruption, Figure 18-11A. The standby UPS activates the switching mechanism when low or no voltage is detected. In the Figure 18-11B, a pair of relay contacts indicates the switching mechanism. Some units use solid-state electronic switches, but even the electronic switches cause a slight interruption in power. Even though this may only take a few milliseconds and go unnoticed in conventional lighting, television, and other pieces of equipment, it is serious enough to lose valuable data during data transactions such as saving, loading, or transferring.

Tech Tip

A UPS should be sized at approximately 130% of the average load, but not less than 110% of the expected maximum load.

Many operating systems have software programs that work with UPS systems. For example, Windows 2000 Server has built-in UPS support designed through a partnership between Microsoft and American Power Conversion (APC). Look at **Figure 18-12.** The **Power Options Properties** dialog box can be used to configure a UPS connected to the server. Once configured, the UPS can display the status of the UPS unit and the status of the commercial power coming into the unit. You can also check low battery and low voltage conditions and monitor the UPS unit from somewhere on the network other than at the server.

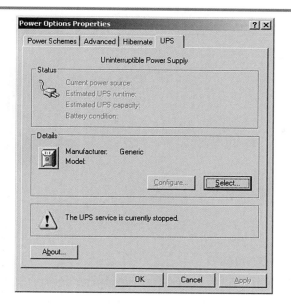

Figure 18-12.
Some operating systems have built-in support for UPS units, which allow the UPS battery and incoming electrical power to be monitored.

Microsoft handles communication with a UPS unit as a service. To start communications with the UPS unit, you may need to open and manually start the UPS service located at **Start | Administrative Tools | Services | Uninterruptible Power Supply**. Look at **Figure 18-13.** Notice that the UPS service can be started, stopped, paused, and resumed.

The UPS communicates with the server through a serial or USB port. The power supply sends alerts to the server or network administrator concerning electrical problems ranging from a total loss of electrical power to low-voltage conditions and the need to replace the existing batteries in the power supply. Most UPS systems are also equipped with RJ-45 and RJ-11 ports to provide surge protection for the network and telephone communications lines. See **Figure 18-14.**

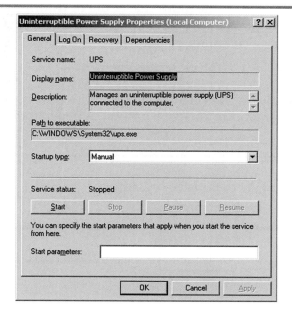

Figure 18-13.
The UPS service can be started, stopped, paused, and resumed.

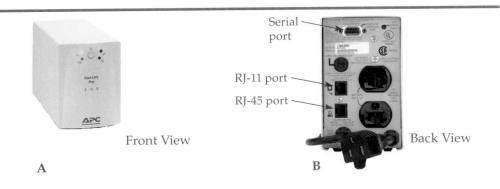

Front View   Back View

A   B

Isolation Transformers

isolation transformer
a device that uses a
transformer to isolate
a circuit from other
circuits emanating
from the same
electrical source.

Dedicated electrical circuits used for supplying computer equipment typically have isolation transformers installed near the source or electrical panel. An
isolation transformer is a device that uses a transformer to isolate a circuit from
other circuits emanating from the same electrical source. For example, electrical
circuits coming from the electrical panel feed regular purpose outlets used for
coffee pots, copiers, paper shredders, pencil sharpeners, and other types of electrical equipment. An isolation transformer isolates the special computer equipment circuit from the other circuits. Transformers are normally used to raise or
lower voltage levels in electrical or electronic systems. Look at **Figure 18-15** to see
how an isolation transformer works.

The transformer consists of a primary, or input, winding and a secondary, or
output, winding. The two windings are not electrically connected. They connect
through a magnetic field produced in the primary winding. The magnetic field
passes through the secondary winding, which in turn produces voltage in the
output that mirrors the input voltage. One of the by-products of using an isolation transformer is its natural ability to suppress voltage spikes. The isolation
transformer isolates the circuit on the output side of the transformer from the
effects of spikes generated on the input side of the transformer. The isolation
transformer will not eliminate very large spikes, but it will suppress the voltage
level, which can damage computer equipment.

generator
a device that creates
and provides
electricity.

Generators

A *generator* is a device that creates and provides electricity. It is used for
computer systems that require 24/7/365 uptime. A generator can maintain power

Figure 18-15.
An isolation transformer
is used to isolate, or
separate, dedicated
power outlets from the
other power outlets in
an electrical distribu-
tion system. Isolation
transformers suppress
electrical voltage spikes.

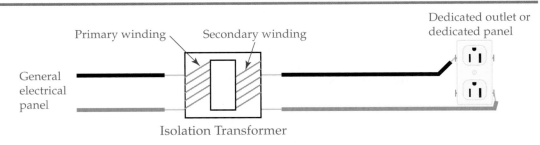

Primary winding   Secondary winding   Dedicated outlet or
dedicated panel

General
electrical
panel

Isolation Transformer

for a few minutes to several hours depending on the size of the unit. Critical systems require one or more generators—for example, a system of computers used in an airfield control tower. If severe weather took the commercial power out, the control tower could still monitor airplanes and provide information to pilots. Other systems that rely on computer technology and require generators are hospitals, police, fire, rescue, security, military, banking, high-rise buildings, and large enterprise businesses.

Protecting Networks from Malware

In Chapter 16—Network Security, you learned that some of the ways a network could be compromised is through malicious code such as Trojan and Macro. However, malicious code cannot only compromise security, it can damage data and cause system failure. An administrator must be on constant guard against computer malicious code, or malware. A network administrator's best protection for the network is a combination of user policies and antivirus software.

Establishing Network System User Policies

All network users need user and system policies that are set in place and enforced. Policies outline how users are to use and not use their workstation and the network. Basically, policies and procedure are a list of dos and don'ts for the user. Typically, these are called an Acceptable Use Policy. An Acceptable Use Policy that helps protect a system against malware may include the following:

- Do not open any e-mail file attachments from unknown sources.
- Do not download any files from unknown sources.
- Update antivirus software on a regular basis.
- Never open e-mail advertisements, chain letters, or junk mail.
- Do not click on icons embedded in files or e-mails.
- Do not open any e-mail attachments with an exe, com, bat, vbs, shs, pif, ovl extension or double extensions, such as GreatPicture.jpg.exe.
- If you doubt the authenticity of an e-mail, check with the indicated source before opening the e-mail, especially an attachment.

Antivirus Software

Antivirus software is typically used to scan files and e-mail for viruses. It can be configured to run in the background and scan files and e-mail as they are accessed. When an antivirus program detects a virus in this mode, it will display a message that says that a virus has been detected. Depending on how the antivirus software is configured, it will either prompt the user for an action to take or alert the user that it has cleaned, deleted, or quarantined the infected file or e-mail. A file is typically quarantined (moved to a quarantine directory) when the antivirus software program is unable to remove the virus. The user can then use the antivirus software program to view the file and delete it.

A user can also scan files and e-mail at anytime by opening the antivirus program and selecting the storage locations or directories to scan. When an antivirus program detects a virus in this mode, it will display a report after it has scanned all of the selected files. Depending on how the antivirus software is configured, it will either display the action taken or allow the user to clean, delete, or

Figure 18-16.
An antivirus scan report typically lists the names of all infected files, the name of the malware with which each file is infected, and the action taken.

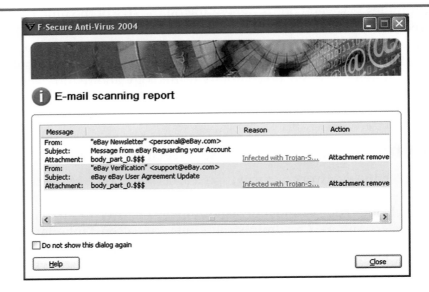

quarantine the infected file or files. **Figure 18-16** shows a report of infected e-mail. Notice that the antivirus software lists the name of the malware with which the file is infected.

Antivirus software should be installed on the server and on each workstation in the network. Antivirus software comes in enterprise versions that allow an administrator to easily install and update antivirus software on all workstations from a single location.

Antivirus software consists of a scan engine and a virus pattern file. The *virus pattern file* is a database of virus signatures, or codes, unique to each known virus. The *scan engine* is a software program that performs the function of reading each file indicated in the scan configuration and checking it against the virus signatures in the virus pattern file. Since antivirus companies continuously update the virus pattern files as new malware is discovered, the virus pattern file should be updated on a regular basis.

An antivirus software program can only remove known viruses. Everyday there are new viruses or variations of old viruses. Virus pattern files are constantly updated to include virus signatures for all known viruses. Sometimes a virus is so new, the latest virus pattern file does not include a virus signature for the new virus. In this case, the administrator must remove the virus manually. Most antivirus sites provide a detailed, step-by-step virus removal process. The process usually involves accessing the system registry and changing values or removing a specific registry key. Normally, you are cautioned never to modify system registry content because doing so could be disastrous. You could change the wrong area of the registry or make a typographical error, which can cause the system to not be able to load the Windows operating system. You should always backup the registry before making any changes to the content.

Most antivirus software packages include additional features such as a firewall, protection against spam, and a popup ad blocker. An antivirus software package that includes additional features is typically referred to as an *antivirus suite*.

Malware

Malicious software is referred to by the term *malware*, which is a contraction of the two-word descriptor. *Malware* is a software program designed to perform

virus pattern file
a database of virus signatures, or codes, unique to each known virus.

scan engine
a software program that performs the function of reading each file indicated in the scan configuration and checking it against the virus signatures in the virus pattern file.

antivirus suite
an antivirus software package that includes additional features.

malware
a software program designed to perform some type of unauthorized activity on a computer.

some type of unauthorized activity on a computer system. The unauthorized activity can be harmless, such as that provided by a joke program, or destructive, such as a program designed to erase all data stored on the system.

The term *malware*, *virus*, and *worm* are often used interchangeably even though this is not technically correct.

Tech Tip

There are many terms used to classify malware, such as *virus*, *worm*, *logic bomb*, *Trojan horse*, *backdoor*, *joke program*, *hoax*, *stealth*, *polymorphic*, and *boot sector*. The exact classification can be confusing when a particular program has a combination of several characteristics. For example, a malware program could be used as an e-mail attachment disguised as a Trojan horse, which is designed to spread throughout the Internet as a worm and infect the boot sector of the hard disk drive in each host. Therefore, we now have a Trojan horse, worm, boot sector virus. This section looks at some of the common classifications of malware based on basic characteristics.

Virus

A *virus* is a type of malware that replicates itself on a computer after infecting it. By strict, technical definition, a virus does not automatically spread itself. A virus is typically introduced into a system, through the use of a disk or disc. It can then be spread by an e-mail attachment. Confusion arises because of the common use of the term *virus* to represent any form of malicious software. The news media is especially fond of the term *virus*.

A virus usually has three phases: infection, replication, and execution. The infection comes from a source such as a floppy disk, CD, e-mail, or network connection such as the Internet. Replication is when the virus duplicates itself to other programs, files, drives, or other computers on a network. If a virus continually replicates itself, it can easily consume all available hard disk drive space. Execution can take many forms, such as harmless, annoying, and destructive.

virus
a type of malware that replicates itself on a computer after infecting it.

Worm

A *worm* penetrates a computer and may reproduce itself. It does, however, infect other computer systems. Technically speaking, a worm is not a true virus, but rather its own destructive program. A virus replicates itself on one computer and infects files on that particular computer. A worm infects files on the infected computer and automatically spreads to other computers. The fact that the worm automatically spreads its own infection is what makes it different from a virus. A classic example of distributing a worm is through the use of e-mail. When a worm is sent as a file attachment, it replicates itself by using the list of contacts in the e-mail database. The people on the contacts list receive an e-mail, which may or may not need to be opened to infect the people on their contacts list. Worms are also referred to as *bacterium virus* by some authorities.

worm
a type of malware that infects files on a computer and automatically spreads to other computers.

Trojan horse

A *Trojan horse* appears as a gift, utility, game, or an e-mail attachment. When the file is opened, the malware is activated. It can cause immediate damage or wait until a later preset date. It can have an outward appearance of a harmless JPEG file or a utility to assist you with some sort of computer operation. You may not even realize that it is a virus. After all, that is the objective of the intruder.

Trojan horse
a type of malware that appears as a gift, a utility, game, or an e-mail attachment.

After a while, things happen. The hard disk drive is erased or filled with duplicates of the original program, leaving no space for any other files to be saved. A Trojan horse can cause most anything imaginable.

Joke program

joke program
a type of malware designed with some sort of joke as the payload.

A *joke program* is designed with some sort of joke as the payload. For example, when the user invokes a certain action, the joke program is triggered. The joke might be designed to display a dialog box on the screen similar to the one in **Figure 18-17.** This joke program was written in Visual Basic. It looks like a real dialog box because it is! Visual Basic uses the same dialog boxes and other features found in the Microsoft operating system family of products.

While this joke appears harmless, there are some that are not. For example, there are joke programs that can render the user's computer unusable. The screen spins before their eyes and cannot be stopped until a secret key combination is used. The prank is harmless, unless the person who receives the joke program never receives the key combination to stop the program. Then the joke program is as destructive as any other malicious software program. Much production time is lost while attempts are made to remove the program. Joke programs are really no joke at all, especially in the world of business.

MBR virus

MBR virus
a type of malware that attacks the master boot record (MBR) of a hard disk drive.

An *MBR virus* is a type of malware that attacks the master boot record (MBR) of a hard disk drive. It is considered extremely destructive. When activated, the virus plants hexadecimal codes in the MBR, rendering the MBR useless. This results in a system boot failure. Damage is usually limited to only the master boot record, which can be rebuilt if backups of system files have been made. Note that this includes the system files, and not just the data files. Many antivirus programs perform MBR backups as part of their installation and require a floppy to save necessary records for rebuilding the system.

One point of particular interest is the **fdisk/mbr** command used to repair the MBR. Certain known virus programs such as the Monkey virus move the location of the MBR. Applying the **fdisk/mbr** command will not remedy the problem and most likely will do more damage.

Logic bomb

logic bomb
a type of malware that remains dormant until a certain event takes place.

A *logic bomb* remains dormant until a certain event takes place. The event could be a date, time, a certain number, word, file name, or even the number of times the virus program is loaded. The idea behind the design of a logic bomb is to wait for a period of time, which allows the virus to spread to other computers before releasing its destructive payload.

Figure 18-17.
Example of a joke program written in Visual Basic. A message box like this could panic a user because it looks like a dialog box produced by the Microsoft operating system.

Backdoor virus

A *backdoor virus* is designed to go undetected and create a backdoor on your system. It is not designed to be directly destructive, but rather to breach security systems. Once the backdoor is created, the computer's file system can be accessed despite standard password and standard security systems. Backdoors are usually associated with network systems, but can also be used on stand-alone PCs.

backdoor virus
a type of malware designed to go undetected and create a backdoor on a computer.

Password virus

A *password virus* is not a virus in the sense that it is designed to destroy a system or replicate itself, but rather, like the backdoor, it is designed to breach security by stealing passwords. The password is stolen and then redirected to another location, possibly on some other network on the Internet to be accessed later. The administrator of the network that houses the stolen password will probably not be aware that the network is being used to store a stolen password. A password virus is closely associated with backdoor viruses and may be used in combination. Both the password and backdoor virus are designed mainly for illegal access to networks. Either can be used to access a private PC when connected to the Internet.

password virus
a type of malware designed to breach security by stealing passwords.

Stealth virus

A *stealth virus* hides from normal detection by incorporating itself into part of a known and usually required program. The signature of a stealth virus is difficult to acquire unless the computer had an antivirus program installed when new. When the antivirus program is installed with a clean PC, it monitors changes in all files, especially those susceptible to a stealth virus. For example, if the program changes in length, it is a sign that a stealth virus has infected it.

stealth virus
a type of malware that hides from normal detection by incorporating itself into part of a known and usually required program.

Polymorphic virus

A *polymorphic virus* is any malware that changes its characteristics or profile as it spreads to resist detection. It can change its program length randomly and also change the location of the files it chooses to infect. By constantly changing its virus profile, it can go undetected by antivirus programs that compare files or programs to the signature or profile of a known virus. The most dangerous viruses use both the polymorphic and stealth characteristics to protect themselves.

polymorphic virus
a type of malware that changes its characteristics or profile as it spreads to resist detection.

Hoax

A *hoax* is as the name implies. It is not a real virus, but a message that is spread about a real or unreal virus. It could be classified as a prank. Hoaxes are not harmless because they can cost money through the loss of production. The hoax may warn of a virus, such as a logic bomb, that will start infecting systems at noon on June 5. People who believe the hoax may shutdown their computer and wait for an antivirus program to clean their computer before resuming work. It may take time before they realize that this is just a hoax and that no virus exists. Much time could be lost, especially if the scenario takes place in the corporate environment involving thousands of computers. As you see, while not incurring harm to the PC and its data or system files, the hoax consumes valuable production time, which cost a company money in the form of labor costs.

hoax
a message that is spread about a real or unreal virus.

Macro virus

A *macro* is a symbol, word, or key that represents a command or list of keystrokes. Most word processing applications, such as Microsoft Word, include a

macro
a symbol, word, or key that represents a command or list of keystrokes.

macro virus
a virus created by the macro feature of a software application program.

macro utility. A *macro virus* is named after the macro utility. A non-virus macro is designed to assist a computer user by converting repetitive tasks into one simple key code or name. Macros can record a set of user keystrokes and then save the set of keystrokes under a file name. When needed, the user can press the keystrokes to invoke the task or set of tasks the keystrokes relate to. For example, a set of keystrokes for creating a company heading to be used repeatedly can easily be created in the Microsoft Word program. A typical heading with the company's name, address, phone number, contact person, and greeting can all be created and saved as a macro. When loaded, the macro automatically recreates the keystrokes producing the desired company name, address, and such. This same method is used to create virus programs. The virus macro is usually distributed as an e-mail attachment.

One way malicious file attachments are hidden from users is by taking advantage of the fact that a file name can be 255 characters long, can use periods in the name, and blank spaces. For example, saving a file with a long string of blank spaces can hide an exe file extension. Look at the following example:

The Report.txt   .exe

The filename in the example is over fifty characters long. If an e-mail was sent to someone with this filename, only the The Report.txt portion of the file name would show in the attachment field at the top of the e-mail. The exe file extension would be hidden from view. The person receiving the e-mail with the attachment would activate the malware when they opened the attachment, unaware that the attachment is an executable program.

Merging Networks

Mergers and acquisitions of different corporations into one new corporation are common in corporate environments. Two competing companies often merge as a way to increase profitability. When two companies merge, many costs can be eliminated. It results in reduced overhead or operational costs, while maintaining the same productivity and customer base. For example, if two competing widget manufacturers merge, they will still have the same, total number of widget customers and have the same combined gross income. However, after the merger, some personnel positions can be reduced or eliminated. The number of high-level management positions can be reduced and departments, such as finance, can be eliminated from one company and absorbed by the other, or one could be greatly reduced. As the departments of the two companies are merged and personnel are reduced, equipment costs and business square footage required can be also reduced or merged. This reduces the total cost of production and thus increases the profits for stockholders and investors.

When the two companies merge, they may also be combining two entirely different networking systems, such as Microsoft Windows and Novell NetWare or Unix/Linux. The networking technicians and administrators are charged with the task of making two different operating systems and networks compatible in the most efficient manner. This is not always an easy challenge. The combining of the two systems may also involve turning the entire network into one network operating system. Combining two diverse network systems into one is often referred to as *system migration*.

system migration
the process of combining two diverse operating systems into one.

The term *migration* is also used to describe converting from one operating system to a newer version of the operating system that uses different technologies.

For example, changing an existing Windows NT Server installation to a Windows 2000 Server installation. Windows 2000 Server uses the Active Directory to maintain the directory structure and provide services. Windows NT does not. Information on the Windows NT system, such as users and groups, must be migrated to the Window 2000 Server Active Directory.

All major network operating systems can be based on TCP/IP communications. This means that all equipment on the network can be identified with IP addresses. When modifying or adding equipment, such as printers to a network, the IP address must be static and not dynamic. It is important that all static addresses be unique.

Also, when providing Dynamic Host Configuration Protocol (DHCP) services from the servers of the two different network systems, the technician must be careful not to duplicate IP address assignments. The two IP address pools should be merged to form one large pool.

There are many aspects to be concerned about when maintaining an existing network. At the technician level, you typically are not responsible for these decisions. The network administrator or administrative team must carry the burden of responsibility for decisions that affect the entire network. The decisions must be made carefully based on an existing body of knowledge. There are many case studies provided at vendor Web sites. Case studies are documented studies of how certain companies proceeded to achieve goals, such as system migration, upgrading, expansion, and new installations. The case studies should be used to serve as a model or as a possible guide to accomplish a similar task. There is little or no need to start your own merger from scratch.

Summary

- A baseline of the system should be established and then routinely collected to predict events that may occur.

- Monitor the network performance and do periodic comparisons between the current indicators and the original baseline to spot trends.

- Software programming flaws are referred to as bugs.

- Patches or fixes are used to correct bugs and security problems as they are discovered.

- Microsoft periodically releases collections of fixes and patches called service packs.

- Software patches, fixes, and service packs should be tested before applying them to the entire network.

- Adding a switch to create two or more virtual LANs (VLANs) can increase an existing network's bandwidth.

- The two types of UPS units are standby and continuous.

- Data integrity is the completeness and accuracy of data stored on the network system.

- Ensuring system integrity requires regular data backups and protecting the data while the server is running.

- The three types of backup are full, incremental, and differential.

- A full backup makes a complete copy of the original data and resets the archive bit.

■ A differential backup copies changed files since the last full backup. It does not reset the archive bit.

■ An incremental backup copies changed files since the last incremental backup and then resets the archive bit.

■ A good antivirus software suite and human resource training regarding computer policies and procedures are the best defenses against malicious software programs.

■ A virus is a self-replicating program that wastes hard disk drive space.

■ A worm is a malicious software program that automatically distributes itself across a network system.

■ A Trojan horse is a malware program that has a deceptive or harmless appearance as a gift.

■ There is no way to completely protect a system from malicious software programs.

■ Caution should be used when making changes to the system registry as part of the steps to remove a virus problem. Always backup the system registry before making any changes.

■ When adding equipment such as printers, use a static IP address, not a dynamic IP address.

Review Questions

Answer the following questions on a separate sheet of paper. Please do not write in this book.

1. Why are networks monitored?

2. Why is a baseline established?

3. What are some types of information gathered for a network baseline?

4. What are some typical events that can be monitored on a server?

5. Problems in software programs are referred to as _____.

6. Software programs installed periodically to solve problems with security or hardware are called _____ or _____.

7. Microsoft periodically releases a major collection of software patches and fixes and refers to them as _____.

8. Why should patches be tested before installing them on a network system?

9. What should you do before installing a new software package on a network system?

10. What should you do before installing a new hardware device on a network server or client?

11. Explain how bandwidth is increased by installing a switch and creating two or more virtual LANs?

12. Which RAID system does not ensure data integrity?

13. What are the three types of system data backups?

14. Which backup system takes the least amount of time to perform for a daily backup?

15. Which backup system takes the most amount of time when restoring data?

16. Which backup system is performed using a series of individual tapes that must be reinstalled in the exact order they were created?

17. When is the archive bit for a file reset?

18. What device is used to protect a server from power blackouts and brownouts?

19. Will a UPS protect against voltage spikes or surges?

20. What are the two major classifications of UPS systems?

21. What is an isolation transformer?

22. All viruses and worms can be referred to as _____.

23. What is the name for a malicious software program that infects files on an infected computer and automatically spreads to other computers?

24. What term is used to describe a malicious software program that appears as a gift, utility, game, or an e-mail attachment?

25. What type of virus is commonly created using a simple word processing program?

26. List seven things that a user can do and not do to help protect a system against malware.

27. What does the term *migration* mean in reference to networking systems?

28. What precaution must be taken when combining two network systems that are both using DHCP?

Network+

Sample Network+ Exam Questions

Answer the following questions on a separate sheet of paper. Please do not write in this book.

1. Which device is required to provide a steady stream of continuous electrical power to a server?
   A. A power strip with surge protection.
   B. A generator rated for at least 12 kW.
   C. A standby UPS system rated for at least 2000 W.
   D. An online UPS system rated for at least 2000 W.

2. You need to design a fault tolerant system for backing up critical data stored on a single network server. The system is in operation five days per week, Monday through Friday. You want to be able to restore the data as fast as possible after a hard disk drive failure. Which fault tolerance system will best meet the requirement of restoring the lost data as quickly as possible?
   A. Perform a full backup Fridays and an incremental backup each day of the week.
   B. Perform a full backup each day of the week.
   C. Perform a full backup Friday and a differential backup each day of the week.
   D. Install a RAID 1 system in the server.

3. You notice that the network has slowed over the last several weeks. While in the server room, you notice that the hard disk drive activity light seems to be constantly blinking. What might be the *most likely* cause of both symptoms, the network speed appearing slower and an increase in hard disk drive activity?
   A. The CPU does not have sufficient processing speed to accommodate network request and activity.
   B. The existing RAM is not of sufficient size to support the amount of network request and activity.
   C. The power supply is heating up, causing a slowdown in processing speed.
   D. A loose or faulty cable is causing too many packets to be dropped thus resulting in increased hard disk drive activity on the server.

4. What should be done before installing an operating system upgrade?
   A. Run **chkdsk** before you install the upgrade to provide more available disk space.
   B. Backup all critical files.
   C. Remove all existing applications before the system upgrade to ensure there will be no software conflicts during the initial boot. Then reinstall the applications.
   D. Stop all support services on the server so that they are not corrupted during the system upgrade.

5. Many more user workstations have been added to your existing Ethernet network since it was first installed. Over a period of time, demand has increased beyond for what the network was originally designed. Which recommendation would best remedy the problem in the shortest amount of time with the least amount of cost?

   A. Install a new server with more RAM, at least two CPUs, and more hard disk space.

   B. Evaluate the system and install a switch that will separate the traffic flow by departments.

   C. Remove the logon and authentication requirement for network users to release more hard disk drive space, decrease CPU demand, and decrease most of the network traffic.

   D. Subdivide the network by department using hubs to block broadcasts from department to department and reduce the overall amount of broadcast across the entire network.

6. What does a file archive bit indicate?

   A. How many times the data has been copied.

   B. The file is not an original.

   C. A backup of the data has been made.

   D. A file accessible to legacy software programs.

7. Which of the following would be required for a cold backup site for the corporate accounting department?

   A. Copies of daily backup tapes.

   B. Equipment matching the corporate headquarters and backup copies of data.

   C. Operating software loaded on backup site computers and copies of data.

   D. Accounting software, operating system software, and backup copies of data loaded on the backup site computers.

8. Which backup technology only copies files that have changed since the last full backup and does not reset the file archive attribute?

   A. Incremental

   B. Differential

   C. Sequential

   D. Sparse

9. Which UPS system provides the best protection for a network server?

   A. Off-line

   B. Standby

   C. Continuous

   D. Generator

10. Which RAID system provides no fault tolerance protection?

   A. RAID 0

   B. RAID 1

   C. RAID 5

   D. RAID 1 and 5

Suggested Laboratory Activities

1. At a vendor Web site, locate and download several case studies concerning system migrations. Study the details of how it was accomplished and make a report to the class.

2. Locate a manufacturer that manufactures switches. Download information on how a switch can affect bandwidth.

3. Do an Internet search to find a Hardware Compatibility List (HCL) that could help you upgrade an existing network operating system to a newer network operating system.

4. Open and explore the Windows 2000 Server UPS support dialog box.

5. Visit a major antivirus Web site, and then download and install a trial basis antivirus suite on a server.

6. Look up the steps required to remove some of the more commonly known or latest viruses.

Interesting Web Sites for More Information

www.adaptec.com

www.antivirus.cai.com

www.apcc.com

www.ca.com/virusinfo/

www.drsolomon.com

www.f-secure.com

www.grisoft.com

www.mcafee.com

www.sans.org

www.symantec.com

www.trendmicro.com

www.vcatch.com

Chapter 18
Laboratory Activity

LANguard Network Security Scanner

After completing this laboratory activity, you will be able to:

■ Install the LANguard Network Security Scanner utility.

■ Determine if the latest patches and service packs are installed on a system.

■ Identify network client and server potential security problems.

■ Inventory workstations, servers, and other network devices.

Introduction

In this laboratory activity, you will download and install a third-part utility called LANguard Network Security Scanner. LANguard is a software utility designed to check for potential security problems and scan for needed maintenance items, such as software patches and service packs. Identifying missing network operating system patches and updates are critical in keeping the network system secure and in working order.

LANguard checks and identifies many security vulnerabilities and recommends corrective actions. The security items include open ports, open shares, passwords that do not meet minimum security requirements, and programs that run automatically, such as Trojans. LANguard also checks for USB devices that might be connected to the network and considered a violation of company security policy, such as wireless adapters, wireless access points, hard drives, cameras, and other devices that use a USB port. As you can see, LANguard is a versatile utility that conducts an in-depth check of the network.

LANguard not only identifies missing patches and service packs, it can be configured to install them from one location. It can also install third-party software patches and updates. LANguard also identifies and inventories most of the devices connected to the network including the type of operating systems installed, MAC and IP addresses, domain names, and user names.

The highlights of LANguard are presented in this laboratory activity. To achieve a more detailed and comprehensive understanding of LANguard, you need to read the user manual, which is in PDF form. It can be downloaded for free from the GFI Web site. Using the LANguard Network Security Scanner utility is very simple. With default configuration settings, you would simply input a range of IP addresses to be scanned and then select the **Scan** button and sit back and watch. Look at the following figure.

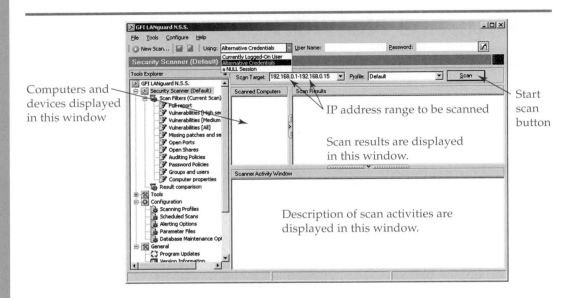

LANguard is set up to scan IP addresses 192.168.0.1 to 192.168.0.15. LANguard can conduct a scan of the network as either an administrator or as a null session. In fact, there are three user types that can be configured before performing a scan: **Currently Logged-On User**, **Alternative Credentials**, and **a NULL Session**. When **a NULL Session** is selected, the results reveal the same vulnerabilities that a hacker might see. The **Currently Logged-On User** option is as implied. You log on to the system as the administrator and then conduct the scan. This reveals detailed information, which requires administrative privileges. The **Alternate Credentials** option allows you to input a user name and password. The results of this type of scan reveal what one of the users assigned to a group in the network might see.

LANguard Network Security Scanner runs a security scan first and then checks for missing patches and service packs, open ports, open shares, auditing policies, password policies, groups and users, and computer properties. The exact sequence is listed on the leftmost pane of the LANguard window under **Tools Explorer**. Computers, servers, and other devices are listed in the **Scanned Computers** (middle) pane. The scan results are displayed in the **Scan Results** (rightmost) pane when the scan is completed. Look at the following figure to see an example of a scan in progress. Take note of the overall progress indicated as a percentage in the **Scanner Activity Window** (bottom) pane.

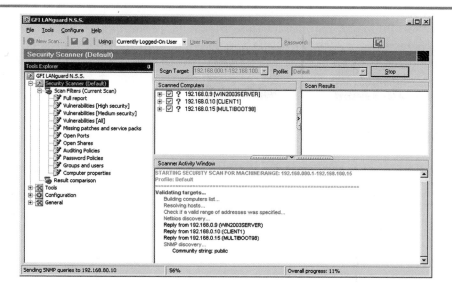

The following figure shows the results of a completed scan. Notice the topics listed in the **Scanned Computers** pane. You can select any of the topics to reveal detailed information. The detailed information displays in the rightmost pane labeled **Scan Results**. In the figure, you can see that 23 patches have been installed on the identified server. You can learn much more about LANguard Network Security Scanner by consulting the manual and installation guide located at the GFI Web site, www.GFI.com.

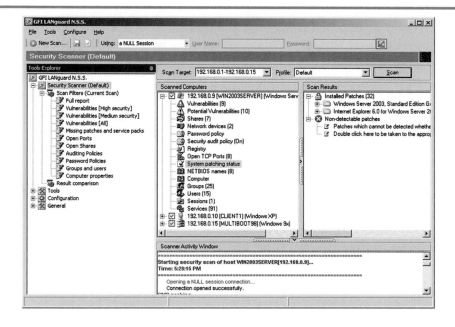

Note:
It is important to note that workstations should not be running a firewall program while performing a scan. Firewalls will interfere with the data collection. In a normal network scenario, there will typically be only one device running a firewall utility. Typically, the designated network gateway will be running firewall software. The network gateway can be a router, a proxy server, or domain server.

Equipment and Materials

- Workstation with Windows XP installed. Can be part of a peer-to-peer or client/server network.
- Administrator user name and password.
- Range of IP addresses you will scan. (Record this information on a separate sheet of paper.)

Note:
Your instructor may supply you with a copy of LANguard Network Security Scanner.

Procedures

1. _____ Gather all required materials and report to your assigned workstation.

2. _____ Boot the workstation and make sure it is working properly.

3. _____ If you already have a copy of LANguard Network Security Scanner, skip to step 6. If not, create a folder in your workstation directory and label it LANguard. This is where you will download the LANguard Network Security Scanner and documentation.

4. _____ Locate the LANguard Network Security Scanner utility by either conducting an Internet search using the keywords *LANguard Network Security Scanner* or going to www.GFI.com and accessing the LANguard Network Security Scanner download page.

5. _____ After locating LANguard Network Security Scanner utility, download the trial copy, the accompanying manuals, and the quick start or installation guide. You will need to refer to the installation guide when installing LANguard on your workstation. Make sure it is the LANguard Network Security Scanner that you download. GFI has many different software utilities that allow you to scan the network and check for vulnerabilities and maintenance items.

6. _____ Read the instructions carefully before installing the LANguard Network Security Scanner utility.

7. _____ After reading the instructions, install the LANguard Network Security Scanner.

8. _____ After installing the LANguard Network Security Scanner, input the range of IP addresses for the network you are scanning, and then scan the network. The utility should run fine with the default settings. The overall progress of the scan will be displayed as a percentage at the bottom of the screen.

9. _____ Run the scanner as a null session and then again as currently logged-on user. Be sure to log on as the system administrator. Explore the tools available in the **Tools Explorer** directory listings.

10. _____ Remove the LANguard Network Security Scanner utility if requested by your instructor.

11. _____ Answer the review questions. You may use the LANguard Network Security Scanner manual to assist you with the review questions.

Review Questions

Answer the following questions on a separate sheet of paper. Please do not write in this book.

1. What scan user types are available in the LANguard Network Security Scanner utility?

2. What kind of information does **a NULL Session** option reveal?

3. What does the **Alternative Credentials** option allow you to do?

4. What type of information can be revealed by the LANguard Network Security Scanner utility?

5. If you did not have a third-party utility such as the LANguard Network Security Scanner, how would you determine which devices needed patches and had security vulnerabilities?

To prevent possible or imminent failure, a network must be constantly monitored.

19

Fundamentals of Troubleshooting the Network

After studying this chapter, you will be able to:

❑ Outline a strategy for troubleshooting network problems.

❑ Explain the CompTIA troubleshooting strategy.

❑ Determine the best course of action to remedy a network problem.

❑ Describe in detail the boot sequence for Microsoft NT–based and Windows 98 workstations.

❑ Determine if the problem is user-, hardware-, or software-generated.

❑ List the most common network problems encountered.

❑ Describe how event logs are used to assist with troubleshooting the network.

❑ Describe common TCP/IP utilities and explain their use as applied to troubleshooting networks.

Network+ Exam—Key Points

Questions concerning troubleshooting consume a major portion of the Network+ Certification exam. You must be familiar with PC repair and support, common TCP/IP troubleshooting commands, Microsoft Recovery Console, and proper troubleshooting procedures. Be sure you can identify the output displays of command-line utilities, such as **tracert**, **ping**, **nslookup**, **netstat**, **nbtstat**, and the various switches used with each.

Many of the questions relating to troubleshooting are scenario-type questions. These questions typically require more thought than the standard fact-recollection questions. Often, exhibits are included in the question, such as typical displays. The test taker must identify the utility that generated the display. To master the troubleshooting portion of the Network+ Certification exam, you need practical experiences. You will have a difficult time solving these types of questions without practical experience.

Key Words and Terms

The following words and terms will become important pieces of your networking vocabulary. Be sure you can define them.

active partition	power-on self-test (POST)
boot partition	protected mode
Hardware Abstract Layer (HAL)	real mode
master boot record (MBR)	system partition
multitasking	

This chapter presents commonly accepted troubleshooting utilities and techniques, and in many ways, serves as a review of all of the areas covered thus far. Troubleshooting requires not only knowledge about the various utilities and tools available, but also their application. Practice using the utilities is essential to become proficient and skilled in troubleshooting.

Troubleshooting a network involves not only knowledge and skills for diagnosing a network problem but also knowledge and skills for diagnosing a PC. If you have not already completed a course in PC support and repair, it is highly recommended you do so. The need will become apparent while studying this chapter.

Much of this chapter is simply a brief review of already mastered knowledge but has been placed in the context of troubleshooting. It is important that you bring together these concepts so that you can apply them in a work setting.

The three most common operating systems you will encounter on a workstation are from Microsoft. They are Windows 98, Windows 2000 Professional, and Windows XP. For this reason, this chapter will use the Microsoft family of operating systems as the model for troubleshooting a system.

Troubleshooting Procedures

This chapter introduces troubleshooting based on CompTIA's recommended troubleshooting procedures. **Figure 19-1** lists the CompTIA procedures for troubleshooting a network problem. These steps are critical. Use them regularly. The following sections discuss each procedure in detail and relate them to real world, network-troubleshooting situations.

Figure 19-1.
CompTIA's network troubleshooting procedures.

CompTIA's Network Troubleshooting Procedures
1. Identify the symptoms and potential causes.
2. Identify the affected area.
3. Establish what has changed.
4. Select the most probable cause.
5. Implement an action plan and solution including potential effects.
6. Test the result.
7. Identify the results and effect of the solution.
8. Document the solution and process.

1. *Establish the Symptoms and Potential Causes*

Troubleshooting a network typically begins with answering a call to the help desk from a system user. The first thing you need to begin the troubleshooting process is an accurate description of the problem. This description is important so that you can clearly establish the symptoms. An accurate description is not always easy to obtain from a user because they do not understand root causes. Users will typically describe problems in vague terms and will not be able to establish any patterned behavior of the problem. Descriptions such as the following are common:

- The computer will not start.
- The computer keeps crashing.
- I can't get to my files.
- The computer is making a funny sound.
- My PageMaker program won't work.
- The printer won't work.

As you can see, most users cannot give you a true, technical description of a problem. Rather, they make general statements about the problem that is readily visible to them. You must be able to communicate with the user without frustration and work with the user through a dialog that will better help you understand the problem, while maintaining a good working relationship with them. Now is not the time to laugh at their description or joke about the terminology they use. They are not technical experts.

As a skilled technician, you can ask the user a series of simple questions to find out more about the nature of the problem. A series of simple questions will typically lead you in the direction of the problem. Some of these questions might include:

- Is this the first time the problem has occurred?
- When did the problem first occur?
- Is anyone else in the department having the same problem?
- Has anyone added new hardware or software to the system?
- What was the last change in the system?

When a problem is first reported by a user or noticed by the IT staff, a series of questions needs to be asked and answered to better understand and isolate the problem. Correctly approaching the problem by analyzing the situation before attempting repairs can save valuable time and frustration. The troubleshooting process *always* begins with questions. Be sure to ask good questions that will lead to a clear understanding of the problem.

Typically in a large networking environment, an incident report is completed by the person at the help desk or when the call is transferred to a technician who will work with the problem. See **Figure 19-2** for a sample of a help desk request form.

2. *Identify the Affected Area*

When diagnosing a network problem, you must first determine the affected area or scope of the problem. The affected area can range from a single PC to the entire network. Remember, a network provides connectivity to local and remote hosts and provides services, file sharing, Internet access, e-mail, and more. There are many areas in a network that can be affected; however, these areas can be

Figure 19-2.
An example of an
incident report form.

Help Desk Ticket Number _____

Caller:_____ Date: _____ Time: _____

Location: _____Phone number/Ext:_____

Department: _____ Tech. Assigned:_____

Complete Description _____

_____

_____

_____

_____

_____

_____

Help Desk Action Taken _____

_____

_____

_____

_____

Followup Date: _____ Time: _____

_____

_____

_____

divided into major areas. A simple LAN can be divided into three major areas: workstation, network infrastructure, and server. A large network can be divided into five major areas:

- Isolated workstation (host).
- Work group.
- Network infrastructure.
- Server(s).
- Outside the LAN (Internet or remote access).

To correct an existing problem, you must first isolate the problem area and then proceed to diagnose and correct the problem. The problem can be thought of in separate areas, such as the workstation, department the network segment belongs to (work group), cabling and equipment (infrastructure), server, and remote access (outside the LAN). See **Figure 19-3.** When you determine the scope of the problem, you are determining how much of the network system is affected. Is the problem limited to a single workstation, several workstations, the entire department, a large section of the network, or the entire network? By determining

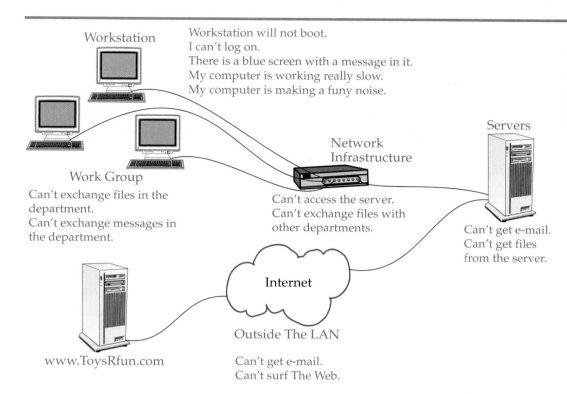

Figure 19-3.
Network problems can be divided into five major areas: workstation, work group, server, network infrastructure, and outside the LAN.

the scope of the problem, you can start a list of probable equipment or software items to inspect. The amount of network area affected will help you determine the device(s) responsible for the problem.

When first establishing the symptoms, you may be isolating the affected areas by applying user information. The user may say, "I can't get to www.toysrfun.com. There's something wrong." You must follow up and ask, "Can you connect to any other sites? Can anyone else near you connect to the site? When did this problem first occur?"

The user's response to the first question starts the troubleshooting process. You do not ask them for the scope of the problem. You ask them a few simple questions. From these questions, you will be able to determine the scope and will have gained valuable information about the problem. The user who is experiencing the problem may or may not have the expertise to help you. For example, they may not know if anyone else is having a problem. By using some of the various TCP/IP utilities available, you should be able to determine the scope of the problem quickly. TCP/IP tools are covered in more detail later in the chapter.

One of the most important items to assist you when troubleshooting a large network system is a detailed site plan that has all of the major networking devices identified. Many times the symptoms will provide you with short lists of devices to check. The map will help you identify the location of these devices.

3. Establish What Has Changed

One of the most important things you must ask yourself and others is, "What has changed?" This question can be asked in many different ways. For example, you may ask one or more of the following questions:

■ What has happened recently that could have caused a problem?

■ Was there a new software package or hardware device installed recently?

■ Was a new service started on the server?

■ Was a new hardware device recently installed?

Any of the answers to these questions could help determine the cause of a network problem. Perhaps, nothing has changed. The system failure may have occurred because of other reasons such as hard disk drive failure, IC failure, and a corrupted system or device file. A device failure and an unrelated event can easily confuse the troubleshooting process. Be aware that coincidences do occur.

4. Select the Most Probable Cause

Troubleshooting requires a series of best guesses. Sometimes the solution to the problem is obvious, and other times the problem can be much more difficult to diagnose. Usually, someone else has encountered the same problem at some other location. There are thousands of documented problems and solutions available for review at no cost on the Internet. Once the symptoms have been clearly established, you can conduct a search using key words related to the symptoms. The search should produce a listing of document abstracts or articles from which to choose.

The operating system vendor typically has a search engine with related technical support articles to help you solve your problem. They will also have free e-mail inquiries and support that you must pay for. Abstracts or articles will typically list possible solutions. Look at **Figure 19-4.** It shows a support document for a connection problem to a NetWare server. It was found by doing a Google search with "Can't Log on to Novell NetWare Server" as the search topic. Notice that the support document shows a list of symptoms, the operating systems to which the problem applies, and a step-by-step fix or possible solution. Although it does not appear in the screen capture because of limited page space, there are several more possible solutions for this problem listed in this support document.

Solutions for problems can come from numerous sources, such as operating system manufacturer Web sites, hardware manufacturer Web sites, and commercial Web sites. Commercial Web sites offer mixed product support, focus groups, organizations, and more. One of the best ways for locating a possible solution to a networking problem is the Google search engine. The World Wide Web has an extensive resource base waiting to be used. Google has a high reliability for hits

Figure 19-4.
Example of a support document. A support document typically contains a list of symptoms, the operating systems the problem applies to, and a step-by-step fix or possible solution.

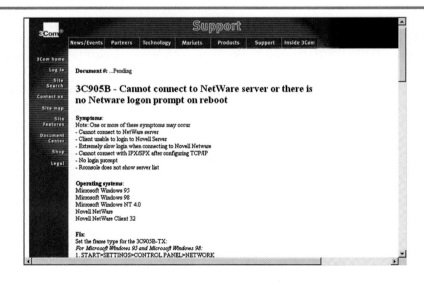

concerning technical information about networks. You will not only receive hits at the manufacturer or software manufacturer Web site, but also from many other reputable sources.

Technical support is also available by your specific operating system. The support can be provided by Web site and telephone. Support may be free or fee-based. Vendor diagnostic software CDs, especially for hardware tests, are also available. As you can see, there are many sources of information, which will lead you to the next step—implement a solution.

5. Implement an Action Plan and Solution Including Potential Effects

When you use the Internet to conduct a search about a probable cause, you will also receive information about possible solutions. You will need to think through the possible solutions to choose the one that is the most appropriate to the situation. This requires experience and a broad knowledge of and familiarity with the network system.

Be sure to document your solution as you progress through the process of diagnosing and implementing possible solutions. Documenting your work as you progress cannot be emphasized enough. Do not trust your memory. Writing down the steps you take is like leaving a trail for you to follow next time and will save you time and effort in the future. For example, if you have made any changes in system configurations, you need to write them down. A technician often loses track of time as they attempt to solve a network problem. As time goes by, they suddenly realize they have made many changes to the system configuration, and may have created additional problems or symptoms caused by those changes made. *Always* record the changes made and return the system configuration to its original settings before trying other changes.

6. Test the Result

This may sound simplistic, but you need to test the fix that you made. Not all problems need to be replicated or tested, but some do. For example, you suspect a new software package recently installed on a workstation has caused a problem, such as preventing the user from accessing a specific printer. When the new software package was removed, the workstation user could access the printer once again. If the new software package really was the culprit, you should be able to duplicate the problem. Try installing the software package on a different workstation and see if the same problem develops. If the problem of connecting to the printer occurs on the new workstation, then the software did actually cause the problem.

If the second workstation did not experience a problem connecting to the printer, then you may have a different problem. For example, a loose network cable connection could cause an intermittent problem with the connection to the printer. You uninstall the suspected software package and the problem clears. While uninstalling the software package, the bad or loose connection reconnects and the printer is accessible once again. If you truly solved a problem, you should be able to duplicate it.

7. Identify the Results and Effect of the Solution

You must realize the full ramification of the corrective action you are taking. Before you apply a possible solution, you need to know the extent of how it will affect the existing system. For example, when all else fails to repair the workstation or server, you can always reinstall the operating system. However, reinstalling the operating system could produce adverse effects on the network or workstation.

You typically have two choices for reinstalling an operating system: perform a clean installation or perform a system upgrade. When you perform a clean installation, all data on the existing partition and possibly on the entire hard disk drive is lost! You must have backups of all data files available, as well as all application software and any required device drivers.

When performing a system upgrade, the original operating system files are replaced. Typically, data files, application software, and original drivers are not required. A bit of caution is necessary when performing an upgrade. You may see messages asking if you want to install the original version of a file or driver or leave the existing file intact. This message occurs when an original system has had patches or service packs installed since the original installation. When performing an upgrade, the existing files are compared to the files about to be installed.

When two file names match but their creation dates differ, the installation process will typically ask, "Do you want to leave the existing file or replace it with the original or older file?" You should answer, "No," and leave the latest copy of the file. The only exception to this rule is if you suspect the latest version is the cause of the system abnormality. In that case, you would install the original operating system as an upgrade and later add each service pack or patch one at a time until you reach the latest, which is the suspected cause of the abnormality. In most cases, a patch or service pack will not have adverse effects on a system. Always check the support documentation at the software vendor's Web site for the latest information when repairing a system.

8. *Document the Solution and Process*

Most network troubleshooting scenarios start with an incident report filed with the help desk. The incident report typically includes an area for the technician to write down the implemented fix for the problem. This documentation will serve as a future reference.

Documentation is an important part of network administration and should be completed thoroughly and in a timely fashion. Documentation will save the company and personnel valuable time and money, and should be considered by all technicians as a professional courtesy, as well as a necessity. Documentation can be shared with other workers especially in a large organization where workers do not get the opportunity to converse on a regular basis. The sharing of documentation can give other workers an insight to problems they may encounter in the future. Documentation also acts as a history of the system. For example, if a network interface card is replaced by another brand or version, you have a record available. The record can be used to update equipment inventories and equipment descriptions.

Knowing what type of equipment is installed on the network is very important, especially when performing a migration to a new operating system. For example, when upgrading from Windows NT 4.0 to Windows Server 2003, an accurate list of system hardware is invaluable. The list can be checked against the Hardware Compatibility List (HCL) to ensure that all hardware used on the existing system is compatible with the new operating system. Without a way to check, a new operating system can be installed and problems may develop. These problems may expend a long period of time troubleshooting only to find out later that a hardware compatibility problem was the issue. Always keep an up-to-date and complete set of documentation for the existing network and associated equipment.

Software programs, such as HelpSTAR, are available to help an administrator keep track of service requests, computer inventory, and history, **Figure 19-5.** HelpSTAR is especially designed for the help desk. It is the only practical way to keep track of help desk calls when working with a large network consisting of many users or when supporting several companies. HelpSTAR can be used to initiate purchase orders and can keep an inventory of all new material and equipment as it is ordered for the network. By controlling the ordering process, the help desk also knows when the equipment needed to repair or replace a device is delivered and can match it to the correct job order. A service request can be initiated by e-mail or telephone request. The request is entered into the HelpSTAR database and assigned a reference number. The technician assigned to the problem can follow up by entering details about the progress and outcome of the incident.

Troubleshooting the Workstation

Troubleshooting the workstation involves a strong knowledge of PC operation, operating system software, and operating system boot sequences. If you are not knowledgeable about PC repair, it is strongly recommended that you take a computer repair course. A computer repair course will not only provide you with a background in PC hardware and repair, it will provide you with basic knowledge about the various desktop operating systems. This section assumes that you have a background in PC hardware and repair and are familiar with the various Microsoft desktop operating systems. It is meant to serve as a review of desktop boot sequences, boot files, and system recovery methods of Windows 98 and Windows NT–based operating systems.

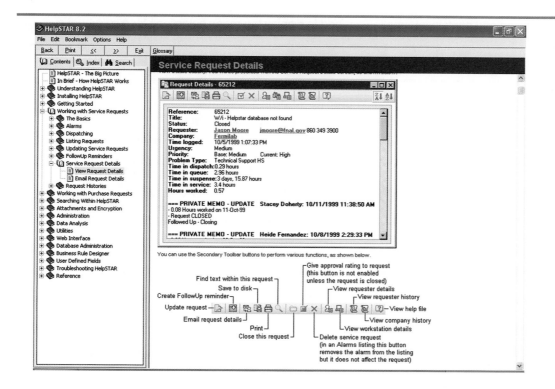

Figure 19-5.
The HelpSTAR program is a valuable tool in managing help desk requests.

The Windows 98 Boot Process

power-on self-test (POST)
a BIOS routine that performs a series of hardware checks to determine if the computer is in minimal working order.

master boot record (MBR)
an area located on the first sector of the hard drive that contains information about the hard disk drive partitions (partition table) and the location of the bootable partition (active partition).

system partition
a partition that contains the required startup files necessary to boot the operating system.

The boot sequence is where many troubleshooting processes begin. You need a good understanding of the boot sequence when determining whether there is a problem with the network system or the PC. The boot sequence is the series of steps that are processed through the CPU until the PC is fully functional. This includes activating the network connection. Remember, the boot process for a typical PC connected to a client/server model network is not complete until the user logs on to the network. **Figure 19-6** outlines the Windows 98 boot process.

All computer systems go through a power-on self-test (POST) when power is applied to the computer. The *power-on self-test (POST)* is controlled by the computer BIOS program. The POST performs a series of hardware checks to determine if the computer is in minimal working order. It is not a detailed assessment of the hardware system. Detailed assessments require third-party diagnostic software and, at times, hardware diagnostic components.

The POST performs checks on the keyboard, mouse, CPU, RAM, and hard disk drive. Additional checks may be performed during the POST by individual components containing their own BIOS, such as video cards and SCSI controller cards.

After successful completion of the POST, the master boot record (MBR) is read. The *master boot record (MBR)* is located on the first sector of the hard drive and contains information about the hard disk drive partitions (partition table) and the location of the bootable partition (active partition). The information from the MBR is loaded into RAM so that it can be quickly and easily accessed anytime during the computer session.

After the MBR has been loaded into RAM, it uses the partition table to locate the system partition and load the other information that is stored in sector 0 (the first sector). A *system partition* contains the required startup files necessary to boot the operating system. Sector 0 contains the startup program code for the operating system.

Tech Tip

In a multiboot operating system environment, the required boot files are located in the system partition and contain information needed to display the menu selections for a multiboot system.

Figure 19-6.
The Windows 98 boot sequence.

Windows 98 Boot Sequence

1. The system BIOS performs the power-on self-test (POST).
2. The BIOS reads the master boot record (MBR) and loads it into memory.
3. The BIOS uses the information in the MBR to locate the system files.
4. The BIOS loads the io.sys file into memory.
5. The io.sys file loads the file allocation table (FAT) into memory and reads boot information from msdos.sys. It also loads drvspace.bin if available.
6. The io.sys file processes config.sys (if it exists).
7. The io.sys file loads command.com.
8. The io.sys file executes the autoexec.bat file (if it exists).
9. The io.sys file executes win.com.
10. The win.com file puts the CPU into protected mode and then loads the Windows kernel, the graphic device interface (GDI), the Explorer shell program, program file libraries, and network support.

During the boot process, the BIOS looks for the boot files in a sequence defined by the BIOS settings. For example, the boot sequence for a PC containing a hard disk drive, floppy disk drive, and CD-ROM drive can be set to check the floppy drive first, the CD-ROM drive next, and hard disk drive last. The exact sequence of the search can be changed to any order, or a device can be eliminated from the boot sequence.

The MBR is independent of the operating system. It is used no matter which operating system is installed on the PC. The MBR is a common target for computer viruses. Corrupting the MBR prevents the computer from booting the operating system.

Real mode

Real mode is the original 8088 processor form of execution. Real mode has limited memory access and can only load and run programs smaller than 640 kB. Note that the limit is in kilobytes, not megabytes, which is common today. Also, real mode cannot perform multitasking. In real mode, only one program can be loaded into memory and executed at any given time. *Multitasking* allows multiple programs to be loaded into memory and the CPU to rapidly switch between the various programs. This rapid switching between programs make it appear to the user that all the programs in memory are running at the same time.

In real mode, the Windows 98 operating system boot process begins with loading the file called io.sys into RAM. The io.sys file is required. The io.sys file is responsible for loading the file allocation table (FAT), reading the msdos.sys file for boot configuration information, and loading drvspace.bin (file compression information) if present.

The familiar Windows logo image is presented on the screen. The Windows logo image may be replaced by a vendor's own design. The system registry values are checked to see if the system completed a proper shut down sequence. If the system was not shut down properly or a system lockup or failure occurred during the use of the last computer session, the ScanDisk utility program is initiated, prompting the user for action. The ScanDisk utility can be run or aborted. The user can select the action to take.

Next, the io.sys file processes the config.sys file, if one exists. The config.sys file is not required and is used only to maintain downward compatibility with legacy DOS programs. Starting with Windows 95, the registry replaced the config.sys file.

After io.sys processes config.sys, it loads command.com and then executes autoexec.bat. The autoexec.bat file is also no longer necessary and has been replaced by the registry. However, the autoexec.bat file, similar to the config.sys file, may be present to provide downward compatibility with legacy programs.

Protected mode

Protected mode allows the computer to access memory above the first 1MB of RAM. While in protected mode, a computer can multitask. Multitasking is not possible without the protected mode feature.

After the autoexec.bat file is processed (if it exists), the win.com file executes. The win.com file is responsible for loading virtual device drivers based on information stored in the system registry and system.ini files. The win.com file puts the CPU into protected mode and then loads the Windows kernel, the graphic device interface (GDI), the Explorer shell program, program file libraries, and network support.

It is important to remember that the computer system initiates real mode before protected mode. It is during real mode that the operating system startup

real mode
the original 8088 processor form of execution. Real mode has limited memory access and can only load and run programs smaller than 640 kB.

multitasking
the process of rapidly switching between various programs that are loaded into memory.

protected mode
a form of CPU execution that allows multiple programs to be loaded into memory and the CPU to rapidly switch between the various programs. It also allows a computer to access memory about the first 1MB of RAM.

can be interrupted and a command-line interface be accessed. The command-line interface allows a user to use various commands recognized by the operating system to diagnose the system.

A user can interrupt the boot process by pressing the [F8] special function key. Pressing the [F8] key will cause a text-based menu, called the Startup Menu, to display a list of options. Through the menu options, the network technician can choose to start the computer with or without loading the network interface card drivers and connecting to the network. See **Figure 19-7** for a list of Windows 98 Startup Menu options and their description.

The Windows NT Family Boot Process

The Windows NT operating system family includes Windows NT (Workstation and Server), Windows 2000 (Professional and Server), Windows Server 2003, and Windows XP (Professional and Home Edition). All four operating systems are often referred to simply as Windows NT operating systems. This makes for some confusion when researching technical problems. Windows NT, Windows 2000, Windows Server 2003, and Windows XP are based on the original NT kernel, so accordingly the boot sequence is similar. **Figure 19-8** outlines the Windows NT family boot process.

After a successful completion of POST, the BIOS locates the active partition and then accesses the master boot record (MBR). The MBR is similar in design and function as described in the previous section. After the MBR is loaded, the next step in the sequence is to load the ntldr file on the hard disk drive and the bootstrap program (boot program). The ntldr is designed to load NT–based operating systems. The ntldr reads the boot.ini file to locate or select the operating system files.

Next, ntldr calls the ntdetect.com program. The ntdetect.com program detects the system hardware information, such as the computer ID, adapters, video information, keyboard, ports, and hard disk drives. The ntdetect.com file passes this

Figure 19-7.
Windows 98 Startup Menu options.

Windows 95/98 Startup Menu Option	Description
Normal	Continues with the startup operation as normal, loading the registry and all startup files.
Logged (\bootlog.txt)	Creates a file called bootlog.txt, which lists the components and files successfully or not successfully loaded during startup.
Safe mode	Starts Windows using only basic system drivers. It bypasses startup files, such as config.sys, autoexec.bat, part of the system.ini, and the registry.
Safe mode with network support	Same as safe mode, but also loads files necessary for network interface card support.
Step-by-step confirmation	Allows user to interact with startup files. The user must confirm each line of the startup file.
Command prompt only	Starts the system with startup files and registry and then displays the command prompt.
Safe mode command prompt only	Starts the system in safe mode and only displays the command prompt. Can be accessed directly by pressing [Shift] [F5].
Previous version of MS-DOS	Starts the system using the previous version of MS-DOS.

Windows NT Family Boot Sequence

1. The system BIOS performs the power-on self-test (POST).
2. The BIOS reads the master boot record (MBR) and loads it into memory.
3. The MBR loads the operating system loader, ntldr.
4. The operating system loader, ntldr, reads the boot.ini file.
5. The operating system loader, ntldr, calls the ntdetect.com program.
6. The ntdetect.com program detects system hardware information and passes it to ntldr.
7. The operating system loader, ntldr, passes system hardware information and control to ntoskrnl.exe.
8. The ntoskrl loads the device drivers, the hal.dll, and initializes the computer settings using the values stored in the system registry.
9. The Winlogon service loads.
10. The user logs onto the system.

Figure 19-8.
The Windows NT family boot sequence.

information back to the ntldr, which in turn passes the information to the NT kernel, ntoskrnl.exe. The ntoskrl.exe program loads the device drivers, initializes the HAL, and initializes the computer settings using the values stored in the system registry. See **Figure 19-9** for a complete list of Windows NT–based boot process files and their description.

The system registry can be accessed and viewed using the Registry Editor (**regedt32**) utility.

Tech Tip

Care should be used when editing the system registry. Changes in the registry can render the computer inoperable. Changes to registry settings are a last resort. Instructions provided by the software manufacturer should be followed carefully. Always make a backup of the system registry before attempting any changes.

Caution

The hardware device drivers are considered part of the operating system for the purpose of this section. Plug and Play is a combination of BIOS, hardware, and operating system. The point at which the BIOS interacts with Plug and Play software varies a great deal. There is no one set standard sequence for loading Plug and Play devices because BIOS designs greatly vary.

The final stage of the boot process is the user logon. When the Winlogon service loads and the user logs on to the system, the entire boot sequence is complete. The user log confirms the authentication of the user. Until the user is authenticated, the user cannot access network system resources. The exception is when starting a workstation in safe mode. When the workstation is started in safe mode, the network connection is not completed. Connecting to the network from safe mode is an option in the Windows Advanced Options Menu. See **Figure 19-10** for a list of Windows NT family menu options.

Often the terms *active partition*, *boot partition*, and *system partition* are used interchangeably, but actually they have different technical meanings. The *active partition* describes the partition that contains the operating system files the

active partition
a partition that contains the operating system files the computer should use to boot.

Figure 19-9.
Window NT family
boot process files.

Boot Process File	Description
boot.ini	A boot initialization file that contains information about operating system locations. It must be located in the root directory of the system partition.
bootsec.dos	The MS-DOS boot sector file used to boot MS-DOS operating systems or Windows 95. It is located in the root directory of the system partition.
hal.dll	The dynamic link library used to support hardware driver commands to hardware devices. The functions of the hal.dll and the NT kernel combine to form the hardware abstract layer.
ntdetect.com	The file used to detect PC hardware installed in the PC. This information is required before loading the operating system kernel. The kernel must know what type of hardware exists to properly load appropriate drivers.
ntldr	The operating system loader program loads the NT operating system or other operating system that is selected. It is located in the root directory of the system partition.
ntoskrnl.exe	This NT executable file is referred to as the kernel or the operating system. It contains the bare essentials of the operating system. It is located at \Winnt\System32\ on the boot partition.
ntbootdd.sys	This file is used for systems with SCSI drives with the SCSI card BIOS disabled. It is located in the root directory of the system partition.
osloader.exe	This file is used for RISC systems. It is similar in function to ntdetect, bootsect.dos, and ntldr combined

Note: The order of files listed illustrates the major steps of the typical NT system startup.

Figure 19-10.
Windows Advanced
Options Menu choices
for NT–based operating
systems.

Windows Advanced Option Menu	Description
Safe Mode	Boot the computer with a minimal set of drivers.
Safe Mode with Networking	Same as safe mode with the addition of initializing network interface card support.
Safe Mode with Command Prompt	Same as safe mode except with the command prompt displayed by loading and running cmd.exe.
Enable Boot Logging	Starts normally and records the loading and initialization of drivers as successful or unsuccessful in a text file called ntbtlog.txt.
Last Known Good Configuration	Loads the most recent settings that worked.
Directory Services Restore Mode	Available only on a Windows domain controller.
Debugging Mode	Allows information to be sent over the COM 2 serial port after running the debug program.
Start Windows Normally	Continues the startup operation as normal, loading the registry and all startup files.
Reboot	Reboot the computer system.

computer should use to boot. A computer system may have many partitions. The BIOS needs to know which partition contains the operating system. In a multi-boot system, two or more partitions may contain an operating system. However, only one partition may be marked as the active partition. For example, a computer may have the Windows XP operating system on the active partition and Windows 98 operating system on another partition. When the computer boots, the BIOS will pass control to the Windows XP operating system because it is on the active partition.

The system partition contains the core files, such as ntldr and ntdetect.com, which are used to start the operating system. In a Windows 98 system, these files are io.sys, msdos.sys, and command.com. They are typically located in the root directory of the active partition, such as the root directory of drive C (C:\). The **boot partition** contains the files needed to operate the computer. These files include device drivers, library files, services, commands, and utilities and are stored in a separate directory, such as \windows and \winnt. **Figure 19-11** shows the Windows NT family arrangement of boot files and operating system files.

Be aware that a single partition can comprise the active partition, boot partition, and system partition. In a multiple partitioned drive, any particular drive can be a combination of boot and system. Typically, drive C is the active partition.

boot partition
a partition that contains the files needed to operate the computer. These files include device drivers, library files, services, commands, and utilities and are stored in a separate directory.

The confusion of system and boot partition terminology comes from the fact that when there is only one partition, the active partition, system partition, and boot partition are all the same partition.

Tech Tip

All Windows NT operating systems have similar NT kernels, but they are not exactly the same. There are many differences in the files that compose each operating system. Therefore, you should only use an Installation CD that corresponds to the version of Windows that is already installed on the computer. For example, only use a Windows XP Professional installation CD with the Windows XP Professional operating system. The same is true in reverse. Do not use the Windows XP Home Edition installation CD for troubleshooting a Windows XP Professional operating system. These CDs have different kernel files and are not compatible. Using the inappropriate CD can cause additional problems.

Hardware Abstract Layer (HAL)

The *Hardware Abstract Layer (HAL)* is a software layer located between the hardware devices and third-party vendor driver software. It is similar to the Application Program Interface (API) associated with Windows 95 and Windows 98 systems. The HAL is created when the hal.dll file is loaded into RAM. HAL

Hardware Abstract Layer (HAL)
a software layer located between the hardware devices and third-party vendor driver software.

System Partition	Boot Partition
boot.ini	ntoskrnl.exe
bootsec.dos	\winnt (contains operating system files)
ntdetect.com	
ntldr	
ntbootdd.sys	

Figure 19-11.
Windows NT–based system partition and boot partition files.

interacts with the kernel to control access to the hardware devices. Through the HAL, the operating system provides a set of interfaces that allow third-party driver software to interact with the hardware. The HAL is used in Windows NT–based operating systems to prevent users from directly accessing hardware devices, such as memory. This helps to prevent many of the most common system lockups and crashes, which are much more common in Microsoft systems based on the earlier DOS structure, such as Windows 95, Windows 98, and Windows Me.

Through the use of the HAL, a programmer can write hardware routines that are device-independent. Device-independent means the program is generic and will work for any standard device, such as a modem, network interface card, and keyboard. The HAL design hides I/O interfaces, software interrupt controllers, and the CPU communication systems. The third-party programmer writes a program that interfaces with the HAL, which in turn interfaces with the system hardware. It allows access to memory, system software IRQs, and other system resources only through the HAL.

An example of a device-independent driver would be similar to printing a document in the Windows operating system environment. When the **Print** command from the program menu is selected, the document prints similar to how it appears on the display. A device-dependent driver, however, must use a set of commands designed for that specific device. Information such as I/O address, port address, memory address, and font and paper type would also need to be sent. This information may require specific hexadecimal codes placed in exact I/O locations in place of generic commands. As you see, HAL makes interaction with hardware devices easier. In this scenario, a user can simply choose **Print** and the HAL layer does the rest.

Dual Boot Systems

Dual boot and multiboot systems are becoming more common as users switch or upgrade from one operating system to a newer one. When users are comfortable with an operating system, they are typically hesitant to change, even if the newer operating system is better. For some users, it is a severe struggle to relearn some common tasks. For example, new operating systems often change the location of features. It is frustrating for the user to locate these features in the new operating system. A common practice is to install the new version as an optional system until the user has been trained on the new version.

Two operating systems on one computer, however, can generate their own unique problems. A common problem is the corruption of the boot.ini file. The boot.ini file provides the user with a choice or selection of operating systems after the POST. An accidental modification of the boot.ini file may prevent any of the operating systems to completely boot or result in only one of the operating systems booting. Look at the contents of a boot.ini file as it appears in Windows XP, **Figure 19-12.** To see the boot.ini file, you must remove the hidden and system file attributes. You can also view the contents of the boot.ini file on a Windows XP system by running the **msconfig** command from the **Run** dialog box located at the **Start** menu. See **Figure 19-13.**

Troubleshooting Windows NT–Based Operating Systems

There are many tools and generally accepted procedures for recovering a failed Windows 98, Windows Me, and NT–based operating system. Some of these utilities are briefly covered in this section. To master these troubleshooting

Windows XP Professional
system files location

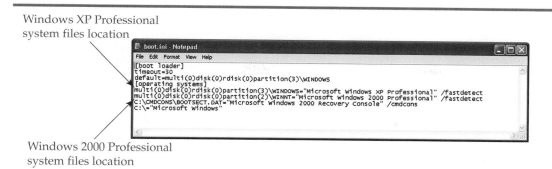

Windows 2000 Professional
system files location

Figure 19-12.
The boot.ini file in a multiboot system contains the locations of system files for each operating system on the computer.

concepts, you need to practice using them. See **Figure 19-14** for a complete list of diagnostic utilities and the Microsoft operating systems in which they are included. This section covers Safe Mode, System Restore utility, Last Known Good Configuration, Recovery Console. It also covers using an emergency repair disk.

Safe Mode

The Safe Mode Startup Menu (Windows 98) or Windows Advanced Options (Windows NT–based operating systems) choice is activated by pressing the special function key [F8] during the system startup. During a safe mode boot, only standard video drivers are loaded and essential drivers that are needed for the keyboard, mouse, and storage devices. No network connection is made. Starting the computer in safe mode can help determine if there is a problem with the PC or with the network. If the system appears to be working correctly with the minimum set of drivers, you may be able to correct the suspected source of the problem. For example, if a recently loaded software or hardware device is suspected of causing the system boot failure, you can delete it from the system and then attempt to boot the computer system normally.

System Restore utility

The System Restore utility does as the name implies. It restores the operating system to a condition established at a previous time. The System Restore utility can be accessed through **Start | All Programs | Accessories | System Tools** even when the computer is running in safe mode. You may use the System Restore utility to roll the operating system back to a previous time when the system was

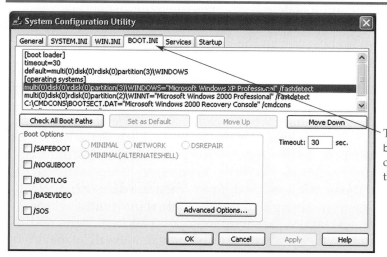

Figure 19-13.
The contents of the boot.ini file can be viewed from the System Configuration Utility.

To view the contents of the boot.ini file in the system configuration utility, select the **BOOT.INI** tab

Figure 19-14.
Common diagnostic
utilities listed by
various Microsoft
operating systems.

Diagnostic Utility	Win 98	Win ME	Win 2000	Win 2003	Win XP
Last Know Good Configuration			XXX	XXX	XXX
System Restore		XXX		XXX	XXX
Driver Roll Back				XXX	XXX
Recovery Console			XXX	XXX	XXX
Safe mode	XXX	XXX	XXX	XXX	XXX
System Configuration Utility (msconfig)		XXX		XXX	XXX
System Information (msinfo32)	XXX	XXX	XXX	XXX	XXX
DirectX Diagnostic Tool			XXX	XXX	XXX
Registry Editor (regedit)*	XXX	XXX	XXX	XXX	XXX
Registry Editor (regedt32)*	XXX	XXX	XXX	XXX	XXX
Net		XXX	XXX	XXX	XXX
Dr. Watson	XXX	XXX	XXX	XXX	XXX
File Signature Verification Utility			XXX	XXX	XXX
Network Diagnostics (netdiag.exe)		XXX	XXX	XXX	XXX

Note: The regedit utility is the preferred registry editor for Windows 98 and Me. The regedt32 utility is the preferred registry editor for all Windows NT-based operating systems.

working perfectly. For example, you may have installed a new pointing device and loaded the required drivers only to find the computer will not boot properly or freezes during the boot process. You can start the computer in safe mode, start the System Restore utility, and select an earlier time. The System Restore utility rolls back the registry but does not change the existing data files, e-mail, and other common file types that you would not want to be removed or changed. The System Restore utility does remove executable files, which means that programs that were installed after the rollback date may be inoperable.

Last Known Good Configuration

The **Last Known Good Configuration** is essential for correcting most boot problems on an NT–based system. Every time an NT–based computer has a successful logon after a completed boot, a backup of the system registry is made. The backup is available for use when a computer fails to boot properly because of some change in the system. The change in the system could be the installation of a new or updated software driver or a change in some hardware configuration. If an error during the boot operation is detected, a message will display such as, "Restore the system to the last known good configuration." When the **Last Known Good Configuration** option is chosen from the Windows Advanced Options menu, the last copy of the registry made during a successful boot replaces the existing registry.

Emergency Repair Disk

All modern operating systems have what is called an emergency repair disk. An emergency repair disk is used to repair or replace items such as the master boot record, file allocation tables, system registry files, or system configuration files. There is one very important aspect to using an emergency repair disk. It can only restore the system to the way it was configured when the disk was made. If the emergency repair disk was made right after the original installation of the operating system, then the system will be set up in its original state. Any modifications to the system since the emergency repair disk was made will be lost if you use the emergency repair disk to repair the system. You may lose the use of any new software systems that were installed, hardware drivers, and system updates or patches. An emergency repair disk should be made whenever changes to the system occur. This will ensure an updated system will be restored rather than the original system. Windows NT (Workstation and Server), Windows 2000 (Professional and Server), Windows Server 2003, and Windows XP Professional allow you to create an emergency repair disk. Also, a third-party vendor, such as Norton Utilities, can be used.

Recovery Console

A powerful, last resort tool is the Recovery Console. When you fail to start the computer with any of the previously stated methods, you may choose to use the Recovery Console feature. Recovery Console is a command-line utility that allows you to issue commands and start a limited number of software utilities that may correct the problem. Recovery Console is not installed during the typical Windows operating system installation. It can be installed by using the installation CD. Once Recovery Console is installed, server services can be started or stopped and files deleted or installed. You may also inspect the system partition(s), repair or replace the master boot record, check the hard disk drive for bad sectors, and more.

See **Figure 19-15** for a list and description of common Recovery Console commands. Also, check Microsoft article Q307654 for more detailed information about the Recovery Console commands available for Windows XP.

Microsoft often includes additional third-party utilities on the operating system installation CD. Microsoft does not provide documentation for these utilities, but most of the utilities contain help files readable either through Internet Explorer or through Word.

Tech Tip

Roll Back Driver

The Roll Back Driver feature is available through the properties dialog box of any hardware device installed on the computer. **Figure 19-16** shows the properties dialog box of a SiS 900 PCI Fast Ethernet Adapter. The dialog box contains buttons that allow you to view driver details, update the driver, roll back the driver, or uninstall the driver.

The Roll Back Driver feature is a better alternative than using System Restore. System Restore rolls back all driver and configuration settings to the selected date. This can result in losing the configuration settings of devices you did not intend to change. The Roll Back Driver feature only reverts to the last driver configured for a specific device. For example, if you suspect a newly installed driver or driver update is causing a problem with a hardware device, you can use the Roll Back Driver feature to return to the previous version of the driver.

Recovery Console Command	Description
chkdsk	Checks the hard disk drive platters for bad sectors.
diskpart	Displays and manages the partitions on a hard disk.
bootcfg	Modifies the boot.ini file.
fixboot	Writes a new boot sector on the system partition.
fixmbr	Repairs the master boot record.

Figure 19-15.
Common Recovery Console commands.

Figure 19-16.
The Rollback Driver feature is available through the properties of each hardware device.

Troubleshooting the Network Infrastructure

The network infrastructure consists of cabling and cabling devices such as hubs, switches, and routers. One of the most commonly encountered infrastructure problems is a loose cable connection. The second most encountered network problem is forgotten user passwords. Many of the tools described in this section were previously explained earlier in the textbook. You may wish to review them at this time.

Windows XP Network Diagnostic Utility

Windows XP and Windows Me include a powerful network diagnostic tool known as Network Diagnostics. The Network Diagnostics utility can be accessed by running **msinfo32** (System Information) from the **Run** dialog box off the **Start** menu. When the System Information window displays, select **Net Diagnostics** from the **Tools** menu, **Figure 19-17.** Selecting the **Scan your system** option in the Network Diagnostics utility automatically runs a series of network diagnostics on the local computer.

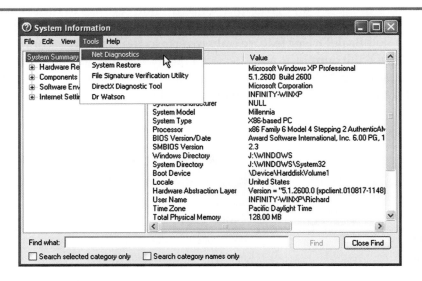

Figure 19-17.
The Network Diagnostics utility can be accessed through the System Information (**msinfo32**) **Tools** menu.

The Network Diagnostic utility combines many of the traditional network utilities into one. It can check network connections, hardware, and services related to networking at one time. Some of the items it automatically checks are the network interface card, telephone modem (if one exists), e-mail, WINS, DNS, and DHCP. As you can readily see, Network Diagnostics is a versatile and powerful tool. Look at **Figure 19-18** to see the results of a typical Network Diagnostics scan performed from a Windows XP workstation.

As you can see, when a particular device passes the test, the word *Pass* in bright green lettering appears to the right side of the device name. If it fails to pass a test, then the word *Failed* is presented in bright red lettering. The Network Diagnostic utility also identifies any items that were scanned but not configured in the workstation by displaying the words *Not Configured* in bright blue.

Network Diagnostics allows the user to customize the objects to scan and save results to a file for viewing at a later date. Saving the results to a file allows a technician or any users to see what the original settings are and compare them to a scan at a later date. A technician could also run a scan of the workstation, save the results to a file, and then transfer the file across the network to a central location. At the central location, it can be added to a network documentation folder containing information about every workstation on the network. This makes network documentation much easier for the technician and the system administrator.

Network Cable Tester

Cables can be analyzed quickly using hardware tools such as a network cable tester. Cable testers go by many different names, such as digital cable analyzer and cable analyzer. A network cable tester performs a series of checks of cable integrity. Cable faults such as opens, shorts, and grounds can be quickly determined and located. Many cable testers can also check for crosstalk, radio interference, EMI, and excessive cable lengths and can determine the exact location of a cable fault. Some of the more expensive models can record troubleshooting data and download information to a computer. The computer can store or print the information.

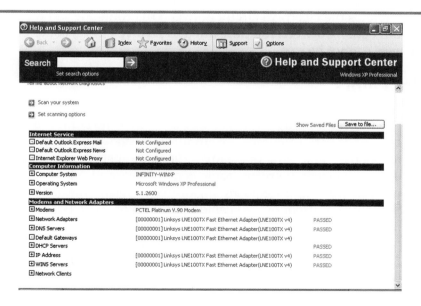

Figure 19-18.
The results of a system scan in Network Diagnostics.

Tone Generator and Tracer

A tone generator and tracer allow a technician to trace the exact location of cable runs inside walls, ceilings, or under the floor or to identify unmarked cables. The tone generator and tracer are separate devices. The tone generator attaches to one end of a cable. It produces an analog or digital signal and transmits it through the cable. The tracer is designed with special circuitry that receives the signal and produces a tone if it is in close range of the cable. When a cable fault is encountered, the signal ends or changes pitch.

The fact that the tone generator and tracer are designed to trace the path of a conductor is an advantage over a cable analyzer, which does not have this feature incorporated. An old term still used by technicians to describe the tone generator and detector is *fox and hound*. The tone generator represents the fox and the tracer represents the hound. In essence, the hound (tracer) follows the fox (signal) until reaching the end of the cable.

Fiber-Optic Cables

Fiber-optic cable requires specialized tools for troubleshooting. Typically, an Optical Time Domain Reflectometer (OTDR), sometimes abbreviated as OTD, is used to locate faults in the fiber-optic cable core. The OTDR uses a light source to detect faults. The light source sends a pulse of light energy through the fiber-optic core. As the light pulse encounters defects in the core and connectors, the light reflects to the source, which also acts as a receiver. The amount of reflected light and the time it took to travel to the fault and back is instantly calculated. The intensity of the reflected and the calculated time indicate the type of flaw and the location in meters where it occurred. A more in-depth explanation of the OTDR can be found in Chapter 3—Fiber-Optic Cable.

NIC Loopback Test

The network interface card loopback test typically employs the use of the **ping** command as part of its diagnostics. A loopback test checks a hardware device's ability to transmit and receive signals. It consists of diagnostic software and an adapter (connector) called a loopback. The loopback is plugged into the back of the network interface card. When the diagnostic software is run, the loopback completes the electrical circuit between the transmit and receive pins. Without a connection between the transmit and receive pins, a complete series of diagnostic tests cannot be run.

Indicator Lights

All network adapters and network devices such as hubs and switches are equipped with indicator lights known as LEDs. There is no standard interpretation of what the light patterns represent. To interpret the exact meaning of the light patterns, such as a flashing or steady light, you must consult the manufacturer's resources. However, there are some general assumptions that can be made. If none of the indicator lights are lit, a cable problem may exist. A cable problem may include a disconnected cable at the hub or switch or a bad connection at either end of the cable. There is also a possibility that the network card is bad, but this is rare. Generally, it is safe to assume network activity when the indicator lights are blinking. The more rapid the lights blink, the higher the rate of activity. A steady light can also indicate high activity.

Network Analyzers

Typically, the entire network infrastructure can be tested using a network analyzer. A network analyzer can be software or a combination of software and hardware. Network analyzers are typically the most expensive of all network testing devices and exceed the budget of IT departments with small LANs. The best network analyzers combine features found in all network testing software and hardware and have elaborate display capabilities.

Protocol Analyzer

Protocol analyzers are used to capture and monitor data frames traveling across the network media. While most protocol analyzers monitor only TCP/IP, more sophisticated protocol analyzers can monitor hundreds of different protocols. Protocol analyzers vary greatly in their features. Protocol analyzers can filter the data they gather by limiting the capture of data to one or more nodes, thus reducing the total amount of data that needs to be analyzed. Many of the more sophisticated protocol analyzers can chart their results in graph form. Protocol analyzers are commonly incorporated into network analyzers.

While protocol analyzers vary somewhat, all have similar functions and characteristics. Look at **Figure 19-19** to view some of these features. Displayed are three viewing panes. The top pane displays the series of frames captured during the brief period the protocol analyzer was running. The frame that is highlighted contains the Simple Mail Transport Protocol (SMTP).

The middle pane displays detailed information about the highlighted frame, such as the IP address of the source and the destination, the time of the capture, and the type of protocol. The bottom pane displays the contents of the packet. The characters on the left side of the bottom pane display the contents as hexadecimal pairs. The ASCII equivalents of the hexadecimal pairs are on the right. Characters outside the range of the basic standard 128 characters are not displayed or are displayed as meaningless symbols. This particular SMTP captured packet contains a plain-text e-mail message. If you look closely at the contents, you will see who sent the message. By scrolling through the pane, the entire contents of the message and the e-mail addresses of the sender and destination are displayed.

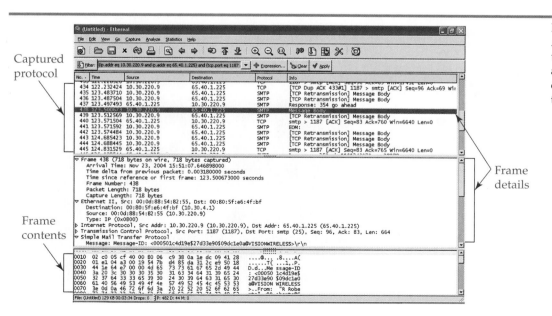

Captured protocol

Frame contents

Frame details

Figure 19-19.
A protocol analyzer is used to capture frames and monitor frames of data traveling across the network media.

A technician in the work place can use a protocol analyzer to analyze network conditions, observe network traffic patterns, and identify bottlenecks. Most protocol analyzers provide features to allow the user to filter packets by source, destination, protocol type, or by most any characteristic of a frame or packet.

Wireless Network Tester/Analyzer

While not all protocol analyzers capture wireless protocols, many do. Wireless network analyzers often require hardware to perform a more in-depth analysis of the wireless infrastructure. Wireless analyzers are often designed to check signal strength and to measure radio and electromagnetic interference.

You may want to visit the Fluke site at www.fluke.com to see the many different types of network test equipment available. It is an excellent way to become more familiar with the various test instruments available for network troubleshooting. Limited trial protocol analyzers are available for download.

A freeware tool for checking a wireless network adapter and wireless access point is Network Stumbler, **Figure 19-20.** The Network Stumbler displays the signal to noise ratio picked up by the wireless device. This can be a very valuable aid if you suspect there might be radio signals (noise) interfering with the wireless network.

Troubleshooting the Server

Server failure is rare, but it does happen. If the correct precautions are taken when installing a server, the server can perform for years without failure. There are several items that can help ensure a long time between system failures and data loss. Some of these items include the following:

- UPS unit.
- RAID system.
- Backup system.
- Virus program.
- Service packs and patches.

Figure 19-20.
The Network Stumbler utility can aid in detecting interference in a wireless network.

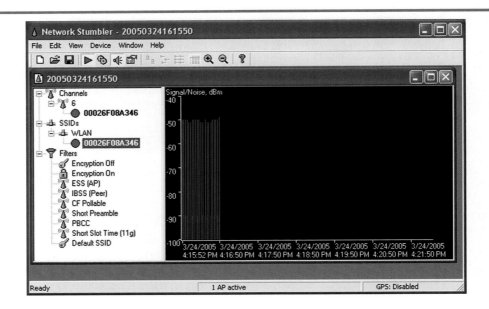

When setting up a server, it is important to install an uninterruptible power supply (UPS). You should also select a fault tolerance type of RAID system that is appropriate for the company budget and for the type of data stored. Because failure is unavoidable, be sure to perform regular data backups or use an off-site storage facility service. Be sure to install patches and service packs as they become available. Install a virus protection software package designed especially for servers. In addition to a virus package for the server, it is also desirable to install a virus protection package on each workstation in the network. Do not overlook laptops used by the organization. The company laptop computers are also frequently attached to the network and can infect the network system. Service packs and patches should be installed on a regular basis as soon as they become available. Remember to test the service pack or patch before installing it on the network server. Many patches and service packs can cause severe problems for the operating system and application programs.

Detailed troubleshooting information covering each of the major network operating systems for servers is beyond the scope of this textbook. For troubleshooting information about a particular server network operating system, you can review the information available at Novell, Microsoft, Sun Micro Systems, IBM, Dell, Red Hat, and Hewlett Packard. Expertise in one or more of these individual server network operating systems may be a certification you wish to obtain after completing this course of study.

Troubleshooting the Most Common Network Problems

After all of the training and certification you receive, you may be disappointed to find that the most common networking problems consume about 90% of a technician's time. In fact, most of the problems will be quite easy to solve with patience. This section presents some of the most common problems encountered. A technician that has a sound knowledge of PC repair and support can easily solve many of these problems.

The User Cannot Log On to the Network/Computer

The most common help desk request is assistance for logging on to the network. Some of the most common user logon problems are listed in **Figure 19-21.** A logon problem is without a doubt the most common problem encountered. Users generate the largest number of network failures, not the hardware or software. Do not blame the user. The fact that users produce so many problems is caused by the lack of user expertise. Users need training, and it is the responsibility of the network administrator to provide that training. The better you train the users, the less you will encounter common problems.

Loose Connections

After logon problems, the most commonly encountered network problem is loose cable connections. I cannot count the times when simply removing and reinserting a cable from a hub, network interface card, or network device has cleared a connection problem. Connections of this type are mechanical connections, not soldered connections. Consequently, they can become loose or slightly oxidized.

Figure 19-21.
Common logon prob-
lems and an explana-
tion of how they
happen.

Logon Problem	Explanation
Wrong password	Users forgot their latest password, which is a very common occurrence, especially if regular changing of passwords is required.
Typographical errors and spelling	Users often misspell their password or perform typographical errors while attempting to log onto the network.
[Caps Lock] key	The [Caps Lock] key is struck accidentally and remains on during the log on procedure. The password is case-sensitive, which results in the user being denied access.
[Num Lock] key	The user accidentally hits the [Num Lock] key, which causes the number pad function to change. If the user has a password that includes numbers and uses the number pad to enter the numbers, password errors will be generated when attempting to log on to the network. This is common with users that do accounting and use the number keypad frequently in their work.
Wrong domain or server	Users may inadvertently attempt to log onto the wrong server or domain when prompted by the logon dialog box. Depending on the network system, some users are capable of logging onto more than one domain controller or server and may have different user names or passwords for each domain controller or server. While not a good practice, this is a real possibility.

Oxidation normally occurs at mechanical connection points where electrical energy flows through the connection. All metal oxidizes at various rates. The oxidation rate is accelerated when electrical energy flows through the connection. The amount of oxidation required for a connection failure need not be detectable by the human eye. There is no need for a heavy layer of electrical corrosion to exist. A microscopic, translucent layer of oxidation can render any electrical connection as an open circuit and produce a system failure.

For example, have you ever had a battery-operated electronic device, such as a television remote control, that quit working? You simply removed and reinserted the batteries, and it began to work. Or, you might have given the remote control a whack or two by hitting it against an object, and it began working. The effect was the same in both cases. The sliding motion of the battery terminals across the remote control contacts removed the microscopic layer of oxidation.

The memory module in **Figure 19-22** illustrates the elusive buildup of oxidation on metallic parts. Oxidation is particularly bothersome on low voltage applications such as computers. A common pencil eraser has been used to remove a small section of oxidation from the memory module contacts that run along the edge. Pay particular attention to how shiny the copper contacts are now that the oxidation has been removed.

Oxidation can occur on any metal connection point in a computer, such as the contact edge of network interface cards and the pins of a CPU or hard disk drive. It can also occur on network cable connections.

The User Cannot Access a Share

There are several reasons a user may not be able to access a network share. First, make sure the user can connect to the share location by using the **ping**

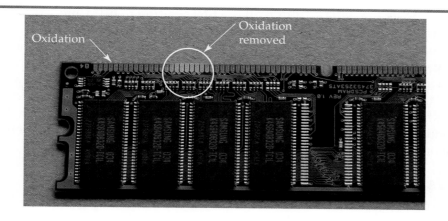

Figure 19-22.
The metal contacts of computer and network hardware are susceptible to oxidation. Note the oxidation on the contacts of this memory module.

command with the IP address of the share location. The inability to access the share location through the **ping** command may indicate a network media problem or network device program. If connectivity with the share location is intact as indicated by the **ping** command, check if the user has rights or permissions to the share. This situation usually happens when the user or the share is new.

The User Cannot Print to Network Printer

Most printer problems are easy to resolve if the printer has worked correctly prior to the problem. Many users do not understand the basics of printer operation, especially if they have never used a networked printer. If a user is experiencing a printer problem, check if electrical power is applied to the printer. This means that the printer is plugged in, the outlet has power, and the printer is turned on. Next, see if the printer will perform a self-test. All printers have a preprogrammed self-test that is activated through the printer menu selections or activated by a button combination or sequence. You may need to check the printer manual for instructions.

If the printer is connected to or managed by the server, check if there is sufficient disk space for the print job(s) that are waiting to print. Check if a print job has stopped because of some other problem, thus preventing any job sent to the printer to be processed. Make sure the person attempting to print has permissions or user rights to print. The user may be trying to use a printer that they do not have permissions to use, especially if they are new. Check if the print server is down. Many of the reasons that a user cannot connect to a network printer would be the same as for connecting to a network share. Check if the DNS or WINS service/server is working.

The Printer is Printing Gibberish

Another common printing problem is a printer that has been working suddenly starts printing gibberish. The following is a list of common printer things to check:

- Does the cache or printer spooler need to be cleaned?
- Is the printer misidentified?
- Did the user attempt to connect to wrong printer?

The print job in the cache or print spooler may have become corrupted or the user may have used wrong printer driver to print to the selected printer. Many times, it is quicker to simply dump all the files waiting to be printed, remove the printer drivers, reinstall the drivers, and reconfigure the printer share. You do not know what has taken place before you arrive at the problem printer. Someone may have already attempted to fix the printer. That person would not typically be a technician, just someone who may have fixed a printer problem before and believes that he or she can most likely fix this one. They may have tried to change many of the settings and even attempted to connect to a different printer share, thus sending all the waiting files to a printer in a different location. Remember your people skills and be patient.

The User Cannot Access the Internet

Internet connection failure on a network can be caused by circumstances inside the local network or outside the local network. The type of physical Internet access used by the network determines the exact method of troubleshooting. For example, if a telephone modem is used to access the Internet through an ISP, then a simple test such as dialing a cell phone number using the Internet connection modem will verify if the modem is working. If a T1 line is used, you may need to call the provider to test the connection from their support office. In general, check if the problem is isolated to the Internet connection equipment, Internet service provider equipment, or a service on the LAN, such as the Internet Connection Sharing (ICS) service.

The User's Computer Has a Virus or Worm

Worms are the real danger for computer networks. Worms are typically distributed by e-mail systems and use the list of e-mail contacts to spread the worm program. A good virus suite of programs that contain e-mail scanning services is essential to protect against worm attacks. There is no way to stop all virus or worm attacks. New viruses and worms are created daily. You can protect the system against known viruses and worms, but you must constantly update the protection package. E-mail alerts from virus protection vendors are sent to users as soon as a new virus or worm is identified.

Most user-reported and inexperienced technician-reported viruses are not really viruses at all. Many suspected viruses are actually the results of users and technicians misconfiguring software or hardware.

Troubleshooting with Event Viewer and System Monitor

In Chapter 18—Maintaining the Network, performance and events monitoring was introduced for system maintenance. These same sources of information are valuable troubleshooting tools. With some of these tools, a complete history of hardware and software installed and removed from the system can be viewed. Reviewing system monitoring tools and logs can help pinpoint causes of system failures. For example, the System Monitor utility can be set up to monitor traffic conditions on the LAN, which will assist the technician when troubleshooting system bottlenecks.

Troubleshooting with TCP/IP Utilities

The TCP/IP protocol suite is the predominant protocol in use. All major operating systems use TCP/IP for communications across the Internet and for

LAN communications. Microsoft NetBEUI and IPX/SPX are rapidly becoming obsolete. An excellent understanding of TCP/IP utilities is mandatory for a network technician. There are several common TCP/IP utilities designed to diagnose problems with the TCP/IP protocol suite. Most of the commands are the same in all operating systems with only a few exceptions. Exceptions will be pointed out when appropriate in this section.

Ping

The **ping** command is designed to send an echo request message to another node using the Internet Control Message Protocol (ICMP). The **ping** command is used to verify that a connection exists between the destination and the source. It is also used to verify that the TCP/IP protocol is configured properly. For example, you can issue the **ping localhost** command to verify that the TCP/IP protocol is configured for the network interface card at the host. Let's look at how you might use the **ping** command to troubleshoot a LAN and beyond. The following sequence will help you troubleshoot a connection on a TCP/IP-based network:

1. Ping the workstation. Use the **ping 127.0.0.1** or **ping localhost** command. This is used to verify that TCP/IP has been configured for the network interface card. Failure indicates a problem with the network adapter.

2. Ping the assigned IP address. If a static address has been assigned to the desktop, then ping that assigned IP address. If DHCP has been selected for an automatic IP address assignment then skip this step.

3. Ping another workstation on the LAN. Ping it by name and by the assigned IP address. If the **ping** command returned successfully when the IP address is used, but fails when the name of the distant workstation is used, it is a good sign of a DNS server or WINS server failure. WINS is used on older Windows systems to resolve NetBIOS computer names to IP addresses.

4. Ping the default network server, gateway, or router. This will verify a complete cable connection from the workstation to the server, gateway, or router. The device you choose will vary according to each network design. Basically, you are verifying a connection to the outer edge of the LAN.

5. Ping the DNS server used for the network. This will verify that the server responsible for resolving domain names to IP addresses is accessible. Ping the DNS server when you are experiencing problems with Internet connections. This may not be necessary for a small network.

6. Ping a distant location, such as Google, AOL, Microsoft, and Novell to verify that connections can be made outside the network. This is especially important when resolving Internet connection problems. You may experience a timeout for the **ping** command. This is because many sites block the **ping** request for security reasons. The **ping** command appears as a probe of the network and may not be successful. You may need to try several well-known sites before one replies to your **ping** request. If either of the last two recommended **ping** steps fail, it may indicate that a DNS server problem exists.

Look at **Figure 19-23.** It shows a list of properties of ICMP that can be blocked. At the top of the list is the **Allow incoming echo request** option. Clearing this option would block any **ping** or **tracert** request from a remote host. The list illustrates how the Windows XP firewall can be configured to stop ICMP probes of a computer or site.

Figure 19-23.
List of ICMP requests
that can be blocked.
Selecting an option will
allow the ICMP request.

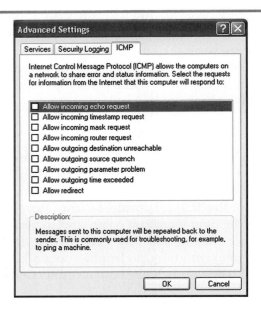

There are some **ping** switches available that can modify its use. The **-n** switch is used to change the number of echo requests to be sent. The **-l** switch may be used to increase the timeout value for special applications. For example, if you are pinging a sight that requires a satellite connection, the time interval may need to be increased because of the inherently slow transmission times associated with satellite communications.

The switches available for the **ping** command can vary according to the operating system used.

Tracert

The **tracert** command traces the route from the workstation to a remote location and displays information about the path taken by the tracert packets. Look at **Figure 19-24** to see a typical **tracert** display. The utility displays a series of hops taken along the route from the workstation to the requested destination. The amount of time taken to reach each router along the route is displayed in milliseconds. The **tracert** command sends a series of **ping** packets from the source to the destination. At each router or gateway encountered along the path, an echo is sent back to the source. Each additional **ping** packet increases the time-to-live to enable the packet to reach the final destination.

There are many third-party tools developed that provide a graphical presentation of a **tracert** result. **Figure 19-25** shows various display formats from a graphical tool called NeoTrace.

The UNIX/Linux **tracert** command is **traceroute**.

Microsoft Windows XP has a **pathping** command, which is a combination of **ping** and **tracert** commands.

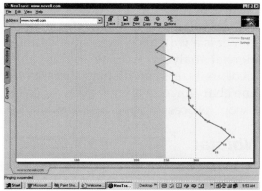

```
  7 x 12 ▼  [□] [▤] [▥] [◈]  [▤] [▤] [A]

Tracing route to novell.com [192.233.80.9]
over a maximum of 30 hops:

  1   247 ms   247 ms   261 ms  tc101.strato.net [209.26.168.249]
  2   248 ms   247 ms   261 ms  tc102-254.strato.net [64.45.197.254]
  3   357 ms   412 ms   247 ms  host193.strato.net [207.30.98.193]
  4   261 ms   261 ms   247 ms  host4.utelfla.com [207.30.208.4]
  5   288 ms   275 ms   275 ms  sl-gw25-atl-9-9.sprintlink.net [144.223.29.9]
  6   261 ms   275 ms   275 ms  sl-bb23-atl-6-0.sprintlink.net [144.232.22.9]
  7   275 ms   288 ms   288 ms  sl-bb25-fw-4-3.sprintlink.net [144.232.20.61]
  8   288 ms   289 ms   274 ms  144.232.11.37
  9   289 ms   288 ms   302 ms  sprint-gw.dlstx.ip.att.net [192.205.32.53]
 10   288 ms   288 ms   302 ms  gbr4-p50.dlstx.ip.att.net [12.123.16.246]
 11   316 ms   329 ms   330 ms  gbr5-p50.dvmco.ip.att.net [12.122.2.102]
 12   329 ms   330 ms   329 ms  gbr3-p60.dvmco.ip.att.net [12.122.5.33]
 13   329 ms   330 ms   329 ms  gar1-p360.slkut.ip.att.net [12.122.2.237]
 14   344 ms   329 ms   330 ms  12.127.106.34
 15   316 ms   330 ms   329 ms  192.94.118.221
 16   330 ms   329 ms   330 ms  192.94.118.2
 17   330 ms   329 ms   330 ms  prv1-mx.provo.novell.com [192.233.80.9]

Trace complete.

C:\WINDOWS>
```

Figure 19-24.
Typical display of the **tracert** command used with a DNS name or IP address. Notice that each router along the path to the destination is listed along with the time in milliseconds it took to reach the router.

Netstat

The **netstat** command displays information about active TCP/IP connections. The **netstat** command displays Ethernet statistics for IP, TCP, ICMP, and UDP for IPv4 and IPv6. When the **netstat** command is used without a switch, it lists all active TCP/IP connections. When used with the **/?** switch, the screen displays a summary of all the common **netstat** switches. **Figure 19-26** displays a list of common **netstat** switches and a description of each.

Figure 19-27 illustrates the **netstat** command issued with the **-e** switch. The cumulative interface statistics for the host's Internet interface card are displayed.

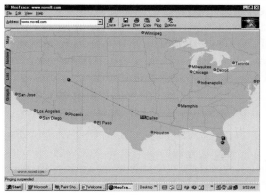

Map Form

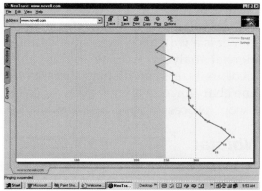

Graph Form

Figure 19-25.
The NeoTrace utility can display traced routes in various formats.

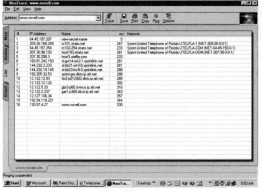

List Form

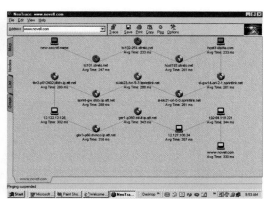

Node Form

Figure 19-26.
Summary of **netstat** switches.

netstat switch	Description
-a	Displays all active TCP and UDP connections.
-e	Displays Ethernet statistics.
-n	Displays active TCP connections expressed numerically.
-p	Displays active connections for a specific protocol.
-s	Displays statistics by protocol.
-r	Displays contents of routing table.
/?	Displays help.

Figure 19-27.
The **netstat -e** command can be used to check if the network is working properly. A high number of discards or errors indicate a problem.

Notice that there are no discards or errors. This indicates that the network is working properly. A high number of discards or errors indicates a problem on the network.

Switches may also be used in combination to create screens containing very specific results. For example, the command **netstat -s -p ip** displays statistics pertaining only to the IP protocol. See **Figure 19-28.**

As you can see, the **netstat** command can be very handy as a troubleshooting tool. It is a way to collect statistics about TCP, IP, UDP, and ICMP. Of course, commercial analyzers display all this same data and more. Commercial network protocol analyzers can display statistics not only as raw data but also in bar, pie, or line chart form in a wide array of colors. These charts can be easily imported into a word processing package to make formal presentations or written reports.

Nbtstat

The **nbtstat** command displays NetBIOS over TCP/IP statistics and information gathered from broadcasts, NetBIOS cache, and WINS services. It can be a

Figure 19-28.
The **netstat -s -p ip** command displays statistics relating to the IP protocol.

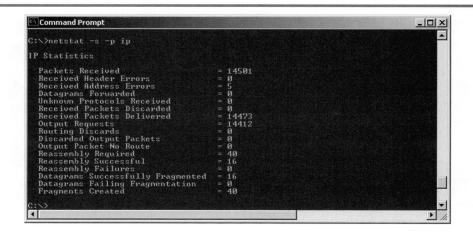

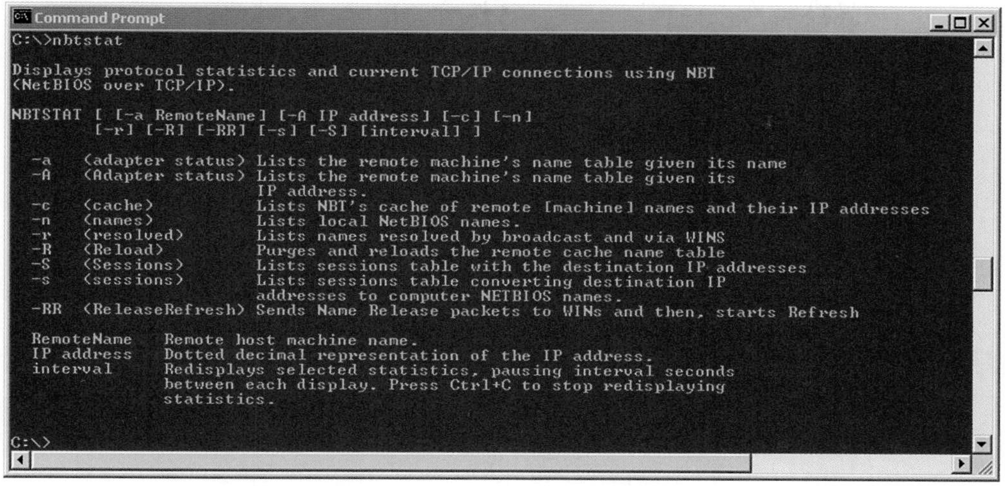

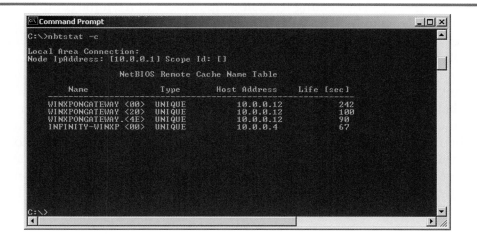

Figure 19-29.
Issuing the **nbtstat** command at the command line displays a list of switches and their descriptions.

very handy tool for verifying that the WINS server is functioning properly or that NetBIOS over TCP has been configured correctly on the network interface card. Look at the list of the common switches in **Figure 19-29** that can be added to the **nbtstat** command to achieve the indicated results.

By applying a switch, you can achieve specific results. For example, by using the **nbtstat -c** command, you will display the contents of the cache of NetBIOS names, their host IP address, and the amount of time in seconds that the information will remain in the name cache. See **Figure 19-30.**

Notice the hexadecimal numbers located inside the greater than and less than symbols (< >) following the NetBIOS names in the list. The NetBIOS name actually contains 16 characters. The sixteenth character is a hexadecimal number used to provide additional programming information about the NetBIOS named device. The hexadecimal number indicates if the device is a server or a workstation or if the name contains blank characters and other program writing concerns. While a user can only enter 15 alphanumeric characters for a NetBIOS name, 16 characters are actually stored for each device. The user is unaware of the sixteenth character. To better illustrate the power of **nbtstat**, some command examples are listed in **Figure 19-31.**

Figure 19-30.
The **nbtstat -c** command displays a list of NetBIOS names that are stored in a cache on the local host.

Figure 19-31.
Some **nbtstat** com-
mands examples.

Command	Description
nbtstat -n	This command displays the contents of the local computer NetBIOS name table.
nbtstat -a station24	This command displays the NetBIOS name table of the computer with the NetBIOS name of *Station24*.
nbtstat -A 192.168.0.23	This command displays the NetBIOS name table of the computer with the IP address 192.168.0.23.

ARP

The **arp** command is used to inspect the contents of the ARP table. The ARP table maps MAC addresses to IP addresses. The **arp** command can be useful when attempting to verify the physical address of another client or device on the local network. Simply ping the IP address or name of the other workstation or device and then check the ARP cache using the **arp -a** command.

IPconfig

Ipconfig is used on the Windows NT family of operating systems to identify the IP configuration of a computer. It reveals the assigned IP address, subnet mask, and default gateway. The equivalent command for Windows 98 and Windows 95 is **winipcfg**. On a UNIX/Linux system, you would use the **ifconfig** command.

Network+ Note:

For the Network+ exam, be sure you can identify which IP configuration tool command is used on each of the following operating systems: Windows NT-based, Windows 95 and 98, and UNIX/Linux.

Nslookup

The **nslookup** command is used to display information about domain name servers. For example, by issuing the **nslookup** command from the command prompt, a reply would identify the name and IP address of the first available DNS server. The **nslookup** command can be used to verify that the DNS server is available and that there is a complete network media path to the DNS server. The dig command is an alternative to **nslookup** but is not supported by Microsoft operating systems.

When **nslookup** is used alone, it returns the network's default DNS server name and location. When used with a domain name, **nslookup** returns information about that domain. See **Figure 19-32**. Notice that **nslookup** was used to gather information about the www.microsoft.com domain. The information returned includes the DNS server name, IP address, and alias. To find out more about the **nslookup** command, issue the **/?** or **help** at the **nslookup** prompt.

Figure 19-32.
When the **nslookup** command is used alone, information about the network's default DNS server is given. When used with a domain name, information about that domain is given.

Summary

- The eight steps to approaching a troubleshooting problem as outlined by CompTIA are as follows.
    1. Identify the symptoms and potential causes.
    2. Identify the affected area.
    3. Establish what has changed.
    4. Select the most probable cause.
    5. Implement an action plan and solution including potential effects.
    6. Test the result.
    7. Identify the results and effect of the solution.
    8. Document the solution and process.

- The five areas of network problems are: workstation, work group, network infrastructure, server(s), and outside the LAN.

- Always ask what has changed recently that could have caused the problem.

- The typical Windows 98 boot sequence is POST, MBR, real mode, protected mode, and then the system logon.

- The typical Windows NT family boot sequence is POST, MBR, ntldr, ntdetect.com, ntldr (again), ntoskrnl.exe, HAL, and then the system logon.

- The HAL is designed to prevent user programs from accessing the hardware directly, thus reducing the most common source of system crashes associated with Windows 95, 98 and Me operating systems.

- The MBR contains information about partitions, files, and operating system file locations and is critical to a successful system boot.

- Real mode allows access to only the first 1 MB of RAM and loads minimal drivers.

- Protected mode allows access to the entire RAM and loads multiple programs into RAM.

- A Microsoft Windows operating system passes through real mode and then protected mode.

- The last step in a boot operation is the logon procedure.

- The startup menu is accessed by pressing [F8] during the boot operation.

- Safe mode allows a computer to boot with a limited number of system drivers and memory space.

- The System Restore utility will roll back the operating system to a previous condition when the system successfully booted.

- The **Last Known Good Configuration** option will attempt to load the operating system using the last good set of registry information.
- Recovery Console is a command-line utility designed for Windows NT–based systems.
- The Network Diagnostic utility is designed for Windows Me and Windows XP systems to diagnose common network problems and causes.
- The Recovery Console is a last resort recovery tool used to recover Windows NT–based computers.
- Windows Me introduced the Network Diagnostics utility, which combines many TCP/IP utilities to perform a quick network diagnoses.
- Ways to ensure network server data integrity and performance involve UPS units, RAID systems, data backups, virus protection, and applying service packs and patches.
- The most common network errors are caused by human interaction with the network system.
- The most common problems involve logon problems and loose network cable connections.

Review Questions

Answer the following questions on a separate sheet of paper. Please do not write in this book.

1. List the eight steps recommended for troubleshooting as outlined by the CompTIA Network+ objectives.

2. What are the five major areas, or locations, of computer/network problems?

3. What is the start of any computer boot process?

4. What controls the sequence of boot device hardware, such as the hard disk drive, floppy disk, and CD-ROM?

5. What Windows software system replaced the need for a config.sys file?

6. Which two files are no longer necessary for completing a Windows 98 boot sequence?

7. What is the name of the Windows NT kernel file?

8. What is the name of the Windows NT file that detects the presence of the system hardware?

9. What is the final stage in the Windows NT boot process?

10. When is the network workstation boot process completed?

11. In a dual boot or multiboot system, which file contains information about the location of each available operating system?

12. In a dual boot or multiboot system, which file displays a menu that allows the user to select an operating system to which to boot?

13. How is the boot.ini file contents displayed in Windows XP?

14. Which special function key is used to access the Safe Mode option on a Windows XP computer?

15. How can you quickly isolate a network interface card from a workstation running Windows 98 without physically removing the card?

16. What are the typical Windows XP startup selections listed in the Windows Advanced Options Menu?

17. Which startup menu selection will start the computer with the system settings at the time of the last successfully completed boot?

18. Explain the term *loopback test*.

19. What is the most common network problem?

20. What are the most common reasons users cannot log on to a network? (List at least three.)

21. What TCP/IP utility is designed to quickly verify a connection between the source and destination?

22. What is the UNIX/Linux command equivalent of the Windows NT **tracert** command?

23. Which TCP/IP utility is designed to list all routers along a path to a distant destination?

24. Which TCP/IP utility will display statistics for IP, TCP, ICMP, and UDP?

Sample Network+ Exam Questions

Answer the following questions on a separate sheet of paper. Please do not write in this book.

1. Which command would you *most likely* use to verify a cable connection to the server on the LAN?
   A. **netstat**
   B. **ping**
   C. **ifconfig**
   D. **tracert**

2. The **Last Known Good Configuration** option is associated with which of the following operating systems? (Select all that apply.)
   A. Windows 95
   B. Windows 98
   C. Windows Me
   D. Windows 2000 Professional
   E. Windows XP Professional
   F. Windows NT Server
   G. Windows XP Home Edition
   H. Windows 2000 Server
   I. Windows Server 2003

3. A Windows 2000 Server fails to complete the boot process. The hard disk drive is formatted with NTFS. You wish to inspect the partition table on the hard disk drive and possibly attempt to repair the MBR. Which utility would be most appropriate for troubleshooting the server?

   A. Run the **fdisk** command using any Windows compatible boot disk.

   B. Run the **diskpart** command after installing the Recovery Console using the installation CD.

   C. Run the **fdisk** command from a DOS system boot disk.

   D. Extract and run the **regedt32** command, which is found on the installation CD.

4. A workstation is connected to an Ethernet 10BaseT network. All workstations are automatically issued IP addresses from the server. While troubleshooting a workstation, you check if a valid IP address has been assigned to the workstation. The first thing you did when you arrived at the workstation was run **winipcfg.exe**. The following dialog box appears. What would be the next appropriate action to take?

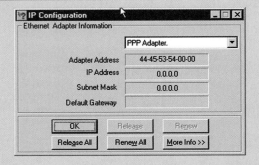

   A. Replace the network interface card.

   B. Open Device Manager and reinstall the network interface card drivers.

   C. Ping the network DHCP server to verify a connection.

   D. Check if the TCP/IP protocol is installed on the computer.

5. Which of the following utilities is responsible for generating the display in the exhibit?

   A. **nbtstat**

   B. **netstat**

   C. **arp**

   D. **ping**

6. Which type of meter would be most appropriate for checking a 100BaseFX cable?

   A. Ohmmeter

   B. Ammeter

   C. Multimeter

   D. TDR meter

7. You have just successfully booted a Linux server. Which command would be the most appropriate to check the server network adapter IP address assignment?

   A. **dig**

   B. **nslookup**

   C. **ifconfig**

   D. **pingpath**

8. Which protocol would you block using a firewall to prevent probes by the **ping** command?

   A. TFTP

   B. ICMP

   C. HTTP

   D. SSL

9. Which command would you issue from the command prompt to view the total number of packets sent and received from a workstation?

   A. **ping -w**

   B. **arp -a**

   C. **tracert -d**

   D. **netstat -e**

10. What is the first thing you should do when troubleshooting a network problem?

   A. Establish what has changed.

   B. Identify the symptoms and potential causes.

   C. Select the most probable cause.

   D. Implement an action plan.

Suggested Laboratory Activities

1. Inspect the boot.ini file using the System Configuration (**msconfig.exe**) utility in Windows XP.

2. Inspect the boot.ini file using Notepad.

3. Make boot disks for Windows 98, Windows 2000, and Windows XP computers.

4. Obtain (download) setup disks for Windows XP.

5. Create an emergency repair disk for a Windows 2000 or 2003 system.

6. Boot an NT–based computer with a non-NT boot disk and observe the error message.

7. Boot an NT–based computer using an NT–based boot disk (one created for that particular system).

8. Practice using boot disks to start operating systems such as Window 98.

9. Practice using Recovery Console in Windows 2000, Windows Server 2003, or Windows XP by using a bootable installation CD.

10. Back up the boot.ini file.

11. Back up a portion of the files on a computer using a storage media such as a CD-R or a tape drive.

12. Inspect the contents of each tab in the System Configuration Utility (**msconf**).

13. Inspect the System Information (**msinfo32**) utility closely, especially available tools and system hardware/software information.

14. Run System Restore and create a restore point in Windows XP.

15. Start the system in safe mode by pressing [F8] during the startup and observe the menu selections closely. Also, try various menu options.

16. Use Help and Support on a Windows Server 2003 computer to research common printer problems.

Interesting Web Sites for More Information

http://developer.novell.com

www.microsoft.com

www.redhat.com/docs/manuals

Chapter 19
Laboratory Activity

Network Diagnostics

After completing this laboratory activity, you will be able to:

■ Access and run the Microsoft Network Diagnostics utility.

■ Modify the Network Diagnostics utility configuration.

■ Identify the tests that can be run through the Network Diagnostics utility.

■ Save the results of a Network Diagnostics scan.

Introduction

In this laboratory activity, you will learn to access, run, and modify the configuration of the Microsoft Network Diagnostic utility on a computer running Windows Server 2003. The Network Diagnostics utility in Windows Server 2003 is similar to the one found in Microsoft XP Professional.

To access Network Diagnostics, you must first start the System Information utility. To start the System Information utility, you simply enter **msinfo32** in the **Run** dialog box off the **Start** menu. After the System Information utility appears, select **Net Diagnostics** from the **Tools** menu as shown in the following figure.

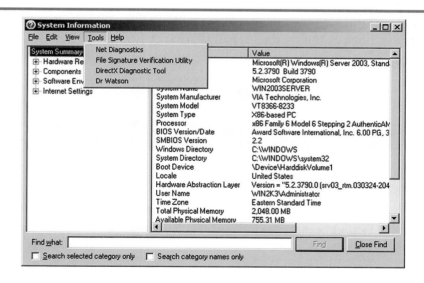

Note:
Another way to access the Net Diagnostics and other utilities is by selecting the **Help and Support** *command off the* **Start** *menu. After the Help and Support window opens, select* **Tools** *and then* **Help and Support Center Tools**. *From the right-side pane of the window, select* **Network Diagnostics**. *As you can see, running* **msinfo32** *from the* **Run** *dialog box is a much simpler way to access the Net Diagnostics utility.*

After selecting **Net Diagnostics**, the Help and Support Center window similar to the one in the following figure will appear. This figure also shows a typical scan result made from a Windows 2003 server. The Net Diagnostics utility scans various items that directly correlate to the network and server. For example, the utility pings the network adaptor to see if it is in working order and configured with the TCP/IP protocol. The utility also pings such items as the DNS, DHCP, and WINS servers as well as the default gateway to verify a connection exists. These services may exist on the server tested or exist on a different server.

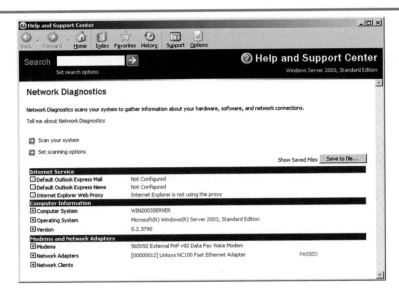

You can configure the Net Diagnostics utility to test all network-related objects or simply a few. Look at the various configuration options shown in the following figure. You will explore the various options during this laboratory activity.

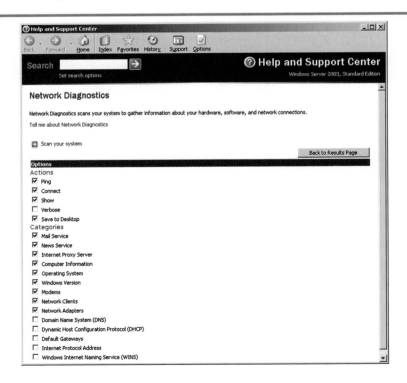

Equipment and Materials

■   Computer with Windows 2003 Server installed.

Note:
You may substitute a Windows XP workstation for this activity with your instructor's permission.

Procedures

1. _____   Report to your assigned workstation.

2. _____   Boot the workstation and make sure it is in working order.

3. _____   Access the Net Diagnostics utility by typing **msinfo32** in the **Run** dialog box located off the **Start** menu. After the System Information window appears, select **Net Diagnostics** located under the **Tools** menu. The Help and Support Center window should open displaying the Network Diagnostics utility.

4. _____   Close the Help and Support window. Now access the utility once more, but this time use the **Start** menu and then select **Help and Support**. After the Help and Support Center window opens, select **Tools** from the right side of the window under **Support Tasks**.

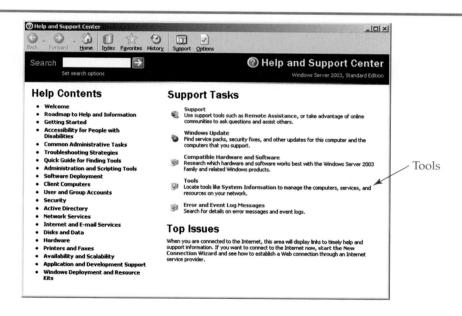

5. _____ After selecting **Tools**, select **Help and Support Center Tools** from the left-side pane of the window. Network Diagnostics will be revealed as an option in both the left-side and right-side panes. Select either option. The following commands will appear: **Scan your system** and **Set scanning options**.

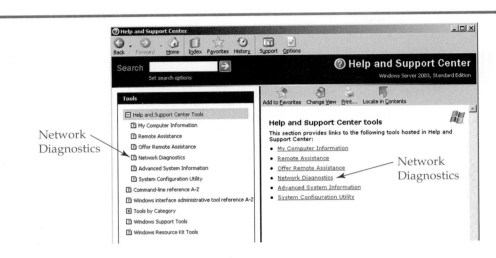

6. _____ Close the Help and Support Center window. Repeat accessing the Network Diagnostics utility using both methods until you feel comfortable with each method.

7. _____ Open the Network Diagnostics utility once more. This time scan the system by selecting **Scan your system** and observe the results.

8. _____ Select **Set scan options** and explore the options available. Record on a separate sheet of paper the options listed under **Actions** and **Categories**.

9. _____ Experiment with the options. Run a scan each time you set different options and then observe the results. Be sure to select the box with a plus sign next to each item listed. The box with a plus sign inside indicates there is more information that can be revealed about the item. For example, if you select the **Network Adapters** box with a plus sign, more information about the network adapter will be revealed.

10. _____ After you have experimented with the scan options, set the scan options so that the information is saved to the desktop. To do this, select **Save to Desktop** under the **Actions** category of **Set scan options**. Scan the system one last time.

11. _____ Close the Help and Support Center window. Look for an icon on the desktop that starts with the name Netdiag and ends with the htm file extension. This is a record of the last scan performed. It was saved to the desktop as an HTML file. When selected, it opens in the Windows Explorer browser window. Select it now by double-clicking the icon. The last scan report should appear. If not, call your instructor for assistance.

12. _____ Close the Network Diagnostics scan report.

13. _____ Answer the review questions.

14. _____ After completing the review questions, return the workstation to the condition specified by your instructor.

Review Questions

Answer the following questions on a separate sheet of paper. Please do not write in this book.

1. What do you type into the **Run** dialog box to access the System Information utility?

2. Where is the Network Diagnostics utility located in the System Information?

3. Identify which items can be tested or verified using the Network Diagnostics utility. Answer with a *Y* for yes *N* for no.

   Verify the server physical address: _____

   Verify the server logical address: _____

   WINS server connection: _____

   DNS server connection: _____

   DHCP server: _____

4. How does the Network Diagnostics utility test the default gateway?

5. How does the Network Diagnostics utility test the assigned IP address of the server's network adapter?

6. What type file extension is used for the Network Diagnostics utility scan when it is saved to the desktop?

7. What does the plus sign inside the box to the left of the list of scanned items indicate?

Not only must a network technician have the knowledge and skills for diagnosing a network, they must be capable of diagnosing a PC.

20

Designing and Installing a New Network

After studying this chapter, you will be able to:

❏ Describe the factors to be considered when designing or modifying a network.

❏ Describe methods used for naming conventions.

❏ Explain the various stages of network design.

❏ Identify and explain terminology used by standards to identify network cable connection locations.

❏ Describe the various facilities used in a telecommunications infrastructure.

❏ Explain and describe the use of a Multi-User Telecommunication Outlet Assembly (MUTOA) and Consolidation Points (CP).

Network+ Exam—Key Points

The Network+ Certification exam does not cover in-depth specifications and standards or installing network systems. CompTIA typically limits their questions to the general installation topics provided in this chapter, such as TIA/EIA standards and architectural design elements. Be sure to know the common horizontal wiring cable standards for distances.

Key Words and Terms

The following words and terms will become important pieces of your networking vocabulary. Be sure you can define them.

backbone

Consolidation Point (CP)

entrance facility

equipment room

horizontal cross connect

horizontal wiring

main entrance

Multi-User Telecommunication Outlet Assembly (MUTOA)

naming convention

patch panel

punch down tool

telecommunications closet

telecommunications room

work area

This chapter introduces the concepts and related knowledge needed to design network systems. At this point in your networking course, you should be capable of designing a small-office/home-office (SOHO) network without assistance. Larger networks can be a challenge. Large network system designers require years of training in all aspects of networking. In fact, some of the many exams required for the Microsoft Certified System Engineer (MCSE) are Designing a Microsoft Windows 2000 Network Infrastructure (70-221), Planning and Maintaining a Microsoft Windows 2003 Infrastructure (70-293), and Designing Windows Server 2003 Active Directory and Network Infrastructure Exam (70-297). All major network operating system software companies have similar examinations.

This chapter is not intended to prepare you for a network design examination, but rather to provide you with a glimpse of design concepts. It also reinforces concepts learned earlier in the textbook.

 **Tech Tip**

Now that you have mastered the basics, you may wish to further your education after finishing this course by taking a course on network design. You can go online and download the requirements from Microsoft and Novell to see what is expected to pass the certification exams.

Needs Assessment and Design

The first step of any design project is conducting a needs assessment. A needs assessment may be a formal or an informal process, depending on the size of the project. For example, a needs assessment for a change in a small-office network can be as simple as interviewing the owner of the company on a one-on-one basis. A complex change involving hundreds or thousands of workstations and thousands of employees may involve a series of formal meetings with numerous personnel. It may also involve several surveys conducted over a long period of time. The scope of the needs assessment is dictated by the complexity of the project.

There are many factors to consider when designing a new network or modifying an existing network. **Figure 20-1** lists the question to consider when designing a network. The questions are organized by physical network structure, security, application, organizational structure, and fault tolerance and data integrity. The following sections take a closer look at some of these questions.

Questions to Consider When Designing a Network

Physical Network Structure
- What type of business is performed at the site?
- What network requirements are necessary for business operations?
- How large is the geographic area to be networked?
- How many users will there be on the network?
- What types of resources are needed (Internet access, printers, and such)?
- Is remote access by sales personnel or other staff required?
- What services will be provided (e-mail, Internet, videoconferencing, and such)?
- How will the Internet be accessed (modem, DSL, ISDN, T-1, fractional T-1, Satellite)?
- What type of electrical concerns should be considered?
- Where will equipment be placed?
- What is the total IT budget?

Security
- What level of security is required?
- What is the physical security level available and required?
- Will there be a firewall?
- Will there be a VPN?
- Will security procedures include passwords, shares, and encryption?

Application
- How much storage will each user require?
- What types of software application packages will be needed on the network (word processing, accounting, spreadsheet, database, and such)?

Organizational Structure
- How do different people, departments, or groups within the organization share data?
- How can services be divided and resources allocated to reduce network traffic?
- Should the security database be divided or replicated to reduce network traffic?

Fault Tolerance and Data Integrity
- How important is data integrity?
- How important is availability?
- What are the data recovery requirements?
- What are the data integrity requirements?
- What are the data redundancy requirements?
- What type of RAID level should be implemented?
- Should server clustering be implemented?
- What data backup method will be implemented?

Figure 20-1.
Questions to consider when designing a network. The questions are organized by physical network structure, security, application, organizational structure, and fault tolerance and data integrity.

Physical Network Structure

The physical structure of the network is typically the first concern of the network designer. The physical structure consists of network devices and media. The number of users and geographic layout determine the type of network equipment and media needed. A company with ten employees working in a small office may only require a single hub to link workstations to the network. Cable length and network traffic will most likely not be an issue. A large company that spans several floors of a building and has many departments will typically need to divide the physical structure of the network into several LANs or virtual LANs to overcome cable length and network traffic issues. This type of network may require a fiber-optic backbone and many switches. A company that spans several geographic locations will need to use leased lines or implement a Virtual Private Network (VPN).

Other factors that determine the type of network equipment and media needed are remote access to and from the network. A company sales force often requires remote access to the network while they are on the road or working from various locations within the state or country. For mobile employees, wireless access may be suitable or a VPN connection over the Internet. For employees working within the company who need to access the Internet, a router, firewall, and high-bandwidth Internet access device will be needed.

Many businesses require a Web site for providing an online catalog of products and supporting online sales. The Web site can be set up as part of the network system or be outsourced to a Web hosting service. When outsourced, the business leases space from the Web hosting service. The Web hosting service provides hardware and software support on 24/7/365 basis. They also provide security.

Setting up a company Web site using a server inside the company network requires expertise and can be quite an expensive proposition. In general, small Web sites should be outsourced to be cost-effective. When a large Web site is required for a business, it may be cheaper for the business to host their own Web site by installing the necessary hardware and software and hiring additional personnel. A simple 56k modem might work well for a low volume site, but if a great deal of business is to be conducted on the company Web site, the company will need a high-volume access method, which can be quite expensive.

Security

Security is vital for a business of any size that provides online sales or remote access. A firewall is typically required to control traffic to and from the network and to block access to sensitive data. The type of firewall to be installed, such as firewall software or a firewall device, must be decided. Cost is typically a main factor when selecting a firewall.

A set of written security procedures needs to be established and enforced. This includes passwords, network access, and the physical security of file server rooms and individual workstations. All personnel using the network need to be trained in security procedures.

Application

The network designer must note the types of applications the network users will need. A company using only a word processing and spreadsheet program will typically not require a large amount of hard drive space on the server for data files. A company using a database program with a large database of customers will require a large amount of hard drive space. A network designer must determine the current size of the database and determine the future size due to growth.

Organizational Structure

Organizational structure questions determine the number of servers, directory structure, location of resources, and partition and volume configuration. It is important to note how the company is organized and how data is shared among departments or work groups. The organizational structure of a company is typically reflected in the organizational structure of the security database, or directory. In a large company, services and resources must be allocated to reduce network traffic. A directory structure must be partitioned to do the same. Server hard drive partitions and volumes should also reflect how the company is organized and how data is shared. This will aid in security and system recovery. A network designer must also establish a naming convention for network objects such as users, printers, servers, and workstations, and must establish the types of attributes that will be filled in for each user. The implementation of TCP/IP and domain names should be planned at this time.

TCP/IP

Typically, all new networks use TCP/IP as their default protocol or they incorporate TCP/IP into their list of protocols bound to the network interface card. How TCP/IP will be implemented for the network requires planning. A network designer must decide if the IP addresses will be statically or dynamically assigned. If dynamically assigned, the designer should determine which server will act as a DHCP server and document the IP address ranges of the IP address pools.

Online purchasing and catalogs are a part of most businesses. An assigned IP address will most likely be required. Also, most companies require a domain name to be used as their URL. To see if a domain name is in use by another company, a simple search can be conducted on the Internet. Most Web hosting sites provide a domain name search for free such as that shown in **Figure 20-2.** After a unique name is found for the company, it can be purchased through a Web host provider. The services of the Web host provider do not need to be used to purchase the name. Originally, URL names were purchased from the U.S. government. As the demand increased, the U.S. government allowed commercial vendors to distribute the names for a reasonable cost—under government supervision, of course. There is a recurring cost each year for the ownership of the name.

Number of servers

A single server may be sufficient for the entire network, or several servers may be required. The overall requirements of the network structure will determine the number of servers needed. Look at the following list of possible server roles:

- File server.
- Application server.
- SQL server.
- Mail server.
- DNS server.

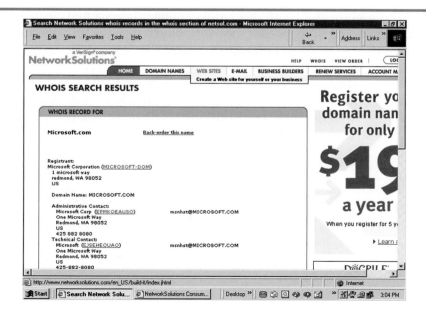

Figure 20-2.
A domain name search engine provided by a domain name registration company.

- DHCP server.
- WINS server.
- RAS server.
- RADIUS server.
- Web server.
- Gateway.
- Firewall.
- Thin client terminal service server.

As you can see, there are many roles and services that may be required by the network server based on the requirements of the business. A server may serve a single role or several. This is typically determined by a server's resources and physical location in the network.

Designing the directory structure

After all the information has been gathered concerning the structure of the organization (people, groups, services, software, workflow through the departments, and business requirements), the directory structure can be designed. There are tools that can help you design the directory structure. These tools are available from all the major network operating system vendors and third-party software developers.

All of the most common directory structure designs are based on the Lightweight Directory Access Protocol (LDAP). The most common directory structures based on LDAP are listed in **Figure 20-3.**

LDAP has become the de facto standard for directory structures. LDAP was designed to overcome limitations of previous proprietary network designs. Many of the network designs were very complex and often required third-party software solutions for integrating different systems into a single unit. With the adoption of TCP/IP as the default internet protocol of all major network systems and LDAP serving as the core for directory structures, networking systems from different companies (Microsoft, Red Hat, Novell, Sun Micro, HP) are becoming easier to integrate.

LDAP is organized into major areas including the root, country, organization, organizational units, objects, and individuals. The exact name used for the components vary somewhat by each network operating system vendor. The name may be different, but the concept and function of the features is the same. See **Figure 20-4.**

The top of the directory structure is the root. Next, is the country container, which is optional in most operating systems. The structure is then divided into organizations followed by organizational units. The last elements are objects. As you can see, the structure closely resembles the Internet hierarchical structure.

Figure 20-3.
Common directory structures based on LDAP.

Operating System	Directory
Microsoft	Active Directory
Novell	eDirectory
IBM	Directory Server
Sun Microsystems	ONE Integrated Server
Open Source UNIX/Linux	OpenLDAP

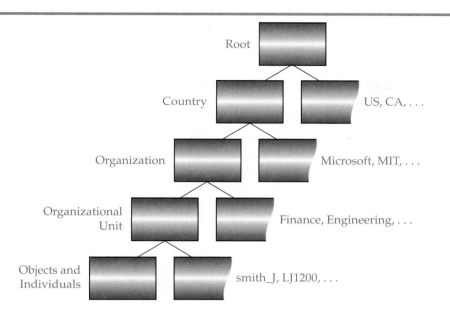

Figure 20-4.
LDAP organization.

This is a very brief summary of the LDAP directory structure. The amount of knowledge required to design a LDAP directory structure requires a complete course. You may wish to take a course on directory design later in your studies.

Microsoft Visio includes templates to assist you in designing an LDAP directory structure.

Directory structure schema

The directory structure is composed of many different objects. Each object has a set of attributes associated with it. Some attributes are mandatory, such as logon name and full name, and some are optional, such as the user's home address and fax number. The network designer should determine and document which attributes are standard for all users.

Naming conventions

A *naming convention* is a standard naming format that is used when providing names for network objects. A naming convention should be established and a written policy developed, especially for a large network installation that involves many technicians. Establishing a naming convention for objects, such as users, printers, and servers, provides for consistency throughout the system. The consistency makes the system much easier to troubleshoot and to locate objects.

A user name may be established in several different ways. Look at the various possible logon names for James Smith:

naming convention
a standard naming format that is used when providing names for network objects.

- Jsmith
- SmithJ
- Smith_J
- J_Smith

The most commonly accepted form is last name, underline, and the initial of the first name. While this is a common form, a more security conscious company may want a different naming convention that would be much harder for an outsider to guess, such as *Smith_J123*. When a user name is easy to guess, the administrator has jeopardized one half of the authentication process. Only the password needs to be guessed to access the network. This is why a more difficult to guess naming convention should be considered. In fact, the user name does not have to resemble the user's name.

A naming convention should be established that describes and identifies all equipment classified as an object. For example, thought should be given to how workstations, servers, printers, and other hardware are named. A workstation may reveal its location as part of the name, such as *WS-23-Bld4Rm110*. The location and the purpose of the server can be incorporated into the server name. This helps a technician to readily identify a server in a large corporate environment or in a network that is spread across the country or world. Some examples of abbreviations that can be incorporated into object names include those in **Figure 20-5.**

A server name may look like the following: *HQServer1*. The server may also follow the LDAP naming convention to match an Internet URL such as HQServer1.ABCcorp.com. This server is located at the headquarters of ABC Corporation and functions as a gateway to the Internet.

Workstations will also have a naming convention based on similar abbreviations. Room numbers, building letters, or departments can be incorporated into the name. A workstation name for the 12th workstation in the accounting department in room 302 of building H might look like *WS12ACCBldH302* or *WS12_ACC_BLD_H302*. It is best to avoid user names to identify workstations because employees often leave the job or are promoted, which means the user name no longer matches the workstation name.

Figure 20-5.
Examples of abbreviations that can be incorporated into object names.

Classification	Name	Abbreviation
Business Locations	California	CA
	Chicago	CHI
	Florida	FL
	Headquarters	HQ
	New York City	NYC
	Operations	OPS
Server Role Identification	Applications	APPS
	Backup server	BKUP
	Domain Name Server	DNS
	File server	File
	Gateway (router, proxy server, firewall, etc.)	GW
	Web server	WEB
Workstation and Printer Identification	Building	BLD
	Color laser printer	CLP
	Department	DPT
	Inkjet	IJ
	LaserJet	LJ
	Room location number	RM
	Workstation	WS

Remember, NetBIOS names are limited to 15 characters. If the network has Windows 95 or Windows 98 workstations, you may want to limit the number of characters in a name to 15; otherwise, name-resolving problems can arise.

It is important that all equipment or nodes on a network have a unique name. Be careful when designing a naming convention. It is easy to duplicate a workstation name while attempting to create a naming convention. For example, if you simply use a workstation number and room number and not a building number, you could easily create duplicate names if more than one building exists. A name such as *WS12RM11*, which uses only a workstation number and room number, can exist in more than one building. This results in duplicate names.

A network designer should consider any naming restrictions. Is there a maximum number of letters that can be used? Are there any illegal symbols that cannot be used? Remember that the network may be remotely connected to another network with a different naming requirement. When creating names for users, workstations, servers, printers, and other network objects, remember to be consistent. Create a maximum length and keep the name compatible with other operating systems.

Figure 20-6 lists typical symbols that should be avoided when creating names. These special symbols are the most universally recognized as reserved symbols and cannot be used by any major operating systems for special names. You will see these symbols not allowed by Microsoft, Novell, Linux, DOS, and LDAP directory structures. The symbols have special meanings that restrict their use and should be avoided.

Some systems have a more extensive set of rules for naming conventions. This can cause some system interoperability problems. For example, some UNIX and Linux systems typically require only lowercase letters be used for object names because object names are case-sensitive.

Partition and volume structure

The partitioning of storage devices is an important aspect to consider. You should not simply create one partition and place all files in it. The design should

Symbol	Description
/	Forward slash
\	Backslash
:	Colon
;	Semicolon
,	Comma
*	Asterisk
[]	Brackets
<>	Greater and lesser than
|	Pipe
+	Plus sign
=	Equal sign
?	Question mark

Figure 20-6.
Typical symbols that should be avoided when creating names in any of the major operating systems.

reflect the work activities of the company and reinforce security. For example, from the security aspect, a separate partition should be used to house the operating system and operating system boot files. The partition can be set up to allow access to administrators. This will add an additional layer of security, in addition to typical user permissions. Work-related partitions can be used to isolate functions such as public shares, artwork, Internet downloads, work groups, departments, or other features unique to the workplace. The partitioning design should be a reflection of the overall directory design.

The hard disk drive system structure must be designed before installing the network operating system. The two main concerns of partition design are security and data recovery. You need to design the partition to limit access and to accommodate system recovery after failure. You can design the system to allow users to access only specific parts of the hard disk drive system. Limiting access strengthens the security of the network from unauthorized users inside and outside of the organization.

The hard disk drive system should be organized to allow efficient system backups. For example, all data storage can be confined to a particular partition or volume. This will allow for easy backup of important system information. The drive that contains the data could easily be mirrored to provide a simple recovery system. The exact design of the partition or volume structure varies and depends mainly on the project administrator.

Fault Tolerance and Data Integrity

The importance a company places on fault tolerance determines the type of RAID system needed or if server mirroring or clustering should be used. Do not forget that fault tolerance also includes the directory structure. A designer should determine how the directory structure is partitioned and replicated, not only to reduce network traffic, but also to provide for redundancy should one of the servers that hosts the directory database fail.

The importance a company places on data integrity determines the type of electrical system that should be employed. Remember, a dedicated electrical system used for computers and networks helps ensure data integrity by eliminating surges produced by office equipment. Many electrical devices produce voltage spikes during normal usage, such as motor-driven devices that use a brush-type motor (kitchen appliances, office machines, vacuum cleaners, drills, saws, and such). The voltage spikes travel along the entire length of the circuit until they encounter suppression devices, such as transformers or voltage suppressers. If sensitive computer equipment share the same circuit as equipment that produce voltage spikes, computer equipment can be damaged.

The isolation of the electrical system starts in the electrical equipment room. Typically, a separate electrical panel is installed to serve only network equipment. The network equipment includes the PCs throughout the facility. The electrical feed from the regular electrical supply to the dedicated network system panel typically uses isolation transformers to separate the two systems. The outlets used for computers are identified by a special color, such as orange.

Network Design Tools

There are many utilities and case studies available to help with network design. You may wish to visit the support Web site of the operating system you will be using and see what utilities and case studies are available. Case studies typically explain how an actual network was designed and implemented. Case studies can provide a valuable insight into how you might approach your design.

Active Directory Sizer

Microsoft offers many software tools that can assist you in planning a network system. One such tool is Active Directory Sizer. Active Directory Sizer is freely available by download from the Microsoft Web site. The tool calculates hard disk drive, RAM, and CPU size based on information provided by the network designer. The network designer simply enters information, such as the number of clients, the number of offices, cities, countries, and the type of software and services that will be provided. After the information is collected, Active Directory Sizer provides an estimate of the size and number of servers required.

Figure 20-7 illustrates how the software program is used to calculate server size. A series of questions are asked. The questions ask about network and server use. After all the information is gathered, the software displays the recommended requirements for the server, such as the number of CPUs, memory size, hard disk size, and RAID configuration.

Approximately 300 utilities are located on the companion disk of the Microsoft Windows 2000 Server Resource kit. Many of the tools are available for free through the Microsoft Support Web site.

Tech Tip

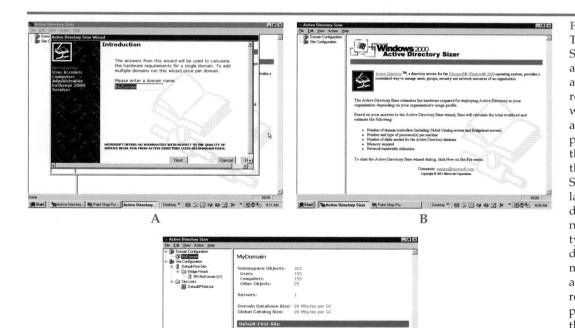

Figure 20-7.
The Active Directory Sizer program provides an estimate of the size and number of servers required. A—The network designer simply answers the questions presented. B—Based on the questions asked, the Active Directory Sizer program calculates the number of domain controllers, number of CPUs and type of CPU, number of disks, and amount of memory needed as well as the bandwidth requirements. C—The program also calculates the number of servers needed and their size.

Figure 20-8.
A network designer can use the Microsoft Visio program to design the physical structure and directory structure of a network.

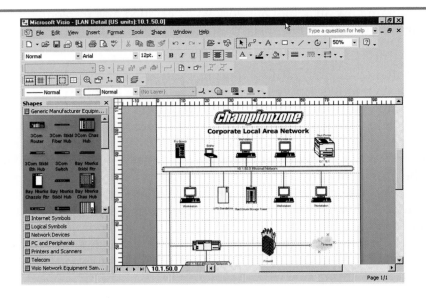

Microsoft Visio

Microsoft Visio is a powerful drawing tool used to diagram network systems and many other types of drawings, such as electrical, mechanical, floor plans, site plans, and LDAP structures. Look at **Figure 20-8.** The Visio drawing program displays a sample drawing of a network. As you can see, various symbols or icons are available to incorporate into the drawing. This drawing is only two dimensional, but Visio is capable of three-dimensional drawings as well. This is an extremely important tool for planning and documenting network systems. The Fluke network analyzer and troubleshooting software can export information into a Visio drawing format. It can present network information in a graphical format. Equipment manufacturers provide stencils and templates of their hardware devices so that they can be incorporated easily into a Visio drawing. The stencils are available for download.

NetworkView

NetworkView can examine existing networks and document hardware located in the network system. It does this by sending out protocol probes to identify hardware IP addresses, names, operating systems, and more. See **Figure 20-9.**

The NetworkView software program identifies all TCP/IP nodes in a network, using DNS, SNMP, TCP ports, and MAC addresses. It draws a high-quality color map of any size that can be printed or saved for future use. This is an extremely valuable tool for identifying existing equipment in a network system that has incomplete or nonexistent drawings.

Developing a Timeline

Installing a network requires teamwork and coordination. A large network requires a planning team with experts from a variety of network skills. Experts might include directory designers, hardware installers, cable installers, security experts, and Web hosting experts. For the team to work together effectively, a timeline needs to be established.

A timeline is a chronological listing of expected project progress and is essential for coordinating all project activities. The timeline must outline the

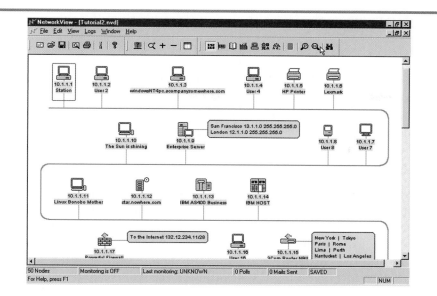

Figure 20-9.
Some software programs, such as NetworkView, can examine an existing network and document hardware located in the network system.

entire project from beginning to end. It is a planning tool that allows everyone involved to chart the pace of the project and to determine if it will be completed on time.

A series of meetings is set up with everyone involved in the project to discuss work accomplished and to identify any problems as they arise. The meetings keep everyone informed and up-to-date, especially if circumstances require a change in the timeline. For example, a delay in delivery of a hardware item, such as a particular router or server, may require a change in the sequence of work activities. If it is expected that the server will be delivered late, the people responsible for installing the operating system and setting up the directory structure need to be notified.

Figure 20-10 shows an example of a timeline. There are several different styles of timelines. The timeline in the example is a GANTT chart. Notice that major tasks are listed. The duration of time they will take to perform is shown in a graphical form in the last column. Bars represent the time period allocated for the particular task. Notice that the time periods of some tasks run concurrently with the time period of other tasks. For example, task 1 and tasks 8, 9, 10 share some of the same time period. This means that while the conduit is installed at the job site, the server can be prepared away from the site.

Figure 20-10.
An example of a timeline for a network installation project.

Installation

Several teams can install various parts of the network simultaneously, especially if the network is large. For example, while the cables are installed, the server operating system and directory structure can be installed. Another team can prepare the workstations using the System Preparation (**sysprep**) tool. The System Preparation tool is used to make an image of a workstation's hard drive. The workstation serves as a model. The image is then used to set up the other workstations.

Implementation

The next step after installing the network is implementation. Implementing and testing the network is a necessity that takes time. Most new, large networks will have problems, even those installed after many hours of planning. There are always oversights. All sections of the network need to be tested to ensure there are no bottlenecks and the hardware and software are in working order. This is typically a combined effort involving the installation crew and business personnel.

Business personnel need to test the system by simulating or actually performing their job duties. Such duties may involve using printers, accessing the Internet, saving data to the backup location, engaging in a network videoconference, exchanging e-mail, and remotely accessing the network. Only after exhaustive testing, can business personnel be sure the network hardware and software has been installed to expectation.

Documentation

Network documentation is vital to an organized approach of network installation and continued service. A set of network documents with specific procedures for security and other areas of network usage must be developed. Rules need to be established to ensure proper use of equipment and software and to prevent the company from being liable for actions taken by its employees.

For example, a set of policies needs to be established that clarify unauthorized use of the network and equipment. Some examples of unauthorized use of the network are gambling, distribution of materials that are not part of the company or part of the user's job, harassment, and unauthorized downloading of graphic material, music, and software. In general, any illegal activity is prohibited. Personal uses of equipment such as printing out your child's homework assignment, club activities, and personal correspondence are typically unauthorized in many companies and are viewed as fraud against the company by the misuse of company time and resources.

Drawings of the finished network system are also needed. Note the word *finished*. Changes occur as the system is built and the finished product can be somewhat different from the original proposal. Too often, a network system is installed, but no drawings of the finished system exist. Only the original proposed plan exists. Changes do occur, especially in complex systems. Documentation of changes that occur after an installation is also important because the network administrator may be the only person who knows how the system was changed.

Training

An essential part of installing a network system is to design a training program for the network users. The user must be trained in password and general security policies and procedures, remote access, e-mail fundamentals, and more.

The amount and depth of the training varies according to the size of the project. For example, a small modification to an existing network may require only a briefing. A large enterprise project may require training for hundreds of employees over a period of days, weeks, or months.

Company procedures and policies need to be developed before conducting training. Policies and procedures need to be disseminated as part of the training. Involving a training professional in the development of the network can be of great assistance in this process. While network designers are really good at their job, they do not often possess the skills and talents necessary to communicate technicalities to a broad audience of users.

Specifications for Network Design

A set of installation standards must be followed when installing a new network or modifying an existing network. A contract for a large installation typically includes a set of specifications or "specs," as they are often referred to in trade jargon. Specifications are used to specify details or expectations about materials and craftsmanship that are acceptable for the installation of the system. The specifications ensure that the owner and the contractor both know exactly what is expected. Without a set of specifications, there can be a wide variation of interpretation between what the owner expects and the finished product. Documentation that includes specifications reduces communication problems.

The overall size of a set of specifications can be quite lengthy. Most specifications make reference to other, previously-established standards to reduce the size of the document. There is no need to call out details about wall outlets and electrical circuits installed for the network. The specifications simply make references to established specifications. For example, a specification in a document might read, "All electrical work shall conform to the minimum standards established by the National Electrical Code (NEC)."

The National Electrical Code (NEC) comprises hundreds of pages of specifications for wire sizes, conduit sizes, supports, and other materials that have been selected by an organization of electrical engineers, contractors, and fire and safety experts. Many years have been dedicated to researching the best minimal standards that are adequate to ensure the safety of personnel.

Tech Tip

This section discusses standard organizations and some standards related to networking. It also defines some architectural design element terms, such as *main entrance way* and *horizontal cross connect*. It is important that you be familiar with these terms because they are often used in the standards related to networking.

Architectural Design Elements

The architectural design elements recognized by ANSI/TIA/EIA are main entrance, entrance facility, equipment room, work area, horizontal cross connect, horizontal wiring, backbone, telecommunications room, and telecommunications closet.

The **main entrance** room is where public or private telecommunications enter the building. In the past, the main entrance room not only housed the telecommunications equipment, but also the air conditioning, heating, electrical, and other

main entrance
the place where public or private telecommunications enter the building.

entrance facility
the room that is used as the entrance location for public or private communication cables.

equipment room
a room that contains the telecommunications equipment for the building such as the Private Branch Exchange (PBX), servers, and telecommunication wiring system terminations.

work area
the place where employees perform their normal office duties.

horizontal cross connect
provides a mechanical means of connecting horizontal cabling systems to other cables or equipment.

horizontal wiring
the section of cable that runs from individual work areas to the telecommunications closet.

equipment. As the building industry evolved, the telecommunications room was eventually established in a separate room in the building. This is preferred due to a multitude of reasons including access, security, and performance issues.

The *entrance facility* is the room that is used as the entrance location for public or private communication cables. The entrance facility may also serve as a telecommunications room or equipment room. Many times, the location depends on the availability of rooms and spaces in an existing building or the design or the complexity of the equipment required.

An *equipment room* contains the telecommunications equipment for the building such as the Private Branch Exchange (PBX), servers, and telecommunication wiring system terminations. Most large companies have their own telephone system referred to as a Private Branch Exchange (PBX). An equipment room may also serve the purpose of an entrance facility or a telecommunications room.

The *work area* is where employees perform their normal office duties. The work area is given this designation because the standards are different for hallways, lounges, reception areas, storage areas, and various other areas in the building. When you consider the various rooms that might be found in a building, designating a work area as a special room is not so unusual. The requirements for a work area are not the same as the requirements for a storage area or a reception area. However, if a reception area were also a work area, there would be different requirements than normal.

A *horizontal cross connect* provides a mechanical means of connecting horizontal cabling systems to other cables or equipment. For example, a horizontal cross connect is used to connect horizontal cabling to a backbone cable or to system equipment such as a router, switch, bridge, or server.

Horizontal wiring refers to the section of cable that runs from individual work areas to the telecommunications closet. The telecommunications closet contains connection equipment for the workstations in the immediate area.

All work areas connect to the telecommunications closet horizontal cross connects. Horizontal cable distance is limited to 90 meters for the horizontal run from the telecommunications outlet to the telecommunications closet. The maximum total distance is 100 meters.

Tech Tip

Some specifications state that the total combined length of patch cables at each end of the horizontal run should be no more than 10 meters. However, as a general rule, a total combined length of patch cables over 10 meters is allowed as long as the maximum total distance for horizontal cabling does not exceed 100 meters.

---

backbone
a cable that is located between the telecommunications closets, equipment rooms, and main entrance facility. The backbone cable connects these areas and does not serve individual workstations. A backbone can run horizontally and vertically through a building.

At least two outlets should be installed in each work area. One outlet is used for voice communications and the other for data communications. Typically, more than two outlets are installed as a combination of network and telephone outlets. Category 3 is the minimum recommendation for telephone wiring, but a higher category of wiring is permitted. The media for data communications can be any one of the following:

- Four pair UTP Category 5 or Category 5e (Category 5e is recommended).
- Two pair STP.
- Two strand, 62.5/125 µm fiber-optic cable.

A network *backbone* is located between the telecommunications closets, equipment rooms, and main entrance facility. The backbone connects these areas

Cable Type	Distance
UTP Category 3 or higher (Category 5 or Category 5e recommended)	800 meters (voice)
	90 meters (data)
STP Category 3 or higher (Category 5 or Category 5e recommended)	800 meters (voice)
	90 meters (data)
Multimode or 62.5/125 fiber-optic cable	2000 meters
Single-mode fiber-optic cable	3000 meters

Figure 20-11.
List of approved cable and distances for backbones.

and does not serve individual workstations. A backbone can run horizontally and vertically through a building. When the backbone is run vertically, it is sometimes referred to as a *riser* and typically passes through 4-inch conduit sleeves in the floor or ceiling. The backbone cannot be run in elevator shafts. See **Figure 20-11** for a list of the approved cable and distances for backbones. Note that when UTP and STP are used for telephone communications (voice), the maximum distance is 800 meters. When UTP and STP are used for networks (data), the backbone is limited to 90 meters.

The most widely accepted material used for network backbones is fiber-optic cable. UTP can be used for small networks, but fiber-optic cable is the preferred media if it can be used within budget restraints. Coaxial cable is not recommended for new installations as either backbone or horizontal wiring. While coaxial has not been prohibited, it is not recommended for new construction.

patch panel
a rack-mounted device with RJ-45 jacks on the front and a matching series of connections on the back. Patch panel cables are used to make connections between the front of the patch panel and equipment. The back of the panel is where the horizontal cable run terminates.

The terminology for network facilities is based on telecommunications standards. The telecommunications industry existed long before computer network technology, and many of the terms used to describe network rooms are actually derived from the communications industry.

Tech Tip

A typical **patch panel** is a rack-mounted device. The device has RJ-45 jacks on the front and a matching series of connections on the back, **Figure 20-12.** Patch panel cables are used for making connections between the front of the patch panel and equipment. The back of the panel is where the horizontal run cable is terminated in the connections.

RJ-45 connections

A   Front View

Punch downs

B   Back View

Figure 20-12.
A patch panel serves as a connection point between cable runs and network equipment.

punch down tool
a tool used for pushing individual twisted pair wires into the connection points on a patch panel.

telecommunications room
a room or enclosed space that houses telecommunications equipment, such as cable termination and cross connect wiring.

telecommunications closet
an enclosed space that houses telecommunication cable termination equipment and is the recognized transition point between the backbone and horizontal wiring.

The individual wires of each cable are pushed into the connections using a punch down tool. A **punch down tool** is specifically made for this type of cable wiring termination, **Figure 20-13.** The connections found on the back of patch panels are also found in outlets and punch down blocks. Patch panels are a convenient way to make connections and provide a means for easily modifying the cable connection system.

The **telecommunications room** is a room or enclosed space that houses telecommunications equipment, such as cable termination and cross connect wiring. The telecommunications room serves as a transition point between backbone and horizontal wiring. Sometimes telecommunications rooms are referred to as equipment rooms. Exact terminology depends on the source of the technology reference. For example, a network technician may use the term *equipment room* while a telecommunications technician may use the term *telecommunications room*. There must be a minimum of one telecommunications room per floor of a multistory building.

The **telecommunications closet** is an enclosed space used to house telecommunications cable termination equipment and is the recognized transition point between the backbone and horizontal wiring. The telecommunications closet is also referred to as telecommunications room. The closet can be a small room about the size of a standard clothes closet, or it can be an enclosed cabinet that is placed securely inside a general office space. The telecommunications closet is typically placed in a centralized section of the work area and is used to house the horizontal cross connect for horizontal wiring. Many companies and organizations use the terms *telecommunication room* and *telecommunication closet* interchangeably.

Standards Organizations

There are many organizations that write standards for communications and network systems. The standards of these organizations are often incorporated into the contract specifications for the new network. Many of the standards appear redundant. However, you must remember that the standards are not just for networks. Many of the standards are written for communications systems such as the telephone system. These communication systems overlap with network communications.

Figure 20-13.
A punch down tool is used to connect twisted pair wire to the connections on the back of patch panels, RJ-45 outlets, and punch down blocks.

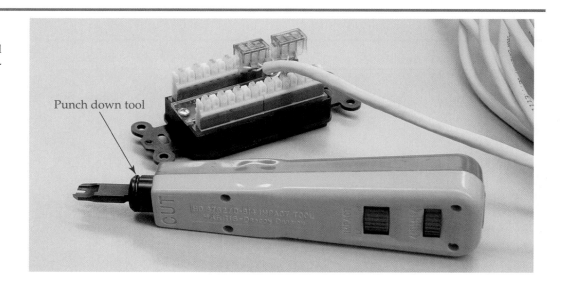

Punch down tool

Many manufacturers publish their own set of standards, often exceeding the standards from the American National Standards Institute (ANSI), Electronic Industries Alliance (EIA), and Telecommunications Industry Association (TIA). This causes a lot of confusion for maximum cable ratings. For example, different manufacturers advertised Category 6 at various rates before it was ever a formal standard published by ANSI/TIA/EIA.

ANSI/TIA/EIA Standards

Network cabling, hardware, and structures are specified by a combination of standards. The set or combination of standards and recommendations used in the United States is a combination of ANSI, TIA, and EIA and is referred to as ANSI/TIA/EIA. While these are separate organizations, they come together for generally agreed on standards for most telecommunications and networking standards such as those in **Figure 20-14.**

Telecommunications is a worldwide media. The ANSI/TIA/EIA is the recognized authority for the United States. ANSI is the main organization responsible for overseeing and distributing information technology standards in the United States. TIA and EIA are international standards. To eliminate redundancy, the TIA, EIA, and ANSI have formed a cooperative group to combine standards that cover common areas. The International Standards Organization (ISO) is the European standards organization with similar responsibilities to the ANSI/TIA/EIA. The OSI model was developed by the ISO. The Canadian Standards Association (CSA) is a Canadian standards organization. As you can see, standards may change from country to country. Also, these are minimum standards. They may be superseded by higher standards written by individual organizations such as bodies of government, the airline industry, and universities.

TIA/EIA 569-A

The 569-A Standard for Telecommunications Pathways and Spaces recommends how communication cables are to be installed. The term *telecommunications* is not limited to data networks, but rather to the broad range of communications, such as telephone and video. It is important to remember when reading standards that many pertain to communications systems and may not meet the requirements of a network standard such as 100BaseT or higher. Keep this in mind when references are made to Category 3 cable systems.

TIA/EIA 568-B

The 568-B standards describe the wiring standards for commercial buildings. They generally define the construction practices and design for the media,

Standard	Description
TIA/EIA 569-A	Commercial Building Standard for Telecommunications Pathways and Spaces
TIA/EIA 568-B.1-2000	Commercial Building Telecommunications Cabling Standard
TIA/EIA 606-A	Administration Standard for Commercial Telecommunications Infrastructure
TIA/EIA 607-A	Commercial Building Grounding and Bonding Requirements for Telecommunications
TSB-75	Additional Horizontal Cabling Practices for Open Offices

Figure 20-14. ANSI/TIA/EIA standards for telecommunications and networking.

connections points, termination, and topology. Note that standards are not solely concerned with network systems, but rather with all types of general telecommunications systems like telephone wiring. You must be careful when reading recommendations because Category 3 is still a recognized UTP for telephone wiring. An example of a 568-B standard is 568-B.1-2000 (Commercial Building Telecommunications Cabling Standard), which was introduced in 2000 when the use of Category 5e cable for network use was approved.

TIA/EIA 606-A

The 606-A Administration Standard for Commercial Telecommunications Infrastructure standard describes telecommunications infrastructure design guidelines. This includes blueprint drawings, PBX records, equipment inventories, identification formats, color coding, and labeling.

TIA/EIA 607-A

The 607-A standard describes the grounding and bonding requirements for telecommunications in commercial buildings. To better understand the concepts relating to grounding and bonding, you must first understand some basic electrical concepts as they relate to personnel safety. Electrical systems carry electrical energy, which can be hazardous to personnel. When electrical equipment fails, electrical current attempts to find the best electrical path to the earth. The path can be provided intentionally for safety reasons or unintentionally, such as through people coming in contact with the equipment. When people provide the path to earth, they can be severely shocked—sometimes fatally.

Electrical cables are insulated to prevent electrical energy from energizing the boxes, conduits, and devices that contain electrical wiring and equipment. To prevent electrical shock, the metal parts of the equipment are grounded and bonded. When equipment is grounded, a better path to earth can be made than through people coming in contact with the equipment.

The term *ground* or *grounding* means the intentional or accidental electrical path between an electronic device and the earth. The TIA/EIA 607-A standard specifies that equipment, such as boxes, conduits, and other metal devices associated with electrical and telecommunications equipment provide protection from electrical shock by grounding. When parts of the system are connected to the grounding system, they are described as bonded. *Bonding* is a term, which means that a short copper jumper joins a non-current carrying device to the electrical system ground. Proper grounding and bonding is a standard safety practice and a requirement by the NEC and the ANSI/TIA/EIA organizations.

The National Electrical Code (NEC) states that there can only be one main system ground for a facility. The main system ground serves as the main ground for all electrical-related equipment, such as the electrical system, telecommunications equipment, and network equipment. A grounding electrode rod is driven into the earth at the main electrical entrance of the building, **Figure 20-15.** A typical grounding electrode is made of copper or copper clad steel approximately 1/2" in diameter and at least 8' long. A grounding conductor made of either bare copper wire or green insulated wire is used to connect all required metallic devices in the system together. The building's steel structural support is also tied to the grounding system. The grounding conductor runs through each section of the facility, including the wiring closets.

TSB-75

The ANSI/TIA/EIA organization released TSB-75 to support the open office space design. TSB-75 is a supplement to the 568-A standard. ANSI/TIA/EIA often

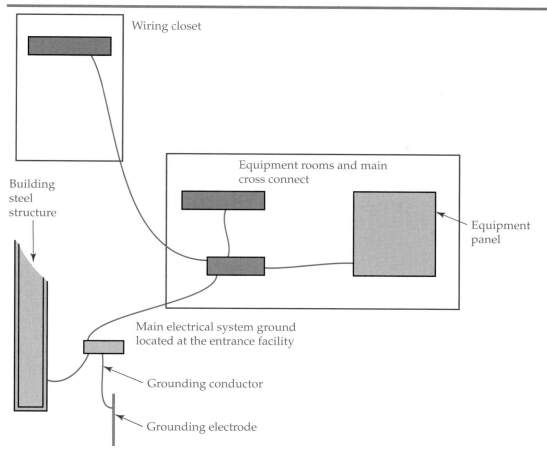

releases updated bulletins to support changes and exceptions to their standards rather than release an entirely new standard.

An open office space design requires a flexible wiring method that can easily support rewiring. Open office areas do not have permanent walls and typically consist of many office cubicles. The cubicle design is easily changed and rearranged. Many building owners who lease space to businesses prefer this style of construction. By using the open office design, the office area can be easily and quickly reorganized for many different types of businesses that require unique office floor arrangements. This type of design can also be used for convention centers and large assembly rooms, which often require the quick rearrangement of the floor plan.

TSB-75 states that a Consolidation Point (CP) and a Multi-User Telecommunications Outlet Assembly (MUTOA) can be used in open office spaces. See **Figure 20-16.** A ***Multi-User Telecommunication Outlet Assembly (MUTOA)*** is a grouping of outlets that serves up to 12 work areas. This is not a cross connect, but rather a prefabricated set of connection points. The MUTOA is mounted on the permanent building structure or on furniture that is permanently secured to the floor. Patch cables run from the MUTOA to the equipment in the work area.

The ***Consolidation Point (CP)*** is a connection to the horizontal wiring system, which in turn feeds to a wall outlet or a MUTOA. A patch cable can be plugged directly into the MUTOA and run directly to the equipment in the work area. Often the patch cable is run through channels in the furniture to conceal the presence of the cabling and ensure office safety. A Consolidation Point uses

Multi-User Telecommunication Outlet Assembly (MUTOA) a grouping of outlets that serves up to 12 work areas.

Consolidation Point (CP) a connection to the horizontal wiring system, which in turn feeds to a wall outlet or a MUTOA.

Figure 20-16.
An open office area has no permanent wall structures. It is designed to be a flexible office area. Several MUTOAs can be placed in the open office area to serve work areas. If the design changes, the MUTOA can be easily rearranged to accommodate the new office layout.

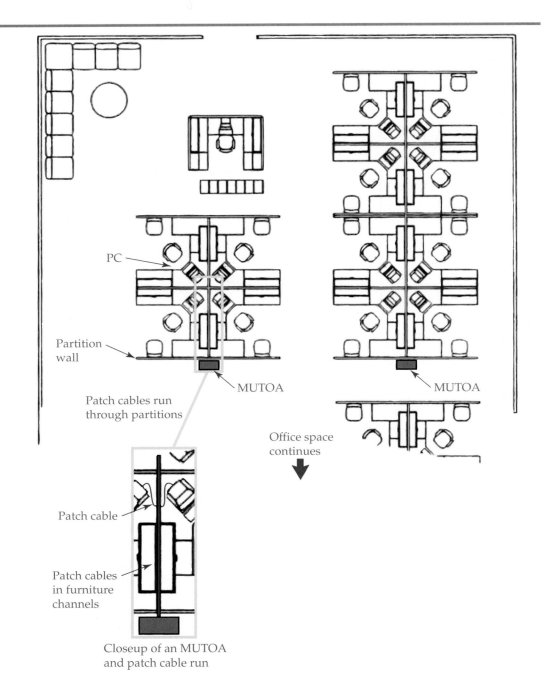

PC

Partition wall

MUTOA

MUTOA

Patch cables run through partitions

Office space continues

Patch cable

Patch cables in furniture channels

Closeup of an MUTOA and patch cable run

twisted pair cable with solid wire rather than stranded wire to connect to the outlets in the work area. See **Figure 20-17** for Consolidation Point and MUTOA systems.

Figure 20-18 lists the maximum recommended distances of cable lengths for horizontal wiring and patch cables. It is important to note that the overall length of the horizontal wiring is limited to 100 meters. The same distances are used for Consolidation Points. Certain equipment manufacturers provide their own distance charts that include modified distances based on conductor characteristics such as diameter. Also note that the maximum length of a patch panel cable that can be used in a telecommunications room is 7 meters.

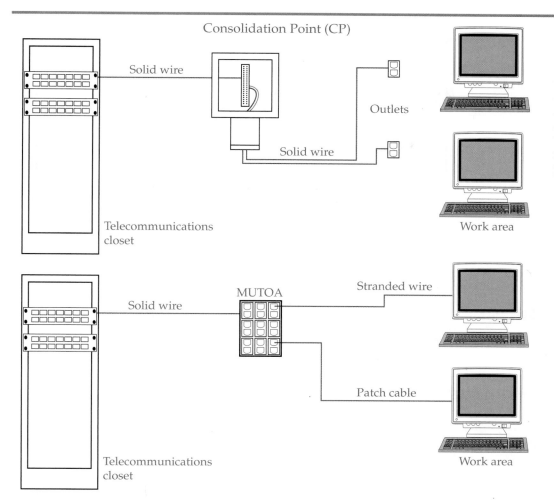

Figure 20-17.
A Consolidation Point (CP) is a central point in a work area that provides connections for outlets in the work areas. An MUTOA is an assembly of outlets in a work area that equipment attaches to using patch cables.

Network and Computer Electrical Requirements

The National Electrical Code (NEC) defines the electrical requirements and standards for networks. NEC specifies standards that ensure safety of personnel and serve as a guide to minimal equipment requirements. The NEC is used as a guide for electricians rather than for a certified network cable installer. Network system construction often requires the expertise of individuals from various backgrounds. All electrical work requires a licensed journeyman electrician and an electrical contractor. Network technicians should never engage in the installation or modifications of electrical systems. Electrical systems are classified in three general groups according to the voltage levels encountered in the system. See **Figure 20-19.**

Telecommunications, video, and signaling systems fall into the less than 50-volt category. Telecommunications wiring must never be in the same conduit or box as any system wiring of a higher voltage rating. The telecommunications must be isolated from the higher voltage systems not only to prevent data corruption and interference, but also to ensure the safety of personnel using network equipment.

The main components of the network wiring system use telecommunications terminology to describe the various sections of the network. Network standards are a subcategory of telecommunications that existed long before networking systems. It is only natural that the existing terminology be used when describing the newer network wiring system.

Tech Tip

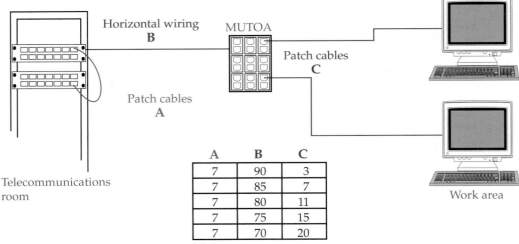

Figure 20-18.
Maximum recommended distances of cable lengths for horizontal wiring and patch cables. Note that the total distance will never exceed 100 meters for the horizontal wiring length. Also note that the longest patch cable on the patch panel is 7 meters.

A	B	C
7	90	3
7	85	7
7	80	11
7	75	15
7	70	20

All distances are in meters

BICSI

The Building Industry Consulting Service International (BICSI) is a worldwide non-profit association dedicated to the education of skilled workers in the telecommunications infrastructure and related fields. The organization is referred to by its acronym, which rhymes with *Dixie*. In recent years, BICSI has expanded their educational offerings in network cabling and is a recognized network cable infrastructure installation authority. The organization offers training programs, publications, and testing for certification.

One such certification is in network cabling, which consists of a written examination and a performance examination. In the performance examination, candidates are required to make connections or splices using several different media, such as Category 5e, fiber-optic cable, and coaxial cable. The performance testing is performed under the direction of a BICSI certified examiner and the cables are sent to the BICSI organization where they are evaluated. Many building specifications state that only certified cable installers be used on the job. If you are interested in becoming a certified network cable installer or simply want to know more, visit the BICSI Web site at www.bicsi.org.

This chapter has presented only a brief introduction to the complexities of network design. Please visit the many recommended Web sites at the end of the chapter to find out more about this interesting aspect of networking technology.

Tech Tip

Network cable installation is typically accomplished by BICSI certified individuals. These individuals have passed the rigid BICSI Certification exam.

Figure 20-19.
The three general groups in which electrical systems are classified. Note that they are classified according to the voltage levels encountered in the system.

Voltage	Application
< 50 volts	Signaling and communications
50 volts–600 volts	Residential and commercial wiring
> 600 volts	High voltage wiring

Summary

- The first part of the design process is to conduct a needs assessment.
- A timeline is a chronological listing of the expected sequence of work from the beginning of the project until the end.
- LDAP is the de facto directory standard used by network operating systems.
- A naming convention must be developed for all objects in the network environment to provide for easy identification.
- An isolated electrical system should always be used for network equipment.
- Specifications are a set of guidelines that explicitly describe the workmanship and materials that will be used in the project.
- Horizontal cable distance is limited to 90 meters for the horizontal run from the telecommunications outlet to the telecommunications room.
- The maximum length of a patch panel cable that can be used in a telecommunications room is seven meters.
- The maximum distance for backbone wiring using UTP as a network medium is 90 meters.
- The maximum distance for a multimode fiber-optic cable backbone is 2000 meters.
- The maximum distance for a single-mode fiber-optic cable backbone is 3000 meters.
- A Consolidation Point (CP) is a location in the horizontal wiring that provides an interconnection, which extends into the work areas.
- A Multi-User Telecommunication Outlet Assembly (MUTOA) is a grouping of outlets that serves up to 12 work areas.
- The main standard used for network design in the U.S. was developed jointly by several organizations and is referred to as ANSI/TIA/EIA.
- International Standards Organization (ISO) is the European standards organization similar to the ANSI/TIA/EIA.
- The Canadian Standards Association (CSA) is similar to ANSI/TIA/EIA.
- All electrical work for a networking system must follow the recommendations of the National Electrical Code (NEC) standards.
- Network technicians should not engage in the installation of electrical devices.
- BICSI is a non-profit organization dedicated to telecommunications installation training and education.

Review Questions

Answer the following questions on a separate sheet of paper. Please do not write in this book.

1. What is the first step in the network design process?
2. What are some of the roles and services provided by a network server? (Name at least 10.)
3. What protocol standard are all major directory structures based on?
4. Why is a naming convention important?

5. List 10 keyboard symbols that should not be used in LDAP computer or other object names.

6. What is the purpose of specifications?

7. What is the name of the room where the telephone system enters the building?

8. Name two other uses or room designations that an equipment room may serve.

9. Where is backbone cable located in a building?

10. What is a work area?

11. Describe a patch panel.

12. What tool is used to connect cable to the back of a patch panel?

13. How many telecommunications rooms must there be in a building?

14. What do the three acronyms ANSI, TIA, EIA represent?

15. What is the difference between a Consolidation Point (CP) and a Multi-User Telecommunication Outlet Assembly (MUTOA)?

16. What is the maximum number of work areas a single MUTAO or Consolidation Point (CP) can serve?

17. Can an MUTOA be mounted behind a drop ceiling?

18. What characteristic does furniture need to allow a MUTOA or Consolidation Point (CP) to be attached to it?

Sample Network+ Exam Questions

Network+

Answer the following questions on a separate sheet of paper. Please do not write in this book.

1. A metal fabrication manufacturer is setting up a production plant that will incorporate network computer-controlled machinery. Welding is commonly used in the metal fabrication process and is automated. Which type of network cable would you recommend be used to counter the effects of EMI generated by the welding equipment on the plant floor.
   A. 24 AWG UTP
   B. 22 AWG UTP
   C. Multimode fiber-optic
   D. 1000BaseT

2. What type of tool is used to make 100BaseT connections at telecommunications outlets in a work area?
   A. RG-58 crimper
   B. Punch down tool
   C. Electric soldering iron
   D. 24 AWG cable stripper

3. What is the maximum length for a horizontal run of UTP cable from a telecommunications closet to the work area outlet?

A. 800 meters

B. 200 meters

C. 100 meters

D. 90 meters

4. What is the maximum length of a patch panel cable that can be used in a telecommunications room?

A. 3 meters

B. 5 meters

C. 7 meters

D. 9 meters

5. What is the maximum distance of a single-mode, fiber-optic backbone?

A. 100 meters

B. 800 meters

C. 2000 meters

D. 3000 meters

6. Which standard describes how to install communication cables in telecommunication pathways and spaces?

A. NEC

B. TIA/EIA 569-A

C. TIA/EIA 606-A

D. TIA/EIA 607-A

7. What is the name of the Ethernet network passive device that is used in an equipment room to serve as a connection between horizontal cable runs and the equipment in the server room?

A. Gateway

B. Router

C. Patch panel

D. MUTOA

8. Which type of cable would *most likely* be used to provide connections between the workstation and the RJ-45 wall jack?

A. Stranded core coaxial cable with a copper shield.

B. UTP with stranded copper wires.

C. STP with solid copper wires.

D. UTP with solid copper wires.

9. Which method provides the best protection from office equipment interference?

A. Double insulate all electrical cables.

B. Use only 120 VAC equipment in the office area.

C. Use insulated RJ-45 connectors on all cable assemblies.

D. Use isolated dedicated electrical circuits for all network equipment.

Network+

10. What is the maximum number of characters that can be used in a NetBIOS name?

   A. 8 plus a 3 letter file extension.

   B. 15

   C. 256

   D. Any number of characters.

Suggested Laboratory Activities

1. Download a free trial version of a network drawing tool such as Microsoft Visio. Use the tool to diagram a small, peer-to-peer network and a client/server network.

2. Download a software program to assist you in estimating the size of server you will need.

3. Design a complete system for your school. The system will include servers, backbones, workstations, and a directory structure.

4. Design a system for an insurance office with five, full-time agents, three secretaries, and one receptionist. The office requires a connection to the corporate headquarters in a neighboring state. Approximately 18 MB of data must be exchanged with the corporate office each 8-hour day. Make a materials list and a bid price for the entire project. You must decide on your own to install a client/server or peer-to-peer network. You will also decide on the software packages to include in the system. Come up with a written proposal and present it to the class.

5. Write an acceptable use policy for the proposed office in activity 4.

6. Write a comparison of features and pricing for three different network operating systems.

7. Visit several Web sites to obtain the latest network installation standards available.

8. Download a copy of the Microsoft network design preparation guide.

9. Download a copy of the Microsoft Active Directory design guide.

Interesting Web Sites for More Information

www.bicsi.org

www.blackbox.com

www.cisco.com/univercd/cc/td/doc/product/icm/icm46/plan/

www.hp.com/rnd/case_studies/index.htm

www.kroneamericas.com

www.microsoft.com/technet/prodtechnol/AD/windows2000

www.siemon.com/standards/ansi_tia_eia.asp#569-A

www.sun.com

Chapter 20
Laboratory Activity

Designing a Small Network

After completing this laboratory activity, you will be able to:

- Construct a simple materials list.
- Design a simple project timeline.
- Prepare a price bid for the installation of a small office network.
- Develop an awareness and appreciation of the skills required in network design.

Introduction

In this laboratory activity, you will design a small network system for a business office. You will decide on the workstations, network equipment, network media, and software. Your assignment is to carefully review the requirements as specified in the scenario and then construct a materials list, which includes prices for materials and labor. You will also figure a profit figure. You will use the Internet for locating project items and obtaining prices.

Equipment and Materials

- Workstation and printer. (The workstation must have Internet access to conduct the research for the project. The printer is necessary to provide equipment specifications and pricing.)

Scenario

RMRoberts Inc. is setting up a brand new office for their Web design business. The office will consist of 16 workstations with the capability of supporting Web design functions. The office will require high-speed Internet access. Each person in the office will require a laser black and white printer. One full-color printer, centrally located, will be shared by all workstations. Each workstation must have installed a Microsoft operating system, Microsoft Office Professional software, and Adobe Studio Professional software. Also needed are office cubical dividers that will accommodate network and electrical wiring. The following figure illustrates what the final office design should look like as well as the office dimensions. Notice that the office has a drop ceiling.

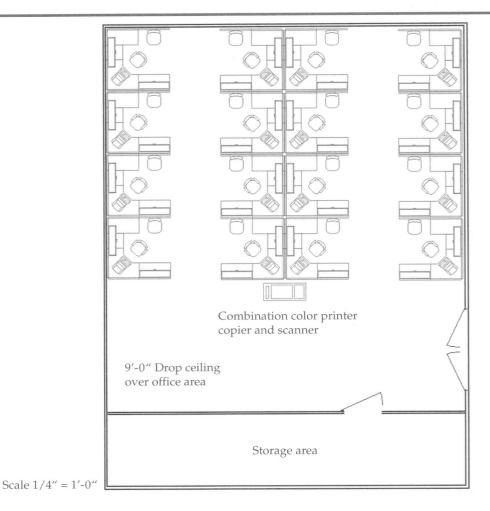

Combination color printer copier and scanner

9'-0" Drop ceiling over office area

Storage area

Scale 1/4" = 1'-0"

In summary, the materials should include the following:

- 16 workstations with a Microsoft operating system and Microsoft Office Professional.
- Adobe Studio Professional for 16 workstations.
- Server(s).
- Network devices, such as hubs, switches, or routers.
- Network media.
- 16 laser black and white printers.
- Office cubicle dividers.

Procedure

1. _____ Select appropriate computer workstations; any needed servers, hubs, switches, or routers; and network medium, such as copper wire, wireless, or fiber-optic, for the network system. Be prepared to defend your selection of materials.

2. _____ Configure a detailed list of materials, material costs, a timeline for installing all hardware and software and providing at least one day of training for all office personnel on the system. A material list must provide the following items:

- Item name.
- Manufacturer.
- Part number.
- Individual price.
- Total cost.
- Shipping cost.
- Approximate delivery time.

3. _____ Submit the following items to your instructor:

- Material list.
- Material costs.
- Labor cost.
- Profit.
- Timeline.
- Hard copy of all prices and equipment specifications found in your Internet search.

4. _____ For extra credit, download a free demo version of VISIO or an equal drawing software program and make a drawing of the network system matching the proposal.

Project Tips

- Keep the system simple; do not try to create an elaborate system.
- Sketch the network system before attempting to start a material list.
- Less expensive materials may not necessarily be the best way to go.
- Feel free to be creative.

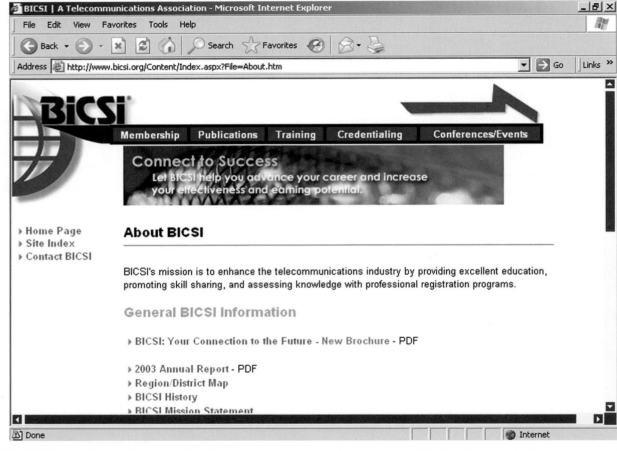

The BICSI organization (www.bicsi.org) provides resources, training, and certification for those in the telecommunications industry. Areas of interest to network designers include voice/data cabling systems, LANs and internetworks, and wireless.

21

Network+ Certification Exam Preparation

After studying this chapter, you will be able to:

❏ Evaluate your level of preparedness for the Network+ Certification Exam.

❏ Identify areas that require additional study.

❏ Develop a strategy for additional study in areas that show a need for improvement.

Network+ Exam—Key Points

You should be able to install, configure, and troubleshoot basic network devices. You should also know the layers of the OSI model and be very familiar with all the common network protocols, especially those associated with the TCP/IP suite.

Be aware that some questions on the Network+ Certification exam will be based on A+ Certification exam knowledge. The CompTIA Web site states that all candidates for the Network+ exam should have the A+ Certification or equivalent knowledge.

Key Words and Terms

The following words and terms will become important pieces of your networking vocabulary. Be sure you can define them.

domains
Network+ Examination Objectives
weight

This chapter is designed to help you prepare for the CompTIA Network+ Certification exam. Test preparation strategies, as well as a practice exam, are included in this chapter. The practice exam simulates what you may encounter when taking the Network+ Certification exam. Every attempt has been made to closely simulate the Network+ Certification exam by covering the exact areas to be tested at the approximate knowledge level required. However, there is no guarantee successfully passing the practice exam will ensure your success on the Network+ Certification exam.

The Network+ Certification Exam

Network+ Examination Objectives
a document that contains the exact areas to be tested on the Network+ Certification exam.

domains
areas of knowledge.

Over 2500 industry representatives were surveyed to determine the critical job areas that would be tested by the Network+ Certification exam. The most common jobs performed by technicians form the basis of the exam. The exact areas to be tested are listed in the *Network+ Examination Objectives* located on CompTIA's Web site. The Network+ Examination Objectives are organized according to *domains,* or areas of knowledge. The Network+ Certification exam covers four domains. Look at **Figure 21-1.** Notice that a certain percentage of exam questions cover each domain.

Network+ Note:

The percentages listed in the Network+ Examination Objectives are approximate and are subject to change.

weight
the amount of points assigned to a test question.

The Network+ Certification exam consists of 72 questions to be completed in 90 minutes. The exam is graded on a scale of 100 to 900. The minimum passing score is 646. Test questions are not equally weighted. This means that each question is worth a different amount of points, or *weight.* This makes it difficult to make a perfect pre-assessment tool, like the practice exam in this chapter.

Figure 21-1.
Network+ domains and the percentage of questions asked in that domain.

Network+ Domain	Percentage of Questions	Approximate Number of Questions
1.0 Media and Topologies	20%	14-15
2.0 Protocols and Standards	20%	14-15
3.0 Network Implementation	25%	18
4.0 Network Support	35%	25
Total	100	72

CompTIA keeps the weights of each question private. However, research has shown that there is a strong correlation between traditional test scores based on percentage and weighted question test scores.

In the introduction of the Network+ Examination Objectives, CompTIA recommends the candidate have a CompTIA A+ Certification before attempting the Network+ exam. The A+ Certification is not required, only recommended. Many of the test items require knowledge that directly correlates to the A+ Certification exam, especially test items relating to troubleshooting and computer hardware. CompTIA assumes that if you are taking the Network+ Certification exam, you have the skills commonly associated with PC support. If you do not have these skills, you will be at a disadvantage when taking the exam.

CompTIA also recommends the candidate have at least nine months work experience in network support or administration before taking the Network+ Certification exam. While there is no substitute for actual job experience, laboratory activities can provide you with much of the same experiences. Be sure to complete all laboratory activities included in this textbook and in the *Networking Fundamentals Laboratory Manual* before attempting the exam. Also, study the questions at the end of the laboratory activities. Many of these questions are designed to help you prepare for the Network+ Certification exam.

Use the *Networking Fundamentals Study Guide* to help reinforce concepts. Many of the concepts presented in the Network+ objectives you will not have an opportunity to work with in the lab. For example, you may not have an opportunity to design and implement an FDDI or a 10GbaseSR network, yet you will need to know the specifications of these standards, such as the types of cable and connectors used and maximum and minimum cable distances. The *Networking Fundamentals Study Guide* will help you organize this information as well as help you review the material required for the Network+ Certification exam. When working in the study guide, feel free to record your own notes and diagrams in the margins or available white space. This will personalize your learning experience. Organizing or presenting the material in your own way will help you gain a better understanding of the material.

Preparation Strategy

The secret to passing the Network+ Certification exam is having an organized strategy for preparation. The most common mistake made by students is to simply sit down and study their notes and available materials. A student needs to approach their exam preparation in the most effective and objective manner. The following is a list of key steps to exam success:

1. Review and analyze the exam objectives.
2. Match exam objective details to resource materials.
3. Identify and practice laboratory activities that match the exam objectives.
4. Take practice exams.
5. Review problem areas.
6. Retake practice exams.
7. Schedule and take the Network+ Certification exam.

Network+ Note:

When preparing for the Network+ Certification exam, be sure to check the CompTIA Web site for the very latest Network+ Examination Objectives. The objectives are constantly revised.

1. Review and Analyze the Exam Objectives

The first step in the exam preparation strategy is to review and analyze the exam objectives outlined by the CompTIA organization. You need to know exactly what areas are required for certification. Networking is an extremely in-depth field that can easily fill volumes of books. The technical vocabulary alone spans over 10,000 words, not including all of the acronyms. Concentrate your studies on what is required for the Network+ Certification exam. For example, if subnet calculations are not required, do not spend hours on this area of networking.

Network+ Note:

Subnet calculations are not required for the Network+ Certification exam, but it is required for other certification exams at the time of this writing. Also, note that changes in the exam composition change on a fairly frequent basis. It is always a good idea to check the CompTIA Web site for the latest content changes.

Too many people spend hours of study on areas that are not on the test. To avoid doing this, make an outline of the current exam objectives and fill in related facts for each objective. A sample of such an outline is presented in **Figure 21-2.**

On your outline, write several facts about each item mentioned in the objective. Use several references for each item in addition to the textbook. A variety of reference materials will greatly enhance your understanding of the topics in the outline.

There are many sites that will provide you with a detailed outline for free or for a small charge. While this would be the most convenient way to make a complete exam outline, it is not a sound learning strategy. To better retain the information, you need to construct your own outline and research the facts and details yourself! This method will better prepare you for the exam. Also, there are many ready-made study guides with inaccurate information and information which is outside the scope of the Network+ Certification exam. Also, much of the material is out of date. You will do a better job by constructing your own outline. This will ensure your study material is up-to-date. Also, by using multiple sources for information, you will be able to avoid inaccurate information.

Network+ Note:

Since CompTIA recommends a Network+ candidate possess CompTIA A+ equivalent knowledge, you may want to download and review the exam objectives for the A+ Certification exams and use it as an outline for review in addition to the Network+ outline.

Figure 21-2.
Sample study guide.

My Study Guide

1.3 Specify the characteristics (For example: speed, length, topology, and cable type) of the following cable standards: 10BaseT, 10BaseFL, 100BaseTX, 100BaseFX, 1000BaseT, 1000BaseCX, 1000BaseSX, 1000BaseLX, 10GBaseSR, 10GBaseLR, and 10BaseER.

10BaseT _____
_____
_____

10BaseFL _____
_____
_____

100BaseTX _____
_____
_____

100BaseFX _____
_____
_____

1000BaseT _____
_____
_____

1000BaseCX _____
_____
_____

1000BaseSX_____
_____
_____

1000BaseLX _____
_____
_____

10GBaseSR _____
_____
_____

10GBaseLR _____
_____
_____

10BaseER _____
_____
_____

2. Match Exam Objective Details to Resource Material

You should gather the material needed to fill in your outline from a variety of reputable sources. Use your textbook, laboratory manual, and the Internet. There are many fine resources available, especially at operating system vendor Web sites. For example, Microsoft has a lot of material on WINS because it was vital to their earlier network operating systems.

When looking up facts for network interoperability, do not simply check the Microsoft Web site to see how to set up communications between a Microsoft system and a Novell system. Check the Novell site as well for their recommendations.

Be careful of some sources, such as chat rooms. Chat rooms that discuss particular exams often have a lot of inaccurate information. It is always better to use only reputable and authoritative Web sites.

3. Identify and Practice Laboratory Activities That Match the Exam Objectives

Many of the objectives, especially those relating to troubleshooting, require the student have knowledge that requires experience. The laboratory activities presented in this textbook and in the *Network Fundamentals Laboratory Manual* are designed to provide you with needed experiences. You may want to review and repeat these laboratory activities. You should be able to perform all of the past laboratory activities with minimal assistance. Laboratory activity experiences are much easier to remember than textbook materials. For example, setting up a DHCP service and troubleshooting one will give you an in-depth knowledge of the subject. This is better than simply memorizing facts.

Remember, a networking profession is not just about passing an exam. You need to know the subject area and be able to perform hands-on activities to obtain and keep a job in the industry. The industry does not pay you to answer questions, but rather to perform duties related to networking.

4. Take Practice Exams

Taking practice exams will help you to identify your areas for needed improvement in both subject matter and test-taking skills. All students have at least one area of weakness that could prevent them from passing the Network+ Certification exam. Take the practice exam included in this chapter and have your instructor score it. After all of the questions are scored, you can do an analysis of in what areas you may need more study or laboratory experiences. Be careful of the many online exams offered for free on the Internet. Only use Web sites that your instructor recommends. Many of the Web sites have questions that do not exactly match the test objectives. Also, some of the provided answers are incorrect.

In general, most of the free practice exams are far more difficult than the actual Network+ Certification exam. Many of the free Internet exam sites are marketing tools to sell you other exams from their Web site. The sample exams may actually be harder than the CompTIA exam. The reason is it enhances their sales if you do poorly on the sample exam. Use caution before purchasing any set of test questions.

5. Review Problem Areas

After taking a practice exam, review your problem areas. Do not simply take a practice exam and study a set of questions. You need to learn the material covered. Review the test items missed. Go back and study these areas. Use the outline you developed from the Network+ Examination Objectives to help you prepare.

6. Retake Practice Exams

Retake practice exams by taking a different practice exam to see if you have improved in the needed areas. Repeat steps four through six until you have gained sufficient confidence that you can pass the commercial Network+ Certification exam.

7. Schedule and Take the Network+ Certification Exam

One of the secrets of preparing for the Network+ Certification exam is to schedule an exact date and time to take the exam. Give yourself a real and practical deadline. Vague deadlines do not work and will actually work against you. By procrastinating, you will lose the knowledge learned for the current exam. For example, do not simply say "I plan to take the exam in the fall." Chances are when fall comes you will still not be prepared and will procrastinate by saying, "I plan to take the exam this winter." Remember, procrastination is not a luxury; in this case, it is your enemy.

Set an exact date. This way you have committed both your financial and mental resources. Use your calendar to pick a date that gives you plenty of time to prepare by taking practice exams and reviewing. Don't wait until you think you are prepared to schedule the exam. In fact, scheduling the exam date should be one of the first things you do if you are serious about certification. Schedule the test, and then set up a specific schedule (timeline) for study.

Network+ Certification Practice Exam

This practice exam is structured according to the domains specified in the Network+ Examination Objectives. Before each set of questions for the domain, the objectives for that domain are listed. After taking the exam, your instructor will score it.

Domain 1—Media and Topologies

■ **1.1 Recognize the following logical or physical network topologies given a diagram, schematic, or description:**

Star, bus, mesh, and ring.

■ **1.2 Specify the main features of 802.2, 802.3, 802.5, 802.11, and FDDI networking technologies, including:**

Speed, access method, topology, and media.

■ **1.3 Specify the characteristics (For example: speed, length, topology, and cable type) of the following cable standards:**

10BaseT, 10BaseFL, 100BaseTX, 100BaseFX, 1000BaseT, 1000BaseCX, 1000BaseSX, 1000BaseLX, 10GBaseSR, 10GBaseLR, and 10BaseER.

■ **1.4 Recognize the following media connectors and describe their uses:**

RJ-11, RJ-45, F-Type, ST, SC, IEEE 1394, LC, and MTRJ.

■ **1.5 Recognized the following media types and describe their uses**:

Category 3, 5, 5e, and 6; UTP; STP; Coaxial cable; SMF optic cable; and MMF optic cable.

■ **1.6 Identify the purposes, features, and functions of the following network components:**

Hubs, switches, bridges, routers, gateways, CSU/DSU, NICs, ISDN adapters, WAPS, modems, transceivers, and firewalls.

■ **1.7 Specify the general characteristics (For example: carrier speed, frequency, transmission type, and topology) of the following wireless technologies:**

802.11 (frequency hopping spread spectrum), 802.11x (direct sequence spread spectrum), infrared, and Bluetooth.

■ **1.8 Identify factors that affect the range and speed of wireless service (For example: interference, antenna type, and environmental factors).**

Domain 1—Practice Exam Questions 1–14

Answer the following questions on a separate sheet of paper. Please do not write in this book.

1. Which network topology provides the most redundancy?

   A. Bus

   B. Token Ring

   C. Mesh

   D. Star

2. What is the maximum transmission speed for a 100BaseTX network?

   A. 10 Mbps

   B. 100 Mbps

   C. 1000 Mbps

   D. 10 Gbps

3. What is 100BaseFX constructed from?

   A. Two pairs of Category 5 cable.

   B. Four pairs of Category 5e cable.

   C. Two strands of multimode or single-mode fiber-optic cable.

   D. One strand of single-mode fiber-optic cable.

4. What is the maximum number of active hubs or repeaters that can be linked in a daisy-chain configuration?

   A. 3

   B. 4

   C. 5

   D. Unlimited

5. What is the frequency band used for IEEE 802.11b devices?

   A. 2.4 MHz

   B. 2.4 GHz

   C. 5 MHz

   D. 5 GHz

6. The central point of a wireless LAN designed to control access to a wired LAN is called the _____.

   A. wireless hub

   B. MAU

   C. Wireless Access Point

   D. Point of Presence

7. Identify the connector type in the following figure.

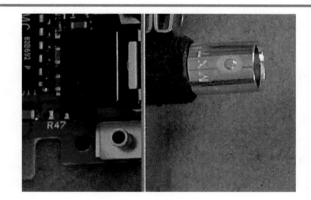

   A. BNC

   B. RJ-45

   C. AUI

   D. ST

8. A T1 line is typically terminated with a(n) _____.

   A. MAU

   B. Cable modem

   C. ISDN adapter

   D. CSU/DSU

9. Which network device bases decisions on MAC addresses?

   A. Passive hub

   B. Bridge

   C. Router

   D. Gateway

10. Wireless LANs use which access method to establish communications?
    A. CSMA/CD
    B. CSMA/CA
    C. Token passing
    D. PPP

11. Which of the following devices can limit Ethernet broadcasts? (Select all that apply).
    A. Hub
    B. Router
    C. Switch
    D. Repeater

12. How many pairs of conductors can be inserted into a standard RJ-45 connector?
    A. One
    B. Two
    C. Three
    D. Four

13. Gigabit Ethernet supports data transfer speeds equal to _____.
    A. 100 Mbps
    B. 1,000 Mbps
    C. 10,000 Mbps
    D. 100,000 Mbps

14. The loss of signal strength through a fiber-optic cable is referred to as _____.
    A. resistance
    B. impedance
    C. attenuation
    D. decibel gain

Domain 2—Protocols and Standards

- **2.1 Identify a MAC address and its parts.**
- **2.2 Identify the seven layers of the OSI model and their functions.**
- **2.3 Identify the OSI layers at which the following network components operate:**
  Hubs, switches, bridges, routers, NICs, WAPs
- **2.4 Differentiate between the following network protocols in terms of routing, addressing schemes, interoperability, and naming conventions:**
  IPX/SPX, NetBEUI, AppleTalk, and TCP/IP.
- **2.5 Identify the components and structure of IP addresses (IPv4 and IPv6) and the required setting for connections across the Internet.**
- **2.6 Identify classful IP ranges and their subnet masks (For example: Class A, B, and C).**
- **2.7 Identify the purpose of subnetting.**

- **2.8 Identify the differences between private and public network addressing schemes.**

- **2.9 Identify and differentiate between the following IP addressing methods:**

  Static, dynamic, and self-assigned (APIPA).

- **2.10 Define the purpose, function, and use of the following protocols used in the TCP/IP suite:**

  TCP, UDP, FTP, SFTP, TFTP, SMTP, HTTP, HTTPS, POP3, Telnet, SSH, ICMP, ARP, NTP, NNTP, SCP, LDAP, IGMP, and LPR.

- **2.11 Define the function of TCP/UDP ports.**

- **2.12 Identify the well-known ports associated with the following commonly used services and protocols.**

  FTP (20), FTP (21), SSH (22), Telnet (23) SMTP (25), DNS (53), TFTP (69), HTTP (80), POP3 (110), NNTP (119), NTP (123), IMAP4 (143), and HTTPS (443).

- **2.13 Identify the purpose of network services and protocols (For example: DNS, NAT, ICS, WINS, SNMP, NFS, Zeroconf, SMB, AFP, and LPD).**

- **2.14 Identify the basic characteristics (For example: speed, capacity, and media) of the following WAN technologies:**

  Packet switching, circuit switching, ISDN, FDDI, T1, T3, OCx, and X.25.

- **2.15 Identify the basic characteristics of the following Internet access technologies:**

  xDSL, Broadband cable, POTS/PSTN, satellite, and wireless.

- **2.16 Define the function of the following remote access protocols and services:**

  RAS, PPP, SLIP, PPPoE, PPTP, VPN, and RDP.

- **2.17 Identify the following security protocols and describe their purpose and function:**

  IPSec, L2TP, SSL, WEP, WPA, and 802.1x.

- **2.18 Identify authentication protocols (For example: CHAP, MS-CHAP, PAP, RADIUS, Kerberos, and EAP).**

Domain 2—Practice Exam Questions 15–29

Answer the following questions on a separate sheet of paper. Please do not write in this book.

15. Which of the following is an example of a MAC address?

    A. 192.168.40.134

    B. AD01:3F:0:23A:B24D:CA7:D43:B3A2

    C. 10111011

    D. A1-D3-CA-12-56-C3

16. Gateway services are associated with which layer(s) of the OSI model?

    A. Physical, data link, and network.

    B. Application, presentation, session, and transport.

    C. Data link and MAC.

    D. Application only.

17. Which protocol is designed to synchronize the time between all clients and servers on a LAN?

    A. FTP

    B. ATM

    C. NTP

    D. ICA

18. Which protocol is used to send an echo request?

    A. FTP

    B. TCP

    C. ICMP

    D. PPP

19. Which protocol is used in a SOHO network when all computers connect to the Internet through one common host and one assigned IP address.

    A. NAT

    B. ATM

    C. PPP

    D. HTTP

20. A new e-mail server is installed on a corporate LAN. For convenience, the corporate executives want to be able to centrally store their e-mail on the server and be able to view their stored e-mails from anywhere in the USA when they travel. Which e-mail protocol will meet the requirements?

    A. FTP

    B. POP

    C. SNMP

    D. IMAP

21. You are in charge of a large LAN containing several hundred workstations. You wish to assign IP addresses automatically when the user logs on to the network. Which service will meet the IP address assignment requirement?

    A. DNS

    B. WINS

    C. TCP/IP

    D. DHCP

22. Which technology would provide the highest data throughput?

    A. ISDN

    B. T1

    C. SONET

    D. DSL

23. Which protocol in the list provides the most secure transactions between a Web site and a user?

A. FTP

B. TCP

C. SSL

D. IP

24. Which port is associated with Web page services?

A. 20

B. 21

C. 80

D. 110

25. Which protocol is designed specifically to support thin client services?

A. ICA

B. TFTP

C. SNMP

D. IPX

26. Which IP Internet standard is IPSec compatible with?

A. IPv4 only.

B. IPv6 only.

C. IPv4 and IPv6.

D. Neither IPv4 nor IPv6.

27. A home-office network based on the Windows 2000 Professional operating system would *most likely* connect to an ISP using a standard V90 telephone modem and which protocol?

A. PPP

B. SLIP

C. TCP

D. IP

28. A virtual LAN would be created using a(n) _____.

A. active hub

B. switch

C. repeater

D. MAU

29. Establishing and terminating a TCP/IP connection are described at the _____ layer of the OSI model.

A. presentation

B. network

C. data link

D. session

Domain 3—Network Implementation

- ▪ **3.1 Identify the basic capabilities (For example: client support, interoperability, authentication, file and print services, application support, and security) of the following server operating systems to access network resources:**

  UNIX/Linux/MAC OS X Server, NetWare, Windows, and AppleShare IP.

- ▪ **3.2 Identify the basic capabilities needed for client workstations to connect to and use network resources (For example: media, network protocols, and peer and server services).**

- ▪ **3.3 Identify the appropriate tool for a given wiring task (For example: wire crimper, media tester/certifier, punch down tool, or tone generator).**

- ▪ **3.4 Given a remote connectivity scenario comprised of a protocol, an authentication scheme, and physical connectivity, configure the connection. Includes connection to the following servers:**

  UNIX/Linux/MAC OS X Server, NetWare, Windows, and AppleShare IP.

- ▪ **3.5 Identify the purpose, benefits, and characteristics of using a firewall.**

- ▪ **3.6 Identify the purpose, benefits, and characteristics of using a proxy service.**

- ▪ **3.7 Given a connectivity scenario, determine the impact on network functionality of a particular security implementation (For example: port blocking/filtering, authentication, and encryption).**

- ▪ **3.8 Identify the main characteristics of VLANs.**

- ▪ **3.9 Identify the main characteristics and purpose of extranets and intranets.**

- ▪ **3.10 Identify the purpose, benefits, and characteristics of using antivirus software.**

- ▪ **3.11 Identify the purpose and characteristics of fault tolerance:**

  Power, link redundancy, storage, and services.

- ▪ **3.12 Identify the purpose and characteristics of disaster recovery:**

  Backup/restore; offsite storage; hot and cold spares; and hot, warm, and cold sites.

Domain 3—Practice Exam Questions 30–47

Answer the following questions on a separate sheet of paper. Please do not write in this book.

30. Which service is used by Windows NT to resolve NetBIOS names to IP addresses?

    A. DHCP

    B. DNS

    C. APIPA

    D. WINS

31. Which RAID level does *not* provide fault tolerance?

    A. RAID 0

    B. RAID 1

    C. RAID 2

    D. RAID 5

32. Which of the following file systems support encryption on a Windows 2000 server?

    A. FAT16

    B. FAT32

    C. NTFS

    D. VFAT

33. What should be done before performing an operating system upgrade? (Select all that apply.)

    A. Perform a backup of all-important data.

    B. Run a virus scan of the hard disk drive.

    C. Remove all current operating system patches.

    D. Reformat the partition where the upgrade is to be installed.

34. What are the main advantages of using a proxy server on a LAN? (Select all that apply.)

    A. The proxy server provides encryption of all data transferred to and from the LAN.

    B. The proxy server improves system performance by saving frequently requested Web pages, thus making access appear quicker.

    C. A proxy server can filter request preventing certain Web sites from being accessed by employees.

    D. A proxy server is used as a database of user names and passwords.

35. The highest level of the UNIX directory structure is _____.

    A. C:

    B. A:

    C. /

    D. /boot

36. Which of the following protocols is native to NetWare 5?

    A. TCP/IP

    B. IPX/SPX

    C. AppleTalk

    D. DLC

37. What is the physical address for the computer indicated in the following figure?

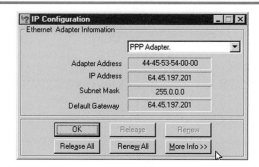

A. 44-45-53-54-00-00

B. 64.45.197.201

C. 255.0.0.0

D. 64.45.197.201

38. What does the acronym NAS represent?

A. Network Access Service

B. Network Attached Storage

C. Novell Access Services

D. Network ASCII Service

39. Which of the following devices would you install on a computer to protect it from unauthorized access from the Internet?

A. Bridge

B. Gateway

C. Firewall

D. Switch

40. Select the correct reason for creating a VLAN?

A. Increase the bandwidth between two or more nodes.

B. Increase the storage capacity of the network.

C. Increase the fault tolerance of the network.

D. Decrease security for transferring data across a public media.

41. When configuring a dial-up connection between a SOHO network and a local ISP, you will be required to have which of the following items? (Select all that apply.)

A. ISP access telephone number.

B. A user name.

C. A password.

D. A Kerberos account.

42. What is the maximum number of characters you can use for a NetBIOS name?

A. 8

B. 15

C. 64

D. 255

43. Which of the following maps computer names (NetBIOS names) to IP addresses?

   A. Lmhosts

   B. Hosts

   C. DNS

   D. DHCP

44. Which of the following passwords is the most secure?

   A. TotalRecall

   B. Pa$$TheBall!

   C. SaraLee

   D. Harmony1

45. The default administrators name for Windows 2000 Server is _____.

   A. Root

   B. Admin

   C. Administrator

   D. SuperUser

46. The default password for Windows 2000 Server is _____.

   A. Root

   B. Admin

   C. Secret

   D. There is no default password.

47. What two ways can IP addresses be issued to clients?

   A. Dynamically and statically

   B. Statically and actively

   C. Confidentially and overtly

   D. Autonomously and statically

Domain 4—Network Support

■ **4.1 Given a troubleshooting scenario, select the appropriate network utility from the following:**

   Tracert/traceroute, ping, arp, netstat, nbstat, ipconfig/ifconfig, winipcfg, and **nslookup/dig.**

■ **4.2 Given output from a network diagnostic utility (For example: those utilities listed in objective 4.1), identify the utility and interpret the output.**

■ **4.3 Given a network scenario, interpret visual indicators (For example: link LEDs and collisions LEDs) to determine the nature of a stated problem.**

■ **4.4 Given a troubleshooting scenario involving a client accessing remote network services, identify the cause of the problem (For example: file services, print services, authentication failure, protocol configuration, physical connectivity, and SOHO router).**

■ **4.5 Given a troubleshooting scenario between a client and the following server environments, identify the cause of a stated problem:**

   UNIX/Linux/MAC OS X Server, NetWare, Windows, and AppleShare IP.

- **4.6 Given a scenario, determine the impact of modifying, adding, or removing network services (For example: DHCP, DNS, and WINS) for network resources and users.**

- **4.7 Given a troubleshooting scenario involving a network with a particular physical topology (For example: bus, star, mesh, or ring) and including a network diagram, identify the network area affected and the cause of the stated failure.**

- **4.8 Given a network-troubleshooting scenario involving an infrastructure (For example: wired or wireless) problem, identify the cause of a stated problem (For example: bad media, interference, network hardware, or environment).**

- **4.9 Given a network problem scenario, select an appropriate course of action based on a logical troubleshooting strategy. This strategy can include the following steps:**

1. Identify the symptoms and potential causes.
2. Identify the affected area.
3. Establish what has changed.
4. Select the most probable cause.
5. Implement an action plan and solution including potential effects.
6. Test the result.
7. Identify the results and effects of the solution.
8. Document the solution and process.

Domain 4—Practice Exam Questions 48–72

Answer the following questions on a separate sheet of paper. Please do not write in this book.

48. Which of the following statements best describes the output display in the following figure.

- A. The WINS server is up and running.
- B. The WINS server is not running.
- C. The DNS server is up and running.
- D. The DNS server is not running.

49. You are troubleshooting a workstation that is having problems connecting to the Internet. The network is a peer-to-peer and is connected to the Internet using a cable modem attached to a workstation serving as the ICS host. Which IP address would you ping to verify the connection to the ICS host?

   A. 169.254.000.100

   B. 192.168.000.001

   C. 127.0.0.1

   D. 127.000.000.255

50. You have just set up and configured a small home office network consisting of three workstations and one printer connected by wireless network adapters. You want to secure the network from access from the Internet. As a solution you set the encryption on each 802.11b network adapter using the passphrase "no_1_home". Which of the following statements best describes your solution based on the required and optional results?

   Required result:

   Protect the network from unauthorized Internet access.

   Optional result:

   Increase overall network performance.

   A. The proposed solution meets the required option only.

   B. The proposed solution meets the optional result only.

   C. The proposed solution meets the required and the optional result.

   D. The proposed solution meets neither the required nor the optional result.

51. A Windows 2000 workstation connected to a client/server network fails to complete the boot sequence and stops with a blank screen displayed. What would be the most appropriate action to take to remedy the problem?

   A. Remove the network adapter and try rebooting the workstation.

   B. Use the Windows 2000 CD to start the recovery console.

   C. Reconfigure the workstation as a member of a peer-to-peer network and try to reboot the workstation.

   D. Shut down the server and leave it offline and then reboot the workstation.

52. What is the term used to describe data signals from one pair of conductors corrupting data signals on another pair of conductors?

   A. Beaconing

   B. Crosstalk

   C. Open loop

   D. Reflection

53. Which of the following tools would be the most appropriate to check a fiber-optic backbone suspected as the cause of a network cabling problem?

   A. Digital multimeter

   B. OTDM

   C. Oscilloscope

   D. Tone generator

54. You want to monitor the packets on a LAN connected to a Window 2000 server. Which utility would be the most appropriate for inspecting packets?

   A. Telnet

   B. Performance Monitor

   C. System Monitor

   D. Network Monitor

55. Users are complaining of very slow downloads in a particular workgroup. What would be the most appropriate course of action to take first?

   A. Ping each of the workstations in the workgroup to ensure each has a good connection.

   B. Run a protocol analyzer and compare the data to the network's baseline data.

   C. Use **netstat** to locate duplicate IP address assignments.

   D. Use a packet sniffer to see if users are using the appropriate user names and passwords, which could cause logon delays.

56. An employee who recently changed their password at the required 90-day company policy calls the help desk complaining that they can no longer log on to the network. What would be the most appropriate course of action to solve this problem?

   A. Check if the user's rights have changed since changing the password.

   B. Have the user check if the [Caps Lock] key is on.

   C. Have the user ping the server to see if a connection can be established.

   D. Have the user do nothing and dispatch a technician to the user's location.

57. Users on a LAN using Windows 2000 Server and Windows 2000 Professional are able to exchange files and communicate with the server, but no one can reach the branch office using the Internet. Which of the following is *most likely* the cause of failure?

   A. The WINS server is down.

   B. The DHCP service has failed.

   C. The lmhosts file is corrupt on the server.

   D. There is insufficient RAM on the server to support Internet access.

58. A network using Windows 2000 Server has recently added 50 workstations. The server response speed seems to be slowing over the last month. You suspect that more RAM is needed in the server. What tool can be used to analyze and determine if sufficient RAM exists on the server?

   A. Network Monitor

   B. Performance Monitor

   C. OTD

   D. Defrag

59. A technician has just installed a new network interface card in a workstation. After completing the installation, the technician cannot successfully log on to the network or see any other devices when browsing the network. Also, when the technician tried to reinstall a driver, he discovered the CD-ROM is no longer working. What is *most likely* the problem?

   A. Faulty network cable.

   B. The wrong protocol has been selected for the network interface card.

   C. There is an IRQ conflict.

   D. This condition is normal until the server recognizes the network card.

60. The time required to transfer a file has greatly increased recently. You suspect that there is an excessive amount of data transfers on the network. Which tool would you use to confirm your suspicions?

   A. Protocol Analyzer

   B. OTD

   C. Network Neighborhood

   D. 1GHz oscilloscope

61. You successfully ping the local host at the workstation but cannot ping anyone else in your workgroup. What is *most likely* the problem?

   A. The TCP/IP protocol is not installed on the workstation.

   B. The DNS server is down.

   C. There is a loose connection at the network interface card or hub.

   D. There is insufficient RAM installed in the workstation.

62. You are checking a network conductor pair using a VOM. What setting is most appropriate to verify there is no short between the two conductors?

   A. Set the meter for amperes and connect the probes to conductor pair.

   B. Set the meter for volts and connect the probes to the conductor pair.

   C. Set the meter for ohms and connect the probes to the conductor pair.

   D. Set the meter for frequency and connect the probes to the conductor pair.

63. There is a weather report indicating the likelihood of severe thunderstorms in your area. What device should be checked to ensure that the network server is protected?

   A. UPS

   B. Gateway

   C. Router

   D. Bridge

64. You suspect that a network connection is preventing a Windows 2000 work-station from booting. What would be the most appropriate action to take?

   A. Connect a protocol analyzer to a different node and monitor the boot process of the suspect workstation.

   B. Open Device Manager on the suspect workstation and verify that the correct drivers are loaded.

   C. Press [F8] during the boot sequence and verify the workstation will start in safe mode.

   D. Use a system boot disk to start the workstation and then open Device Manager to verify the network interface card settings.

65. You suspect that you have a corrupt MBR on a Windows 2000 server that is using NTFS. What is the most appropriate remedy for this problem in the list of choices and is typically the last resort?

   A. Start the server using Recovery Console and then enter the **fixmbr** command.

   B. Start the server using a DOS system disk and then type **fdisk/mbr** at the command prompt.

   C. The server's hard disk drive will need to be replaced and the system files reinstalled.

   D. Perform a system restore from a backup tape, which will automatically repair the MBR.

66. What information is revealed when issuing an **ipconfig** command at the command prompt of a Windows 2000 workstation on a TCP/IP Ethernet network? (Select all that apply.)

   A. The physical address.

   B. The IP address.

   C. The workgroup name.

   D. The subnet mask.

67. What does the error message "NTLDR is missing" indicate?

   A. The network adapter card failed to load.

   B. The NT boot loader failed to load.

   C. The kernel is corrupt.

   D. The system failed the POST.

68. What is the most appropriate utility or tool to use to check a network cable connection from a workstation to a local server?

   A. **ping**

   B. **ipconfig**

   C. VOM

   D. Packet Sniffer

69. Which of the following items can cause interference with a wireless LAN?

   A. Wireless telephones.

   B. Kitchen appliances.

   C. Office equipment.

   D. All the above.

70. A user calls the help desk and says that they cannot print to the network printer. What is the first thing that should be done?

   A. Have the user check if the printer is online.

   B. Have the user check if the correct driver is installed.

   C. Have the user open the printer control console and see if the buffer cache is full.

   D. Have the user reboot their desktop computer.

71. A technician wishes to trace a UTP cable to check for a suspected break and to verify its path through the building. Which tool would be most appropriate?

   A. Tone generator and tracer

   B. ODT

   C. Protocol analyzer

   D. Multimeter

72. What would be the most appropriate last step in a troubleshooting process?

   A. Identify the affected area.

   B. Establish what has changed.

   C. Test the result.

   D. Document the solution and process.

Scoring the Exam

Copy the following chart onto a separate sheet of paper and fill in your scores. Please do not write in this book.

Network+ Domain	Your Score (Number of correctly answered questions.)
1.0 Media and Topologies	
2.0 Protocols and Standards	
3.0 Network Implementation	
4.0 Network Support	

As a quick indication of how you performed on the practice exam, refer to the chart in **Figure 21-3.** The chart shows the total number of questions in each domain and the number of correctly answered questions you need to have to score an 80% in that domain.

Network+ Domain	Total Number of Questions on the Exam	Approximate Number of Correct Questions Needed to Score an 80%
1.0 Media and Topologies	14	11
2.0 Protocols and Standards	15	12
3.0 Network Implementation	18	15
4.0 Network Support	25	20
Total Number of Questions	72	58

Figure 21-3.
Use this chart to determine if you have achieved an 80% on the practice exam or in any of the domains.

Summary

- The Network+ Certification exam contains 72 questions.
- The maximum time allotted to answer the questions on the Network+ Certification exam is 90 minutes.
- Questions on the Network+ Certification exam are not equally weighted.
- Review and analyze all of the Network+ Certification exam objectives to become familiar with the knowledge areas for which the exam tests.
- Match the Network+ Certification exam objective details to quality resource material.
- Identify and practice laboratory activities that match the Network+ Certification exam objectives.
- Take practice exams, and then identify and review problem areas.

Review Questions

1. How many questions are on the Network+ Certification exam?
2. What is the minimum passing score for the Network+ Certification exam?
3. Outline the key points for an exam preparation strategy.
4. Why do you *not* want to depend on a commercial study guide to prepare for the exam?
5. What other CompTIA exam objectives may you want to use to help you prepare for the Network+ Certification exam?

Suggested Laboratory Activities

1. Download the latest Network+ Examination Objectives.
2. Make a test objective outline and fill it in with facts and details about each objective.
3. Locate several Web sites that contain free test materials and evaluate the contents by comparing them to the current test objectives.

Interesting Web Sites for More Information

www.CompTIA.org

http://support.microsoft.com

Network+ Note:

Visit the Microsoft's support Web site and look at articles that will reinforce knowledge required for the Network+ Certification exam. Examples are installing Windows 98, 2000, and XP. Pay particular attention to problems associated with the installation process. Also, look at the articles on troubleshooting tools provided by the operating systems. I have no doubt that many of these articles are a source of information for the Network+ Certification exam.

22

Employment in the Field of Networking Technology

After studying this chapter, you will be able to:

- ❏ Discuss a variety of networking technology careers.
- ❏ Define entrepreneur and entrepreneurship.
- ❏ Identify career information sources.
- ❏ Discuss networking technology educational and certification requirements.
- ❏ Identify advanced training options.
- ❏ Outline ideas for a successful job search.
- ❏ Conduct a job search.
- ❏ Write a resume.
- ❏ Identify appropriate interview skills.

Key Words and Terms

The following words and terms will become important pieces of your networking vocabulary. Be sure you can define them.

A+ Certification
business plan
Certified Document Imaging Architect (CDIA+) Certification
Certified Linux Engineer (CLE) Certification
Certified Linux Professional (CLP) Certification
Certified Novell Administrator (CNA) Certification
Certified Novell Engineer (CNE) Certification
consultant
engineer
entrepreneur
i-Net+ Certification
Linux+ Certification
Master CNE (MCNE) Certification
Microsoft Certified Database Administrator (MCDA) Certification
Microsoft Certified Desktop Support Specialist (MCDST) Certification

Microsoft Certified Professional (MCP) Certification
Microsoft Certified Solution Developer (MCSD) Certification
Microsoft Certified Systems Administrator (MCSA) Certification
Microsoft Certified Systems Engineer (MCSE) Certification
Microsoft Office Specialist Certification
Network+ Certification
network administrator
network support specialist
programmer
programmer analyst
Server+ Certification
systems analyst
Web administrator
Web master

By successfully reaching this point in the textbook, you have probably decided whether or not to pursue a career in networking technology. Networking is one of the fastest growing fields within the scope of computer technology. If you have the interest, desire, and ability, you can find a rewarding career in this field.

Networking technology and related computer careers require a basic education in computers and networking as well as an advanced education in the field. Technology is constantly and rapidly advancing. New ideas become a reality every day. As a professional in the computer and networking field, this condition requires that you continually update your skills through education and training. Commercial certification in one or more technology fields is often required.

To keep up with the rapid changes in networking technology, you must form some type of action plan. Your personal plan must include strategies for gaining initial training and certifications and keeping up-to-date with the changes and ways to expand your opportunities for advancement in the IT field. This chapter discusses ways to gain employment and ways to advance your career. While you may already be employed in a networking, computer repair, or related job, this chapter may help you better define your career goals. First, let's look at some of the many different job titles in the IT industry.

Information Technology Industry Careers

There are a wide variety of careers in the Information Technology (IT) industry. Networking is the key to communications and information systems that play a major role in business and daily life. Think of all the things that depend on computer networks. For example, in the communications industry, the telephone, television, and radio are linked to computer networks. The manufacturing industry uses computers and networks to design, build, inventory, and sell most everything it makes. Computer networks control robots and automate assembly lines. In the business world, computer networks make communication and business nearly instantaneous. Sharing information within and among companies and the Internet is commonplace. Tasks that, in the past, required days and months to complete are now completed automatically or within minutes. The entire banking industry relies on networked computers to calculate and track money exchanges and post records of interest, earnings, and mortgage statistics.

Network technology has also saturated the field of medicine. Surgeons can perform laser and other surgery through long-distance communication over networks. Patient medical records are computerized, **Figure 22-1.** MRI scans can be transmitted instantly across a network to any distant city where other doctors can evaluate them. Architects and engineers rely on computers to design structures around the world. Not only can they design the structure, they can take a virtual tour of the design from another country. Law enforcement, the military, and other governmental units are completely reliant on networking technology.

The need for people with networking and other computer skills has rapidly grown and will continue to grow. There are many careers in the IT industry from which to choose. Some of the many careers include the following:

- Network support specialist.
- Network administrator.
- Systems analyst.
- Consultant.
- Technical sales person.
- Web administrator.
- Programmer.
- Software engineer.
- Entrepreneur.

Figure 22-1.
Network technology plays a vital role in the field of medicine. It can provide access to patient records from remote locations, allow surgeons to perform laser and other surgery through long-distance communication, and transport MRI scans to any distant city where other doctors can evaluate them.

network support specialist
a person who assists users and customers whenever they encounter a network problem.

network administrator
a person who is responsible for the installation, configuration, and maintenance of a LAN, MAN, or WAN. The network administrator may perform many routine administrative functions, such as work with hardware and software vendors, make written and oral presentations about the state of the network, interview prospective employees, evaluate employee performance, coordinate work schedules, order supplies and equipment, and prepare budget reports.

Network Support Specialist

As a *network support specialist*, you will be required to assist users and customers whenever they encounter a network problem. Often a network support specialist is assigned to a help desk. They answer phone calls, interpret user problems, and recommend solutions or dispatch a technician to the user's workstation. Most of the support they provide is through the Remote Assistance utility found in Windows XP, **Figure 22-2.** As a part of the role of support, network support specialists are often required to conduct training sessions for employees and write training manuals.

The network support specialist works under the direct supervision of a network administrator. The network administrator usually directs their activities for the day. Many network administrators started their careers as a network support specialist.

A network support specialist must continuously learn about new software packages, operating systems, hardware, security, data storage, and other key elements of networking. New hardware systems, software packages, service packs, and security software are continuously developed. You will need to expand your knowledge in these areas to make yourself even more valuable as a network support specialist and to advance to network administrator.

Network Administrator

Most of the maintenance of a network system is the result of constant monitoring by the network administrator. While the network support specialist may perform routine maintenance under the direction of the network administrator, the *network administrator* is ultimately responsible for the type, extent, and frequency of maintenance to be performed. The network administrator is also responsible for the installation, configuration, and maintenance of the LAN, MAN, or WAN. The network administrator may perform many routine administrative functions, such as work with hardware and software vendors, make written and oral presentations about the state of the network, interview prospective employees, evaluate employee performance, coordinate work schedules,

Figure 22-2.
A network support specialist is often assigned to a help desk. The network support specialist interprets user problems and can often provide support over the phone or remotely through a program like Remote Assistance.

order supplies and equipment, and prepare budget reports. It is not unusual for this type of position to require a four-year degree in addition to one to two years of technical training.

Systems Analyst

A *systems analyst* is responsible for performing analysis, evaluations, and recommendations of business software systems. Typically, a systems analyst is involved in designing system, and making recommendations rather than involved in day-to-day network operations. Many systems analysts work for hardware and software companies. The system analyst works with clients by analyzing the client's business and making recommendations. They also design network systems to match the needs of the client's business.

Some systems analysts work as independent contractors or consultants. As a consultant, they sell their expertise as a service to organizations that are about to install, upgrade, or modify a network system. A systems analyst must be able to communicate effectively with programmers, technicians, and engineers and with clients who have limited computer system knowledge. A systems analyst position usually requires a four-year degree.

systems analyst
a person who is responsible for performing analysis, evaluations, and recommendations of business software systems.

Consultant

Consulting is another growing business in the IT industry. A *consultant* works with clients on projects and makes recommendations based on their expertise, **Figure 22-3.** The specific job they do often depends on what work is needed. The client pays the consultant for their expertise. When the job is completed, the consultant is free to move on to a new job and client.

In the case of networking technology, it is recommended that a person become a consultant after extensive experience in the field. Consultants are expected to have answers to many challenges or be able to quickly research them. Consultants usually have vast expertise in one or more particular aspects of the IT industry, such as security, systems analysis, network design, and electronic marketing.

consultant
a person who works with clients on projects and makes recommendations based on their expertise.

Figure 22-3.
Companies often rely on consultants for their expertise in certain areas.

Technical Sales Person

Technical sales persons are always needed in the industry to represent manufacturers of hardware and software. Many technical sales persons originally started out as a technician and then moved into sales. They are typically required to obtain a certification directly related to their field of expertise. For example, if they are representing a virus software manufacturer, they will be required to obtain certification on the product they are selling. A person seeking a career in technical sales needs an outgoing personality, must be well groomed, and must like working with the people. This is not a job for the shy or timid.

Web Administrator

Web administrator
a person who is responsible for converting written documents into HTML or other Web page programming languages. This job requires not only training in Web page programming language, but also some expertise in networking technology.

Web administration requires the design and implementation of Web pages for external and internal Web sites. The *Web administrator* is responsible for converting written documents into HTML or other Web page programming languages. The Web administrator usually has to interpret the needs of their employer and convert those needs into Web page presentations. This job requires not only training in Web page programming language, but also some expertise in networking technology. The Web administrator also requires some expertise in art and graphics. The use of digital camera, video equipment, and video editing is beneficial.

Web master
a person who has expertise in Web programming languages, such as HTML, XML, JAVA, JavaScript, ActiveX, and familiarity with server management of IIS, Apache, UNIX, Linux, and Microsoft systems.

The Web page is often referred to as the front end while the support of the Web server is referred to as the back end. There are many self-taught Web page designers, but to land a job as a Web master, administrator, or technician, requires additional training in networking and server technology. A *Web master* requires expertise in Web programming languages, such as HTML, XML, JAVA, JavaScript, ActiveX, and familiarity with server management of IIS, Apache, UNIX, Linux, and Microsoft systems. Certification is typically a minimal job requirement.

Programmer

programmer
a person who writes, tests, and modifies software programs.

A *programmer* writes, tests, and modifies software programs. There are many programming languages to choose from. Many programmers master more than one programming language. Some programming languages you may wish to learn are C, C++, Prolog, or JAVA. C and C++ are excellent programming languages to learn and a recommended starting point. Prolog is a programming language used in artificial intelligence and machines. One of the hottest programming languages today is JAVA. Other languages to consider are Visual Basic and XML.

programmer analyst
a person who analyzes the needs of a client and performs the actual programming.

Programmers vary from self-taught to college-educated. When college-educated, programmers are generally referred to as *software engineers*. Many programmers work as independent contractors who program or modify custom software packages to meet the needs of their employer. Some programmers are known as programmer analysts. A *programmer analyst* not only analyzes the needs of a client, they perform the actual programming.

Engineer

engineer
a degreed professional who possesses high-level skills necessary to solve problems related to their field of expertise.

An *engineer* is a degreed professional who possesses high-level skills necessary to solve problems related to their field of expertise. To become an engineer requires a four- to six-year college degree. Engineers require a strong background in mathematics, sciences, programming languages, and hardware. They typically advance to management and project leadership positions. If an engineer exhibits leadership skills, they often become the chief technology officer. Engineers command the highest salaries in the IT job market. Those that earn a doctoral degree often become university engineering and computer science faculty. They also are employed in the area of research where they are responsible for developing new technologies.

Entrepreneur

An *entrepreneur* owns and operates a business. They usually start with an idea for filling a gap in the marketplace or identify where a new product or service is needed. An entrepreneur then develops a business plan. A business plan is required if the entrepreneur is seeking financial support to open a business. The **business plan** outlines the goals for the business and includes action plans and a timetable for meeting those goals. A good business plan is vital if the business is to succeed.

In addition to a sound business plan, a successful entrepreneur possesses certain skills. The successful entrepreneur has knowledge of the product or service they plan to provide. This knowledge allows the entrepreneur to make smart business decisions. The successful entrepreneur also has sound management skills. These skills allow the entrepreneur to manage resources such as money, time, and employees. Resource management is critical to success. A serious lack of skill in any of these areas can lead to certain failure. The ability to think creatively and wisely are also important skills for the successful entrepreneur. These skills allow the entrepreneur to control the business and move it in the right direction.

Entrepreneurial opportunities are vast in the IT industry. Along with the growth of networking technologies and products, growth has occurred in servicing these technologies and products. Installing, maintaining, and servicing networks is and will continue to be a sound business in the industry.

entrepreneur
a person who owns and operates a business.

business plan
a document that outlines the goals for the business and includes an action plan and a timetable for meeting those goals.

Career Information Sources

The major source of job information, especially information regarding computer and networking, is the Internet. There are many Internet sources that contain job listings and job matching services. Nearly every college and school is connected to such a service. Another reference is the *Occupational Outlook Handbook*. It is published by the United States Department of Labor and Bureau of Statistics. Most school, community college, university, and public libraries have copies of this book. The book can be ordered from the Bureau of Labor Statistics Web site, www.bls.gov.

School guidance counselors and local labor market offices are other outstanding sources of career information. They can help you find information on particular careers and colleges, training programs, and military service opportunities. School guidance counselors and those who work in local labor market offices are typically well informed and ready to help you in your search for a job or training. Colleges typically have job information and placement services available.

A great amount of career and training information can be found on the Internet. Many company Web sites list employment opportunities, the required skills and educational level, and a brief job description. The following are some careers you may find listed when searching these sources:

- Entry-level network technician.
- Network support professional.
- Network installer.
- Technical sales and marketing professional.
- Technical writer.
- Technical forensics technician.
- Customer service representative.
- Network security specialist.
- Internet Web site developer.
- Internet systems administrator.
- Database specialist.
- Network hardware specialist.
- Network engineer.
- Systems engineer.
- Software engineer.
- Programmer.
- Analyst.
- Chief information officer.
- Telecommunications data specialist.
- Voice over IP (VoIP) engineer.
- Data communications engineer.
- ATM engineer.
- Web master.
- Web developer.
- Industrial control engineer.
- IT consultant.
- Service/help desk technician.
- Network training specialist.

Each person must ask themselves several questions before going further into the computer arena. Have you enjoyed this *Networking Fundamentals* textbook and the laboratory activities? The fact that you are studying networking shows that you already have a good background in computers and an interest in this area. Have you taken other classes in computer technology? Did you like them? Did you do well? Give serious thought to your responses to the above questions and guidelines. Your responses may be the start you need to achieve success in your chosen area.

General and College Education

The educational requirements for jobs in the IT industry vary. A minimum of a high school education is a solid foundation on which to build. Very few high school graduates enter the networking industry directly, but are required to receive specialized education in the training programs maintained by large companies. Specialized training may also come from formal training in colleges, technical schools, or the military. To gain promotions, however, advanced certifications and degrees are required.

College is an excellent option for advancement in the IT industry. A college education in one of the IT industry-related fields is an excellent choice because of the demand for people with a college degree. A college education not only provides a technical background, it provides preparation for the business world through other classes, such as technical writing, business writing, and speech. These classes teach basic communication skills, which are in as much demand as IT skills.

If you plan to advance in the IT business world, you must master the skills of communication, both written and spoken. You may need to give presentations not only inside your company but also to prospective clients for your company. You need to have the skills to prepare written materials and make a presentation that represents your company. For example, your company may be bidding to contract a networking project for a firm. During the process of meeting with the firm, a single person or a team representing your company will need to make a presentation to attempt to gain the firm's business. If you plan to be a part of the team, you will need the very best communication skills. A technician who knows the technology and who can communicate well is always in demand, **Figure 22-4.**

There is a direct correlation between salary and education level. Research proves that as your education level increases, so does your salary level. While the direct correlation indicates college as a major factor, additional certifications are also a direct influence. An ideal situation for a person planning a future in the IT industry is to combine technology education with another field of study such as business administration, finance, mathematics, science, electronics, medical, law enforcement, or any other field of interest. A combination of degrees greatly increases a person's employability.

Many community colleges offer courses and programs in the network technology field. You can take specific courses to learn more about a particular

Figure 22-4.
Strong communication skills are a plus. In the networking field you may be required to present a product or a system plan to a group of clients. The ability to communicate technical information to those without your expertise is crucial.

subject without going for a degree. Alternative pathways, such as military training, are also available to gain a particular skill. The military offers many specialized areas of study in the networking and computer technology fields. The opportunities for education and for gaining valuable work experiences are excellent in the military.

As an IT professional, you will be learning the rest of your life. The total knowledgebase related to the computer industry is never ending. Part of what you know now may be obsolete in just a few months. If you did not learn another thing from this point forward, your skills would be obsolete within a year or two.

Certification

Certification is a way to advance your knowledge and career in the IT industry. Many companies require certifications prior to hiring. Certification combined with work experience can prove your abilities to an employer. It can also advance your career inside a company and provide you with job security.

Obtaining your CompTIA Network+ and A+ Certification should just be the beginning for you. Once you've obtained these certifications, you should immediately start advancing toward other CompTIA certifications, such as the i-Net+ or Security+. You may also wish to start in an area of specialization, such as network administration, and obtain the Microsoft Certified Systems Administrator (MCSA).

Novell, Microsoft, Cisco Systems, and many other companies offer certification exams in many advanced fields of study. Two of the most demanding certifications offered by Microsoft are the Microsoft Certified Systems Engineer (MCSE) and Microsoft Certified Systems Administrator (MCSA). Training preparation for these examinations can be delivered in many ways such as private schools, public schools, special seminars, online educational programs, and self-study. The main point is you must pass a series of exams to prove your knowledge and competency before you receive the certification. Let's look at some of these certifications.

CompTIA Certifications

The CompTIA organization offers certification in many areas. Receiving an A+ and Network+ Certification from CompTIA is just the beginning. Some advanced CompTIA certifications that you may elect to pursue are CDIA+, i-Net+, Server+, and Linux+.

A+

A+ Certification
a CompTIA certification that proves a person's knowledge of PC repair, hardware, and software.

The *A+ Certification* is designed to test a person's knowledge regarding PC repair, hardware, and software. The candidate is tested in two areas of PC technology. These areas are represented by two exams: Core Hardware and Operating System Technologies. The A+ Core Hardware exam tests the candidate's knowledge of PC hardware. This includes installation, configuration, and troubleshooting. The A+ Operating System Technologies exam covers the installation, configuration, and troubleshooting of Microsoft operating systems, such as Windows 2000 and Windows XP. The A+ Certification is universally recognized and is a good place to start your certification process.

Network+

Network+ Certification
a CompTIA certification that measures the skills necessary for an entry-level network support technician.

The *Network+ Certification* measures the skills necessary for an entry-level network support technician. The typical Network+ candidate should have the A+ Certification or equivalent knowledge. It is also recommended that the candidate have at least nine months of network experience before attempting the exam. The exam is vender-neutral and is recognized by leading network organizations such as Microsoft, Novell, Compaq, Cisco, 3Com, and Lotus as part of their certification process. It is an excellent starting point for students interested in further certification in the networking technology areas.

CDIA+

The *Certified Document Imaging Architect (CDIA+) Certification* is a good advanced certification to obtain. A great need exists for persons with expertise in computer imaging. The CDIA+ Certification exam tests a candidate's knowledge of imaging systems such as scanners, displays, printers, graphic file types, file conversion, and image enhancement. Typical questions include those about storage systems, transmission speeds across networks, and PC performance as related to imaging. There is a tremendous need for technicians who can quickly and efficiently convert text pages and illustrations into formats recognized by various computer systems.

Certified Document Imaging Architect (CDIA+) Certification a CompTIA certification that proves a person's knowledge of imaging systems such as scanners, displays, printers, graphic file types, file conversion, and image enhancement.

i-Net+

The *i-Net+ Certification* is designed to prove competency working with Web services. It includes Internet communications and installing, updating, and modifying Web pages. The i-Net+ Certification exam tests areas concerned with Internet system administration, application development, security, e-commerce, and Web site design. The competencies are similar to those in the Network+ Certification but have a greater emphasis on Internet communication, construction, and design. Do not confuse the contents of the i-Net+ Certification with Web design graphics. Web design graphics is more like commercial art. It involves the presentation of the display, sequencing, and general artwork related to Web design. The i-Net+ Certification is concerned with equipment, protocols, networking systems, and hardware requirements related to Web services. Many of the skills and knowledge you gain from the Network+ Certification overlap the requirements for the i-Net+ Certification. Obtaining the Network+ Certification before you pursue the i-Net+ Certification will give you a head start on your studies.

i-Net+ Certification a CompTIA certification that proves a person's competency working with Web services, including Internet communications and installing, updating, and modifying Web pages.

Server+

The *Server+ Certification* tests a person's knowledge of network server hardware and software. The candidate is tested on installing, configuring, diagnosing, and troubleshooting network server hardware and network operating systems. The exam requires an in-depth knowledge of protocols, backup system standards, and system security.

Server+ Certification a CompTIA certification that proves a person's knowledge of network server hardware and software.

Linux+

The *Linux+ Certification* tests a person's knowledge of the Linux operating system. It covers installing, configuring, administering, and troubleshooting the Linux operating system for a single PC and a network server.

Linux+ Certification a CompTIA certification that proves a person's knowledge of the Linux operating system.

Microsoft Certifications

As with all computer technology, the certifications available rapidly change; however, Microsoft has a rich offering of various certifications at this time. These include the Microsoft Office Specialist, Microsoft Certified Professional (MCP), Microsoft Certified Desktop Support Specialist (MCDST), Microsoft Certified Systems Engineer (MCSE), Microsoft Certified Systems Administrator (MCSA),

Microsoft Office Specialist Certification
a Microsoft certification that proves a person's knowledge of Microsoft Office software such as Word, Excel, Access, PowerPoint, Project, and Outlook. There are presently four levels of Microsoft Office Specialist Certification: Specialist, Expert, Master, and Master Instructor.

Microsoft Certified Professional (MCP) Certification
a Microsoft certification that proves a person's expertise in one or more of the Microsoft software products.

Microsoft Certified Desktop Support Specialist (MCDST) Certification
a Microsoft certification that measures the skills required to work as a help desk technician or a customer support representative.

Microsoft Certified Systems Engineer (MCSE) Certification
a Microsoft certification that proves a person's ability to design, implement, maintain, upgrade, troubleshoot, and administer a Microsoft network system based on the Windows 2000 or 2003 platform.

Microsoft Certified Solution Developer (MCSD), and Microsoft Certified Database Administrator (MCDA).

These certifications usually require the candidate to pass two or more exams. Many of the required exams apply toward more than one certification. For example, the Managing and Maintaining a Windows Server 2003 Environment exam can be used toward multiple certifications.

The series of Microsoft certifications are designed to support the Microsoft family of products. There are many specialized exams designed for testing knowledge about Microsoft operating systems such as Windows Server 2003, Windows 2000, and Windows XP, as well as specialized areas, such as security. Many of the skills required for these exams can help toward certification in other areas or companies.

Microsoft Office Specialist

The *Microsoft Office Specialist Certification* is based on knowledge of Microsoft Office software such as Word, Excel, Access, PowerPoint, Project, and Outlook. There are presently four levels of certification: Specialist, Expert, Master, and Master Instructor.

MCP

The *Microsoft Certified Professional (MCP) Certification* is designed to prove expertise in one or more of the Microsoft software products. An MCP Certification is commonly earned while in the pursuit of a more advanced certification such as the Microsoft Certified Systems Engineer (MCSE).

MCDST

The *Microsoft Certified Desktop Support Specialist (MCDST) Certification* measures the skills required to work as a help desk technician or a customer support representative. The certification requires the candidate to take two certification exams: *Supporting Users and Troubleshooting a Microsoft Windows XP Operating System* and *Supporting Users and Troubleshooting Desktop Applications on a Microsoft XP Operating System.*

Supporting Users and Troubleshooting a Microsoft Windows XP Operating System measures skills relating to hardware and operating system issues. Many of the skills required for this exam match the skill set associated with the CompTIA A+ exams. In fact, Microsoft will automatically award credit toward the MCDST if you already have the A+ Certification.

The Troubleshooting Desktop Applications on a Microsoft XP Operating System exam measures skills associated with supporting Microsoft Office products, Outlook Express, and Internet Explorer. You do not need an in-depth knowledge of Microsoft software applications, but you should be able to answer the most common questions typically encountered while working at a help desk.

MCSE

Microsoft currently has two tracks for the Microsoft Certified Systems Engineer (MCSE) Certification: MCSE on Windows 2000 and MCSE on Windows Server 2003. The *Microsoft Certified Systems Engineer (MCSE) Certification* is based on the ability of the candidate to design, implement, maintain, upgrade, troubleshoot, and administer a Microsoft network system based on the Windows 2000 or 2003 platform.

Some job titles associated with this certification are systems engineer, technical support engineer, systems analyst, network analyst, and technical consultant. Do not be confused by the term *engineer* as related to this certification.

You do not have to go to college to be a Microsoft systems engineer. The term *engineer* better describes the duties associated with the certification and job function. Any of the certifications that include the term *engineer* require many hours of training both in the classroom and on the job. It is not unusual for a person seeking this certification to dedicate one to two years of constant study.

The MCSE consists of a series of exams leading to certification. A typical candidate is a person who has been working with large network systems that comprise 200 to 26,000 users spread over 5 to 150 physical locations. The MCSE Certification requires expertise in a network operating system, network design, and administration. Requirements for the MCSE Certification are constantly changing. Check the Microsoft Web site for up-to-date information.

MCSA

The *Microsoft Certified Systems Administrator (MCSA) Certification* was developed to answer a demand for a less stringent certification than the MCSE. The MCSA is designed for a person who will be responsible for the day-to-day operations of a middle-to-large network rather than responsible for its design and installation.

The MCSA does not require as many exams as the MCSE. At this time, the MCSE requires six core exams and one elective compared to three core exams and one elective required for the MCSA. Many students opt to obtain the MCSA before going on to achieve the MCSE. This is a preferred practice since the same course work can be applied to the MCSE. This method allows a student to obtain the impressive MCSA Certification in a much shorter time than the MCSE.

MCSD

The *Microsoft Certified Solution Developer (MCSD) Certification* is awarded to individuals that prove they have the ability to design, implement, and administer business solutions using Microsoft Office or Back Office products. In other words, they can set up a network system for a business based on individual business needs, select the appropriate software packages, design the hardware requirements or specifications, and install and maintain the system.

MCDA

The *Microsoft Certified Database Administrator (MCDA) Certification* is earned by proving expertise knowledge of SQL Server databases. The MCDA Certification proves expertise in creating, maintaining, optimizing, installing, and managing SQL Server databases. Companies, government, educational institutions, and research facilities use database technology extensively.

Novell Certifications

Novell offers several certifications, such as the Certified Novell Administrator (CNA), Certified Novell Engineer (CNE), Master CNE, Certified Linux Professional (CLP), and Certified Linux Engineer (CLE). The certifications are tied directly to Novell software products such as Netware, SuSE Linux, ZENworks, and GroupWise. The Novell certifications are similar to the Microsoft certifications; however, they are designed to support Novell network products.

CNA

The *Certified Novell Administrator (CNA) Certification* proves that a person has the skills necessary to set up and manage workstations and to

Microsoft Certified Systems Administrator (MCSA) Certification a Microsoft certification that proves a person's ability to handle the day-to-day operations of a middle-to-large network based on the Windows 2000 or 2003 platform.

Microsoft Certified Solution Developer (MCSD) Certification a Microsoft certification that proves a person's ability to design, implement, and administer business solutions using Microsoft Office or Back Office products.

Microsoft Certified Database Administrator (MCDA) Certification a Microsoft certification that proves a person's expertise in creating, maintaining, optimizing, installing, and managing SQL Server databases.

Certified Novell Administrator (CNA) Certification a Novell certification that proves a person's ability to set up and manage workstations and to manage network system resources such as files, printers, and software in a NetWare network.

Certified Novell Engineer (CNE) Certification
a Novell certification that proves a person's ability in LAN design, development, implementation, support, and troubleshooting of a NetWare network.

Master CNE Certification
an advanced Novell certification with an emphasis on project management and design and expertise in areas such as ZENworks desktop management, Integrating eDirectory and Active Directory, GroupWise Administration, and eDirectory Tools and Diagnostics.

Certified Linux Professional (CLP) Certification
a Novell certification designed for people who wish to prove they have mastered the basic skills required to manage and maintain a SuSE Linux server. The skills measured are installing a SuSE Linux server, managing users and groups, troubleshooting the SuSE Linux file system, and managing and compiling the Linux kernel.

Certified Linux Engineer (CLE) Certification
a Novell certification that verifies expertise in both Linux and Novell services that run on a Linux system.

manage network system resources such as files, printers, and software. It also shows that the person understands how to monitor network performance and provide remote access to the network.

CNE

The *Certified Novell Engineer (CNE) Certification* is designed to prove expertise in a wider range of network applications than the CNA. A person must prove expertise in LAN design, development, implementation, support, and troubleshooting. The CNE may specialize in one particular area of network software such as Netware 6.5 or may certify in multiple software packages, such as NetWare 6 and NetWare 6.5.

Master CNE

The *Master CNE (MCNE) Certification* is an advanced certification. It is the next logical step after achieving the CNE. Additional skills are required with an emphasis on project management and design. The CompTIA Project+ Certification can be used toward fulfilling the requirements of Master CNE certification. Other required areas of expertise include any of the following: ZENworks desktop management, Integrating eDirectory and Active Directory, GroupWise Administration, and eDirectory Tools and Diagnostics.

CLP

The *Certified Linux Professional (CLP) Certification* is designed for people who wish to prove they have mastered the basic skills required to manage and maintain a SuSE Linux server. The skills measured are installing a SuSE Linux server, managing users and groups, troubleshooting the SuSE Linux file system, and managing and compiling the Linux kernel. The CLP covers three Novell Linux training areas: Novell Linux Fundamentals, Novell Linux Administration, and Advanced Linux Administration. The certification is based on passing one practical exam that measures the skills reflected in the three training areas.

CLE

The *Certified Linux Engineer (CLE) Certification* is the next logical certification to pursue after achieving the CLP. The Novell CLE Certification verifies expertise in both Linux and Novell services that run on a Linux system. A broad background in Linux administration is recommended before attempting the examination. This certification also requires a practical exam rather than a written exam to prove competency. To see a complete listing of available courses and text objectives, visit the Novell Web site (www.novell.com).

Cisco Certifications and Training

Cisco Corporation provides training opportunities at high schools and colleges all over the world. Cisco sites are referred to as *academies*. The major emphasis of Cisco academies is network design, implementation, and troubleshooting Cisco products.

The academy courses are designed as a combination of lecture, textbook, online learning, and hands-on laboratory activities. Some of the certifications available through Cisco are Cisco Certified Network Associate (CCNA), Cisco Certified Design Associate (CCDA), Cisco Certified Network Professional (CCNP), Cisco Certified Design Professional (CCDP), and Cisco Certified Internetworking Expert (CCIE). The certificates are obtained by passing specific Cisco exams. You can locate the Cisco academy nearest you by visiting the Cisco Web site (www.cisco.com).

Other Certifications

Many other companies have certification programs besides the ones outlined in the previous sections. Some of the companies are 3Com, Nortel Networks, HP, Red Hat, Sun, and Oracle. More in-depth information about these certifications, such as exam outlines, study materials, available training, and prerequisites, can be found at the related Web sites. There are over 120 different companies and organizations that offer certifications and over 300 different types of certifications you may earn.

As you see, many different certifications are available for network technicians and IT professionals. You can customize your credentials by selecting different areas of certification similar to the way a university student selects classes for credit toward a degree. In fact, a collection of certifications is similar to a university degree.

Two certifications that are highly recommended are Security+ and i-Net+. Demand for Web server support personnel and network security experts are on the rise. The Network+ Certification and A+ Certification are accepted as an elective requirement toward Microsoft's MCSA Certification. Novell recommends having Network+ Certification before pursuing the Novell CNA Certification. Some colleges award credit toward degree programs based on receiving the Network+ Certification.

After finishing this course, it is suggested that you select a field of specialization. This is the time to begin training for a specialized certification. Some of the most common areas are Microsoft, Novell, and Cisco. These three areas have shown the most interest of students at the entry level of training. You may wish to seek employment before beginning your next level of training or train concurrently while employed. Going to school while employed is an excellent method to reinforce what you are learning in the classroom. Remember, there is a direct correlation between knowledge and earnings. The more you know, the more you will earn.

The IT industry demands continuing education of its workforce. If you choose to join the IT workforce, be prepared for a lifetime of study, **Figure 22-5.** New systems, protocols, software, and hardware are constantly introduced. Often, these new systems demand new expertise to install, service, and repair. You can continue your education in a number of ways, such as attending college classes, seminars, online courses, and self-study.

Figure 22-5.
To keep up with the continuously changing field of computer and networking technology requires constant study. Some of the ways you can continue your education include attending college classes and seminars, taking online courses, and self-study.

You may choose to not pursue other certifications. This decision is acceptable, as long as you keep your skills up-to-date. You should subscribe to and read professional journals and newsletters in your area of work. Take as many courses as you possibly can, such as digital electronics, to enhance your networking and computer skills. If you are not willing to continuously hone your skills, then you should choose another, less demanding field of employment.

Employment

The three most important factors that will determine your ability to secure a position with an employer are your job history, technical expertise, and the job interview. A job history tells the employer a lot about your future as an employee, even if your job experiences are unrelated to the job for which you are applying. You may be just entering the IT profession. A solid recommendation from a past employer can make the difference. For example, if an applicant that has worked part-time while going to school for the last two years and a recommendation from their past employer shows a proven record of dependability, this applicant will have a great advantage. A person with no work experience leaves this aspect unknown.

Your training and work experiences show your technical expertise. If you have prior technical experiences, this can prove to be a real asset. Another applicant may have no technical employment history or any technical training. They may have simple, informal experience with a computer at home or have helped friends, and now they believe they can handle the job. This is where your training, experience, and especially your certification will put you ahead of the others.

Job Search Ideas

Finding a job can be a time-consuming and difficult task. The Internet has the most up-to-date information regarding job searches, **Figure 22-6.** Almost every computer-related site has a section devoted to career opportunities. You can often complete an application online. There are also many Web sites that will allow you to post a resume.

Figure 22-6.
The Internet is a great place to conduct a job search. Many sites have search engines that allow you to enter a key word, job category, and location.

The Occupational Outlook Handbook also has tips for conducting a job search. Start by talking to your parents, neighbors, teachers, or guidance counselors. These people may know of job openings that have not been advertised. Read the classified advertisements in the newspaper, especially in the Sunday editions. Look through the Yellow Pages to generate a list of local companies and their addresses and phone numbers. Companies are grouped according to industry in the Yellow Pages. City, county, or state employment services may provide useful job leads. Private employment agencies might also provide leads, but they often charge a fee for a job placement.

Preparing a Resume

On your resume, list all education and training you have had, in sequence. List all past employment in sequence, and do not leave gaps in your employment history. If you stopped your employment to go to school, note that in your employment history. This makes sure your potential employer does not think you are leaving out an employer who may not give you a good reference. List any reason there may be a break in employment. Additional material that is optional as a part of your resume can include copies of certifications, transcripts, or letters of reference. Typically, these will be verified if listed on your resume. However, you should not include more than a few. Employers do not appreciate lengthy resumes.

A quality resume can make a great impression on an employer. Use a good-quality paper and a cover sheet. Make sure you have produced a good-quality copy that they may keep. Never give a potential employer a poor-quality copy.

Preparing for the Interview

Always learn about the company or prospective employer before your interview. Many times, this information is available on their Web site. This preparation lets the employer know you are truly interested in their company and that you possess the personal initiative to research and learn.

Have a specific job or jobs in mind, generally at an entry level. Most companies do not begin a new employee in a high-level job until they have proven themselves to the company. Remember, in the networking field, you will have access to the highest level of company security. A trustworthy employee is critical.

Review your qualifications for the job. Make sure your qualifications match those desired by the employer. Do not waste their time or yours by interviewing for a job that is far beyond your level. If you are an entry-level technician, do not apply for a job as a network administrator.

Prepare to answer broad questions about yourself. It is wise to practice interviewing with someone who has knowledge about job interviews. A family member or friend who regularly does hiring for a company can be a great help, even if the hiring they do is not related to the field you are seeking. Practicing the interview will help you learn to control your natural nervousness.

Arrive at least 15 minutes prior to the scheduled time of your interview, **Figure 22-7.** Showing up late for your interview does not enhance your prospects for the job. It displays a lack of care for the job, the company, and the interviewer. Locate the building in advance. Do a practice run so that you will know exactly how to get to the interview and how long it will take to get there. Consider the traffic conditions for that time of day. If you really want that job interview, have a backup plan in case you have difficulty with your transportation.

Figure 22-7.
Always arrive fifteen minutes early for your interview. To predict how long it will take you to arrive at the interview site, make a practice run a day or two before. Be sure to take into consideration traffic conditions for the time of day in which your interview is scheduled.

Personal Appearance at the Interview

Additionally, you should pay close attention to your physical appearance during a job interview. While some IT jobs allow a casual appearance, most employers require a more professional appearance. This is because technicians and administrators often interact with professionals within the company. Your neat, clean, and professional appearance shows that you take pride in yourself and in your work. Even if the job is a backroom position, always dress appropriately for the interview. Blue jeans and a T-shirt are never appropriate.

Regardless of the job conditions, men and women should always dress up rather than down for an interview. This does not mean that you should dress for a party. Rather, you should dress for a formal business setting. Men should always wear a tie and a long-sleeved shirt at minimum. If you wear a suit or jacket, make sure it matches and fits properly. Women should always wear a suit or business dress.

Smile and use a firm handshake when you introduce yourself, **Figure 22-8.** This shows your confidence. Do not chew gum, eat candy, or smoke at anytime when you are on the company premises. This is not a social visit. You may encounter your prospective supervisor on the premises prior to your interview.

Figure 22-8.
A neat appearance, good posture, and a firm handshake demonstrate self-confidence.

Information to Bring to an Interview

The common information required at an interview is social security number, driver's license number, and a copy of your resume. It is customary to provide references for a job. The quality of your references can mean a great deal to the employer. Some good references are teachers, past employers, supervisors, and fellow employees. Your immediate past employer will be required as a reference. Friends, family, and clergy are not considered good references. Get permission from people before using them as a reference. For each reference, provide a name, address, telephone number, and occupation. Also, note if the person was a past supervisor. You will need this information to complete a job application, which is typically required by the company's personnel office.

Job applications vary in appearance but are designed to gather the same basic information. Besides routine information, such as your name, address, telephone number, and social security number, the application typically requires information about your references and work experiences. It will also ask about police records or any crimes for which you may have been convicted. The application may ask you to write a short summary of why you want the position.

Whenever you fill out the job application, do your best to make it legible to the reader, **Figure 22-9.** If the job application is not legible, it will often not be reviewed very thoroughly. In fact, some personnel offices will not even consider an application that is barely legible.

The Job Interview

The major factor that determines your employment is the job interview. The job interview gives the employer a chance to assess the applicant both physically and mentally with a series of questions. The way a question is answered is at times more important than the answer itself. For example, if an employer asks: "What would you do if you could not fix a networking problem?" The way you answer that question may tell the employer about your character, your confidence, and your ability to work with others. The employer is looking for certain traits in the individual they are about to hire. Some common traits are honesty, confidence, dependability, and the ability to work well as a team member.

Employers many not ask the same questions for all applicants and will usually not be direct about the qualities they are seeking. The employer will ask questions to probe for the character and job-related qualities they want. For

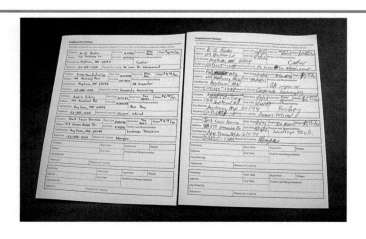

Figure 22-9.
Write neatly (application on the left) when filling out employment applications or taking tests. Poor penmanship (application on the right) can make a bad impression on a prospective employer.

example, an employer may ask you to describe a time you had a problem with a fellow employee and how it was resolved. The answer to this question can tell an experienced interviewer a lot about the character of the applicant.

Once you have secured an interview, it is important to be prepared for it. Read the following tips for a successful interview:

- *Answer all questions to the best of your ability.* If you do not know the answer, simply say so. Do not try to devise an answer. Express your willingness to learn any new topics with which you may not be familiar. The person conducting the interview is an expert. You will not fool them by trying to invent an answer, and you will find they will most likely respect your honesty.

- *Use proper English and avoid slang.*

- *Speak slowly and concisely.*

- *Use good manners.* Always address the persons who are conducting the interview as "Sir" and "Ma'am." Even if you are personally acquainted with your prospective employer, treat the employer as if he or she is a total stranger. Do not become complacent or presume you have the job, and never use foul language, even in a joking manner.

- *Convey a sense of cooperation and enthusiasm.* Your body language will convey a lot about your personality. Keep smiling.

- *Remember that the interview has not ended when you start asking questions.* As a matter of fact, your questions can reveal even more to the employer. Asking about how many breaks you will get during a day will send an undesirable message to the employer. It is not necessary to ask questions, especially if the interview has been thorough. However, do not hesitate if you believe there is pertinent information you must know. You may ask questions about the position and the organization, but limit your questions to operation conditions, which you do not understand. Many issues regarding a job can be obtained prior to the interview, especially if the information was posted.

- *Do not ask questions regarding salary unless it has not been covered in a job posting, uncovered through your research, or discussed by your prospective employer during the interview.* Remember, if you are offered the job, there will be additional time at that point to make any clarifications or salary negotiations that you believe are necessary.

Testing at the Interview

Employer testing is a common part of the job interview process. An employer can tell a lot about your technical knowledge and communication skills through a test, especially if handwritten answers are required. Written responses reveal a lot about an interviewee. When taking a test, be sure you understand all test directions. Verbally confirm them if you have questions before you begin. The following guidelines will improve your testing success:

- Read each question carefully.

- Write legibly and clearly. If your handwriting is poor, print instead.

- Budget your time wisely. Don't dwell on one question.

Summary

- An entrepreneur owns and operates a business.
- There is a direct correlation between the amount of education for a specific field and salary.
- An information technology (IT) position requires continuous education.
- Recognized commercial certifications are a way of proving your expertise in most computer technology specialties, including networking.
- After gaining A+ Certification, Network+ Certification is a good, next step in your career advancement.
- CompTIA offers A+, Network+, CDIA+, i-Net+, Server+, Linux+, and many other certifications.
- Microsoft, Novell, Cisco, and many other companies offer certifications based on exams regarding their respective operating systems and hardware.
- You must continually update your knowledge and skills to remain vital to the IT industry.
- When you are interviewing for a new job, be sure to take your social security card, driver's license, and a good-quality copy of your resume.
- Be able to produce a list of at least three references, including addresses and phone numbers. Make sure you obtain permission to use associates as references.
- Be on time or 15 minutes early for your interview.
- Smile and show a sense of cooperation and enthusiasm when interviewed.
- If tested by the prospective employer, be sure you understand the instructions for the test.

Review Questions

Answer the following questions on a separate sheet of paper. Please do not write in this book.

1. What is an entrepreneur?
2. Where can objective data be accessed for labor statistical information in the USA?
3. What is the correlation between education and salaries?
4. What certification do you think you wish to achieve next?
5. Why is continuing education required for an employee in the IT industry?
6. What items should you bring to a job interview?
7. What is needed concerning your references?

Suggested Laboratory Activities

1. Research a possible certification you wish to obtain. List the approximate amount of time required to prepare for the certification. Find out the cost of the certification's exam(s). Download a list of the knowledge areas, or objectives, required for the certification. Identify the type of job associated with the certification.

2. Research college degree IT programs. List the classes required for the degree. Identify and describe the type of job associated with the degree. What is the total cost of earning the degree, credit hour cost, and approximate textbook costs? What will the approximate cost be to support yourself during the period of time you are in school? This cost includes rent, utilities, food, clothing, transportation, and miscellaneous expenses.

3. Prepare a resume.

4. Use the Internet to locate three different job descriptions for each of the following jobs. Make a composite of job skills required for each.
   - Web master.
   - Network administrator.
   - Network support specialist.
   - Help desk technician.

Interesting Web Sites for More Information

www.bls.gov

www.computerworld.com/careertopics/careers

www.networkmagazine.com

www.techcareers.informationweek.com

Appendix A

List of Acronyms

A

ABR	Available Bit Rate
ACL	access control list
ADC	Analog to Digital Converter
AH	Authentication Header
ANSI	American National Standards Institute
APIPA	Automatic Private IP Addressing
ARP	Address Resolution Protocol
ASCII	American Standard Code for Information Interchange
ATM	Asynchronous Transfer Mode

B

BAP	Bandwidth Allocation Protocol
BCD	Binary Coded Decimal
BDC	backup domain controller
B-ISDN	Broadband ISDN
BOOTP	Bootstrap Protocol
BRI-ISDN	Basic Rate ISDN

C

CA	Certificate Authority

CBR	Constant Bit Rate
CERN	*Centre European pour la Recherche Nucléaire*
CHAP	Challenge Handshake Authentication Protocol
CIR	Committed Information Rate
CP	Consolidation Point
CRC	Cyclic Redundancy Check
CSMA/CD	Carrier Sense Multiple Access with Collision Detection
CSU/DSU	Channel Service Unit/Data Service Unit
CUPS	Common UNIX Printing System

D

DAC	Digital to Analog Converter
DHCP	Dynamic Host Configuration Protocol
DMA	Direct Memory Access
DNS	Domain Name System
DoS	Denial of Service
DSL	Digital Subscriber Line

E

EAP	Extensible Authentication Protocol

EBCDIC	Extended Binary Coded Decimal Interchange Code
EIA	Electronic Industry Association
ELFEXT	Equal Level Far-End Crosstalk
ESP	Encapsulation Security Payload

F

FDDI	Fiber Distributed Data Interface
FEXT	Far-End Crosstalk
FIR	Fast Infrared
FLSM	Fixed Length Subnet Mask
FQDN	Fully Qualified Domain Name

G

GUI	graphical user interface

H

HAL	Hardware Abstract Layer
HTML	Hypertext Markup Language
HTTP	Hypertext Transfer Protocol

I

I/O	Input/Output
ICANN	Internet Corporation for Assigned Names and Numbers

IEEE	Institute of Electrical and Electronic Engineers
IGMP	Internet Group Management Protocol
IMAP	Internet Message Access Protocol
InterNIC	Integrated Network Information Center
IPSec	IP Security
IPX/SPX	Internetwork Packet Exchange/Sequenced Packet Exchange
IRQ	interrupt request
ISDN	Integrated Services Digital Network
ISO	International Organization for Standardization
ITU	International Telecommunications Union

L

L2F	Layer 2 Forwarding
L2TP	Layer 2 Tunneling Protocol
LAN	local area network
LDR	Logical Disk Manager
LEC	Local Exchange Carrier
LSA	Local Security Authority
LUN	logical unit number

M

MAC	media access code *also* Media Access Control
MAN	metropolitan area network
MAU	multistation access unit

MBR	master boot record	
MIME	Multipurpose Internet Mail Extensions	
MITM	man in the middle	
MLPPP	Multilink Point-to-Point Protocol	
MPEG	Moving Picture Experts Group	
MS-CHAP	Microsoft Challenge Handshake Authentication Protocol	
MSR	Microsoft Reserved	
MUTOA	Multi-User Telecommunication Outlet Assembly	

N

NAS	network attached storage
NAT	Network Address Translation
NetBEUI	NetBIOS Enhanced User Interface
NetBIOS	Network Basic Input/Output System
NEXT	Near-End Crosstalk
NIC	network interface card
NNTP	Network News Transfer Protocol
NOS	network operating system
NRZ	non-return to zero
NSS	Novell Storage Services
NTFS	New Technology File System
NWFS	NetWare File System

O

OFDM	orthogonal frequency-division multiplexing

OSI	Open Systems Interconnection
OSPF	Open Shortest Path First
OTDR	Optical Time Domain Reflectometer

P

PAP	Password Authentication Protocol
PDC	primary domain controller
PoP	Point of Presence
POP	Post Office Protocol
POST	power-on self-test
POTS	Plain Old Telephone Service
PPP	Point-to-Point Protocol
PPPoE	Point-to-Point Protocol over Ethernet
PPTP	Point-to-Point Tunneling Protocol
PRI-ISDN	Primary Rate ISDN
PSTN	Public Switched Telephone Network
PVC	permanent virtual circuit

Q

QoS	Quality of Service

R

RADIUS	Remote Authentication Dial-In User Service
RAID	Redundant Array of Independent Disks
RARP	Reverse Address Resolution Protocol
RDP	Remote Desktop Protocol
RIP	Routing Information Protocol

S

SAM	Security Account Manager
SAN	storage area network
SAP	Service Access Point
SCP	Secure Copy Protocol
SCSI	Small Computer Systems Interface
SFTP	Secure File Transfer Protocol
S-HTTP	Secure HTTP
SLIP	Serial Line Internet Protocol
SMTP	Simple Mail Transfer Protocol
SNMP	Simple Network Management Protocol
SOAP	Simple Object Access Protocol
SSH	Secure Shell
SSL	Secure Sockets Layer

T

TCP/IP	Transmission Control Protocol/Internet Protocol
TFTP	Trivial File Transfer Protocol
TIA	Telecommunications Industry Association
TTL	Time to Live

U

UBR	Unspecified Bit Rate
UL	Underwriters Laboratories
UNC	Universal Naming Convention
UPS	Uninterruptible Power Supply
URL	Uniform Resource Locator

V

VBR	Variable Bit Rate
VBR-NRT	Variable Bit Rate-Non Real Time
VBR-RT	Variable Bit Rate-Real Time
VLAN	Virtual LAN
VLSM	Variable Length Subnet Mask
VoIP	Voice over IP
VPN	Virtual Private Network

W

W3C	World Wide Web Consortium
WAN	wide area network
WAP	Wireless Access Point
WAP	Wireless Application Protocol
WEP	Wired Equivalent Privacy
WINS	Windows Internet Naming Service
WPA	Wi-Fi Protected Access

X

XML	Extensible Markup Language

Appendix B

Binary Math

Binary math accurately represents digital circuitry. In digital electronics, a circuit is either on or off or a voltage condition is high or low. For example, a digital circuit may have two distinct conditions: 5 volts present or 0 volts present.

Binary math uses only two numbers, *1* and *0*, to represent an infinite range of numbers. The binary number system accomplishes this in basically the same way the decimal number system does, by placing numbers into discrete, digit positions. The decimal number system fills these digits with values 0 through 9. The first digit position is commonly referred to as the 1s. The maximum value that can be entered here is 9. The second digit position must therefore be the 10s. If the maximum value of 9 is entered in both the 10s position and the 1s position, the resulting number is 99. The third position must therefore be the 100s position, and so on. Notice in **Figure B-1** that each of the positions can be expressed as an exponent of the base 10.

Digit Positions	1s	10s	100s	1,000s	10,000s	100,000s
Exponent	10^0	10^1	10^2	10^3	10^4	10^5
Range	0–9	10–90	100–900	1,000–9,000	10,000–90,000	100,000–900,000

Figure B-1.

Note:
The Range row indicates the allowed values for each digit position.

Look at the number 2,753. This number includes the following:

2-1000s	2,000
7-100s	700
5-10s	50
3-1s	3
Total	2,753

Add the values together for a total of 2,753.

Binary numbers are expressed in similar fashion. However, instead of each digit position being 10 times greater than the position before it, the value of each position is double that of the position before it. See **Figure B-2.**

Digit Positions	1s	2s	4s	8s	16s	32s	64s	128s
Exponent	2^0	2^1	2^2	2^3	2^4	2^5	2^6	2^7
Value	0	2	4	8	16	32	64	128

Figure B-2.

Look at the binary number 101011000001 and compare it to **Figure B-3.** This number includes the following:

1-2048s	2,048
0-1024s	0
1-512s	512
0-256s	0
1-128s	128
1-64s	64
0-32s	0
0-16s	0
0-8s	0
0-4s	0
0-2s	0
1-1s	1
Total	2,753

2048s	1024s	512s	256s	128s	64s	32s	16s	8s	4s	2s	1s
1	0	1	0	1	1	0	0	0	0	0	1

Figure B-3.

As you can see, 101011000001 is the binary equivalent of 2,753.

To convert a binary number to a decimal number, simply record the decimal value assigned to the position for each position that contains a binary number 1 and then add the decimal numbers together. See **Figure B-4.**

32s	16s	8s	4s	2s	1s
1	0	1	1	0	1
32	+	8 +	4	+	1 = 45

Figure B-4.

When converting a decimal number to a binary number, you simply reverse the previous operation. For example, to convert the decimal number 178 to a binary number, you must divide it by a series of "powers of 2." The "powers of 2" are 1, 2, 4, 8, 16, 32, 64, 128, 256, 512, 1,024, 2,048, and so on.

To convert the decimal number 178 to binary, start by finding the largest "power of 2" that does not exceed 178. The largest "power of 2" value that does not exceed 178 is 128 (2^7). Place a 1 in the binary number position that represents 128. See **Figure B-5.**

Powers of 2	128	64	32	16	8	4	2	1
Binary Digit	1							

Figure B-5.

Subtracting 128 from 178 leaves 50. The number 50 is less than 64 (2^6), the next smaller "power of 2." Therefore, you must insert a 0 in the 64s position, as seen in **Figure B-6.**

Powers of 2	128	64	32	16	8	4	2	1
Binary Digit	1	0						

Figure B-6.

Next, 50 is larger than 32 (2^5), so place a 1 in the 32s position. See **Figure B-7.**

Powers of 2	128	64	32	16	8	4	2	1
Binary Digit	1	0	1					

Figure B-7.

Now, subtract 32 from 50. The difference is 18. The number 18 is larger than 16 (2^4), so place a 1 in the 16s position. See **Figure B-8.**

Powers of 2	128	64	32	16	8	4	2	1
Binary Digit	1	0	1	1				

Figure B-8.

Subtracting 16 from 18 leaves 2. The number 2 is smaller than the next two "powers of 2," 8 (2^3) and 4 (2^2). That means the next two positions in the binary number are both 0s. See **Figure B-9.**

Powers of 2	128	64	32	16	8	4	2	1
Binary Digit	1	0	1	1	0	0		

Figure B-9.

The next "power of 2" is 2 (2^1), and the number remaining from the last step is also 2. Therefore, a 1 goes into the 2s position. See **Figure B-10.**

Powers of 2	128	64	32	16	8	4	2	1
Binary Digit	1	0	1	1	0	0	1	

Figure B-10.

There are no decimal numbers remaining, so the 1s position should be filled with a 0. The binary equivalent of the decimal number 178 is 10110010, **Figure B-11.**

Powers of 2	128	64	32	16	8	4	2	1
Binary Digit	1	0	1	1	0	0	1	0

Figure B-11.

Appendix C

Number Conversion Table			
Decimal	Binary	Octal	Hexadecimal
0	000000	0	0
1	000001	1	1
2	000010	2	2
3	000011	3	3
4	000100	4	4
5	000101	5	5
6	000110	6	6
7	000111	7	7
8	001000	10	8
9	001001	11	9
10	001010	12	A
11	001011	13	B
12	001100	14	C
13	001101	15	D
14	001110	16	E
15	001111	17	F
16	010000	20	10
17	010001	21	11
18	010010	22	12
19	010011	23	13
20	010100	24	14
21	010101	25	15
22	010110	26	16
23	010111	27	17
24	011000	30	18
25	011001	31	19
26	011010	32	1A
27	011011	33	1B
28	011100	34	1C
29	011101	35	1D
30	011110	36	1E

Number Conversion Table (continued)			
Decimal	Binary	Octal	Hexadecimal
31	011111	37	1F
32	100000	40	20
33	100001	41	21
34	100010	42	22
35	100011	43	23
36	100100	44	24
37	100101	45	25
38	100110	46	26
39	100111	47	27
40	101000	50	28
41	101001	51	29
42	101010	52	2A
43	101011	53	2B
44	101100	54	2C
45	101101	55	2D
46	101110	56	2E
47	101111	57	2F
48	110000	60	30
49	110001	61	31
50	110010	62	32
51	110011	63	33
52	110100	64	34
53	110101	65	35
54	110110	66	36
55	110111	67	37
56	111000	70	38
57	111001	71	39
58	111010	72	3A
59	111011	73	3B
60	111100	74	3C
61	111101	75	3D
62	111110	76	3E
63	111111	77	3F

Appendix D

Table of Standard ASCII Characters				(Continued)			(Continued)	
0	NUL	Null	43	+		86	V	
1	SOH	Start of header	44	,		87	W	
2	STX	Start of text	45	-		88	X	
3	ETX	End of text	46	.		89	Y	
4	EOT	End of transmission	47	/		90	Z	
5	ENQ	Enquiry	48	0		91	[	
6	ACK	Acknowledgment	49	1		92	\	
7	BEL	Bell	50	2		93	]	
8	BS	Backspace	51	3		94	^	
9	HT	Horizontal tab	52	4		95	_	
10	LF	Line feed	53	5		96	`	
11	VT	Vertical tab	54	6		97	a	
12	FF	Form feed	55	7		98	b	
13	CR	Carriage return	56	8		99	c	
14	SO	Shift out	57	9		100	d	
15	SI	Shift in	58	:		101	e	
16	DLE	Data link escape	59	;		102	f	
17	DC1	Device control 1	60	<		103	g	
18	DC2	Device control 2	61	=		104	h	
19	DC3	Device control 3	62	>		105	i	
20	DC4	Device control 4	63	?		106	j	
21	NAK	Negative acknowledgment	64	@		107	k	
22	SYN	Synchronous idle	65	A		108	l	
23	ETB	End of transmit block	66	B		109	m	
24	CAN	Cancel	67	C		110	n	
25	EM	End of medium	68	D		111	o	
26	SUB	Substitute	69	E		112	p	
27	ESC	Escape	70	F		113	q	
28	FS	File separator	71	G		114	r	
29	GS	Group separator	72	H		115	s	
30	RS	Record separator	73	I		116	t	
31	US	Unit separator	74	J		117	u	
32	SP	Space	75	K		118	v	
33	!		76	L		119	w	
34	"		77	M		120	x	
35	#		78	N		121	y	
36	$		79	O		122	z	
37	%		80	P		123	{	
38	&		81	Q		124			
39	'		82	R		125	}	
40	(		83	S		126	~	
41	)		84	T		127	DEL	
42	*		85	U				

Appendix E

Protocol Family Encapsulations

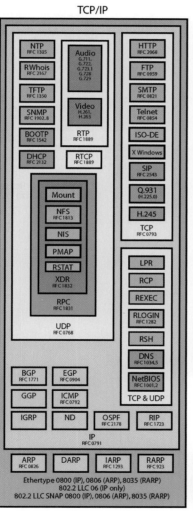
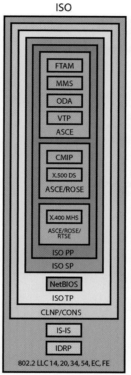
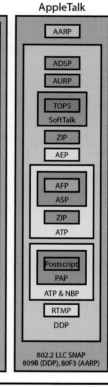
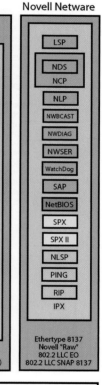

WildPackets, Inc.
Reproduced with Permission

AARP	AppleTalk Address Resolution Protocol
ADSP	AppleTalk Data Stream Protocol
AEP	AppleTalk Echo Protocol
AFP	AppleTalk Filing Protocol
ARP	Address Resolution Protocol
ASCE	Association Control Service Element
ASP	AppleTalk Session Protocol
ATP	AppleTalk Transaction Protocol
AURP	AppleTalk Update Routing Protocol
BGP	Border Gateway Protocol
BOOTP	Boot Protocol
CLNP	Connectionless Network Protocol
CMIP	Common Management Information Protocol
CONS	Connection Oriented Network Protocol
DARP	Dynamic Address Resolution Protocol
DDP	Datagram Delivery Protocol
DHCP	Dynamic Host Configuration Protocol
EGP	Exterior Gateway Protocol
FTAM	File Transfer Access and Management
FTP	File Transfer Protocol
GGP	Gateway to Gateway Protocol
HTTP	HyperText Transfer Protocol
IARP	Inverse Address Resolution Protocol
ICMP	Internet Control Message Protocol
IDRP	Interdomain Routing Protocol
IGRP	Interior Gateway Routing Protocol
IP	Internet Protocol
IPX	Internet Packet Exchange
IS-IS	Intermediate System to Intermediate System Routing
ISO PP	Presentation Protocol
ISO SP	Session Protocol
ISO TP	Transport Protocol
ISO-DE	Development Environment
LPR	Remote Print
LSP	NetWare Lite Sideband Protocol
MMS	Manufacturing Message Service
Mount	Mount Server
NBP	Name Binding Protocol
NCP	NetWare Core Protocol
ND	Network Disk
NDS	NetWare Directory Service
NetBEUI	NetBIOS Enhanced User Interface
NetBIOS	Network Basic Input/Output System
NFS	Network File System
NIS	Network Information Services
NLP	NetWare Lite Protocol
NLSP	NetWare Link State Protocol
NTP	Network Time Protocol
NWBCAST	NetWare Broadcast Message Notification
NWDIAG	NetWare Diagnostic Support Protocol
NWSER	NetWare Serialization Protocol
ODA	Office Document Architecture
OSPF	Open Shortest Path First
PAP	Printer Access Protocol
PING	Packet Internet Groper
PMAP	Port Mapper
RARP	Reverse Address Resolution Protocol
RCP	Remote Copy
REXEC	Remote Execution
RIP	Routing Information Protocol
RLOGIN	Remote Login
ROSE	Remote Operations Service Element
RPC	Remote Procedure Call
RSH	Remote Shell
RSTAT	Remote Statistics
RTCP	RTP Control Protocol
RTMP	Routing Table Maintenance Protocol
RTP	Real-time Transport Protocol
RTSE	Reliable Transfer Service Element
RWhois	Remote Whois
SAP	Service Advertisement Protocol
SIP	Session Initialization Protocol
SMB	Server Message Block
SMTP	Simple Mail Transfer Protocol
SNMP	Simple Network Management Protocol
SPX	Sequenced Packet Exchange
TCP	Transmission Control Protocol
Telnet	Telecommunications Network Protocol
TFTP	Trivial File Transfer Protocol
TOPS	Transcendental Operating System
UDP	User Datagram Protocol
VTP	Virtual Terminal Protocol
WatchDog	Station Keep Alive
X Windows	Graphical Windows
X.400 MHS	Message Handling System
X.500 DS	Directory Services
XDR	Exchange Data Representative Protocol
ZIP	Zone Information Protocol

Glossary

1000BaseCX: an IEEE 802.3 classification that specifies the use of Category 5 cable, a data rate of 1000 Mbps, a maximum segment length of 25 m, a minimum segment length of .6 m, and a star topology.

1000BaseT: an IEEE 802.3 classification that specifies the use of Category 5 cable, a data rate of 1000 Mbps, a maximum segment length of 100 m, a minimum segment length of .6 m, and a star topology.

100BaseT4: an IEEE 802.3 classification that specifies the use of Category 3, 4, and 5 cable, a data rate of 100 Mbps, a maximum segment length of 100 m, a minimum segment length of .6 m, and a star topology.

100BaseTX: an IEEE 802.3 classification that specifies the use of Category 5 cable, a data rate of 100 Mbps, a maximum segment length of 100 m, a minimum segment length of .6 m, and a star topology.

10Base2: an IEEE 802.3 classification that specifies the use of RG-58 cable, a data rate of 10 Mbps, a maximum segment length of 185 m, a minimum segment length of .5 m, and a bus topology.

10Base5: an IEEE 802.3 classification that specifies the use of RG-8 cable, a data rate of 10 Mbps, a maximum segment length of 500 m, a minimum segment length of 2.5 m, and a bus topology.

10BaseT: an IEEE 802.3 classification that specifies the use of Category 3, 4, or 5 cable, a data rate of 10 Mbps, a maximum segment length of 100 m, a minimum segment length of .6 m, and a star topology.

A

A+ Certification: a CompTIA certification that proves a person's knowledge of PC repair, hardware, and software.

access control list (ACL): a list of users, groups, and permissions that are associated with a resource.

access method: a method of gaining access to the network media.

access port: a connection into which network cables plug.

active hub: a network device that acts as a central connection point for network cables and regenerates digital signals like a repeater. The active hub can also determine whether a packet should remain in the isolated section of the network or pass the packet through the hub to another section of the network.

active monitor: a computer that is responsible for monitoring the necessary administrative functions associated with Token Ring technology.

active partition: a partition that contains the operating system files the computer should use to boot.

ad hoc mode: a wireless network that does not contain a Wireless Access Point (WAP).

Address Resolution Protocol (ARP): a service that maps a MAC address to an IP address.

administrative server: a server that is used to administer network security and activities.

American National Standards Institute (ANSI): a private, nonprofit organization that does not develop standards, but rather prompts voluntary conformity and standardization.

American Standard Code for Information Interchange (ASCII): a character code that uses eight bits to represent alphanumeric characters.

amplifier: an electronic device designed to raise a signal's amplitude.

amplitude: the maximum voltage, or height, of an electronic signal.

analog signal: an electronic signal that varies in values.

Analog to Digital Converter (ADC): a computer chip designed to change an analog signal to a digital signal.

antivirus suite: an antivirus software package that includes additional features.

AppleTalk: a suite of protocols developed to support the Macintosh computer when configured as a network. AppleTalk uses CSMA/CA as its access method.

application gateway: a hardware device or software package that provides security for specific applications by accepting traffic based on the exact match of the application permitted.

application layer: the highest layer of the OSI model that starts network data communications using a program such as e-mail or a Web browser.

archive bit: a file attribute that identifies if a file's contents has changed since the last full or incremental backup.

asymmetric-key encryption: a key encryption method in which the sender and the receiver use two different keys: a private key and a public key.

Asynchronous Transfer Mode (ATM): a protocol especially designed for carrying audio, video, and multimedia. It can support a bandwidth of 622 Mbps.

asynchronous transmission: a type of transmission in which a digital signal is not synchronized with a reference signal.

ATM (Asynchronous Transfer Mode): a protocol designed especially for transmitting data, voice, and video.

attenuation: the loss of signal strength.

auditing: a service that tracks the events, use, and access of network resources and writes these actions to a log.

authentication: the process of identifying a user and ensuring the user is who they claim to be. Also, the process of verifying a user's identity and allowing the user access to the network.

Authentication Header (AH): an IPSec protocol that provides authentication and anti-replay but does not encrypt the data payload inside the packet.

automatic private IP addressing (APIPA): a feature that assigns an IP address from a special set of IP addresses that have been set aside by IANA when a DHCP server cannot be reached.

Available Bit Rate (ABR): an ATM data transfer classification most appropriate for file transfer. It uses the available bit rate associated with the networking medium. The speed of a file transfer using the ABR classification is affected by the amount of traffic on the network system.

average utilization: the average amount of traffic throughout the network in the time period monitored.

AWG rating: a rating that describes the size of a conductor's diameter.

B

backbone: a cable that is located between the telecommunications closets, equipment rooms, and main entrance facility. The backbone cable connects these areas and does not serve individual workstations. A backbone can run horizontally and vertically through a building. Also, a cable that serves as a common path and often employs high-speed network cable such as fiber-optic.

backdoor: a software access port to a computer a Trojan horse has infected.

backdoor virus: a type of malware designed to go undetected and create a backdoor on a computer.

backplane: a simple motherboard designed with minimal components. It typically serves as an interface of all major components in the system.

backup domain controllers (BDC): a domain controller that stores backup copies of the user account and security database. A BDC can perform authentication and can be promoted to a PDC in case the PDC fails.

bandwidth: a measurement of the network media's ability to carry data.

Bandwidth Allocation Protocol (BAP): a protocol that can change the number of lines or channels used in communication according to current bandwidth.

Baseband: a method of transmitting data in the form of a digital signal, using the entire bandwidth of a cable.

baseline: a measurement of performance characteristics that can be used at a later date to objectively determine if the network or server is performing satisfactory.

basic disk: the term applied to the old system of hard disk drive configuration, which includes creating primary and extended partitions.

Basic Rate ISDN (BRI-ISDN): a category of ISDN that consists of three conductors: two B channels, referred to as bearer channels, and one D channel, referred to as the delta channel.

beaconing: a recovery process used when a hardware failure occurs on the ring.

Binary Coded Decimal (BCD): a binary number format where each number is represented as a four-digit binary code.

biometrics: the science of using the unique physical features of a person to confirm their identification for authentication purposes.

bipolar digital signal: a digital signal that fluctuates between a positive five-volt level and a negative five-volt level.

bit rate: the number of bits used to represent the amplitude of an analog signal.

blackout: a total loss of electrical energy, or voltage.

blade server: a powerful server that is extremely thin. It is designed to be mounted in a small space with other blade servers.

boot partition: a partition that contains the files needed to operate the computer. These files include device drivers, library files, services, commands, and utilities and are stored in a separate directory.

bootloader: a program that starts the operating system load process.

Bootstrap Protocol (BOOTP): a service that uses a centralized database of the MAC addresses and IP assignments of all devices on the network and assigns the appropriate IP address to a host when it boots.

bridge: a network device that divides the network into smaller segments, reducing the chance of collisions.

Broadband: a method of transmitting data in the form of several analog signals at the same time.

Broadband ISDN (B-ISDN): a category of ISDN that is designed to carry multiple frequencies.

broadcast frame: a frame intended for every computer on the network to read.

broadcast storm: a condition in which a network is flooded with a continuous number of collisions and rebroadcasts.

brouter: a network device that combines router and bridge functions.

brownout: a partial loss of electrical energy, or voltage, in which the voltage level is lower than normal or average levels.

browser-based management utility: a management utility typically associated with, but not limited to, a thin client. The thin client connects to the network device and then runs scripts hosted on the network device, not on the thin client.

buffer: a small amount of memory used to temporarily store data.

bug: a software program error.

bus topology: a topology that uses a single cable or conductor to connect all nodes on a network.

business plan: a document that outlines the goals for the business and includes an action plan and a timetable for meeting those goals.

C

Carrier Sense Multiple Access with Collision Detection (CSMA/CD): a set of rules that define how two or more devices will access network media and how they respond after a collision occurs.

carrier wave: an electromagnetic wave of a set frequency that is used to carry data.

cell: the area that is served by a radio access tower.

cellular technology: a technology based on radio waves connecting to designated areas referred to as cells.

centralized administration: the methodology used to administer a client/server network.

CERN: an organization that is responsible for the original development of the World Wide Web.

Certificate Authority (CA): a service that contains a security list of users authorized to access the private key owner's messages, using a public key.

Certified Document Imaging Architect (CDIA+) Certification: a CompTIA certification that proves a person's knowledge of imaging systems such as scanners, displays, printers, graphic file types, file conversion, and image enhancement.

Certified Linux Engineer (CLE) Certification: a Novell certification that verifies expertise in both Linux and Novell services that run on a Linux system.

Certified Linux Professional (CLP) Certification: a Novell certification designed for people who wish to prove they have mastered the basic skills required to manage and maintain a SuSE Linux server. The skills measured are installing a SuSE Linux server, managing users and groups, troubleshooting the SuSE Linux file system, and managing and compiling the Linux kernel.

Certified Novell Administrator (CNA) Certification: a Novell certification that proves a person's ability to set up and manage workstations and to manage network system resources such as files, printers, and software in a NetWare network.

Certified Novell Engineer (CNE) Certification: a Novell certification that proves a person's ability in LAN design, development, implementation, support, and troubleshooting of a NetWare network.

Challenge Handshake Authentication Protocol (CHAP): an authentication protocol that sends an encrypted string of characters that represent the user name and password. It does not send the actual user name and password.

channel: the bandwidth of a carrier wave.

Channel Service Unit/Data Service Unit (CSU/DSU): electronic circuitry that converts signals from the LAN to signals that can be carried by the T1 line, and vice versa.

ciphertext: encrypted data.

circuit switching: a type of transmission which establishes a permanent connection between two points for the duration of the data transfer period.

circuit-level gateway: a hardware device or software package that monitors a connection until it is successfully established between the destination and source host.

Class A network: a TCP/IP network classification that supports up to 16 million hosts on each of 127 networks.

Class B network: a TCP/IP network classification that supports up to 65,000 hosts on each of 16,000 networks.

Class C network: a TCP/IP network classification that supports 254 hosts on each of 2 million networks.

client/server network: a type of network model that consists of computers connected via a network to one or more servers.

cluster: a group of servers that share the network demand.

clustering: a technology that allows up to eight servers to be connected and act as a single server.

coaxial cable: a cable consisting of a copper core conductor surrounded by an insulator.

codec (compressor/decompressor): software, hardware, or a combination of software and hardware that compress and decompress video and audio information.

cold spare: any compatible drive that is in storage and used to replace a failed drive.

collision domain: a section of a network where collisions occur.

command prompt: a text-based interface where commands are typed and entered.

command syntax: the correct manner and arrangement in which a command is to be typed.

Committed Information Rate (CIR): a guaranteed bandwidth a commercial carrier will provide a subscriber.

Common UNIX Printing System (CUPS): a printing system that supports network printing using the Internet Printing Protocol (IPP).

complex trust relationship: a trust relationship in which more than two domains have a full-trust relationship.

concentrator: a network device that serves as a central connection point for network cables.

connectionless communication: a type of connection in which data is transmitted to the destination without first establishing a connection.

connectionless-oriented: a type of network connection in which there is neither a direct connection nor a consistent connection between the destination and the source.

connection-oriented: a type of network connection that maintains a constant and consistent connection between the source and the destination.

connection-oriented communication: a type of connection in which communication is first established between the source and destination computers before data is transmitted.

console: the keyboard and monitor located at the NetWare server.

Consolidation Point (CP): a connection to the horizontal wiring system, which in turn feeds to a wall outlet or a MUTOA.

Constant Bit Rate (CBR): a steady stream of ATM cells moving at a predictable rate. CBR is best used for video conferencing, telephone conversations, television broadcast, or anywhere real-time data transfer is required.

consultant: a person who works with clients on projects and makes recommendations based on their expertise.

context: the location of an object in the NDS tree.

contiguous namespace: a namespace in which the location must use the root domain name as part of its URL or as its complete name.

continuous UPS: a UPS unit that provides a steady supply of electrical energy at all times, even when there is no electrical problem.

copper core cable: an electrical cable that consists of a copper wire surrounded by plastic or synthetic insulation.

cracker: anyone who gains access to a computer system without authorization and with the intent to do harm or play pranks.

crossed pair: a wiring fault in which two pairs within a twisted pair cable have switched positions at the opposite end of the cable.

crosstalk: interference that comes from neighboring conductors inside a wire's insulating jacket.

cryptology: the science of encoding data.

Cyclic Redundancy Check (CRC): a sophisticated data integrity check that uses complicated mathematical algorithms to determine if one or more bits are corrupt.

D

daemon: a program that runs in the background and waits for a client to request its services.

data encryption: the encoding of data based on a mathematical formula, which converts the original data symbol into another symbol.

data link layer: the layer of the OSI model that describes how raw data is packaged for transfer from one network interface card to another.

database server: a server that contains data files and software programs that query the data.

datagram: a short message block that can be sent to a particular computer, sent to a group of computers, or broadcast to all computers connected to the media.

de facto standard: a standard developed because of its widely accepted use by industry.

decentralized administration: the methodology used to administer a peer-to-peer network.

decibel (dB): a unit of measurement used to express the relationship of power between two electrical forces.

dedicated server: a server that serves a single function.

default gateway address: the address of the computer that provides a connection to the Internet.

demodulation: the process of separating a data signal from a carrier wave.

Denial of Service (DoS): denying access to a server by overloading the server with bogus requests.

desktop environment: a software package that includes many applications (X clients) and a windows manager. The desktop environment provides a common look among X clients.

device file: a file that serves as a means of communication between a driver and a device.

dial-up networking: a network system in which a dial-up connection is used to access a remote server.

differential backup: a backup method that backs up all data that has been changed since the last full backup.

digital certificate: a file that commonly contains data such as a user's name and e-mail address, the public key value assigned to the user, the validity period of the public key, and issuing authority identifier information.

digital encoding: the conversion of data into a digital pattern acceptable to the networking media.

digital signal: an electronic signal that has discrete values.

Digital Subscriber Line (DSL): a leased line dedicated to networking that uses multiple frequencies as separate channels on the existing telephone local loop. The multiple channels combine to carry more data than the original telephone modem design.

Digital to Analog Converter (DAC): a computer chip designed to change a digital signal to an analog signal.

Direct Memory Access (DMA) channel: a circuit that allows devices to communicate and transfer data to and from RAM without the need of CPU intervention.

direct sequencing: a spread spectrum technique that transmits data on multiple channels sequentially.

directional: the ability of an antenna to focus or aim an electromagnetic signal in a particular direction.

disaster recovery: the restoration of a system to normal operation after a disaster has occurred.

disjointed namespace: a domain name that is a part of the Active Directory, but does not follow the contiguous namespace requirement.

disk mirroring: the act of writing the same information to two hard disk drives at the same time.

disk quota: the amount of disk space assigned to specific users.

disk striping: a storage technique that divides data into sections and writes the data sections across several hard disk drives at the same time.

diskless station: a workstation that does not have hard disk drives. It can also mean a workstation that does not have a floppy disk drive.

dispersion: the distortion of a light wave as it reflects off the cladding of a fiber-optic cable.

distinguished name: a naming system that uniquely identifies the location of the object in the Active Directory structure.

domain: a logical grouping of users and equipment as defined by the network administrator.

Domain Name System (DNS): a system that associates a host or domain name with an IP address, making it easy to identify and find hosts and networks.

domains: areas of knowledge.

dotted decimal notation: an IP address displayed in decimal form, not binary.

driver: a software program that allows a PC to communicate with and to transfer data to and from computer hardware.

dumb terminal: a type of computer that consists of only a display and keyboard. The keyboard contains the electronic components needed to support communication between the mainframe and the terminal. A dumb terminal does not contain a CPU.

duplexing: a technique of placing each mirrored hard drive on a separate hard disk drive controller.

dynamic addressing: automatically assigning IP addresses.

dynamic disk: a new system of disk configuration and management introduced with NTFS5.0. Dynamic Disks are managed by the system as a group and allow for the creation of five different volume types: simple, spanned, mirrored, RAID-5, and striped.

Dynamic Host Configuration Protocol (DHCP): a service that assigns IP addresses automatically to the hosts on a network.

dynamic IP address table: a table of IP addresses generated by software programs that communicate with nearby routers.

dynamic IP assignment: an IP address that is issued automatically, typically when the computer boots and joins the network.

E

eDirectory: a database of network resources similar to Microsoft's Active Directory.

EFI partition: a partition on a GPT disk that contains all programs required to boot the computer in the same way the BIOS boots the computer.

electrical spike: a very short burst of abnormally high voltage, typically caused by electrical equipment.

electrical surge: a higher-than-normal voltage level, typically caused by lightning.

electromagnetic wave: a form of energy that behaves like a wave and can travel through a vacuum.

Electronic Industry Association (EIA): a trade association that is concerned with radio communications.

encapsulation: the process of adding information to a segment that identifies such things as the source address, the destination address, the end of the segment, and the size of the segment.

Encapsulation Security Payload (ESP): an IPSec protocol that provides authentication and anti-replay and encrypts the payload data inside the packet.

encryption: the method of using an algorithm to encode data.

engineer: a degreed professional who possesses high-level skills necessary to solve problems related to their field of expertise.

entrance facility: the room that is used as the entrance location for public or private communication cables.

entrepreneur: a person who owns and operates a business.

Equal Level Far-End Crosstalk (ELFEXT): a measurement calculated by subtracting the effects of attenuation from the FEXT.

equipment room: a room that contains the telecommunications equipment for the building such as the Private Branch exchange (PBX), servers, and telecommunication wiring system terminations.

error correction: a RAID technique that uses traditional error checking code (ECC) or parity.

explicit rights: rights that are specifically granted to a trustee, and are not granted through inheritance.

Extended Binary Coded Decimal Interchange Code (EBCDIC): an IBM character code similar to ASCII.

Extensible Authentication Protocol (EAP): an authentication protocol used on both wired and wireless network systems.

Extensible Markup Language (XML): a markup language based on the same principles as HTML, with the added ability to create custom tags.

extranet: a network that allows internal and external access to Web pages by authorized personnel.

extrinsic losses: signal losses caused by physical factors outside the normal core such as splices, connectors, and bends in the core.

F

fabric switch: a switch designed especially for Fibre Channel networking.

Far-End Crosstalk (FEXT): a measurement of reflective loss at the far end, or output end, of the cable.

fault tolerance: the ability to recover from a hard disk or hard disk controller failure without the loss of stored data. When applied to a network infrastructure, fault tolerance means the ability to continue operation during a system hardware or software error.

Fiber Distributed Data Interface (FDDI): a standard developed by ANSI that employs fiber-optic cable over great distances. The distances can reach as far as 40 kilometers in length and support data speeds of 100 Mbps or higher. FDDI is used mainly as a backbone for large network systems such as a MAN or a WAN.

fiber-optic cable: a type of cable that uses a glass or clear plastic core rather than copper.

Fibre Channel: a high-speed access method that typically uses fiber-optic cable as media.

file access protocol: provides the means for a client to access networked files and print services.

file server: a server that stores data files that can be accessed by a client.

FIR (Fast Infrared): a protocol used for transmitting data from laptop computers to PCs without the use of cables.

firewall: a hardware device or software package that passes or blocks data packets as they enter and exit the network system.

firmware: the combination of the BIOS chip and the software program within the chip.

Fixed Length Subnet Mask (FLSM): a subnet mask that is expressed in standard eight-bit or one-byte values. An FLSM has subnets that are equal in range and have an equal number of hosts.

forest: a collection of domain trees that share a common Active Directory database.

frame: a packet that is encapsulated with information needed to travel the Internet.

Frame Relay: a packet switching protocol that typically uses leased lines such as T1 to carry data over long distances.

frame size average: the average of all frame sizes during the time period monitored.

frame size peak: a record of the largest frame size recorded during the time period monitored.

frequency: the number of cycles of an electronic signal that occur in 1 second, measured in Hertz (Hz).

frequency hopping: a spread spectrum technique that transmits data on multiple channels simultaneously.

Fresenel reflection loss: a type of signal loss that commonly occurs at connection points in fiber-optic cabling and is due to the refraction property differences in the core material, the connector materials used for sealing the connector, and air.

full backup: a backup method that backs up all designated data.

full-duplex: a communication mode in which bi-directional communication occurs simultaneously.

Fully Qualified Domain name (FQDN): a combination of a host name and a domain name.

fusion splice: the joining of two fiber-optic cores using heat to fuse, or melt, the materials together.

G

gateway: a network device that connects networks that use different protocols.

gateway service: a service that translates communications between networks that use different network operating systems.

generator: a device that creates and provides electricity.

geosynchronous orbit: an orbit in which a satellite's rotational speed is synchronized with the earth's rotational speed, making the satellite appear to be in a stationary position.

global security policy: a security policy that affects domain users.

graded-index multimode fiber-optic cable: a type of multimode fiber-optic cable that is designed with a varying grade of core material. The core is designed with maximum light conduction at its center and gradually diminished light conduction toward its cladding.

graphical user interface (GUI): a pictorial representation of commands and computer hardware that allow a user to access resources and programs with a click of a mouse button.

ground: an occurrence of a conductor connecting to the earth through a continuous path.

group account: a collection of users that typically share a common job-oriented goal or similar function.

H

H.323 standard: a telecommunications standard for audio, video, and data communications using IP or packet-type networks defined by the International Telecommunication Union (ITU).

half-duplex: a communication mode in which bi-directional communication occurs in one direction at a time.

Hardware Abstract Layer (HAL): a software layer located between the hardware devices and third-party vendor driver software.

hashing: a technique that relies on an algorithm or an encryption device based on mathematical algorithms for guessing a password.

head-end: the local area distribution point of a Cable television service provider.

hierarchical star: a network topology that is created when two or more star topologies are merged using network devices such as hubs, switches, or routers.

hoax: a message that is spread about a real or unreal virus.

horizontal cross connect: provides a mechanical means of connecting horizontal cabling systems to other cables or equipment.

horizontal wiring: the section of cable that runs from individual work areas to the telecommunications closet.

host: a name for a computer, printer, or network device associated with an IP address on a TCP/IP network.

hot spare: a backup component that can automatically replace the failed system component without the intervention of a technician.

hot swapping: the process of removing system components without shutting down the system.

hot-swap technology: allows a component to be removed or installed while the system is running.

HTML tag: an instruction for how text and graphics should appear when displayed in a Web browser.

hub: a network device that provides a common electrical connection to all nodes in a star topology.

hybrid topology: a physical arrangement that is a mixture of star, bus, and ring topologies.

hyperlink: a link to another Web page or to area on the same Web page.

Hypertext Markup Language (HTML): a programming language used to create Web pages.

Hypertext Transfer Protocol (HTTP): a protocol designed for communications between a Web browser and a Web Server.

I

impedance: the opposition to alternating current (AC) in a conductor.

incremental backup: a backup method that backs up only data that has changed since the previous backup.

industrial field bus technology: a combination of computer network devices and industrial controller devices sharing the same bus in the industrial environment.

i-Net+ Certification: a CompTIA certification that proves a person's competency working with Web services, including Internet communications and installing, updating, and modifying Web pages.

infrastructure mode: a wireless network that contains one or more Wireless Access Points (WAPs).

inherited rights: rights that are granted to a trustee through inheritance.

inode: a table entry that contains information such as permissions, file size, name of the owner, the time stamps of the file's creation, modification, and last access, and a pointer to where the file is stored.

Input/Output (I/O) port: a small amount of memory assigned to a device that temporarily holds small amounts of data and is used to transfer data between two locations.

Institute of Electrical and Electronic Engineers (IEEE): a professional organization that continually develops standards for the networking and communications industry.

Integrated Network Information Center (InterNIC): a branch of the United States government that was responsible for regulating the Internet, overseeing the issue of domain names, and assigning IP addresses to them.

Integrated Services Digital Network (ISDN): a long distance connection technology that provides a means of a fully digital transmission over channels that are capable of speeds of up to 64 kbps each.

interactive logon: an authentication process in which the user is verified and given access to the Active Directory.

interference: an undesired electromagnetic signal imposed on a desired signal that distorts or corrupts the desired signal.

International Organization for Standardization (ISO): an organization interested in the standardization of computer equipment.

International Telecommunications Union (ITU): an organization responsible for standardizing communications on an international level. The ITU regulates radio, television, and satellite frequencies and standards and telephony standards.

Internet: a collection of interconnected networks from all around the world.

Internet Corporation for Assigned Names and Numbers (ICANN): a company that manages domain name registrations by allocating domain name registration to select, private companies.

Internet Group Management Protocol (IGMP): a TCP/IP protocol designed to support multicasting. IGMP informs a multicast router of the names of the multicast group to which a host belongs.

Internet Message Access Protocol (IMAP): a sophisticated mail access protocol that can manipulate e-mail while it is on the server. It also allows a user to access their e-mail and then leave the e-mail on the server.

interrupt request (IRQ): a circuit that communicates with the CPU.

intranet: a private network that provides Web page distribution to a specific group of users within a LAN.

IP Security (IPSec): a protocol that secures IP packets across the Internet. It works with IPv4 and IPv6 and supports two modes of encryption: transport and tunnel mode.

IPv6: an Internet addressing scheme that uses 128 bits to represent an IP address. The 128 bits are divided into 8 units of 16 bits. These units can be represented as a 4-digit hexadecimal number separated by colons.

IPX internal network number: a unique hexadecimal number that serves to identify a NetWare server on the network.

IPX/SPX (Internetwork Packet Exchange/Sequenced Packet Exchange): a routable protocol, developed by Novell, that controls how packets of data are delivered and routed between nodes and LANs.

ISM band: the band of radio frequencies associated with industrial, scientific, and medical devices.

isolation transformer: a device that uses a transformer to isolate a circuit from other circuits emanating from the same electrical source.

J

JAVA-based management utility: a management utility based on the Sun Microsystems JAVA programming language.

jitter: small staggers or hesitations in the delivery sequence of audio or video data caused by latency or missing packets.

joke program: a type of malware designed with some sort of joke as the payload.

journal file: a log of all file activity on a journaling file system.

journaling file system: a file system that relies on a journaling file to attain a faster boot time, use less system memory, and prevent file corruption.

K

Kerberos: a security authentication system that provides both authentication and encryption services. It uses a two-way method of authentication.

kernel: the core of an operating system.

key: an algorithm used to encode data.

L

latency: the amount of time it takes a signal to travel from its source to its destination.

layer 1 device: a network device that makes no decision about where a packet is sent.

layer 2 device: a network device that makes decisions about where a packet is sent based on a MAC address or a logical name.

Layer 2 Forwarding (L2F): a remote access protocol, similar to PPTP, designed to enhance security and to make use of a virtually private network using the public Internet. It was developed by Cisco Systems and then released for use by other venders. L2F requires special equipment on the host side to support data transfer whereas PPTP does not.

Layer 2 Tunneling Protocol (L2TP): a tunneling protocol that has the best features of both L2F and PPTP protocols. It uses IPSec to encrypt the contents of the encapsulated PPP protocol.

layer 3 device: a network device that makes a decision about where a packet is sent based on a protocol such as the Internet protocol.

lightning arrestor: a special piece of electrical equipment designed to dampen the effects of an electrical surge caused by lightning.

Linux+ Certification: a CompTIA certification that proves a person's knowledge of the Linux operating system.

loading coil: a device used to amplify voice signals.

local area network (LAN): a network that is usually confined to a single building and is managed by a single entity such as a company.

Local Central Office: the location where the customer's telephone lines connect to the switchgear. The Local Central Office connects to long distance carriers to provide long distance access to individual residencies.

Local Exchange Carrier (LEC): a local carrier. It is often made of one or more Local Central Offices.

local loop: the section of wiring between customer premises and the Local Central Office.

Local Security Authority (LSA): a service that validates local and remote logons and generates a security access token.

local security policy: a security policy that affects local users.

LocalTalk: the network media in an AppleTalk network.

logic bomb: a type of malware that remains dormant until a certain event takes place.

Logical Disk Manager (LDM) partition: a partition on a GPT disk that contains information about dynamic volumes and is created during the conversion from NTFS4.0 to NTFS.0. There can be more than one LDM partition in a system.

logical identification: a name that uniquely identifies a computer on the network.

logical unit number (LUN): a numbering scheme to identify SCSI devices attached to an extender card.

logon right: the ability to log on to the network.

loopback test: a signal that tests the network interface card to ensure that it is functioning properly.

M

MAC (media access code) address: a six-byte hexadecimal number, such as 00 C0 12 2B 14 C5, that uniquely identifies a network card.

macro: a symbol, word, or key that represents a command or list of keystrokes.

macro virus: a virus created by the macro feature of a software application program.

magnetic induction: an electrical phenomenon in which the magnetic field encircling a current carrying conductor induces a current flow in a conductor of close proximity.

mail filter: filters, or blocks, unwanted e-mail messages.

mail gateway: a special software and computer used to connect two normally incompatible e-mail systems.

main entrance: the place where public or private telecommunications enter the building.

malware: a software program designed to perform some type of unauthorized activity on a computer.

man in the middle (MITM): a method of intercepting a network transmission, reading it, and then placing it back on route to its intended destination.

Manchester encoding: an encoding scheme that is characterized by a digital pulse transitioning during the midpoint of the timing period.

master boot record (MBR): an area located on the first sector of the hard drive that contains information about the hard disk drive partitions (partition table) and the location of the bootable partition (active partition).

Master CNE Certification: an advanced Novell certification with an emphasis on project management and design and expertise in areas such as ZENworks desktop management, Integrating eDirectory and Active Directory, GroupWise Administration, and eDirectory Tools and Diagnostics.

MBR virus: a type of malware that attacks the master boot record (MBR) of a hard disk drive.

media: a general term that identifies the material used to transport packets.

Media Access Control (MAC) filter: a feature that allows or restricts WAP access based on the MAC address of a wireless network card.

member server: a server that belongs to a domain but does not serve as a domain controller. It authenticates users through another server that acts as the PDC or a BDC.

memory address assignment: a large block of memory assigned to a device and is used to transfer data between two locations.

mesh topology: a physical arrangement in which each node on the network connects to every other node on the network.

metropolitan area network (MAN): a network, under one management, that consists of two or more LANs connected with private communication lines within the same geographic area, such as a city or a university campus.

Microsoft Certified Database Administrator (MCDA) Certification: a Microsoft certification that proves a person's expertise in creating, maintaining, optimizing, installing, and managing SQL Server databases.

Microsoft Certified Desktop Support Specialist (MCDST) Certification: a Microsoft certification that measures the skills required to work as a help desk technician or a customer support representative.

Microsoft Certified Professional (MCP) Certification: a Microsoft certification that proves a person's expertise in one or more of the Microsoft software products.

Microsoft Certified Solution Developer (MCSD) Certification: a Microsoft certification that proves a person's ability to design, implement, and administer business solutions using Microsoft Office or Back Office products.

Microsoft Certified Systems Administrator (MCSA) Certification: a Microsoft certification that proves a person's ability to handle the day-to-day operations of a middle-to-large network based on the Windows 2000 or 2003 platform.

Microsoft Certified Systems Engineer (MCSE) Certification: a Microsoft certification that proves a person's ability to design, implement, maintain, upgrade, troubleshoot, and administer a Microsoft network system based on the Windows 2000 or 2003 platform.

Microsoft Challenge Handshake Authentication Protocol (MS-CHAP): an enhanced version of CHAP that encrypts the user name, password, and data. MS-CHAP must be used with Microsoft operating systems.

Microsoft Office Specialist Certification: a Microsoft certification that proves a person's knowledge of Microsoft Office software such as Word, Excel, Access, PowerPoint, Project, and Outlook. There are presently four levels of Microsoft Office Specialist Certification: Specialist, Expert, Master, and Master Instructor.

Microsoft Reserved (MSR) partition: a partition on a disk that reserves disk space for use by system components.

modulation: the process of mixing a data signal with a carrier wave.

module: a small program, such as a hardware driver or kernel enhancement.

monitor contention: the process in which an active monitor is selected.

mount point: the location in the directory structure where a device is mounted or inserted.

MPEG (Moving Picture Experts Group): an industry standard developed to ensure compatibility between different cameras, displays, and other multimedia equipment.

multicast frame: a frame intended for a preselected number of computers, such as a specific workgroup.

multicasting: a process of sending the same data packet to a group of clients identified by one IP address.

Multilink Point-to-Point Protocol (MLPPP): a remote access protocol that can combine two or more physical links in such a way that they act as one, thus increasing the supported bandwidth. This protocol works with ISDN lines, PSTN lines, and X.25 technology.

multimaster replication: a type of security database replication in which all domain controllers store a copy of the Active Directory database. Changes to the database can be made at any DC in the domain. When changes to the database are made, the changes are replicated to the other domain controllers.

multimode fiber-optic cable: a type of fiber-optic cable that has a large core diameter and is susceptible to attenuation due to dispersion.

Multipurpose Internet Mail Extensions (MIME): a protocol that encodes additional information known as mail attachments to e-mail protocols that normally could not transfer attachments such as graphics.

multistation access unit (MAU): a network device that allows for the quick connection and disconnection of Token Ring cables while maintaining the logic of the ring topology.

multitasking: the process of rapidly switching between various programs that are loaded into memory.

Multi-User Telecommunication Outlet Assembly (MUTOA): a grouping of outlets that serves up to 12 work areas.

N

namespace: the label that identifies a unique location in a structure such as the Internet.

naming convention: a standard naming format that is used when providing names for network objects.

Near-End Crosstalk (NEXT): a measurement of the reflected loss at the near-end, or input end, of a cable.

NetBEUI (NetBIOS Enhanced User Interface): a protocol that is an enhanced version of NetBIOS.

NetBIOS (Network Basic Input/Output System): software that allows a computer to communicate with other computers and provides basic services for data transfer.

NetWare File System (NWFS): NetWare's traditional file system.

network: an interconnected collection of computers, computer-related devices, and communications devices.

Network Address Translation (NAT): a protocol that translates private network addresses into an assigned Internet address, and vice versa. In other words, it allows an unregistered private network address to communicate with a legally registered IP address.

network administrator: a person who controls access to the network and to its shares. This person may also be responsible for the installation, configuration, and maintenance of a LAN, MAN, or WAN. The network administrator may perform many routine administrative functions, such as work with hardware and software vendors, make written and oral presentations about the state of the network, interview prospective employees, evaluate employee performance, coordinate work schedules, order supplies and equipment, and prepare budget reports.

network attached storage (NAS): a device or collection of devices used to provide storage for network data.

network authentication: an authentication process that occurs when a user accesses a resource.

network interface card (NIC): a network device that contains the electronic components to send and receive a digital signal.

network layer: the layer of the OSI model that is responsible for routing packets from one network to another using the IP addressing format.

network media: a general term for all forms of pathways that support network communications.

Network News Transfer Protocol (NNTP): a TCP/IP protocol that is designed to distribute news messages to NNTP clients and NNTP servers across the Internet.

network operating system (NOS): software that provides a communication system between computers, printers, and other intelligent hardware that exist on the network.

network share: a resource on the network that is shared among assigned users.

network support specialist: a person who assists users and customers whenever they encounter a network problem.

network topology: the physical arrangement of computers, computer-related devices, communication devices, and cabling in a network.

Network+ Certification: a CompTIA certification that measures the skills necessary for an entry-level network support technician.

Network+ Examination Objectives: a document that contains the exact areas to be tested on the Network+ Certification exam.

network-centric database: a single database that serves an entire network.

New Technology File System (NTFS): the native file format for Windows NT and Windows 2000/2003.

newsgroup: news articles that are arranged in a group or category on an NNTP server. Newsgroups are also referred to as discussion groups.

node: any device attached to the network that is capable of processing and forwarding data.

noise: electromagnetic interference.

non-return to zero (NRZ): a digital signal that fluctuates between a high voltage level and a low voltage level and never returns to zero volts for any measurable period of time.

Novell Storage Services (NSS): NetWare's enhanced files system.

NTFS permissions: permissions assigned to directories and files. NTFS permissions are effective locally and over the network.

O

object rights: rights that allow a trustee certain types of access to an object.

octet: an 8-bit (or one byte) binary value.

omni-directional: the ability of an antenna to transmit electromagnetic signals in all directions.

one's compliment: a digital-based mathematical calculation.

one-way trust relationship: a trust relationship in which one domain is the trusted domain and the other is the trusting domain. The trusting domain allows the trusted domain to access its resources, but the trusted domain does not allow the trusting domain to access it resources.

open: a wiring fault in which there is a break in a circuit path.

Open Shortest Path First (OSPF): a routing protocol that only exchanges the most recently changed information in the routing table.

Open Systems Interconnection (OSI) model: a model that describes how hardware and software should work together to form a network communications system.

Optical Time Domain Reflectometer (OTDR): a meter used for testing and troubleshooting long runs of fiber-optic cable. The OTDR conducts measurements based on the principle of attenuation.

organizational unit: a container that holds objects or other organizational units and is used to organize a network into manageable units.

orthogonal frequency-division multiplexing (OFDM): a transmission technique that transmits data over different channels within an assigned frequency range. Each channel is broadcast separately and is referred to as multiplexed. It can achieve a data rate as high as 54 Mbps.

overbooking: when the combined amount of declared NSS volume sizes is greater than the size of the storage pool.

P

packet filter: a hardware device or software package that accepts or rejects packets based on a set of rules set up by the network administrator.

packet sniffer: a network monitoring utility that captures data packets as they travel across a network.

packet switching: a type of transmission which does not use a permanent connection between two points for the duration of the data transfer period. Packets may travel different routes to the same destination.

parallel processing: processing a program through more than one CPU simultaneously.

parity check: a method of verifying the integrity of transmitted data.

passive hub: a network device that acts only as a central connection point for network cables.

pass-through authentication: the ability to access all resources throughout the entire network system with only a single logon.

Password Authentication Protocol (PAP): an authentication protocol that sends a user name and password in plain text format.

password virus: a type of malware designed to breach security by stealing passwords.

patch panel: a rack-mounted device with RJ-45 jacks on the front and a matching series of connections on the back. Patch panel cables are used to make connections between the front of the patch panel and equipment. The back of the panel is where the horizontal cable run terminates.

peak utilization: the highest level of utilization experienced by the network.

peer-to-peer network: a network in which all computers are considered as peers or equals.

permanent virtual circuit (PVC): a type of connection that behaves like a hard-wired connection between the destination and source.

permission: the ability to access a network share.

phishing: a method of Internet fraud, which involves using e-mail to steal a person's identity as well as other sensitive information, such as financial.

physical layer: the lowest layer of the OSI model, which focuses on the physical characteristics of a network such as cabling and connectors.

piconet: a Bluetooth network.

Plain Old Telephone Service (POTS): an older telephone technology that uses twisted pair cabling and analog signals rather than digital.

plenum-rated: a rating given to a cable that has a special type of insulation that will not give off toxic gases should the cable be consumed by fire.

Point of Presence (PoP): the point where the telephone company line connects to the subscriber line.

Point-to-Point Protocol (PPP): a remote access protocol that enables a PC to connect to a remote network using a serial line connection, typically through a telephone line. PPP is a synchronous protocol that supports multiple protocols such as IPX and AppleTalk.

Point-to-Point Protocol over Ethernet (PPPoE): a remote access protocol that provides one or more host on an Ethernet network the ability to establish an individual PPP connection with an ISP. PPPoE frames the PPP protocol so that the PPP frame can travel over an Ethernet network.

Point-to-Point Tunneling protocol (PPTP): a remote access protocol that is an enhanced version of PPP. It is designed to enhance security and to make use of a virtually private network using the public Internet.

polymorphic virus: a type of malware that changes its characteristics or profile as it spreads to resist detection.

port: a number that represents a logical connection and matches a service with a computer.

port number: a number that is associated with the TCP/IP protocol and is used to create a virtual connection between two computers running TCP/IP.

Post Office Protocol (POP): a mail access protocol designed to access a mail server and download messages to the e-mail client.

power conditioning: the process of eliminating spikes and any type of variation in the desired AC signal pattern.

power on self-test (POST): a quick initial check of the major components in a computer, such as memory, disk drives, keyboard, mouse, and monitor, to be sure that a minimum working system is available.

power-on self-test (POST): a BIOS routine that performs a series of hardware checks to determine if the computer is in minimal working order.

presentation layer: the layer of the OSI model that ensures character code recognition.

primary domain controller (PDC): a domain controller that hosts the user and security database and manages user access to the network. The chief function of the primary domain controller is to authenticate domain users as they log on to the network.

Primary Rate ISDN (PRI-ISDN): a category of ISDN that consists of twenty-three B channels and one D channel. It has a total data rate of 1.544 Mbps.

print server: a server that coordinates printing activities between clients and printers.

programmer: a person who writes, tests, and modifies software programs.

programmer analyst: a person who analyzes the needs of a client and performs the actual programming.

propagation delay: the time it takes data to be transmitted from the earth and satellite.

protected mode: a form of CPU execution that allows multiple programs to be loaded into memory and the CPU to rapidly switch between the various programs. It also allows a computer to access memory about the first 1MB of RAM.

protocol: a group of computer programs that handle packet formatting and control data transmission. Also, a set of rules governing communication between devices on a network.

protocol analyzer: a special tool used to inspect protocol activity and contents.

proxy server: a firewall component that is typically installed on a server and resides between Internet servers and the LAN hosts.

Public Switched Telephone Network (PSTN): an older telephone technology that uses twisted pair cabling and analog signals rather than digital.

punch down tool: a tool used for pushing individual twisted pair wires into the connection points on a patch panel.

Q

Quality of Service (QoS): a protocol that gives time-sensitive packets, such as those carrying telephone conversations, a higher priority than data packets.

R

radio interference: interference that matches the frequency of the carrier wave.

radio waves: electromagnetic waves with a frequency range of 10 kHz to 3,000,000 MHZ.

real mode: the original 8088 processor form of execution. Real mode has limited memory access and can only load and run programs smaller than 640 kB.

receiver: an electronic device that receives a modulated signal and demodulates it.

redirector: a software package that determines where a client request for a resource should be routed.

Redundant Array of Independent Disks (RAID): a system that can provide fault tolerance, increase disk speed when large blocks of data need to be constantly accessed, or both.

reflected loss: the amount of signal reflected from the far end of the cable.

registrar: a select, private company that is assigned a pool of IP addresses from ICANN and handles domain registration.

remote access server: a server that is accessed through dial-up networking.

Remote Authentication Dial-In User Service (RADIUS): a service that allows remote access servers to authenticate to a central server.

Remote Desktop Protocol (RDP): a presentation protocol that allows Windows XP Professional computers to communicate directly with Windows-based clients.

repeater: a network device that regenerates a weak signal into its original strength and form.

replay attack: a method of using information, such as an IP and MAC address, that had previously been copied and stored during a network transmission to establish an unauthorized connection with the destination.

resistance: the opposition to direct current (DC) in a conductor.

resolver: a software program located on a host that queries a DNS server to resolve a host name to an IP address.

Reverse Address Resolution Protocol (RARP): a service that finds the MAC address of a host when the IP address is known.

reversed pair: a wiring fault in which conductors within a twisted pair cable pair reverse their connection at the opposite end of the cable.

RG-8: a thick, rigid cable also known as thicknet and used in a 10Base5 network.

RG-58: a thin, flexible cable also known as thinnet and used in a 10Base2 network.

ring polling: a ring poll conducted by the active monitor that identifies if a computer has logged on or off the ring.

ring purge: removing a defective token and replacing it with a new one.

ring topology: a physical arrangement that consists of a single cable that runs continuously from node to node.

routable protocol: a protocol capable of delivering a packet to a network different from the source network.

router: a network device that navigates packets across large networks, such as the Internet, using the most efficient route.

Routing Information Protocol (RIP): a routing protocol that periodically exchanges an entire table of routing information with nearby routers.

routing table: a database of IP addresses and computer names used to direct packet flow to each side of a router.

S

Samba: a free software package that allows UNIX and Linux systems to share files and printers with Windows-based clients.

sampling frequency: how often the amplitude of an analog signal is measured per second.

sampling rate: the number of times a signal is sampled during a specific period of time.

scan engine: a software program that performs the function of reading each file indicated in the scan configuration and checking it against the virus signatures in the virus pattern file.

scattering: the loss of signal strength due to impurities in the core material.

screen: an instance of a running program, utility, or command on a NetWare server.

Secure Copy Protocol (SCP): a protocol that provides a secure way of transferring files between computers. It is the replacement for **rcp**.

Secure File Transfer Protocol (SFTP): a secure version of FTP, which encrypts the user name, password, and data to provide the highest level of security compared to FTP and TFTP.

Secure HTTP (S-HTTP): an Internet security protocol that secures individual messages between the client and server rather than the connection.

Secure Shell (SSH): a protocol that provides secure network services over an insecure network medium such as the Internet.

Secure Sockets Layer (SSL): an Internet security protocol developed by Netscape to ensure secure transmissions between Web servers and individuals using the Internet for such purposes as credit card transactions.

security access token: contains the user's security identifier (SID) number and the security identifier number of any groups the user belongs to.

Security Account Manager (SAM): a service that maintains the security account database in a Windows NT domain.

security policy: a blanket policy that secures resources on the network.

segment: raw data that is divided into a smaller unit.

segmenting: the act of dividing a network into smaller sections to avoid collisions.

Serial Line Internet Protocol (SLIP): a remote access protocol that enables a PC to connect to a remote network using a serial line connection, typically through a telephone line. SLIP is an asynchronous protocol that supports only IP.

server: a computer that provides services to networked computers or clients.

Server+ Certification: a CompTIA certification that proves a person's knowledge of network server hardware and software.

server-centric database: a database that is used for a single server.

servertop: the GUI on a NetWare server.

Service Access Point (SAP): a combination of a MAC address and a number that identifies a protocol and its service. It provides a means of direct communication between protocols at different layers of the OSI model and between the source and destination.

service pack: a collection of patches and fixes.

session: communication that is limited between two particular computers. Also, a logical connection with the Linux computer.

session layer: the layer of the OSI model that establishes a connection between two computers and provides security based on computer and user name recognition.

share permissions: permissions assigned to a network share. Common share permissions are full control, change, and read.

share-level security: a type of security that provides password protection and minimal share permissions to network shares. Share-level security applies only to shares that are accessed over the network.

shell: a user interface that interprets and carries out commands of the user similar to the way the DOS command interpreter (command.com) interprets and carries out commands.

short: a wiring fault which results in electrical current flow taking a shorter path.

Simple Mail Transfer Protocol (SMTP): a protocol that is part of the TCP/IP protocol suite and is designed to transfer plain text e-mail messages from an e-mail client to a mail server and from a mail server to a mail server.

Simple Network Management Protocol (SNMP): a protocol that is part of the TCP/IP suite and is designed to allow an administrator to manage and monitor network devices and services from a single location.

Simple Object Access Protocol (SOAP): an XML message format that allows a client to freely interact with a Web page on the Web server, rather than download it.

simple volume: a dynamic disk volume that exists on a single drive.

simplex: a communication mode in which communications occur in one direction only.

single-mode fiber-optic cable: a type of fiber-optic cable that has a small core diameter. Dispersion is limited in this type of cable because the core is designed to closely match the wavelength of the light signal. Single-mode fiber-optic cable can carry light farther than multimode fiber-optic cable.

Small Computer Systems Interface (SCSI): a computer bus technology that allows you to connect multiple devices to a single controller.

smart card: a technology that incorporates a special card into the security system, which is used in conjunction with a personal identification number (PIN).

snap-in: a tool or utility that is added to the Microsoft Management Console (MMC).

social engineering: the manipulation of personnel by the use of deceit to gain security information.

socket: a port number combined with an IP address.

software upgrade: a major improvement or enhancement to existing software programs.

spam: an unwanted e-mail, such as advertisement.

spammer: a person who engages in distributing unsolicited e-mail or sending e-mail with some sort of advertisement as a probe.

spamming: the distribution of unsolicited e-mail.

spanned volume: a single dynamic disk volume that spans many drives.

split pair: a wiring fault in which one conductor from each of two pairs within a twisted pair cable have switched positions at the opposite end of the cable.

spoofing: a method of fooling the destination by using an IP address other than the true IP address of a source to create a fake identity.

spread spectrum: a transmission technique that uses multiple channels to transmit data either simultaneously or sequentially.

stand-alone server: a server that is not part of a domain. Security on a stand-alone server is handled locally and all security information is stored in the local database.

standard: a set of recommendations or practices presented by an organization that help define specific aspects about a technology.

standby monitor: a computer on a Token Ring network that can become an active monitor if the current active monitor fails or logs off the network.

standby UPS: a UPS unit that waits until there is a disruption in commercial electricity before it takes over the responsibility of supplying electrical energy.

star topology: a physical configuration in which cables running from each node in a network connect to a single point, such as a hub.

static addressing: assigning an IP address manually.

static IP address table: a table of IP addresses entered manually.

static IP assignment: an IP address that is entered manually for each host on a network.

station address: an IPX/SPX-based address that consists of only a network number and node address.

stealth virus: a type of malware that hides from normal detection by incorporating itself into part of a known and usually required program.

step-index multimode fiber-optic cable: a general multimode fiber-optic cable that does not counter dispersion.

storage area network (SAN): a separate, high-speed network used to provide a storage facility for one or more networks.

storage pools: collections of storage devices with logical volumes.

striped volume: a dynamic disk volume that increases the read/write access speed by spreading data across multiple hard disk drives.

subdomain: any level domain located beneath the secondary domain. Also called a lower-level domain.

subnet mask: a number similar to an IP address that is used to determine in which subnet a particular IP address belongs.

subnetting: dividing a network into subnetworks, or subnets.

subnetwork: a network within a network.

sunspots: magnetic energy storms that occur at the surface of the sun.

swap file: hard drive disk space that supplements the RAM.

switch: a network device that filters network traffic or creates subnetworks from a larger network.

symmetric-key encryption: a key encryption method in which the sender and receiver use the same key.

synchronous transmission: a type of transmission in which a digital signal is synchronized with a reference signal to ensure proper timing.

system migration: the process of combining two diverse operating systems into one.

system partition: a partition that contains the required startup files necessary to boot the operating system.

systems analyst: a person who is responsible for performing analysis, evaluations, and recommendations of business software systems.

T

tar ball: a compressed file containing one or more software programs.

T-carrier: a leased line that follows one of the standards known as T1, fractional T1, T2, or T3. The T-carrier is a dedicated, permanent connection that is capable of providing a high bandwidth.

TCP/IP (Transmission Control Protocol/Internet/Protocol): a protocol developed for UNIX to communicate over the Internet.

telecommunications closet: an enclosed space that houses telecommunication cable termination equipment and is the recognized transition point between the backbone and horizontal wiring.

Telecommunications Industry Association (TIA): an organization that is concerned with fiber optics, user equipment, network devices, wireless communications, and satellite communications.

telecommunications room: a room or enclosed space that houses telecommunications equipment, such as cable termination and cross connect wiring.

terminating resistor: an electronic device, employed in a coaxial-type bus topology, that absorbs transmitted signals, preventing the signals from deflecting and the data from distorting.

thin client: a computer that relies on a thin client server's processing power and memory.

thin client server: a server that provides applications and processing power to a thin client.

thin server: a server that has only the hardware and software needed to support and run a specific function, such as Web services, print services, and file services.

time period: the rate of recurrence of an expected signal level change.

Time to Live (TTL): a maximum time allowed for an electronic signal to circulate a network.

token: a short binary code that is passed to computers on a ring topology and in some cases a bus topology. Also, the name given to the packet that exchanges session information between the source and destination.

transmitter: an electronic device that generates a carrier wave and modulates data signals into the carrier wave.

transport layer: the layer of the OSI model that ensures reliable data by sequencing packets and reassembling them into their correct order.

transport mode: an IPSec mode that encrypts only the payload portion of a packet.

tree: a network topology that is created when two or more star topologies are connected by a common backbone. A tree topology is also defined as a hierarchical star by some network manufacturers. Also, a collection of domains that share a common root domain name and Active Directory database.

Trivial File Transfer Protocol (TFTP): a lightweight version of FTP, which does not require a user name and password because it uses UDP packets for transferring data. It allows a client to transfer files, but not to view the directory listing at the FTP site.

Trojan horse: a type of malware that appears as a gift, a utility, game, or an e-mail attachment.

trunk lines: the hundreds of pairs of twisted pair cable or fiber-optic cable that connect Local Central Offices.

trust relationship: a relationship between domains, which allows users from one domain to access resources on another domain in which they do not have a user account. The user can then access resources on another domain without having to log on to that domain. The user, however, must have permissions to access the resources.

trustee: a user or group that is granted rights to an object. Also, a user, group, or container object that is granted rights to directories and files.

tunnel mode: an IPSec mode that encrypts the payload and the header information of a packet.

twisted pair: a type of cable consisting of four pairs of twisted conductors.

two-way trust relationship: a trust relationship in which both domains are designated as a trusted domain and a trusting domain. A two-way trust relationship allows both domains in the trust relationship to share its resources with the other.

typeful name: the name of an object preceded by the object's abbreviation, such as CN, O, and OU, and an equal sign (=).

typeless name: the name of an object followed by the path to the top of the directory. Each object in the path is separated by a dot (.).

U

unbounded media: an unrestricted path for network transmissions.

Underwriters Laboratories (UL): a nonprofit organization that tests products and materials against safety standards.

Unicode: a character code that uses 16 bits to represent individual characters.

Uniform Resource Locator (URL): a user-friendly name, such as http://support.microsoft.com/, which allows users to access Web-based resources.

Uninterruptible Power Supply (UPS): a device that ensures constant and consistent network performance, by supplying electrical energy in case of a power failure or blackout.

unipolar digital signal: a digital signal that fluctuates between a positive five-volt and zero-volt level.

Universal Naming Convention (UNC): a standard naming convention used by NetBIOS.

Unspecified Bit Rate (UBR): an ATM data transfer classification that does not guarantee any speed or meet requirements of any special application such as multimedia or telephony. This classification is typically applied inside TCP/IP frames.

upconverter: a radio transmitter integrated into a satellite dish.

user-level security: a type of security that requires a user to authenticate through a security database to access a share.

Variable Bit Rate (VBR): a cell rate that automatically adjusts to support time sensitive data. It uses multiplexing techniques to provide a minimum CBR for time sensitive audio and video transmissions while controlling the data rate of not time sensitive data, such as text or plain e-mail.

V

Variable Bit Rate-Non Real Time (VBR-NRT): an ATM data transfer classification that allows cells to move at a variable rate. The rate of movement depends on the type of data contained in each cell.

Variable Bit Rate-Real Time (VBR-RT): an ATM data transfer classification that allows cells to move at a variable rate to support real-time audio and video transfers.

Variable Length Subnet Mask (VLSM): a subnet mask that is not expressed in standard eight-bit or one-byte values. A VLSM occurs when a subnet is further divided into smaller subnets, which are not equal to the original subnet in length or in the number of hosts.

virtual circuit: a connection-oriented communication.

Virtual LAN (VLAN): a LAN that is created using software and hardware techniques to connect workstations on separate network segments as though they were on the same segment.

virtual network: a logical network within a LAN.

Virtual Private Network (VPN): a simulated, independent network created by software over a public network.

virus: a type of malware that replicates itself on a computer after infecting it.

virus pattern file: a database of virus signatures, or codes, unique to each known virus.

Voice over IP (VoIP): an Internet telephony protocol that relies on existing TCP/IP technology and existing TCP/IP networking equipment.

W

wavelength: the total distance an electromagnetic or light wave travels during one full cycle.

Web administrator: a person who is responsible for converting written documents into HTML or other Web page programming languages. This job requires not only training in Web page programming language, but also some expertise in networking technology.

Web browser: a software program that permits a user to navigate the World Wide Web and display and interpret Web pages.

Web master: a person who has expertise in Web programming languages, such as HTML, XML, JAVA, JavaScript, ActiveX, and familiarity with server management of IIS, Apache, UNIX, Linux, and Microsoft systems.

Web server: a server configured to provide Web services, such as Web pages, file transfer, and HTML-based e-mail.

Web site: a location on the World Wide Web, which contains a collection of Web pages and files. A Web site is owned and managed by an individual, company, or organization.

Web-based management utility: A management utility based on existing open source standards such as HTTP and XML.

weighted: the amount of points assigned to a test question.

wide area network (WAN): a network that consists of a large number of networks and PCs connected with private and public communications lines throughout many geographic areas.

Wi-Fi: a term coined by the Wi-Fi Alliance that refers to wireless network products used in the 802.11b category.

Wi-Fi Protected Access (WPA): a standard developed to ensure the safe exchange of data between a wireless network and a portable Wi-Fi device, such as a cell phone.

windowing: the process of deciding the maximum size of each segment and the amount of segments that will be sent before requiring an acknowledgment.

Windows Internet Naming Service (WINS): a service that resolves NetBIOS names to IP addresses.

windows manager: an X client that controls the display. In other words, it provides the GUI such as the icons, boxes, and buttons.

Wired Equivalent Privacy (WEP): a data encryption protocol that makes a wireless network as secure as a wired network.

Wireless Access Point (WAP): a wireless network device that provides a connection between a wireless network and a cable-based network and controls the flow of all packets on the wireless network.

Wireless Application Protocol (WAP): a protocol that combines the authentication method with encryption.

wireless topology: a physical arrangement that uses infrared light or radio transmission as a means of communication between nodes.

work area: the place where employees perform their normal office duties.

workgroup: a group of computers that share resources such as files and hardware.

working group: a standard not fully developed and adopted as an official standard recognized by IEEE.

World Wide Web consortium (W3C): an organization that focuses on the development of common protocols used for Internet communications.

worm: a software program that can spread easily and rapidly to many different computers.

worm: a type of malware that infects files on a computer and automatically spreads to other computers.

X

X client: an application or a windows manager.

X server: a program that communicates with the computer hardware, such as the keyboard, mouse, and monitor.

X Windows: a system that provides a GUI for UNIX and Linux systems.

X.25: a protocol that uses analog signals to transmit data across long distances.

Z

Zero Configuration (Zeroconf): a standard developed by IEEE which recommends how to design a device that automatically detects other devices on the same network or on a nearby network segment without the need of intervention by an administrator or a DHCP or DNS server.

Index